Handbook of Experimental Pharmacology

Volume 146

Springer-Verlag Berlin Heidelberg GmbH

Fibrinolytics and Antifibrinolytics

Contributors

F. Bachmann, D.C. Berridge, C. Bode, H. Bounameaux,
F.J. Castellino, J.H. Chesebro, A.C. Chiu, D. Collen,
G.J. del Zoppo, W. Dietrich, P. Donner, J.J. Emeis, L. Flohé,
V. Fuster, R. Gallo, W.A. Günzler, V. Gurewich, E. Haber,
S.B. Hawley, R. Hayes, T. Herren, J.L. Hoover-Plow, J. Horrow,
T. Kooistra, E.K.O. Kruithof, H.R. Lijnen, V.J. Marder,
L.A. Miles, K. Peter, E.F. Plow, V.A. Ploplis, A. Redlitz,
M.S. Runge, C.M. Samama, M.M. Samama, A.A. Sasahara,
W.-D. Schleunig, G.V.R.K. Sharma, O. Taby, E.J. Topol,
M. Verstraete, S. Xue

Editor:

F. Bachmann

 Springer

Professor emeritus
F. BACHMANN
FRAT – Hôpital Nestlé 2077
CHUV
CH-1011 Lausanne
Switzerland
e-mail: Fedor.Bachmann@dmed.unil.ch

Private address:
Chemin Praz Mandry 20
CH-1052 Le Mont
Switzerland

With 47 Figures and 50 Tables

ISBN 978-3-642-63029-3   ISBN 978-3-642-56637-0 (eBook)
DOI 10.1007/978-3-642-56637-0

Library of Congress Cataloging-in-Publication Data
Fibrinolytics and antifibrinolytics / contributors, F. Bachmann . . . [et al.]; editor, Fedor Bachmann.
     p.   cm. – (Handbook of experimental pharmacology; v. 146)
   Includes bibliographical references and index.

   1. Thrombolytic therapy.   2. Fibrinolytic agents.   3. Fibrinolysis.   4. Antifibrinolytic agents.
I. Bachmann, Fedor.   II. Series.
   [DNLM: 1. Fibrinolytic Agents.   2. Antifibrinolytic Agents. QV 190 F4431 2001]
QP905 .H3   vol. 146
[RC694.3]
615′.1s – dc21
[615′.718]

                                                                            00-059471

© Springer-Verlag Berlin Heidelberg 2001
Originally published by Springer-Verlag Berlin Heidelberg New York in 2001
Softcover reprint of the hardcover 1st edition 2001

The use of general descriptive names, registered names, etc. in this publication does not imply, even in the absence of a specific statement, that such names are exempt from the relevant protective laws and regulations and free for general use.

Product liability: The publishers cannot guarantee the accuracy of any information about dosage and application contained in this book. In every individual case the user must check such information by consulting the relevant literature.

Cover design: design & production GmbH, Heidelberg
Typesetting: Best-set Typesetter Ltd., Hong Kong

Preface

In 1978, Fritz Markwardt edited *Fibrinolytics and Antifibrinolytics*, vol. 46 of the *Handbook of Experimental Pharmacology* series. Since then tremendous strides have been made in our understanding of the pathophysiology of the fibrinolytic system, of myocardial infarction and of stroke.

I am very grateful to the contributors to this new edition of *Fibrinolytics and Antifibrinolytics*, all outstanding scientists and/or clinicians in their field. They have succeeded in presenting an up-to-date account of basic aspects of the fibrinolytic system, of both older and new thrombolytic agents, and have analyzed in a balanced fashion the results of first to third generation thrombolytic drugs in clinical trials.

The book is arranged in three major sections. The first one deals with the molecular biology, biochemistry, physiology and pharmacology of the plasminogen–plasmin enzyme system. The progress that has been made in this field since 1978 is staggering. Methods taken from molecular biology have been used successfully for cloning the zymogen plasminogen, the two principal activators and three major inhibitors of the system, many receptors for plasmin(ogen) and its activators, and some modulators of this clot-dissolving system. For many compounds the crystal structure has been uncovered. Numerous regulators of the system, particularly cytokines, growth factors and hormones, have been identified and their action and signal transduction elucidated. This section is of particular interest to both basic scientists and pharmacologists.

The large second section on the clinical use of thrombolytic agents is subdivided into three subsections. In the first of these subsections, a great deal of space is devoted to the use of thrombolytic agents for patients suffering from acute myocardial infarction. Analysis encompasses all major megatrials up to early 2000. Given that the formation of an occlusive thrombotic blood clot is a very dynamic process, which involves platelets and activated clotting factors (particularly thrombin), ample space is devoted to the discussion of adjuvant therapy with antiplatelet and antithrombotic agents. This part of the book will be of great interest to cardiologists. The second subsection provides a competent discussion of thrombolytic therapy of massive pulmonary embolisms, deep venous thrombosis, peripheral arterial occlusion and thromboembolic cerebrovascular disease. It should make for interesting reading for internists, angiologists, vascular surgeons and neurologists. The third subsection deals mostly with thrombolytic agents that are still in development and with compounds capable of stimulating the endogenous production and release of tissue-type plasminogen activator.

The last major section of the volume deals with the pharmacology and clinical use of antifibrinolytic agents and will interest surgeons, internists and obstetricians. I have personally edited many of these chapters in order to prevent significant overlaps. Several chapters that had been submitted earlier by the authors have been updated to reflect the state of the art in early 2000.

This book is dedicated to my loving wife Edith who – understanding the demands on time that editing this book made – laid aside many projects for common activities in recent years. Special thanks are due to Ms. Doris Walker and her collaborators of Springer-Verlag for their uncomplicated and fruitful collaboration.

Lausanne, June 2000   Fedor Bachmann

List of Contributors

BACHMANN, F., Department of Medicine, University of Lausanne Medical Centre, FRAT – Hôpital Nestlé 2077, CHUV, CH-1011 Lausanne, Switzerland Private address: Chemin Praz Mandry 20, CH-1052 Le Mont, Switzerland
e-mail: Fedor.Bachmann@dmed.unil.ch

BERRIDGE, D.C., Department of Vascular and Endovascular Surgery, First Floor, Block A, Lincoln Wing, St. James's University Hospital, Leeds, LS9 7TF, United Kingdom
e-mail: David Berridge@GW.SJUH.NORTHY.NHS.UK

BODE, C., Medizinische Universitätsklinik Freiburg, Abteilung Innere Medizin III, Kardiologie und Angiologie, Hugstetter Strasse, D-79106 Freiburg, Germany
e-mail: bode@mm31.ukl.uni-freiburg.de

BOUNAMEAUX, H., University of Geneva School of Medicine, Division of Angiology and Hemostasis, Department of Internal Medicine, University Hospital of Geneva, CH-1211 Geneva 14, Switzerland
e-mail: bounamea@cmu.unige.ch

CASTELLINO, F.J., Department of Chemistry and Biochemistry, University of Notre Dame, Notre Dame, IN 46556, USA
e-mail: castellino.1@nd.edu

CHESEBRO, J.H., Cardiovascular Institute, Mount Sinai Medical Center, Box 1030, One Gustave L. Levy Place, New York, NY 10029-6574, USA
e-mail: james.chesebro@mssm.edu

CHIU, A.C., SMDC Health System Section of Cardiology 407 East Third Street Duluth, MN 55805
e-mail: achiu@smdc.org

Present address: Section of Cardiology, SMDC Health System,
400 East Third Street, Duluth, MN 55805, USA
e-mail: achiu@smdc.org

COLLEN, D., Centre for Molecular and Vascular Biology,
University of Leuven, Campus Gasthuisberg, O&N, Herestraat 49,
B-3000 Leuven, Belgium
e-mail: desire.collen@med.kuleuven.ac.be

DEL ZOPPO, G.J., Department of Molecular and Experimental Medicine,
The Scripps Research Institute, 10550 North Torrey Pines Road,
MEM-132, La Jolla, CA 92037, USA
e-mail: grgdl.zop@hermes.scripps.edu

DIETRICH, W., Deutsches Herzzentrum, Institut für Anaesthesiologie,
Lothstrasse 11, D-80335 München, Germany

DONNER, P., Preclinical Drug Research, Schering Research Laboratories,
Müllerstr. 178, D-13342 Berlin, Germany

EMEIS, J.J., Gaubius Laboratory, TNO-Prevention and Health, P.O. Box 2215,
NL-2301 CE Leiden, The Netherlands
e-mail: jj.emeis@pg.tno.nl

FLOHÉ, L., Department of Biochemistry, TU Braunschweig,
Mascheroder Weg 1, D-38124 Braunschweig, Germany
e-mail: lfl@gbf.de

FUSTER, V., Cardiovascular Institute, Mount Sinai Medical Center, Box 1030,
One Gustave L. Levy Place, New York, NY 10029-6574, USA

GALLO, R., Cardiovascular Institute, Mount Sinai Medical Center, Box 1030,
One Gustave L. Levy Place, New York, NY 10029-6574, USA

GÜNZLER, W.A., Grünenthal GmbH, Postfach 50 04 44, D-52088 Aachen,
Germany

GUREWICH, V., Harvard Medical School, Vascular Research Laboratory,
21–27 Burlington/BIDMC, POB 15709, Boston, MA 02215, USA
e-mail: vgurevic@caregroup.harvard.edu
and: Institute for the Prevention of Cardiovascular Disease,
Beth Israel-Deaconess Medical Center, Boston, MA, USA

HABER, E., Center for the Prevention of Cardiovascular Disease, Division of
Biological Sciences, Harvard School of Public Health and Harvard
Medical School, 677 Huntington Avenue, Boston, MA 02115, USA

HAWLEY, S.B., Department of Vascular Biology,
The Scripps Research Institute, 10550 N. Torrey Pines Road, La Jolla,
CA 92037, USA

HAYES, R., Cardiovascular Institute, Mount Sinai Medical Center, Box 1030,
One Gustave L. Levy Place, New York, NY 10029-6574, USA

HERREN, T., Joseph J. Jacobs Center for Thrombosis and Vascular Biology,
Cleveland Clinic Foundation/NB50, 9500 Euclid Avenue, Cleveland,
OH 44139, USA

HOOVER-PLOW, J.L., Joseph J. Jacobs Center for Thrombosis and Vascular
Biology, Cleveland Clinic Foundation/NB50, 9500 Euclid Avenue,
Cleveland, OH 44139, USA

HORROW, J., IBEX Technologies Corporation, 5 Great Valley Parkway,
Suite 300, Molvern, PA 19355, USA

KOOISTRA, T., Gaubius Laboratory, TNO-Prevention and Health,
P.O. Box 2215, NL-2301 CE Leiden, The Netherlands
e-mail: T.Kooistra@pg.tno.nl

KRUITHOF, E.K.O., Research Laboratories,
Division of Angiology and Hemostatis, University Hospital,
CH-1211 Geneva, Switzerland
e-mail: Egbert.Kruithof@hcuge.ch

LIJNEN, H.R., Center for Molecular and Vascular Biology,
Campus Gasthuisberg, O&N, Herestraat 49, B-3000 Leuven,
Belgium
e-mail: roger.lijnen@med.kuleuven.ac.be

MARDER, V.J., Orthopaedic Hospital/UCLA, Vascular Medicine Program,
2400 S. Flower Street, Los Angeles, CA 90007, USA
e-mail: vmarder@laoh.ucla.edu

MILES, L.A., Department of Vascular Biology,
The Scripps Research Institute, 10550 N. Torrey Pines Road, La Jolla,
CA 92037, USA

PETER, K., Medizinische Universitätsklinik Freiburg,
Abteilung Innere Medizin III, Kardiologie und Angiologie,
Hugstetter Strasse, D-79106 Freiburg, Germany

PLOW, E.F., Joseph J. Jacobs Center for Thrombosis and Vascular Biology,
Cleveland Clinic Foundation/NB50, 9500 Euclid Avenue, Cleveland,
OH 44139, USA
e-mail: plowe@ccf.org

PLOPLIS, V.A., Department of Chemistry and Biochemistry,
    University of Notre Dame, Notre Dame, IN 46556, USA

REDLITZ, A., Joseph J. Jacobs Center for Thrombosis and Vascular Biology,
    Cleveland Clinic Foundation/FF20, 9500 Euclid Avenue, Cleveland,
    OH 44139, USA

RUNGE, M.S., Division of Cardiology and Sealy Center for Molecular
    Cardiology, University of Texas Medical Branch at Galveston,
    5.106 John Sealy Hospital, 301 University Boulevard, Galveston,
    TX 77555-0553, USA

SAMAMA, C.M., Départment d'Anesthésie-Réanimation, Hôpital Avicenne,
    125 route de Stalingrad, F-93009 Bobigny Cedex, France
    e-mail: cmsamama@invivio.edu

SAMAMA, M.M., Hôpital Hotel Dieu, Place du Parvis de Notre Dame,
    F-75004 Paris, France
    e-mail: mmsamama@aol.com

SASAHARA, A.A., Cardiovascular Division and the Departments of Medicine,
    Brigham and Women's Hospital and Harvard Medical School,
    1115 Beacon Street, Unit 12, Newton, MA 02461-1154, USA
    e-mail: arthur.a.sasahara@abbott.com

SCHLEUNIG, W.-D., Preclinical Drug Research, Schering Research
    Laboratories, Müllerstr. 178, D-13342 Berlin, Germany
    e-mail: Wolfdieter.Schleuning@Schering.de

SHARMA, G.V.R.K., the Departments of Medicine, Brigham and Women's
    Hospital and Harvard Medical School, and the Cardiology Section,
    Veterans Affairs Medical Center, West Roxbury, MA 02132, USA

TABY, O., Hôpital Hotel Dieu, Place du Parvis de Notre Dame, F-75004
    Paris, France

TOPOL, E.J., Department of Cardiology, Desk F-25, J.J. Jacobs Center for
    Thrombosis and Vascular Biology, Cleveland Clinic Foundation,
    9500 Euclid Avenue, Cleveland, OH 44195, USA
    e-mail: topole@ccf.org

VERSTRAETE, M., Centre for Molecular and Vascular Biology, University of
    Leuven, O&N, Herestraat 49, B-3000 Leuven, Belgium
    e-mail: marc.verstraete@med.kuleuven.ac.be

XUE, S., Department of Vascular Biology, The Scripps Research Institute,
    10550 N. Torrey Pines Road, La Jolla, CA 92037, USA

Contents

CHAPTER 4

Urinary-type Plasminogen Activator (uPA)
W.A. GÜNZLER and L. FLOHÉ. With 2 Figures ....

CHAPTER 5

The Inhibitors of the Fibrinolytic System

CHAPTER 6

Assembly of the Plasminogen System on Cell Surfaces

Section II: Clinical Use of Thrombolytic Agents
In Acute Myocardial Infarction

CHAPTER 7

Streptokinase and Anisoylated Lys-Plasminogen Streptokinase Activator Complex

CHAPTER 9

Urokinase and Single-Chain Urokinase-Type Plasminogen Activator (Pro-urokinase)

CHAPTER 11

Conjunctive Therapy to Reduce the Occurrence of Coronary Reocclusion After Thrombolytic Treatment of AMI

Thrombolytic Treatment of Other Clinical Thromboembolic Conditions

CHAPTER 17

Desmodus rotundus (Common Vampire Bat)
Salivary Plasminogen Activator
W.-D. Schleuning and P. Donner. With 5 Figures. 451

CHAPTER 18

Thrombus-Targeting of Plasminogen Activators
C. BODE, K. PETER, M.S. RUNGE, and E. HABER ....

CHAPTER 19

The Hunt for the Ideal Thrombolytic Agent:
Mutants of tPA and uPA, Chimera of Both Molecules, Fibrolase
M. VERSTRAETE. With 3 Figures   493

CHAPTER 20

Agents which Increase Synthesis and Release of Tissue-Type Plasminogen Activator

Section I
Molecular Biology, Biochemistry, Physiology, and Pharmacology of the Plasminogen-Plasmin Enzyme System

CHAPTER 1

The Fibrinolytic System and Thrombolytic Agents

F. BACHMANN

A. The Fibrinolytic System

The formation of a haemostatic thrombus is a useful defence mechanism for the closure of vascular lesions. However, undesirable thrombi are also formed in closed vessels, e.g. over atherosclerotic plaques or after rupture of such plaques. It has long been assumed that the primary function of the fibrinolytic system consists of dissolving such thrombi, a task to which it often does not measure up. Indeed, repeat venography in patients with deep venous thrombosis treated with heparin and oral anticoagulation alone often shows only minimal resorption of the venous thrombus (DUROUX et al. 1991). The situation is somewhat more favourable for arterial thrombi. The classical work of DEWOOD et al. (1980) has demonstrated that coronary thrombi undergo thrombolysis in the absence of thrombolytic therapy. While thrombotic lesions were present in 87% of patients undergoing coronary angiography 1–4 h after the onset of symptoms, this figure was only 68% in patients examined 6–12 h after start of symptoms ($p < 0.01$). In the UPET study the spontaneous recanalisation of pulmonary emboli was quite remarkable. Seven days after the embolic event, pulmonary angiography no longer showed a difference between control patients and those who had been treated with urokinase (UROKINASE PULMONARY EMBOLISM TRIAL 1970).

Figure 1 is a simplified scheme of the fibrinolytic system. The zymogen plasminogen is activated to its active enzyme form, plasmin, by two plasminogen activators – tissue-type plasminogen activator (tPA) and urinary-type plasminogen activator (uPA). Plasminogen and its activation is discussed in Chap. 2, tPA in Chap. 3 and uPA in Chaps. 4 and 9. The endothelial lining of blood vessels synthesises constitutively tPA and has storage sites in vesicles for release (LEVIN and LOSKUTOFF 1982; EMEIS et al. 1997; PARMER et al. 1997; ROSNOBLET et al. 1999). The coagulation of blood generates substances which trigger the release of tPA, such as bradykinin, which is formed when the contact system of the blood coagulation system is activated, and thrombin, the enzyme which brings about clotting of fibrinogen (HANSS and COLLEN 1987; LEVIN and SANTELL 1988). Endothelial cells also produce uPA, although this cell is not the major source for this zymogen (CAMOIN et al. 1998). It is not

4

4

4 F. BACHMANN

THE PLASMINOGEN-PLASMIN SYSTEM

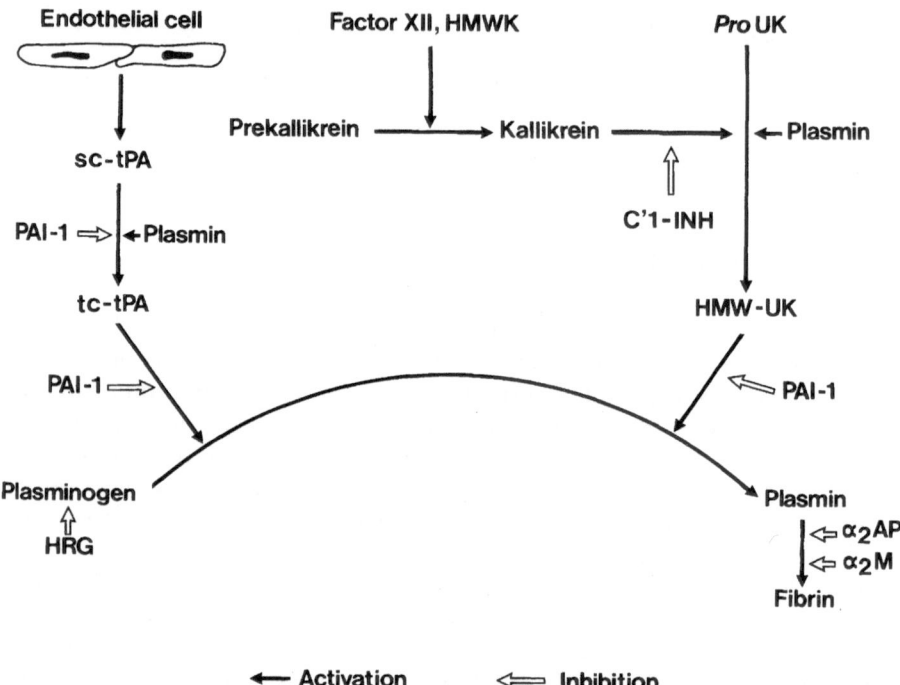

Fig. 1. Scheme of the fibrinolytic system. tPA, tissue-type plasminogen activator; UK, urokinase; sc, single-chain; tc, two-chain; HRG, histidin-rich glycoprotein; HMWK, high-molecuclar-weight kininogen; C'1-INH, C'1 inhibitor; PAI-1, plasminogen activator inhibitor type 1; α₂AP, α₂-antiplasmin; α₂M, α₂-macroglobulin

clearly established what role the contact system plays. Activated factor XII has been shown to convert plasminogen directly to plasmin (SCHOUSBOE et al. 1999). A third plasminogen activator has also been described, but was not characterised in detail (BINNEMA et al. 1990). The generation of kallikrein by the contact phase of coagulation leads to the conversion of single-chain uPA (sc-uPA, also pro-urokinase) to its two-chain form (tc-uPA, also denominated HMW-urokinase) (ICHINOSE et al. 1986; HAUERT et al. 1989).

The two principal inhibitors in plasma are plasminogen activator inhibitor type-1 (PAI-1) and α₂-antiplasmin. These are discussed in Chap. 5. The role of histidine-rich glycoprotein is not clearly established. It can bind to lysine-binding sites (LBS) on plasminogen and thus diminish the amount of free plasminogen in the blood (LIJNEN et al. 1980). C'1 inhibitor is the most important inhibitor of the contact phase of coagulation. C'1 deficiency results in hereditary angioneurotic oedema. During acute episodes some activation of the fibrinolytic system may occur (CUGNO et al. 1993). The molar concentration of α_2-antiplasmin is about $1\,\mu mol/l$, that of plasminogen $2\,\mu mol/l$. During thrombolytic therapy with non-fibrinspecific agents most of the plasminogen is converted to plasmin and rapidly inhibited by α_2-antiplasmin (WADA et al. 1989; WILLIAMS 1989). The formed complex is removed from the circulation. In this situation, where there is still excess of plasmin but exhaustion of α_2-antiplasmin, α_2-macroglubulin acts as a scavenger inhibitor in inhibiting free plasmin, albeit at a slower rate.

The plasminogen-plasmin system is also involved in embryogenesis, cell migration, wound healing and spread of tumour cells. This is not discussed in this chapter. The reader is referred to some recent references covering these subjects (DEAR and MEDCALF 1998; CHAPMAN 1997; CARROLL and BINDER 1999; LOSKUTOFF et al. 1999; RIFKIN et al. 1999).

B. Mechanisms which Lead to the Lysis of a Thrombus

There is minimal fibrinolytic activity in the normal circulating blood. tPA is a very poor activator of plasminogen in the absence of fibrin (CAMIOLO et al. 1971) and sc-uPA has virtually no enzymatic activity (PANNELL and GUREWICH 1987; PETERSEN et al. 1988; LIJNEN et al. 1989). Even when the levels of tPA increase some 20- to 100-fold, such as after strenuous exercise, after the intravenous injection of DDAVP to healthy volunteers, or in an occluded region during a venous stasis test, only trace amounts of plasmin form because plasminogen activation is negligible at these still physiological tPA concentrations (ARAI et al. 1990; WEISS et al. 1998) This does not apply during the treatment of acute myocardial infarction (AMI) with recombinant tPA, where circulating plasma concentrations of tPA are achieved which are approximately 1000–5000 times higher than those observed in normal plasma (LUCORE et al. 1992).

Obviously, the fibrinolytic system is geared to remove fibrin from the circulation but its main function probably is to prevent excessive fibrin accumulation. Indeed, at the very earliest stages of fibrin formation, tPA and plasminogen bind to the forming fibrin strands (THORSEN et al. 1972). Once small amounts of tPA and plasminogen are bound to fibrin in the form of a ternary complex, the catalytic efficiency of tPA for plasminogen is several hundred times higher than in the absence of fibrin (reviewed by FEARS 1989). Plasmin generation will cause proteolytic cleavage of fibrin which starts at the C-terminal portion of the α-chain of fibrin and produces new C-terminal lysyl residues (SUENSON et al. 1984; TRAN-THANG et al. 1984; HARPEL et al. 1985; DE VRIES et al. 1990). Partially digested fibrin binds up to ten times more Glu-plasminogen than native, undegraded fibrin (TRAN-THANG et al. 1986). sc-uPA binds with high affinity to plasminogen and appears to activate selectively Glu-plasminogen that is bound to C-terminal lysines in the partially degraded fibrin

(Lenich et al. 1991; Longstaff et al. 1992). Trace amounts of plasmin also activate the single chain form of uPA to its enzymatically active two-chain form (Lijnen et al. 1987).

Over de-endothelialised lesions of the vessel wall, the contact system probably becomes activated locally by interaction of Factor XII with fibrillar subendothelial material. Activation of the contact phase of coagulation generates Factor XIIa that is able directly to activate plasminogen (Schousboe et al. 1999) and kallikrein which can convert sc-uPA to tc tPA (Ichinose et al. 1986; Hauert et al. 1989). Thus each reaction leads to further events, all increasing the efficiency of fibrin breakdown and at the same time restricting the process to places where fibrin has been formed. During the generation of thrombin, some of it will bind to thrombomodulin, a membrane glycoprotein. The complex thrombin/thrombomodulin initiates the protein C pathway, resulting in inhibition of coagulation (reviewed by Esmon 1995), and activates the zymogen TAFI (thrombin-activatable fibrinolysis inhibitor; procarboxypeptidase U or B; reviewed by Nesheim 1998). Active TAFI cleaves C-terminal lysine and thus down-regulates the binding of plasminogen to these lysyl residues, resulting in inhibition of fibrinolysis (described in Chap. 5).

Global circulating (a rare observation unless thrombolytic agents are administered to patients at high doses) and local fibrinolytic activity is modulated by serpins. sc-tPA, tc-tPA and tc-uPA are efficiently inhibited by PAI-1, which is nearly always present in molar excess over tPA (Chandler et al. 1995). About 80% of blood PAI-1 is located in the platelets and promptly released upon platelet activation (Kruithof et al. 1986). Although most of the platelet PAI-1 is of the inactive type, there is still enough active PAI-1 in platelets to stabilise platelet rich thrombi, as occur in the arterial circulation where the blood pressure is high and premature lysis of a haemostatic plug is undesirable.

Plasmin bound to fibrin is partly protected from inactivation by α_2-antiplasmin. This assures that fibrinolysis proceeds on the surface of a clot. Free plasmin, that spills over into the general circulation, however, is rapidly inactivated by α_2-antiplasmin (Wiman and Collen 1978).

Two types of receptors play important roles in the regulation of fibrinolytic activity. Activation receptors localise PAs on cell surfaces and greatly enhance plasminogen activation. Clearance receptors continuously remove free PAs and PA/serpin complexes from the circulation (discussed in Chap. 6). Many of the components of the plasminogen/plasmin system are highly regulated (Grant and Medcalf 1990) (see also Chap. 20). This is especially the case for tPA, PAI-1 and the uPAR. These constituents of the fibrinolytic system are up- or downregulated by hormones, growth factors, cytokines and also by oxidised LDL and lp(a).

Thus the fibrinolytic system is a highly modulated enzyme system and nature has taken many steps to limit its action in time and space, but has also provided the necessary feedback loops which enhance the fibrinolytic system on the local level.

C. Pathophysiology of the Fibrinolytic System

I. Decreased Fibrinolytic Activity and Deep Venous Thrombosis

Elevated levels of PAI-1 have been found to be associated with deep vein thrombosis (DVT) and arterial thrombotic events in patients with atherosclerosis, particularly in coronary heart disease, myocardial infarction and stroke. However, the question whether PAI-1 is a pathogenic factor favouring thromboembolic disease or merely a marker of disease has not been clearly answered to this date. Unlike thrombophilia caused by antithrombin III, protein C or protein S deficiency, very few families have been described with genetically increased PAI-1 levels. In most instances, a high PAI-1 level appears to be an acquired condition associated with an inflammatory state, with hypertriglyceridaemia, insulin resistance, obesity or pregnancy. PAI-1 belongs to the acute-phase proteins (JUHAN-VAGUE et al. 1985; KLUFT et al. 1985) and may be elevated for 2–4 months after an episode of deep vein thrombosis (JUHAN-VAGUE et al. 1984; JANSSON et al. 1989).

1. Familial Thrombophilia with High PAI-1 Levels

I have found only four families in which it was clearly demonstrated that high PAI-1 levels were transmitted through at least two generations and were associated with idiopathic thrombosis. Many of these families were restudied several years after the first report had been published (JOHANSSON et al. 1978; BUESSECKER et al. 1993; JØRGENSEN et al. 1982; JØRGENSEN and BONNEVIE-NIELSEN 1987; PATRASSI et al. 1992; ANGLÉS-CANO et al. 1993; GLUECK et al. 1993); reviewed by BACHMANN 1995.

2. Acquired Association of Hypofibrinolysis and Deep Venous Thrombosis

There have been many studies on the association of idiopathic DVT with a decreased basal fibrinolytic activity, mostly due to elevated levels of PAI-1 or to a deficient release of tPA after a venous occlusion test or a DDAVP-infusion (reviewed by (JUHAN-VAGUE et al. 1995; WIMAN 1999). Some 20 case control studies in patients with DVT have been published (literature review and references by BACHMANN 1995). Among the descriptive studies, one is particularly interesting because it dealt specifically with the question of the mechanisms implicated in hypofibrinolysis before and after venous occlusion. It also established correlations with other risk factors (JUHAN-VAGUE et al. 1987).

These studies are very heterogeneous. In many reports the diagnosis of DVT was not confirmed by phlebography, some studies included all cases with

DVT, others only 'young' patients (mostly under the age of 40 years) and/or only subjects with no underlying disease predisposing to DVT. In several studies the interval between the acute episode and the execution of the fibrinolytic studies was not mentioned or the studies were made within the first two months of the acute disease, when PAI-1 levels may still be increased due to the inflammatory reaction occurring in DVT. In several studies some patients were on oral anticoagulation, others not. Methodologies to measure hypofibrinolysis varied greatly and only studies from 1985 on reported results of PAI-1 activity and antigen. Reference intervals were often arbitrarily chosen. Taken together, several conclusions can be drawn:

1. The incidence of hypofibrinolysis in patients with idiopathic DVT varies from a few percent to 40%. Most likely the true figure will be below 10%.
2. The two principal reasons for a low basal fibrinolytic activity or a deficient increase after stimulation with a venous occlusion test or a DDAVP-infusion are high PAI-1 levels in about 80% of patients and deficient release of tPA from the endothelial cells in about 20%. Much rarer (<1%) are dys- and hypoplasminogenaemias, and dysfibrinogenaemias associated with poor binding of tPA or plasminogen to fibrin.
3. Familial DVT associated with familial hypofibrinolysis is very rare. Although up to 10% individuals with hypofibrinolysis are detected who also report a positive family history of DVT, it is often found subsequently that there is no correlation between test results and clinical events in a particular family.

One truly randomised prospective trial and two prospective studies of patients who had already had an episode of DVT merit further discussion. The primary incidence rate of DVT was investigated in physicians participating in the Physicians Health Study. This randomised double-blind, placebo-controlled trial of aspirin in the prevention of cardiovascular disease and of β-carotene in the prevention of cancer provided a unique opportunity to assess whether baseline PAI-1 and tPA levels are predictive of future DVT. Serum samples were prospectively collected from 14,916 participating physicians. Fifty-five individuals developed DVT during a mean follow-up period of 60 months, and were matched with an equal number of subjects for age, sex and smoking habits (RIDKER et al. 1992). There was no difference in the mean PAI-1 and tPA antigen levels between those physicians with and those without thrombotic event. It appears therefore that these fibrinolytic parameters do not predict the occurrence of a first future DVT.

The first prospective study of patients who had already had an episode of DVT was reported by KORNINGER et al. (1984) In 121 patients with a history of DVT and/or pulmonary embolism a venous occlusion test was performed at the earliest 4 months (median 19 months) after the last thromboembolic episode when oral anticoagulation had been discontinued. The main criterion for the presence or absence of a hypofibrinolytic state was a euglobulin lysis time (ELT) of >60min and <59min respectively after venous occlusion.

Approximately 5 years later (57 ± 19 months) patients were re-examined. The investigator who questioned the patients, evaluated medical records and performed a physical examination was unaware of the previous laboratory results. The recurrence rate for thromboembolic disease during this follow-up period was 4.8% per year in 76 patients with ELTs of <59 min after stimulation, but 10.3% in 48 patients with ELTs over 60 min ($p < 0.05$). A similar recent study in patients with idiopathic deep venous thrombosis demonstrated that increased PAI-1 levels predict the development of another thrombosis. However, the difference in plasma PAI-1 concentration between the group with a subsequent event as compared to the event-free group was small. Thus, these data are of limited prognostic value, even if a statistically significant difference had been obtained in the two groups (SCHULMAN et al. 2000).

Several studies have searched for predictive markers for the development of postoperative thrombosis. Among all the biological data evaluated only age and body mass index (CLAYTON et al. 1976) and elevated baseline PAI-1 levels (MELLBRING et al. 1985; ROCHA et al. 1988; ERIKSSON et al. 1989; PRINS and HIRSH 1991) were predictive markers.

In conclusion, for patients with an established episode of DVT the evidence for an association between elevated PAI-1 levels and the risk of experiencing a further episode of a thromboembolic event is weak. There is no evidence so far for a predictive value of increased PAI-1 levels for the development of a first episode of DVT. Families with a proven association between a hypofibrinolytic state and DVT are very rare. The incidence of postoperative thrombosis after orthopaedic surgery appears to be increased in patients with elevated PAI-1 levels.

II. Impaired Fibrinolysis and Coronary Heart Disease

After the first classical demonstration that increased PAI-1 levels in patients under the age of 45 years were associated with a significant risk for a subsequent event (HAMSTEN et al. 1987), many further studies have shown increased PAI-1 and tPA antigen levels in patients with atherosclerosis (CORTELLARO et al. 1993; SALOMAA et al. 1995), angina pectoris (THOMPSON et al. 1995; JUHAN-VAGUE et al. 1996; HELD et al. 1997a,b; LEHMANN et al. 1999; HOFFMEISTER et al. 1999), AMI (RIDKER et al. 1993) and peripheral arterial disease (SMITH et al. 1995; reviewed recently by WIMAN 1999). Multivariate analyses revealed that PAI-1 antigen and activity were related to insulin resistance but not to other classical risk factors for AMI. tPA antigen was related to insulin resistance, to markers of inflammation and endothelial damage (JUHAN-VAGUE et al. 1996). Increased PAI-1 levels also are predictive of poor outcome in patients with AMI undergoing thrombolytic therapy (BARBASH et al. 1989; SINKOVIC 1998). About 4–12 h after the administration of SK, tPA, or uPA there is in addition a massive rebound of PAI-1 levels, independent of the agent used (GENSER et al. 1998). It is not clear whether this rebound is related to outcome.

D. Thrombolytic Therapy

I. Basic Considerations

1. The Proteolytic State and Bleeding Complications

Therapeutic doses of SK, uPA and, to a lesser extent, of tPA result in degradation of fibrinogen, Factor V, Factor VIII and von Willebrand factor and consumption of α_2-antiplasmin (COLLEN et al. 1986; TRACY et al. 1997) (see Chap. 7, Sect. A.II). Ex vivo studies in man and animal experiments revealed that this effect is greatly attenuated or absent in the presence of highly fibrin-specific agents, such as staphylokinase and vampire bat plasminogen activator (VANDERSCHUEREN et al. 1997; MELLOTT et al. 1992; MUSCHICK et al. 1993) (see also Chaps. 16 and 17). In addition, activation of the complement system takes place (OREN et al. 1998). This effect is more marked with SK than with tPA (HOFFMEISTER et al. 1998) and the activation of kallikrein may be one of the reasons why the coagulation cascade is activated.

The question as to whether avoidance of a proteolytic state results in less bleeding has not been answered unambiguously. The comparison of SK vs tPA infusion in animals suggested that tPA produced less bleeding (AGNELLI et al. 1985; WEITZ 1995) but in the large multicentre trials of patients with AMI this observation could not be confirmed. It even appears that the more fibrin specific tPA causes a higher incidence of cerebral haemorrhage (THE GUSTO INVESTIGATORS 1993).

2. Thrombin Generation and Reocclusion

All classical thrombolytic agents activate prothrombin to thrombin (EISENBERG et al. 1988; WEITZ et al. 1988; GULBA et al. 1991; SEITZ et al. 1993; GRANGER et al. 1998; HOFFMEISTER et al. 1998) as demonstrated by the increase of thrombin-antithrombin complexes and of fibrinopeptide A. Thrombin may also be found on non-dissolved thrombi and is a risk factor for reocclusion (GULBA et al. 1990). While free circulating thrombin is readily inactivated by antithrombin, particularly in the presence of heparin, thrombin bound to fibrin or the subendothelial matrix is poorly accessible to the action of antithrombin/heparin (WEITZ et al. 1990, 1998). Heparin attenuates the extent of circulating thrombin-antithrombin complexes but is not able to suppress these completely (RAPOLD et al. 1992). An intensive search is under way to develop drugs which are direct thrombin inhibitors and are able to neutralise effectively clot-bound thrombin (see Chap. 11).

3. Delay to Treatment

Numerous studies have shown that the earlier thrombolytic treatment is started after onset of symptoms, the better the results (discussed in extenso in Chap. 7, Sect. B.I.3.d).

4. Platelet Activation

Platelets play an important role in the natural history of AMI. Intravascular platelet activation may lead to the extension of the primary platelet plug forming over injured endothelium and platelet-rich thrombi are more resistant to lysis by thrombolytic agents. Platelet-aggregating effects of thrombolytic agents have been described as measured by increased generation of prostacyclin and of thromboxane (KERINS et al. 1989), β-thromboglobulin release (UDVARDY et al. 1992), P-selectin expression (YUSUF et al. 1996; KAWANO et al. 1998; SEREBRUANY and GURBEL 1999), exposure of the Gp IIb/IIIa receptor leading to increased binding of fibrinogen (RABHI-SABILE and PIDARD 1995; BIHOUR et al. 1995; RABHI-SABILE et al. 1998) and stimulation of aggregation by immuncomplexes arising when SK reacts with high anti-SK antibodies (VAUGHAN et al. 1991). The effect of staphylokinase on platelet function in vitro was negligible (ABDELOUAHED et al. 1997). However, antiaggregating effects due to inhibition of platelet aggregation by fibrin(ogen) breakdown products (GOUIN et al. 1992; PARISE et al. 1993; LEBRAZI et al. 1995) and redistribution of GP Ib from the platelet surface to the canalicular system (LU et al. 1993) have also been noted. In at least one study, platelet aggregability was reduced for the first hours after thrombolytic therapy and increased above baseline values several hours later (GURBEL et al. 1998). Some of the contradictory results are due to methodological differences (BLOCKMANS et al. 1996). Plasmin can exert its effect in an unimpeded fashion when it acts on washed platelets. In plasma, free plasmin is rapidly inhibited by α_2-antiplasmin.

Clinical evidence for a beneficial effect of aspirin on the outcome of AMI was first demonstrated in THE ISIS-2 (SECOND INTERNATIONAL STUDY OF INFARCT SURVIVAL) COLLABORATIVE GROUP (1988). The 35-day vascular mortality in 4300 patients having received placebo treatment was 13.2%; in another 4300 subjects an infusion of 1.5 Mio units of SK was administered and 35-day mortality was 10.4%, a reduction of 28%. Results in patients receiving only 160 mg/day of aspirin were similar (10.7%, a reduction of 23%). From 1988 onwards aspirin was therefore added as adjuvant drug in the large majority of studies on thrombolytic treatment of AMI. More effective drugs inhibiting exposed Gp IIb/IIIa receptor have since been developed, particularly a humanised monoclonal antibody (abciximab) directed against this fibrinogen receptor. The efficacy of antiplatelet agents as an adjuvant therapy in AMI trials is discussed in Chap. 11.

5. Contraindications for Thrombolytic Therapy

A list of the contraindications is found in Table 3 of Chap. 7.

6. Fibrin Specificity

SOBEL (1999) has proposed the following nomenclature with reference to fibrin specificity:

Fibrin binding   The binding of a protein or other moiety to fibrin.
Fibrin selectivity*   A PA's increased enzymatic activity in the presence of
                          fibrin as compared with activity in its absence.
Clot selectivity   The clinically relevant effect of fibrin selectivity, i.e. the
                          extent to which a lytic drug activates clot-bound plas-
                          minogen to form plasmin, as opposed to acting on cir-
                          culating plasminogen.

Classifying the thrombolytic agents currently on the market or tested in clinical trials it can be stated that fibrin binding is greatest with TNK-tPA > alteplase = monteplase = pamiteplase > reteplase = lanoteplase (these two mutants have deletions in their finger domain) = vampire bat PA (binding takes place exclusively via the finger region) > APSAC > SK, uPA and staphylokinase, which do not bind to fibrin. Fibrin selectivity is highest for staphylokinase and the vampire bat PA. Both of these two PAs do not convert plasminogen to plasmin in the absence of fibrin; followed by TNK-tPA $\geq$ alteplase > monteplase = pamiteplase = reteplase. APSAC has some fibrin selectivity in vitro. sc-uPA, tc-uPA and SK have no fibrin selectivity. Considering clot selectivity the ranking order is identical as for fibrin selectivity with the exception of sc-uPA which binds to Glu-plasminogen at C-terminal lysines in partially digested clots. It is still an open questions as to whether fibrin selective agents produce less bleeding (AGNELLI et al. 1985; WEITZ 1995).

7. Antigenicity

SK is highly antigenic and most persons have anti-SK antibodies in their plasma due to previous streptococcal infections (BACHMANN 1968; OJALVO et al. 1999). Since thrombolytic therapy with SK often results in up to 100 times higher anti-SK titres, repeat treatment with this PA is better avoided. In patients treated with SK, moderately high anti-SK-titres have been demonstrated up to 4 years after such therapy (see Chap. 7, Sect. B.I.3.a). The antigenicity of APSAK was reported to be lower in animal experiments that that of SK. In man, APSAC probably is as antigenic as SK. The incidence and titres of preformed anti-staphylokinase antibodies is lower, but treatment with staphylokinase also results in an increase of anti-staphylokinase antibodies (DECLERCK et al. 1994). Great efforts have been undertaken to reduce the antigenicity of staphylokinase by mutagenesis (COLLEN et al. 1996a,b). Sc-uPA, tc-uPA and also LMW-urokinase, as well as recombinant tPA, are not antigenic. It remains to be determined whether mutants of tPA evoke immunological responses in man.

8. Treatment of Elderly Patients

While mortality in patients with AMI increases with age, there is no scientifically valid argument not to use thrombolytic agents in elderly patients. The

---

* In many publications it is designated as fibrin specificity.

benefit of thrombolytic therapy in these patients is at least as good as in younger patients (GOTTLIEB et al. 1997; BARAKAT et al. 1999). Elderly patients, however, present more frequent contraindications to thrombolytic therapy such as arterial hypertension.

E. Milestones in Thrombolytic Therapy of Acute Myocardial Infarction

I. First Generation Agents

The first non-randomised clinical trial of treating AMI with SK were performed by the St. Louis team in the late 1950s (FLETCHER et al. 1958; FLETCHER and SHERRY 1960). A few years later HMW-urokinase was also developed as a thrombolytic agent for AMI by the same group (FLETCHER et al. 1965) and by SASAHARA (UROKINASE PULMONARY EMBOLISM TRIAL 1970) (see Chap. 12) for pulmonary embolism. By the 1970s SK was already widely used for the treatment of deep venous thrombosis (see Chap. 13). Around 1980 it was increasingly realised that the pathogenesis of coronary occlusion was mainly due to thrombi, based on the pioneering study of DEWOOD et al. (1980) but also on the demonstration by RENTROP et al. (1979, 1981) of rapid recanalisation of coronary arteries upon intracoronary infusions of SK (see also Chap. 7).

II. Second Generation Agents

In the early to mid 1980s, tPA and sc-uPA were cloned (PENNICA et al. 1983; VERDE et al. 1984; HOLMES et al. 1985) and APSAC (anistreplase) was bio-chemically synthesised (SMITH et al. 1981). All three agents were rapidly tested in man (Chaps. 4, 7, 8 and 9). This period also saw the beginning of the randomised megatrials, investigating different thrombolytic regimens starting with THE ISIS-2 (SECOND INTERNATIONAL STUDY OF INFARCT SURVIVAL) COLLABORATIVE GROUP (1988). SK was compared with tPA, with APSAC and with sc-uPA (Chaps. 7, 9 and 10). The first trials in peripheral arterial occlusions were started (Chap. 14) and t-PA was used in the treatment of pulmonary embolism (Chap. 12). Table 1 lists some of the salient features of first and second generation agents. Table 2 in Chap. 19 gives similar information on tPA mutants.

III. Third Generation Agents

The 1990s saw a panoply of megatrials comparing various dosage schedules of different thrombolytic agents such as the GUSTO trial comparing SK and tPA (Chap. 11). It was also a period of great human inventiveness. Hundreds of mutants of tPA and of uPA, of chimeras consisting of domains of tPA, uPA and plasminogen were created and first tested in vivo and later in animals with the sobering experience that is was hard to create better agents than nature

Table 1. Features of some thrombolytic agents

Source	Streptokinase	Anistreplase	tc-uPA	sc-uPA	Staphylokinase	Vampire bat PA (DSPA α1)
	Streptococcal culture	SK + Lys-plasminogen	Mammmalian tissue culture	Mammalian tissue culture	Staphylococcal tissue culture	Mammalian tissue culture
Molecular mass, kDa	47	131	32 and 54	54	17	52
Fibrin binding	0	(+)	0	0	0	+
Fibrin selectivity	0	(+)	0	0	+++	+++
Clot selectivity	0	(+)	0	+	+++	+++
Plasma half-clearance rate (min)	~20	~30	7	7	α-Phase 6 β-Phase 37	>60
Antigenicity	Yes	Yes	No	No	Yes	?

Characteristics of alteplase and mutants of tPA are listed in Table 2, Chap. 19.
DSPA, *Desmodus rotundus* plasminogen activator.

had already developed in hundreds of thousands of years of evolution (see Chap. 19). In the search for agents with still higher binding to fibrin, bifunctional agents were created consisting of a PA linked to antifibrin – or antiplatelet antibodies (see Chap. 18) and the highly fibrin-selective agents staphylokinase and vampire bat PA were cloned and tested in animals and then in man (Chaps. 16 and 17). New tPA mutants with longer half-lives, such as reteplase and TNK-tPA, were investigated. Cardiologists explored increasingly the value of adjuvant therapy with thrombin inhibitors, such as hirudin or argatroban and with GpIIb/IIIa receptor inhibitors such as the humanised monoclonal antibody abciximab and synthetic inhibitors (Chap. 11). Several large clinical trials were started in patients with stroke (Chap. 15). Antifibrinolytic agents were increasingly used in patients with open heart surgery, orthotopic liver transplantation, uncontrolled menorrhagia and in a small series of promyelocytic leukaemia (see Chap. 21).

IV. The Future of Thrombolytic Therapy

In the new millennium we will see several mutants of tPA, of staphylokinase and of the vampire bat PA more thoroughly investigated. There is a tendency to use lesser amounts of PA and to combine thrombolytic agents with antiplatelet and/or antithrombin inhibitors. Only large and well designed clinical trials will be able to give answers as to the optimal treatment of AMI.

List of Abbreviations and Acronyms

AMI	acute myocardial infarction
APSAC	anisoylated plasminogen streptokinase activator complex
DDAVP	1-deamino-8-D-arginine vasopressin
GISSI	Gruppo Italiano per lo Studio della Streptochinasi nell'Infarto miocardico
GUSTO	Global Utilization of Streptokinase and Tissue plasminogen activator for Occluded coronary arteries
HMW	high molecular weight
ISIS	International Study of Infarct Survival
LBS	lysine binding site
LMW	low molecular weight
PAI-1	plasminogen activator inhibitor type-1
SK	streptokinase
tPA	tissue-type plasminogen activator
sc-tPA	single chain tPA
TAFI	thrombin activatable fibrinolysis inhibitor
tc-tPA	two-chain tPA
TNK-tPA	triple mutant of tPA
uPA	urinary-type (or urokinase-type) plasminogen activator, also called urokinase

sc-uPA   single-chain uPA, also called pro-urokinase
tc-uPA   two-chain uPA, also called urokinase

References

Abdelouahed M, Hatmi M, Helft G, Emadi S, Elalamy I, Samama MM (1997) Comparative effects of recombinant staphylokinase and streptokinase on platelet aggregation. Thromb Haemost 77:815–817

Agnelli G, Buchanan MR, Fernandez F, Boneu B, Van Ryn J, Hirsh J, Collen D (1985) A comparison of the thrombolytic and hemorrhagic effects of tissue-type plasminogen activator and streptokinase in rabbits. Circulation 72:178–182

Anglés-Cano E, Gris JC, Loyau S, Schved JF (1993) Familial association of high levels of histidine-rich glycoprotein and plasminogen activator inhibitor-1 with venous thromboembolism. J Lab Clin Med 121:646–653

Arai M, Yorifuji H, Ikematsu S, Nagasawa H, Fujimaki M, Fukutake K, Katsumura T, Ishii T, Iwane H (1990) Influences of strenuous exercise (triathlon) on blood coagulation and fibrinolytic system. Thromb Res 57:465–471

Bachmann F (1968) Development of antibodies against perorally and rectally administered streptokinase in man. J Lab Clin Med 72:228–238

Bachmann F (1995) The role of plasminogen activator inhibitor type 1 (PAI-1) in the clinical setting, including deep vein thrombosis. In: Glas-Greenwalt P (ed) Fibrinolysis in disease: molecular and hemovascular aspects of fibrinolysis, CRC Press, Boca Raton, pp 79–86

Barakat K, Wilkinson P, Deaner A, Fluck D, Ranjadayalan K, Timmis A (1999) How should age affect management of acute myocardial infarction? A prospective cohort study. Lancet 353:955–959

Barbash GI, Hod H, Roth A, Miller HI, Rath S, Zahav YH, Modan M, Zivelin A, Laniado S, Seligsohn U (1989) Correlation of baseline plasminogen activator inhibitor activity with patency of the infarct artery after thrombolytic therapy in acute myocardial infarction. Am J Cardiol 64:1231–1235

Bihour C, Durrieu-Jaïs C, Besse P, Nurden P, Nurden AT (1995) Flow cytometry reveals activated GP IIb-IIIa complexes on platelets from patients undergoing thrombolytic therapy after acute myocardial infarction. Blood Coagul Fibrinolysis 6:395–410

Binnema DJ, Dooijewaard G, Iersel JJL, Turion PNC, Kluft C (1990) The contact-system dependent plasminogen activator from human plasma: Identification and characterization. Thromb Haemost 64:390–397

Blockmans D, Deckmyn H, Van den Hove L, Vermylen J (1996) The effect of plasmin on platelet function. Platelets 7:139–148

Buessecker F, Reinartz J, Schirmer U, Kramer MD (1993) Enzyme-linked immunosorbent assays for plasminogen activators. J Immunol Methods 162:193–200

Camiolo SM, Thorsen S, Astrup T (1971) Fibrinogenolysis and fibrinolysis with tissue plasminogen activator, urokinase, streptokinase-activated human globulin, and plasmin. Proc Soc Exp Biol Med 138:277–280

Camoin L, Pannell R, Anfosso F, Lefevre JP, Sampol J, Gurewich V, Dignat-George F (1998) Evidence for the expression of urokinase-type plasminogen activator by human venous endothelial cells in vivo. Thromb Haemost 80:961–967

Carroll VA, Binder BR (1999) The role of the plasminogen activation system in cancer. Semin Thromb Hemost 25:183–197

Chandler WL, Levy WC, Stratton JR (1995) The circulatory regulation of TPA and UPA secretion, clearance, and inhibition during exercise and during the infusion of isoproterenol and phenylephrine. Circulation 92:2984–2994

Chapman HA (1997) Plasminogen activators, integrins, and the coordinated regulation of cell adhesion and migration. Curr Opin Cell Biol 9:714–724

Clayton JK, Anderson JA, MacNicol GP (1976) Preoperative prediction of postoperative deep vein thrombosis. Br Med J 2:910–912

Collen D, Bernaerts R, Declerck P, De Cock F, Demarsin E, Jenné S, Laroche Y, Lijnen HR, Silence K, Verstreken M (1996a) Recombinant staphylokinase variants with altered immunoreactivity. I: Construction and characterization. Circulation 94:197–206

Collen D, Bounameaux H, De Cock F, Lijnen HR, Verstraete M (1986) Analysis of coagulation and fibrinolysis during intravenous infusion of recombinant human tissue-type plasminogen activator in patients with acute myocardial infarction. Circulation 73:511–517

Collen D, Moreau H, Stockx L, Vanderschueren S (1996b) Recombinant staphylokinase variants with altered immunoreactivity. 2. Thrombolytic properties and antibody induction. Circulation 94:207–216

Cortellaro M, Cofrancesco E, Boschetti C, Mussoni L, Donati MB, Cardillo M, Catalano M, Gabrielli L, Lombardi B, Specchia G, Tavazzi L, Tremoli E, Pozzoli E, Turri M (1993) Increased fibrin turnover and high PAI-1 activity as predictors of ischemic events in atherosclerotic patients: A case-control study. Arterioscler Thromb 13:1412–1417

Cugno M, Hack CE, de Boer JP, Eerenberg AJM, Agostini A, Cicardi M (1993) Generation of plasmin during acute attacks of hereditary angioedema. J Lab Clin Med 121:38–43

De Vries C, Veerman H, Koornneef E, Pannekoek H (1990) Tissue-type plasminogen activator and its substrate Glu-plasminogen share common binding sites in limited plasmin-digested fibrin. J Biol Chem 265:13547–13552

Dear AE, Medcalf RL (1998) The urokinase-type-plasminogen-activator receptor (CD87) is a pleiotropic molecule. Eur J Biochem 252:185–193

Declerck PJ, Vanderschueren S, Billiet J, Moreau H, Collen D (1994) Prevalence and induction of circulating antibodies against recombinant staphylokinase. Thromb Haemost 71:129–133

DeWood MA, Spores J, Notske R, Mouser LT, Burroughs R, Golden MS, Lang HT (1980) Prevalence of total coronary occlusion during the early hours of transmural myocardial infarction. N Engl J Med 303:897–902

Duroux P, Ninet J, Bachet Ph, Prandoni P, Ruol A, Vigo M, Barret A, Mericq O, Boneu B, Janvier G, Girard Ph, Laprevote-Heully MC, Sourou P, Robert D, Chagny M, Nenci G, Agnelli G, d'Addato M, Palumbo H, Bensaid J, Gouffault J, Leborgne P, Le Hellocco A, Ducreux JC, Tempelhoff G, Sala-Planell E, Rosendo-Carrera A, Torres-Gomez A, Blettery B, Bachmann F (1991) A randomised trial of subcutaneous low molecular weight heparin (CY 216) compared with intravenous unfractionated heparin in the treatment of deep vein thrombosis. A collaborative European multicentre study. Thromb Haemost 65:251–256

Eisenberg PR, Miletich JP, Sobel BE, Jaffe AS (1988) Differential effects of activation of prothrombin by streptokinase compared with urokinase and tissue-type plasminogen activator (t-PA). Thromb Res 50:707–717

Emeis JJ, van den Eijnden-Schrauwen Y, Van den Hoogen CM, de Priester W, Westmuckett A, Lupu F (1997) An endothelial storage granule for tissue-type plasminogen activator. J Cell Biol 139:245–256

Eriksson BI, Eriksson E, Gyzander E, Teger-Nilsson A-C, Risberg B (1989) Thrombosis after hip replacement – relationship to the fibrinolytic system. Acta Orthop Scand 60:159–163

Esmon CT (1995) Thrombomodulin as a model of molecular mechanisms that modulate protease specificity and function at the vessel surface. FASEB J 9:946–955

Fears R (1989) Binding of plasminogen activators to fibrin: characterization and pharmacological consequences. Biochem J 261:313–324

Fletcher AP, Alkjaersig N, Smyrniotis FE, Sherry S (1958) Treatment of patients suffering from early myocardial infarction with massive and prolonged streptokinase therapy. Trans Assoc Am Physicians 71:287–296

Fletcher AP, Sherry S (1960) Thrombolytic therapy for coronary heart disease. Circulation 22:619–626

Fletcher AP, Alkjaersig N, Sherry S, Genton E, Hirsh J, Bachmann F (1965) The development of urokinase as a thrombolytic agent: maintenance of a sustained thrombolytic state in man by its intravenous infusion. J Lab Clin Med 65:713–731

Genser N, Lechleitner P, Maier J, Dienstl F, Artner-Dworzak E, Puschendorf B, Mair J (1998) Rebound increase of plasminogen activator inhibitor type I after cessation of thrombolytic treatment for acute myocardial infarction is independent of type of plasminogen activator used. Clin Chem 44:209–214

Glueck CJ, Glueck HI, Mieczkowski L, Tracy T, Speirs J, Stroop D (1993) Familial high plasminogen activator inhibitor with hypofibrinolysis, a new pathophysiologic cause of osteonecrosis? Thromb Haemost 69:460–465

Gottlieb S, Goldbourt U, Boyko V, Barbash G, Mandelzweig L, Reicher-Reiss H, Stern S, Behar S, for the SPRINT and Thrombolytic Survey Groups (1997) Improved outcome of elderly patients ($\geq$75 years of age) with acute myocardial infarction from 1981–1983 to 1992–1994 in Israel. Circulation 95:342–350

Gouin I, Lecompte T, Morel M-C, Lebrazi J, Modderman PW, Kaplan C, Samama MM (1992) In vitro effect of plasmin on human platelet function in plasma. Inhibition of aggregation caused by fibrinogenolysis. Circulation 85:935–941

Granger CB, Becker R, Tracy RP, Califf RM, Topol EJ, Pieper KS, Ross AM, Roth S, Lambrew C, Bovill EG, for the GUSTO-I Hemostasis Substudy Group (1998) Thrombin generation, inhibition and clinical outcomes in patients with acute myocardial infarction treated with thrombolytic therapy and heparin: results from the GUSTO-I Trial. J Am Coll Cardiol 31:497–505

Grant PJ, Medcalf RL (1990) Hormonal regulation of haemostasis and the molecular biology of the fibrinolytic system. Clin Sci 78:3–11

Gulba DC, Bode C, Topp J, Höpp H-W, Westhoff-Bleck M, Rafflenbeul W, Lichtlen PR (1990) Die Häufigkeit von Residualthromben nach erfolgreicher Thrombolysetherapie bei akutem Herzinfarkt und ihre Bedeutung für die Rate früher Reokklusionen. Ein Bericht von der multizentrischen Dosisfindungsstudie zur Thrombolysetherapie mit Urokinase-präaktivierter natürlicher Prourokinase (TCL 598). Z Kardiol 79:279–285

Gulba DC, Barthels M, Westhoff-Bleck M, Jost S, Rafflenbeul W, Daniel WG, Hecker H, Lichtlen PR (1991) Increased thrombin levels during thrombolytic therapy in acute myocardial infarction. Relevance for the success of therapy. Circulation 83:937–944

Gurbel PA, Serebruany VL, Shustov AR, Bahr RD, Carpo C, Ohman EM, Topol EJ, for the GUSTO-III Investigators (1998) Effects of reteplase and alteplase on platelet aggregation and major receptor expression during the first 24 hours of acute myocardial infarction treatment. J Am Coll Cardiol 31:1466–1473

Gurewich V, Pannell R (1987) Fibrin binding and zymogenic properties of single-chain urokinase (pro-urokinase). Semin Thromb Hemost 13:146–150

Hamsten A, De Faire U, Walldius G, Dahlen G, Szamosi A, Landou C, Blombäck M, Wiman B (1987) Plasminogen activator inhibitor in plasma: risk factor for recurrent myocardial infarction. Lancet 2:3–9

Hanss M, Collen D (1987) Secretion of tissue-type plasminogen activator and plasminogen activator inhibitor by cultured human endothelial cells: modulation by thrombin, endotoxin, and histamine. J Lab Clin Med 109:97–104

Harpel PC, Chang T, Verderber E (1985) Tissue plasminogen activator and urokinase mediate the binding of Glu-plasminogen to plasma fibrin I. Evidence for new binding sites in plasmin-degraded fibrin I. J Biol Chem 260:4432–4440

Hauert J, Nicoloso G, Schleuning WD, Bachmann F, Schapira M (1989) Plasminogen activators in dextran sulfate-activated euglobulin fractions: a molecular analysis of factor XII- and prekallikrein-dependent fibrinolysis. Blood 73:994–999

Held C, Hjemdahl P, Rehnqvist N, Wallén NH, Björkander I, Eriksson SV, Forslund L, Wiman B (1997a) Fibrinolytic variables and cardiovascular prognosis in patients with stable angina pectoris treated with verapamil or metoprolol. Results from the Angina Prognosis Study In Stockholm. Circulation 95:2380–2386

Held C, Hjemdahl P, Rehnqvist N, Wallén NH, Forslund L, Björkander I, Angelin B, Wiman B (1997b) Haemostatic markers, inflammatory parameters and lipids in male and female patients in the Angina Prognosis Study In Stockholm (APSIS). A comparison with healthy controls. J Intern Med 241:59–69

Hoffmeister HM, Szabo S, Kastner C, Beyer ME, Helber U, Kazmaier S, Wendel HP, Heller W, Seipel L (1998) Thrombolytic therapy in acute myocardial infarction: comparison of procoagulant effects of streptokinase and alteplase regimens with focus on the kallikrein system and plasmin. Circulation 98:2527–2533

Hoffmeister HM, Jur M, Helber U, Fischer M, Heller W, Seipel L (1999) Correlation between coronary morphology and molecular markers of fibrinolysis in unstable angina pectoris. Atherosclerosis 144:151–157

Holmes WE, Pennica D, Blaber M, Rey MW, Günzler WA, Steffens GJ, Heyneker HL (1985) Cloning and expression of the gene for pro-urokinase in Escherichia coli. Biotechnology 3:923–929

Ichinose A, Fujikawa K, Suyama T (1986) The activation of pro-urokinase by plasma kallikrein and its inactivation by thrombin. J Biol Chem 261:3486–3489

Jansson J-H, Norberg B, Nilsson TK (1989) Impact of acute phase on concentrations of tissue plasminogen activator and plasminogen activator inhibitor in plasma after deep-vein thrombosis or open-heart surgery. Clin Chem 35(7):1544–1546

Johansson L, Hedner U, Nilsson IM (1978) A family with thromboembolic disease associated with deficient fibrinolytic activity in vessel wall. Acta Med Scand 203:477–480

Juhan-Vague I, Moerman B, De Cock F, Aillaud MF, Collen D (1984) Plasma levels of a specific inhibitor of tissue-type plasminogen activator (and urokinase) in normal and pathological conditions. Thromb Res 33:523–530

Juhan-Vague I, Aillaud MF, De Cock F, Philip Joet C, Arnaud C, Serradimigni A, Collen D (1985) The fast-acting inhibitor of tissue-type plasminogen activator is an acute phase reactant protein. In: Davidson JF, Donati MB, Coccheri S (eds) Progress in Fibrinolysis, Vol. 7, Churchill-Livingstone, Edinburgh, pp 146–149

Juhan-Vague I, Valadier J, Alessi MC, Aillaud MF, Ansaldi J, Philip Joet C, Holvoet P, Serradimigni A, Collen D (1987) Deficient t-PA release and elevated PA inhibitor levels in patients with spontaneous or recurrent deep venous thrombosis. Thromb Haemost 57:67–72

Juhan-Vague I, Alessi MC, Declerck PJ (1995) Pathophysiology of fibrinolysis. Baillieres Clin Haematol 8:329–343

Juhan-Vague I, Pyke SDM, Alessi MC, Jespersen J, Haverkate F, Thompson SG, on behalf of the ECAT Study Group (1996) Fibrinolytic factors and the risk of myocardial infarction or sudden death in patients with angina pectoris. Circulation 94:2057–2063

Jørgensen M, Mortensen JZ, Madsen AG, Thorsen S, Jacobsen B (1982) A family with reduced plasminogen activator activity in blood associated with recurrent venous thrombosis. Scand J Haematol 29:217–223

Jørgensen M, Bonnevie-Nielsen V (1987) Increased concentration of the fast-acting plasminogen activator inhibitor in plasma associated with familial venous thrombosis. Br J Haematol 65:175–180

Kawano K, Aoki I, Aoki N, Homori M, Maki A, Hioki Y, Hasumura Y, Terano A, Arai T, Mizuno H, Ishikawa K (1998) Human platelet activation by thrombolytic agents: effects of tissue-type plasminogen activator and urokinase on platelet surface P-selectin expression. Am Heart J 135:268–271

Kerins DM, Roy L, FitzGerald GA, Fitzgerald DJ (1989) Platelet and vascular function during coronary thrombolysis with tissue-type plasminogen activator. Circulation 80:1718–1725

Kluft C, Verheijen JH, Jie AFH, Rijken DC, Preston FE, Sue Ling HM, Jespersen J, Aasen AO (1985) The postoperative fibrinolytic shutdown: a rapidly reverting acute phase pattern for the fast-acting inhibitor of tissue-type plasminogen activator after trauma. Scand J Clin Lab Invest 45:605–610

Korninger C, Lechner K, Niessner H, Gossinger H, Kundi M (1984) Impaired fibrinolytic capacity predisposes for recurrence of venous thrombosis. Thromb Haemost 52:127–130

Kruithof EKO, Tran-Thang C, Bachmann F (1986) Studies on the release of a plasminogen activator inhibitor by human platelets. Thromb Haemost 55:201–205

Lebrazi J, Abdelouahed M, Mirshahi M, Samama MM, Lecompte T (1995) Streptokinase and APSAC inhibit platelet aggregation in vitro by fibrinogenolysis: effect of plasma fibrinogen degradation products X and E. Fibrinolysis 9:113–119

Lehmann KG, Gonzales E, Tri BD, Vaziri ND (1999) Systemic and translesional activation of coagulation, fibrinolytic, and inhibitory systems in candidates for coronary angioplasty: basal state and effect of successful dilation. Am Heart J 137:274–283

Lenich C, Pannell R, Gurewich V (1991) The effect of the carboxy-terminal lysine of urokinase on the catalysis of plasminogen activation. Thromb Res 64:69–80

Levin EG, Loskutoff DJ (1982) Cultured bovine endothelial cells produce both urokinase and tissue-type plasminogen activators. J Cell Biol 94:631–636

Levin EG, Santell L (1988) Stimulation and desensitization of tissue plasminogen activator release from human endothelial cells. J Biol Chem 263:9360–9365

Lijnen HR, Hoylaerts M, Collen D (1980) Isolation and characterization of a human plasma protein with affinity for the lysine binding sites in plasminogen. J Biol Chem 255:10214–10222

Lijnen HR, Van Hoef B, Collen D (1987) Activation with plasmin of two-chain urokinase-type plasminogen activator derived from single-chain urokinase-type plasminogen activator by treatment with thrombin. Eur J Biochem 169:359–364

Lijnen HR, Van Hoef B, De Cock F, Collen D (1989) The mechanism of plasminogen activation and fibrin dissolution by single chain urokinase-type plasminogen activator in a plasma milieu in vitro. Blood 73:1864–1872

Longstaff C, Clough AM, Gaffney PJ (1992) Kinetics of plasmin activation of single chain urinary-type plasminogen activator (scu-PA) and demonstration of a high affinity interaction between scu-PA and plasminogen. J Biol Chem 267:173–179

Loskutoff DJ, Curriden SA, Hu G, Deng G (1999) Regulation of cell adhesion by PAI-1. Acta Pathol Microbiol Immunol Scand 107:54–61

Lu H, Soria C, Soria J, De Romeuf C, Perrot J-Y, Tenza D, Garcia I, Caen JP, Martin Cramer E (1993) Reversible translocation of glycoprotein Ib in plasmin-treated platelets: Consequences for platelet function. Eur J Clin Invest 23:785–793

Lucore CL, Soule HR, Hagan RR, Dorian CJ, Sobel BE (1992) Quantification of free tissue-type plasminogen activator in plasma from patients undergoing coronary thrombolysis. Coron Artery Dis 3:831–838

Mellbring G, Dahlgren S, Wiman B, Sunnegardh O (1985) Relationship between preoperative status of the fibrinolytic system and occurrence of deep vein thrombosis after major abdominal surgery. Thromb Res 39:157–163

Mellott MJ, Stabilito II, Holahan MA, Cuca GC, Wang S, Li P, Barrett JS, Lynch JJ, Gardell SJ (1992) Vampire bat salivary plasminogen activator promotes rapid and sustained reperfusion without concomitant systemic plasminogen activation in a canine model of arterial thrombosis. Arterioscler Thromb 12:212–221

Muschick P, Zeggert D, Donner P, Witt W (1993) Thrombolytic properties of *Desmodus* (vampire bat) salivary plasminogen activator DSPA$_{a1}$, alteplase and streptokinase following intravenous bolus injection in a rabbit model of carotid artery thrombosis. Fibrinolysis 7:284–290

Nesheim M (1998) Fibrinolysis and the plasma carboxypeptidase. Curr Opin Hematol 5:309–313

Ojalvo AG, Pozo L, Labarta V, Torréns I (1999) Prevalence of circulating antibodies against a streptokinase C-terminal peptide in normal blood donors. Biochem Biophys Res Commun 263:454–459

Oren S, Maslovsky I, Schlesinger M, Reisin L (1998) Complement activation in patients with acute myocardial infarction treated with streptokinase. Am J Med Sci 315: 24–29

Pannell R, Gurewich V (1987) Activation of plasminogen by single-chain urokinase or by two-chain urokinase – A demonstration that single-chain urokinase has a low catalytic activity (pro-urokinase). Blood 69:22–26

Parise P, Hauert J, Iorio A, Callegari P, Nenci GG (1993) Fibrinogen degradation products generation is the major determinant of platelet inhibition induced by plasminogen activators in platelet-rich plasma. Fibrinolysis 7:379–385

Parmer RJ, Mahata M, Mahata S, Sebald MT, O'Connor DT, Miles LA (1997) Tissue plasminogen activator (t-PA) is targeted to the regulated secretory pathway. Catecholamine storage vesicles as a reservoir for the rapid release of t-PA. J Biol Chem 272:1976–1982

Patrassi GM, Sartori MT, Saggiorato G, Boeri G, Girolami A (1992) Familial thrombophilia associated with high levels of plasminogen activator inhibitor. Fibrinolysis 6:99–103

Pennica D, Holmes WE, Kohr WJ, Harkins RN, Vehar GA, Ward CA, Bennett WF, Yelverton E, Seeburg PH, Heynecker HL, Goeddel DV, Collen D (1983) Cloning and expression of human tissue-type plasminogen activator cDNA in E. coli. Nature 301:214–221

Petersen LC, Lund LR, Nielsen LS, Danø K, Skriver L (1988) One-chain urokinase-type plasminogen activator from human sarcoma cells is a proenzyme with little or no intrinsic activity. J Biol Chem 263:11189–11195

Prins MH, Hirsh J (1991) A critical review of the evidence supporting a relationship between impaired fibrinolytic activity and venous thromboembolism. Arch Intern Med 151:1721–1731

Rabhi-Sabile S, Pidard D (1995) Exposure of human platelets to plasmin results in the expression of irreversibly active fibrinogen receptors. Thromb Haemost 73:693–701

Rabhi-Sabile S, De Romeuf C, Pidard D (1998) On the mechanism of plasmin-induced aggregation of human platelets: implication of secreted von Willebrand factor. Thromb Haemost 79:1191–1198

Rapold HJ, de Bono D, Arnold AER, Arnout J, De Cock F, Collen D, Verstraete M, for the European Cooperative Study Group (1992) Plasma fibrinopeptide A levels in patients with acute myocardial infarction treated with alteplase. Correlation with concomitant heparin, coronary artery patency, and recurrent ischemia. Circulation 85:928–934

Rentrop KP, Blanke H, Karsch KR, Wiegand V, Köstering H, Oster H, Leitz K (1979) Acute myocardial infarction: intracoronary application of nitroglycerin and streptokinase. Clin Cardiol 2:354–363

Rentrop P, Blanke H, Karsch KR, Kaiser H, Köstering H, Leitz K (1981) Selective intracoronary thrombolysis in acute myocardial infarction and unstable angina pectoris. Circulation 63:307–317

Ridker PM, Vaughan DE, Stampfer MJ, Manson JE, Shen C, Newcomer LM, Goldhaber SZ, Hennekens CH (1992) Baseline fibrinolytic state and the risk of future venous thrombosis. A prospective study of endogenous tissue-type plasminogen activator and plasminogen activator inhibitor. Circulation 85:1822–1827

Ridker PM, Vaughan DE, Stampfer MJ, Manson JE, Hennekens CH (1993) Endogenous tissue-type plasminogen activator and risk of myocardial infarction. Lancet 341:1165–1168

Rifkin DB, Mazzieri R, Munger JS, Noguera I, Sung J (1999) Proteolytic control of growth factor availability. Acta Pathol Microbiol Immunol Scand 107:80–85

Rocha E, Alfaro MJ, Páramo JA, Cañadell JM (1988) Preoperative identification of patients at high risk of deep venous thrombosis despite prophylaxis in total hip replacement. Thromb Haemost 59:93–95

Rosnoblet C, Vischer UM, Gerard RD, Irminger J-C, Halban PA, Kruithof EKO (1999) Storage of tissue-type plasminogen activator in Weibel-Palade bodies of human endothelial cells. Arterioscler Thromb Vasc Biol 19:1796–1803

Salomaa V, Stinson V, Kark JD, Folsom AR, Davis CE, Wu KK (1995) Association of fibrinolytic parameters with early atherosclerosis. The ARIC study. Circulation 91:284–290

Schousboe I, Feddersen K, Røjkjaer R (1999) Factor XIIa is a kinetically favorable plasminogen activator. Thromb Haemost 82:1041–1046

Schulman S, Wiman B, for the Duration of Anticoagulation (DURAC) Trial Study Group (2000) The significance of hypofibrinolysis for the risk of recurrence of venous thromboembolism. Thromb Haemost 84:(In Press)

Seitz R, Pelzer H, Immel A, Egbring R (1993) Prothrombin activation by thrombolytic agents. Fibrinolysis 7:109–115

Serebruany VL, Gurbel PA (1999) Effect of thrombolytic therapy on platelet expression and plasma concentration of PECAM-1 (CD31) in patients with acute myocardial infarction. Arterioscler Thromb Vasc Biol 19:153–158

Sinkovic A (1998) Pretreatment plasminogen activator inhibitor-1 (PAI-1) levels and the outcome of thrombolysis with streptokinase in patients with acute myocardial infarction. Am Heart J 136:406–411

Smith FB, Lee AJ, Rumley A, Fowkes FGR, Lowe GDO (1995) Tissue-plasminogen activator, plasminogen activator inhibitor and risk of peripheral arterial disease. Atherosclerosis 115:35–43

Smith RAG, Dupe RJ, English PD, Green J (1981) Fibrinolysis with acyl-enzymes: A new approach to thrombolytic therapy. Nature 290:505–508

Sobel BE (1999) The language of lysis: a proposal to standardize the nomenclature. J Am Coll Cardiol 34:1226–1227

Suenson E, Lützen O, Thorsen S (1984) Initial plasmin-degradation of fibrin as the basis of a positive feedback mechanism in fibrinolysis. Eur J Biochem 140:513–522

The GUSTO Investigators (1993) An international randomized trial comparing four thrombolytic strategies for acute myocardial infarction. N Engl J Med 329: 673–682; erratum in Topol EJ, Califf RM, Lee KL (1994) More on the GUSTO trial. N Engl J Med 331:277–278

The ISIS-2 (Second International Study of Infarct Survival) Collaborative Group (1988) Randomised trial of intravenous streptokinase, oral aspirin, both, or neither among 17 187 cases of suspected acute myocardial infarction: ISIS-2. Lancet 2:349–360

Thompson SG, Kienast J, Pyke SDM, Haverkate F, van de Loo JCW (1995) Hemostatic factors and the risk of myocardial infarction or sudden death in patients with angina pectoris. N Engl J Med 332:635–641

Thorsen S, Glas-Greenwalt P, Astrup T (1972) Differences in the binding to fibrin of urokinase and tissue plasminogen activator. Thromb Diath Haemorrh 28:65–74

Tracy RP, Rubin DZ, Mann KG, Bovill EG, Rand M, Geffken D, Tracy PB (1997) Thrombolytic therapy and proteolysis of factor V. J Am Coll Cardiol 30:716–724

Tran-Thang C, Kruithof EKO, Bachmann F (1984) Tissue-type plasminogen activator increases the binding of Glu-plasminogen to clots. J Clin Invest 74:2009–2016

Tran-Thang C, Kruithof EKO, Atkinson J, Bachmann F (1986) High-affinity binding sites for human Glu-plasminogen unveiled by limited plasmic degradation of human fibrin. Eur J Biochem 160:599–604

Udvardy M, Harsfalvi J, Boda Z, Rak K (1992) Beta thromboglobulin and increased platelet activation after streptokinase treatment of acute myocardial infarction. Am J Cardiol 70:837–838

Urokinase pulmonary embolism trial (1970) Phase 1 results: a cooperative study. JAMA 214:2163–2172

Vanderschueren S, Dens J, Kerdsinchai P, Desmet W, Vrolix M, De Man F, Van den Heuvel P, Hermans L, Collen D, Van de Werf F (1997) Randomized coronary patency trial of double-bolus recombinant staphylokinase versus front-loaded alteplase in acute myocardial infarction. Am Heart J 134:213–219

Vaughan DE, Van Houtte E, Declerck PJ, Collen D (1991) Streptokinase-induced platelet aggregation. Prevalence and mechanism. Circulation 84:84–91

Verde P, Stoppelli MP, Galeffi P, Di Nocera P, Blasi F (1984) Identification and primary sequence of an unspliced human urokinase poly(A)$^+$ RNA. Proc Natl Acad Sci USA 81:4727–4731

Wada K, Takahashi H, Tatewaki W, Takizawa S, Shibata A (1989) Plasmin-α_2-plasmin inhibitor complex in plasma of patients with thromboembolic diseases. Thromb Res 56:661–665

Weiss C, Seitel G, Bärtsch P (1998) Coagulation and fibrinolysis after moderate and very heavy exercise in healthy male subjects. Med Sci Sports Exerc 30:246–251

Weitz JI, Cruickshank MK, Thong B, Leslie B, Levine MN, Ginsberg J, Eckhardt T (1988) Human tissue-type plasminogen activator releases fibrinopeptides A and B from fibrinogen. J Clin Invest 82:1700–1707

Weitz JI, Hudoba M, Massel D, Maraganore J, Hirsh J (1990) Clot-bound thrombin is protected from inhibition by heparin- antithrombin III but is susceptible to inactivation by antithrombin III-independent inhibitors. J Clin Invest 86:385–391

Weitz JI (1995) Limited fibrin specificity of tissue-type plasminogen activator and its potential link to bleeding. J Vasc Interv Radiol 6 [Suppl S]:19S–23S

Weitz JI, Leslie B, Hudoba M (1998) Thrombin binds to soluble fibrin degradation products where it is protected from inhibition by heparin-antithrombin but susceptible to inactivation by antithrombin-independent inhibitors. Circulation 97:544–552

Williams EC (1989) Plasma α_2-antiplasmin activity. Role in the evaluation and management of fibrinolytic states and other bleeding disorders. Ann Intern Med 149:1769–1772

Wiman B, Collen D (1978) On the kinetics of the reaction between human antiplasmin and plasmin. Eur J Biochem 84:573–578

Wiman B (1999) Predictive value of fibrinolytic factors in coronary heart diseas. Scand J Clin Lab Invest 59 [Suppl 230]:23–31

Yusuf SW, Sanderson H, Heptinstall S, Nurden AT, Wenham PW, Hopkinson BR (1996) ADP-induced P-selectin expression on platelets as a predictor of successful thrombolysis. Blood Coagul Fibrinolysis 7:266–269

CHAPTER 2
Plasminogen and Streptokinase

F.J. CASTELLINO and V.A. PLOPLIS

A. Primary Structure of Human Plasminogen

The primary structure of human plasminogen (HPg), diagrammed in Fig. 1, has been deduced from the nucleotide sequence of the cDNA (FORSGREN et al. 1987) and genomic DNA (PETERSEN et al. 1990) that encode this protein, and has been directly determined by amino acid sequence analysis (WIMAN 1973; WIMAN and WALLEN 1975; WIMAN 1977; SOTTRUP-JENSEN et al. 1978). HPg is synthesized as an 810-residue single polypeptide chain. A 19-residue leader peptide is excised during secretion, producing the mature form of HPg, which contains 791 amino acid residues (FORSGREN et al. 1987). The only other known processing steps involved in production of plasma HPg are N- and O-linked glycosylation (HAYES and CASTELLINO 1979a,b), and phosphorylation (WANG et al. 1997).

Plasmin (HPm) is formed from HPg as a result of activator-catalyzed cleavage of the Arg^{561}-Val^{562} peptide bond in the zymogen (ROBBINS et al. 1967). The resulting Glu^1-Pm contains a heavy chain of 561 amino acid residues, originating from the amino-terminus of HPg, disulfide-linked to a light chain of 230 amino acid residues. This latter chain, containing the carboxyl-terminus of HPg, is homologous to serine proteases, such as trypsin and elastase. The heavy and light polypeptide chains of HPm are covalently linked by two disulfide bonds. One such bond bridges Cys^{548} of the heavy chain and Cys^{666} of the light chain, and another links Cys^{558} of the heavy chain and Cys^{566} of the light chain. A second functionally-significant hydrolytic reaction, catalyzed by HPm, occurs in HPg between residues Lys^{77} and Lys^{78}, with additional assorted peptide bond cleavages within this 77-amino acid polypeptide, particularly at Lys^{62} and Arg^{68} (HORREVOETS et al. 1995). Hydrolysis of this peptide bond from the amino-terminus of Glu^1-Pg, and/or the amino-terminus of the heavy chain of Glu^1-Pm, provides Lys^{78}-Pg and Lys^{78}-Pm, respectively (VIOLAND and CASTELLINO 1976).

The catalytic triad of amino acids that define serine proteases is entirely present within HPm, and involves His^{603}, Asp^{646}, and Ser^{741}. The crystal structure of the Ser741Ala mutant of the catalytic domain has been solved at 2.0 Å resolution. It revealed a deformed catalytic triad and a blocked S1

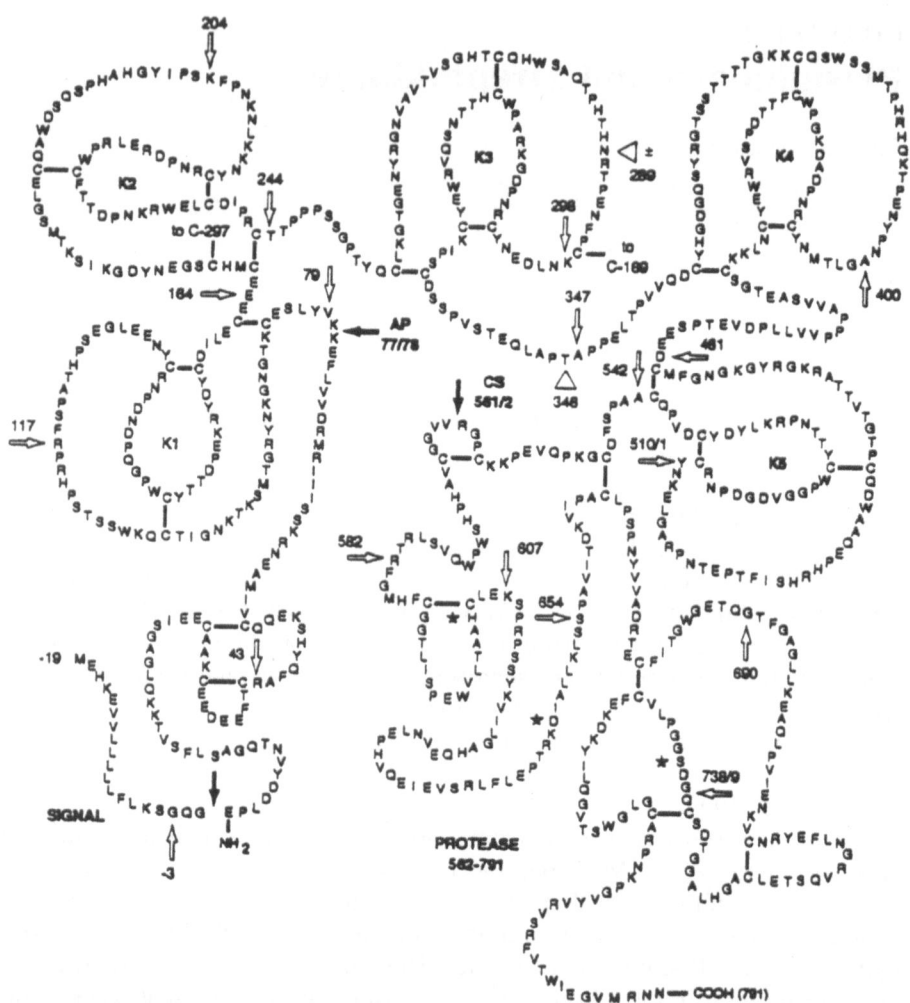

Fig. 1. Primary structure of human plasminogen. The cleavage sites of the signal peptide between residues –1 and 1, necessary for proper maturation of the plasma protein, and that between residues 77 and 78 required for release of the activation peptide (AP) resulting in the transformation of Glu1-plasminogen (Glu[1]-Pg) to Lys[78]-plasminogen (Lys[78]-Pg), respectively of Glu[1]-plasmin (Glu[1]-Pm) to Lys[78]-plasmin (Lys[78]-Pm), and that between residues 561 and 562 required for activation of HPg to HPm (CS), are indicated by *filled arrows*. Positions of introns in the gene sequence are represented by *unfilled arrows*. The locations of the N-linked oligosaccharide at sequence position 289 and the O-linked glycan at position 346 are indicated by triangles. Members of the catalytic triad of plasmin consisting of His[603], Asp[646], and Ser[741] are displayed (*). Disulfide bonds are depicted by *heavy bars*

specificity pocket (WANG et al. 2000). One consensus Asn-linked glycosylation sequence, Asn[289]-Arg-Thr, is present, which in human plasma contains biantennary, bisialylated glycan in approximately one-half of the HPg molecules (HAYES and CASTELLINO 1979a). HPg also contains one site containing O-linked glycan at Thr[346] (HAYES and CASTELLINO 1979b) that is occupied on all HPg molecules. Other minor O-linked glycosylation sites may exist at Ser^{249} (PIRIE-SHEPHERD et al. 1997) and Ser^{339} (HORTIN 1990). The two N-linked glycoforms of HPg can be resolved on Sepharose-lysine affinity chromatography columns (BROCKWAY and CASTELLINO 1972).

The heavy chain of HPm consists of a series of repeating homologous triple-disulfide-linked peptide regions, ca. 80 amino acid residues in length, termed kringles (SOTTRUP-JENSEN et al. 1978). Five such repeats are present within the latent HPm heavy chain, namely residues Cys^{84} to Cys^{162}, Cys^{166} to Cys^{243}, Cys^{256} to Cys^{333}, Cys^{358} to Cys^{435}, and Cys^{462} to Cys^{541}. These modules are present in several other clotting and fibrinolytic proteins, such as prothrombin (MAGNUSSON et al. 1975), factor XII (MCMULLEN and FUJIKAWA 1985), tPA (PENNICA et al. 1983), uPA (STEFFENS et al. 1982), and apolipoprotein(a) (MCLEAN et al. 1987). Some of these kringles are responsible for interactions with regulators of these proteins. Additionally, a major phosphorylation site was identified at residue Ser^{578} (WANG et al. 1997).

The functions of kringles in HPg are primarily involved with mediation of protein-protein interactions, such as those between fibrin(ogen) and HPg (SUENSON and THORSEN 1981), with binding of HPg to mammalian (MILES et al. 1988) and bacterial (BERGE and SJORBRING 1993; DÍCOSTA and BOYLE 1998; SJOBRING et al. 1998) cell surfaces, and with interactions of HPg and small molecule activation effectors, such as Cl^- and ω-amino acids (URANO et al. 1987a,b). Kringles in HPg that display interactions with effector molecules are kringles 1, 2, 4, and 5. $K3_{Pg}$ has not been directly implicated in the functioning of HPg. The interactions of HPg and HPm with fibrinogen and fibrin, as well as with cell surfaces, are inhibited by lysine and analogues. $K1_{Pg}$ and $K4_{Pg}$ contain the strongest of its ω-amino acid binding sites (MENHART et al. 1991, 1993; SEHL et al. 1990), while $K5_{Pg}$ exhibits somewhat weaker interactions with these ligands (MCCANCE et al. 1994; CHANG et al. 1998). $K2_{Pg}$ shows even weaker binding with this class of agents (MARTI et al. 1999).

Based upon site-directed mutagenesis studies with isolated kringle domains, the nature of the binding sites of kringles for lysine-type ligands has been identified. The anionic site that coordinates the amino group of the ligand consists of Asp residues in homologous locations to Asp^{54} and Asp^{56} (using $K4_{Pg}$ as the reference structure and numbering from the first Cys residue of the kringle) (HOOVER et al. 1993; MCCANCE et al. 1994; CHANG et al. 1998). The cationic donor sites of these ligands in HPg kringles are the residues homologous to Arg^{69} of $K4_{Pg}$ (HOOVER et al. 1993; MCCANCE et al. 1994; CHANG et al. 1998). In addition to these ion-pair contacts, there are a number of interactions between the methylene backbone of the ligand and certain hydrophobic residues in the kringle binding sites. The most relevant hydrophobic residues are those homologous to Trp^{60} and Trp^{70} of $K4_{Pg}$ (HOOVER et al. 1993; MCCANCE et al. 1994; CHANG et al. 1998). These results correlate well with predictions from models of the ligand/kringle complexes based on crystal structures of EACA/$K1_{Pg}$ (MATHEWS et al. 1996), EACA/$K4_{Pg}$ (WU et al. 1991), and $K5_{Pg}$ (CHANG et al. 1998).

B. Gene Organization of Human Plasminogen

The cDNA encoding HPg has been cloned and sequenced (Forsgren et al. 1987), as has its genomic DNA (Petersen et al. 1990). The gene for HPg, which has been mapped to chromosome 6q26–6q27 (Murray et al. 1987), encompasses 52.5 kb. The HPg coding sequence includes a 57 bp signal sequence and a total of 2373 nucleotides for the mature protein. Nineteen exons (Fig. 1), with a size range of 75–387 bp, and 18 introns, of type I, type II, and type O (Sharp 1981) are contained within the HPg gene. The first exon (amino acids –19 to –3) comprises most of the signal sequence of the protein, while exons II and III (amino acids –3 to 43 and 44 to 79, respectively) code for the amino-terminal peptide that is liberated consequent to activation of HPg to HPm. Each of the five kringles (amino acids 79–461) is encoded by two exons (exons IV–XIII). Exon XIV (amino acids 542–582) contains the Arg^{561}-Val^{562} peptide bond that is cleaved by HPg activators, the two cysteine residues (Cys^{548} and Cys^{558}) on the heavy chain which form disulfide bonds with the light chain, as well as Cys^{566} on the light chain which forms a disulfide bond with Cys^{558}. This latter bond appears to be essential for the substrate specificity of the Pg-molecule (Linde et al. 1998). Exons XV and XVI (amino acids 583–607 and 608–654) consist of the coding regions for the HPm active site residues His^{603} and Asp^{646}. A stretch of amino acids (655–689) is contained in exon XVII, within which exists the partner (Cys^{666}) of Cys^{548} that covalently stabilizes the two-chain structure of HPm. Exon XVIII is translated into a sequence of amino acids (690–738) which contains a disulfide loop of unknown functional significance. However, the additional Cys^{737} residue in this exon, which pairs with Cys^{765}, is located proximal to the active center residue Ser^{741}, and may be of importance for the specific functioning of the active site through its importance to the folding of this region of the molecule. Finally, this active site serine residue of HPm (Ser^{741}) is contained in exon XIX, beginning at Gly^{739}, and terminates at an undetermined location 3′ of the translation stop sequence.

The regulatory portions of the HPg gene, contained in the nucleotide sequences 5′ and 3′ of the coding region, have been partly identified (Petersen et al. 1990). Transcription control nucleotide sequences 5′ of the signal sequence-initiating methionine codon include TATAA promoter elements and the forward and reverse CCAAT proximal upstream promoter element boxes. Nucleotide sequences (CTGGGA), found in several acute phase reactant proteins, e.g., fibrinogen (Fowlkes et al. 1984), human haptoglobin (Maeda 1985), α_1-antitrypsin (Long et al. 1984), and transferrin (Adrian et al. 1986), are also located upstream of the Met^{-19} signal-initiation codon of the HPg. Two 5′-sequences, which appear to be recognition sites for hepatocyte-enriched HNF-1 and a general nuclear factor, AP-3, located within the nucleotide span 2.5 kb upstream of the translational initiation signal regulate transcriptional activation and liver specificity of the HPg gene (Meroni et al. 1996). These two transcription factors act synergistically in regulating the transcription of the HPg gene. Several other putative regulatory transcription sites, such as for transcription factors IL-6, AP-1, DBP, C/EBP, GATA, LF/A-1, and CREB, are present in the 5′ flanking region (reviewed by KIDA et al. 1997). Regulatory GC boxes were not observed in the 5′ region of the HPg gene. In the 3′ noncoding region of the cDNA, a primary mRNA polyadenylation recognition sequence, AATAAA, is found 46 bp upstream from the poly(A) tail, and a second polyadenylation recognition unit, CTTTG (BERGET 1984), is positioned 13 bp downstream from the above primary consensus sequence. In addition, another AATAAA primary polyadenylation signal was found downstream of the first, as well as a secondary polyadenylation recognition unit, CATTG, 43 bp downstream from this second recognition site. The YGTGTTYY consensus sequence that is also needed for efficient polyadenylation of mRNA (MCLAUCHLAN et al. 1985) is present 32 bp downstream of this latter alternative polyadenylation site in the HPg cDNA. This is close to the location of this sequence (within 24–30 bp of the AATAAA sequence) found in a variety of other mammalian gene structures (MCLAUCHLAN et al. 1985). No such sequence is found within this distance from the first polyadenylation signal.

C. The Activation of Human Plasminogen

I. General Considerations

The activation of HPg results in formation of the serine protease, HPm, an enzyme potentially involved in many physiological and pathological pathways, the most notable of which is clearance by proteolytic degradation of the fibrin clot. Regulation of the generation and localization of HPm occurs by nature of the presence of a variety of agents present in cells, on cell surfaces, and in the fluid phase. A summary of these events is provided in Fig. 2.

HPm is produced from HPg as a consequence of cleavage of a single peptide bond, Arg^{561}-Val^{562}, in the zymogen. This event is catalyzed by agents termed plasminogen activators, which normally have very limited specificity. One activator of this type, tissue-type plasminogen activator (tPA), preferentially activates the subpopulation of HPg that is bound to the clot, thereby directing HPm toward clot dissolution. While residing on the clot surface, HPm and tPA are resistant to reaction with, and subsequent inactivation by, their major inhibitors, α_2-plasmin inhibitor (α_2PI) and α_2-macroglobulin (α_2M), for HPm (COLLEN 1980), and plasminogen activator inhibitor type 1 (PAI-1) for tPA (KRUITHOF et al. 1993). After lysis of the thrombus, HPm and tPA are released from the clot and efficiently inactivated by the above circulating inhibitors.

Activation of HPg also occurs on cell surfaces, and is based in the localized presence of another type of plasminogen activator, two-chain urinary-type plasminogen activator (tc-uPA) (VASSALLI et al. 1985), on normal (BAJPAI and BAKER 1985; STOPPELLI et al. 1985; LAZARUS and JENSEN 1991) and neoplastic (STOPPELLI et al. 1986; BOYD et al. 1988; NIELSEN et al. 1988; DUGGAN et al. 1995) cells (see also Chap 5). Binding of tc-uPA to these types

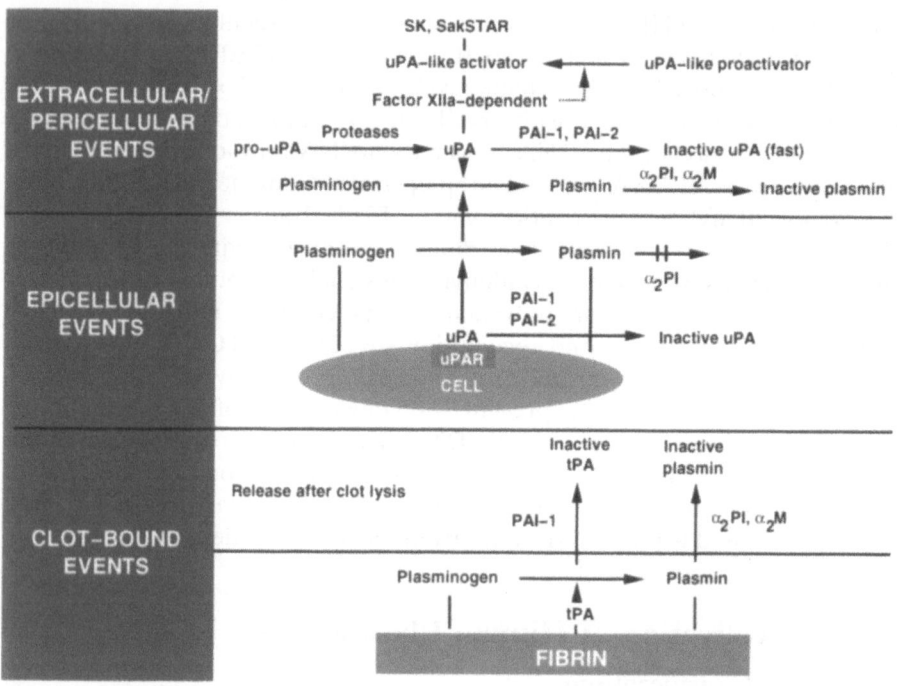

Fig. 2. The interactions involved in the activation of human plasminogen (HPg). Top: Solution phase activation of HPg readily occurs with two-chain urokinase (tc-uPA). uPA bound to its receptor (uPAR) can also catalyze this activation. The presence of circulating inhibitors of both uPA, viz., plasminogen activator inhibitor-1 (PAI-1) and plasminogen activator inhibitor-2 (PAI-2), and of plasmin (HPm), viz., α_2-plasmin inhibitor (α_2PI) and α_2-macroglobulin (α_2M), probably limit the physiological effectiveness of true extracellular activation pathways, but may allow effective pericellular activation of HPg to be important, wherein local concentrations of HPm may be sufficiently high to overcome local levels of inhibitors. Solution phase activation of HPg may also occur directly from proteases, i.e., kallikrein, that evolve from the contact phase of blood coagulation, or indirectly from these and other proteases by way of activation of sc-uPA (single-chain urokinase), or from activation of a uPA-related proactivator. The HPm formed in solution may be responsible for various proteolytic events. Exogenously added streptokinase (SK) or staphylokinase (SakSTAR), which are used in thrombolytic therapy, also activate HPg through indirect mechanisms by first complexing with HPg and HPm to form the actual HPg activator. In these cases, SK or SakSTAR is added in sufficient amounts to allow levels of HPm to form that overcome the circulating concentration of a$_2$PI. Middle: HPg is also activated on various cell surfaces by uPA bound to its cellular receptor (uPAR). The HPm thus formed is protected from inactivation by a$_2$PI, and probably functions directly or indirectly in extracellular matrix degradation, perhaps locally after dissociation from the cell surface. The cell-bound uPA appears to be susceptible to inactivation by PAI-1 and PAI-2, but at a some what slower rate than free uPA. Bottom: Activation of HPg also occurs on the surface of the blood clot. Here, the clot-bound activator, tissue-type plasminogen activator (tPA), activates fibrin-bound HPg. The HPm formed on the clot surface is resistant to inhibition by a$_2$PI, as is clot-bound tPA toward PAI-1. After dissolution of the clot and release of HPm and tPA, these proteases are inhibited by their fast-acting circulating inhibitors

of cells occurs via a specific cellular receptor, uPAR (BEHRENDT et al. 1990), which has been cloned and its nucleotide sequence determined (CASEY et al. 1994; WANG et al. 1995). That this interaction results in functional consequences for the cells is evidenced by the fact that binding of uPA to uPAR initiates a transmembrane signal, as a result of phosphorylation on a Try residue of a 38kDa protein in ovarian cancer cells (DUMLER et al. 1993), or serine phosphorylation of 47kDa and 55kDa proteins in human epithelial cells (BUSSO et al. 1994). These activities of phosphotyrosine kinase and/or protein kinase C lead to a signal transduction pathway, which activates nuclear transcription factors (DUMLER et al. 1994; reviewed in DEAR and MEDCALF 1998). Cells enriched in surface-bound tc-uPA promote HPm formation from the HPg that is also bound to normal and carcinoma cell surfaces (HAJJAR et al. 1986; PLOW et al. 1986; MILES and PLOW 1987; MILES et al. 1988), perhaps through the HPg binding proteins, α-enolase (HAMANOUE et al. 1994; REDLITZ et al. 1995), actin (DUDANI and GANZ 1997), annexin II (reviewed in HAJJAR and KRISHNAN 1999), and/or a cytokeratin 8-like protein (HEMBROUGH et al. 1995, 1996). α-Enolase was also found to be a Pg binding protein on the surface of pathogenic streptococci (PANCHOLI and FISCHETTI 1998). This epicellular HPm is also resistant to inactivation by α_2PI (MILES and PLOW 1988), and receptor-bound tc-uPA is slightly more resistant to PAI-1 and PAI-2 than is solution phase tc-uPA (ELLIS et al. 1990). These cell-associated fibrinolysis mechanisms provide a basis for the proposed roles of HPg activators in generating a protease employed for normal processes involving cell migration in tissue remodeling. These include macrophage invasion in inflammation (UNKELESS et al. 1974; WOHLWEND et al. 1987), mammary cell involution after lactation (OSSOWSKI et al. 1979), breakdown of the follicular wall for ovulation (REICH et al. 1985), trophoblast invasion into the endometrium during embryogenesis (STRICKLAND et al. 1976; SAPPINO et al. 1989), angiogenesis (GROSS et al. 1983), and keratinocyte accumulation after wound healing (MORIOKA et al. 1991). Furthermore, tc-uPA and uPAR, likely through formation of epicellular HPm, have been implicated in pathological processes of cell migration that are involved in tumor cell growth and invasion of surrounding tissue, perhaps leading to metastasis (OSSOWSKI et al. 1979; HEARING et al. 1988; DE VRIES et al. 1994; FAZIOLI and BLASI 1994; STAHL and MUELLER 1994). These concepts are supported by the ability of HPm to degrade extracellular matrix proteins, such as proteoglycans (EDMONDS-ALT et al. 1980), fibronectin (JILEK 1977), laminin (SCHLECHTE et al. 1989, 1990; GOLDFINGER et al. 1998), fibulin (SASAKI et al. 1996), tenascin-C (GUNDERSSEN et al. 1997), perlecan (WHITELOCK et al. 1996), and type IV collagen (MACKAY et al. 1990), either directly, or indirectly through activation of the metalloprotease zymogens, stromelysin and procollagenase (STRICKLIN et al. 1977; HE et al. 1989). As a result of degradation of the extracellular matrix, cell migration into surrounding areas is enhanced.

Mechanisms for solution-phase activation of HPg also exist, providing HPm that may play roles as an extracellular or pericellular protease. However, the physiological relevance of extracellular proteolytic mechanisms involving HPm are uncertain because of the presence of relatively high levels of α_2PI in plasma, which should rapidly inactivate circulating HPm. This inhibition may be attenuated by localized high concentrations of HPm, which may overcome local levels of α_2PI, and/or by compartmentalization in areas of low inhibitor concentration. When present, solution-phase HPm may participate in such diverse processes as zymogen conversions in the classic (AGOSTONI et al. 1994) and alternate complement pathway (BRADE et al. 1974), in the contact phase of blood coagulation (COCHRANE et al. 1974), in proinsulin to insulin conversion (VIRGI et al. 1980), in bradykinin generation from kininogen (HABAL et al. 1976), and in proteolytic destruction of other plasma proteins (MIRSKY et al. 1959; JANEWAY et al. 1968; PIZZO et al. 1972; OMAR and MANN 1987; FEDERICI et al. 1993; OHKURA et al. 1998). Solution-phase activation of HPg is catalyzed by soluble tc-uPA (KLUFT et al. 1981) and tc-uPA bound to uPAR (ELLIS et al. 1991). Additionally, tPA released from the vasculature by stimuli such as venous occlusion also provides a low level of activation of plasma HPg (RIJKEN et al. 1980). In vitro, HPg is also directly activated to HPm by several plasma proteases involved in the contact activation of the clotting cascade, among which are factor XIa (MANDLE and KAPLAN 1979), factor XIIa (GOLDSMITH et al. 1978; SCHOUSBOE et al. 1999), and kallikrein (COLMAN 1980). Indirect activation of HPg by plasma proteases, e.g., HPm (ELLIS et al. 1987; URANO et al. 1988a) and kallikrein (ICHINOSE et al. 1986; HAUERT et al. 1989), by means of their abilities to convert sc-uPA to tc-uPA can also occur. Finally, a factor XIIa-dependent mechanism for HPg activation might exist (OGSTON et al. 1969), which is based upon the kallikrein-catalyzed activation of a HPg proactivator that shares some homology with uPA (BINNEMA et al. 1990).

Products released from HPg as a result of activation, and/or of limited proteolysis of HPg or HPm, may also play functional roles in other processes. As one example, proteases such as elastase are capable of in vitro proteolysis of very specific peptide bond that liberate intact kringle domains or combinations of kringles (SOTTRUP-JENSEN et al. 1978). Of great recent interest is the discovery that angiostatin, an inhibitor of angiogenesis, is structurally related to the kringle 1–3, or 1–4 domain region of HPg, and that this fragment of HPg, but not the intact protein, blocked neovascularization in a Lewis lung carcinoma model (O'REILLY et al. 1994) and inhibited capillary endothelial cell proliferation (CAO et al. 1996, 1999). Recent studies have shown that angiostatin can be generated, in vitro, by matrix metalloproteases (MMP), viz., MMP-3 (LIJNEN et al. 1998), MMP-7 and MMP-9 (PATTERSON et al. 1997), and MMP-12 (DONG et al. 1997); serine proteases, e.g., urokinase, in the presence of free sulfhydryl donors (GATELY et al. 1997); and protein disulfide isomerase combined with thioredoxin (STATHAKIS et al. 1997). The in vivo mechanism(s) and the relative contribution of host and tumor derived factors in angiostatin generation is still unknown. Speculation on its mode of activity in inhibiting angiogenesis involves mechanisms associated with increased endothelial cell apoptosis and activation of focal adhesion kinase (CLAESSON-

WELSH et al. 1998), as well as arrest of mitosis (GRISCELLI et al. 1998). Other kringle structures have been shown to exhibit angiostatin-like activity, such as kringle 2 of prothrombin (LEE et al. 1998). Amino acid sequence alignment of this kringle with $K1_{Pg}$ indicates that each kringle has six conserved cysteine residues. Interestingly, it has been shown that isolated $K5_{Pg}$ also exerts a strong angiostatin-like activity, to the extent that this property is reflected in its ability to inhibit cell growth and/or cell migration (CAO et al. 1997; JI et al. 1998). Thus, HPg and other kringle-containing proteins that share structural similarities with the functionally active site(s) of angiostatin may act as substrates for the proteolytic generation of functionally relevant fragments exhibiting potent angiogenic inhibitor activities.

II. Activation of Human Plasminogen by Two-chain Urokinase and Tissue-type Plasminogen Activator

1. Mechanism of Activation of Human Plasminogen

The obligatory step in conversion of HPg to HPm is the cleavage of the $Arg^{561}Val^{562}$ peptide bond in the zymogen, which alone is sufficient to generate active HPm (ROBBINS et al. 1967). All HPg activators catalyze this peptide bond cleavage. Both positive and negative regulators of this activation exist, which play important roles in activation of this zymogen-enzyme system.

The overall scheme for conversion of circulating Glu^1-Pg to the final product, Lys^{78}-Pm, is shown in Scheme 1 (VIOLAND and CASTELLINO 1976). Here, circulating Glu^1-Pg is activated to Glu^1-Pm by activator (A)-catalyzed cleavage of the Arg^{561}-Val^{562} peptide bond in Glu^1-Pg. This step most likely provides the first molecules of plasmin (Pm). This reaction is very slow in the presence of Cl⁻, where Glu^1-Pg exists in a compact (T-state) and a poorly activatable form (VIOLAND et al. 1975, 1978; URANO et al. 1987a,b, 1988b). However, ω-amino acids, such as EACA, enhance the rate of this reaction when Cl⁻ is present because of the "loosening" (R-state) of the Cl⁻-dependent conformation of Glu^1-Pg (VIOLAND et al. 1978; URANO et al. 1987a,b), leading

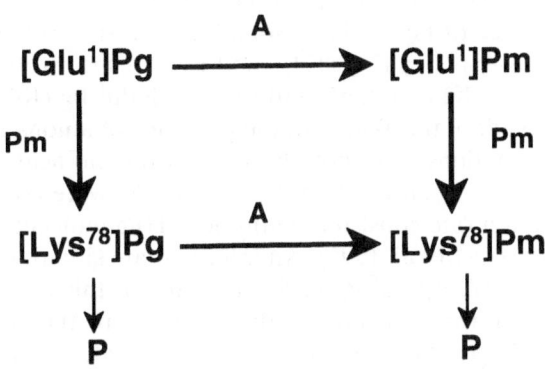

to more facile access of the activator to peptide bond cleavage at the required locale. The initial Glu^1-Pm formed catalyzes cleavage of a 77-amino acid peptide from the amino-terminus of the Glu^1-Pm heavy chain, providing Lys^{78}-Pm, or from the amino terminus of Glu^1-Pg (VIOLAND and CASTELLINO 1976; GONZALEZ-GRONOW et al. 1977), yielding Lys^{78}-Pg. The activator cannot catalyze cleavage of this latter peptide bond at a rate sufficiently significant to be involved in the activation mechanism (VIOLAND and CASTELLINO 1976; GONZALEZ-GRONOW et al. 1977).

2. Positive and Negative Activation Effectors

The activation rate of Lys^{78}-Pg is substantially faster than that of Glu^1-Pg (CLAEYS and VERMYLEN 1974), regardless of the presence of Cl^- and/or EACA, and its activation rate is approximately the same as that of Glu^1-Pg when this latter protein resides in its R-state. This latter conformational state only occurs in the absence of anions, in the presence of low concentrations of weakly bound anions, such as acetate, or when saturating levels of ω-amino acids are present (which reverses the anion-induced T-state of Glu^1-Pg). Thus, Cl^- is a negative effector of Glu^1-Pg activation and ω-amino acid analogues are positive effectors of Glu^1-Pg activation in the presence of Cl^-, or other Glu^1-Pg-bound monovalent anions (URANO et al. 1987a,b, 1988b). From these considerations, it is clear that these regulatory events in HPg activation are dependent upon the presence of the 77-residue activation peptide in Glu^1-Pg which, when cleaved from the HPg molecule, results in Lys^{78}-Pg. This form of HPg does not have the ability to undergo transformation by anions into the relatively poorly activatable T-state. Thus, Lys^{78}-Pg loses its ability to be regulated by these positive and negative conformation and activation effectors.

The molecular basis of the reciprocal effects of anions and EACA on Glu^1-Pg structure and activation has been investigated further by generation of r-Glu^1-Pg variants in which the EACA binding site in individual kringles was greatly diminished by alteration of a critical Asp residue in each to Asn (HOOVER et al. 1993; MCCANCE et al. 1994). Referring to the Glu^1-Pg amino acid sequence of Fig. 1, these homologous Asp residues are located at Asp^{139}, Asp^{413}, and Asp^{518} in $K1_{Pg}$, $K4_{Pg}$, and $K5_{Pg}$, respectively. The mutants, r-$[Asp^{139}Asn]Glu^1$-Pg ($K1_{Pg}$ mutant) and r-$[Asp^{413}]Glu^1$-Pg ($K4_{Pg}$ mutant), were no longer able to adopt the T-state in the presence of anions, such as Cl^-, and, thus, their conformations were not altered by ω-amino acids, such as EACA (MCCANCE and CASTELLINO 1995). These variants were concomitantly activated by tc-uPA at a rate consistent with those HPg molecules that existed in the R-state (MENHART et al. 1995). An intermediate situation was found with the mutant r-$[Asp^{518}Asn]Glu^1$-Pg ($K5_{Pg}$ mutant). In this case, Cl^- placed this variant in a conformation intermediate between the T- and R-states (MCCANCE and CASTELLINO 1995). The activation rate of this zymogen in the presence of Cl⁻, and the effect of EACA on this rate, were also situated between those of HPg in the T-and R-states (MENHART et al. 1995).

These studies point to complex structure-activation relationships in Glu^1-Pg. We propose that, in the presence of Cl⁻, Lys side chains in the activation peptide, and perhaps in other regions of HPg, interact with the ω-amino acid binding sites of $K1_{Pg}$, $K4_{Pg}$, and, to a lesser extent, $K5_{Pg}$. Several activation peptide candidate residues have been identified as possible donors to the lysine binding sites of the kringle domains that assist in maintaining the T conformation (HORREVOETS et al. 1995). These interactions lead to conformational transformations that result in formation of the poorly activated compact T-conformation of Glu^1-Pg. Addition of ω-amino acids displace these intramolecular interactions and leads to the more readily activatable, conformationally-expanded, and more internally flexible R-state of Glu^1-Pg. Since Lys^{78}-Pg does not contain the activation peptide, these same intramolecular protein interactions are not favorable. Thus, this latter form of HPg cannot transform to the T-state, always exists in the R-state, and is activated at a high rate by tc-uPA.

Fibrin, and to a lesser degree fibrinogen, is also a potent positive effector of HPg activation with tPA, due to the ability of tPA, HPg, and HPm to interact and localize at the clot surface, as discussed above (see also Chap. 3). In fact, the basis of the efficacy of interventional clot-lysis therapy with tPA in the early stages of clot-related myocardial infarction is due to efficient activation of HPg by tPA in the presence of fibrin. Solution-phase tPA catalyzes HPg activation rather inefficiently compared with tc-uPA. However, when fibrin is present, a large increase in the rate of tPA-catalyzed HPg activation occurs, primarily as a result of a large decrease in the Km of the activation reaction (HOYLAERTS et al. 1982). Fibrin does not significantly stimulate the activation rate of the tc-uPA-catalyzed activation of HPg. In cases of fibrin-stimulation of HPg activation by tPA, virgin fibrin is not as effective an activation potentiator as is fibrin that is partially digested by HPm. Since these latter forms of fibrin contain COOH-terminal Lys and Arg residues, Pg and tPA would be more tightly bound to the clot when these residues are liberated than is the case in undigested fibrin. That these considerations may have more general relevance is evidenced by the fact the HPm binds very efficiently to its major plasma inhibitor, α_2PI, a protein which contains a COOH-terminal Lys residue, but this inhibition is much less effective after treatment of the inhibitor with carboxypeptidase B (HORTIN et al. 1989).

III. Activation of Human Plasminogen by Single-chain Urokinase

The activation of HPg by sc-uPA represents a special case. Debates have occurred as to whether sc-uPA is a zymogen with a low level (<1%) of HPg activator activity (PANNELL and GUREWICH 1987), or whether it is a true inactive zymogen, with some initial HPg activator activity observed as a result of low levels of tc-uPA contaminant, and a consequent feedback effect of the

HPm produced, converting sc-uPA to the potent HPg activator, tc-uPA (Petersen et al. 1988; Urano et al. 1988a; see also Chap. 4). An attempt to resolve this issue was forwarded with construction of a mutant of sc-uPA that was not capable of being converted into its two-chain form, and the finding that this mutant possessed a low level (ca. 1%) of the HPg activator activity of tc-uPA. Similarly, examination of the activation rate of an HPg mutant in which the active site Ser^{740} was altered (to Ala), the activation of which should not lead to feedback conversion of sc-uPA to tc-uPA by the generated inactive [Ser^{740}Ala]r-Pm, showed that the mutant HPg was activated (measured by gel monitoring of the cleavage of the Arg^{561}-Val^{562} peptide bond in the mutant HPg) at a rate consistent with a low level of inherent activity of sc-uPA, viz., <1% of the rate of activation of [Ser^{740}Ala]r-Pm by tc-uPA (Lijnen et al. 1990).

Thus, the overall mechanism for the activation of Glu^{1}-Pg by sc-uPA is suggested to involve: (a) a low rate of conversion of Glu^{1}-Pg to Glu^{1}-Pm by the inherent sc-uPA activity (or by the presence of a small amount of tc-uPA if present); (b) feedback activation by the initial Glu^{1}-Pm of sc-uPA to tc-uPA, a step that would accelerate this activation; and (c) similar feedback conversion by Glu^{1}-Pm of Glu^{1}-Pg to the more readily activatable Lys^{78}-Pg, which also results in a much higher rate of activation of the newly generated Lys^{78}-Pg. During this process Glu^{1}-Pm is also converted to Lys^{78}-Pm. The fibrin specificity observed for this activation might involve a fibrin-stimulated conversion of sc-uPA to tc-uPA.

IV. Activation of Human Plasminogen by Streptokinase

1. Structural and Activation Features of Streptokinases

SK, a protein secreted into the growth medium in cultures of several strains of hemolytic *streptococci*, is able to activate HPg, as well as plasminogens from some other mammalian plasmas. Interestingly, SK does not display hydrolytic activity. Therefore, the most intriguing biochemical questions surrounding SK involve first the nature of the mechanism whereby a protein without the inherent required activity nonetheless activates HPg, and second the reasons for its species selectivity. Amino acid sequence investigations of a group C streptokinase, isolated from a human host (HSKc), establishes that this protein consists of 415 amino acid residues contained in a single polypeptide chain (Fig. 3). No disulfide bonds are present. Employing amino acid homology relationships, it has been proposed that SK evolved as two independent domains, each related to serine proteases, such as bovine trypsin and *Streptomyces griseus* protease, but containing critical amino acid substitutions, especially at the active center His of the catalytic triad, that render the protein enzymatically ineffectual (Jackson and Tang 1982). NMR investigations have led to the conclusion that three autonomous domains exist in HSKc. These modules are predicted to encompass amino acid residues, 1–146, 147–287, and 293–380 (Parrado et al. 1996).

```
      -26                  -20                       -10
HSKc:  M K N Y L S M G M F A L L F A L T F G T V N S V Q A
HSKa:  M K N Y L S M G M F A L L F A L T F G T V K P V Q A
HSKg:  M K N Y L S F G M F A L L F A L T F G T V N S V Q A

       +1                      10                      20
HSKc:  I A G P E W L L D R P S V N N S Q L V V S V A G T
HSKc:  I A G P E W L L D R P S V N N S Q L V V S V A G T
HSKa:  I A G Y E W L L D R P S V N N S Q L V V S V A G T
HSKg   I A G P E W L L D R P S V N N S Q L V V S V A G T
ESKc:                          N N Y A K P I Y K V G T

                       30                40                      50
HSKc:  V E G T N Q D I S L K F F E I D L T S R P A H G G
HSKc:  V E G T N Q D I S L K F F E I D L T S R P A H G G
HSKa;  V E G T N Q E I S L K F F E I D L T S R P A H G G
HSKg:  V E G T N Q D I S L K F F E I D L T S R P A H G G
ESKc:  Y Q P T D D T V F N S K D Y Q D T ? G L Y L T

                       60                      70
HSKc:  K T E Q G L S P K S K P F A T D S G A M S H K L E
HSKc:  K T E Q G L S P K S K P F A T D S G A M S H K L E
HSKa:  K T E Q G L S P K S K P F A T D K G A M S H K L E
HSKg:  K T E Q G L S P K S K L F A T D S G A M P H K L E

                       80                90                      100
HSKc:  K A D L L K A I Q E Q L I A N V H S N D D Y F E V
HSKc:  K A D L L K A I Q E Q L I A N V H S N D D Y F E V
HSKa:  K A D L L K A I Q E Q L I A N V H S N D G Y F E V
HSKg:  K A D L L K A I Q E Q L I A N V H S N D D Y F E V

                       110               120
HSKc:  I D F A S D A T I T D R N G K V Y F A D K D G S V
HSKc:  I D F A S D A T I T D R N G K V Y F A D K D G S V
HSKa:  I D F A S D A T I T D R N G K V Y F A D K D D S V
HSKg;  I D F A S D A T I T D R N G G V Y F A D K D G S V

                       130               140               150
HSKc:  T L P T Q P V Q E F L L S G H V R V R P Y K E K P
HSKc:  T L P T Q P V Q E F L L S G H V R V R P Y K E K P
HSKa:  T L P T Q P V Q E F L L S G H V R V K P Y Q P K A
HSKg:  T L P I Q P V Q E F L L K G H V R V R P Y K E K P

                       160               170
HSKc:  I Q N Q A K S V D V E Y T V Q F T P L N P D D D F
HSKc:  I Q N Q A K S V D V E Y T V Q F T P LD N P D D D F
HSKa:  V H N S A E R V N V N Y E V S F V S E T G D L D F
HSKg:  V Q N Q A K S V D V E Y T V Q F T P L N P D D D F
```

Fig. 3. Amino acid sequences of group A (HSKa), group C (HSKc), and group G (HSKg) streptokinases, isolated from human hosts, and a group C SK (ESKc) isolated from an equine host (ESKc). The amino acid sequence of a group C SK, obtained directly from the protein (HSKc) (JACKSON and TANG 1982), is compared to the amino acid sequence as deduced from the sequence of the SK gene, obtained from this same *streptococcol* strain (HSKc) (MALKE et al. 1985). These sequences are compared to those of group A (HSKa) and group G (HSKg) streptokinases, deduced from the sequence of the gene encoding this protein (HUANG et al. 1989), and of ESKc, obtained from protein sequencing. Amino acid residues present in HSKa, HSKg, and ESKc, as well as those determined directly on the protein, that differ from those deduced from the gene sequence of HSKc are presented in *bold and underlined type*

The gene for an HSKc from *Streptococcus equisimilis* H46 A cells, which is the same strain of bacteria used in past studies to purify the protein, has been cloned and sequenced (MALKE and FERRETTI 1984; MALKE et al. 1985). This gene consists of a 26-residue leader sequence and a mature protein containing 414 amino acid residues. The single amino acid extension found in the

```
                180                              190                          200
HSKc:  R  P  G  L  K  D  T  K  L  L  K  T  L  A  I  G  D  T  I  T  S  Q  E  L  L
HSKc:  R  P  G  L  K  D_L T  K  L  L  K  T  L  A  I  G  D  T  I  T  S  Q  E  L  L
HSKa:  T  P  L  L  R  N  Q  Y  H  L  T  T  L  A  V  G  D  S  L  S  S  Q  E  L  A
HSKg:  R  P  A  L  K  D  T  K  L  L  K  T  L  A  I  G  D  T  I  T  S  Q  E  L  L

                                 210                          220
HSKc:  A  Q  A  Q  S  I  L  N  K  N  H  P  G  Y  I  I  Y  E  R  D  S  S  I  V  T
HSKc:  A  Q  A  Q  S  I  L  N  K  N  H  P  G  Y  T  I  Y  E  R  D  S  S  I  V  T
HSKa:  A  I  A  Q  F  I  L  S  K  K  H  P  D  Y  I  I  T  K  R  D  S  S  I  V  T
HSKg:  A  Q  A  Q  S  I  L  N  K  N  H  P  G  Y  T  I  Y  E  R  D  S  S  I  V  T

                230                              240                          250
HSKc:  H  D  N  D  I  F  R  T  I  L  P  M  D  Q  E  F  T  Y  R  V  K  N  R  E  Q
HSKc:  H  D  N  D  I  F  R  T  I  L  P  M  D  Q  E  F  T  Y  R  V  K  N  R  E  Q
HSKa:  H  D  N  D  I  F  R  T  I  L  P  M  D  Q  E  F  T  Y  H  I  K  D  R  E  Q
HSKg:  H  D  N  D  I  F  R  T  I  L  P  M  D  Q  E  F  T  Y  H  V  K  N  R  E  Q

                                 260                          270
HSKc:  A  Y  R  I  N  K  K  S  G  L  N  E  E  I  N  N  T  D  L  I  S  *  E  K  Y
HSKc:  A  Y  R  I  N  K  K  S  G  L  N  E  E  I  N  N  T  D  L  I  S  L  E  Y  K
HSKa:  A  Y  K  A  N  S  K  T  G  I  E  E  K  T  N  N  T  D  L  I  S  *  E  K  Y
HSKg:  A  Y  R  I  N  K  K  S  G  L  N  E  E  I  N  N  T  D  L  I  S  *  E  K  Y

                280                              290                          300
HSKc:  Y  V  L  K  K  G  E  K  P  Y  D  P  F  D  R  S  H  L  K  L  F  T  I  K  Y
HSKc:  Y  V  L  K  K  G  E  K  P  Y  D  P  F  D  R  S  H  L  K  L  F  T  I  K  Y
HSKa:  Y  V  L  K  K  G  E  K  P  Y  D  P  F  D  R  S  H  L  K  L  F  T  I  N  Y
HSKg:  Y  V  L  K  K  G  E  K  P  Y  D  P  F  D  R  S  H  L  K  L  F  T  I  K  Y

                                 310                          320
HSKc:  V  D  V  D  T  N  E  L  L  K  S  E  Q  L  L  T  A  S  E  R  N  L  D  F  R
HSKc:  V  D  V  D  T  N  E  L  L  K  S  E  Q  L  L  T  A  S  E  R  N  L  D  F  R
HSKa:  V  D  V  N  T  N  K  L  L  K  S  E  Q  L  L  T  A  S  E  R  N  L  D  F  R
HSKg:  V  D  V  N  T  N  E  L  L  K  S  E  Q  L  L  T  A  S  E  R  N  L  D  F  R

                330                              340                          350
HSKc:  D  L  Y  D  P  R  D  K  A  K  L  L  Y  N  N  L  D  A  F  G  I  M  D  Y  T
HSKc:  D  L  Y  D  P  R  D  K  A  K  L  L  Y  N  N  L  D  A  F  G  I  M  D  Y  T
HSKa:  D  L  Y  D  P  R  D  K  A  K  L  L  Y  N  N  L  D  A  F  G  I  M  D  Y  T
HSKg:  D  L  Y  D  P  R  D  K  A  K  L  L  Y  N  N  L  D  I  F  G  I  M  D  Y  T

                                 360                          370
HSKc:  L  T  G  K  V  E  D  N  H  D  D  T  N  R  I  I  T  V  Y  M  G  K  R  P  E
HSKc:  L  T  G  K  V  E  D  N  H  D  D  T  N  R  I  I  T  V  Y  M  G  K  R  P  E
HSKa:  L  T  G  K  V  E  D  N  H  D  D  T  N  R  I  I  T  V  Y  M  G  K  R  P  E
HSKg:  L  T  G  K  V  E  D  N  H  D  D  T  N  R  I  I  T  V  Y  M  G  K  R  P  E

                380                              390                          400
HSKc:  G  E  N  A  S  Y  H  L  A  Y  D  K  D  R  Y  T  E  E  E  R  E  V  Y  S  Y
HSKc:  G  E  N  A  S  Y  H  L  A  Y  D  K  D  R  Y  T  E  E  E  R  E  V  Y  S  Y
HSKa:  G  E  N  A  S  Y  H  L  A  Y  D  K  D  R  Y  T  E  E  E  R  E  V  Y  S  Y
HSKg:  G  E  N  A  S  Y  H  L  A  Y  D  K  D  R  Y  T  E  E  E  R  E  V  Y  S  Y

                                 410
HSKc:  L  R  Y  T  G  T  P  I  P  D  N  P  N  D  K
HSKc:  L  R  Y  T  G  T  P  I  P  D  N  P  D  D  K
HSKa:  L  R  Y  T  G  T  P  I  P  D  N  P  K  D  K
HSKg:  L  R  Y  T  G  T  P  I  P  D  N  P  N  D  K
```

Fig. 3. *Continued*

protein sequence is due to the identification of a Leu^{272} that was not observed in the gene (Fig. 3). Double Leu/Asp residues found at positions 169 and 181 in the protein sequence have been resolved as Leu and Asp, respectively, in the gene sequence. Other differences found in the protein and gene sequences of HSKc include a $Tyr^{274}Lys^{275}$ in the protein for a $Lys^{274}Tyr^{275}$ in the gene and an Asp^{413} in the protein for an Asn^{413} in the gene. However, these apparent differences may not be real, and could be accounted for by an inadvertent error in reporting the former, since the two amino acids are simply inverted, and a deamidation in the latter. The other difference, a Thr^{215} in the protein in place of an Ile^{215} deduced from the gene sequence, is seemingly a disparity between the two sequences.

In order to investigate possible strain heterogeneities in the SK produced, SK genes from other streptococcol strains have been cloned and sequenced. In one example, the nucleotide sequence of a human host-derived group A streptokinase (HSKa), from *Streptococcus pyogenes* NZ131 M type 49 cells (HUANG et al. 1989), has been reported and its amino acid sequence deduced (Fig. 3). As was the case for HSKc, this gene also encoded 440 amino acids, including a 26-amino acid leader sequence. A total of 62 amino acids were found to be different in the deduced amino acid sequences of HSKa and HSKc, with only 17 of these classified as non-conserved alterations (HUANG et al. 1989).

Additionally, the nucleotide sequence of a human-derived streptokinase from a group G *streptococcus* (HSKg), strain 19908, has been reported (WALTER et al. 1989), and its amino acid sequence deduced (Fig. 3). This gene is the same size as the other HSK genes and possesses few amino acid sequence differences from HSKc. This comparative sequence analysis is useful in that any sequence differences found among the various HSK proteins cannot be responsible for the functioning of HSK in HPg activation, unless compensatory multiple changes occur, since these proteins are fully active. However, these same differences probably account for the immunological variation found in HSK among different strains of *streptococci* (DILLON and WANNAMAKER 1965; GERLACH and KOHLER 1977).

A high degree of conservation and linkage relationships of the SK gene have been discovered in a variety of pathogenic *streptococci* (FRANK et al. 1995). Using DNA probes from the HSKc gene cloned from *Streptococcus equisimilis* H46 A cells, it was found that the HSKa gene content and gene order (dexB-abc-Irp-skc-orfl-re) are preserved in *Streptococcus pyogenes* strains A374, NA131, and SF130/13, as well as in the HSKg gene from strain G19908. A similar conclusion was made when this analysis was extended to the equine (E) SKc gene from *Streptococcus equisimilis*, strain 87–542-W. Such similarities were not found when genes from strains of bovine (B)-host derived *Streptococcus uberis* were similarly probed. These latter strains did not produce an SK-like HPg activator.

2. Mechanism of Activation of Human Plasminogen by Human Host-derived Streptokinase

The mechanism of activation of HPg by HSKc involves two major steps: (1) formation of a plasminogen activator, and (2) activation of plasminogen. The sequence of reactions in the first step, summarized below, result in generation of protein-protein complexes that serve as HPg activators.

$$SK + HPg \longleftrightarrow SK\text{-}HPg \longrightarrow SK\text{-}HPg^* \longrightarrow SK\text{-}HPg' \longrightarrow SK^\wedge\text{-}HPm$$
$$\uparrow$$
$$SK + HPm$$

Initially, a stoichiometric complex of SK and HPg forms (SK-HPg), within which a conformational rearrangement of the HPg takes place allowing an active site to form in the HPg moiety of the complex (SK-HPg*) (SCHICK and CASTELLINO 1974). This active site in HPg* is sensitive to inhibition by Cl⁻ and stimulation by fibrin(ogen) (CHIBBER et al. 1985, 1986; CHIBBER and CASTELLINO 1986). With time, another complex (SK-HPg') forms, which possesses a diminished ability to be regulated by Cl⁻ and fibrin(ogen), finally yielding the most stable of these complexes, SK^-HPm (CHIBBER et al. 1986). In this latter case, SK undergoes degradation at both its amino-terminal (BROCKWAY and CASTELLINO 1974) and carboxyl-terminal regions (SIEFRING and CASTELLINO 1976), leading to several forms of SK, globally designated here as SK^. This latter complex is also formed from SK and HPm.

In the second step of the activation, the SK-HPg*, SK-HPg', and SK^-HPm complexes become catalytic activators of remaining HPg, as shown with SK-HPg' being a more potent activator than SK-HPm (DAVIDSON et al. 1990).

$$
\begin{array}{c}
SK\text{-}HPg^* \\
SK\text{-}HPg' \\
HPg \xrightarrow{\;\;SK^\wedge\text{-}HPm\;\;} HPm
\end{array}
$$

Detailed kinetic characteristics of SK-HPg* have not been performed because of the inherent instability of this complex. HPm, alone, cannot directly catalyze activation of HPg (VIOLAND et al. 1975; GONZALEZ-GRONOW et al. 1977), and none of the activator complexes are able to remove the 77-residue activation peptide from the amino terminus of Glu^1-Pg. At this second stage of the activation, the mechanism is essentially the same as that described for tcu-PA and tPA, above in terms of the peptide bonds cleaved in HPg, and the enzyme species that catalyze their cleavages.

The basis for the species selectivity of HSKc in activation of mammalian plasminogens (WULF and MERTZ 1969; McKEE et al. 1971; SUMMARIA et al. 1974) is as yet unclear. In early investigations with rabbit plasminogen, a zymogen that is weakly activated by HSKc (SODETZ et al. 1972), we found that the HSKc in the complex was rapidly degraded to inactive forms (SCHICK and CASTELLINO 1973). Thus, while a complex formed, the HSKc moiety within the complex was unstable. These are the only biochemical studies to date that shed some light on this question. Recent investigations with SK isolated from streptococcal strains obtained from other mammalian hosts extend knowledge somewhat further on this subject. A strain (*Streptococcus equisimilis*, group C, strain 87–542-W) of equine *streptococcus*, produced an ESKc (MCCOY et al. 1991) that was purified and partially cloned (NOWICKI et al. 1994). Like HSKc, the ESKc did interact with HPg, but the human zymogen was not activated by ESKc (NOWICKI et al. 1994). A similar situation was found with activation of equine plasminogen by HSKc. Thus, species specificity is found with both components of this activation system. A partial amino acid sequence has been forwarded for ESKc, and this is provided in Fig. 3. When compared to the HSK sequences in the highly conserved amino-terminal region, ESKc exhibits few similarities with these other human host-derived streptococcal strains, an observation confirmed by the finding that DNA probes complementary to regions of HSKc did not hybridize with a variety of strains of ESK (NOWICKI et al. 1994). These data show that this activator can vary widely in primary structure among different strains of *streptococci* that produce this protein, and amino acid and concomitant conformational properties of the SK could be tailored to activate the plasminogen present in the mammal that is infected by the bacteria.

3. Functional Regions of Human Plasminogen and Human-derived Streptokinase that Mediate Activation

The loci within HPg that interact with HSKc have been reduced to elements within the latent protease chain of HPg. In this regard, it has been found that the isolated protease chain of HPm formed a functional complex with HSKc (SUMMARIA and ROBBINS 1976), and that a proteolytic fragment of HPg, consisting of only the $K5_{Pg}$ domain covalently linked to the latent protease chain (Val^{442}-Pg), was activated by HSKc at a rate similar to that of the intact protein (POWELL and CASTELLINO 1980). In further refining the regions of the HPm protease chain wherein HSKc functional binding resides, it is useful to compare the amino acid sequence of the HPm light chain with those of bovine (B) and porcine (P) Pm, the latter two of which do not form functional complexes with HSKc. The amino acid sequences of the protease chains of these three plasminogens are very similar, and unless single base changes alter their reactivities with HSKc, then a search of amino acid clusters in BPg and PPg that differ from HPg might suggest areas of the protease chain that should be further investigated with regard to functional SK binding. Three such areas can be initially identified to be of possible importance in this regard, viz., a region surrounding Arg^{582} another in the area of Asn^{625} of the human zymogen, and an additional locus near its COOH-terminus.

Studies of the sites within HSKc that possibly interact in a functional manner with HPg and HPm have yielded valuable information. Initial work on this topic demonstrated that a fragment of HSKc comprising amino acid residues Ser^{60}-Lys^{387} was fully functional in HPg activation, and that much more extensive COOH-terminal degradation was not well tolerated in terms of maintenance of a functional HSKc (Schick and Castellino 1973; Brockway and Castellino 1974; Siefring and Castellino 1976). Later work extended these conclusions as a result of the demonstration that the peptide regions, 91–414, 127–414, and 158–414 were not effective activators of HPg (Young et al. 1995). Additional probing of the functions of various fragments of HSKc demonstrated that the peptide region Ser^{60}-Lys^{333} was essential for minimal SK activator activity, and that the sequence encompassing Ala^{334}-Lys^{387} was needed for high affinity interactions with HPm. It was also proposed that the 59-residue amino-terminal polypeptide Ile^{1}-Lys^{59} stabilized the proper conformation of HSKc (Shi et al. 1994; Young et al. 1998; Nihalani et al. 1998; Fay and Bokka 1998), and enhanced the ability of streptokinase fragment 56–414 to activate HPg (Nihalani et al. 1998). Other fragments of HSKc have been shown to possess tight binding sites to HPg and/or HPm. Binding, and even slow generation of an active site in stoichiometric complexes of HPg, was found with an r-HSKc fragment composed of residues Val^{143}-Lys^{386} (Rodriguez et al. 1994), while tight binding without such active site generation was observed with an r-HSKc fragment, Val^{143}-Lys^{293} (Rodriguez et al. 1995). Another tight HPg binding fragment of r-HSKc was later discovered, and consisted of residues, Arg^{244}-Thr^{352}. This fragment, while not itself serving as an activator of HPg, nonetheless inhibited the activation of HPg by full-length r-HSKc. This suggests that it constitutes a necessary binding locale of HSKc to HPg (Reed et al. 1995). Analysis of point mutations in HSKc suggest that residues around positions 13–20 of SK are important for production of amidolytic activity with HPg. Residues of HSKc proximal to 364–374 are not only required for this same amidolytic activity, but also for generating plasminogen activator activity with plasmin(ogen) (Fay and Bokka 1998; Chaudhary et al. 1999). Many of these concepts have been confirmed as a result of examination of the crystal structure of HSKc complexed with the catalytic chain of HPm (Wang et al. 1998).

V. Activation of Human Plasminogen by Staphylokinase

A variation of the above mechanism with HSKc occurs with another bacterial HPg activator, SakSTAR (see also Chap. 16). The amino acid sequence of this protein, as deduced from the SakSTAR gene, has been determined, and shows the presence of 163 amino acids, which includes a 27-amino acid signal sequence (Sako et al. 1983; Sako and Tsuchida 1983; Collen et al. 1992b). This DNA has been expressed in several bacterial systems (Sako 1985; Behnke and Gerlach 1987; Collen et al. 1992b). Like SK, SakSTAR does not possess protease activity and functions by first forming a stoichiometric complex with HPg (Kowalska-Loth and Zakrzewski 1975), but a major difference with SK is grounded in the observation that SakSTAR shows fibrin-stimulated activation of HPg (Sakai et al. 1989), a finding that has encouraged studies with SakSTAR regarding its ability to efficiently function as a potential in vivo thrombolytic agent (Collen et al. 1992a; Collen and

LIJNEN 1994). Endothelial cells also enhance the plasminogen activator activity of SakSTAR (UESHIMA et al. 1996). Binding of SakSTAR to HPg appears similar to that of HSKc, with available evidence showing that the $K1_{Pg}$-$K4_{Pg}$ region of HPg is not necessary for the binding of either agent, and that there are distinct, but partially overlapping binding sites on HPg for HSKc and SakSTAR (RODRIGUEZ et al. 1995). A determinant on HPm of SakSTAR binding is Arg^{719} (JESPERS et al. 1998), a residue that has also been found to function similarly for HSKc binding to HPm (DAWSON et al. 1994).

The mechanism of activation of HPg with SakSTAR is similar to that of HSKc, but it appears as though the initial complex of SakSTAR-HPg does not develop plasminogen activator activity. Such activity exists in SakSTAR-HPm that is formed from SakSTAR-HPg in the presence of other HPg activators, i.e., tPA. (COLLEN et al. 1993; GRELLA and CASTELLINO 1997). A reduction in size of the SakSTAR in the complex with HPm also occurs, associated with cleavage of the Lys^{10}-Lys^{11} peptide bond (SCHLOTT et al. 1997), but is not needed to induce clot lysis in the presence of HPg (UESHIMA et al. 1993). Then, like tcu-PA, tPA, and the SK-HPg and SK-HPm complexes, SakSTAR-HPm functions as an HPg activator (LIJNEN et al. 1994).

D. Spontaneous and Induced Phenotypes of Plasminogen Deficiency

The recent availability of mice with a targeted deficiency for plasminogen has proved to be a valuable resource for directly investigating the role of plasminogen in a number of biological processes. Plasminogen-deficient mice, Pg(–/–), survive embryonic development, attain adulthood, and are fertile. However, they are predisposed to severe thrombosis and spontaneously develop thrombotic lesions in a number of organs as well as gastrointestinal tract ulcerations and rectal prolapse (BUGGE et al. 1995; PLOPLIS et al. 1995). Reconstitution with murine plasminogen was shown to normalize the thrombolytic potential and significantly resolve endogenous fibrin deposits, indicating that the major physiologic pathway of fibrinolysis is the plasminogen system (LIJNEN et al. 1996). Another spontaneous phenotype in Pg(–/–) mice, palpebral and bulbar conjunctivitis, was shown to be strain-sensitive, occurring at higher frequency and severity in the C57Bl/6J strain relative to that of 129/Black Swiss (DREW et al. 1998). The conjunctivitis that develops is grossly and histologically indistinguishable to that observed in plasminogen-deficient humans (MINGER et al. 1997; SCHUSTER et al. 1997).

Other induced phenotypes in Pg(–/–) mice have been observed from in vivo analyses. Pg(–/–) mice demonstrate impaired skin and corneal epithelial wound healing which is normalized in Pg(–/–) crosses with fibrinogen (Fg)-deficient mice, Pg(–/–)/Fg(–/–), suggesting that plasmin plays an important role in mediating resolution of provisional fibrin matrices associated with the wound healing process (RØMER et al. 1996; BUGGE et al. 1996; KAO et al.

1998). Pg(–/–) mice also display enhanced fibrin-mediated glomerular injury as demonstrated in a model of crescentic glomerulonephritis (KITCHING et al. 1997). Additionally, cell migration associated with the inflammatory response has also been shown to be compromised in Pg(–/–) mice when challenged with an inflammatory mediator, as well as in a model of transplant arteriosclerosis (PLOPLIS et al. 1998; MOONS et al. 1998). As a result, media necrosis, breakdown of the elastic laminae, and adventitial remodeling were more pronounced in arterial grafts in Pg(+/+) mice than in Pg(–/–) mice. While a number of these studies potentially implicate fibrin as mediating these events, studies that have examined the effect of plasminogen on excitotoxin-induced neurodegeneration indicated that Pg(–/–) mice display resistance to neurodegeneration, while wild-type mice were sensitive. Pg(–/–)/Fg(–/–) mice demonstrated a similar resistance as that seen in Pg(–/–) mice, thus implicating a substrate for plasmin other than fibrin in mediating excitotoxin-induced neurodegeneration (TSIRKA et al. 1997). Additional studies identified this target protein as laminin, and its degradation by plasmin led to disruption of the interaction between extracellular matrix protein and neurons ultimately resulting in cell death (CHEN et al. 1997). Further investigations utilizing Pg(–/–) mice should prove to be a valuable approach towards elucidating the in vivo contribution of plasminogen in a number of physiological and pathophysiological processes.

List of Abbreviations

HPg	any molecular form of human plasminogen
HPm	any molecular form of human plasmin that is cleaved at the Arg^{561}-Val^{562} peptide bond
Glu^1-Pg	human plasminogen containing Glu^1 as the amino-terminal amino acid
Lys^{78}-Pg	human plasminogen containing Lys^{78} as the amino-terminal amino acid
Glu^1-Pm	human plasmin containing Glu^1 as the amino-terminal amino acid of the plasmin heavy chain
Lys^{78}-Pm	human plasmin containing Lys^{78} as the amino-terminal amino acid of the plasmin heavy chain
$K1_{Pg}$	the kringle 1 module of human plasminogen, which consists of amino acid residues Cys^{84}-Cys^{162}
$K2_{Pg}$	the kringle 2 region of human plasminogen, containing amino acid residues Cys^{166}-Cys^{243}
$K3_{Pg}$	the kringle 3 region of human plasminogen, spanning amino acid residues Cys^{256}-Cys^{333}
$K4_{Pg}$	the kringle 4 module of human plasminogen, which includes amino acid residues Cys^{358}-Cys^{435}
$K5_{Pg}$	the kringle 5 module of human plasminogen, which contains amino acid residues Cys^{462}-Cys^{541}

tPA	tissue-type plasminogen activator
$K2_{tPA}$	the kringle 2 module of human tissue-type plasminogen activator, which consists of amino acid residues Cys^{180}-Cys^{261}
sc-uPA	single-chain urokinase-type plasminogen activator
tc-uPA	two-chain urokinase-type plasminogen activator obtained by cleavage of the Lys^{158}-Ile^{159} peptide bond in the single chain form (a low molecular weight variant of tc-uPA exists in which the peptide bond Lys^{135}-Lys^{136} has been additionally cleaved, resulting in the loss of the amino-terminal 135 amino acid residues from the enzyme
uPAR	urokinase-type plasminogen activator receptor
SKc	a group C streptokinase
HSKc	a group C streptokinase obtained from a human host
SKa	a group A streptokinase
HSKa	a group A streptokinase obtained from a human host
ESK	equine host-derived streptokinase
SakSTAR	staphylokinase
α_2PI	α_2-plasmin inhibitor (used synonymously with α_2-antiplasmin)
α_2M	α_2-macroglobulin
PAI-1	plasminogen activator inhibitor-1
r	recombinant

References

Adrian GS, Korinek BW, Bowman GH, Yang F (1986) The human transferrin gene: 5' region contains conserved sequences which match the control elements regulated by heavy metals, glucocorticoids and acute phase reaction. Gene 49:167–175

Agostoni A, Gardinali M, Frangi D, Cafaro C, Conciato L, Sponzilli C, Salvioni A, Cugno M, Cicardi M (1994) Activation of complement and kinin systems after thrombolytic therapy in patients with acute myocardial infarction. A comparison between streptokinase and recombinant tissue-type plasminogen activator. Circulation 90:2666–2670

Bajpai A, Baker JB (1985) Cryptic urokinase binding sites on human foreskin fibroblasts. Biochem Biophys Res Commun 133:475–482

Behnke D, Gerlach D (1987) Cloning and expression in *Escherichia coli, Bacillus subtilis,* and *Streptococcus sanguis* of a gene for staphylokinase-a bacterial plasminogen activator. Molec Gen Genet 210:528–534

Behrendt N, Rønne E, Ploug M, Petri T, Lober D, Nielsen LS, Schleuning W-D, Blasi F, Appella E, Danø K (1990) The human receptor for urokinase plasminogen activator. NH₂-terminal amino acid sequence and glycosylation variants. J Biol Chem 265:6453–6460

Berge A, Sjöbring U (1993) PAM, a novel plasminogen-binding protein from *Streptococcus pyogenes.* J Biol Chem 268:25417–25424

Berget SM (1984) Are U4 small nuclear ribonucleoproteins involved in polyadenylation? Nature 309:179–182

Binnema DJ, Dooijewaard G, van Iersel JJL, Turion PNC, Kluft C (1990) The contact-system dependent plasminogen activator from human plasma: Identification and characterization. Thromb Haemostas 64:390–397

Boyd D, Florent G, Kim P, Brattain M (1988) Determination of the levels of urokinase and its receptor in human colon carcinoma cell lines. Cancer Res 48:3112–3116

Brade V, Nicholson A, Bitter-Suermann D, Hadding V (1974) Formation of the C-3 cleaving properdin enzyme on zymosen. J Immunol 113:1735–1743

Brockway WJ, Castellino FJ (1972) Measurement of the binding of antifibrinolytic amino acids to various plasminogens. Arch Biochem Biophys 151:194–199

Brockway WJ, Castellino FJ (1974) A characterization of native streptokinase and altered streptokinase isolated from a human plasminogen activator complex. Biochemistry 13:2063–2070

Bugge TH, Flick MJ, Daugherty CC, Degen JL (1995) Plasminogen deficiency causes severe thrombosis but is compatible with development and reproduction. Genes Dev 9:794–807

Bugge TH, Kombrinck KW, Flick MJ, Daugherty CC, Danton MJ, Degen JL (1996) Loss of fibrinogen rescues mice from the pleiotropic effects of plasminogen deficiency. Cell 15:709–719

Busso N, Masur SK, Lazega D, Waxman S, Ossowski L (1994) Induction of cell migration by pro-urokinase binding to its receptor: possible mechanism for signal transduction in human epithelial cells. J Cell Biol 126:259–270

Cao R, Wu H-L, Veitonmäki N, Linden P, Farnebo J, Shi G-Y, Cao Y (1999) Suppression of angiogenesis and tumor growth by the inhibitor K1-5 generated by plasmin-mediated proteolysis. Proc Natl Acad Sci USA 96:5728–5733

Cao Y, Chen A, An SSA, Ji RW, Davidson D, Llinás M (1997) Kringle 5 of plasminogen is a novel inhibitor of endothelial cell growth. J Biol Chem 272:22924–22928

Cao YH, Ji RW, Davidson D, Schaller J, Marti D, Söhndel S, McCance SG, O'Reilly MS, Llinás M, Folkman J (1996) Kringle domains of human angiostatin. Characterization of the anti-proliferative activity on endothelial cells. J Biol Chem 271: 29461–29467

Casey JR, Petranka JG, Kottra J, Fleenor DE, Rosse WF (1994) The structure of the urokinase-type plasminogen activator receptor gene. Blood 84:1151–1156

Chaudhary A, Vasudha S, Rajagopal K, Komath SS, Garg N, Yadav M, Mande SC, Sahni G (1999) Function of the central domain of streptokinase in substrate plasminogen docking and processing revealed by site-directed mutagenesis. Protein Sci 8:2791–2805

Chen ZL, Strickland S (1997) Neuronal death in the hippocampus is promoted by plasmin-catalyzed degradation of laminin. Cell 91:917–925

Chibber BAK, Castellino FJ (1986) Regulation of the streptokinase-mediated activation of human plasminogen by fibrinogen and chloride ions. J Biol Chem 261:5289–5295

Chibber BAK, Morris JP, Castellino FJ (1985) Effects of human fibrinogen and its cleavage products on activation of human plasminogen by streptokinase. Biochemistry 24:3429–3434

Chibber BAK, Radek JT, Morris JP, Castellino FJ (1986) Rapid formation of an anion sensitive active site in stoichiometric complexes of streptokinase and human [Glu^1]plasminogen. Proc Natl Acad Sci USA 83:1237–1241

Claesson-Welsh L, Welsh M, Ito N, Anand-Apte B, Soker S, Zetter B, O'Reilly M, Folkman J (1998) Angiostatin induces endothelial cell apoptosis and activation of focal adhesion kinase independently of the integrin-binding motif RGD. Proc Natl Acad Sci USA 95:5579–5583

Claeys H, Vermylen J (1974) Physicochemical and proenzyme properties of amino-terminal glutamic acid and amino-terminal lysine human plasminogen. Biochim Biophys Acta 342:351–359

Cochrane CG, Revak SD, Wuepper WG (1974) Activation of Hageman factor in solid and fluid phases. J Exp Med 138:1564–1583

Collen D (1980) On the regulation and control of fibrinolysis. Thromb Haemostas 43:77–89

Collen D, De Cock F, Vanlinhout I, Declerck PJ, Lijnen HR, Stassen JM (1992a) Comparative thrombolytic and immunogenic properties of staphylokinase and streptokinase. Fibrinolysis 6:232–242

Collen D, Lijnen HR (1994) Staphylokinase, a fibrin-specific plasminogen activator with therapeutic potential? Blood 84:680–686

Collen D, Schlott B, Engelborghs Y, Van Hoef B, Hartmann M, Lijnen HR, Behnke D (1993) On the mechanism of the activation of human plasminogen by recombinant staphylokinase. J Biol Chem 268:8284–8289

Collen D, Zhao ZA, Holvoet P, Marynen P (1992b) Primary structure and gene structure of staphylokinase. Fibrinolysis 6:226–231

Colman RW (1980) Activation of plasminogen by human plasma kallikrein. Biochem Biophys Res Commun 35:273–279

Davidson DJ, Higgins DL, Castellino FJ (1990) Plasminogen activator activities of equimolar complexes of streptokinase with variant recombinant plasminogens. Biochemistry 29:3585–3590

DíCosta SS, Boyle MDP (1998) Interaction of a group A *Streptococcus* within human plasma results in assembly of a surface plasminogen activator that contributes to occupancy of surface plasmin-binding structures. Microb Pathog 24:341–349

Dawson KM, Marshall JM, Raper RH, Gilbert RJ, Ponting CP (1994) Substitution of arginine 719 for glutamic acid in human plasminogen substantially reduces its affinity for streptokinase Biochemistry 33:12042–12047

Dear AE, Medcalf RL (1998) The urokinase-type-plasminogen-activator receptor (CD87) is a pleiotropic molecule. Eur J Biochem 252:185–193

De Vries TJ, Quax PHA, Denijn M, Verheijen JH, Verspaget HW, Weidle UH, Ruiter DJ, Vanmuijen GNP, Verrijp KN (1994) Plasminogen activators, their inhibitors, and urokinase receptor emerge in late stages of melanocytic tumor progression. Am J Pathol 144:70–81

Dillon HC, Wannamaker LW (1965) Physical and immunological differences among streptokinases. J Exp Med 121:351–360

Dong Z, Kumar R, Yang X, Fidler IJ (1997) Macrophage-derived metalloelastase is responsible for the generation of angiostatin in Lewis lung carcinoma. Cell 88: 801–810

Drew AF, Kaufman AH, Kombrinck KW, Danton MJ, Daugherty CC, Degen JL, Bugge TH (1998) Ligneous conjunctivitis in plasminogen deficient mice. Blood 91: 1616–1624

Dudani AK, Ganz PR (1996) Endothelial cell surface actin serves as a binding site for plasminogen, tissue plasminogen activator and lipoprotein(a). Br J Haematol 95:168–178

Duggan C, Maguire T, McDermott E, O'Higgins N, Fennelly JJ, Duffy MJ (1995) Urokinase plasminogen activator and urokinase plasminogen activator receptor in breast cancer. Int J Cancer 61:597–600

Dumler I, Petri T, Schleuning W-D (1993) Interaction of urokinase-type plasminogen activator (u-PA) with its cellular receptor (u-PAR) induces phosphorylation on tyrosine of a 38 kDa protein. FEBS Lett 322:37–40

Dumler I, Petri T, Schleuning W-D (1994) Induction of c-fos gene expression by urokinase-type plasminogen activator in human ovarian cancer cells. FEBS Lett 343:103–106

Edmonds-Alt X, Quisquater E, Vaes G (1980) Proteoglycan- and fibrin-degrading neutral proteinase activities of Lewis lung carcinoma cells. Eur J Cancer 16: 1257–1261

Ellis V, Behrendt N, Danø K (1991) Plasminogen activation by receptor-bound urokinase. Kinetic study with both cell-associated and isolated receptor. J Biol Chem 266:12572–12578

Ellis V, Scully MF, Kakkar VV (1987) Plasminogen activation by single-chain urokinase in functional isolation. J Biol Chem 262:14998–15003

Ellis V, Wun T-C, Behrendt N, Rønne E, Danø K (1990) Inhibition of receptor-bound urokinase by plasminogen activator inhibitors. J Biol Chem 265:9904–9908

Fazioli F, Blasi F (1994) Urokinase-type plasminogen activator and its receptor. New targets for anti-metastatic therapy. Trends Pharmacol Sci 15:25–29

Fay WP, Bokka LV (1998) Functional analysis of the amino- and carboxyl-termini of streptokinase. Thromb Haemost 79:985–991

Federici AB, Berkowitz SD, Lattuada A, Mannucci PM (1993) Degradation of von
    Willebrand factor in patients with acquired clinical conditions in which there is
    heightened proteolysis. Blood 81:720–725
Forsgren M, Raden B, Israelsson M, Larsson K, Hedén L-O (1987) Molecular cloning
    and characterization of a full-length cDNA clone for human plasminogen. FEBS
    Lett 213:254–260
Fowlkes DM, Mullis MT, Comeau CM, Crabtree GR (1984) Potential basis for regu-
    lation of the coordinately expressed fibrinogen genes: homology in the 5' flanking
    regions. Proc Natl Acad Sci USA 81:2313–2316
Frank C, Steiner K, Malke H (1995) Conservation of the organization of the streptok-
    inase gene region among pathogenic streptococci. Med Microbiol Immunol
    184:139–146
Gately S, Twardowski P, Stack MS, Cundiff DL, Grella D, Castellino FJ, Enghild J,
    Kwaan HC, Lee F, Kramer RA, Volpert O, Bouck N, Soff GA (1997) The mecha-
    nism of cancer-mediated conversion of plasminogen to the angiogenesis inhibitor
    angiostatin. Proc Natl Acad Sci U S A 94:10868–10872
Gerlach D, Kohler W (1977) Studies of the heterogeneity of streptokinase of different
    origin. Zbl Bakt Hyg I Abt Orig 238:336–349
Goldfinger LE, Stack MS, Jones JC (1998) Processing of laminin-5 and its functional
    consequences: role of plasmin and tissue-type plasminogen activator. J Cell Biol
    141:255–265
Goldsmith GH, Saito H, Ratnoff OD (1978) The activation of plasminogen by
    Hageman factor (factor XII) and Hageman factor fragments. J Clin Invest
    62:54–60
Gonzalez-Gronow M, Violand BN, Castellino FJ (1977) Purification and some prop-
    erties of the glu- and lys- human plasmin heavy chains. J Biol Chem 252:2175–
    2177
Grella DK, Castellino FJ (1997) Activation of human plasminogen by staphylokinase.
    Direct evidence that preformed plasmin is necessary for activation to occur. Blood
    89:1585–1589
Griscelli F, Li H, Bennaceur-Griscelli A, Soria J, Opolon P, Soria C, Perricaudet M, Yeh
    P, Lu H (1998) Angiostatin gene transfer: inhibition of tumor growth in vivo by
    blockage of endothelial cell proliferation associated with a mitosis arrest. Proc
    Natl Acad Sci USA 95:6367–6372
Gross JL, Moscatelli D, Rifkin DB (1983) Increased capillary endothelial cell protease
    activity in response to angiogenic stimuli in vitro. Proc Natl Acad Sci USA 80:
    2623–2627
Gundersen D, Tran-Thang C, Sordat B, Mourali F, Rüegg C (1997) Plasmin-induced
    proteolysis of tenascin-C. Modulation by T lymphocyte-derived urokinase-type
    plasminogen activator and effect on T lymphocyte adhesion, activation, and cell
    clustering. J Immunol 158:1051–1060
Habal FM, Burrowes CE, Movat HZ (1976) Generation of kinin by plasma kallikrein
    and plasmin and the effect of α_1-antitrypsin and antithrombin III on the kinino-
    genases. Adv Exp Med Biol 70:23–36
Hajjar KA, Krishnan S (1999) Annexin II: a mediator of the plasmin/plasminogen
    activator system. Trends Cardiovasc Med 9:128–138
Hajjar KA, Harpel PC, Jaffe EA, Nachman RL (1986) Binding of plasminogen to cul-
    tured human endothelial cells. J Biol Chem 261:11656–11662
Hauert J, Nicoloso G, Schleuning WD, Bachmann F, Schapira M (1989) Plasminogen
    activators in dextran sulfate-activated euglobulin fractions: a molecular analysis of
    factor XII- and prekallikrein- dependent fibrinolysis. Blood 73:994–999
Hayes ML, Castellino FJ (1979a) Carbohydrate of human plasminogen variants. II.
    Structure of the asparagine-linked oligosaccharide unit. J Biol Chem 254:
    8772–8776
Hayes ML, Castellino FJ (1979b) Carbohydrate of human plasminogen variants. III.
    Structure of the O-glycosidically-linked oligosaccharide unit. J Biol Chem 254:
    8777–8780

He C, Wilhelm SM, Pentland AP, Marmer BL, Grant GA, Eisen AZ, Goldberg GI (1989) Tissue cooperation in a proteolytic cascade activating human interstitial collagenase. Proc Natl Acad Sci USA 86:2632–2636

Hearing V, Law L, Corti A, Appella E, Blasi F (1988) Modulation of metastatic potential by surface urokinase of murine melanoma cells. Cancer Res 48:1270–1278

Hembrough TA, Li L, Gonias SL (1996) Cell-surface cytokeratin 8 is the major plasminogen receptor on breast cancer cells and is required for the accelerated activation of cell-associated plasminogen by tissue-type plasminogen activator. J Biol Chem 271:25684–25691

Hembrough TA, Vasudevan J, Allietta MM, Glass WF, Gonias SL (1995) A cytokeratin 8-like protein with plasminogen-binding activity is present on the external surfaces of hepatocytes, HepG2 cells and breast carcinoma cell lines. J Cell Sci 108: 1071–1082

Hoover GJ, Menhart N, Martin A, Warder S, Castellino FJ (1993) Amino acids of the recombinant kringle 1 domain of human plasminogen that stabilize its interaction with ω-amino acids. Biochemistry 32:10936–10943

Horrevoets AJG, Smilde AE, Fredenburgh JC, Pannekoek H, Nesheim ME (1995) The activation-resistant conformation of recombinant human plasminogen is stabilized by basic residues in the amino-terminal hinge region. J Biol Chem 270:15770–15776

Hortin GL (1990) Isolation of glycopeptides containing O-linked oligosaccharides by lectin affinity chromatography on jacalin-agarose. Anal Biochem 191:262–267

Hortin GL, Trimpe BL, Fok KF (1989) Plasmin's peptide-binding specificity: Characterization of ligand sites in α_2-antiplasmin. Thromb Res 54:621–632

Hoylaerts M, Rijken DC, Lijnen HR, Collen D (1982) Kinetics of the activation of plasminogen by human tissue plasminogen activator. Role of fibrin. J Biol Chem 257: 2912–2919

Huang T-T, Malke H, Ferretti JJ (1989) The streptokinase gene of group A streptococci: cloning, expression in *Escherichia coli*, and sequence analysis. Mol Microbiology 3:197–205

Ichinose A, Fujikawa K, Suyama T (1986) The activation of pro-urokinase by plasma kallikrein and its activation by thrombin. J Biol Chem 261:3486–3489

Jackson KW, Tang J (1982) Complete amino acid sequence of streptokinase and its homology with serine proteases. Biochemistry 21:6620–6625

Janeway CA, Merler E, Rosen FS, Salmon S, Crain JO (1968) Intravenous *gamma*-globulin: Mechanism of gamma-globulin fragments in normal and agamma-globulinemic persons. New Eng J Med 278:919–923

Jespers L, VanHerzeele N, Lijnen HR, VanHoef B, DeMaeyer M, Collen D, Lasters I (1998) Arginine 719 in human plasminogen mediates formation of the staphylokinase:plasmin activator complex. Biochemistry 37:6380–6386

Ji WR, Barrientos LG, Llinás M, Gray H, Villarreal X, Deford ME, Castellino FJ, Kramer RA, Trail PA (1998) Selective inhibition by kringle 5 of human plasminogen on endothelial cell migration, an important process in angiogenesis. Biochem Biophys Res Commun 247:414–419

Jilek F (1977) Cold-insoluble globulin III. Cyanogen bromide and plasminolysis fragments containing a label introduced by transamidation. Hoppe-Seyler's Z Physiol Chem 358:1165–1168

Kao WW, Kao CW, Kaufman AH, Kombrinck KW, Converse RL, Good WV, Bugge TH, Degen JL (1998) Healing of corneal epithelial defects in plasminogen- and fibrinogen-deficient mice. Invest Ophthalmol Vis Sci 39:502–508

Kida M, Wakabayashi S, Ichinose A (1997) Expression and induction by IL-6 of the normal and variant genes for human plasminogen. Biochem Biophys Res Commun 230:129–132

Kitching AR, Holdsworth SR, Ploplis VA, Plow EF, Collen D, Carmeliet P, Tipping PG (1997) Plasminogen and plasminogen activators protect against renal injury in crescentic glomerularnephritis J Exp Med 185:963–968

Kluft C, Wijngaards G, Jie AFH (1981) The factor XII-independent plasminogen pro-activator system includes urokinase-related activators. Thromb Haemostas 46: 343–350

Kowalska-Loth B, Zakrzewski K (1975) The activation by staphylokinase of human plasminogen. Acta Biochim Pol 22:327–339

Kruithof EKO, Ransijn A, Bachmann F (1983) Inhibition of tissue plasminogen activator by human plasma. In: Davidson JF, Bachmann F, Bouvier CA, Kruithof EKO (eds) Progress in Fibrinolysis, vol 6. Churchill Livingstone, Edinburgh, pp 362–366

Lazarus GS, Jensen PJ (1991) Plasminogen activators in epithelial biology. Sem Thromb Haemostas 17:210–216

Lee TH, Rhim TY, Kim SS (1998) Prothrombin kringle-2 has a growth inhibitory activity against basic fibroblast growth factor-stimulated capillary endothelial cells. J Biol Chem 273:28805–28812

Lijnen HR, Carmeliet P, Bouche A, Moons L, Ploplis VA, Plow EF, Collen D (1996) Restoration of thrombolytic potential in plasminogen-deficient mice after bolus administration of plasminogen. Blood 88:870–876

Lijnen HR, De Cook F, Van Hoef B, Schlott B, Collen D (1994) Characterization of the interaction between plasminogen and staphylokinase. Eur J Biochem 224: 143–149

Lijnen HR, Ugwu F, Bini A, Collen D (1998) Generation of an angiostatin-like fragment from plasminogen by stromelysin-1 (MMP-3). Biochemistry 37:4699–4702

Lijnen HR, Van Hoef B, Nelles L, Collen D (1990) Plasminogen activation with single-chain urokinase-type plasminogen activator (scu-PA). Studies with active site mutagenized plasminogen (Ser^{740} → Ala) and plasmin-resistant scu-PA (Lys^{158} → Glu). J Biol Chem 265:5232–5236

Linde V, Nielsen LS, Foster DC, Petersen LC (1998) Elimination of the Cys^{558}-Cys^{566} bond in Lys^{78}-plasminogen. Effect on activation and fibrin interaction. Eur J Biochem 251:472–479

Long GL, Chandra T, Woo SLC, Davie EW, Kurachi K (1984) Complete sequence of the cDNA for human α_1-antitrypsin and the gene for the S variant. Biochemistry 23:4828–4837

Mackay AR, Corbitt RH, Hartzler JL, Thorgeirsson UP (1990) Basement membrane type IV collagen degradation: Evidence for the involvement of a proteolytic cascade independent of metalloproteinases. Cancer Res 50:5997–6001

Maeda M (1985) Nucleotide sequence of the haptoglobin and haptoglobin-related gene pair. J Biol Chem 260:6698–6709

Magnusson S, Petersen TE, Sottrup-Jensen L, Claeys H (1975) Complete primary structure of prothrombin: Isolation and reactivity of ten carboxylated glutamic residues and regulation of prothrombin activation by thrombin. In: Reich E, Rifkin DB, Shaw E (eds) Proteases and biological control. Cold Spring Harbor Laboratories, Cold Spring Harbor, NY, pp 123–149

Malke H, Ferretti JJ (1984) Streptokinase: cloning, expression, and excretion by *Escherichia coli*. Proc Natl Acad Sci USA 81:3557–3561

Malke H, Roe B, Ferretti JJ (1985) Nucleotide sequence of the streptokinase gene from *Streptococcus equisimilis* H46A. Gene 34:357–362

Mandle RJ, Kaplan AP (1979) Generation of fibrinolytic activity by the interaction of activated factor XI and plasminogen. Blood 54:850–861

Marti DN, Schaller J, Llinás M (1999) Solution structure and dynamics of the plasminogen kringle 2-AMCHA complex: 3_1-helix in homologous domains. Biochemistry 38:15741–15755

Mathews II, Vanderhoff-Hanaver P, Castellino FJ, Tulinsky A (1996) Crystal structures of the recombinant kringle 1 domain of human plasminogen in complexes with the ligands ε-aminocaproic acid and *trans*-4-(aminomethyl)cyclohexane-1-carboxylic acid. Biochemistry 35:2567–2576

McCance SG, Castellino FJ (1995) Contributions of individual kringle domains toward maintenance of the chloride-induced tight conformation of human glutamic acid-1 plasminogen. Biochemistry 34:9581–9586

McCance SG, Menhart N, Castellino FJ (1994) Amino acid residues of the kringle-4 and kringle-5 domains of human plasminogen that stabilize their interactions with omega-amino acid ligands. J Biol Chem 269:32405–32410

McCoy HE, Border CC, Lottenberg R (1991) Streptokinases produced by pathological group C *Streptococci* demonstrate species-specific plasminogen activation. J Inf Dis 164:515–521

McKee PA, Lemmon WB, Hampton JW (1971) Streptokinase and urokinase activation of human chimpanzee and baboon plasminogen. Thromb Diath Haemorrh 26: 512–522

McLauchlan J, Gaffney D, Whitton JL, Clements JB (1985) The consensus sequence YGTGTTYY located downstream from the AATAAA signal is required for efficient formation of mRNA 3' termini. Nucleic Acids Res 13:1347–1368

McLean JW, Tomlinson JE, Kuang W-J, Eaton DL, Chen EY, Gless GM, Scanu AM, Lawn RM (1987) cDNA sequence of human apolipoprotein(a) is homologous to plasminogen. Nature 330:132–137

McMullen BA, Fujikawa K (1985) Amino acid sequence of the heavy chain of human α-factor XIIa (activated Hageman factor). J Biol Chem 260:5328–5341

Menhart N, Hoover GJ, McCance SG, Castellino FJ (1995) Roles of individual kringle domains in the functioning of positive and negative effectors of human plasminogen activation. Biochemistry 34:1482–1488

Menhart N, McCance SG, Sehl LC, Castellino FJ (1993) Functional independence of the kringle 4 and kringle 5 regions of human plasminogen. Biochemistry 32: 8799–8806

Menhart N, Sehl LC, Kelley RF, Castellino FJ (1991) Construction, expression and purification of recombinant kringle 1 of human plasminogen and analysis of its interaction with ω-amino acids. Biochemistry 30:1948–1957

Meroni G, Buraggi G, Mantovani R, Taramelli R (1996) Motifs resembling hepatocyte nuclear factor 1 and activator protein 3 mediate the tissue specificity of the human plasminogen gene. Eur J Biochem 236:373–382

Miles LA, Dahlberg CM, Plow EF (1988) The cell binding domains of plasminogen and their function in plasma. J Biol Chem 263:11928–11934

Miles LA, Plow EF (1987) Receptor mediated binding of the fibrinolytic components, plasminogen and urokinase to peripheral blood cells. Thromb Haemostas 58: 936–942

Miles LA, Plow EF (1988) Plasminogen receptors: ubiquitous sites for cellular regulation of fibrinolysis. Fibrinolysis 2:61–71

Minger AM, Heimburger N, Zeitler P, Kreth HW, Schuster V (1997) Homozygous type I plasminogen deficiency. Semin Thromb Hemost 23:259–269

Mirsky IA, Perisutti G, Davis NC (1959) The destruction of glucagon, adrenocorticotropin and somatotropin by human blood plasma. J Clin Invest 38:14–20

Moons L, Shi C, Ploplis V, Plow E, Haber E, Collen D, Carmeliet P (1998) Reduced transplant arteriosclerosis in plasminogen deficient mice. J Clin Invest 102: 1788–1797

Morioka S, Lazarus GS, Baird JL, Jensen PJ (1991) Migrating keratinocytes express urokinase-type plasminogen activator. J Invest Dermatol 88:418–423

Murray JC, Buetow KH, Donovan M, Hornung S, Motulsky AG, Disteche C, Dyer K, Swisshelm K, Anderson J, Giblet E, Sadler E, Eddy R, Shows TB (1987) Linkage disequilibrium of plasminogen polymorphisms and assignment of the gene to human chromosome 6q26–6q27. Am J Hum Genet 40:338–350

Nakajima K, Hamanoue M, Takemoto N, Hattori T, Kato K, Kohsaka S (1994) Plasminogen binds specifically to α-enolase on rat neuronal plasma membrane. J Neurochem 63:2048–2057

Nielsen LS, Kellerman GM, Behrendt N, Picone R, Danø K, Blasi F (1988) A 55 000–65 000 Mr receptor for urokinase-type plasminogen activator. Identification in human tumor cell lines and partial purification. J Biol Chem 263:2358–2363

Nihalani D, Kumar R, Rajagopal K, Sahni G (1998) Role of the amino-terminal region of streptokinase in the generation of a fully functional plasminogen activator complex probed with synthetic peptides. Protein Sci 7:637–648

Nowicki ST, Minningwenz D, Johnston KH, Lottenberg R (1994) Characterization of a novel streptokinase produced by *Streptococcus equisimilis* of non-human origin. Thromb Haemostas 72:595–603

O'Reilly MS, Holmgren L, Sing Y, Chen C, Rosenthal RA, Moses M, Lane WS, Cao Y, Sage EH, Folkman J (1994) Angiostatin: a novel angiogenesis inhibitor that mediates suppression of metastasis by a Lewis lung carcinoma. Cell 79:315–328

Ogston D, Ogston DM, Ratnoff OD, Forbes CD (1969) Studies on a complex mechanism for the activation of plasminogen by kaolin and by chloroform: the participation of Hageman factor and additional cofactors. J Clin Invest 48:1786–1801

Ohkura N, Hagihara Y, Yoshimura T, Goto Y, Kato H (1998) Plasmin can reduce the function of human β2 glycoprotein I by cleaving domain V into a nicked form. Blood 91:4173–4179

Omar MN, Mann KG (1987) Inactivation of factor Va by plasmin. J Biol Chem 262: 9750–9755

Ossowski L, Biegel D, Reich E (1979) Mammary plasminogen activator: Correlation with involution hormonal modulation and comparison between normal and neoplastic tissue. Cell 16:929–940

Pancholi V, Fischetti VA (1998) α-enolase, a novel strong plasmin(ogen) binding protein on the surface of pathogenic streptococci. J Biol Chem 273:14503–14515

Pannell R, Gurewich V (1987) The activation of plasminogen by single-chain urokinase or by two-chain urokinase-a demonstration that single-chain urokinase has a low catalytic activity (pro-urokinase). Blood 67:22

Parrado J, Conejero-Lara F, Smith RAG, Marshall JM, Ponting CP, Dobson CM (1996) The domain organization of streptokinase: Nuclear magnetic resonance, circular dichroism, and functional characterization of proteolytic fragments. Protein Sci 5:693–704

Patterson BC, Sang QA (1997) Angiostatin-converting enzyme activities of human matrilysin (MMP-7) and gelatinase B/type IV collagenase (MMP-9). J Biol Chem 272:28823–28825

Pennica D, Holmes WE, Kohr WJ, Harkins RN, Vehar GA, Ward CA, Bennett WF, Yelverton E, Seeburg PH, Heyneker HL, Goeddel DV, Collen D (1983) Cloning and expression of human tissue-type plasminogen activator cDNA in *E. coli.* Nature 301:214–221

Petersen LC, Lund LR, Nielsen LS, Danø K, Skriver L (1988) One-chain urokinase-type plasminogen activator from human sarcoma cells is a proenzyme with little or no intrinsic activity. J Biol Chem 263:11189–11195

Petersen TE, Martzen MR, Ichinose A, Davie EW (1990) Characterization of the gene for human plasminogen, a key proenzyme in the fibrinolytic system. J Biol Chem 265:6104–6111

Pizzo SV, Schwartz ML, Hill RL, McKee PA (1972) The effect of plasmin on the subunit structure of human fibrinogen. J Biol Chem 247:636–645

Ploplis VA, Carmeliet P, Vazirzadeh S, Van Vlaenderen I, Moons L, Plow EF, Collen D (1995) Effects of disruption of the plasminogen gene on thrombosis, growth and health in mice. Circulation 92:2585–2593

Ploplis VA, French EL, Carmeliet P, Collen D, Plow EF (1998) Plasminogen deficiency differentially affects recruitment of inflammatory cell populations in mice. Blood 91:2005–2009

Plow EF, Freany F, Plescia J, Miles LA (1986) The plasminogen system and cell surfaces: Evidence for plasminogen and urokinase receptors on the same cell type. J Cell Biol 103:2411–2430

Powell JR, Castellino FJ (1980) Activation of neo-plasminogen-Val$_{442}$ by urokinase and streptokinase and a kinetic characterization of neo-plasmin-Val$_{442}$. J Biol Chem 255:5329–5335

Pirie-Shepherd SR, Stevens RD, Andol NL, Enghild JJ, Pizzo SV (1997) Evidence for a novel O-linked sialylated trisaccharide on Ser-248 of human plasminogen 2. J Biol Chem 272:7408–7411

Redlitz A, Fowler BJ, Plow EF, Miles LA (1995) The role of an enolase-related molecule in plasminogen binding to cells. Eur J Biochem 227:407–415

Reed GL, Lin L-F, Parhami-Seren B, Kussie P (1995) Identification of a plasminogen-binding region in streptokinase that is necessary for creation of a functional streptokinase-plasminogen complex. Biochemistry 34:10266–10271

Reich R, Miskin R, Tsafriri A (1985) Follicular plasminogen activator: Involvement in ovulation. Endocrinology 116:516–521

Rijken DC, Wijngaards G, Welbergen J (1980) Relationship between tissue plasminogen activator and activators in blood and vessel wall. Thromb Res 18:815–830

Robbins KC, Summaria L, Hsieh B, Shah RJ (1967) The peptide chains of human plasmin. Mechanism of activation of human plasminogen to plasmin. J Biol Chem 242:2333–2342

Rodríguez P, Collen D, Lijnen HR (1995) Binding of streptokinase and staphylokinase to plasminogen. Fibrinolysis 9:298–303

Rodríguez P, Fuentes D, Muñoz E, Rivero D, Orta D, Alburquerque S, Perez S, Besada V, Herrera L (1994) The streptokinase domain responsible for plasminogen binding. Fibrinolysis 8:276–285

Rodríguez P, Fuentes P, Barro M, Alvarez JG, Muñoz E, Collen D, Lijnen HR (1995) Structural domains of streptokinase involved in the interaction with plasminogen. Eur J Biochem 229:83–90

Rømer J, Bugge TH, Pyke C, Lund LR, Flick MJ, Degen JL, Danø K (1996) Impaired wound healing in mice with a disrupted plasminogen gene. Nat Med 2:287–292

Sakai M, Watanuki M, Matsuo O (1989) Mechanism of fibrin-specific fibrinolysis by staphylokinase: Participation of α_2-plasmin inhibitor. Biochem Biophys Res Commun 162:830–837

Sako T (1985) Overproduction of staphylokinase in Escherichia coli and its characterization. Eur J Biochem 149:557–563

Sako T, Sawaki S, Sakurai T, Ito S, Yoshizawa Y, Kondo I (1983) Cloning and expression of the staphylokinase gene of Staphylococcus aureus in E. coli. Molec Gen Genet 190:271–277

Sako T, Tsuchida N (1983) Nucleotide sequence of the staphylokinase gene from Staphylococcus aureus. Nucleic Acids Res 11:7679–7693

Sappino AP, Huarte J, Belin D, Vassalli J-D (1989) Plasminogen activators in tissue remodeling and invasion: mRNA localization in mouse ovaries and implanting embryos. J Cell Biol 109:2471–2479

Sasaki T, Mann K, Murphy G, Chu M-L, Timpl R (1996) Different susceptibilities of fibulin-1 and fibulin-2 to cleavage by matrix metalloproteinases and other tissue proteases. Eur J Biochem 240:427–434

Schick LA, Castellino FJ (1973) Interaction of streptokinase and rabbit plasminogen. Biochemistry 12:4315–4321

Schick LA, Castellino FJ (1974) Direct evidence for the generation of an active site in the plasminogen moiety of the streptokinase-human plasminogen activator complex. Biochem Biophys Res Comm 57:47–54

Schlechte W, Brattain M, Boyd D (1990) Invasion of extracellular matrix by cultured colon cancer cells: Dependence on urokinase receptor display. Cancer Commun 2:173–179

Schlechte W, Murano G, Boyd D (1989) Examination of the role of the urokinase receptor in human colon cancer mediated laminin degradation. Cancer Res 49:6064–6069

Schlott B, Guhrs KH, Hartmann M, Rocker A, Collen D (1997) Staphylokinase requires NH_2-terminal proteolysis for plasminogen activation. J Biol Chem 272:6067–6072

Schousboe I, Feddersen K, Røjkjaer R (1999) Factor XIIa is a kinetically favorable plasminogen activator. Thromb Haemost 82:1041–1046

Schuster V, Mingers AM, Seidenspinner S, Nussgens Z, Pukrop T, Kreth HW (1997) Homozygous mutations in the plasminogen gene of two unrelated girls with ligneous conjunctivitis. Blood 90:958–966

Sehl LC, Castellino FJ (1990) Thermodynamic properties of the binding of α-, ω-amino acids to the isolated kringle 4 region of human plasminogen as determined by high sensitivity titration calorimetry. J Biol Chem 265:5482–5486

Sharp PA (1981) Speculations on RNA splicing. Cell 23:643–646

Shi G-Y, Chang B-I, Chen S-M, Wu D-H, Wu H-L (1994) Function of streptokinase fragments in plasminogen activation. Biochem J 304:235–241

Siefring GE, Castellino FJ (1976) The interaction of streptokinase and plasminogen. Isolation and characterization of a streptokinase degradation product. J Biol Chem 251:3913–3921

Sodetz JM, Brockway WJ, Castellino FJ (1972) Multiplicity of rabbit plasminogen. Physical characterization. Biochemistry 11:4451–4458

Sottrup-Jensen L, Claeys H, Zajdel M, Petersen TE, Magnusson S (1978) The primary structure of human plasminogen: isolation of two lysine-binding fragments and one "mini" plasminogen (MW, 38000) by elastase-catalyzed-specific limited proteolysis. Prog Chem Fibrinolysis and Thrombolysis 3:191–209

Stahl A, Mueller BM (1994) Binding of urokinase to its receptor promotes migration and invasion of human melanoma cells in vitro. Cancer Res 54:3066–3071

Stathakis P, Fitzgerald M, Matthias LJ, Chesterman CN, Hogg PJ (1997) Generation of angiostatin by reduction and proteolysis of plasmin. Catalysis by a plasmin reductase secreted by cultured cell. J Biol Chem 272:20641–20645

Steffens GJ, Günzler WA, Ötting F, Frankus E, Flohé L (1982) The complete amino acid sequence of low molecular mass urokinase from human urine. Hoppe-Seyler's Z Physiol Chem 363:1043–1058

Stoppelli MP, Corti A, Stoffientini A, Cassani G, Blasi F, Assoian RK (1985) Differentiation-enhanced binding of the amino-terminal fragment of human urokinase plasminogen activator to a specific receptor on U937 monocytes. Proc Natl Acad Sci USA 82:4939–4943

Stoppelli MP, Tacchetti C, Cubellis MV, Corti A, Hearing VJ, Cassani G, Appella E, Blasi F (1986) Autocrine saturation of prourokinase receptors on human A431 cells. Cell 45:675–684

Strickland S, Reich E, Sherman MZ (1976) Plasminogen activator in early embryogenesis: Enzyme production by trophoblast and parietal endotherm. Cell 9:231–240

Stricklin GP, Bauer EA, Jeffrey JJ, Eisen A (1977) Human skin collagenase: Isolation of precursor and active forms from both fibroblast and organ cultures. Biochemistry 16:1607–1615

Suenson E, Thorsen S (1981) Secondary-site binding of glu-plasmin, lys-plasmin and miniplasmin to fibrin. Biochem J 197:619–628

Summaria L, Arzadon L, Bernabe P, Robbins KC (1974) The interaction of streptokinase with human, cat, dog, and rabbit plasminogens. J Biol Chem 15:4670–4679

Summaria L, Robbins KC (1976) Isolation of a human-plasmin-derived functionally active, light (B) chain capable of forming with streptokinase an equimolar light (B) chain- streptokinase complex with plasminogen activating activity. J Biol Chem 251:5810–5813

Tsirka SE, Bugge TH, Degen JL, Strickland S (1997) Neuronal death in the central nervous system demonstrates a non-fibrin substrate for plasmin. Proc Natl Acad Sci USA 94:9779–9781

Ueshima S, Silence K, Collen D, Lijnen HR (1993) Molecular conversions of recombinant staphylokinase during plasminogen activation in purified systems and in human plasma. Thromb Haemostas 70:495–499

Ueshima S, Okada K, Matsumoto H, Takaishi T, Fukao H, Matsuo O (1996) Effects of endothelial cells on activity of staphylokinase. Blood Coagulat Fibrinol 7:522–529

Unkeless JC, Gordon S, Reich E (1974) Secretion of plasminogen activator by stimulated macrophages. J Exp Med 139:834–850

Urano T, Chibber BAK, Castellino FJ (1987a) The reciprocal effects of ε-aminohexanoic acid and chloride ion on the activation of human [Glu¹]plasminogen by human urokinase. Proc Natl Acad Sci USA 84:4031–4034

Urano T, de Serrano VS, Chibber BAK, Castellino FJ (1987b) The control of the urokinase-catalyzed activation of human glutamic acid 1-plasminogen by positive and negative effectors. J Biol Chem 262:15959–15964

Urano T, de Serrano VS, Gaffney PJ, Castellino FJ (1988a) The activation of human [Glu¹] plasminogen by human single-chain urokinase. Arch Biochem Biophys 264:222–230

Urano T, de Serrano VS, Gaffney PJ, Castellino FJ (1988b) Effectors of the activation of human [Glu¹]plasminogen by human tissue plasminogen activator. Biochemistry 27:6522–6528

Vassalli J-D, Baccino D, Belin D (1985) A cellular binding site for the Mr 55000 form of the human plasminogen activator, urokinase. J Cell Biol 100:86–92

Violand BN, Byrne R, Castellino FJ (1978) The effect of α-ω-amino acids on human plasminogen structure and activation. J Biol Chem 253:5395–5401

Violand BN, Castellino FJ (1976) Mechanism of urokinase-catalyzed activation of human plasminogen. J Biol Chem 251:3906–3912

Violand BN, Sodetz JM, Castellino FJ (1975) The effect of ε-aminocaproic acid on the gross conformation of plasminogen and plasmin. Arch Biochem Biophys 170: 300–305

Virgi MAG, Vassalli JD, Estensen RD, Reich E (1980) Plasminogen activator of islets of Langerhans: modulation by glucose and correlation with insulin production. Proc Natl Acad Sci USA 77:875–879

Walter F, Siegel M, Malke H (1989) Nucleotide sequence of a streptokinase gene from a group G *Streptococcus*. Nucleic Acids Res 17:1261

Wang H, Prorok M, Bretthauer RK, Castellino FJ (1997) Serine-578 is a major phosphorylation locus in human plasma plasminogen. Biochemistry 36:8100–8106

Wang X, Lin X, Loy JA, Tang J, Zhang XC (1998) Crystal structure of the catalytic domain of human plasmin complexed with streptokinase. Science 281:1662–1665

Wang X, Terzyan S, Tang J, Loy JA, Lin X, Zhang XC (2000) Human plasminogen catalytic domain undergoes an unusual conformational change upon activation. J Mol Biol 295:903–914

Wang Y, Dang JJ, Johnson LK, Selhamer JJ, Doe WF (1995) Structure of the human urokinase receptor gene and its similarity to CD59 and the ly-6 family. Eur J Biochem 227:116–122

Whitelock JM, Murdoch AD, Iozzo RV, Underwood PA (1996) The degradation of human endothelial cell-derived perlecan and release of bound basic fibroblast growth factor by stromelysin, collagenase, plasmin, and heparanases. J Biol Chem 271:10079–10086

Wiman B (1973) Primary structure of peptides released during activation of human plasminogen by urokinase. Eur J Biochem 39:1–9

Wiman B (1977) The primary structure of the (light) chain of human plasmin. Eur J Biochem 76:129–137

Wiman B, Wallén P (1975) Amino-acid sequence of the cyanogen-bromide fragment from human plasminogen that forms the linkage between the plasmin chains. Eur J Biochem 58:539–547

Wistedt AC, Kotarsky H, Marti D, Ringdahl U, Castellino FJ, Schaller J, Sjöbring U (1998) Kringle 2 mediates high affinity binding of plasminogen to a defined natural sequence in *streptococcal* surface protein PAM. J Biol Chem 273:24420–24424

Wohlwend A, Belin D, Vassali J-D (1987) Plasminogen activator-specific inhibitors produced by human monocytes/macrophages. J Exp Med 165:320–339

Wu T-P, Padmanabhan K, Tulinsky A, Mulichak AM (1991) The refined structure of the ε-aminocaproic acid complex of human plasminogen kringle 4. Biochemistry 30:10589–10594

Wulf RJ, Mertz ET (1969) Studies on plasminogen. VIII. Species specificity of streptokinase. Can J Biochem 47:927–931
Young K-C, Shi G-Y, Chang Y-F, Chang B-I, Chang L-C, Lai M-D, Chuang W-J, Wu
H-L (1995) Interaction of plasminogen and streptokinase-studies with truncated
streptokinase peptides. J Biol Chem 270:29601–29606
Young K-C, Shi G-Y, Wu D-H, Chang L-C, Chang B-I, Ou C-P, Wu H-L (1998) Plasminogen activation by streptokinase via a unique mechanism. J Biol Chem 273:
3110–3116

CHAPTER 3

Tissue-type Plasminogen Activator (tPA)

F. Bachmann

A. Milestones in Tissue-type Plasminogen Activator (tPA) Research

At the turn of this century, the Belgian physiologist Nolf (1908) observed that under certain experimental conditions the lysis of a blood clot could be obtained and introduced the concept of the "fourth state of blood coagulation," namely the dissolution of thrombi. During the same period Loeb in Germany was studying the process of wound healing. He observed that epithelial cells were able to liquefy the fibrinous wound scab and thus assured their progression into the scab. He concluded that epithelial cells were able to produce "peptonizing enzymes" (Loeb 1904; Fleisher and Loeb 1915). In 1936, the Russian investigator Yudin (1936, 1937) observed that cadaver blood was fluid and caused, after transfusion into man, a fibrinolytic state. In the 1930s and 1940s, MacFarlane and Biggs in Oxford were able to demonstrate, using a dilute plasma clot lysis assay, that several pathophysiological conditions triggered a release of fibrinolytic activity in man, such as surgery, trauma, physical exercise, mental stress, or the intravenous injection of adrenaline (MacFarlane 1937; MacFarlane and Biggs 1946; Biggs et al. 1947). During the same time period, Christensen and MacLeod (1945) discovered the zymogen plasminogen in human serum that could be activated by streptokinase (SK) to form active plasmin, and Astrup and collaborators started their seminal work on the fibrinolytic system at the Carlsberg Foundation Research Institute in Copenhagen. Astrup and Permin (1947) were the first investigators who clearly demonstrated that there exist two different plasminogen activators in mammalians, tPA and urokinase. Astrup and Müllertz (1952) also developed a sensitive test to measure the activity of tPA, the plasminogen-enriched fibrin plate test.

tPA proved to be a very difficult enzyme to purify since it was tightly bound to particulate matter and could not be eluted into solution by physiological buffers. Astrup and Stage (1952) finally succeeded in solubilizing tPA using chaotropic agents, such as 2mol/l potassium thiocyanate (KSCN). However, during subsequent purification steps, exchange of 2mol/l KSCN for more physiological solutions resulted in the precipitation of tPA, and the best preparations in Astrup's lab probably were at most tenfold purified. Support for the tissue localization of tPA came from Todd's laboratory. Todd (1959, 1964) developed a histochemical method for the assay of tPA, consisting of thin tissue slices which were then covered with a solution of plasminogen-rich fibrinogen and thrombin, generating a thin fibrin film which permitted one to follow the dissolution of the fibrin. Another method that was widely used in these years is the euglobulin lysis time. It derives from an observation by Milstone (1941) that the plasma euglobulin fraction contains a "lytic factor." In this test diluted plasma is acidified, centrifuged, and the precipitate resuspended in buffer and clotted with thrombin. Increased concentrations of tPA shorten the euglobulin lysis time.

Progress in the purification of tPA was achieved when 5 mol/l urea or pH 4.2 high salt acetic acid buffers were used for the extraction of tPA from pig heart, followed by ammonium sulphate fractionation, precipitation at low ionic strength with 3 mmol/l Zn^{++} and gelfiltration on Sephadex G-100/G-200. This method resulted in an approximately 1700-fold purification compared to the crude extract (Bachmann et al. 1964); in subsequent work on the purification of tPA this extraction procedure was widely used. In the 1970s, Aoki and von Kaulla (1971) reported on the purification of a "vascular activator" extracted from the vascular tree of human cadavers and demonstrated that it was biochemically and immunologically different from human urokinase. These findings were confirmed by Binder et al. (1979), but even then it was not clear whether this vascular activator was identical to tPA. The same year Rijken et al. (1979) were able to produce the first really pure tPA preparation and two years later Rijken and Collen (1981) produced larger amounts of human tPA, using a melanoma cell line. The cDNA structure of tPA was revealed at the 6th International Congress on Fibrinolysis in Lausanne by Diane Pennica and reported a few months later in the journal Nature (Pennica et al. 1983) (see also Chap. 8). At this congress agreement was reached that the vascular plasminogen activator was identical to tPA.

The great interest in using tPA for therapeutic purposes originates to a large extent from the original observations of Thorsen et al. (1972) who demonstrated that tPA, but not urokinase, binds to fibrin. The first trials using tPA for the treatment of acute myocardial infarction started in 1983 (see Chap. 8). In the mid-1990s the first tPA knock-out was produced in mice (Carmeliet et al. 1994) and the crystal structure of the protease domain of tPA elucidated (Lamba et al. 1996).

B. Sources of tPA

tPA, a serine protease of 68 kDa (synonyms: tissue plasminogen activator, vascular plasminogen activator, extrinsic activator; EC 3.4.21.68), is one of the two physiological plasminogen activators present in human blood. This enzyme exerts its effect primarily in the vascular system and is the principal agent for the dissolution of thrombi via activation of clot-bound plasminogen to plasmin, whereas the urinary-type plasminogen activator (uPA, urokinase) is reported to play a role in cell migration and tissue remodeling (see Chaps. 4, 6, and 9).

tPA has been isolated from many tissues (reviewed in BACHMANN and KRUITHOF 1984; DANØ et al. 1985; COLLEN et al. 1989). Its principal site of synthesis is the endothelial cell (LEVIN and LOSKUTOFF 1982) but other cells have been shown to produce tPA, such as monocytes (HART et al. 1989), megakaryocytes (BRISSON-JEANNEAU et al. 1990a,b), mesothelial cells (VAN HINSBERGH et al. 1990; LATRON et al. 1991), mast cells (BANKL et al. 1999), vascular smooth muscle cells (PAPADAKI et al. 1998), cardiac fibroblast cells (TYAGI et al. 1998), and neuronal cells (WANG et al. 1998). Substantially more tPA is produced by several cell types when these are grown on flexible membranes which are subjected to repetitive stretch (IBA et al. 1991; TYAGI et al. 1998) or when arterial shear stresses of 14–28 dynes/cm^2 are applied to the cultures (DIAMOND et al. 1989, 1990; PAPADAKI et al. 1998). The increased expression of tPA by shear stress appears to be mediated by transforming growth factor β_1 (TGF-β_1) (UEBA et al. 1997).

In tissues and cell cultures, tPA is present in very small quantities and is bound firmly to particulate cell constituents. Therefore, isolation and purification of tPA from tissues has been difficult. Highly purified tPA preparations have been obtained from pig heart (COLE and BACHMANN 1977; WALLÉN et al. 1982), human uterus (RIJKEN et al. 1979), and from vascular perfusion fluids (BINDER et al. 1979). Real progress in the characterization of tPA was made when it was discovered that larger quantities of this enzyme could be obtained from certain cell cultures, such as a human melanoma cell line (RIJKEN and COLLEN 1981) and HeLa cells (WALLER and SCHLEUNING 1985; KRUITHOF et al. 1985).

In normal plasma the antigen concentration of tPA is about 5 µg/l, which corresponds to about 70 pmol/l (RIJKEN et al. 1983; STALDER et al. 1985; HOLVOET et al. 1985; NICOLOSO et al. 1988; TAKADA and TAKADA 1989). Most of the tPA is present in a complex with its primary inhibitor in plasma, the plasminogen activator inhibitor type 1 (PAI-1) (STALDER et al. 1985; BOOTH et al. 1987). Even when blood is collected in a buffer/anticoagulant mixture which brings about immediate acidification to approximately pH 5, thus preventing the complexing of still free tPA to PAI-1 (NILSSON and MELLBRING 1989), the amount of free tPA activity rarely exceeds 0.5 units/ml of plasma (equivalent to 1 µg/l since pure tPA has a specific activity of about 500 000–700 000 units/mg (GAFFNEY and CURTIS 1987).

C. Metabolism of tPA

Free, as well as complexed, tPA is rapidly bound to a variety of receptors on endothelial cells and hepatocytes and removed from the circulation (reviewed by Camani and Kruithof 1994). In normal individuals the half-life of tPA is about 3–4 min (Seifried et al. 1988; Tanswell et al. 1989; Huber et al. 1991). Interestingly, α-phase t/2 clearance of tPA was faster in subjects with low PAI-1 levels (3.5 min) vs high PAI-1 (5.3 min; $p = 0.006$). Clearance of tPA, PAI-1 and of the tPA/PAI-1 complex was best fit by a two-compartment pharmaco-kinetic model, and was greater for active tPA (t/2α of 2.4 min) than for tPA/PAI-1 complexes (t/2α of 5.0 min; $p = 0.006$) (Chandler et al. 1997). The slower clearance of tPA in the presence of high concentrations of PAI-1, and the slower clearance of the tPA/PAI-1 complex well explain why high levels of PAI-1 activity are associated with high tPA antigen concentrations in some patients.

Due to its clearance by the liver and the dependence of clearance on liver blood flow (Bounameaux et al. 1986; de Boer et al. 1992; Van Griensven et al. 1996) the half-life may be considerably prolonged in patients with hepatic cirrhosis (Fletcher et al. 1964; Huber et al. 1991; Lasierra et al. 1991).

D. Structure and Biochemistry of tPA

I. Gene Structure of tPA

1. Translated Regions

The fairly large gene for tPA comprises 32.7 kb and is located on bands p12–q11.2 on chromosome 8 (Benham et al. 1984; Triputti et al. 1986; Verheijen et al. 1986b). The 14 exons code for domains which are commonly encountered in serine proteases (Browne et al. 1985; Fisher et al. 1985; Friezner Degen et al. 1986). Figure 1 indicates the position of the interven-ing introns. The leader sequence of the pre-pro-tPA is encoded by the first exon, the signal peptide by the second, and the pro-sequence by the third exon. The finger region and the epidermal growth factor (EGF) domain are each coded by one exon, the two kringles by two exons each, as is the case for plas-minogen. The catalytic domain is encoded by the remaining five exons. There is considerable similarity of gene organisation and of exon-intron junction types (Patthy 1990) between tPA and the urokinase gene but the homologies of nucleotides and of amino acids between these two plasminogen activators is only about 40% (Friezner Degen et al. 1986).

The complete cDNA comprises 2530 bp (Pennica et al. 1983; Edlund et al. 1983; Fisher et al. 1985). The glycine^{-3} residue which represents probably the real NH$_2$-terminal of tPA (Wallén et al. 1983; Berg and Grinnell 1991) is preceded by 32 amino acids, of which approximately 20 constitute the hydrophobic signal peptide involved in the secretion of tPA.

2. 5′ and 3′ Flanking Regions

Over 9500 bp have been sequenced in the 5′ flanking region. Common tran-scription elements, such as three TATA boxes were identified; a CAAT box is

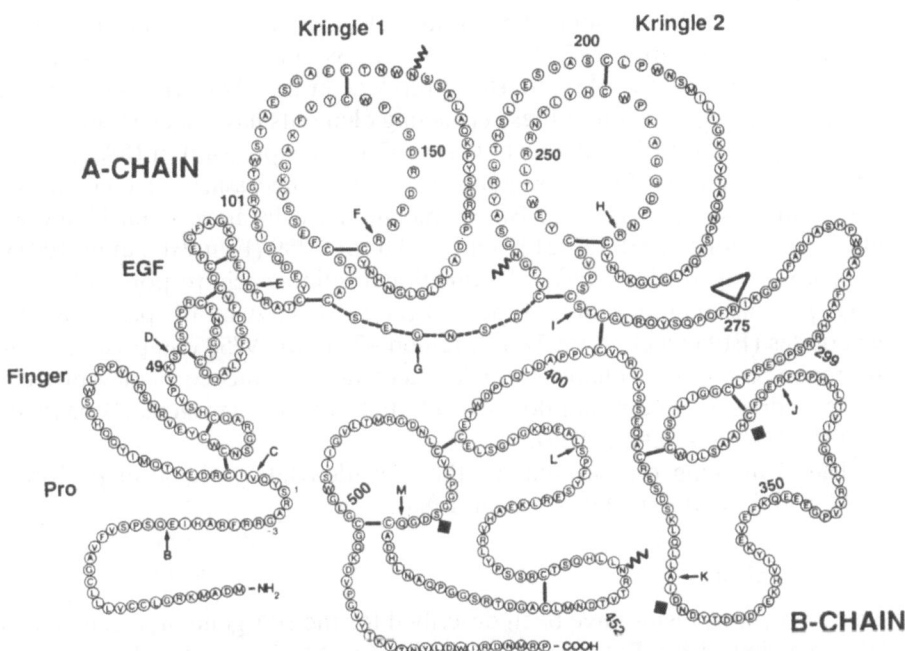

Fig. 1. Model of the secondary structure of the tPA precursor protein, including signal peptide and pro-sequence. The *solid black bars* indicate the potential disulfide bridges. The *arrows* (B–M) indicate the position of the individual introns in the protein. The cleavage site, dividing the heavy and the light chain is indicated with the *triangle*. The three amino acids His[322], Asp[371], and Ser[478], which make up the catalytic site, are marked with a *solid square*. *Zigzag lines* indicated the position of the three glycosylation sites. (Reproduced, slightly modified, with permission from Ny et al. 1984)

situated at position −112 to −116 respectively. Controversy exists concerning the transcription initiation site of the tPA gene. Two independent studies demonstrated the same start site 22 bp downstream of a TATAAA consensus sequence (FISHER et al. 1985; FRIEZNER DEGEN et al. 1986) but two other reports located the major transcription initiation start site 110 bp further downstream at a sequence homologous to the start sites identified in the mouse and rat tPA gene (HENDERSON and SLEIGH 1992; COSTA et al. 1998). Thus, two 5' untranslated regions (5'UTR) of mRNA have been found, the longer 5'UTR having 209, the shorter one 99 nucleotides (COSTA et al. 1998). DNase I protection analysis of the tPA gene promoter in human endothelial and phorbol stimulated HeLa cells revealed several protected regions: a CRE-like/TRE-like element between positions −102 to −115. This element is particularly interesting since it binds preferentially the cAMP-responsive element-binding protein (CREB). Mice with a CREB deletion exhibit defective long-term memory (BOURTCHULADZE et al. 1994), as do mice with a deletion of the tPA gene (HUANG et al. 1996). One could speculate therefore that the defects seen in CREB[−/−] mice are due to a defective regulation of the tPA gene. A CTF/NF-1-like element is located at position −92 to −77, three Sp1 binding sites at positions −72 to −66, −48 to −39, and +60 to +74, and three GC/GT boxes between −43 and +68 (Medcalf et al. 1990; Arts et al. 1997; Costa et al. 1998). Several of these *cis*-acting elements have been shown to be involved in constitutive and phorbol ester induced expression of tPA mRNA. Higher upstream two further regulatory elements, necessary for constitutive expression of the tPA gene in Bowes melanoma cells, have been identified between positions −2288 to −2129 and −2390 to −2289 (Fujiwara et al. 1994). Far upstream at −7.3 kb is located a functional retinoic acid response element (RARE), consisting of a direct repeat of the GGGTCA motif spaced by five nucleotides (Bulens et al. 1995). The region −7145 to −9758 comprises a multihormone responsive enhancer which is activated by glucocorticoids, progesterone, androgens, and mineralocorticoids, but not by oestrogens (Bulens et al. 1997; Merchiers et al. 1999).

The 3′ flanking region contains the polyadenylation signal at positions 32 688–32 693 (Friezner Degen et al. 1986).

3. Polymorphisms

Several polymorphisms have been described for the tPA gene or, respectively, cDNA (reviewed by Tishkoff et al. 1996). An Alu insertion/deletion polymorphism was more widely studied as to its physiopathological significance. With the exception of one case-control study which found that homozygosity for the insertion was associated with twice as many cases of AMI as was homozygosity for the deletion (van der Bom et al. 1997), other clinical studies found no correlation of the Alu insertion polymorphism with AMI or stroke (Iacoviello et al. 1996; Ridker et al. 1997; Steeds et al. 1998). No correlation was also found for the EcoRI restriction fragment length polymorphism and allograft vasculopathy in heart transplant recipients (Benza et al. 1998).

II. Protein Structure of tPA

1. Primary Amino Acid Sequence

The mature protein exists in two forms of different length. The N-terminal of Bowes melanoma cell tPA exhibits either glycine or serine (Wallén et al. 1983; Jörnvall et al. 1983; Berg and Grinnell 1991) but recombinant tPA nearly always has serine as the N-terminal amino acid (Pennica et al. 1983). In this chapter we will use the numbering based on the Ser-terminus of recombinant tPA, because of the very extensive literature which exists on structure-function relationships of this form of tPA. The wild-type version of tPA is a glycoprotein, composed of 527 amino acids. It is secreted as a single-chain molecule (sc-tPA) but can be easily converted to the two-chain form (tc-tPA) by plasmin-mediated cleavage of the Arg^{275}-Ile^{276} peptide bond. Surprisingly, the single-chain form is not a zymogen but a protease which, in the presence of fibrin, is nearly as active as the two-chain form, i.e., it is characterized by an abnormally low zymogenicity (RÅNBY et al. 1982; RIJKEN et al. 1982). The unusually high catalytic activity of sc-tPA may result from the absence of interactions present in typical zymogens, that stabilize an inactive conformation of the zymogen, or the presence of interactions, absent in typical zymogens, that stabilize an active conformation of the sc-tPA. Based on the crystal structure of tc-tPA and modeling of the sc-tPA, TACHIAS and MADISON (1996) concluded that in sc-tPA the interaction between Lys[156] and Asp[194] (chymotrypsin numbering, corresponding tPA numbering: 429 and 477) forms an important salt bridge which stabilizes the active conformation of sc-tPA. In sc-uPA this salt bridge does not exist because Glu[144], which is a His residue in tPA (uPA/tPA numbering 417), provides an alternative, electrostatic interaction for Lys[156]. Mutation of His[144] of tPA to an acidic residue, as in u-PA, selectively suppressed the activity of sc-tPA and thereby increased the zymogenicity by a factor of 16. Likewise, replacement of Lys[156] by Tyr increased the zymogenicity of sc-tPA 47-fold (TACHIAS and MADISON 1997b). The fully solvent-exposed, hydrophobic region comprising amino acids 420–423 of tPA which forms a surface loop near one edge of the active site of tPA constitutes an important secondary site for the interaction of tPA with plasminogen in the absence of fibrin (see below: enzyme kinetics in the presence of fibrin) (KE et al. 1997).

The N-terminal portion, also called A-chain, contains several domains typical of serine proteases (Fig. 1), such as the finger- and epidermal growth factor (EGF)-domains, and two kringle domains. The catalytic portion of tPA, also called B-chain, is homologous to the proteolytic domain of other serine proteases.

a) The Finger Domain

The finger domain extends from residues 6 to 43 and has a 34% homology with the first finger of bovine fibronectin (PETERSEN et al. 1983). It is involved in the binding of tPA to fibrin, since a mutant which contains only the finger region and the proteolytic domain bind to fibrin (VAN ZONNEVELD et al. 1986; VERHEIJEN et al. 1986a). Urokinase, which does not contain this domain, does not bind to fibrin and a degraded form of tPA which has lost an N-terminal fragment of 12 kDa by proteolytic degradation bound less well to fibrin than wild type tPA (BÁNYAI et al. 1983). The solution structure of the finger domain, determined by ¹H-NMR, shows striking similarity to that of the seventh type 1 repeat of human fibronectin (DOWNING et al. 1992).

b) The Epidermal Growth Factor Domain

The sequence 44 to 92 exhibits homology with the murine and human EGF (SAVAGE et al. 1973). Mutants lacking this region exhibit diminished clearance of tPA from the circulation and it thus appears that epitopes on the EGF are recognized by tPA receptors in the liver (see below). ¹H-NMR demonstrated that the EGF module of tPA closely matches the EGF consensus module structure, previously determined (SMITH et al. 1994).

c) The Kringle Domains

tPA contains two kringles, which have both been isolated in high purity by chemical or recombinant methodologies (CLEARY et al. 1989; WILHELM et al. 1990; VLAHOS et al. 1991). The majority of workers in this field have not found a biological function for kringle 1 (VAN ZONNEVELD et al. 1986; VERHEIJEN et al. 1986a; DE MUNK et al. 1989). In contrast, kringle 2 has affinity for lysine, ω-amino acids, such as ε-amino caproic acid (EACA) and fibrin (VAN ZONNEVELD et al. 1986; VERHEIJEN et al. 1986a). The affinity for the binding of EACA (a model compound for C-terminal lysine residues) and of N-acetyllysine methyl ester (a model compound for intrachain lysine residues) is about equal, suggesting that tPA does not prefer C-terminal lysine residues (as plasminogen does) for binding. Intact tPA and a variant consisting only of kringle 2 and the protease domains were found to bind to fibrinogen CNBr fragment FCB-2, the very fragment that also binds plasminogen and acts as a stimulator of tPA-catalysed plasminogen activation. In both cases, binding was completely inhibited by EACA, pointing to the involvement of a lysine binding site in this interaction. The binding of the finger domain appears to involve a binding site which is different from that exposed on the FCB-2 and cannot be blocked by EACA (DE MUNK et al. 1989).

The structure-function relationships between kringle 2 and ω-amino acids and lysine have been studied in detail using ^{1}H-NMR spectroscopy (BYEON et al. 1991, 1995), microcalorimetry (KELLEY et al. 1991), crystallography (DE VOS et al. 1992), and site directed mutagenesis of residues thought to be critical for the interaction of kringle 2 with ligands (WEENING-VERHOEFF et al. 1990; COLLEN et al. 1990; KELLEY et al. 1991; TULINSKY 1991; DE SERRANO et al. 1992). The crystal structure of kringle 2 resembles that of plasminogen kringle 4; however, there are differences in the lysine binding pocket. The core of kringle 2 is formed by a hydrophobic cluster of three tryptophan residues in positions 25, 63, and 74 surrounded by aromatic and hydrophobic side chains which form, at the surface of the kringle, a hydrophobic grove. Ligand binding appears to rely mostly on the integrity of Trp[63] and Trp[74] and aromaticity at Tyr[76]. With regard to aromatic amino acids, kringle folding is most dependent on Tyr[9], Trp[25], Tyr[52], Trp[63], and Tyr[76] (CHANG et al. 1997, 1999). Mutation of the critical amino acids Lys[33], Asp[55], Asp[57], or Trp[72] results in markedly diminished binding to lysine-Sepharose and/or interaction with EACA (DE VOS et al. 1992; DE SERRANO et al. 1992). Attempts to improve the efficiency of therapeutically used recombinant tPA by modifications which might increase the affinity for fibrin, such as the replacement of amino acids in kringle 1 or 2 with arginine residues in positions corresponding to those found in the plasminogen kringles at positions 34 and 71, yielded indeed molecules which, in vitro, had a slightly superior catalytic efficiency for plasminogen activation, but were not significantly better as a thrombolytic agent, in vivo, than wild-type tPA (COLLEN et al. 1990).

d) The Protease (Catalytic) Domain

The catalytic domain of tPA is made up of 230 amino acids whose sequence shows homology with that of the other serine proteases. It contains in the active site the three amino acids His[322], Asp[371], and Ser[478].

2. Crystal Structure of the Protease Domain

The 2.3-Å crystal structure of the recombinant tc-tPA, consisting of a 17 amino acid A-chain and 252 residue B-chain revealed that the catalytic domain of tPA is a roughly spherical molecule of a radius of 25 Å. The molecule consists of two six-stranded barrel-like sub-domains held together by three *trans*-domain straps (LAMBA et al. 1996). It display strong structural similarity with other trypsin-like serine proteases, in particular with thrombin. The active site cleft is shaped and narrowed by four surface loops. The 37-loop around Arg[299] (chymotrypsin (c) numbering c37B) exhibits five additional residues (RRSPG) compared with chymotrypsinogen. It projects out of the molecular surface as a β-hairpin and is of fundamental importance for the interaction with PAI-1 and also for the fibrin specificity of tPA (MADISON et al. 1989, 1990; BENNETT et al. 1991; PAONI et al. 1992, 1993a; TACHIAS and MADISON 1997a). The 60-loop around Arg[327] (c60B) exhibits some similarity to, but is shorter than the corresponding loop in thrombin. Further loops are found around Ser[381] (the 110-loop; c110B) and Gly[465] (the 186-loop; c186D).

Analysis of the crystal structure of the catalytic domain of sc-tPA in a covalent complex with Glu-Gly-Arg-chloromethylketone revealed that the Lys[429] side chain (c156) is bound in the Ile[176] (c16) pocket and forms an asymmetric salt bridge with Asp[477] (c194). This explains the low zymogenicity of sc-tPA (RENATUS et al. 1997; STUBBS et al. 1998) (see above).

3. Glycosylation of tPA

The tPA protein has four potential glycosylation sites of which three are occupied in type 1 (Asn[117], Asn[184], and Asn[448]) and two in type II tPA (Asn[117] and Asn[448]). The M_r of type II tPA is about 3000 Da smaller (RÅNBY 1982b). In natural tPA, as well as in recombinant tPA expressed in Chinese hamster ovary cells, residue 117 is predominantly N-glycosylated with oligomannose-type structures; residues 184 and 448 are predominantly associated with complex-type structures in recombinant tPA and tPA isolated from fibroblast cell lines, but with both complex- and oligomannose-type structures when isolated from melanoma cells (PAREKH et al. 1989; SPELLMAN et al. 1989). Glycosylation at residue 184 influences the biological properties of tPA. The second order rate constant for the plasmin-mediated conversion of sc-tPA to tc-tPA for the type II form is about twice that for type I, and the type II form also has a lower affinity for lysine (and fibrin) and a lower fibrinolytic activity (reviewed in DWEK 1995; RUDD et al. 1995).

Like urokinase, tPA has an O-linked fucose attached to Thr^{61} in the EGF domain (Harris et al. 1991). Addition of fucose or the enzymatic removal of α-fucose reduced the binding of tPA to HepG2 cells (Hajjar and Reynolds 1994). This suggests that the EGF-associated O-linked α-fucose mediates binding to, and degradation of tPA by, hepatocytes.

III. Enzymatic Properties of tPA

1. Enzyme Kinetics in the Absence and Presence of Fibrin

tPA is a very specific serine protease and its only known substrate is plasminogen where it cleaves the single Arg^{561}-Val^{562} peptide bond. It is a very inefficient activator of plasminogen in the absence of fibrin but in its presence the activation of plasminogen is greatly potentiated (Camiolo et al. 1971). This is explicable by the assembly of plasminogen and of tPA on the fibrin surface. Since the original observation of Thorsen, that fibrin binds tPA, but not urokinase (Thorsen et al. 1972), many investigators have observed this phenomenon (Binder et al. 1979; Aasted 1980; Radcliffe and Heinze 1981; Allen and Pepper 1981; Rånby 1982a; Kruithof and Bachmann 1982; Tran-Thang et al. 1984a; Rijken and Groeneveld 1986; reviewed by Fears 1989). Most agree that binding of sc-tPA and of tc-tPA is roughly comparable, although the single chain form may bind slightly better (Higgins and Vehar 1987). The $K_{d}s$ for binding of tPA to fibrin clots, in the absence of plasminogen, range from 140 nmol/l to 400 nmol/l (Rijken et al. 1982; Rånby et al. 1982; Kruithof and Bachmann 1982; Higgins and Vehar 1987; Larsen et al. 1988; Bosma et al. 1988). In the presence of plasminogen the affinity of tPA to fibrin increases about 20-fold (K_{d} of 20 nmol/l) (Rånby et al. 1982). Two explanations are given for this observation: (1) the formation of a ternary complex comprising tPA, plasminogen, and fibrin, (2) binding of tPA to plasminogen which, upon binding to fibrin, has taken on an open conformation. The isolated A-chain of tPA was indeed shown to bind to Glu- and mini-plasminogen with a K_{d} of 100 nmol/l (Geppert and Binder 1992).

The kinetic parameters which have been reported for the activation of plasminogen by tPA show great variations. This is due to a multitude of factors including variations in the different tPA and plasminogen preparations and in the concentrations of substrate used and, for studies using fibrin or fibrin derivatives, the nature of the fibrin stimulator used and the non-Michaelis-Menten behavior of enzyme kinetics.

In the absence of fibrin, K_{m} values for the activation of Glu-plasminogen by tPA from 9 μmol/l (Rånby 1982b), 65 μmol/l (Hoylaerts et al. 1982) to slightly over 100 μmol/l (Rijken et al. 1982) have been found, using plasminogen concentrations around 0.5–3 μmol/l, i.e., corresponding to those found in human plasma (2 μmol/l). When studying plasminogen activation by tPA at very low substrate concentrations, such as may occur in the tissues where the plasminogen-plasmin system plays a key role in the facilitation of cell migration, much lower K_ms are found which approximate those encountered when studying plasminogen activation by tPA in the presence of fibrin (NIEUWEN-HUIZEN et al. 1988; GEPPERT and BINDER 1992). In general, K_ms are three to four times lower with tc-tPA than with the sc-tPA when the activation of Glu-plasminogen is investigated in the absence of fibrin. This difference disappears when fibrin is present and K_m values typically are two orders of magnitude smaller with only moderate increases of k_{cat}. Values for K_m in the presence of fibrin range from $0.16\,\mu mol/l$ (HOYLAERTS et al. 1982) to $1.1\,\mu mol/l$ (RIJKEN et al. 1982) and k_{cat} values from $0.1\cdot sec^{-1}$ (HOYLAERTS et al. 1982) to $1.1\cdot sec^{-1}$ (RIJKEN et al. 1982). However, as mentioned above, several authors found non-linear enzyme kinetics, particularly in the presence of fibrin (RÅNBY 1982b; NORRMAN et al. 1985; GEPPERT and BINDER 1992). This behavior is apparently not due to conversion of sc-tPA to tc-tPA and/or the conversion of Glu- to Lys-plasminogen, since non-Michaelis-Menten kinetics are also found when activating Lys-plasminogen by tc-tPA (GEPPERT and BINDER 1992).

Norrman concluded that there were two phases in the activation of Glu-plasminogen by tPA in the presence of fibrin; the K_m in the initial phase was $1.05\,\mu mol/l$ and k_{cat} $0.15\,sec^{-1}$. In the later phase K_m decreased to $0.07\,\mu mol/l$ and k_{cat} remained unchanged at $0.14\,sec^{-1}$ (NORRMAN et al. 1985). This change of K_m can be explained by the generation of new, high affinity binding sites for plasminogen and tPA in partially digested fibrin exposing new C-terminal lysine residues (RIJKEN et al. 1979; SUENSON et al. 1984; TRAN-THANG et al. 1984b, 1986; HARPEL et al. 1985; DE VRIES et al. 1990).

Non-Michaelis-Menten kinetics for a single substrate enzyme can be due to a modifier mechanism. When such a mechanism is found in a purified system initially consisting of only the enzyme (tPA) and its substrate (plasminogen) an allosteric site, in the enzyme, for the substrate or the product can be assumed. The allosteric site will become occupied at higher plasminogen concentrations, thereby shifting the enzyme to an altered conformational state (GEPPERT and BINDER 1992).

2. Differences Between sc-tPA and tc-tPA

In the absence of fibrin, sc-tPA has about a 3- to 4-fold lower catalytic efficiency than the tc-tPA to cleave synthetic substrates, such as H-D-isoleucyl-L-prolyl-L-arginine-p-nitroanilide (S-2288) (RÅNBY et al. 1982; RIJKEN et al. 1982; LOSCALZO 1988; ANDREASEN et al. 1991). Tc-tPA is also more rapidly inhibited by DFP and by serpins, particularly PAI-1 (SPRENGERS and KLUFT 1987; KRUITHOF 1988). In the presence of fibrin this difference disappears. Sc-tPA has a higher affinity for fibrin and for endothelial cells (TATE et al. 1987; HUSAIN et al. 1989; AERTS et al. 1989; HIGGINS et al. 1990) but the number of low affinity binding sites on fibrin was reported to be higher for tc-tPA (HUSAIN et al. 1989). Taken together these observations suggest that sc-tPA undergoes a conformational change on binding to fibrin or when cleaved at position Arg^{275}.

3. Localization of the Epitopes in Fibrin that Enhance the Rate of Plasminogen Activation by tPA

Two sites in the fibrin molecule, referred to as fibrin-specific epitopes, have been identified which accelerate plasminogen activation, one in the cyanogen bromide cleavage fragment FCB-2 which encompasses the sequence Aα(148–160) chain (Voskuilen et al. 1987), and the other in the fragment FCB-5 (Schielen et al. 1991; Yonekawa et al. 1992). The FCB-5 fragment consists of fibrin(ogen) chain fragments $\gamma^{311–336}$ and $\gamma^{337–379}$ which are linked by a disulfide bond. Sequences $\alpha^{148–160}$ and $\gamma^{312–324}$ are buried in fibrinogen and become exposed during the conversion of fibrinogen into fibrin as demonstrated with monoclonal antibodies directed against these peptides (Schielen et al. 1989, 1991). The term fibrin-specific epitope is not entirely appropriate, since Mosesson et al. (1998) have shown that the critical event for their exposure is not the release of fibrinopeptides A and B but rather the polymerization process of fibrin(ogen) strands. Indeed, exposure of these sites could be obtained when fibrinogen was crosslinked (polymerized) by Factor XIIIa without inducing clotting of fibrinogen in the classical sense (Fig. 2). The CNB-2 fragment which encompasses Aα(148–160) binds plasminogen and tPA, whereas the $\gamma^{311–379}$ sequence binds tPA. It is therefore probable that epitope exposure, binding of tPA and plasminogen, and the enhancement of the rate of plasminogen activation by tPA, are all linked together.

IV. Interactions of tPA with Serpins

The physiologically important, primary inhibitor of sc- and of tc-tPA is PAI-1. PAI-1 is produced by endothelial cells, adipocytes, and megacaryocytes (Loskutoff et al. 1983; Morange et al. 1999; Madoiwa et al. 1999). In human plasma there exists normally a slight molar excess of PAI-1 over tPA; therefore most of the tPA molecule is present in a complex with PAI-1. Table 1 lists the second order rate constants for the inhibition of tPA by several serpins. Sc-tPA is only efficiently inhibited by PAI-1 and the second order rate constants of other serpins do not reach values which are likely to be of biological significance for the inhibition of tPA. An exception may be the recently discovered mammary tumor suppressor gene maspin. It inhibits tPA relatively efficiently and may represent a regulatory mechanism in the control of mammary tumor cell motility and invasion (Sheng et al. 1998). tPA was also reported to interact with α2-macroglobulin, but in this complex tPA remains active (Ieko et al. 1997).

By 1990 Madison et al. (1990) had already demonstrated that replacement of the positively charged amino acids Arg^{298} and Arg^{299} in the 37-loop (see Sect. D.II.2) by Glu resulted in mutant tPAs that associated poorly with PAI-1. A similar result was obtained when Arg^{304}, situated on the edge of the active site cleft, was converted to Glu (Madison et al. 1989). The authors predicted that the tPA sequence 298–302 interacts with the negatively charged

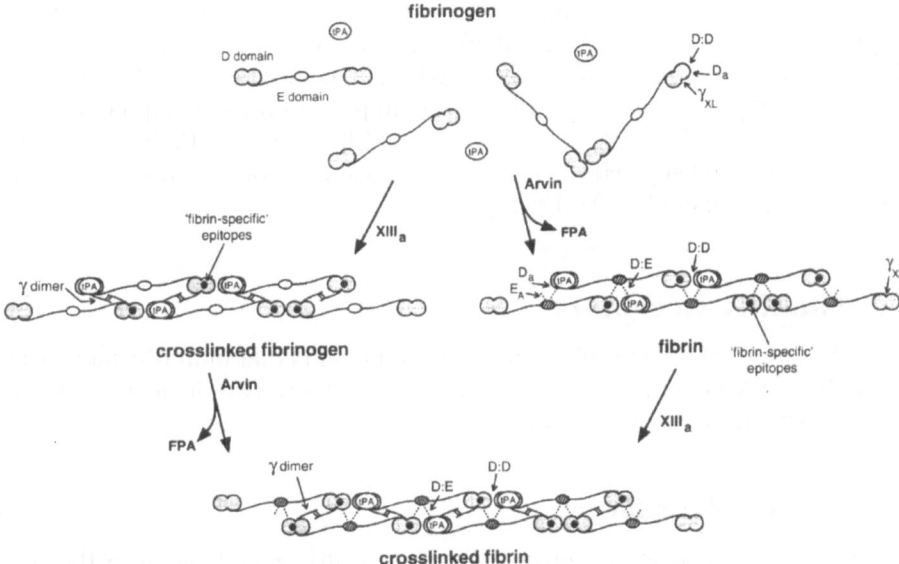

Fig. 2. Diagram of the formation of crosslinked fibrinogen (*left hand side*), respectively of fibrin (*right hand side*). In the presence of Factor XIIIa the XL sites on the γ-chain of fibrinogen are crosslinked, forming γ-dimers). The E-domain is not activated in this reaction (*empty circles*) and does not participate in the polymer formation. Crosslinking of the D-domains on the γ-chains leads to exposure of fibrin-specific epitopes (*solid circles*); tPA binding to these epitopes is shown. Upon treatment of the fibrinogen polymer with arvin (*right hand side*), fibrinopeptides A (FPA) are cleaved and the activated E-domain (*cross-hatched circles*) now establishes the DDE polymer structure. The D:E interaction that takes place between the E_A and D_a sites during fibrin assembly is indicated, as is the site of the D:D self-association reaction. This reaction also generates fibrin-specific epitopes. Crosslinking between D-domains however takes only place when Factor XIIIa is present. (Taken from MOSESSON et al. 1998, with permission)

Table 1. Second order rate constants for the inhibition of tPA by several serpins (Taken from NORDENHEM and WIMAN 1998; SHENG et al. 1998)

	Second order rate constant of inhibition in $M^{-1} \cdot sec^{-1}$	
	sc-tPA	tc-tPA
PAI-1	10^7	3×10^7
PAI-2	9×10^2	2×10^5
Protease nexin	1.5×10^3	3×10^4
Maspin	10^6	
Protein C inhibitor	10^3	10^3
C1-Inhibitor	1.3	
α_2-Plasmin inhibitor	13	

PAI-1 sequence 350–355 (Glu-Glu-Ile-Met-Asp). Replacement of all three arginines in positions 298, 299, and 304 resulted in a mutant which was inhibited 120000 times less rapidly by PAI-1 than wild-type tPA (Tachias and Madison 1997a). These observations lead, in part, to the development of the new tPA mutant TNK-tPA (tenecteplase; see Chap. 19, Sect. B.VII) where the sequence 296–299 (Lys-His-Arg-Arg) is replaced by four alanines (Bennett et al. 1991; Paoni et al. 1992, 1993b).

E. The tPA Receptors

A wide variety of structurally unrelated components that bind tPA have been described. tPA binding moieties can be separated into two distinct functional groups: activation receptors and clearance receptors.

I. Activation Receptors

As a rule, the first group localizes tPA on a cell surface and enhances the activation of plasminogen by tPA (see also Chap. 6, Sect. C). This class comprises 42-kDa annexin II (Hajjar et al. 1994; Hajjar and Krishnan 1999; Kang et al. 1999; Fitzpatrick et al. 2000), 45-kDa actin (Dudani and Ganz 1996), heparan sulfate and chondroitin sulfate-like proteoglycans (Bohm et al. 1996), cytokeratin 8 and 18 (Hembrough et al. 1996; Kralovich et al. 1998), and tubulin (Beebe et al. 1990). Several, as yet poorly defined, tPA receptors have been identified on endothelial cells and it is not clear yet whether some of these are identical with each other. Beebe et al. (1989) described a receptor on endothelial cells that interacts with the finger domain of tPA; a peptide corresponding to residues 7–17 of tPA inhibited the binding of tPA to human umbilical endothelial cells (HUVEC). In addition to actin, Dudani et al. (1996) have described two further 37-kDa and 45-kDa tPA-binding proteins on HUVECs that are distinct from actin. Lysine inhibited the binding to these three receptors. Cheng et al. (1996) have described a 55-kDa protein that binds to a sequence in the 37-loop comprising residues Ala-Lys-His-Arg-Arg-Ser-Pro-Gly-Glu-Arg, the sequence in the B-chain that also binds to PAI-1. However tPA bound to this 55-kDa protein maintained its activity and was protected from inactivation by PAI-1. Lysine analogues did not inhibit the interaction of the 55-kDa receptor with tPA. Still another 20-kDa tPA-binding protein was characterized and purified from HUVECs by Fukao and Matsuo (1998). This protein did not interact with plasminogen, but binding was inhibited by ω-amino acids, suggesting that the lysine-binding site of tPA is involved. The purified protein did not interact with antibodies directed against annexin II or α-enolase. In the presence of this receptor plasminogen activation by tPA was enhanced 90-fold. A receptor that interacts with the B-chain of tPA and potentiates plasminogen activation about 100-fold has recently been described on vascular smooth muscle cells (Ellis and WHAWELL 1997; WERNER et al. 1999). This receptor appears to be distinct from any of the other known tPA receptors.

II. Clearance Receptors

As described in Sect. C, tPA is rapidly cleared from the circulation. Receptor-mediated clearance of tPA and its subsequent degradation is an important regulatory mechanism to control the plasma concentration of tPA and to remove inactive tPA/PAI-1 complexes from the circulation. The liver is the principal organ involved in tPA clearance. At least two different clearance mechanisms exist: liver endothelial cells and Kupffer cells bind tPA via the mannose receptor, whereas parenchymal hepatocytes bind tPA via the low density lipoprotein receptor-related protein (LRP)/α_2-macroglobulin receptor. A secondary role is probably played by some types of endothelial cells in the general circulation. Endothelial cells synthesize PAI-1 that may be bound to the cell surface via vitronectin and capture free tPA, which can then be internalized by the LRP pathway. This process is inhibited in the presence of the 39-kDa receptor-associated protein (RAP) (MULDER et al. 1997).

1. The Mannose Receptor

Early studies had shown that the uptake of tPA by liver endothelial and by Kupffer cells was inhibited by ovalbumin and that tPA mutants lacking the high-mannose glycosylation site at Arg[117] (see Sect. D.II.3) were more slowly cleared from the circulation of experimental animals (reviewed by CAMANI and KRUITHOF 1994; REDLITZ and PLOW 1995). This led to the identification of the mannose receptor (TAYLOR et al. 1990) as a major tPA receptor by OTTER et al. (1991). Binding of ligands to the mannose receptor is pH- and Ca^{++}-dependent. Binding of tPA to the mannose receptor is inhibited by mannose-albumin, mannan, D-mannose and L-fucose and particularly by cluster mannosides of the composition M_6L_5 (OTTER et al. 1991; NOORMAN et al. 1997; BIESSEN et al. 1997). tPA has a much higher affinity (K_D of 1–4nmol/l) for the mannose receptor than other high-mannose-type containing glycoproteins such as ribonuclease B, β-glucuronidase, and ovalbumin (K_D of 60–600nmol/l) (NOORMAN et al. 1998).

Two mutants of tPA are in clinical use where the Asn[117] residue in kringle 1 has been replaced by Gln: tenecteplase (TNK-tPA) and lanoteplase (nPA, SUN 9216) (see Chap. 19, Sect. B). In pamiteplase (YM 866) kringle 1 has been deleted. All three mutants exhibit considerably longer half-lives than wild-type tPA. At present it is not known whether the fucose residue at Thr[61] in the EGF-domain also binds to the mannose receptor (fucose inhibits the binding of tPA to the mannose receptor in a purified system) or is taken up by the previously described fucose receptor (LEHRMAN and HILL 1986). Mutants lacking the finger- and EGF domains, such as reteplase (rPA), also exhibit a longer plasma clearance rate. However at least two groups found that inhibition of the mannose receptor and of the LRP pathways of clearance resulted in at least a tenfold decrease of tPA clearance and estimated that clearance due to fucose amounted to 5% of total clearance at most (NARITA et al. 1995; BIESSEN et al. 1997).

2. The LDL Receptor-Related Protein (LRP, α_2-Macroglobulin Receptor)

The other major pathway for tPA clearance involves the α_2-macroglobulin receptor/LRP. The 600-kDa LRP, the largest plasma membrane protein ever described, comprises 4525 amino acids. It mediates the clearance of apolipoprotein E-enriched chylomicron remnants, toxins, cytokines, complexes of α_2-macroglobulin with proteases from all subclasses, and of free tPA and tPA/PAI-1 and uPA/PAI-1 complexes (reviewed by ANDREASEN et al. 1994). Complex formation of tPA with PAI-1 increases the rate of clearance by LRP by at least one order of magnitude compared to that of free tPA (WING et al. 1991; CAMANI et al. 1994). The binding site(s) for tPA/PAI-1 and for uPA/PAI-1 complexes are situated in the second and forth complement-type domain cluster of LRP (MOESTRUP et al. 1993; WILLNOW et al. 1994; HORN et al. 1997; NEELS et al. 1999). RAP inhibits endocytosis of most ligands to LRP (HERZ et al. 1991; BU et al. 1992; ORTH et al. 1992; CAMANI et al. 1994).

Other LRP-like multiligand receptors such as the 600-kDa glycoprotein 330 and the 130-kDa VLDL receptor are also able to mediate the clearance of free and of PAI-1 complexed tPA (ANDREASEN et al. 1994; MOESTRUP 1994; KASZA et al. 1997; MIKHAILENKO et al. 1999). For efficient uptake and clearance of several ligands LRP works in cooperative fashion with co-receptors. Examples of such receptors are the uPAR for the LRP-mediated degradation of uPA/PAI-1 complexes (KOUNNAS et al. 1993; NYKJAER et al. 1994), and glycosaminoglycans for the degradation of lipoprotein lipase (KOUNNAS et al. 1995) and thrombin/protease nexin complexes (KNAUER et al. 1997). CAMANI et al. (2000) recently found that some monocyte-like cell lines that expressed LRP on the cell surface were unable to degrade tPA and postulated the existence of a co-receptor for efficient degradation of tPA by LRP.

F. Pathophysiology of tPA

I. Intravascular Function of tPA

During the formation of a hemostatic plug biochemical mechanisms are already initiated to limit the extent of the hemostatic process and to reestablish normal blood flow. To a large extent this is accomplished by localized activation of the plasminogen-plasmin enzyme system, also called the fibrinolytic system. To accomplish healing of a vascular lesion without compromising the stability of the hemostatic plug too early and to limit the activation of the plasminogen system to the injured area, a finely tuned dynamic mechanism is necessary consisting of plasminogen, tPA, uPA, and the fibrinolytic inhibitors PAI-1 and α_2-antiplasmin. Endothelial cells produce both tPA and PAI-1. Since tPA and plasminogen bind to fibrin and plasminogen activation is greatly enhanced when these two components assemble on fibrin, it is generally accepted that the dissolution of a thrombus is initiated by the release of endothelial tPA. If the dynamic balance between pro- and antifibrinolytic constituents is upset, bleeding may occur if there is an excess of fibrinolytic activity or a deficiency of one of the fibrinolytic inhibitors. Conversely, it has been assumed that an abnormal release of tPA (KORNINGER et al. 1984; STALDER et al. 1985; JUHAN-VAGUE et al. 1987; GRIMAUDO et al. 1992; reviewed by DECLERCK et al. 1994) or an excess of PAI-1 favors the development of thrombosis. However, although an association of a defective release of tPA or increased levels of PAI-1 with postoperative or idiopathic thrombosis has been convincingly demonstrated, prospective studies on the role of a deficient fibrinolytic system as the cause of thromboembolic disease are lacking (reviewed by PRINS and HIRSH 1991; DECLERCK et al. 1994; BACHMANN 1995).

A lifelong hemorrhagic disorder associated with enhanced fibrinolysis due to an isolated increase of circulating tPA has been described in only two cases (AZNAR et al. 1984; BOOTH et al. 1983). Hereditary deficiency of either the α_2-plasmin inhibitor (AOKI et al. 1979, 1980) or PAI-1 (DIÉVAL et al. 1991; FAY et al. 1992, 1997; LEE et al. 1993) may also result in a lifelong bleeding disorder. Bleeding is presumably caused by premature lysis of hemostatic plugs at sites of vascular trauma. It may occur after initial hemostasis following surgery but may also be manifest spontaneously, for example, by epistaxis or ecchymoses.

Congenital deficiencies of tPA or uPA have not been described. Until publication of the seminal work of the Leuven group it was assumed that such a deficiency constituted a lethal condition. CARMELIET et al. (1994) reported on the successful deletion of the tPA and uPA gene in mice. Unexpectedly, mice with a disrupted tPA gene developed normally and did not exhibit any macroscopic abnormality. Microscopic examination revealed mild glomerulonephritis. In a plasma clot lysis assay fibrinolytic activity was reduced compared to wild-type animals. After endotoxin-induced thrombogenic stimulation tPA$^{-/-}$ mice did exhibit more extended thrombotic lesions than tPA$^{+/+}$ mice. Mice with a deleted uPA gene exhibited a slightly increased thrombotic susceptibility with occasional spontaneous intravascular fibrin deposition. In both deficiency states the abnormalities found were rather mild but mice with a double deficiency of tPA and uPA suffered from severe spontaneous thrombosis (CARMELIET et al. 1994; CARMELIET and COLLEN 1996). It must therefore be concluded that, in mice, there exists a certain redundancy of biological functions with respect to plasminogen activators. On the other hand it is obvious that a complete knock-out of the fibrinolytic activators leads to severe disease. It is not known to what extent these results apply to human pathophysiology.

Support for the importance of continuous release of tPA from the vasculature comes from a recent study in rabbits. WAUGH et al. (1999) cloned the human tPA gene into an adenoviral vector under the control of the RSV promoter. Three days after transfection of HUVEC cultures with the adenovirus tPA levels in the medium were three to four times higher than in controls. To determine whether these levels of local tPA production would be adequate to limit thrombus formation, the construct was tested in an in vivo model of arterial thrombosis. The common femoral artery in rabbits was divided, resutured, and constriction applied to produce a thrombogenic stimulus. Control and mock-virus treated rabbits and a group that was treated with the conventional intravenous infusion of tPA (0.15 mg/kg bolus, followed by 0.75 mg/kg given over 30 min, then 0.5 mg/kg over 60 min) all developed obstructive thrombosis. The injection of 5×10^9 plaque-forming units into the isolated femoral arterial segment greatly diminished thrombus formation and obstructive thrombosis was not encountered ($p < 0.01$). These experiments demonstrate that, in rabbits, the continuous release of tPA successfully counteracted the harsh thrombogenic stimulus, whereas the infusion of tPA as used for the treatment of acute myocardial infarction was ineffective.

A very large literature exists on the predictive value of tPA antigen and PAI as a risk factor for acute myocardial infarction and stroke. As stated above, high tPA antigen concentrations are generally associated with and due to high PAI-1 levels. This is explained by the lower clearance rate of the tPA/PAI-1 complex (CHANDLER et al. 1997). By multivariate analysis high tPA antigen levels were shown to be significantly associated with PAI-1 activity, hypertriglyceridemia, the type of angina, multivessel disease, and hypercholesterolemia (GEPPERT et al. 1998). Since this chapter does not deal with PAI-1, I will only list some references on this subject, from over 200 articles published on this particular aspect of the topic: (RIDKER et al. 1994; THOMPSON et al. 1995; RIDKER 1997, 1999; VAN DER BOM et al. 1997; SMITH et al. 1998; THÖGERSEN et al. 1998; JANSSON et al. 1998; CARTER et al. 1998; MACKO et al. 1999).

The regulation of the tPA expression is discussed in Chap. 20.

II. Other Functions of tPA

Proteolysis generated by the action of tPA upon plasminogen has been described in many physiological processes, such as ovulation, embryo implantation and embryogenesis, mammary involution, etc. However, mice with single deficiencies of tPA or uPA were found to have normal reproduction, although mice with a combined deficiency of tPA and uPA were significantly less fertile. This implies a functionally redundant mechanism for plasmin formation during gonadotropin-induced ovulation (LEONARDSSON et al. 1995).

By in situ hybridization tPA mRNA was localized in neuro-ectodermal structures, neural crest-, mesoderm-, and endoderm-derived tissues. Analysis of β-gal expression of transgenic mice and rat embryos bearing a lacZ reporter gene under the control of the human tPA promoter revealed blue staining in areas of active cell migration, particularly of neural crest-derived tissues, suggesting a role for tPA in tissue remodeling during neuronal development (THEURING et al. 1995). Neuronal plasticity is important for many brain functions and tPA has been shown to be involved in remodeling and elongation of axons (LOCHNER et al. 1998; MÜLLER and GRIESINGER 1998; SEEDS et al. 1999) and hippocampal long-term potentiation and learning (FREY et al. 1996; HUANG et al. 1996; BARANES et al. 1998; MADANI et al. 1999; ZUO et al. 2000). Indeed, tPA$^{-/-}$ mice were slower learners in the water maze test. (It consists of a circular pool of 160 cm diameter; water is made opaque by the addition of skimmed milk. At one position a 15 cm diameter circular platform is submerged 1.5 cm below the water levels. The swimming mice have to find this hidden platform).

Two studies found that tPA also exerts an anti-inflammatory effect (STRINGER et al. 1997; KRAPOHL et al. 1998) and stimulates endothelial cell proliferation (WELLING et al. 1996). The latter effect was not related to tPA's proteolytic activity, since it was not inhibited in the presence of aprotinin.

The development of tPA as a thrombolytic agents is discussed in Chap. 8.

List of Symbols and Abbreviations

AP1	activator protein 1 binding site
AP2	activator protein 2 binding site
CRE	cAMP responsive element
CREB	cAMP responsive element binding-protein
EACA	ε-amino caproic acid
EGF	epidermal growth factor
ERK	extracellular signal regulated kinase
HMW	high molecular weight
HUVEC	human umbilical endothelial cell
LMW	low molecular weight
LRP	low density lipoprotein receptor-related protein (α_2-macroglobulin receptor)
MMP	metalloproteinase
NF-κB	nuclear transcription factor, first found in κ-light chain of B lymphocytes
NMR	nuclear magnetic resonance
PAI-1	plasminogen activator inhibitor type-1
RAP	receptor-associated protein
RARE	retinoic acid responsive element
SK	streptokinase
Sp1	specific promoter-1 binding site
tPA	tissue-type plasminogen activator
sc-tPA	single chain tPA
tc-tPA	two chain tPA
uPA	urinary-type (or urokinase-type) plasminogen activator, also called urokinase

sc-uPA	single chain uPA, also called pro-urokinase
tc-uPA	two chain uPA, also called urokinase
TGFβ	transforming growth factor β
UTR	untranslated region

References

Aasted B (1980) Purification and characterization of human vascular plasminogen activator. Biochim Biophys Acta 621:241–254

Aerts RJ, Gillis K, Pannekoek H (1989) Single-chain and two-chain tissue-type plasminogen activator (t- PA) bind differently to cultured human endothelial cells. Thromb Haemost 62:699–703

Allen RA, Pepper DS (1981) Isolation and properties of human vascular plasminogen activator. Thromb Haemost 45:43–50

Andreasen PA, Petersen LC, Danø K (1991) Diversity in catalytic properties of single chain and two chain tissue-type plasminogen activator. Fibrinolysis 5:207–215

Andreasen PA, Sottrup-Jensen L, Kjøller L, Nykjær A, Moestrup SK, Petersen CM, Gliemann J (1994) Receptor-mediated endocytosis of plasminogen activators and activator/inhibitor complexes. FEBS Lett 338:239–245

Aoki N, Saito H, Kamiya T, Koie K, Sakata Y, Kobakura M (1979) Congenital deficiency of α_2-plasmin inhibitor associated with severe hemorrhagic tendency. J Clin Invest 63:877–884

Aoki N, Sakata Y, Matsuda M, Tateno K (1980) Fibrinolytic states in a patient with congenital deficiency of α_2-plasmin inhibitor. Blood 55:483–488

Aoki N, von Kaulla KN (1971) Dissimilarity of human vascular plasminogen activator and human urokinase. J Lab Clin Med 78354:354–362

Arts J, Herr I, Lansink M, Angel P, Kooistra T (1997) Cell-type specific DNA-protein interactions at the tissue-type plasminogen activator promoter in human endothelial and HeLa cells in vivo and in vitro. Nucleic Acids Res 25:311–317

Astrup T, Müllertz S (1952) The fibrin plate method for estimating fibrinolytic activity. Arch Biochem Biophys 40:346–356

Astrup T, Permin PM (1947) Fibrinolysis in the animal organism. Nature 159:681–683

Astrup T, Stage A (1952) Isolation of a soluble fibrinolytic activator from animal tissue. Nature 170:929–930

Aznar J, Estellés A, Vila V, Regañón E, España F, Villa P (1984) Inherited fibrinolytic disorder due to an enhanced plasminogen activator level. Thromb Haemost 52: 196–200

Bachmann F (1995) The role of plasminogen activator inhibitor type 1 (PAI-1) in the clinical setting, including deep vein thrombosis. In: Glas-Greenwalt P (ed) Fibrinolysis in Disease. Molecular and Hemovascular Aspects of Fibrinolysis, CRC Press, Boca Raton, pp 79–86

Bachmann F, Fletcher AP, Alkjaersig N, Sherry S (1964) Partial purification and properties of the plasminogen activator from the pig heart. Biochemistry 3:1578–1585

Bachmann F, Kruithof EKO (1984) Tissue plasminogen activator: chemical and physiological aspects. Semin Thromb Hemost 10:6–17

Bankl HC, Grossschmidt K, Pikula B, Bankl H, Lechner K, Valent P (1999) Mast cells are augmented in deep vein thrombosis and express a profibrinolytic phenotype. Hum Pathol 30:188–194

Bányai L, Váradi A, Patthy L (1983) Common evolutionary origin of the fibrin-binding structures of fibronectin and tissue-type plasminogen activator. FEBS Lett 163: 37–41

Baranes D, Lederfein D, Huang Y-Y, Chen M, Bailey CH, Kandel ER (1998) Tissue plasminogen activator contributes to the late phase of LTP and to synaptic growth in the hippocampal mossy fiber pathway. Neuron 21:813–825

Beebe DP, Miles LA, Plow EF (1989) A linear amino acid sequence involved in the interaction of tPA with its endothelial cell receptor. Blood 74:2034–2037

Beebe DP, Wood LL, Moos M (1990) Characterization of tissue plasminogen activator binding proteins isolated from endothelial cells and other cell types. Thromb Res 59:339–350

Benham FJ, Spurr N, Povey S, Brinton BT, Goodfellow PN, Solomon E, Harris TJR (1984) Assignment of tissue-type plasminogen activator to chromosome 8 in man and identification of a common restriction length polymorphism within the gene. Mol Biol Med 2:251–259

Bennett WF, Paoni NF, Keyt BA, Botstein D, Jones AJS, Presta L, Wurm FM, Zoller MJ (1991) High resolution analysis of functional determinants on human tissue-type plasminogen activator. J Biol Chem 266:5191–5201

Benza RL, Grenett HE, Bourge RC, Kirklin JK, Naftel DC, Castro PF, McGiffin DC, George JF, Booyse FM (1998) Gene polymorphisms for plasminogen activator inhibitor-1/tissue plasminogen activator and development of allograft coronary artery disease. Circulation 98:2248–2254

Berg DT, Grinnell BW (1991) Signal and propeptide processing of human tissue plasminogen activator: activity of a pro-tPA derivative. Biochem Biophys Res Commun 179:1289–1296

Biessen EAL, van Teijlingen M, Vietsch H, Barrett-Bergshoeff MM, Bijsterbosch MK, Rijken DC, Van Berkel TJC, Kuiper J (1997) Antagonists of the mannose receptor and the LDL receptor-related protein dramatically delay the clearance of tissue plasminogen activator. Circulation 95:46–52

Biggs R, MacFarlane RG, Pilling G (1947) Observations on fibrinolysis. Experimental activity produced by exercise or adrenalin. Lancet 1:402–405

Binder BR, Spragg J, Austen KF (1979) Purification and characterization of human vascular plasminogen activator derived from blood vessel perfusates. J Biol Chem 254:1998–2003

Bohm T, Geiger M, Binder BR (1996) Isolation and characterization of tissue-type plasminogen activator-binding proteoglycans from human umbilical vein endothelial cells. Arterioscler Thromb Vasc Biol 16:665–672

Booth NA, Bennett B, Wijngaards G, Grieve JHK (1983) A new life-long hemorrhagic disorder due to excess plasminogen activator. Blood 61:267–275

Booth NA, Walker E, Maughan R, Bennett B (1987) Plasminogen activator in normal subjects after exercise and venous occlusion: tPA circulates as complexes with C1 inhibitor and PAI-1. Blood 69:1354–1362

Bosma PJ, Rijken DC, Nieuwenhuizen W (1988) Binding of tissue-type plasminogen activator to fibrinogen fragments. Eur J Biochem 172:399–404

Bounameaux H, Stassen JM, Seghers C, Collen D (1986) Influence of fibrin and liver blood flow an the turnover and the systemic fibrinogenolytic effects of recombinant human tissue-type plasminogen activator in rabbits. Blood 67:1493–1497

Bourtchuladze R, Frenguelli B, Blendy J, Cioffi D, Schutz G, Silva AJ (1994) Deficient long-term memory in mice with a targeted mutation of the cAMP-responsive element-binding protein. Cell 79:59–68

Brisson-Jeanneau C, Nelles L, Rouer E, Sultan Y, Benarous R (1990a) Tissue-plasminogen activator RNA detected in megakaryocytes by in situ hybridization and biotinylated probe. Histochemistry 95:23–26

Brisson-Jeanneau C, Solberg LA Jr, Sultan Y (1990b) Presence of functionally active tissue plasminogen activator in human CFU-M derived megakaryocytes in vitro. Fibrinolysis 4:107–115

Browne MJ, Tyrell AWR, Chapman CG, Carey JE, Glover DM, Grosveld FG, Dodd I, Robinson JH (1985) Isolation of a human tissue-type plasminogen activator genomic DNA clone and its expression in mouse L cells. Gene 33:279–284

Bu G, Williams S, Strickland DK, Schwartz AL (1992) Low density lipoprotein receptor-related protein/α2-macroglobulin receptor is an hepatic receptor for tissue-type plasminogen activator. Proc Natl Acad Sci USA 89:7427–7431

Bulens F, Ibañez-Tallon I, van Acker P, de Vriese A, Nelles L, Belayew A, Collen D (1995) Retinoic acid induction of human tissue-type plasminogen activator gene expression via a direct repeat element (DR5) located at –7 kilobases. J Biol Chem 270:7167–7175

Bulens F, Merchiers P, Ibañez-Tallon I, de Vriese A, Nelles L, Claessens F, Belayew A, Collen D (1997) Identification of a multihormone responsive enhancer far upstream from the human tissue-type plasminogen activator gene. J Biol Chem 272:663–671

Byeon I-JL, Kelley RF, Llinás M (1991) Kringle-2 domain of the tissue-type plasminogen activator. ^{1}H-NMR assignments and secondary structure. Eur J Biochem 197:155–165

Byeon I-JL, Kelley RF, Mulkerrin MG, An SSA, Llinás M (1995) Ligand binding to the tissue-type plasminogen activator kringle 2 domain: structural characterization by ^{1}H-NMR. Biochemistry 34:2739–2750

Camani C, Bachmann F, Kruithof EKO (1994) The role of plasminogen activator inhibitor type 1 in the clearance of tissue-type plasminogen activator by rat hepatoma cells. J Biol Chem 269:5770–5775

Camani C, Gavin O, Kruithof EKO (2000) Cellular degradation of free and inhibitor-bound tissue-type plasminogen activator – requirement for a co-receptor? Thromb Haemost 83:290–296

Camani C, Kruithof EKO (1994) Clearance receptors for tissue-type plasminogen. Int J Hematol 60:97–109

Camiolo SM, Thorsen S, Astrup T (1971) Fibrinogenolysis and fibrinolysis with tissue plasminogen activator, urokinase, streptokinase-activated human globulin, and plasmin. Proc Soc Exp Biol Med 138:277–280

Carmeliet P, Collen D (1996) Targeted gene manipulation and transfer of the plasminogen and coagulation systems in mice. Fibrinolysis 10:195–213

Carmeliet P, Schoonjans L, Kieckens L, Ream B, Degen J, Bronson R, De Vos R, van den Oord JJ, Collen D, Mulligan RC (1994) Physiological consequences of loss of plasminogen activator gene function in mice. Nature 368:419–424

Carter AM, Catto AJ, Grant PJ (1998) Determinants of tPA antigen and associations with coronary artery disease and acute cerebrovascular disease. Thromb Haemost 80:632–636

Chandler WL, Alessi MC, Aillaud MF, Henderson P, Vague P, Juhan-Vague I (1997) Clearance of tissue plasminogen activator (TPA) and TPA/plasminogen activator inhibitor type 1 (PAI-1) complex: relationship to elevated TPA antigen in patients with high PAI-1 activity levels. Circulation 96:761–768

Chang Y, Nilsen SL, Castellino FJ (1999) Functional and structural consequences of aromatic residue substitutions within the kringle-2 domain of tissue-type plasminogen activator. J Pept Res 53:656–664

Chang Y, Zajicek J, Castellino FJ (1997) Role of tryptophan-63 of the kringle 2 domain of tissue-type plasminogen activator in its thermal stability, folding, and ligand binding properties. Biochemistry 36:7652–7663

Cheng X-F, Pohl G, Back O, Wallén P (1996) Characterization of receptors interacting specifically with the B-chain of tissue plasminogen activator on endothelial cells. Fibrinolysis 10:167–175

Christensen LR, MacLeod CM (1945) A proteolytic enzyme of serum. Characterization, activation, and reaction with inhibitors. J Gen Physiol 28:559–583

Cleary S, Mulkerrin MG, Kelley RF (1989) Purification and characterization of tissue plasminogen activator kringle-2 domain expressed in *Escherichia coli*. Biochemistry 28:1884–1891

Cole ER, Bachmann F (1977) Purification and properties of a plasminogen activator from pig heart. J Biol Chem 252:3729–3737

Collen D, Lijnen HR, Bulens F, Vandamme AM, Tulinsky A, Nelles L (1990) Biochemical and functional characterization of human tissue-type plasminogen activator variants with mutagenized kringle domains. J Biol Chem 265:12184–12191

Collen D, Lijnen HR, Todd PA, Goa KL (1989) Tissue-type plasminogen activator. A review of its pharmacology and therapeutic use as a thrombolytic agent. Drugs 38:346–388

Costa M, Shen Y, Maurer F, Medcalf RL (1998) Transcriptional regulation of the tissue-type plasminogen-activator gene in human endothelial cells: identification of nuclear factors that recognise functional elements in the tissue-type plasminogen-activator gene promoter. Eur J Biochem 258:123–131

Danø K, Andreasen PA, Grøndahl-Hansen J, Kristensen P, Nielsen LS, Skriver L (1985) Plasminogen activators, tissue degradation, and cancer. Adv Cancer Res 44:139–266

de Boer A, Kluft C, Kroon JM, Kasper FJ, Schoemaker HC, Pruis J, Breimer DD, Soons PA, Emeis JJ, Cohen AF (1992) Liver blood flow as a major determinant of the clearance of recombinant human tissue-type plasminogen activator. Thromb Haemost 67:83–87

de Munk GAW, Caspers MPM, Chang GTG, Pouwels PH, Enger-Valk BE, Verheijen JH (1989) Binding of tissue-type plasminogen activator to lysine, lysine analogues, and fibrin fragments. Biochemistry 28:7318–7325

de Serrano VS, Sehl LC, Castellino FJ (1992) Direct identification of lysine-33 as the principal cationic center of the ω-amino acid binding site of the recombinant kringle 2 domain of tissue-type plasminogen activator. Arch Biochem Biophys 292:206–212

de Vos AM, Ultsch MH, Kelley RF, Padmanabhan K, Tulinsky A, Westbrook ML, Kossiakoff AA (1992) Crystal structure of the kringle 2 domain of tissue plasminogen activator at 2.4-Å resolution. Biochemistry 31:270–279

De Vries C, Veerman H, Koornneef E, Pannekoek H (1990) Tissue-type plasminogen activator and its substrate Glu-plasminogen share common binding sites in limited plasmin-digested fibrin. J Biol Chem 265:13547–13552

Declerck PJ, Juhan-Vague I, Felez J, Wiman B (1994) Pathophysiology of fibrinolysis. J Intern Med 236:425–432

Diamond SL, Eskin SG, McIntire LV (1989) Fluid flow stimulates tissue plasminogen activator secretion by cultured human endothelial cells. Science 243:1483–1485

Diamond SL, Sharefkin JB, Dieffenbach C, Frasier-Scott K, McIntire LV, Eskin SG (1990) Tissue plasminogen activator messenger RNA levels increase in cultured human endothelial cells exposed to laminar shear stress. J Cell Physiol 143:364–371

Diéval J, Nguyen G, Gross S, Delobel J, Kruithof EKO (1991) A lifelong bleeding disorder associated with a deficiency of plasminogen activator inhibitor type 1. Blood 77:528–532

Downing AK, Driscoll PC, Harvey TS, Dudgeon TJ, Smith BO, Baron M, Campbell ID (1992) Solution structure of the fibrin binding finger domain of tissue-type plasminogen activator determined by ^{1}H nuclear magnetic resonance. J Mol Biol 225:821–833

Dudani AK, Ganz PR (1996) Endothelial cell surface actin serves as a binding site for plasminogen, tissue plasminogen activator and lipoprotein(a). Br J Haematol 95:168–178

Dwek RA (1995) Glycobiology: "towards understanding the function of sugars". Biochem Soc Trans 23:1–25

Edlund T, Ny T, Rånby M, Hedén LO, Palm G, Holmgren E, Josephson S (1983) Isolation of cDNA sequences coding for a part of human tissue plasminogen activator. Proc Natl Acad Sci USA 80:349–352

Ellis V, Whawell SA (1997) Vascular smooth muscle cells potentiate plasmin generation by both urokinase and tissue plasminogen activator-dependent mechanisms: evidence for a specific tissue-type plasminogen activator receptor on these cells. Blood 90:2312–2322

Fay WP, Parker AC, Condrey LR, Shapiro AD (1997) Human plasminogen activator inhibitor-1 (PAI-1) deficiency: characterization of a large kindred with a null mutation in the PAI-1 gene. Blood 90:204–208

Fay WP, Shapiro AD, Shih JL, Schleef RR, Ginsburg D (1992) Complete deficiency of plasminogen-activator inhibitor type 1 due to a frame-shift mutation. N Engl J Med 327:1729–1733

Fears R (1989) Binding of plasminogen activators to fibrin: characterization and pharmacological consequences. Biochem J 261:313–324

Fisher R, Waller EK, Grossi G, Thompson D, Tizard R, Schleuning W-D (1985) Isolation and characterization of the human tissue-type plasminogen activator structural gene including its 5' flanking region. J Biol Chem 260:11223–11230

Fitzpatrick SL, Kassam G, Choi K-S, Kang H-M, Fogg DK, Waisman DM (2000) Regulation of plasmin activity by annexin II tetramer. Biochemistry 39:1021–1028

Fleisher MS, Loeb L (1915) On tissue fibrinolysins. J Biol Chem 21:477–501

Fletcher AP, Biederman O, Moore D, Alkjaersig N, Sherry S (1964) Abnormal plasminogen-plasmin system activity (fibrinolysis) in patients with hepatic cirrhosis: its cause and consequences. J Clin Invest 43:681–695

Frey U, Muller M, Kuhl D (1996) A different form of long-lasting potentiation revealed in tissue plasminogen activator mutant mice. J Neurosci 16:2057–2063

Friezner Degen JF, Rajput B, Reich E (1986) The human tissue plasminogen activator gene. J Biol Chem 261:6972–6985

Fujiwara J, Kimura T, Ayusawa D, Oishi M (1994) A novel regulatory sequence affecting the constitutive expression of tissue plasminogen activator (tPA) gene in human melanoma (Bowes) cells. J Biol Chem 269:18558–18562

Fukao H, Matsuo O (1998) Analysis of tissue-type plasminogen activator receptor (t-PAR) in human endothelial cells. Semin Thromb Hemost 24:269–273

Gaffney PJ, Curtis AD (1987) A collaborative study to establish the Second International Standard for tissue plasminogen activator (tPA). Thromb Haemost 59: 1085–1087

Geppert A, Graf S, Beckmann R, Hornykewycz S, Schuster E, Binder BR, Huber K (1998) Concentration of endogenous tPA antigen in coronary artery disease: relation to thrombotic events, aspirin treatment, hyperlipidemia, and multivessel disease. Arterioscler Thromb Vasc Biol 18:1634–1642

Geppert AG, Binder BR (1992) Allosteric regulation of tPA-mediated plasminogen activation by a modifier mechanism: evidence for a binding site for plasminogen on the tPA A-chain. Arch Biochem Biophys 297:205–212

Grimaudo V, Bachmann F, Hauert J, Christe M-A, Kruithof EKO (1992) Hypofibrinolysis in patients with a history of idiopathic deep vein thrombosis and/or pulmonary embolism. Thromb Haemost 67:397–401

Hajjar KA, Jacovina AT, Chacko J (1994) An endothelial cell receptor for plasminogen/tissue plasminogen activator. I. Identity with annexin II. J Biol Chem 269: 21191–21197

Hajjar KA, Krishnan S (1999) Annexin II: a mediator of the plasmin/plasminogen activator system. Trends Cardiovasc Med 9:128–138

Hajjar KA, Reynolds CM (1994) α-Fucose-mediated binding and degradation of tissue-type plasminogen activator by HepG2 cells. J Clin Invest 93:703–710

Harpel PC, Chang T, Verderber E (1985) Tissue plasminogen activator and urokinase mediate the binding of Glu-plasminogen to plasma fibrin I. Evidence for new binding sites in plasmin-degraded fibrin I. J Biol Chem 260:4432–4440

Harris RJ, Leonard CK, Guzzetta AW, Spellman MW (1991) Tissue plasminogen activator has an O-linked fucose attached to threonine-61 in the epidermal growth factor domain. Biochemistry 30:2311–2314

Hart PH, Vitti GF, Burgess DR, Singleton DK, Hamilton JA (1989) Human monocytes can produce tissue-type plasminogen activator. J Exp Med 169:1509–1514

Hembrough TA, Kralovich KR, Li L, Gonias SL (1996) Cytokeratin 8 released by breast carcinoma cells in vitro binds plasminogen and tissue-type plasminogen activator and promotes plasminogen activation. Biochem J 317:763–769

Henderson BR, Sleigh MJ (1992) TATA box-independent transcription of the human tissue plasminogen activator gene initiates within a sequence conserved in related genes. FEBS Lett 309:130–134

Herz J, Goldstein JL, Strickland DK, Ho YK, Brown MS (1991) 39-kDa protein modulates binding of ligands to low density lipoprotein receptor-related protein/α2-macroglobulin receptor. J Biol Chem 266:21232–21238

Higgins D, Vehar GA (1987) Interaction of one-chain and two-chain tissue plasminogen activator with intact and plasmin-degraded fibrin. Biochemistry 26:7786–7791

Higgins DL, Lamb MC, Young SL, Powers DB, Anderson S (1990) The effect of the one-chain to two-chain conversion in tissue plasminogen activator: characterization of mutations at position 275. Thromb Res 57:527–539

Holvoet P, Cleemput H, Collen D (1985) Assay of human tissue-type plasminogen activator (tPA) with an enzyme-linked immunosorbent assay (ELISA) based on three murine monoclonal antibodies to tPA. Thromb Haemost 54:684–687

Horn IR, van den Berg BM, van der Meijden PZ, Pannekoek H, van Zonneveld AJ (1997) Molecular analysis of ligand binding to the second cluster of complement-type repeats of the low density lipoprotein receptor-related protein. Evidence for an allosteric component in receptor-associated protein-mediated inhibition of ligand binding. J Biol Chem 272:13608–13613

Hoylaerts M, Rijken DC, Lijnen HR, Collen D (1982) Kinetics of the activation of plasminogen by human tissue plasminogen activator. Role of fibrin. J Biol Chem 257:2912–2919

Huang Y-Y, Bach ME, Lipp HP, Zhuo M, Wolfer DP, Hawkins RD, Schoonjans L, Kandel ER, Godfraind J-M, Mulligan R, Collen D, Carmeliet P (1996) Mice lacking the gene encoding tissue-type plasminogen activator show a selective interference with late-phase long-term potentiation in both Schaffer collateral and mossy fiber pathways. Proc Natl Acad Sci USA 93:8699–8704

Huber K, Kirchheimer JC, Korninger C, Binder BR (1991) Hepatic synthesis and clearance of components of the fibrinolytic system in healthy volunteers and in patients with different stages of liver cirrhosis. Thromb Res 62:491–500

Husain SS, Hasan AAK, Budzynski AZ (1989) Differences between binding of one-chain and two-chain tissue plasminogen activators to non-cross-linked and cross-linked fibrin clots. Blood 74:999–1006

Iacoviello L, Di Castelnuovo A, de Knijff P, D'Orazio A, Amore C, Kluft C, Donati MB (1996) Alu-repeat polymorphism in the tissue-type plasminogen activator (tPA) gene, tPA levels and risk of familial myocardial infarction (AMI). Fibrinolysis 10 [Suppl 2]:13–16

Iba T, Shin T, Sonoda T, Rosales O, Sumpio BE (1991) Stimulation of endothelial secretion of tissue-type plasminogen activator by repetitive stretch. J Surg Res 50:457–460

Ieko M, Sawada K, Yasukouchi T, Sakurama S, Tohma Y, Shiroshita K, Kurosawa S, Ohmoto A, Kohno M, Satoh M, Koike T (1997) Protection by α2-macroglobulin of tissue plasminogen activator against inhibition by plasminogen activator inhibitor-1. Br J Haematol 97:214–218

Jansson J-H, Nilsson TK, Johnson O (1998) von Willebrand factor, tissue plasminogen activator, and dehydroepiandrosterone sulphate predict cardiovascular death in a 10 year follow up of survivors of acute myocardial infarction. Heart 80:334–337

Juhan-Vague I, Valadier J, Alessi MC, Aillaud MF, Ansaldi J, Philip Joet C, Holvoet P, Serradimigni A, Collen D (1987) Deficient tPA release and elevated PA inhibitor levels in patients with spontaneous or recurrent deep venous thrombosis. Thromb Haemost 57:67–72

Jörnvall H, Pohl G, Bergsdorf N, Wallén P (1983) Differential proteolysis and evidence for a residue exchange in tissue plasminogen activator suggest possible association between two types of protein microheterogeneity. FEBS Lett 156:47–50

Kang H-M, Choi K-S, Kassam G, Fitzpatrick SL, Kwon M, Waisman DM (1999) Role of annexin II tetramer in plasminogen activation. Trends Cardiovasc Med 9:92–102

Kasza A, Petersen HH, Heegaard CW, Oka K, Christensen A, Dubin A, Chan L, Andreasen PA (1997) Specificity of serine proteinase/serpin complex binding to very-low-density lipoprotein receptor and α2-macroglobulin receptor/low-density-lipoprotein-receptor-related protein. Eur J Biochem 248:270–281

Ke S-H, Tachias K, Lamba D, Bode W, Madison EL (1997) Identification of a hydrophobic exosite on tissue type plasminogen activator that modulates specificity for plasminogen. J Biol Chem 272:1811–1816

Kelley RF, DeVos AM, Cleary S (1991) Thermodynamics of ligand binding and denaturation for His64 mutants of tissue plasminogen activator kringle-2 domain. Proteins: Struct Funct Genet 11:35–44

Knauer MF, Kridel SJ, Hawley SB, Knauer DJ (1997) The efficient catabolism of thrombin-protease nexin 1 complexes is a synergistic mechanism that requires both the LDL receptor-related protein and cell surface heparins. J Biol Chem 272:29039–29045

Korninger C, Lechner K, Niessner H, Gossinger H, Kundi M (1984) Impaired fibrinolytic capacity predisposes for recurrence of venous thrombosis. Thromb Haemost 52:127–130

Kounnas MZ, Chappell DA, Wong H, Argraves WS, Strickland DK (1995) The cellular internalization and degradation of hepatic lipase is mediated by low density lipoprotein receptor-related protein and requires cell surface proteoglycans. J Biol Chem 270:9307–9312

Kounnas MZ, Henkin J, Argraves WS, Strickland DK (1993) Low density lipoprotein receptor-related protein/α_2-macroglobulin receptor mediates cellular uptake of pro-urokinase. J Biol Chem 268:21862–21867

Kralovich KR, Li L, Hembrough TA, Webb DJ, Karns LR, Gonias SL (1998) Characterization of the binding sites for plasminogen and tissue-type plasminogen activator in cytokeratin 8 and cytokeratin 18. J Protein Chem 17:845–854

Krapohl BD, Siemionow M, Zins JE (1998) Effect of tissue-plasminogen activator on leukocyte-endothelial interactions at the microcirculatory level. Plast Reconstr Surg 102:2388–2394

Kruithof EKO (1988) Inhibitors of plasminogen activators. In: Kluft C (ed) Tissue-Type Plasminogen Activator (tPA). Physiological and Clinical Aspects. Vol I, CRC Press, Boca Raton, FL, pp 189–210

Kruithof EKO, Bachmann F (1982) Studies on the binding of tissue plasminogen activator to fibrinogen and fibrin. In: Henschen A, Graeff H, Lottspeich A (eds) Fibrinogen – recent biomedical and medical aspects. Walter de Gruyter, Berlin, pp 377–387

Kruithof EKO, Schleuning WD, Bachmann F (1985) Human tissue-type plasminogen activator. Production in continuous serum-free culture and rapid purification. Biochem J 226:631–636

Lamba D, Bauer M, Huber R, Fischer S, Rudolph R, Kohnert U, Bode W (1996) The 2.3 Å crystal structure of the catalytic domain of recombinant two-chain human tissue-type plasminogen activator. J Mol Biol 258:117–135

Larsen GR, Henson H, Blue Y (1988) Variants of human tissue-type plasminogen activator. Fibrin binding, fibrinolytic, and fibrinogenolytic characterization of genetic variants lacking the fibronectin finger-like and/or the epidermal growth factor domains. J Biol Chem 263:1023–1029

Lasierra J, Aza MJ, Viladés E, Poblet S, Barrao F, Bayon E, Gonzalez J (1991) Tissue plasminogen activator and plasminogen activator inhibitor in patients with liver cirrhosis. Fibrinolysis 5:117–120

Latron Y, Alessi MC, George F, Anfosso F, Poncelet P, Juhan-Vague I (1991) Characterization of epitheloid cells from human omentum: comparison with endothelial cells from umbilical veins. Thromb Haemost 66:361–367

Lee MH, Vosburgh E, Anderson K, McDonagh J (1993) Deficiency of plasma plasminogen activator inhibitor 1 results in hyperfibrinolytic bleeding. Blood 81:2357–2362

Lehrman MA, Hill RL (1986) The binding of fucose-containing glycoproteins by hepatic lectins. J Biol Chem 261:7419–7425

Leonardsson G, Peng X-R, Liu K, Nordström L, Carmeliet P, Mulligan R, Collen D, Ny T (1995) Ovulation efficiency is reduced in mice that lack plasminogen activator gene function: functional redundancy among physiological plasminogen activators. Proc Natl Acad Sci USA 92:12446–12450

Levin EG, Loskutoff DJ (1982) Cultured bovine endothelial cells produce both urokinase and tissue-type plasminogen activators. J Cell Biol 94:631–636

Lochner JE, Kingma M, Kuhn S, Meliza CD, Cutler B, Scalettar BA (1998) Real-time imaging of the axonal transport of granules containing a tissue plasminogen activator/green fluorescent protein hybrid. Mol Biol Cell 9:2463–2476

Loeb L (1904) Versuche über einige Bedingungen der Blutgerinnung, insbesondere über die Spezifität der in den Geweben vorhandenen Coaguline. Virchows Arch Pathol Anat Physiol 176:10–47

Loscalzo J (1988) Structural and kinetic comparison of recombinant human single- and two-chain tissue plasminogen activator. J Clin Invest 82:1391–1397

Loskutoff DJ, van Mourik JA, Erickson LA, Lawrence D (1983) Detection of an unusually stable fibrinolytic inhibitor produced by bovine endothelial cells. Proc Natl Acad Sci USA 80:2956–2960

MacFarlane RG (1937) Fibrinolysis following operation. Lancet 1:10–12

MacFarlane RG, Biggs R (1946) Observations on fibrinolysis. Spontaneous activity associated with surgical operations, trauma, etc. Lancet 2:862–864

Macko RF, Kittner SJ, Epstein A, Cox DK, Wozniak MA, Wityk RJ, Stern BJ, Sloan MA, Sherwin R, Price TR, McCarter RJ, Johnson CJ, Earley CJ, Buchholz DW, Stolley PD (1999) Elevated tissue plasminogen activator antigen and stroke risk: The Stroke Prevention In Young Women Study. Stroke 30:7–11

Madani R, Hulo S, Toni N, Madani H, Steimer T, Muller D, Vassalli J-D (1999) Enhanced hippocampal long-term potentiation and learning by increased neuronal expression of tissue-type plasminogen activator in transgenic mice. EMBO J 18:3007–3012

Madison EL, Goldsmith EJ, Gerard RD, Gething MJH, Sambrook JF (1989) Serpin-resistant mutants of human tissue-type plasminogen activator. Nature 339:721–724

Madison EL, Goldsmith EJ, Gerard RD, Gething MJH, Sambrook JF, Bassel-Duby RS (1990) Amino acid residue that affect interaction of tissue-type plasminogen activator with plasminogen activator inhibitor 1. Proc Natl Acad Sci USA 87:3530–3533

Madoiwa S, Komatsu N, Mimuro J, Kimura K, Matsuda M, Sakata Y (1999) Developmental expression of plasminogen activator inhibitor-1 associated with thrombopoietin-dependent megakaryocytic differentiation. Blood 94:475–482

Medcalf RL, Rüegg M, Schleuning WD (1990) A DNA motif related to the cAMP-responsive element and an exon-located activator protein-2 binding site in the human tissue- type plasminogen activator gene promoter cooperate in basal expression and convey activation by phorbol ester and cAMP. J Biol Chem 265:14618–14626

Merchiers P, Bulens F, de Vriese A, Collen D, Belayew A (1999) Involvement of Sp1 in basal and retinoic acid induced transcription of the human tissue-type plasminogen activator gene. FEBS Lett 456:149–154

Mikhailenko I, Considine M, Argraves KM, Loukinov D, Hyman BT, Strickland DK (1999) Functional domains of the very low density lipoprotein receptor: molecular analysis of ligand binding and acid-dependent ligand dissociation mechanisms. J Cell Sci 112:3269–3281

Milstone JH (1941) A factor in normal human plasma which participates in streptococcal fibrinolysis. J Immunol 42:109–116

Moestrup SK (1994) The α_2-macroglobulin receptor and epithelial glycoprotein-330: two giant receptors mediating endocytosis of multiple ligands. Biochim Biophys Acta 1197:197–213

Moestrup SK, Holtet TL, Etzerodt M, Thøgersen HC, Nykjær A, Andreasen PA, Rasmussen HH, Sottrup-Jensen L, Gliemann J (1993) α_2-macroglobulin-proteinase complexes, plasminogen activator inhibitor type-1-plasminogen activator complexes, and receptor-associated protein bind to a region of the α_2-macroglobulin receptor containing a cluster of eight complement-type repeats. J Biol Chem 268:13691–13696

Morange PE, Alessi MC, Verdier M, Casanova D, Magalon G, Juhan-Vague I (1999)
    PAI-1 produced ex vivo by human adipose tissue is relevant to PAI-1 blood level.
    Arterioscler Thromb Vasc Biol 19:1361–1365
Mosesson MW, Siebenlist KR, Voskuilen M, Nieuwenhuizen W (1998) Evaluation of
    the factors contributing to fibrin-dependent plasminogen activation. Thromb
    Haemost 79:796–801
Mulder M, Kohnert U, Fischer S, van Hinsbergh VW, Verheijen JH (1997) The inter-
    action of recombinant tissue type plasminogen activator and recombinant plas-
    minogen activator (r-PA/BM 06.022) with human endothelial cells. Blood Coagul
    Fibrinolysis 8:124–133
Müller CM, Griesinger CB (1998) Tissue plasminogen activator mediates reverse occlu-
    sion plasticity in visual cortex. Nat Neurosci 1:47–53
Narita M, Bu G, Herz J, Schwartz AL (1995) Two receptor systems are involved in the
    plasma clearance of tissue-type plasminogen activator (tPA) in vivo. J Clin Invest
    96:1164–1168
Neels JG, Van den Berg BMM, Lookene A, Olivecrona G, Pannekoek H, Van Zon-
    neveld A-J (1999) The second and fourth cluster of class A cysteine-rich repeats
    of the low density lipoprotein receptor-related protein share ligand-binding
    properties. J Biol Chem 274:31305–31311
Nicoloso G, Hauert J, Kruithof EKO, Van Melle G, Bachmann F (1988) Fibrinolysis in
    normal subjects-comparison between plasminogen activator inhibitor and other
    components of the fibrinolytic system. Thromb Haemost 59:299–303
Nieuwenhuizen W, Voskuilen M, Vermond A, Hoegee-de Nobel B, Traas DW (1988)
    The influence of fibrin(ogen) fragments on the kinetic parameters of the tissue-
    type plasminogen-activator-mediated activation of different forms of plasmino-
    gen. Eur J Biochem 174:163–169
Nilsson TK, Mellbring G (1989) Impact of immediate acidification of blood on mea-
    surement of plasma tissue plasminogen activator (tPA) activity in surgical patients.
    Clin Chem 35:1999
Nolf P (1908) Contributions à l'étude de la coagulation du sang. V. La fibrinolyse. Arch
    Int Physiol 6:306–359
Noorman F, Barrett-Bergshoeff MM, Biessen EA, van de Bilt E, van Berkel TJ, Rijken
    DC (1997) Cluster mannosides can inhibit mannose receptor-mediated tissue-type
    plasminogen activator degradation by both rat and human cells. Hepatology
    26:1303–1310
Noorman F, Barrett-Bergshoeff MM, Rijken DC (1998) Role of carbohydrate and
    protein in the binding of tissue-type plasminogen activator to the human mannose
    receptor. Eur J Biochem 251:107–113
Nordenhem A, Wiman B (1998) Tissue plasminogen activator (tPA) antigen in plasma:
    correlation with different tPA/inhibitor complexes. Scand J Clin Lab Invest
    58:475–483
Norrman B, Wallén P, Rånby M (1985) Fibrinolysis mediated by tissue plasminogen
    activator. Disclosure of a kinetic transition. Eur J Biochem 149:193–200
Ny T, Elgh F, Lund B (1984) The structure of the human tissue-type plasminogen acti-
    vator gene: correlation of intron and exon structures to functional and structural
    domains. Proc Natl Acad Sci USA 81:5355–5359
Nykjaer A, Kjøller L, Cohen RL, Lawrence DA, Garni-Wagner BA, Todd RF, III, Van
    Zonneveld A-J, Gliemann J, Andreasen PA (1994) Regions involved in binding of
    urokinase-type-1 inhibitor complex and pro-urokinase to the endocytic α_2-
    macroglobulin receptor/low density lipoprotein receptor-related protein. Evi-
    dence that the urokinase receptor protects pro-urokinase against binding to the
    endocytic receptor. J Biol Chem 269:25668–25676
Orth K, Madison EL, Gething M-J, Sambrook JF, Herz J (1992) Complexes of
    tissue-type plasminogen activator and its serpin inhibitor plasminogen-activator
    inhibitor type 1 are internalized by means of the low density lipoprotein receptor-
    related protein/α_2-macroglobulin receptor. Proc Natl Acad Sci USA 89:7422–
    7426

Otter M, Barrett-Bergshoeff MM, Rijken DC (1991) Binding of tissue-type plasminogen activator by the mannose receptor. J Biol Chem 266:13931–13935

Paoni NF, Chow AM, Peña LC, Keyt BA, Zoller MJ, Bennett WF (1993a) Making
tissue-type plasminogen activator more fibrin specific. Protein Eng 6:529–534

Paoni NF, Keyt BA, Refino CJ, Chow AM, Nguyen HV, Berleau LT, Badillo J, Peña LC,
Brady K, Wurm FM, Ogez J, Bennett WF (1993b) A slow clearing, fibrin-specific,
PAI-1 resistant variant of tPA (T103N, KHRR 296–299 AAAA). Thromb
Haemost 70:307–312

Paoni NF, Refino CJ, Brady K, Peña LC, Nguyen HV, Kerr EM, Johnson AC, Wurm
FM, van Reis R, Botstein D, Bennett WF (1992) Involvement of residues 296–299
in the enzymatic activity of tissue-type plasminogen activator. Protein Eng 5:
259–266

Papadaki M, Ruef J, Nguyen KT, Li F, Patterson C, Eskin SG, McIntire LV, Runge MS
(1998) Differential regulation of protease activated receptor-1 and tissue plasminogen activator expression by shear stress in vascular smooth muscle cells. Circ
Res 83:1027–1034

Parekh RB, Dwek RA, Thomas JR, Opdenakker G, Rademacher TW (1989) Cell-type-
specific and site-specific N-glycosylation of type I and type II human tissue plasminogen activator. Biochemistry 28:7644–7662

Patthy L (1990) Evolutionary Assembly of Blood Coagulation Proteins. Semin Thromb
Hemost 16:245–259

Pennica D, Holmes WE, Kohr WJ, Harkins RN, Vehar GA, Ward CA, Bennett WF,
Yelverton E, Seeburg PH, Heynecker HL, Goeddel DV, Collen D (1983) Cloning
and expression of human tissue-type plasminogen activator cDNA in E. coli.
Nature 301:214–221

Petersen TE, Thøgersen HC, Skorstengaard K, Vite-Pedersen K, Sahl P, Sottrup-Jensen
L, Magnussen S (1983) Partial primary structure of bovine plasma fibronectin:
Three types of internal homology. Proc Natl Acad Sci USA 80:137–141

Prins MH, Hirsh J (1991) A critical review of the evidence supporting a relationship
between impaired fibrinolytic activity and venous thromboembolism. Arch Intern
Med 151:1721–1731

Radcliffe R, Heinze T (1981) Stimulation of tissue plasminogen activator by denatured
proteins and fibrin clots: A possible additional role for plasminogen activator?
Arch Biochem Biophys 211:750–761

Rånby M (1982a) Tissue plasminogen activator: isolation, enzymatic properties and
assay procedure. Thesis. Umea, Sweden, Umea University Medical Dissertations:
New Series No. 90

Rånby M (1982b) Studies on the kinetics of plasminogen activation by tissue plasminogen activator. Biochim Biophys Acta 704:461–469

Rånby M, Bergsdorf N, Nilsson T (1982) Enzymatic properties of the one- and two-
chain form of tissue plasminogen activator. Thromb Res 27:175–183

Redlitz A, Plow EF (1995) Receptors for plasminogen and tPA: an update. Baillieres
Clin Haematol 8:313–327

Renatus M, Engh RA, Stubbs MT, Huber R, Fischer S, Kohnert U, Bode W (1997)
Lysine 156 promotes the anomalous proenzyme activity of tPA; X-ray crystal structure of single chain human tPA. EMBO J 16:4797–4805

Ridker PM (1997) Intrinsic fibrinolytic capacity and systemic inflammation: novel risk
factors for arterial thrombotic disease. Haemostasis 27 [Suppl 1]:2–11

Ridker PM (1999) Evaluating novel cardiovascular risk factors: can we better predict
heart attacks? Ann Intern Med 130:933–937

Ridker PM, Baker MT, Hennekens CH, Stampfer MJ, Vaughan DE (1997) Alu-repeat
polymorphism in the gene coding for tissue-type plasminogen activator (tPA) and
risks of myocardial infarction among middle-aged men. Arterioscler Thromb Vasc
Biol 17:1687–1690

Ridker PM, Hennekens CH, Stampfer MJ, Manson JE, Vaughan DE (1994) Prospective study of endogenous tissue plasminogen activator and risk of stroke. Lancet
343:940–943

Rijken DC, Collen D (1981) Purification and characterization of the plasminogen activator secreted by human melanoma cells in culture. J Biol Chem 256:7035–7041

Rijken DC, Groeneveld E (1986) Isolation and functional characterization of the heavy and light chains of human tissue-type plasminogen activator. J Biol Chem 261:3098–3102

Rijken DC, Hoylaerts M, Collen D (1982) Fibrinolytic properties of one-chain and two-chain human extrinsic (tissue-type) plasminogen activator. J Biol Chem 257:2920–2925

Rijken DC, Juhan-Vague I, De Cock F, Collen D (1983) Measurement of human tissue-type plasminogen activator by a two-site immunoradiometric assay. J Lab Clin Med 101:274–284

Rijken DC, Wijngaards G, Zaal De Jong M, Welbergen J (1979) Purification and partial characterization of plasminogen activator from human uterine tissue. Biochim Biophys Acta 580:140–153

Rudd PM, Woods RJ, Wormald MR, Opdenakker G, Downing AK, Campbell ID, Dwek RA (1995) The effects of variable glycosylation on the functional activities of ribonuclease, plasminogen and tissue plasminogen activator. Biochim Biophys Acta 1248:1–10

Savage CR Jr, Hash JH, Cohen S (1973) Epidermal growth factor: Location of disulfide bonds. J Biol Chem 248:7669–7672

Schielen WJG, Adams HPHM, van Leuven K, Voskuilen M, Tesser GI, Nieuwenhuizen W (1991) The sequence γ-(312–324) is a fibrin-specific epitope. Blood 77:2169–2173

Schielen WJG, Voskuilen M, Tesser GI, Nieuwenhuizen W (1989) The sequence Aα-(148–160) in fibrin, but not in fibrinogen, is accessible to monoclonal antibodies. Proc Natl Acad Sci USA 86:8951–8954

Seeds NW, Basham ME, Haffke SP (1999) Neuronal migration is retarded in mice lacking the tissue plasminogen activator gene. Proc Natl Acad Sci USA 96:14118–14123

Seifried E, Tanswell P, Rijken DC, Barrett-Bergshoeff MM, Su CAPF, Kluft C (1988) Pharmacokinetics of antigen and activity of recombinant tissue-type plasminogen activator after infusion in healthy volunteers. Arzneimittelforschung 38:418–422

Sheng S, Truong B, Fredrickson D, Wu R, Pardee AB, Sager R (1998) Tissue-type plasminogen activator is a target of the tumor suppressor gene maspin. Proc Natl Acad Sci USA 95:499–504

Smith BO, Downing AK, Dudgeon TJ, Cunningham M, Driscoll PC, Campbell ID (1994) Secondary structure of fibronectin type 1 and epidermal growth factor modules from tissue-type plasminogen activator by nuclear magnetic resonance. Biochemistry 33:2422–2429

Smith FB, Rumley A, Lee AJ, Leng GC, Fowkes FGR, Lowe GDO (1998) Haemostatic factors and prediction of ischaemic heart disease and stroke in claudicants. Br J Haematol 100:758–763

Spellman MW, Basa LJ, Leonard CK, Chakel JA, O'Connor JV, Wilson S, Van Halbeek H (1989) Carbohydrate structures of human tissue plasminogen activator expressed in Chinese hamster ovary cells. J Biol Chem 264(24):14100–14111

Sprengers ED, Kluft C (1987) Plasminogen activator inhibitors. Blood 69:381–387

Stalder M, Hauert J, Kruithof EKO, Bachmann F (1985) Release of vascular plasminogen activator (v-PA) after venous stasis: electrophoretic-zymographic analysis of free and complexed v-PA. Br J Haematol 61:169–176

Steeds R, Adams M, Smith P, Channer K, Samani NJ (1998) Distribution of tissue plasminogen activator insertion/deletion polymorphism in myocardial infarction and control subjects. Thromb Haemost 79:980–984

Stringer KA, Bose SK, Mccord JM (1997) Antiinflammatory activity of tissue plasminogen activator in the carrageenan rat footpad model. Free Radical Biol Med 22:985–988

Stubbs MT, Renatus M, Bode W (1998) An active zymogen: unravelling the mystery of tissue-type plasminogen activator. Biol Chem 379:95–103

Suenson E, Lützen O, Thorsen S (1984) Initial plasmin-degradation of fibrin as the basis of a positive feedback mechanism in fibrinolysis. Eur J Biochem 140:513–522

Tachias K, Madison EL (1996) Converting tissue-type plasminogen activator into a zymogen. J Biol Chem 271:28749–28752

Tachias K, Madison EL (1997a) Converting tissue type plasminogen activator into a zymogen – important role of Lys[156]. J Biol Chem 272:28–31

Tachias K, Madison EL (1997b) Variants of tissue-type plasminogen activator that display extraordinary resistance to inhibition by the serpin plasminogen activator inhibitor type 1. J Biol Chem 272:14580–14585

Takada Y, Takada A (1989) Plasma levels of tPA, free PAI-1 and a complex of tPA with PAI-1 in human males and females at various ages. Thromb Res 55:601–609

Tanswell P, Seifried E, Su PCAF, Feuerer W, Rijken DC (1989) Pharmacokinetics and systemic effects of tissue-type plasminogen activator in normal subjects. Clin Pharmacol Ther 46:155–162

Tate KM, Higgins DL, Holmes WE, Winkler ME, Heyneker HL, Vehar GA (1987) Functional role of proteolytic cleavage at arginine-275 of human tissue plasminogen activator as assessed by site-directed mutagenesis. Biochemistry 26:338–343

Taylor ME, Conary JT, Lennartz MR, Stahl PD, Drickamer K (1990) Primary structure of the mannose receptor contains multiple motifs resembling carbohydrate-recognition domains. J Biol Chem 265:12156–12162

Theuring F, Aguzzi A, Kropp C, Wohn K-D, Hoffmann S, Schleuning W-D (1995) Analysis of the human tissue-type plasminogen activator gene promoter activity during embryogenesis of transgenic mice and rats. Fibrinolysis 9:277–287

Thompson SG, Kienast J, Pyke SDM, Haverkate F, van de Loo JCW (1995) Hemostatic factors and the risk of myocardial infarction or sudden death in patients with angina pectoris. N Engl J Med 332:635–641

Thorsen S, Glas-Greenwalt P, Astrup T (1972) Differences in the binding to fibrin of urokinase and tissue plasminogen activator. Thromb Diath Haemorrh 28:65–74

Thögersen AM, Jansson J-H, Boman K, Nilsson TK, Weinehall L, Huhtasaari F, Hallmans G (1998) High plasminogen activator inhibitor and tissue plasminogen activator levels in plasma precede a first acute myocardial infarction in both men and women. Evidence for the fibrinolytic system as an independent primary risk factor. Circulation 98:2241–2247

Tishkoff SA, Ruano G, Kidd JR, Kidd KK (1996) Distribution and frequency of a polymorphic Alu insertion at the plasminogen activator locus in humans. Hum Genet 97:759–764

Todd AS (1959) The histological localization of fibrinolysin activator. J Pathol Bacteriol 78:281

Todd AS (1964) Localization of fibrinolytic activity in tissues. Br Med Bull 20:210–212

Tran-Thang C, Kruithof EKO, Bachmann F (1984a) Tissue-type plasminogen activator increases the binding of Glu-plasminogen to clots. J Clin Invest 74:2009–2016

Tran-Thang C, Kruithof EKO, Bachmann F (1984b) The mechanism of in vitro clot lysis induced by vascular plasminogen activator. Blood 63:1331–1337

Tran-Thang C, Kruithof EKO, Atkinson J, Bachmann F (1986) High-affinity binding sites for human Glu-plasminogen unveiled by limited plasmic degradation of human fibrin. Eur J Biochem 160:599–604

Triputti P, Blasi F, Ny T, Emanuel BS, Letosfsky J, Croce CM (1986) Tissue-type plasminogen activator gene is on chromosome 8. Cytogenet Cell Genet 42:24–28

Tulinsky A (1991) The structures of domains of blood proteins. Thromb Haemost 66:16–31

Tyagi SC, Lewis K, Pikes D, Marcello A, Mujumdar VS, Smiley LM, Moore CK (1998) Stretch-induced membrane type matrix metalloproteinase and tissue plasminogen activator in cardiac fibroblast cells. J Cell Physiol 176:374–382

Ueba H, Kawakami M, Yaginuma T (1997) Shear stress as an inhibitor of vascular smooth muscle cell proliferation. Role of transforming growth factor-β1 and tissue-type plasminogen activator. Arterioscler Thromb Vasc Biol 17:1512–1516

van der Bom JG, de Knijff P, Haverkate F, Bots ML, Meijer P, de Jong PTVM, Hofman
    A, Kluft C, Grobbee DE (1997) Tissue plasminogen activator and risk of myocar-
    dial infarction. The Rotterdam Study. Circulation 95:2623–2627
Van Griensven JMT, Huisman LGM, Stuurman T, Dooijewaard G, Kroon R,
    Schoemaker RC, Kluft K, Cohen AF (1996) Effects of increased liver blood flow
    on the kinetics and dynamics of recombinant tissue-type plasminogen activator.
    Clin Pharmacol Ther 60:504–511
Van Hinsbergh VWM, Kooistra T, Scheffer MA, van Bockel JH, van Muijen GNP
    (1990) Characterization and fibrinolytic properties of human omental tissue
    mesothelial cells. Comparison with endothelial cells. Blood 75:1490–1497
Van Zonneveld A-J, Veerman H, Pannekoek H (1986) On the interaction of the finger-
    and kringle 2-domain of tissue-type plasminogen activator with fibrin. Inhibition
    of kringle-2 binding to fibrin by ε-aminocaproic acid. J Biol Chem 261:14214–
    14218
Verheijen JH, Caspers MPM, Chang GTG, de Munk GAW, Pouwels PH, Enger-Valk
    BE (1986a) Involvement of finger domain and kringle-2 domain of tissue-type
    plasminogen activator in fibrin binding and stimulation of activity by fibrin.
    EMBO J 5:3525–3530
Verheijen JH, Visse R, Wijnen JT, Chang GT, Kluft C, Meera-Khan P (1986b) Assign-
    ment of the human tissue-type plasminogen activator gene (PLAT) to chromo-
    some 8. Hum Genet 72:153–156
Vlahos CJ, Wilhelm OG, Hassell T, Jaskunas SR, Bang NU (1991) Disulfide pairing of
    the recombinant kringle-2 domain of tissue plasminogen activator produced in
    Escherichia coli. J Biol Chem 266:10070–10072
Voskuilen M, Vermond A, Veeneman GH, van Boom JH, Klasen EA, Zegers ND,
    Nieuwenhuizen W (1987) Fibrinogen lysine residue Aα157 plays a crucial role in
    the fibrin-induced acceleration of plasminogen activation, catalyzed by tissue-type
    plasminogen activator. J Biol Chem 262:5944–5946
Waller EK, Schleuning WD (1985) Induction of fibrinolytic activity in HeLa cells by
    phorbol myristate acetate. Tissue-type plasminogen activator antigen and mRNA
    augmentation require intermediate protein biosynthesis. J Biol Chem 260:6354–
    6360
Wallén P, Bergsdorf N, Rånby M (1982) Purification and identification of two structural
    variants of porcine tissue plasminogen activator by affinity adsorption on fibrin.
    Biochim Biophys Acta 719:318–328
Wallén P, Pohl G, Bergsdorf N, Rånby M, Ny T, Jörnvall H (1983) Structural charac-
    terization of tissue plasminogen activator purified by immunosorbent chromatog-
    raphy. In: Davidson JF, Bachmann F, Bouvier CA, Kruithof EKO (eds) Progress
    in Fibrinolysis, vol 6. Churchill-Livingstone, Edinburgh, pp 338–343
Wang Y, Hand AR, Wang Y-H, Mina M, Gillies C, Peng T, Cone RE, O'Rourke J (1998)
    Functional and morphologic evidence of the presence of tissue-plasminogen acti-
    vator in vascular nerves: implications for a neurologic control of vessel wall fibri-
    nolysis and rigidity. J Neurosci Res 53:443–453
Waugh JM, Kattash M, Li J, Yuksel E, Kuo MD, Lussier M, Weinfeld AB, Saxena R,
    Rabinovsky ED, Thung S, Woo SLC, Shenaq SM (1999) Gene therapy to promote
    thromboresistance: local overexpression of tissue plasminogen activator to
    prevent arterial thrombosis in an in vivo rabbit model. Proc Natl Acad Sci USA
    96:1065–1070
Weening-Verhoeff EJD, Quax PHA, Van Leeuwen RTJ, Rehberg EF, Marotti KR, Ver-
    heijen JH (1990) Involvement of aspartic and glutamic residues in kringle-2 of
    tissue-type plasminogen activator in lysine binding, fibrin binding and stimulation
    of activity as revealed by chemical modification and oligonucleotide-directed
    mutagenesis. Protein Eng 4:191–198
Welling TH, Huber TS, Messina LM, Stanley JC (1996) Tissue plasminogen activator
    increases canine endothelial cell proliferation rate through a plasmin-
    independent, receptor-mediated mechanism. J Surg Res 66:36–42

Werner F, Razzaq TM, Ellis V (1999) Tissue plasminogen activator binds to human vascular smooth muscle cells by a novel mechanism. Evidence for a reciprocal linkage between inhibition of catalytic activity and cellular binding. J Biol Chem 274:21555–21561

Wilhelm OG, Jaskunas SR, Vlahos CJ, Bang NU (1990) Functional properties of the recombinant kringle-2 domain of tissue plasminogen activator produced in *Escherichia coli*. J Biol Chem 265:14606–14611

Willnow TE, Orth K, Herz J (1994) Molecular dissection of ligand binding sites on the low density lipoprotein receptor-related protein. J Biol Chem 269:15827–15832

Wing LR, Hawksworth GM, Bennett B, Booth NA (1991) Clearance of tPA, PAI-1, and tPA-PAI-1 complex in an isolated perfused rat liver system. J Lab Clin Med 117:109–114

Yonekawa O, Voskuilen M, Nieuwenhuizen W (1992) Localization in the fibrinogen γ-chain of a new site that is involved in the acceleration of the tissue-type plasminogen activator-catalysed activation of plasminogen. Biochem J 283:187–191

Yudin S (1936) Transfusion of cadaver blood. JAMA 106:997–999

Yudin S (1937) Transfusion of stored cadaver blood. Lancet 2:361–366

Zhuo M, Holtzman DM, Li Y, Osaka H, DeMaro J, Jacquin M, Bu G (2000) Role of tissue plasminogen activator receptor LRP in hippocampal long-term potentiation. J Neurosci 20:542–549

CHAPTER 4
Urinary-type Plasminogen Activator (uPA)

W.A. GÜNZLER and L. FLOHÉ

A. Milestones in uPA Research

Although the proteolytic activity of urine was recognized more than 100 years ago, its fibrinolytic activity was first investigated by MACFARLANE and PILLING in 1947. WILLIAMS (1951) and ASTRUP and STERŃDORFF (1952) attributed the activity to a "kinase" which activates plasminogen. SOBEL et al. (1952) coined the name "urokinase" for this plasminogen activator.

Early attempts to isolate, quantitate, and characterize urokinase from human urine (PLOUG and KJELDGAARD 1957) were intensified in the 1960s due to commercial interests in the potential clinical use of urokinase as a thrombolytic agent. Since the mid-1960s urokinase has been used clinically in the treatment of thromboembolic diseases, e.g., pulmonary embolism, myocardial infarction, peripheral thrombosis. Lack of antigenicity turned out as its principal advantage (cf. Chaps. 5 and 12). Urokinase in crystallized form was first obtained by LESUK et al. (1965). High and low molecular weight forms of urokinase were purified and characterized by WHITE et al. (1966). The glycoprotein nature of urokinase was later described by ZUBAIROV et al. (1974). Further insight into the structure of uPA and evidence for interdependency of uPA forms was provided by elucidation of its amino acid and corresponding DNA sequences. Amino acid sequences of low and high molecular weight urokinases were reported in 1982 (STEFFENS et al. 1982; GÜNZLER et al. 1982b). These data revealed that full-length uPA is a multi-domain protein consisting of an epidermal growth factor-like domain, a kringle, and a serine proteinase domain. The amino acid sequence was confirmed by cDNA-sequencing (HOLMES et al. 1985; NAGAI et al. 1985) and provided the basis for production of recombinant urokinase (GÜNZLER et al. 1985) and pro-urokinase in *E. coli*. The structure of the human uPA gene was reported in 1985 (RICCIO et al. 1985). A recent report on the crystal structure of the proteinase domain (SPRAGGON et al. 1995) demonstrates that the enzyme has an S1 specificity pocket similar to that of trypsin. At six positions insertions of extra residues in loop regions create unique surface areas. One of these loop regions is characteristic of a small subset of serine proteases such as tPA, factor XII, and complement factor I.

The proenzyme of urokinase was shown to be activated by plasmin or plasminogen in the presence of fibrin by Bernik and Oller (1973). These observations led to the concept of clot-specific thrombolysis by pro-urokinase (Gurewich et al. 1984; Hanbücken et al. 1987) which was soon scrutinized by clinical trials (Flohé 1985; Van de Werf et al. 1986a,b; PRIMI Trial Study Group 1989; Bär et al. 1997; Tebbe et al. 1998) (cf. Chaps. 5 and 12).

Despite the well-documented fibrinolytic activity of uPA, its physiological role in the hemostatic balance remains ambiguous and additional or alternative biological functions of uPA are being discussed. A role for uPA in cell migration and tissue remodeling is suggested by the observation that uPA is able to bind specifically to a uPA receptor (uPAR) on cell surfaces (Stoppelli et al. 1985; Vasalli et al. 1985) by means of a peptide loop within its epidermal growth factor domain (Appella et al. 1987). The uPA-deficient mouse model (Carmeliet et al. 1994) revealed a whole network of functions in which uPA appears to be directly or indirectly involved. This will be discussed at the end of this chapter.

B. Nomenclature of uPA Forms

The names of biomolecules usually refer to their source and function. The historical, still widely used name "urokinase" has yet to be rated as misleading, since uPA is not a phosphorylating "kinase" but a serine proteinase. The preferred generic term "urinary-type (or urokinase-type) plasminogen activator" (uPA) better complies with this principle and allows differentiation from other plasminogen activators, e.g., tissue-type plasminogen activator (tPA). Chemical Abstracts registers urokinase under the term "kinase (enzyme-activating)-uro" [CAS no: 9039–53–6] and the International Union of Biochemistry under EC 3.4.21.31.

According to the nomenclature of serine proteinases the precursors of mature uPA which are produced within or are secreted from the cell were named prepro-urokinase and pro-urokinase, respectively. However, due to doubts about its proenzyme character, the International Committee on Thrombosis and Haemostasis recommends the name "single-chain urokinase-type plasminogen activator" (sc-uPA) for pro-urokinase. The mature serine proteinase urokinase, formed after activating cleavage is termed "two-chain urokinase-type plasminogen activator" (tc-uPA). Limited proteolysis generates various forms of sc-uPA and tc-uPA which, according to their size, are termed "high and low molecular weight (HMW and LMW)" sc-uPA or tc-uPA. If not stated otherwise, the name uPA refers to the human-type glycosylated protein. Full-length (Ser^1-Leu^{411}) unglycosylated human-type single-chain uPA from recombinant bacteria is termed "saruplase" by International Nonproprietary Name. The prefix "r" is often used as a general term for recombinant uPA forms, e.g., "rsc-uPA."

C. Sources of uPA

uPA is produced by a large variety of cells, the most significant producers being epithelial cells, e.g., of the kidney (SAPPINO et al. 1991) and cells involved in migratory processes. The latter type of cells usually express uPAR concomitantly.

The concentration of uPA in human urine is in the range of 40–80 μg/l, of which 25% is sc-uPA and 75% tc-uPA, a part of which is bound to an inhibitor (STUMP et al. 1986a). Sc-uPA is also present in human plasma in the range of 2–10 μg/l. It has been proposed that venous endothelial cells may be a source of the uPA intrinsic to blood (CAMOIN et al. 1998).

uPA is fairly rapidly cleared from the circulation. Negative correlation of uPA plasma concentration with liver blood flow suggests prevalence of hepatic elimination (VAN GRIENSVEN et al. 1997). The dominant initial plasma half-life of unglycosylated sc-uPA was found to be slightly shorter than that of glycosylated tc-uPA (about 9 vs 12 min) (POEPPELMEIER et al. 1996). Differential hepatic elimination mechanisms, either via the α_2-macroglobulin receptor/low density lipprotein receptor-related protein or via the asialoglycoprotein receptor, have been proposed for unglycosylated and glycosylated uPA (VAN DER KAADEN et al. 1998).

Formerly, human urine was the common source of tc-uPA, while human kidney cell culture was used for production of glycosylated sc-uPA. More recently, recombinant cells have been employed as a source for uPA. Initially, the uPA gene had been cloned in prokaryotes, e.g., *E. coli* or *B. subtilis*. Recombinant K12-type *E. coli* equipped with specially designed high expression vectors express high levels of unglycosylated uPA (BRIGELIUS-FLOHÉ et al. 1992). The expressed uPA protein accumulates rapidly within the cell in the form of inactive inclusion bodies, which can be refolded in vitro into its active conformation. In contrast, recombinant eukaryotic cells allow excretion of the folded, glycosylated uPA product into the culture medium. The uPA gene has further been cloned in microbial and mammalian eukaryotic cells, e.g., yeast or streptomyces and Chinese hamster ovary or myeloma cells, respectively. Transgenic mice which carried a hybrid casein-uPA gene secreted active uPA in their milk in high concentrations (MEADE et al. 1990).

D. Assay of uPA

The quality of tc-uPA for use as a drug substance is defined by a monograph within the European Pharmacopoeia (2nd suppl. 1993) where test methods and requirements for identity, purity, and potency of urokinase from human urine are laid down.

In vitro tests for quantitation of uPA either make use of the enzymatic activity or the antigen nature of the protein or both. The activity of tc-uPA and, after activation by plasmin, of sc-uPA can be reliably assayed by chromogenic substrate tests. Suitable chromogenic substrates for the assay of "amidolytic" activity, e.g., pyro-Glu-Gly-Arg-pNA, are oligopeptides which mimic the cleavage site of plasminogen (FRIBERGER 1982). Release of the chromophore p-nitroaniline by hydrolysis allows one to monitor continuously the enzyme activity. This test system is independent of calibration by a reference standard and allows one to express enzyme units in katal (mol/sec) in compliance with the SI system. The sensitivity of the chromogenic substrate test is increased if uPA-dependent plasminogen activation is assayed by use of a chromogenic substrate for plasmin.

Fibrinolytic test systems, e.g., clot lysis test (PLOUG and KJELDGAARD 1957), fibrin plate test (BRAKMAN 1967), or fibrin agar plate test (SCHUMACHER and SCHILL 1972), include plasminogen activation and require calibration by an appropriate reference standard. A glycosylated HMW tc-uPA International Reference Standard (87/594) is supplied by the National Institute for Biological Standards and Control, Blanche Lane, South Mimms, Potters Bar, Hertfordshire EN6 3QG, U.K., and allows one to express the activity of tc-uPA in International Units (IU). Pure tc-uPA has a specific activity of about 100 000 IU/mg.

Measurement of uPA activities in biological materials requires taking precautions in order to avoid in vitro alterations of sample, e.g., inhibition of tc-uPA activity or activation of sc-uPA. It is also often desirable to distinguish between different plasminogen activators. Stabilization of plasma samples by aprotinin and benzamidine and a modified chromogenic test for tc-uPA and sc-uPA, which includes specificity-conferring immuno binding, has been proposed (GÜNZLER et al. 1990). A selective chromogenic assay for sc-uPA in plasma in which sc-uPA is activated by thermolysine has also been described (KEBABIAN and HENKIN 1992).

Since the uPA content in certain malignant tumors is a prognostic marker, e.g., in breast cancer (JÄNICKE et al. 1994), determination of uPA antigen in tissue cytosols or extracts meets increasing interest. While commercial test kits for uPA antigen are available, the best choice of extraction methods for tissue samples and standardization of uPA antigen are still under debate (ROMAIN et al. 1995; BENRAAD et al. 1996).

Animal models suitable for in vivo studies of thrombolytic efficacy of uPA in arterial systems were reviewed by BUSH and SHEBUSKI (1990). A most common animal model of venous thrombolysis uses experimental jugular vein thrombosis in rabbits (COLLEN et al. 1983). Small animal thrombosis models, e.g., a pulmonary embolism model in adult hamsters, have been proposed for easy performance if a great number of experiments is to be conducted (STASSEN et al. 1991).

E. Biochemistry of uPA

I. The uPA Gene

The uPA gene is localized on chromosome 10 (RAJPUT et al. 1985). It is organized into 11 exons and is 6.3 kb long (RICCIO et al. 1985). The intron-exon organization of the uPA gene supports the hypothesis of a modular evolution of the serine protease genes. A high degree of homology attests to the close relationship of uPA with tPA. Despite some single base differences, the cDNA (HOLMES et al. 1985; NAGAI et al. 1985) and poly(A)RNA (VERDE et al. 1984) sequences from different human cell lines and the genomic DNA sequence comply with the amino acid sequence of uPA isolated from urine. However, either Ile or Met have been identified at position 194 of the protein, as well as a Pro/Leu polymorphism at position 121 (CONNE et al. 1997).

The 5′ flanking regulatory region of the human uPA gene exhibits a typical TATA box 25 bp upstream, and four GC rich sequences in the first 200 bp upstream of the transcription initiation site. The latter are potential binding sites for the transcription factor SP1. Two putative AP-2 binding sites close to the transcription initiation site are responsible for protein kinase A-dependent induction of uPA expression. Two NF-κB elements at −1580 bp and −1865 bp have been identified in the human promoter. The latter site acts as a repressor, as do two further negative regulatory sites, one situated between −1824 bp and −1572 bp, and the other involving an enhancer-dependent and cell-specific silencing region between −660 bp and −536 bp. The uPA promoter is strongly enhanced by an enhancer located about 2000 bp upstream of the transcription initiation site. This enhancer consists of an upstream PEA/AP-1 A site and a downstream AP-1B site. All three sites are required for induction of uPA gene transcription by a variety of extracellular stimuli. Synergism between PEA/AP-1 A and AP-1B depends on the integrity of an intervening COM site which contains several sequences for binding of urokinase enhancer factors (DE CESARE et al. 1996, 1997; BERTHELSEN et al. 1998; reviewed and literature references in BESSER et al. 1997).

II. The uPA Structure

As mentioned above, uPA exists in different forms. The amino acid structure of prepro-urokinase has been derived from the cDNA sequence (HOLMES et al. 1985; NAGAI et al. 1985). The amino terminal signal peptide, which consists of 20 predominantly hydrophobic amino acids, is cleaved off during the process of secretion.

Sc-uPA, consists of a single peptide chain of 411 amino acids. The amino acid sequence of human sc-uPA derived from the amino acid sequence of urinary tc-uPA (STEFFENS et al. 1982; GÜNZLER et al. 1982b) and from the DNA sequences of the uPA gene and the cDNA is given in Fig. 1. The cysteine residues form 12 disulfide bridges. The molecular mass of sc-uPA based on its amino acid sequence is 46344 Da, and about 54 kDa in the glycosylated form. Three characteristic domains can be distinguished. The amino-terminal domain including three disulfide bridges resembles the structure of epidermal growth factor and is termed "EGF domain." It contains a fucosyl binding site at Thr[18] (BUKO et al. 1991) and the peptide loop 13–30 responsible for binding to the uPAR (APPELLA et al. 1987). The adjacent domain also containing three disulfides is homologous to "kringle" domains common to many serine

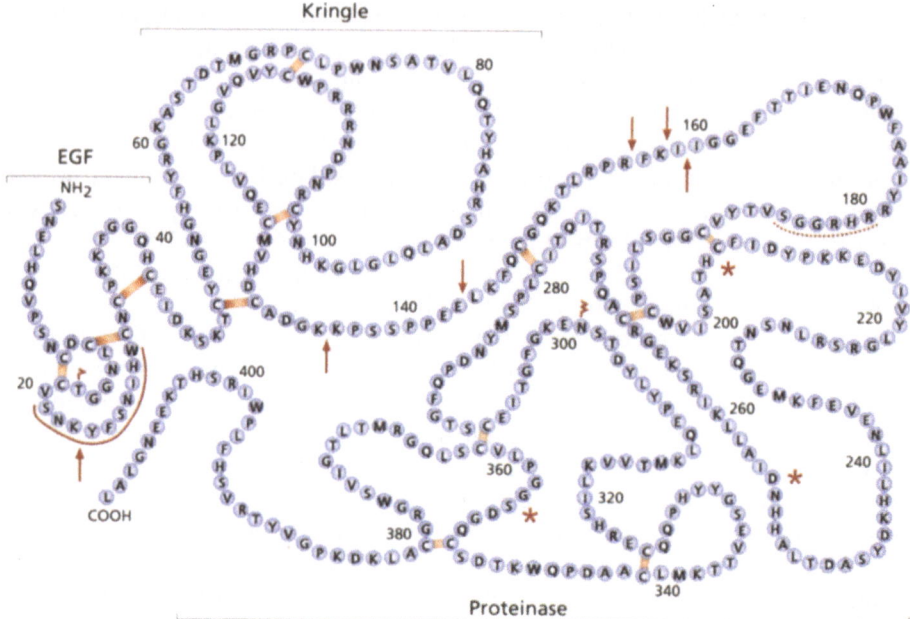

Fig. 1. The schematic presentation shows the amino acid sequence of sc-uPA in single letter code (modified from Holmes et al. 1985). Disulfide bridges are given as *solid connections* between cysteine residues (C). EGF, kringle and proteinase domains are indicated. Annexed *zig-zag lines* mark glycosylation sites. *Asterisks* indicate active site amino acids of the serine proteinase. *Arrows* mark peptide cleavage sites which appear of relevance for uPA function (see Table 1). Cleavage of the Lys^{158}-Ile^{159} bond converts sc-uPA into the fully active serine proteinase tc-uPA. The peptide loop conferring uPAR binding specificy (positions 20–30) and the region responsible for binding of PAI-1 (positions 179–184) are marked by *solid and stippled lines*, respectively

proteinases. Dominance of positively charged amino acids suggests that the kringle may interact with negatively charged entities, e.g., heparin. A "connecting peptide" links the kringle domain to the "proteinase domain" which contains the amino acids of the typical serine proteinase active site, His^{204}, Asp^{255}, and Ser^{356}, and a glycosylation site at Asn^{302}.

Cleavage of the peptide bond Lys^{158}-Ile^{159} transforms sc-uPA into the mature serine proteinase tc-uPA. The amino terminal A-chain of HMW tc-uPA containing the EGF and kringle domains remains connected to the B-chain, i.e., the proteinase domain, by the disulfide bond Cys^{148}-Cys^{279}.

Peptide cleavage of the Lys^{135}-Lys^{136} bond between kringle domain and the inter-chain disulfide bond results in the loss of the amino terminal EGF and kringle domains, also termed amino terminal fragment (ATF), from the proteinase. The resulting LMW uPA has a molecular mass of about 33 kDa and its truncated A-chain consists of only 21 amino acids (Lys^{136}-Arg^{156}) (Günzler et al. 1982a; Schaller et al. 1983). From the medium of a human lung adenocarcinoma cell line a LMW sc-uPA was characterized which starts at Leu[144] (STUMP et al. 1986b).

The conformation of uPA and of its domains has been studied by means of small angle neutron scattering (MANGEL et al. 1991), one- and two-dimensional NMR (BOGUSKY et al. 1989; LI et al. 1991; NOWAK et al. 1994) and X-ray crystallography (SPRAGGON et al. 1995). The domains of uPA are folded essentially as in other serine proteinases with few significantly different details. The uPA molecule is highly asymmetric; the domains resemble differently shaped beads on a string. EGF, kringle, and the three subdomains within the proteinase domain behave autonomously and thus can be considered essentially as independent functional units. The amino terminal region of the B-chain is extremely resistant to unfolding. Binding of a pseudosubstrate inhibitor, Glu-Gly-Arg-chloromethyl ketone, stabilizes the proteinase subdomain. High flexibility was observed within the uPAR binding region. The interaction of Lys[300] in the flexible loop region 297–313 with Asp[355] was shown to stabilize the active proteinase conformation of sc-uPA (SUN et al. 1997, 1998). The positively charged cluster of residues 179–184 in the surface loop known as variable region 1 interacts with the main physiological inhibitor of tc-uPA, plasminogen activator inhibitor type-1 (PAI-1) (ADAMS et al. 1991), which, as a pseudosubstrate, forms an inactive adduct by acylation of the active site serine residue of the proteinase (EGELUND et al. 1998).

III. Non-protein Components of uPA

The composition of the carbohydrate side chain at Asn[302] depends on the producing cell type. In uPA derived from recombinant cells a higher carbohydrate content than in urinary uPA and a different sialic acid contents were observed (SARUBBI et al. 1989). The N-glycosidic carbohydrate moiety of uPA from urine was found to be heterogeneous and rather uncommon (BERGWERFF et al. 1995). It consists mainly of two- and three-antennal chains containing terminal N-acetylglactosamine(β1–4)-N-acetylglucosamine elements. The latter residues partially carry a fucose residue at O3 and the former are mainly sulfated at O4. According to cDNA-derived amino acid sequences the glycosylation site in porcine uPA is located in the kringle domain (NAGAMINE et al. 1984), while murine uPA lacks any N-glycosylation site (BELIN et al. 1985). These differences make it unlikely that N-glycosylation of uPA is of basic functional significance.

Fucosylation of a distinct threonine residue (Thr[18]) has been observed first in uPA (BUKO et al. 1991) but might be a common feature of EGF-like structures. It has been implicated in cell proliferation and mitogenicity of glycosylated uPA (RABBANI et al. 1992).

There are two phosphorylation sites, on Ser[138] and Ser[303], which modulate proadhesive ability of uPA without interference of its binding to uPAR (FRANCO et al. 1997). Phosphorylation of tyrosine residues within uPA from urine or produced by HT-1080 cells has been reported (BARLATI et al. 1991); the site of modification is, however, not specified.

IV. Proteolytic Conversions of uPA

The significance of activation of sc-uPA by proteolytic action was recognized by BERNIK and OLLER (1973). The activation by the product of uPA action, i.e., plasmin, was recognized as a particularly key step in the fibrinolytic action of sc-uPA. Meanwhile, it has become evident that the activation site of sc-uPA is also prone to cleavage by several other serine proteinases, cysteine proteinases, and even a metalloproteinase (Table 1). While plasma kallikrein and factor XIIa may contribute to fibrinolysis, cathepsins are thought to participate in the uPA activation of cell surface-related plasminogen activation.

Cleavage of sc-uPA by thrombin at the Arg^{156}-Phe^{157} bond results in an essentially inactive tc-uPA (ICHINOSE et al. 1986; GUREWICH et al. 1987), since the amino group released does not appropriately interact with the proteinase active site. The inactivation is accelerated by thrombomodulin (DE MUNK et al. 1991). This inactive tc-uPA form can be activated by release of the amino terminal dipeptide of its B-chain by cathepsin C (NAULAND and RIJKEN 1994) or, although slowly, by plasmin (LIJNEN et al. 1987). Cleavage at the Ile^{159}-Ile^{160} bond results in an irreversibly inactive tc-uPA (SCHMITT et al. 1989; LEARMONTH et al. 1992).

Cleavage within the uPAR binding region, e.g., of the Lys^{23}-Tyr^{24} bond, abolishes uPA receptor binding and therefore modulates cell-surface uPA

Table 1. Proteinases acting on individual cleavage sites (see Fig. 1) within sc-uPA being of relevance for uPA function

Cleavage site	Proteinase	Reference
Lys^{158}-Ile^{159}	Plasmin, trypsin	BERNIK and OLLER 1973
	Serum kallikrein	ICHINOSE et al. 1986
	Glandular kallikrein	LIST et al. 2000
	Factor XIIa	ICHINOSE et al. 1986
	Nerve growth factor-γ	WOLF et al. 1993
	Mast cell tryptase	STACK and JOHNSON 1994
	Cathepsin B	KOBAYASHI et al. 1991
	Cathepsin L	SCHMITT et al. 1992
	Cathepsin G (with heparin)	DRAG and PETERSEN 1994
	Hu TSP-1	BRUNNER et al. 1990
	Thermolysin	MARCOTTE and HENKIN 1993
Arg^{156}-Phe^{157}	Thrombin	ICHINOSE et al. 1986
Ile^{159}-Ile^{160}	Granulocyte elastase	SCHMITT et al. 1989
	Cathepsin G	LEARMONTH et al. 1992
Lys^{135}-Lys^{136}	Urokinase, plasmin	SCHALLER et al. 1983
Glu^{143}-Leu^{144}	Pump 1 metalloproteinase	MARCOTTE et al. 1992a
	Metalloproteinase 3	UGWU et al. 1998
Lys^{23}-Tyr^{24}	Plasmin	MARCOTTE et al. 1992b
	Cathepsin G (with heparin)	LEARMONTH et al. 1992

activity (MARCOTTE et al. 1992b; LEARMONTH et al. 1992). A B-type carboxypeptidase like carboxypeptidase U or thrombin-activable fibrinolysis inhibitor, which was recently identified in plasma (EATON et al. 1991; WANG et al. 1994; BAJZAR et al. 1995) might be responsible for the lack of carboxy-terminal Lys of the A-chain in tc-uPA from urine (GÜNZLER et al. 1982b).

V. uPA Activity

uPA activates the pro-enzyme plasminogen to the active serine proteinase plasmin by cleavage of the single Arg^{561}-Val^{562} bond (cf. Chap. 2). Almost all proteolytic activities ascribed to uPA are finally due to the plasmin generated this way. Direct cleavage by tc-uPA of fibronectin (GOLD et al. 1989), processing of pro-hepatocyte growth factor (NALDINI et al. 1995) and of basic fibroblast growth factor (ODEKON et al. 1992), as well as promotion of axon growth (MUIR et al. 1998) are a few exceptions.

On a molar basis HMW and LMW tc-uPA show similar activities in chromogenic substrate tests (PHILO and GAFFNEY 1981). However, higher efficacy was reported for HMW tc-uPA in clot lysis and fibrin plate tests (SAMAMA et al. 1982). No difference in inhibition rate by α_2-antiplasmin or in plasma was observed between HMW and LMW tc-uPA (SAMAMA et al. 1982). Unglycosylated tc-uPA was faster inhibited in plasma than glycosylated tc-uPA, while kinetic constants with chromogenic substrate were the same (LENICH et al. 1992).

Whether sc-uPA is an active entity has long been a matter of debate. Like a pro-enzyme, it is essentially resistant to inhibition, i.e., its activity persists in plasma (PANNELL and GUREWICH 1986) and it does not form an inactive adduct with PAI-1. It is also accepted now that sc-uPA exhibits only very little (<1/1000) activity compared to tc-uPA in a chromogenic test system. However, in fibrinolytic or thrombolytic test systems, which include the activation of plasminogen, sc-uPA produces considerable activity in vivo and in vitro. The lytic activity of sc-uPA is, in part, due to the reciprocal activation of sc-uPA by plasmin and of plasminogen by tc-uPA (PETERSEN 1997). The evidence for a genuine plasminogen-activating activity of sc-uPA, possibly sufficient to trigger the lytic effect, has been provided by the activity of the Lys^{158}Glu sc-uPA mutant, which is no longer cleavable by plasmin (NELLES et al. 1987) but can activate plasminogen. Furthermore, the plasminogen mutant Ser^{740}Ala, the active site of which is inoperative, was still cleaved at the Arg^{561}-Val^{562} site by sc-uPA (LIJNEN et al. 1990). This cleavage was most effective if the mutant plasminogen was bound to partially degraded fibrin, suggesting that for fibrin-bound plasminogen, sc-uPA might be a sufficient activator on its own (FLEURY et al. 1993).

It is evident from studies in vitro and in vivo that clot lysis by sc-uPA in contrast to that by tc-uPA can occur fibrin-specifically. This means that clot lysis proceeds without major systemic effects induced by plasmatic plasmin generation. The way by which sc-uPA acts fibrin-specifically has not been definitely established. Since sc-uPA does not bind to fibrin, this is not a matter of local binding. Instead, a probable explanation (Fig. 2) considers the prevalence of sc-uPA activity towards fibrin-bound vs free plasmatic Glu-plasminogen (GUREWICH et al. 1992). This concept agrees with the above-mentioned in vitro experiments of plasminogen activation (FLEURY et al. 1993), with the frequent observation of a concentration-dependent lag-phase in sc-uPA clot lysis in vitro, and with the claim of synergistic thrombolytic actions of tPA and sc-uPA (GUREWICH and PANNELL 1986; COLLEN et al. 1987). Accordingly, initial fibrin degradation, e.g., by tPA, generates free carboxy terminal lysine residues in fibrin fragment E. Glu-plasminogen binds to these residues and thereby undergoes a conformational change which renders it highly susceptible to activating cleavage by sc-uPA. Plasmin, thus generated, attacks fibrin and, simultaneously, produces tc-uPA from sc-uPA, thereby enhancing the fibrinolytic process (Fig. 2b). In contrast, in plasma, Glu-plasminogen is not efficiently activated by sc-uPA and mature proteinases such as tc-uPA and plasmin, eventually generated, are under control of plasmatic inhibitors (Fig. 2a). Therefore, as long as the inhibitor capacity is adequate, plasminaemia and, consequently, systemic fibrinogen breakdown is prevented.

A conformational change of Glu-plasminogen upon interaction with lysine analogues similar to that upon binding to fibrin could explain the observation of a rapid conversion of sc-uPA to tc-uPA in plasma containing tranexamic acid (GÜNZLER et al. 1990). When sc-uPA and tranexamic acid were coadministered to rabbits in a pulmonary embolism model, fibrinolysis was antagonized, possibly by interfering with the fibrin binding of plasminogen, while a profound fibrinogen degradation indicated a fulminant systemic plasminogen activation (SCHNEIDER 1990).

Interaction of sc-uPA with uPAR was reported to increase its activity, possibly by reversible change of its conformation (HIGAZI et al. 1995). Unglycosylated sc-uPA was activated faster by plasmin and was more active in clot lysis than glycosylated sc-uPA (LENICH et al. 1992). Functional properties of HMW and LMW sc-uPA were found to be very similar (LIJNEN et al. 1988).

F. Physiology of uPA

The physiological role of uPA within the plasminogen-plasmin system is still not completely understood. There appear to be two main aspects of plasmin-dependent uPA functions:

1. Maintenance of intravascular or intratubular patency
2. Tissue remodelling involved in cell migration processes

The former function is suggested by secretion of uPA into, and presence of free uPA in, the vascular or tubular fluids. The latter function depends on localization of uPA specifically on the cell surface, e.g., by uPAR binding (cf.

(a)

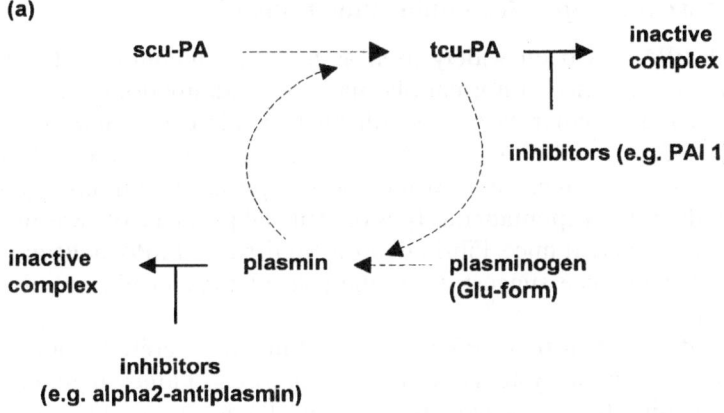

(b)

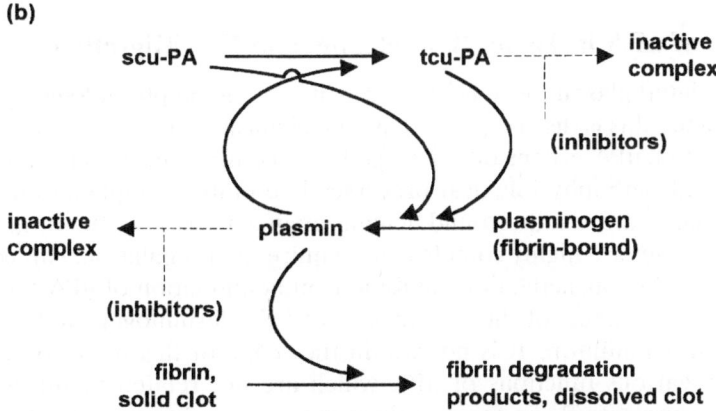

Fig. 2a,b. Explanation of clot-specific action of sc-uPA. **a** Processes in plasma: negligible plasminogen activation by sc-uPA; proteinases (tc-uPA, plasmin) under inhibitor control (cf. Chap. 5). **b** Processes at the clot: activation by sc-uPA of plasminogen bound to fibrin fragment E; positive feedback activation of sc-uPA by plasmin to tc-uPA; inefficient local inhibition; efficient degradation of the fibrin clot

Chap. 6). Vitronectin might contribute to the localization of uPA activity by binding of soluble uPAR/uPA complexes (CHAVAKIS et al. 1998). A special type of uPA binding seems to be involved in the uptake into human platelets (GUREWICH et al. 1995; JIANG et al. 1996). In this instance, HMW uPA appears securely fixed to the platelet membrane while its activity is conserved. The multiligand receptors α_2-macroglobulin receptor/low density lipoprotein receptor-associated protein and epithelial glycoprotein 330 are reported to be responsible for internalization of several types of uPA complexes (CONESE and BLASI 1995; VAN DER KAADEN et al. 1998).

I. Intravascular and Intratubular Functions of uPA

Although uPA has been widely used as a useful human-type thrombolytic agent and its presence in normal plasma is beyond any doubt, its physiological role in the vascular system is still unclear. First experiences with mice devoid of a functional uPA gene (CARMELIET et al. 1994) indicated that uPA, like tPA, might protect veins from fibrin deposits. In pulmonary arteries, however, the rate of spontaneous lysis of artificial plasma clots was not altered in uPA gene-deficient mice. Fibrin deposits in tissues of uPA-deficient, but not in tPA-deficient mice suggest a specific role of uPA in fibrinolysis (BUGGE et al. 1996).

Since it is known that uPA is produced in, and secreted from, epithelial cells lining the tubular system of different organs, e.g., kidney, mammary gland, and male genital tract, it seems obvious that uPA is responsible for maintaining their patency. The regulation of the exocrine process and the structural determinants of uPA secretion are still unclear.

II. Role of uPA in Tissue Remodelling and Cell Migration

Cell-associated plasminogen activation is another main physiological function of uPA. Since this is discussed as a basic mechanism of tissue remodelling and cell migration, uPA is considered a significant component of important physiological and pathophysiological processes like embryo implantation, migration of inflammatory cells, wound healing, angiogenesis, and tumor metastasis. These processes are highly complex and require tight regulation. This concerns not only expression, activation, inhibition, and elimination of uPA but also of the other components of the system, e.g., uPAR, plasminogen, and plasminogen activator inhibitors. It is not within the scope of this book to discuss in detail the various functions of uPA which are not related to thrombolysis. The significance of uPA in some of these processes is, however, questioned by experience with mice deficient in functional uPA gene (CARMELIET et al. 1994). uPA-deficient mice appeared normal at birth, suggesting no pivotal role for uPA in embryonic development. The finding that macrophage function was not affected by uPA deficiency indicates that some uPA functions may, in part, be taken over by other enzymes (biological redundancy). However, uPA deficient mice are predisposed to staphylococcal infections (SHAPIRO et al. 1997). They also exhibit some resistance to induction of primary cutaneous melanoma, suggesting that uPA is involved in malignant progression (SHAPIRO et al. 1996). uPA was shown to induce smooth muscle cell migration. The EGF-region was essential for the chemotactic effect of uPA, suggesting a pivotal role of uPAR binding (STEPANOVA et al. 1997). Arterial injury in uPA-deficient mice revealed that uPA plays a significant, but uPAR-independent, role in vascular wound healing and arterial intima formation (CARMELIET et al. 1997a, 1998).

Mice deficient in apolipoprotein A develop early atherosclerotic lesions, thinning of the aortic wall, and fragmentation of elastic membranes, resulting in microaneurysms of the aorta. Doubly deficient mice in apoE and uPA were protected against media destruction, probably due to reduced plasmin-dependent activation of pro-metalloproteases (CARMELIET et al. 1997b, 1998). The induction of vascular smooth muscle proliferation by uPA appears to follow a different pathway since it was found to be independent of uPAR-binding and of its proteinase activity (KANSE et al. 1997).

uPA is able to cleave directly fibronectin (MONZON-BORDONABA et al. 1997), soluble latent membrane-type matrix metalloproteinase (MMP-2) (KAZES et al. 1998), as well as several proforms of growth factors. The N-terminal fragment inhibits $\alpha 4\beta 1$ integrin-mediated lymphocyte adhesion independently of its catalytic activity (OLIVIER et al. 1999).

Binding of uPA to its receptor induces a series of signal transmission events (reviewed by DEAR and MEDCALF 1998 and by BESSER et al. 1997), such as activation of the Jak/Stat pathway (DUMLER et al. 1998), of the early response genes *fos*, *jun*, and *myc* (RABBANI et al. 1997), of tyrosine kinase-dependent glucose transporters (ANICHINI et al. 1997), and extracellular signal-regulated kinases ERK1 and ERK2 (NGUYEN et al. 1998; KONAKOVA et al. 1998), to cite just a few examples. uPA, therefore, is a multifunctional protein whose many roles in cell migration and tissue remodelling still await further elucidation.

List of Abbreviations

AP-1   activator protein 1 binding site
AP-2   activator protein 2 binding site
ATF   amino-terminal fragment (of uPA)
COM   co-operation mediating (DNA sequences)
EGF   epidermal growth factor
ERK   extracellular signal regulated kinase
HMW   high molecular weight
LMW   low molecular weight
MMP   metalloproteinase
NF-κB   nuclear transcription factor, first found in κ-light chain of B lymphocytes
PAI-1   plasminogen activator inhibitor type-1
PEA   polyoma early promoter activator
Sp1   specific promoter-1 binding site
uPA   urinary-type (or urokinase-type) plasminogen activator
sc-uPA single chain uPA, also called pro-urokinase
tc-uPA two chain uPA, also called urokinase

References

Adams DS, Griffin LA, Nachajko WR, Reddy VB, Wei C (1991) A synthetic DNA encoding a modified human urokinase resistant to inhibition by serum plasminogen activator inhibitor. J Biol Chem 266:8476–8482

Anichini E, Zamperini A, Chevanne M, Caldini R, Pucci M, Fibbi G, Del Rosso M (1997) Interaction of urokinase-type plasminogen activator with its receptor rapidly induces activation of glucose transporters. Biochemistry 36:3076–3083

Appella E, Robinson EA, Ullrich SJ, Stoppelli MP, Corti A, Cassani G, Blasi F (1987) The receptor binding sequence of urokinase. A biological function for the growth-factor module of proteases. J Biol Chem 262:4437–4440

Astrup T, Sterndorff I (1952) An activator of plasminogen in normal urine. Proc Soc Exp Biol Med 81:675–678

Bär FW, Meyer J, Vermeer F, Michels R, Charbonnier B, Haerten K, Spiecker M, Macaya C, Hanssen M, Heras M, Boland JP, Morice MC, Dunn FG, Uebis R, Hamm C, Ayzenberg O, Strupp G, Withagen AJ, Klein W, Windeler J, Hopkins G, Barth H, von Fisenne MJ, for the SESAM Study Group (1997) Comparison of saruplase and alteplase in acute myocardial infarction. SESAM Study Group. Am J Cardiol 79:727–732

Bajzar L, Manuel R, Nesheim ME (1995) Purification and characterization of TAFI, a thrombin-activable fibrinolysis inhibitor. J Biol Chem 271:16603–16608

Barlati S, Paracini F, Belotti D, DePetro G (1991) Tyrosine phosphorylation of human urokinase-type plasminogen activator. FEBS Lett 281:137–140

Belin D, Vassalli JD, Combépin C, Godeau F, Nagamine Y, Reich E, Kocher HP, Duvoisin RM (1985) Cloning nucleotide sequencing and expression of cDNAs encoding mouse urokinase-type plasminogen activator. Eur J Biochem 148: 225–232

Benraad TJ, Geurts-Moespot J, Grøndahl-Hansen J, Schmitt M, Heuvel JJTM, De Witte JH, Foekens JA, Leake RE, Brünner N, Sweep CGJ (1996) Immunoassays (ELISA) of urokinase-type plasminogen activator (uPA):report of an EORTC/BIOMED-1 workshop. Eur J Cancer 32 A:1371–1381

Bergwerff AA, Van Oostrum J, Kamerling JP, Vliegenthart JF (1995) The major N-linked carbohydrate chains from human urokinase. The occurrence of 4–0-sulfated, (alpha2–6)-sialylated or (alpha 1–3)-fucosylated N-acetylgalactosamine (beta 1–4)-N-acetylglucosamine elements. Eur J Biochem 228:1009–1019

Bernik MB, Oller EP (1973) Increased plasminogen activator (urokinase) in tissue culture after fibrin deposition. J Clin Invest 52:823–834

Berthelsen J, Zappavigna V, Mavilio F, Blasi F (1998) Prep1, a novel functional partner of Pbx proteins. EMBO J 17:1423–1433

Besser D, Verde P, Nagamine Y, Blasi F (1997) Signal transduction and the u-PA/u-PAR system. Fibrinolysis 10:215–237

Bogusky MJ, Dobson CM, Smith RAG (1989) Reversible independent unfolding of the domains of urokinase monitored by ^{1}H-NMR. Biochemistry 28:6728–6735

Brakman P (1967) Fibrinolysis. A standardized fibrin plate method and a fibrinolytic assay of plasminogen. Scripta Medica 8, Scheltema & Holkema NV, Amsterdam

Brigelius-Flohé R, Steffens G, Straßburger W, Flohé L (1992) High expression vectors for the production of recombinant single-chain urinary plasminogen activator from Escherichia coli. Appl Microbiol Biotechnol 36:640–649

Brunner G, Simon MM, Kramer MD (1990) Activation of prourokinase by the human T cell-associated serine proteinase Hu TSP-1. FEBS Lett 260:141–144

Bugge TH, Flick MJ, Danton MJ, Daugherty CC, Rømer J, Danø K, Carmeliet P, Collen D, Degen J (1996) Urokinase-type plasminogen activator is effective in fibrin clearance in the absence of its receptor or tissue-type plasminogen activator. Proc Natl Acad Sci USA 93:5899–5904

Buko AM, Kentzer EJ, Petros A, Menon G, Zuiderweg ERP, Sarin VK (1991) Characterization of a posttranslational fucosylation in the growth factor domain of urinary plasminogen activator. Proc Natl Acad Sci USA 88:3992–3996

Bush LR, Shebuski RJ (1990) In vivo models of arterial thrombosis and thrombolysis. FASEB J 4:3087–3098

Camoin L, Pannell R, Anfosso F, Lefevre JP, Sampol J, Gurewich V, Dignat-George F (1998) Evidence for the expression of urokinase-type plasminogen activator by human endothelial cells in vivo. Thromb Haemost 80:961–967

Carmeliet P, Moons L, Dewerchin M, Rosenberg S, Herbert JM, Lupu F, Collen D (1998) Receptor-independent role of urokinase-type plasminogen activator in pericellular plasmin and matrix metalloproteinase proteolysis during vascular wound healing in mice. J Cell Biol 140:233–245

Carmeliet P, Moons L, Herbert J-M, Crawley J, Lupu F, Lijnen R, Collen D (1997a) Urokinase but not tissue type plasminogen activator mediates arterial neointima formation in mice. Circ Res 81:829–839

Carmeliet P, Moons L, Lijnen R, Baes M, Lemaître V, Tipping P, Drew A, Eeckhout Y, Shapiro S, Lupu F, Collen D (1997b) Urokinase-generated plasmin activates matrix metalloproteinases during aneurysm formation. Nat Genet 17:439–444

Carmeliet P, Schoonjans L, Kieckens L, Ream B, Degen J, Bronson R, De Vos R, van den Oord JJ, Collen D, Mulligan RC (1994) Physiological consequences of loss of plasminogen activator gene function in mice. Nature 368:419–424

Chavakis T, Kanse SM, Yutzy B, Lijnen HR, Preissner KT (1988) Vitronectin concentrates proteolytic activity on the cell surface and extracellular matrix by trapping soluble urokinase receptor-urokinase complexes. Blood 91:2305–2312

Collen D, Stassen JM, De Cock F (1987) Synergistic effect of thrombolysis of sequential infusion of tissue-type plasminogen activator (tPA), single-chain urokinase-type plasminogen activator (sc-uPA) and urokinase in the rabbit jugular vein thrombosis model. Thromb Haemostas 58:943–946

Collen D, Stassen JM, Verstraete M (1983) Thrombolysis with extrinsic (tissue-type) plasminogen activator in rabbits with experimental jugular vein thrombosis. Effect of molecular form and dose of activator, age of the thrombus, and route of administration. J Clin Invest 71:368–376

Conese M, Blasi F (1995) Urokinase/urokinase receptor system: Internalization/degradation of urokinase-serpin complexes: mechanism and regulation. Biol Chem Hoppe Seyler 376:143–155

Conne B, Berczy M, Belin D (1997) Detection of polymorphisms in the human urokinase-type plasminogen activator gene. Thromb Haemost 77:434–435

De Cesare D, Palazzolo M, Blasi F (1996) Functional characterization of COM, a DNA region required for cooperation between AP-1 sites in urokinase gene transcription. Oncogene 13:2551–2562

De Cesare D, Palazzolo M, Berthelsen J, Blasi F (1997) Characterization of UEF-4,a DNA-binding protein required for transcriptional synergism between two AP-1 sites in the human urokinase enhancer. J Biol Chem 272; 23921–23929

Dear AE, Medcalf RL (1998) The urokinase-type-plasminogen-activator receptor (CD87) is a pleiotropic molecule. Eur J Biochem 252:185–193

De Munk GAW, Groeneveld E, Rijken DC (1991) Acceleration of the thrombin inactivation of single chain urokinase-type plasminogen activator (pro-urokinase) by thrombomodulin. J Clin Invest 88:1680–1684

Drag B, Petersen LC (1994) Activation of pro-urokinase by cathepsin G in the presence of glucosaminoglycans. Fibrinolysis 8:192–199

Dumler I, Weis A, Mayboroda OA, Maasch C, Jerke U, Haller H, Gulba DC (1998) The Jak/Stat pathway and urokinase receptor signaling in human aortic vascular smooth muscle cells. J Biol Chem 273:315–321

Eaton DL, Malloy BE, Tsai SP, Henzel W, Drayna D (1991) Isolation, molecular cloning and partial characterization of a novel carboxypeptidase B from human plasma. J Biol Chem 266:21833–21838

Egelund R, Rodenburg KW, Andreasen PA, Rasmussen MS, Guldberg RE, Petersen TE (1998) An ester bond linking a fragment of a serine proteinase to its serpin inhibitor. Biochemistry 37:6375–6379

Fleury V, Lijnen HR, Angelés-Cano E (1993) Mechanism of the enhanced intrinsic activity of single-chain urokinase-type plasminogen activator during ongoing fibrinolysis. J Biol Chem 268:18554–18559

Flohé L (1985) Single-chain urokinase-type plasminogen activators: New hopes for clot-specific lysis. Eur Heart J 6:905–908

Franco P, Iaccarino C, Chiaradonna F, Brandazza A, Iavarone C, Mastronicola MR, Nolli ML, Stoppelli MP (1997) Phosphorylation of human pro-urokinase on Ser138/303 impairs its receptor-dependent ability to promote myelomonocytic adherence and motility. J Cell Biol 137:779–791

Friberger P (1982) Chromogenic peptide substrates. Scan J Clin Lab Invest 42 [Suppl 162]:15–98

Gold LI, Schwimmer R, Quingley JP (1989) Human plasma fibrinectin as a substrate for human urokinase. Biochem J 262:529–534

Günzler WA, Beier H, Flohé L (1990) Activity and antigen of saruplase and two chain urokinase-related plasminogen activator are stabilized by a combination of aprotinin and benzamidine in citrated plasma. Fibrinolysis 4 [Suppl 2]:145–147

Günzler WA, Cramer J, Frankus E, Friderichs E, Giertz H, Hennies H-H, Henninger W, Kim SMA, Ötting F, Schneider J, Steffens GJ, Straßburger W, Wollmer A, Flohé L (1985) Chemical, enzymological and pharmacological equivalence of urokinases isolated from genetically transformed bacteria and human urine. Arzneim Forsch/Drug Res 35(I):652–662

Günzler WA, Steffens GJ, Ötting F, Buse G, Flohé L (1982a) Structural relationship between high and low molecular mass urokinase. Hoppe Seyler's Z Physiol Chem 363:133–141

Günzler WA, Steffens GJ, Ötting F, Kim SMA, Frankus E, Flohé L (1982b) The primary structure of high molecular mass urokinase from human urine. The complete amino acid sequence of the A chain. Hoppe Seyler's Z Physiol Chem 363:1155–1165

Gurewich V, Johnstone MT, Pannell R (1995) The selective uptake of high molecular weight urokinase-type plasminogen activator by human platelets. Fibrinolysis 9:188–195

Gurewich V, Liu JN, Pannell R (1992) The intrinsic lysis concept. Ann NY Acad Sci 667:224–232

Gurewich V, Pannell R (1986) A comparative study of the efficacy and specificity of tissue plasminogen activator and pro-urokinase. Demonstration of synergism and of different thresholds of non-selectivity. Thromb Res 44:217–228

Gurewich V, Pannell R (1987) Inactivation of single-chain urokinase (pro-urokinase) by thrombin and thrombin-like enzymes: relevance of the findings to the interpretation of fibrin-binding experiments. Blood 69:769–772

Gurewich V, Pannell R, Louie S, Kelley P, Suddith RL, Greenlee R (1984) Effective and fibrin-specific clot lysis by a zymogen precursor form of urokinase (pro-urokinase) J Clin Invest 73:1731–1739

Hanbücken FW, Schneider J, Günzler WA, Friderichs E, Giertz H, Flohé L (1987) Selective fibrinolytic activity of recombinant non-glycosylated human pro-urokinase (single-chain urokinase-type plasminogen activator) from bacteria. Arzneim Forsch/Drug Res 37(II):993–997

Higazi AAR, Cohen RL, Henkin J, Kniss D, Schwartz BS, Cines DB (1995) Enhancement of the enzymatic activity of single-chain urokinase plasminogen activator by soluble urokinase receptor. J Biol Chem 270:17375–17380

Holmes WE, Pennica D, Blaber M, Rey MW, Günzler WA, Steffens J, Heyneker HL (1985) Cloning and Expression of the gene for pro-urokinase in Escherichia coli. Bio/Technology 3:923–929

Ichinose A, Fujikawa K, Suyama T (1986) The activation of pro-urokinase by kallikrein and its inactivation by thrombin. J Biol Chem 261:3486–3489

Jänicke F, Pache L, Schmitt M, Ulm K, Thomssen C, Prechtl A, Graeff H (1994) Both the cytosols and detergent extracts of breast cancer tissues are suited to evaluate the prognostic impact of the urokinase-type plasminogen activator and its inhibitor, plasminogen activator inhibitor type 1. Cancer Res 54:2527–2530

Jiang YP, Pannell R, Liu JN, Gurewich V (1996) Evidence for a novel binding protein to urokinase-type plasminogen activator in platelet membranes. Blood 87:2775–2781

Kazes I, Delarue F, Hagège J, Bouzhir-Sima L, Rondeau E, Sraer J-D, Nguyen G (1998) Soluble latent membrane-type 1 matrix metalloprotease secreted by human mesangial cells is activated by urokinase. Kidney Int 54:1976–1984

Kebabian PR and Henkin J (1992) A chromogenic enzymatic assay capable of detecting prourokinase-like material in plasma. Thromb Res 65:401–407

Kobayshi H, Schmitt M, Goretzki L, Chucholowski N, Calvete J, Kramer M, Günzler WA, Jänicke F, Graeff H (1991) Cathepsin B efficiently activates the soluble and the tumour cell receptor-bound form of the proenzyme urokinase-type plasminogen activator (Pro-uPA). J Biol Chem 266:5147–5152

Konakova M, Hucho F, Schleuning WD (1998) Downstream targets of urokinase-type plasminogen-activator-mediated signal transduction. Eur J Biochem 253:421–429

Learmonth MP, Li W, Namiranian S, Kakkar VV, Scully MF (1992) Modulation of the cell binding property of single chain urokinase-type plasminogen activator by neutrophil cathepsin G. Fibrinolysis 6 [Suppl 4]:113–116

Lenich C, Pannell R, Henkin J, Gurewich V (1992) The influence of glycosylation on the catalytic and fibrinolytic properties of pro-urokinase. Thromb Haemostas 68:539–544

Lesuk A, Terminiello L, Traver JH (1965) Crystalline human urokinase: some properties. Science 147:880–881

Li X, Smith RAG, Dobson CM (1992) Sequential ^{1}H NMR assignment and secondary structure of the kringle domain from urokinase. Biochemistry 31:9562–9571

Lijnen HR, Nelles L, Holmes WE, Collen D (1988) Biochemical and thrombolytic properties of a low molecular weight form (comprising Leu 144 through Leu 411) of recombinant single-chain urokinase-type plasminogen activator. J Biol Chem 263:5594–5598

Lijnen HR, Van Hoef B, Collen D (1987) Activation with plasmin of two-chain urokinase-type plasminogen activator derived from single-chain urokinase-type plasminogen activator. Eur J Biochem 169:359–364

Lijnen HR, Van Hoef B, Nelles L, Collen D (1990) Plasminogen activation with single-chain urokinase-type plasminogen activator scu-PA. Studies with active site mutagenized plasminogen (Ser^{740},Ala) and plasmin-resistant scu-PA (Lys^{158},Glu). J Biol Chem 265:5232–5236

List K, Jensen ON, Bugge TH, Lund LR, Ploug M, Danø K, Behrendt N (2000) Plasminogen-independent initiation of the pro-urokinase activation cascade in vivo. Activation of pro-urokinase by glandular kallikrein (mGK-6) in plasminogen-deficient mice. Biochemistry 39:508–515

MacFarlane RG, Pilling J (1947) Fibrinolytic activity of normal urine. Nature 159:779

Mangel WF, Lin B, Ramakrishnan (1991) Conformation of one and two-chain high molecular weight urokinase analyzed by small-angle neutron scattering and vacuum ultraviolet circular dichroism. J Biol Chem 266:9408–9412

Marcotte PA, Henkin J, Credo RB, Badylak SF (1992b) A-chain isozymes of recombinant and natural urokinases: preparation, characterization, and their biochemical and fibrinolytic properties. Fibrinolysis 6:69–78

Marcotte P, Henkin J (1993) Characterization of the activation of pro-urokinase by thermolysin. Biochim Biophys Acta 1161:105–112

Marcotte PA, Kozan IM, Dorwin SA, Ryan JM (1992a) The matrix metalloproteinase pump-1 catalyzes formation of low molecular weight (pro)urokinase in cultures of normal human kidney cells. J Biol Chem 267:13803–13806

Meade H, Gates L, Lacy E, Lonberg N (1990) Bovine alpha S1-casein gene direct high level expression of active human urokinase in mouse milk. Bio/Technology 8:443–446

Monzon-Bordonaba F, Wang CL, Feinberg RF (1997) Fibronectinase activity in cultured human trophoblasts is mediated by urokinase-type plasminogen activator. Am J Obstet Gynecol 176:58–65

Muir E, Du JS, Fok-Seang J, Smith-Thomas LC, Housden ES, Rogers J, Fawcett JW (1998) Increased axon growth through astrocyte cell lines transfected with urokinase. Glia 23:24–34

Nagai M, Hiramatsu R, Kaneda T, Hayasuke N, Arimura H, Nishida M, Suyama T (1985) Molecular cloning of cDNA coding for human preprourokinase. Gene 36:183–188

Nagamine Y, Pearson D, Altus MS, Reich E (1984) cDNA and gene nucleotide sequence of porcine plasminogen activator. Nucleic Acids Res 12:9525–9541

Naldini L, Vigna E, Bardelli A, Follenzi A, Galimi F, Comoglio PM (1995) Biological activation of pro-HGF (hepatocyte growth factor) by urokinase is controlled by a stoichiometric reaction. J Biol Chem 270:603–611

Nauland U, Rijken DC (1994) Activation of thrombin-inactivated single-chain urokinase-type plasminogen activator by dipeptidyl peptidase I (cathepsin C). Eur J Biochem 223:497–501

Nelles L, Lijnen HR, Collen D, Holmes WE (1987) Characterization of recombinant human single chain urokinase-type plasminogen activator mutants produced by site-specific mutagenesis of Lys 158. J Biol Chem 262:5682–5689

Nguyen DH, Hussaini IM, Gonias SL (1998) Binding of urokinase-type plasminogen activator to its receptor in MCF-7 cells activates extracellular signal-regulated kinase 1 and 2 which is required for increased cellular motility. J Biol Chem 273:8502–8507

Nowak UK, Cooper A, Saunders D, Smith RAG, Dobson CM (1994) Unfolding studies of the proteinase domain of urokinase-type plasminogen activator: the existence of partly folded states and stable subdomains. Biochemistry 33:2951–2960

Odekon LE, Sato Y, Rifkin DB (1992) Urokinase-type plasminogen activator mediates basic fibroblast growth factor-induced bovine endothelial cell migration independent of its proteolytic activity. J Cell Physiol 150:258–263

Olivier P, Bieler G, Müller KM, Hauzenberger D, Rüegg C (1999) Urokinase-type plasminogen activator inhibits $\alpha 4 \; \beta 1$ integrin-mediated T lymphocyte adhesion to fibronectin independently of its catalytic activity. Eur J Immunol 29:3196–3209

Pannell R, Gurewich V (1986) Pro-urokinase – a study of its stability in plasma and of a mechanism for its selective fibrinolytic effect. Blood 67:1215–1223

Petersen LC (1997) Kinetics of reciprocal pro-urokinase/plasminogen activation. Stimulation by a template formed by the urokinase receptor bound to poly(D-lysine). Eur J Biochem 245:316–323

Philo RD, Gaffney PJ (1981) Assay methodology for urokinase: its use in assessing the composition of mixtures of high- and low-molecular weight urokinase. Thromb Res 21:81–88

Ploug J, Kjeldgaard NO (1957) Urokinase. An activator of plasminogen from human urine. Biochim Biophys Acta 24:278–289

Poeppelmeier J, Beier H, Carlsson J, Günzler WA, Meierhenrich R, Hopkins GR, Tebbe U (1996) Comparison of the pharmacokinetics and effects on the hemostatic system of saruplase and urokinase in patients with acute myocardial infarction. J Thromb Thrombol 3:385–390

PRIMI trial study group (1989) Randomized double-blind trial of recombinant pro-urokinase against streptokinase in acute myocardial infarction. The Lancet 1989 I:863–867

Rabbani SA, Gladu J, Mazar AP, Henkin J, Goltzman D (1997) Induction in human osteoblastic cells (SaOS2) of the early response genes fos, jun, and myc by the amino terminal fragment (ATF) of urokinase. J Cell Physiol 172:137–145

Rabbani SA, Mazar AP, Bernier SM, Haq M, Bolivar I, Henkin J, Golzman D (1992) Structural requirements for the growth factor activity of the amino-terminal domain of urokinase. J Biol Chem 267:14151–14156

Rajput B, Degen SF, Reich E, Waller EK, Axelrod J, Eddy RL, Shows TB (1985) Chromosomal locations of human tissue plasminogen activator and urokinase genes. Science 230:672–674

Riccio A, Grimaldi G, Verde P, Sebastio G, Boast S, Blasi F (1985) The human urokinase plasminogen activator gene and its promoter. Nucleic Acids Res 13:2759–2771

Romain S, Spyratos F, Laine-Bidron C, Bouchet C, Guirou O, Martin PM, Oglobine J, Magdelenat H (1995) Comparative study of four extraction procedures for urokinase type plasminogen activator and plasminogen activator inhibitor-1 in breast cancer tissues. Eur J Clin Chem Clin Biochem 33:603–608

Samama M, Castel M, Matsuo O, Hoylaerts M, Lijnen HR (1982) Comparative study of the activity of high and low molecular weight urokinase in the presence of fibrin. Thromb Haemostas 47:36–40

Sappino AP, Huarte J, Vassalli JD, Belin D (1991) Sites of synthesis of urokinase and tissue-type plasminogen activators in the murine kidney. J Clin Invest 87:962–970

Sarubbi E, Nolli ML, Robbiati F, Soffientini A, Parenti F, Cassani G (1989) The differential glycosylation of human pro-urokinase from various recombinant mammalian cell lines does not affect activity and binding to PAI-1. Thromb Haemostas 62:927–933

Schaller J, Nick H, Rickli E, Gillessen D, Lergier W, Studer RO (1983) The two-chain structure of human low molecular weight urinary urokinase, evidenced by sequence analysis. In: Davidson JF, Bachmann F, Bouvier CA, Kruithof EKO (eds) Progress in Fibrinolysis, vol 6. Churchill Livingstone, Edinburgh pp 348–352

Schmitt M, Kanayama N, Henschen A, Hollrieder A, Hafter R, Gulba D, Jänicke F, Graeff H (1989) Elastase released from human granulocytes stimulated with N-formyl-chemostatic peptide prevents activation of tumour cell prourokinase (pro-uPA). FEBS Lett 255:83–88

Schmitt M, Goretzki L, Chucholowski N, Jänicke F, Graeff H (1992) Effective activation of the proenzyme form of the urokinase-type plasminogen activator (pro-uPA) by the cysteine protease cathepsin L. Fibrinolysis 6 [Suppl 2]:122

Schneider J (1990) Interactions of saruplase with acetylsalicylic acid, heparin, glyceryl trinitrate, tranexamic acid and aprotinin in a pulmonary thrombosis model. Arzneim Forsch/Drug Res 40(II):1180–1184

Schumacher GF, Schill WB (1972) Radial diffusion in gel for micro determination of enzymes. 2. Plasminogen activator, elastase, and nonspecific proteinases. Anal Biochem 48:9–26

Shapiro RL, Duquette JG, Nunes I, Roses DF, Harris MN, Wilson EL, Rifkin DB (1997) Urokinase-type plasminogen activator-deficient mice are predisposed to staphylococcal botryomycosis, pleuritis, and effacement of lymphoid follicles. Am J Pathol 150:359–369

Shapiro RL, Duquette JG, Roses DF, Nunes I, Harris MN, Kamino H, Wilson EL, Rifkin DB (1996) Induction of primary cutaneous melanocytic neoplasms in urokinase-type plasminogen activator (uPA)-deficient and wild-type mice – cellular blue nevi invade but do not progress to malignant melanoma in uPA-deficient animals. Cancer Res 56:3597–3604

Sobel GW, Mohler SR, Jones NW, Dowdy ABC, Guest MM (1952) Urokinase: an activator of plasma profibrinolysin extracted from urine. Am J Physiol 171:768–769

Spraggon G, Phillips C, Nowak UK, Ponting CP, Saunders D, Dobson CM, Stuart DI, Jones EY (1995) The crystal structure of the catalytic domain of human urokinase-type plasminogen activator. Structure 3:681–691

Stack MS, Johnson DA (1994) Human mast cell tryptase activates single-chain urinary-type plasminogen activator. J Biol Chem 269:9416–9419

Stassen JM, Lijnen HR, Kieckens L, Collen D (1991) Small animal thrombosis models for the evaluation of thrombolytic agents. Circulation 83 [Suppl IV]:IV-65–IV-72

Steffens GJ, Günzler WA, Ötting F, Frankus E, Flohé L (1982) The complete amino acid sequence of low molecular mass urokinase from human urine. Hoppe-Seyler's Z Physiol Chem 363:1043–1058

Stepanova V, Bobik A, Bibilashvily R, Belogurov A, Rybalkin I, Domogatsky S, Little PJ, Goncharova E, Tkachuk V (1997) Urokinase plasminogen activator induces smooth muscle cell migration: key role of growth factor-like domain. FEBS Lett 414:471–474

Stoppelli MP, Corti A, Soffientini A, Cassani G, Blasi F, Assoian RK (1985) Differentiation-enhanced binding of the amino-terminal fragment of human urokinase plasminogen activator to a specific receptor on U 937 monocytes. Proc Natl Acad Sci USA 82:4939–4943

Stump DC, Thienpont M, Collen D (1986a) Urokinase-related proteins in human urine. J Biol Chem 261:1267–1273

Stump DC, Lijnen HR, Collen D (1986b) Purification and characterization of a novel low molecular weight form of single-chain urokinase-type plasminogen activator. J Biol Chem 261:17120–17126

Sun Z, Jiang Y, Ma Z, Wu H, Liu BF, Xue Y, Tang W, Chen Y, Li C, Zhu D, Gurewich V, Liu JN (1997) Identification of a flexible loop region (297–313) of urokinase-type plasminogen activator, which helps determine its catalytic activity. J Biol Chem 19;272:23818–23823

Sun Z, Liu BF, Chen Y, Gurewich V, Zhu D, Liu JN (1998) Analysis of the forces which stabilize the active conformation of urokinase-type plasminogen activator. Biochemistry 37:2935–2940

Tebbe U, Michels R, Adgey J, Boland J, Caspi A, Charbonnier B, Windeler J, Barth H, Groves R, Hopkins GR, Fennell W, Betriu A, Ruda M, Mlczoch J, for the Comparison Trial of Saruplase and Streptokinase (COMPASS) Investigators (1998) Randomized, double-blind study comparing saruplase with streptokinase therapy in acute myocardial infarction: the COMPASS Equivalence Trial. J Am Coll Cardiol 31:487–493

Ugwu F, Van Hoef B, Bini A, Collen D, Lijnen HR (1998) Proteolytic cleavage of urokinase-type plasminogen activator by stromelysin-1 (MMP-3). Biochemistry 19;37:7231–7236

van der Kaaden ME, Rijken DC, van Berkel TJC, Kuiper J (1998) Plasma clearance of urokinase-type plasminogen activator. Fibrinolysis Proteolysis 12:251–258

Van de Werf F, Nobuhara M, Collen D (1986a) Coronary thrombolysis with human single-chain urokinase-type plasminogen activator (pro-urokinase) in patients with acute myocardial infarction. Ann Internal Med 104:345–346

Van de Werf F, Vanhaeke J, deGeest H, Verstraete MN, Collen D (1986b) Coronary thrombolysis with recombinant single-chain urokinase-type plasminogen activator in patients with acute myocardial infarction. Circulation 74:1066–1070

van Griensven JM, Koster RW, Hopkins GR, Beier H, Günzler WA, Kroon R, Schoemaker RC, Cohen AF (1997) Effect of changes in liver blood flow on the pharmacokinetics of saruplase in patients with acute myocardial infarction. Thromb Haemost 78:1015–1020

Vassalli JD, Baccino D, Belin D (1985) A cellular binding site for the M_r 55000 form of the human plasminogen activator, urokinase. J Cell Biol 100:86–92

Verde P, Stoppelli MP, Galeffi P, DiNocera P, Blasi F (1984) Identification and primary sequence of an unspliced human urokinase poly (A)⁺RNA. Proc Natl Acad Sci USA 81:4727–4731

Wang W, Hendriks DF, Scharpé SS (1994) Carboxypeptidase U, a plasma carboxypeptidase with high affinity for plasminogen. J Biol Chem 269:15937–15944

White WF, Barlow GH, Mozen MM (1966) The isolation and characterization of plasminogen activators (urokinase) from human urine. Biochemistry 5:2160–2169

Williams JRB (1951) The fibrinolytic activity of urine. Brit J Exp Path 32:530–537

Wolf BB, Vasudevan J, Henkin J, Gonias SL (1993) Nerve growth factor-γ activates soluble and receptor-bound single chain urokinase-type plasminogen activator. J Biol Chem 268:16327–16331

Zubairov DM, Asadullin MG, Zinkevich OD, Chenborisova GSh, Timerbaev VN (1974) The isolation and some properties of human urokinase (in Russian). Biokhimiya 39:378–383

CHAPTER 5

The Inhibitors of the Fibrinolytic System

E.K.O. KRUITHOF

A. Introduction

Studies in transgenic mice deficient in plasminogen, tissue-type plasminogen activator (tPA), urinary-type plasminogen activator (uPA), or plasminogen activator inhibitor type-1 (PAI-1) have confirmed the major role of the PA system in fibrin clot surveillance, wound healing, tissue remodeling, cell migration, and arterial neointima formation (CARMELIET and COLLEN 1997; PLOPLIS 1995; MOONS 1998; BUGGE 1998).

To avoid excessive proteolysis and tissue damage, a precise, coordinated regulation of the PA system is required. This is achieved by many regulatory mechanisms including: (a) inhibition by specific inhibitors; (b) binding of plasminogen, PAs, and their inhibitors to fibrin, to extracellular matrix proteins, or to cell surface receptors; (c) modulation of PA and PAI gene expression; and (d) clearance of free and PAI-bound PAs via specific receptors. Inhibitors are of paramount importance for the regulation of the fibrinolytic system (Fig. 1). The principal inhibitor of plasmin is α_2-antiplasmin (A2AP), whereas the plasminogen activator inhibitors PAI-1 (LOSKUTOFF 1991; JUHAN-VAGUE and ALESSI 1993; ANDREASEN 1997; VAN MEIJER and PANNEKOEK 1995) and PAI-2 (KRUITHOF et al. 1995) are the principal inhibitors of tPA and uPA. These inhibitors are all members of the serpin superfamily of proteins which form irreversible complexes with their target proteases (POTEMPA et al. 1994; LAWRENCE 1997; WHISSTOCK et al. 1998). Lipoprotein(a), histidine-rich glycoprotein, A2AP, and carboxypeptidase interfere with the binding of plasminogen to fibrin and cell surface receptors and may reduce the rate of plasminogen activation. The aim of the present review is to summarize recent findings concerning the properties of these inhibitors of the fibrinolytic system (Table 1).

B. α_2-Antiplasmin

I. Gene and Protein Structure

The A2AP gene (PLI) is located on chromosome 17p13 (KATO et al. 1993) (Table 2). Mature A2AP has 464 amino acids (TONE et al. 1987; HOLMES et al.

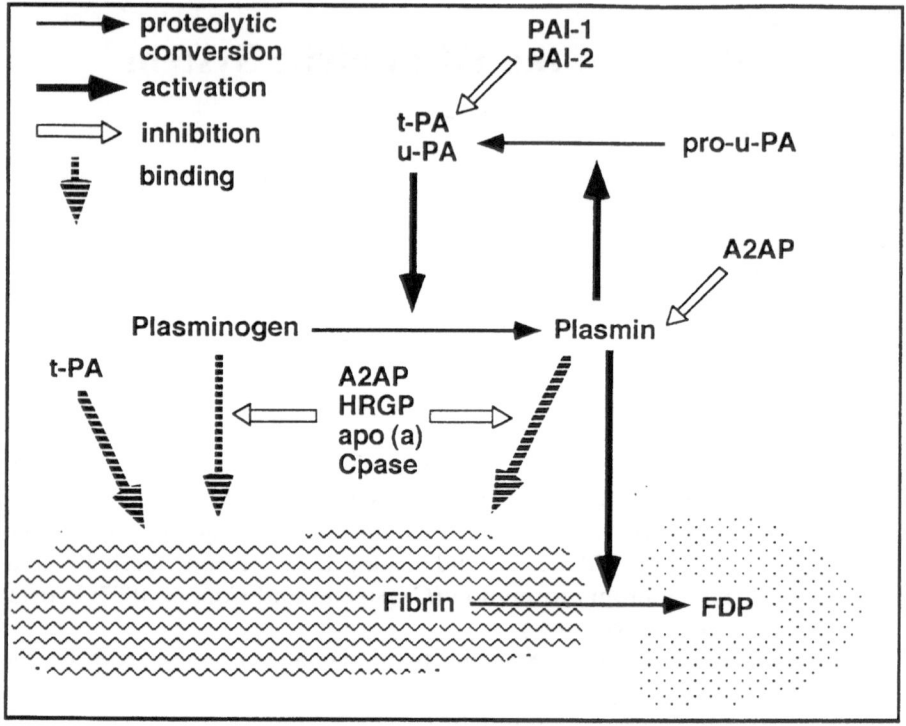

Fig. 1. Simplified model of fibrinolysis. Plasminogen is activated by tissue-type plasminogen activator (tPA) or urokinase (uPA). Active uPA is generated from pro-uPA by the action of plasmin and of kallikrein. The binding of plasminogen and of tPA to fibrin greatly accelerates the rate of plasminogen activation. Inhibitors modulate fibrinolytic activity at the level of plasmin activity (α_2-antiplasmin, A2AP) and of PA activity (PAI-1 and PAI-2). The binding of plasminogen to fibrin is inhibited by A2AP, by histidine-rich glycoprotein (HGRP), by apolipoprotein(a), and by caboxypeptidases (Cpase), which removes C-terminal lysines that bind plasminogen

1987) and a C-terminal region much longer than that of other serpins. It has been purified from plasma (AOKI et al. 1993) in which it is present at 69 µg/ml (~1 µmol/l). A2AP has three interaction sites which are important for its activity. The reactive center is located at Arg^{376}–Met^{377}. The C-terminal region has Lys^{448} and Lys^{464} that bind to the lysine binding site of plasmin(ogen) (SUGIYAMA et al. 1988) and interfere with the binding of plasminogen to fibrin or cell surfaces and reduce the rate of plasminogen activation. A transglutamination site (Glu^2) mediates cross-linking by factor XIII of A2AP to fibrin (JANSEN et al. 1987). Several degradation products of A2AP have been described in plasma that have lost the plasminogen binding site or the transglutamination site. Mature A2AP is secreted by liver cells with amino-terminal Met-Glu (KOYAMA et al. 1994). However, in plasma part of the A2AP

Table 1. Inhibitors of the fibrinolytic system

Inhibitor	Effect on fibrinolysis
A2AP	Primary inhibitor of plasmin, is crosslinked by factor XIII to fibrin, inhibits the binding of plasminogen to fibrin and cell surfaces
PAI-1	Principal inhibitor of tPA and uPA
PAI-2	Efficient inhibitor of uPA and two-chain tPA, mainly intracellular
HRGP	Inhibits the binding of plasminogen to fibrin and cell surfaces
Lp(a)	Contains the plasminogen-like apolipoprotein(a), which inhibits the binding of plasminogen to fibrin and cell surfaces
Cpase B	Cleaves off C-terminal lysines from fibrin or cell surface proteins, which bind plasminogen

A2AP, α_2-antiplasmin; PAI, plasminogen activator inhibitor (type 1 or type 2); HRGP, histidine-rich glycoprotein; Lp(a), lipoprotein(a); Cpase B, carboxypeptidase B.

Table 2. Basic information on the principal serpin inhibitors of fibrinolysis

	A2AP	PAI-1	PAI-2
Chromosomal location	17p13	7q22.1–22.3	18q22.1
cDNA size (kbp)	2.3	3.2 and 2.2	1.9
Molecular weight (kDa)	67	52	47 and 60
Number of amino acids	464	379	415
Reactive center	Arg-Met	Arg-Met	Arg-Ser
N-linked glycosylation sites	4	3	3
Plasma concentration (nmol/l)	1000	0.1–1	<0.1[a]
Heparin binding	No	Yes	No

[a] In pregnancy plasma PAI-2 may increase up to 5 nmol/l.

molecules have lost the first 12 amino acids containing the transglutamination site. Furthermore one third of plasma A2AP has lost the plasminogen binding site (CLEMMENSEN et al. 1981). A2AP is not only produced by hepatocytes, but also in the kidney in epithelial cells lining the convoluted portion of proximal tubules; its accumulation is under androgen control (MENOUD et al. 1996). Moderate amounts of A2AP mRNA are detected in other murine tissues such as muscle, intestine, central nervous system, and placenta.

II. Target Enzyme Specificity

In human plasma A2AP inhibits free plasmin activity with an estimated half life of 0.1 s (calculated from an A2AP concentration of 1 μmol/l and a $k_1 = 10^7 M^{-1} s^{-1}$) (Table 3). Fibrin bound plasmin has both its active site and its lysine binding site occupied and therefore reacts much more slowly with A2AP than

Table 3. Second order rate constants (in $M^{-1}s^{-1}$) of A2AP, PAI-1, and PAI-2 with respect to plasmin and the PAs

	A2AP	PAI-1	PAI-2
Plasmin	10^7	7×10^5	10^2
sc-tPA	$\sim10^2$	10^7	10^3
tc-tPA	$\sim10^2$	10^7	2×10^5
tc-uPA	$\sim10^2$	10^7	10^6

free plasmin (RAKOCZI et al. 1978). However, crosslinking of A2AP to fibrin enhances the efficacy of the inhibitor towards fibrin-bound plasmin. The inhibitory effect of factor XIII on clot lysis was shown to be due to cross-linking of A2AP to fibrin rather than to fibrin-fibrin cross-linking (JANSEN et al. 1987). Also, A2AP reduces the rate of plasminogen activation by interfering with the binding of plasminogen to fibrin and to cell surfaces. Compared to PAI-1 or PAI-2, A2AP is a poor inhibitor of tPA and uPA.

III. α_2-Antiplasmin Deficiencies

Congenital autosomal recessive A2AP deficiencies have been described. Patients with homozygous deficiencies are characterized by easy bruising, post-traumatic rebleeding (typically after cessation of initial hemorrhage), hematoma, hemarthrosis, and excessive bleeding after dental extraction (SAITO 1988). Subjects with heterozygous deficiencies may have a mild bleeding tendency, but are often asymptomatic. A2AP Enschede has normal plasma antigen concentrations, but an absent activity. The molecular defect is due to an insertion of an alanine in the reactive center loop, resulting in a loss of inhibitory activity (KLUFT et al. 1987). Two A2AP mutants (Nara and Okinawa) have low plasma concentrations due to an impaired secretion by hepatocytes (MIURA et al. 1989a,b).

Acquired A2AP deficiencies may be due to a deficient A2AP biosynthesis, which may occur in severe liver disease, or to excessive plasminogen activation during thrombolytic therapy or during disseminated intravascular coagulation.

C. Plasminogen Activator Inhibitor Type 1

I. Gene Structure

The PAI-1 gene (PLANH1) is located on chromosome 7q22 (KLINGER et al. 1987). Several regulatory elements were identified in the PAI-1 promoter that may mediate the response to phorbol ester, cAMP, transforming growth factor-β (TGFβ), interleukin 1, or glucocorticoids (VAN ZONNEVELD et al. 1988;

KEETON et al. 1991; WESTERHAUSEN et al. 1991; BRUZDZINSKI et al. 1993; DAWSON et al. 1993; KNUDSEN et al. 1994; CHANG and GOLDBERG 1995; DONG et al. 1996). A heterogeneity (4G/5G) is found at position 675 upstream of the transcription initiation site. In the general population the 4G and 5G alleles occur at almost equal frequency. Individuals homozygous for the 4G allele at this position have 30% higher PAI-1 levels (DAWSON et al. 1991; GRANT 1997). The genotype related differences in PAI-1 levels may be due to genotype-specific responses to PAI-1-inducing factors. Thus, the response of the PAI-1 gene to interleukin-1 or triglycerides may depend on the 4G/5G phenotype (DAWSON et al. 1993; ERIKSSON et al. 1998a). Several studies suggest that environmental factors such as smoking, or patient characteristics such as body mass index, triglycerides levels, gender, or heterogeneities in other genes such as the angiotensin-converting enzyme gene may be more important than genotype differences in determining plasma levels of PAI-1 (MARGAGLIONE et al. 1998b; BURZOTTA et al. 1998; HENRY et al. 1998). The contribution of the 4G genotype to cardiovascular disease is still controversial. Several studies observed an association of the 4G genotype with cardiovascular disease (ERIKSSON et al. 1995; IWAI et al. 1998; MARGAGLIONE et al. 1998a), whereas other studies did not (YE et al. 1995; CATTO et al. 1997; GRANT 1997; RIDKER et al. 1997). In Pima Indians with type 2 diabetes, the presence of the 4G allele of the PAI-1 gene was associated with a higher risk of diabetic retinopathy (NAGI et al. 1997). In stroke patients, PAI-1 level but not 4G/5G genotype was associated with early mortality (CATTO et al. 1997). In one study, the PAI-1 genotype affected the phenotypic expression of thrombophilia in protein S deficient individuals (ZÖLLER et al. 1998). The observation of a higher frequency of the 4G polymorphism in centenarians than in a young healthy control population (MANNUCCI et al. 1997) suggests that a higher level of PAI-1, despite its association with impaired fibrinolysis, is compatible with successful aging.

II. Protein Structure

Human PAI-1 is a 50 kDa glycoprotein of 379 amino acids, having its reactive center at the Arg^{348}-Met^{349} bond. Its crystal structure has been determined (SHARP et al. 1999). It has been purified from conditioned media of eukaryotic cells and has been expressed in *E. coli* (REILLY et al. 1994). PAI-1 is released from cells in an active form, but at physiological pH and 37°C it rapidly (half-life 2 h) loses its activity (LEVIN and SANTELL 1987). The resulting "latent" form can be reactivated by treatment with chaotropic agents (HEKMAN and LOSKUTOFF 1985). The active form of PAI-1 is stabilized by storage at lower pH and temperature and by adding vitronectin (LINDAHL et al. 1989a). A third form of PAI-1, the substrate form, is cleaved after incubation with PAs (DE CLERCK et al. 1992; SANCHO et al. 1995). Peptides corresponding to the reactive center loop convert the inhibitory form of PAI-1 into a substrate form (EITZMAN et al. 1995). The three dimensional structure of latent PAI-1 is known and shows that in latent PAI-1 the reactive center loop is partially buried in β-sheet A

(MOTTONEN et al. 1992). The crystal structure of a complex of a reactive center loop peptide with PAI-1 shows that this peptide binds in the space between β-strands 3 A and 5 A and is able to prevent insertion of the reactive-center loop into β-sheet A, thereby abolishing the ability of the serpin to inactivate irreversibly its target enzyme (XUE et al. 1998).

III. Target Enzyme Specificity of PAI-1

PAI-1 reacts very efficiently with two-chain uPA and tPA with rate constants above $10^7 M^{-1} s^{-1}$ (Table 3) but also with trypsin (7×10^6), plasmin (7×10^5), thrombin (9×10^2), factor Xa (2×10^4), factor XIa (2×10^5), plasma kallikrein (6×10^4), and factor XIIa (2×10^4) (THORSEN et al. 1988; HEKMAN and LOSKUTOFF 1988; BERRETTINI et al. 1989). Heparin and vitronectin enhance the rate of inhibition by PAI-1 of thrombin to 6×10^4 and 2×10^5, respectively (VAN MEIJER et al. 1997) and that of factor Xa to 5×10^4 and 2×10^5, respectively (URANO et al. 1996). In addition to enhancing the rate of inhibition, both cofactors increase the apparent stoichiometry of the PAI-1-thrombin interaction. Therefore, thrombin will efficiently inactivate PAI-1 in the presence of either vitronectin or heparin, unless a sufficient excess of the inhibitor is present.

IV. Deficiencies, Knockout, and Overexpression in Mice

Functional deficiencies of PAI-1 are associated with a delayed type bleeding. In one patient a complete PAI-1 deficiency was due to a frame-shift mutation (FAY et al. 1992), whereas in other patients PAI-1 was undetectable in plasma but normal in platelets (DIÉVAL et al. 1991; LEE et al. 1993). In a large kindred with PAI-1 deficiency, 19 individuals heterozygous for the PAI-1 null allele and 7 homozygous individuals with complete PAI-1 deficiency were identified (FAY et al. 1997). Clinical manifestations of PAI-1 deficiency were restricted to abnormal bleeding, which was observed only after trauma or surgery in homozygous affected individuals. Fibrinolysis inhibitors, including ε-aminocaproic acid and tranexamic acid, were effective in treating and preventing bleeding episodes. Other than abnormal bleeding, no significant developmental or other abnormalities were observed in homozygous PAI-1-deficient individuals. In mice, functional disruption of the PAI-1 gene did not result in a bleeding tendency, but accelerated the rate of spontaneous clot lysis. Furthermore, PAI-1-deficient mice were resistant to the development of venous thrombosis after injection of endotoxin into the footpad (CARMELIET et al. 1993). In contrast, overexpression of PAI-1 in mice resulted in an increased risk of thrombotic complications (ERICKSON et al. 1990).

V. Role of PAI-1 in Cell Migration

Malignant keratinocytes transplanted into a PAI-deficient mouse model prevented local invasion and tumor vascularization (BAJOU et al. 1998). In contrast, a model of arterial wound healing after electric injury revealed that PAI-1$^{(-/-)}$ smooth muscle cells, originating from the uninjured borders, migrated more rapidly into the necrotic center of the arterial wound than wild-type smooth muscle cells (CARMELIET et al. 1997). These in vivo studies suggest that PAI-1 plays a complex role in cell migration, essential under certain circumstances and inhibitory under other circumstances. The mechanism of migration inhibition by PAI-1 appears to be related to the binding of PAI-1 to vitronectin, because cell migration is enhanced by the presence of vitronectin and requires the expression of the vitronectin receptor (α_v integrins) and because active but not latent PAI-1 inhibits the interaction of vitronectin with α_v (STEFANSSON and LAWRENCE 1996; KJØLLER 1997; GERMER et al. 1998). Formation of a complex between PAI-1 and plasminogen activators results in loss of PAI-1 affinity for vitronectin and restores cell migration. In contrast, PAI-I can also stimulate cell migration on vitronectin, presumably by inhibiting uPA/uPAR-mediated cell adhesion to vitronectin and thereby releasing the inhibition of cell migration induced by uPA (STAHL and MUELLER 1997; WALTZ et al. 1997). Cells may also directly adhere to immobilized PAI-1. This requires the presence of urokinase/uPAR at the cell surface (PLANUS et al. 1977). $\alpha_v\beta_3$ Integrin might be the transmembrane molecule that physically connects the complex of PAI-1, urokinase, and uPAR to the cytoskeleton. Taken together, these results suggest that adhesion and dissociation of cells are both required for cell migration and that this process is regulated by the complex interactions between PAI-1, uPA, uPAR, vitronectin, and α_v integrins (reviewed by LOSKUTOFF et al. 1999).

VI. Distribution of PAI-1 In Vivo

PAI-1 is normally detected in human plasma and in human platelets. Ninety percent of PAI-1 antigen, but only 50% of PAI activity in human blood is associated with platelets and released upon platelet activation (ERICKSON et al. 1984; KRUITHOF et al. 1987b). The discrepancy is due to the fact that part of platelet PAI-1 is in the latent form (DE CLERCK et al. 1988). The source of PAI-1 in plasma is not known but liver and endothelial cells are most likely to contribute. Other cell types that may contribute are smooth muscle cells (LUPU et al. 1993), monocytes/macrophages and adipocytes (ALESSI et al. 1997; LOSKUTOFF and SAMAD 1998; ERIKSSON et al. 1998b).

An early report observed that fibrin-bound tPA is protected from plasma PAI-1 (KRUITHOF et al. 1984). However, during clot formation in vivo, PAI-1 is released from platelets and may be incorporated in the clots (STRINGER et al. 1994). This increases the resistance of such clots for tPA-mediated clot lysis (ZHU et al. 1999). Inhibition of PAI-1 by monoclonal antibodies facilitated spontaneous clot lysis and reduced the extension of experimental thrombosis (LEVI et al. 1992; BIEMOND et al. 1995b). Similar results were obtained using a low-molecular-weight inhibitor of PAI-1 (FRIEDRICH et al. 1997).

VII. Gene Regulation of PAI-1

PAI-1 is expressed by a majority of cell types in culture. A common finding is that PAI-1 expression by cells in culture is much higher than that of their normal counterparts in vivo. This suggests that the process of cell proliferation by itself is a powerful inducer of PAI-1 biosynthesis. PAI-1 expression may be further increased by a variety of agents such as TGF-β, tumor necrosis factor (TNF), interleukin-1, retinoic acid, endotoxin, phorbol ester, insulin, angiotensin II, and glucocorticoids (MEDCALF et al. 1988a,b; LOSKUTOFF 1991; ZELEZNA et al. 1992; ANFOSSO et al. 1993; SPENCER-GREEN 1994; MEDINA et al. 1989; PRESTA et al. 1990; SAWDEY et al. 1989; VAN HINSBERGH et al. 1988; HEATON et al. 1992). Interleukin-6 had little or no effect on PAI-1 expression by hepatoma cells (BERGONZELLI and KRUITHOF 1991), but markedly enhanced the interleukin-1-mediated induction of PAI-1 (HEALY and GELEHRTER 1994) and induced an increase of plasma PAI-1 in vivo (see below). Cyclic AMP and prostaglandin E2 inhibited basal PAI-1 expression or the induction of PAI-1 (BERGONZELLI et al. 1992; THALACKER and NILSEN-HAMILTON 1992; SANTELL and LEVIN 1988). Furthermore, in human endothelial cells and in human hepatoma or fibrosarcoma cells, PAI-1 expression is increased by hyperthermia (BERGONZELLI and KRUITHOF 1991; ANG and DAWES 1994).

In mice, injection of endotoxin in vivo induced a markedly increased expression of PAI-1 in renal endothelial cells, which suggests a role for PAI-1 in the pathogenesis of renal disease (KEETON et al. 1993).

VIII. PAI-1 as a Marker for Vascular and Arterial Disease

The relation between elevated PAI-1 and vascular and arterial disease has been demonstrated in a large number of studies (HAMSTEN et al. 1987; MUNKVAD et al. 1990; JUHAN-VAGUE and ALESSI 1993). This led to the idea that elevated PAI-1 is a risk factor for (i.e., causally related to) the occurrence of thromboembolic complications and myocardial infarction. However, PAI-1 concentrations are correlated with risk factors of the metabolic syndrome of insulin resistance which are also associated with atherothrombotic disease, like triglycerides (JUHAN-VAGUE and ALESSI 1993; JUHAN-VAGUE et al. 1993), with markers of the acute phase response, like fibrinogen and C-reactive protein (THOMPSON et al. 1995), and with hypertension (URANO et al. 1993; GLUECK et al. 1994). The intricate relation of PAI-1 with these risk markers (or factors) will make it difficult to determine whether PAI-1 by itself has an independent causal effect for thromboembolic disease (JUHAN-VAGUE and ALESSI 1993). Perhaps studies with PAI-1 deficient mice may ultimately help to settle whether PAI-1 is a risk factor or a risk marker.

IX. PAI-1 and the Acute Phase Response

There are several indications for PAI-1 being an acute phase protein. In humans, PAI-1 concentrations increase in the early postoperative phase

(JANSSON et al. 1989; AILLAUD et al. 1985). Plasma from the postoperative period was able to augment PAI-1 biosynthesis twofold in human umbilical vein endothelial cells, which suggests the presence of a PAI-1 inducing factor in postoperative plasma (KASSIS et al. 1992). In baboons, injection of interleukin-6, a mediator of the acute phase response, induced a 30-fold increase of PAI-1, but also a fourfold increase of tPA (MESTRIES et al. 1994; KRUITHOF et al. 1997). The maximal response was observed after 8h, whereas C-reactive protein, an early acute phase protein, had its maximum only after one or two days. Taken together, these studies suggest that PAI-1 and tPA are very early acute phase proteins.

X. PAI-1 and Septicemia

Extremely high plasma levels of PAI-1 (up to 6000ng/ml) have been found in patients with septicemia and septic shock and predict a poor outcome (PRALONG et al. 1989; HESSELVIK et al. 1989; LORENTE et al. 1993; MESTERS et al. 1996; KORNELISSE et al. 1996; VERVLOET et al. 1998). The induction of PAI-1 is most likely mediated by endotoxin, because injection of endotoxin in humans leads first to an increase of tPA antigen (maximum after 4h) and later an increase of PAI-1, resulting in a net reduction of fibrinolytic activity (SUFFREDINI et al. 1989). Similar results were obtained in chimpanzees; in these animals, the administration of pentoxifylline or anti-TNF antibodies, but not anti-interleukin-6 antibodies, strongly attenuated the increase of tPA and PAI-1 (BIEMOND et al. 1995a). This result suggests that TNF is a more potent in vivo inducer of PAI-1 than interleukin-6. The finding that T-686 – an inhibitor of PAI-1 – had a protective effect in a septicemia model in mice suggests that the high levels of PAI-1 in septicemia contribute to the risk of complications (MURAKAMI et al. 1997).

D. Plasminogen Activator Inhibitor Type 2

I. Gene and Protein Structure

The PAI-2 gene (PLANH2) is situated on chromosome 18q22 within a serpin cluster that also contains the genes for squamous cell carcinoma antigens 1 and 2, maspin gene, and cytoplasmic antiproteinase 2 (SILVERMAN et al. 1998) and is located at 600kbp centromeric of the bcl-2 proto-oncogene, a major regulator of apoptosis (SCHNEIDER et al. 1995). The PAI-2 promoter contains several regulatory elements that may mediate the phorbol ester, cAMP, and retinoic acid responses of the gene and also transcriptional silencers (COUSIN et al. 1991; SCHUSTER et al. 1994; ANTALIS et al. 1993, 1996; DEAR et al. 1996, 1997; MAHONY et al. 1998).

PAI-2 mRNA encodes a protein of 415 amino acids, having its reactive center at the Arg^{380}-Thr^{381} bond. The protein has been purified from human placenta (ÅSTEDT et al. 1985) or from monocyte-like cells (KRUITHOF et al.

1986) and has been produced by recombinant DNA techniques in CHO cells, in bacteria, and in yeast (Mikus et al. 1993; Steven et al. 1991). Its crystal structure has been solved at 2.0 Å resolution (Harrop et al. 1999). There are two forms of PAI-2, an intracellular nonglycosylated 47 kDa, pI 5.0 form and a secreted, glycosylated 60 kDa, pI 4.4 form of PAI-2 (Kruithof et al. 1986; Genton et al. 1987; Wohlwend et al. 1987), having similar inhibition characteristics (Mikus et al. 1993). In most cell types studied, the majority of PAI-2 is non-glycosylated and intracellular (Genton et al. 1987; Wohlwend et al. 1987; Medcalf et al. 1988a; Ny et al. 1989). PAI-2 has no classical aminoterminal signal sequence (Ye et al. 1988) and both intracellular and secreted PAI-2 are produced from the same mRNA and the same start codon (Belin et al. 1989). Increasing the hydrophobicity of α-helices A and B results in improved PAI-2 secretion (Von Heijne et al. 1991; Belin 1993, 1996). Under certain conditions relatively large amounts of the 47 kDa form of PAI-2 have been detected in cell-conditioned medium, e.g., after prolonged exposure of U-937 cells to phorbol ester, when cell viability is poor (Kruithof et al. 1986). Cell death (tissue necrosis or apoptosis) may thus be one route enabling the 47 kDa form of PAI-2 to reach the extracellular environment. The possibility of other mechanisms leading to the release of 47 kDa PAI-2 from viable cells cannot be excluded, but hard data are lacking at present. Glycosylated PAI-2 has been detected in pregnancy plasma (Kruithof et al. 1987a; Lecander and Åstedt 1987) implying a regular in vivo mechanism for PAI-2 secretion. Recent reports suggested that the distribution of the secreted and intracellular forms PAI-1 is modulated by serum (Ritchie and Booth 1998). The presence of transglutamination sites provides a mechanism for the cross-linking of PAI-2 to other proteins (Jensen et al. 1994a).

II. Target Enzyme Specificity

PAI-2 efficiently inhibits tc-uPA and tc-tPA, but poorly sc-tPA (Table 3). Compared to PAI-1, it reacts 10 times less rapidly with tc-uPA and 50 and 10000 times less rapidly with two-chain and single-chain tPA, respectively (Kruithof et al. 1986; Thorsen et al. 1988a; Mikus et al. 1993).

III. Distribution of PAI-2 In Vivo

PAI-2 has been consistently detected in pregnancy plasma (see below), in placenta, and in plasma from patients with myelomonocytic leukemias (Scherrer et al. 1991). It has furthermore been detected in human monocytes and macrophages (Vassalli et al. 1984; Chapman and Stone 1985).

PAI-2 is the predominant epidermal PA inhibitor being found throughout the normal epidermis, especially in the granular layers (Chen et al. 1993; Gissler et al. 1993; Lyons-Giordano et al. 1994). The finding that PAI-2 becomes incorporated into the cornified envelope during terminal differentiation of the keratinocyte suggests that PAI-2 has a role in skin formation (Jensen et al. 1995).

IV. Gene Regulation of PAI-2

In general, basal levels of PAI-2 levels expressed by cells in culture are low. However, PAI-2 biosynthesis can be induced by many factors. PAI-2 expression is regulated both at the level of gene transcription and at the level of mRNA stability. The interaction between various signal transduction pathways provides for additional possibilities for regulating PAI-2 expression. The observation that PAI-2 is induced by inflammatory agents suggests that PAI-2 is involved in inflammation. Thus, lipopolysaccharide, phorbol ester, TNF, and interleukin-1 induce PAI-2 in monocytes, in fibroblasts, in mesothelial cells, and in endothelial cells, whereas glucocorticoids and TGF-β may reduce PAI-2 expression. cAMP stimulates PAI-2 production and potentiates the effect of phorbol ester in monocyte-like cells, whereas in fibroblast-like cells and in melanoma cells it antagonizes the phorbol ester- or TNF-mediated induction of PAI-2. In monocyte-like cells retinoic acid may potentiate the phorbol ester-mediated induction of PAI-2 (for a recent extensive review and references see: KRUITHOF et al. 1995).

V. The Physiological Role of PAI-2

The physiological role of PAI-2 still has to be established. Secreted PAI-2 may regulate PA activity near the cell surface and in the pericellular space. Intracellular PAI-2 may also contribute to PA inhibition in the extracellular space. In monocytes/macrophages suitable stimulation may increase PAI-2 up to several percent of total cell-associated protein, which provides a local storage pool of inhibitory activity. Release of these large amounts of PAI-2 during cell death (or by as yet unidentified alternative mechanisms) could block local PA-dependent proteolytic activity and contribute to tissue encapsulation as was seen with PAI-2 transfected tumor cells injected into mice (LAUG et al. 1993; MUELLER et al. 1995).

Some reports suggested a role for PAI-2 as an inhibitor of TNF-mediated cytolysis or apoptosis (KUMAR and BAGLIONI 1991; DICKINSON et al. 1995) and of the cytopathic effects of α-virus infection (ANTALIS et al. 1998). The observation of a cleaved form of PAI-2 in apoptotic cells and in keratinocytes suggest the presence of an intracellular target enzyme (JENSEN et al. 1994b; RISSE et al. 2000).

E. Plasminogen Activator Inhibitors in Pregnancy

Pregnancy is accompanied by increased plasma concentrations of PAI-1 (tenfold) and of PAI-2, from below detection limit, (<5 ng/ml) to a maximum of 250 ng/ml at term (KRUITHOF et al. 1987; LINDOFF et al. 1993). However, increases of tPA and uPA are also consistently observed during pregnancy. Pregnancy plasma contains mainly 60 kDa, but also some 47 kDa PAI-2 (LECANDER and ÅSTEDT 1987; KRUITHOF et al. 1987). PAI-1 protein and mRNA was detected in epithelial cells of the chorioamniotic membrane and PAI-2 in placental villous syncytiotrophoblasts (ÅSTEDT et al. 1986; FEINBERG et al. 1989; FAZLEABAS et al. 1991). PAI-2 concentrations remain elevated for several days postpartum, while PAI-1 concentrations return to normal within a few hours.

Pre-eclampsia is a multisystemic obstetric disease of unknown etiology that is commonly associated with fibrin deposition, occlusive lesions in placental vasculature, and intrauterine fetal growth retardation. Pre-eclampsia and intrauterine growth retardation show clear modifications with respect to normal pregnancies of the same gestational age and are characterized by increased plasma PAI-1 and tPA levels, and decreased plasma PAI-2 and uPA as compared to uncomplicated pregnancies (DE BOER et al. 1988; GILABERT et al. 1990, 1994; REITH et al. 1993; KOH et al. 1993; LINDOFF and ÅSTEDT 1994; SCHJETLEIN et al. 1997). However, in patients with intrauterine growth retardation without pre-eclampsia, PAI-2 levels are low and PAI-1 levels normal, whereas in patients with pre-eclampsia without intrauterine growth retardation, PAI-2 levels are normal and PAI-1 levels elevated. These results suggest that elevated PAI-1 is a predictive marker for pre-eclampsia (BALLEGEER et al. 1989) and that reduced PAI-2 is a marker for a decreased placental function and intrauterine growth retardation (ESTELLÉS et al. 1989; GILABERT et al. 1994). By using the antigen ratio of PAI-1 to PAI-2 a better discrimination between pre-eclampsia and normal pregnancy was obtained, the ratio in normal pregnancy being on average 0.6 while that in pre-eclampsia was 2.5 (REITH et al. 1993).

The mechanisms leading to the high levels of PAI-1 or low levels of PAI-2 in pre-eclampsia are not yet clear. For PAI-1 they may be related to changes in placental circulation and in local cytokine expression. A high resistance in the placental circulation was associated with elevated plasma PAI-1 activity levels but not with low PAI-2 antigen levels (HE et al. 1995). An increase of TNF antigen in pre-eclamptic placenta was also observed, which was associated with an increase in PAI-1 mRNA not only in syncytiotrophoblast and infarction areas, but also in fibroblasts and in some endothelial cells of fetal vessels in placentas from pre-eclamptic patients (ESTELLÉS et al. 1998). The decrease in PAI-2 levels in pre-eclampsia plasma is correlated with a decrease in PAI-2 mRNA in the placenta (GRANCHA et al. 1996). However, as TNF normally increases PAI-2 gene transcription, it is unlikely that the modifications of PAI-2 during pre-eclampsia are due to this cytokine.

F. Therapeutic Applications of PAI-1 and PAI-2

PAI-1 or PAI-2 have a potential therapeutic benefit in clinical conditions where plasminogen activation needs to be limited, e.g., hemorrhagic disease, excessive proteolysis, or to inhibit undesired cell migration such as in tumor invasion. In an experimental animal model of arthritis, tranexamic acid, an inhibitor of plasminogen activation, exhibited significant anti-osteoarthritic activity (BUTLER et al. 1983). Animal studies on the therapeutic use of PAI-1 or PAI-2 are rare, but the availability of recombinant proteins in sufficient quantities should enable such studies.

I. PAI-1

The utility of PAI-1 has been investigated in a rabbit model of tPA-mediated thrombolysis. Injection of recombinant PAI-1 at the end of tPA infusion resulted in complete reversal, within 5 min, of the prolongation of the bleeding time, and in a disappearance of the bleeding tendency (VAUGHAN et al. 1989). In a rabbit ear model of blood loss after treatment with tPA, administration of PAI-1 was more efficient in blocking tPA-induced blood loss, than ε-aminocaproic acid (RACANELLI et al. 1992). Also, in a rat dorsal flap model, topical administration of PAI-1 was more efficient in reducing blood loss than topical thrombin (REILLY et al. 1994).

II. PAI-2

The potential use of PAI-2 as an inhibitor of tumor cell migration and metastasis was investigated using PAI-2 transfected cells. In nude mice, tumors derived from PAI-2 transfected human fibrosarcoma cells displayed a thickened collagenous-like encapsulation which was absent in tumors formed from sham-transfected cells (LAUG et al. 1993). Also, human melanoma cells stably transfected with PAI-2 and injected subcutaneously in SCID mice were surrounded by a thick peritumoral capsule and had significantly less metastases in lung and lymph nodes than mock transfected control melanoma cells (MUELLER et al. 1995). Furthermore, administration of PAI-2 resulted in a decrease of the number of pulmonary metastases after injection of mammary carcinoma cells in rats (EVANS and LIN 1995). Thus, PAI-2 may be of potential use as an inhibitor of metastasis, or in other situations where inhibition of cell migration is desirable, such as inflammation or angiogenesis.

G. Other Protease Inhibitors of the Fibrinolytic System

Several other protease inhibitors, besides A2AP, PAI-1, and PAI-2, were reported to inhibit the fibrinolytic system. In normal plasma the activity half-life of added plasmin is approximately 0.1 s, whereas in A2AP depleted plasma it is approximately 30 s, mainly because of inhibition by α_2-macroglobulin. This implies that A2AP is the principal inhibitor of plasmin, but that α_2-macroglobulin may have an accessory role in limiting plasmin activity.

When added in excess to plasma, tPA and uPA form complexes with α_2-macroglobulin, A2AP, α_1-antitrypsin, C1'-inhibitor, activated protein C inhibitor, and antithrombin III (KRUITHOF et al. 1984; RIJKEN et al. 1983). However, the rate of inhibition of uPA (half-life 20 min) or of tPA (half-life

90min) in PAI-1 depleted plasma was much lower than the rate of in vivo clearance via the liver, which is in the order of a few minutes. It is thus unlikely that these inhibitors regulate PA activity in the blood circulation.

In plasma uPA is present mainly in a proenzyme form, which can be activated by plasmin, thereby forming a positive feed-back mechanism for acceleration of plasminogen activation. Kallikrein is another plasmatic protein that may activate pro-uPA (ICHINOSE et al. 1986; HAUERT et al. 1989). C1'-inhibitor being the principal inhibitor of kallikrein in plasma may potentially limit kallikrein-mediated pro-uPA activation. Indeed, in patients with hereditary angioedema, due to a deficiency of C1'-inhibitor, increased concentrations of plasmin-A2AP complexes, but also of A2AP have been observed, which suggests an activation of the fibrinolytic system (NILSSON and BACK 1985; DONALDSON and HARRISON 1982). To what extent this phenomenon is due to an enhanced activation of pro-uPA or to an activation of blood coagulation with reactive activation of the fibrinolytic system remains to be established.

H. Inhibitors of Plasminogen Binding to Fibrin or Cell Surfaces

Activation of plasminogen by tPA proceeds at a very slow rate in solution. The binding of plasminogen and tPA to fibrin increases the rate of plasminogen activation by several orders of magnitude. Likewise, binding of plasminogen and tPA or uPA to receptors at a cell surface increases the rate of plasminogen activation. Specific receptors for tPA (annexin II) and uPA (the uPAR) have been identified and are well characterized. Specific plasminogen receptors have been postulated, but plasminogen may also bind to any protein with a C-terminal lysine as well as to gangliosides (PLOW et al. 1995; see Chap. 6). Furthermore, plasmin-mediated cleavage of proteins after lysines may create additional plasminogen binding sites. Several proteins are known to interfere with the binding of plasminogen to fibrin or to the cell surface and thereby may decrease the rate of plasminogen activation. A2AP is one of these and has been described above. A few other examples are listed below.

I. Histidine-rich Glycoprotein

Histidine-rich glycoprotein (HRGP) is a monomeric 75kDa protein that is present in plasma at concentrations of approximately 1.5μmol/l. It has repetitive sequences (consensus Gly-His-His-Pro-His), which are similar to a histidine-rich region of high molecular weight kininogen (KOIDE et al. 1986). An elevation of HRGP was found in a family with thrombophilia having no other hemostatic deficiencies that may explain the familial thrombophilia (ENGESSER et al. 1987). HRGP binds to the same lysine binding site of plasminogen as A2AP with a similar affinity of 1μmol/l (LIJNEN et al. 1980; HOLVOET et al. 1986). To what extent variations of HRGP are clinically relevant is not clear. In one study the rate of plasmin generation in normal plasma (HRGP level of 100%) was found to be the same as that in plasmas with HGRP levels between 160% and 280% (ANGLES-CANO et al. 1993).

II. Lipoprotein(a)

Apolipoprotein(a) is a component of lipoprotein(a) and is disulfide-linked to apolipoprotein B-100. It is an independent risk factor for myocardial infarction. Strong evidence for a causative role for apo(a) in the development of arterial disease was obtained with transgenic mice expressing human apo(a), which developed aortic lesions after 3.5 months of an atherogenic diet (LAWN 1994). The apo(a) gene contains at least 19 different alleles varying in length between 48 kb and 190 kb, partially impacting on the plasma levels of lipoprotein(a) (SCANU 1992). A gene cluster was identified on chromosome 6q27-27, which also contained, besides apo(a), plasminogen and two apo(a)-related genes of which one has a domain structure similar to apo(a) and is transcribed in the liver (LINDAHL et al. 1989b; FRANK et al. 1988; BYRNE et al. 1994). Three related genes reside on chromosomes 2 and 4.

Apo(a) has a marked size polymorphism from 300 kDa to 800 kDa, which is related to the number of repeats which are similar to the kringle 4 of plasminogen (MCLEAN et al. 1987; SCANU 1992). Several studies suggested that lipoprotein(a) exerts its pathogenic effect via an inhibition of the plasminogen activator system. Lipoprotein(a) has an affinity for lysine and may compete with plasminogen and tPA for fibrin binding. As a result, fibrin-dependent plasminogen activation by tPA may be attenuated (LOSCALZO et al. 1990; EDELBERG and PIZZO 1994). Plasmin treatment of immobilized fibrin induced an increase in lipoprotein(a) binding, in the same way as plasminogen binding is increased by plasmin-pretreatment (HARPEL et al. 1989). Lipoprotein(a) may also inhibit plasminogen activation at the surface of endothelial cells, of platelets, and of monocyte-like cells (PLOW et al. 1995; MILES et al. 1989; HAJJAR et al. 1989; GONZALEZ-GRONOW et al. 1989; EZRATTY et al. 1993). Furthermore, tPA may bind to apo(a) and apo B-100 and lose its ability to activate plasminogen (SIMON et al. 1991). These results suggest that lipoprotein(a) has an antifibrinolytic effect which may explain its relation to atherothrombotic disease. However, several studies found no (LU et al. 1990; HALVORSEN et al. 1992) or an enhancing (MAO and TUCCI 1990) effect of apo(a) on tPA-induced thrombolysis in a plasma milieu. Also, in atherosclerotic lesions no evidence was found for a displacement of plasminogen by apo(a) (SMITH and CROSBIE 1991). The causes for these discrepancies are not clear, but may be due to donor dependent differences in the fibrin affinity of apo(a) (BAS LEERINK et al. 1992); e.g., a single point mutation (Trp to Arg) in position 72 of kringle 4-37 was associated with a defect in lysine binding (SCANU et al. 1994). Also, the relative concentration of apo(a) with respect to plasminogen may be critically important. In one study, apo(a) inhibited fibrin-dependent plasminogen activation by tPA in a purified system, but had no effect on plasminogen activation in plasma even at high apo(a) concentrations (LU et al. 1990).

In mice, expression of human apo(a) reduced plasminogen activation in vivo and inhibited the activation of TGF-β (GRAINGER et al. 1994). These findings suggest an alternative mechanism for the contribution of apo(a) to the development of atherothrombosis.

III. Carboxypeptidases

A significant part of plasminogen binding to fibrin or cell surfaces is mediated by C-terminal lysines, which are potential substrates for basic plasma carboxypeptidases (see also Chaps. 3 and 6). Plasmin-mediated cleavage of cell-surface proteins increases the density of C-terminal lysines and thus of plasminogen binding sites. A recent study addressed the question whether carboxypeptidases, at their physiologic plasma concentrations may regulate plasminogen binding sites (REDLITZ et al. 1995). Cells incubated in plasma consistently exhibited reduced plasminogen binding; inhibition of carboxypeptidase activity abolished the effect of plasma. However, even prolonged incubation with carboxypeptidase did not completely abolish plasminogen binding, indicating that binding is partially due to interaction with internal lysines or gangliosides (PLOW et al. 1995; REDLITZ 1996). Furthermore, plasma carboxypeptidase B reduced the rate of whole blood clot lysis induced by tPA. Further studies showed that carboxypeptidase B circulates in plasma as a precursor (procarboxypeptidase B) that is activated by thrombin and plasmin (SCHATTEMAN et al. 2000). The rate of activation of procarboxypeptidase B by thrombin is increased a thousandfold by the presence of thrombomodulin (SAKHAROV et al. 1997; BAJZAR et al. 1998). Taken together, these results suggest a model for activation of plasminogen at the cell surface, which has positive and negative feedback mechanisms:

1. Cell-bound PA efficiently activates plasminogen provided it is bound to cells – mostly via C-terminal lysines of cell-surface proteins. The formed plasmin cleaves cell-surface proteins and thereby generates additional plasminogen binding sites.
2. Thrombomodulin-bound thrombin A activates procarboxypeptidase B at the cell surface, which results in the removal of C-terminal lysines and thereby reduces plasminogen binding.

I. Conclusions

The large amount of information gathered in recent years underscores the importance of inhibitors for the regulation of plasminogen activator activity. In the context of fibrinolysis, i.e., the removal of intravascular blood clots, the principal inhibitors are A2AP, PAI-1, and carboxypeptidase B. PAI-2 seems to be more relevant for the regulation of extravascular plasminogen activator activity and not for regular fibrinolysis: its normal plasma concentration is much lower than that of the more efficient inhibitor PAI-1 and it is a relatively inefficient inhibitor of sc-tPA, the form found in the circulation. The role (if any) of the other inhibitors described above for the regulation of physiological fibrinolysis remains to be established.

Acknowledgments. This work was supported by a Program Grant of the "Association Vaud-Genève" and by the Swiss National Fund for Scientific Research (grant no: 31-050645.97).

List of Abbreviations

A2AP	α_2 Antiplasmin (also designated as α_2 Plasmin Inhibitor)
Apo(a)	apolipoprotein(a)
CHO	Chinese hamster ovary (cells)
HRGP	histidine-rich glycoprotein
PA	plasminogen activator
PAI-1	plasminogen activator inhibitor type-1
PAI-2	plasminogen activator inhibitor type-2
pro-uPA	pro-urokinase, sc form of uPA
sc	single-chain
tc	two-chain
tPA	tissue-type plasminogen activator
TGF	transforming growth factor
TNF	tumor necrosis factor
uPA	urokinase-type (also urinary-type) plasminogen activator
uPAR	uPA receptor

References

Aillaud MF, Juhan-Vague I, Alessi MC, Marecal M, Vinson MF, Arnaud C, Vague P, Collen D (1985) Increased PA-inhibitor levels in the postoperative period – no cause-effect relation with increased cortisol. Thromb Haemost 54:466–468

Alessi MC, Peiretti F, Morange P, Henry M, Nalbone G, Juhan-Vague I (1997) Production of plasminogen activator inhibitor 1 by human adipose tissue: possible link between visceral fat accumulation and vascular disease. Diabetes 46:860–867

Andreasen PA, Kjøller L, Christensen L, Duffy MJ (1997) The urokinase-type plasminogen activator system in cancer metastasis: a review. Int J Cancer 72:1–22

Anfosso F, Chomiki N, Alessi MC, Vague P, Juhan-Vague I (1993) Plasminogen activator inhibitor-1 synthesis in the human hepatoma cell line Hep G2. Metformin inhibits the stimulating effect of insulin. J Clin Invest 91:2185–2193

Ang C, Dawes J (1994) The effects of hyperthermia on human endothelial monolayers: modulation of thrombotic potential and permeability. Blood Coagul Fibrinolysis 5:193–199

Angles-Cano E, Gris JC, Loyau S, Schved JF (1993) Familial association of high levels of histidine-rich glycoprotein and plasminogen activator inhibitor-1 with venous thromboembolism. J Lab Clin Med 121:646–653

Antalis TM, Godbolt D, Donnan KD, Stringer BW (1993) Southwestern blot mapping of potential regulatory proteins binding to the DNA encoding plasminogen activator inhibitor type 2. Gene 134:201–208

Antalis TM, Costelloe E, Muddiman J, Ogbourne S, Donnan K (1996) Regulation of the plasminogen activator inhibitor type-2 gene in monocytes: localization of an upstream transcriptional silencer. Blood 88:3686–3697

Antalis TM, La Linn M, Donnan K, Mateo L, Gardner J, Dickinson JL, Buttigieg K, Suhrbier A (1998) The serine proteinase inhibitor (serpin) plasminogen activation inhibitor type 2 protects against viral cytopathic effects by constitutive interferon α/β priming. J Exp Med 187:1799–1811

Aoki N, Sumi Y, Miura O, Hirosawa S (1993) Human α_2-plasmin inhibitor. Methods Enzymol 223:185–197

Åstedt B, Lecander I, Brodin T, Lundblad A, Löw K (1985) Purification of a specific placental plasminogen activator inhibitor by monoclonal antibody and its complex formation with plasminogen activator. Thromb Haemost 53:122–125

Åstedt B, Hägerstrand I, Lecander I (1986) Cellular localisation in placenta of placental type plasminogen activator inhibitor. Thromb Haemost 56:63–65

Bajou K, Noel A, Gerard RD, Masson V, Brunner N, Holst-Hansen C, Skobe M, Fusenig NE, Carmeliet P, Collen D, Foidart JM (1998) Absence of host plasminogen activator inhibitor 1 prevents cancer invasion and vascularization. Nat Med 4:923–928

Bajzar L, Nesheim M, Morser J, Tracy PB (1998) Both cellular and soluble forms of thrombomodulin inhibit fibrinolysis by potentiating the activation of thrombin-activable fibrinolysis inhibitor. J Biol Chem 273:2792–2798

Ballegeer V, Spitz B, Kieckens L, Moreau H, Van Assche A, Collen D (1989) Predictive value of increased plasma levels of fibronectin in gestational hypertension. Am J Obstet Gynecol 161:432–436

Bas Leerink C, Duif PF, Gimpel JA, Kortlandt W, Bouma BN, van Rijn HJ (1992) Lysine-binding heterogeneity of Lp(a): consequences for fibrin binding and inhibition of plasminogen activation. Thromb Haemost 68:185–188

Belin D, Wohlwend A, Schleuning WD, Kruithof EKO, Vassalli JD (1989) Facultative polypeptide translocation allows a single mRNA to encode the secreted and cytosolic forms of plasminogen activators inhibitor 2. EMBO J 8:3287–3294

Belin D (1993) Biology and facultative secretion of plasminogen activator inhibitor-2. Thromb Haemost 70:144–147

Belin D, Bost S, Vassalli JD, Strub K (1996) A two-step recognition of signal sequences determines the translocation efficiency of proteins. EMBO J 15:468–478

Bergonzelli GE, Kruithof EKO (1991) Induction of plasminogen activator inhibitor 1 biosynthesis by hyperthermia. J Cell Physiol 148:306–313

Bergonzelli GE, Kruithof EKO, Medcalf RL (1992) Transcriptional antagonism of phorbol ester-mediated induction of plasminogen activator inhibitor types 1 and 2 by cyclic adenosine 3′,5′-monophosphate. Endocrinology 131:1467–1472

Berrettini M, Schleef RR, España F, Loskutoff DJ, Griffin JH (1989) Interaction of type 1 plasminogen activator inhibitor with the enzymes of the contact activation system. J Biol Chem 264:11738–11743

Biemond BJ, Levi M, ten Cate H, van der Poll T, Büller HR, Hack CE, ten Cate JW (1995a) Plasminogen activator and plasminogen activator inhibitor I release during experimental endotoxaemia in chimpanzees: effect of interventions in the cytokine and coagulation cascades. Clin Sci (Colch) 88:587–594

Biemond BJ, Levi M, Coronel R, Janse MJ, ten Cate JW, Pannekoek H (1995b) Thrombolysis and reocclusion in experimental jugular vein and coronary artery thrombosis. Effects of a plasminogen activator inhibitor type 1-neutralizing monoclonal antibody. Circulation 91:1175–1181

Bruzdzinski CJ, Johnson MR, Goble CA, Winograd SS, Gelehrter TD (1993) Mechanism of glucocorticoid induction of the rat plasminogen activator inhibitor-1 gene in HTC rat hepatoma cells: identification of cis-acting regulatory elements. Mol Endocrinol 7:1169–1177

Bugge TH, Lund LR, Kombrinck KK, Nielsen BS, Holmbäck K, Drew AF, Flick MJ, Witte DP, Danø K, Degen JL (1998) Reduced metastasis of Polyoma virus middle T antigen-induced mammary cancer in plasminogen-deficient mice. Oncogene 16:3097–3104

Burzotta F, Di Castelnuovo A, Amore C, D'Orazio A, Di Bitondo R, Donati MB, Iacoviello L (1998) 4G/5G promoter PAI-1 gene polymorphism is associated with plasmatic PAI-1 activity in Italians: a model of gene-environment interaction. Thromb Haemost 79:354–358

Butler M, Colombo C, Hickman L, O'Byrne E, Steele R, Steinetz B, Quintavalla J, Yokoyama N (1983) A new model of osteoarthritis in rabbits III. Evaluation of the anti-osteoarthritic effects of selected drugs administered intraarticularly. Arthritis Rheum 26:1380–1386

Byrne CD, Schwartz K, Meer K, Cheng JF, Lawn RM (1994) The human apolipoprotein(a)/plasminogen gene cluster contains a novel homologue transcribed in liver. Arterioscler Thromb 14:534–541

Carmeliet P, Stassen JM, Schoonjans L, Ream B, van den Oord JJ, De Mol M, Mulligan RC, Collen D (1993) Plasminogen activator inhibitor-1 gene-deficient mice. II. Effects on hemostasis, thrombosis, and thrombolysis. J Clin Invest 92:2756–2760

Carmeliet P, Collen D (1997) Molecular genetics of the fibrinolytic and coagulation systems in haemostasis, thrombogenesis, restenosis and atherosclerosis. Curr Opin Lipidol 8:118–125

Carmeliet P, Moons L, Lijnen R, Janssens S, Lupu F, Collen D, Gerard RD (1997) Inhibitory role of plasminogen activator inhibitor-1 in arterial wound healing and neointima formation: a gene targeting and gene transfer study in mice. Circulation 96:3180–3191

Catto AJ, Carter AM, Stickland M, Bamford JM, Davies JA, Grant PJ (1997) Plasminogen activator inhibitor-1 (PAI-1) 4G/5G promoter polymorphism and levels in subjects with cerebrovascular disease. Thromb Haemost 77:730–734

Chang E, Goldberg H (1995) Requirements for transforming growth factor-β regulation of the pro-$\alpha 2$(I) collagen and plasminogen activator inhibitor-1 promoters. J Biol Chem 270:4473–4477

Chapman HA, Stone OL (1985) A fibrinolytic inhibitor of human alveolar macrophages. Induction with endotoxin. Am Rev Respir Dis 132:569–575

Chen CS, Lyons-Giordano B, Lazarus GS, Jensen PJ (1993) Differential expression of plasminogen activators and their inhibitors in an organotypic skin coculture system. J Cell Sci 106:45–53

Clemmensen I, Thorsen S, Mullertz S, Petersen LC (1981) Properties of three different molecular forms of the alpha 2 plasmin inhibitor. Eur J Biochem 120:105–112

Cousin E, Medcalf RL, Bergonzelli GE, Kruithof EKO (1991) Regulatory elements involved in constitutive and phorbol ester-inducible expression of the plasminogen activator inhibitor type 2 gene promoter. Nucleic Acids Res 19:3881–3886

Dawson S, Hamsten A, Wiman B, Henney A, Humphries S (1991) Genetic variation at the plasminogen activator inhibitor-1 locus is associated with altered levels of plasma plasminogen activator inhibitor-1 activity. Arterioscler Thromb 11:183–190

Dawson SJ, Wiman B, Hamsten A, Green F, Humphries S, Henney AM (1993) The two allele sequences of a common polymorphism in the promoter of the plasminogen activator inhibitor-1 (PAI-1) gene respond differently to interleukin-1 in HepG2 cells. J Biol Chem 268:10739–10745

Dear AE, Shen Y, Rüegg M, Medcalf RL (1996) Molecular mechanisms governing tumor-necrosis-factor-mediated regulation of plasminogen-activator inhibitor type-2 gene expression. Eur J Biochem 241:93–100

Dear AE, Costa M, Medcalf RL (1997) Urokinase-mediated transactivation of the plasminogen activator inhibitor type 2 (PAI-2) gene promoter in HT-1080 cells utilises AP-1 binding sites and potentiates phorbol ester-mediated induction of endogenous PAI-2 mRNA. FEBS Lett 402:265–272

De Boer K, Lecander I, ten Cate JW, Borm JJJ, Treffers PE (1988) Placental-type plasminogen activator inhibitor in preeclampsia. Am J Obstet Gynecol 158:518–522

De Clerck PJ, Alessi MC, Verstreken M, Kruithof EKO, Juhan-Vague I, Collen D (1988) Measurement of plasminogen activator inhibitor 1 in biologic fluids with a murine monoclonal antibody-based enzyme-linked immunosorbent assay. Blood 71:220–225

De Clerck PJ, De Mol M, Vaughan DE, Collen D (1992) Identification of a conformationally distinct form of plasminogen activator inhibitor-1, acting as a noninhibitory substrate for tissue-type plasminogen activator. J Biol Chem 267:11693–11696

Dickinson JL, Bates EJ, Ferrante A, Antalis TM (1995) Plasminogen activator inhibitor type 2 Inhibits tumor necrosis factor alpha-induced apoptosis. Evidence for an alternate function. J Biol Chem 270:27894–27904

Diéval J, Nguyen G, Gross S, Delobel J, Kruithof EKO (1991) A lifelong bleeding disorder associated with a deficiency of plasminogen activator inhibitor type 1. Blood 77:528–532

Donaldson VH, Harrison RA (1982) Complexes between C1-inhibitor, kallikrein, high molecular weight kininogen, plasma thromboplastin antecedent, and plasmin in normal human plasma and hereditary angioneurotic edema plasmas containing dysmorphic C1-inhibitors: role of cold activation. Blood 60:121–129

Dong G, Schulick AH, DeYoung MB, Dichek DA (1996) Identification of a cis-acting sequence in the human plasminogen activator inhibitor type-1 gene that mediates transforming growth factor-β_1 responsiveness in endothelium in vivo. J Biol Chem 271:29969–29977

Edelberg J, Pizzo SV (1994) Lipoprotein (a) regulates plasmin generation and inhibition. Chem Phys Lipids 67–68:363–368

Eitzman DT, Fay WP, Lawrence DA, Francis-Chmura AM, Shore JD, Olson ST, Ginsburg D (1995) Peptide-mediated inactivation of recombinant and platelet plasminogen activator inhibitor-1 in vitro. J Clin Invest 95:2416–2420

Engesser L, Kluft C, Briet E, Brommer EJ (1987) Familial elevation of plasma histidine-rich glycoprotein in a family with thrombophilia. Br J Haematol 67:355–358

Erickson LA, Ginsberg MH, Loskutoff DJ (1984) Detection and partial characterization of an inhibitor of plasminogen activator in human platelets. J Clin Invest 74:1465–1472

Erickson LA, Fici GJ, Lund JE, Boyle TP, Polites HG, Marotti KR (1990) Development of venous occlusions in mice transgenic for the plasminogen activator inhibitor-1 gene. Nature 346:74–76

Eriksson P, Kallin B, van 't Hooft FM, Båvenholm P, Hamsten A (1995) Allele-specific increase in basal transcription of the plasminogen-activator inhibitor 1 gene is associated with myocardial infarction. Proc Natl Acad Sci USA 92:1851–1855

Eriksson P, Nilsson L, Karpe F, Hamsten A (1998a) Very-low-density lipoprotein response element in the promoter region of the human plasminogen activator inhibitor-1 gene implicated in the impaired fibrinolysis of hypertriglyceridemia. Arterioscler Thromb Vasc Biol 18:20–26

Eriksson P, Reynisdottir S, Lönnqvist F, Stemme V, Hamsten A, Arner P (1998b) Adipose tissue secretion of plasminogen activator inhibitor-1 in non-obese and obese individuals. Diabetologia 41:65–71

Estellés A, Gilabert J, Aznar J, Loskutoff DJ, Schleef RR (1989) Changes in the plasma levels of type 1 and type 2 plasminogen activator inhibitors in normal pregnancy and in patients with severe preeclampsia. Blood 74:1332–1338

Estellés A, Gilabert J, Grancha S, Yamamoto K, Thinnes T, España F, Aznar J, Loskutoff DJ (1998) Abnormal expression of type 1 plasminogen activator inhibitor and tissue factor in severe preeclampsia. Thromb Haemost 79:500–508

Evans DM, Lin PL (1995) Suppression of pulmonary metastases of rat mammary cancer by recombinant urokinase plasminogen activator inhibitor. Am Surg 61:692–696

Ezratty A, Simon DI, Loscalzo J (1993) Lipoprotein(a) binds to human platelets and attenuates plasminogen binding and activation. Biochemistry 32:4628–4633

Fay WP, Shapiro AD, Shih JL, Schleef RR, Ginsburg D (1992) Brief report: complete deficiency of plasminogen-activator inhibitor type 1 due to a frame-shift mutation. N Engl J Med 327:1729–1733

Fay WP, Parker AC, Condrey LR, Shapiro AD (1997) Human plasminogen activator inhibitor-1 (PAI-1) deficiency: characterization of a large kindred with a null mutation in the PAI-1 gene. Blood 90:204–208

Fazleabas AT, Everly SL, Lottenberg R (1991) Immunological and molecular characterization of plasminogen activator inhibitors 1 and 2 in baboon (Papio anubis) placental tissues. Biol Reprod 45:49–56

Feinberg RF, Kao LC, Haimowitz JE, Queenan JT, Wun TC, Strauss JF, Kliman HJ (1989) Plasminogen activator inhibitor types 1 and 2 in human trophoblasts PAI-1 is an immunocytochemical marker of invading trophoblasts. Lab Invest 61:20–26

Frank SL, Klisak I, Sparkes RS, Mohandas T, Tomlinson JE, McLean JW, Lawn RM, Lusis AJ (1988) The apolipoprotein(a) gene resides on human chromosome 6q26–27, in close proximity to the homologous gene for plasminogen. Hum Genet 79:352–356

Friederich PW, Levi M, Biemond BJ, Charlton P, Templeton D, van Zonneveld AJ, Bevan P, Pannekoek H, ten Cate JW (1997) Novel low-molecular-weight inhibitor of PAI-1 (XR5118) promotes endogenous fibrinolysis and reduces postthrombolysis thrombus growth in rabbits. Circulation 96:916–921

Genton C, Kruithof EKO, Schleuning WD (1987) Phorbol ester induces the biosynthesis of glycosylated and non-glycosylated plasminogen activator inhibitor 2 in high excess over urokinase-type plasminogen activator in human U-937 lymphoma cells. J Cell Biol 104:705–712

Germer M, Kanse SM, Kirkegaard T, Kjøller L, Felding-Habermann B, Goodman S, Preissner KT (1998) Kinetic analysis of integrin-dependent cell adhesion on vitronectin–the inhibitory potential of plasminogen activator inhibitor-1 and RGD peptides. Eur J Biochem 253:669–674

Gilabert J, Estellés A, Ridocci F, España F, Aznar J, Galbis M (1990) Clinical and haemostatic parameters in the HELLP syndrome: relevance of plasminogen activator inhibitors. Gynecol Obstet Invest 30:81–86

Gilabert J, Estellés A, Ayuso MJ, España F, Chirivella M, Grancha S, Mico JM, Aznar J (1994) Evaluation of plasminogen activators and plasminogen activator inhibitors in plasma and amniotic fluid in pregnancies complicated with intrauterine fetal growth retardation. Gynecol Obstet Invest 38:157–162

Gissler HM, Frank R, Kramer MD (1993) Immunohistochemical characterization of the plasminogen activator system in psoriatic epidermis. Br J Dermatol 128:612–618

Glueck CJ, Glueck HI, Hamer T, Speirs J, Tracy T, Stroop D (1994) Beta blockers, Lp(a), hypertension, and reduced basal fibrinolytic activity. Am J Med Sci 307:317–324

Gonzalez-Gronow M, Edelberg JM, Pizzo SV (1989) Further characterization of the cellular plasminogen binding site: evidence that plasminogen 2 and lipoprotein a compete for the same site. Biochemistry 28:2374–2377

Grainger DJ, Kemp PR, Liu AC, Lawn RM, Metcalfe JC (1994) Activation of transforming growth factor-beta is inhibited in transgenic apolipoprotein(a) mice. Nature 370:460–462

Grancha S, Estellés A, Gilabert J, Chirivella M, España F, Aznar J (1996) Decreased expression of PAI-2 mRNA and protein in pregnancies complicated with intrauterine fetal growth retardation. Thromb Haemost 76:761–767

Grant PJ (1997) Polymorphisms of coagulation/fibrinolysis genes: gene environment interactions and vascular risk. Prostaglandins Leukot Essent Fatty Acids 7:473–477

Hajjar KA, Gavish D, Breslow JL, Nachman RL (1989) Lipoprotein(a) modulation of endothelial cell surface fibrinolysis and its potential role in atherosclerosis. Nature 339:303–305

Halvorsen S, Skjønsberg OH, Berg K, Ruyter R, Godal HC (1992) Does Lp(a) lipoprotein inhibit the fibrinolytic system? Thromb Res 68:223–232

Hamsten A, de Faire U, Walldius G, Dahlen G, Szamosi A, Landou C, Blombäck M, Wiman B (1987) Plasminogen activator inhibitor in plasma: risk factor for recurrent myocardial infarction. Lancet 2:3–9

Harpel PC, Gordon BR, Parker TS (1989) Plasmin catalyzes binding of lipoprotein (a) to immobilized fibrinogen and fibrin. Proc Natl Acad Sci USA 86:3847–3851

Harrop SJ, Jankova L, Coles M, Jardine D, Whittaker JS, Gould AR, Meister A, King GC, Mabbutt BC, Curmi PMG (1999) The crystal structure of plasminogen activator inhibitor 2 at 2.0 Å resolution: implications for serpin function. Structure Fold Des 7:43–54

Hauert J, Nicoloso G, Schleuning WD, Bachmann F, Schapira M (1989) Plasminogen activators in dextran sulfate-activated euglobulin fractions: a molecular analysis of factor XII- and prekallikrein-dependent fibrinolysis. Blood 73:994–999

He S, Bremme K, Blombäck M (1995) Increased blood flow resistance in placental circulation and levels of plasminogen activator inhibitors types 1 and 2 in severe preeclampsia. Blood Coagul Fibrinolysis 6:703–708

Healy AM, Gelehrter TD (1994) Induction of plasminogen activator inhibitor-1 in HepG2 human hepatoma cells by mediators of the acute phase response. J Biol Chem 269:19095–19100

Heaton JH, Kathju S, Gelehrter TD (1992) Transcriptional and posttranscriptional regulation of type 1 plasminogen activator inhibitor and tissue-type plasminogen activator gene expression in HTC rat hepatoma cells by glucocorticoids and cyclic nucleotides. Mol Endocrinol 6:53–60

Hekman CM, Loskutoff DJ (1985) Endothelial cells produce a latent inhibitor of plasminogen activators that can be activated by denaturants. J Biol Chem 260:11581–11587

Hekman CM, Loskutoff DJ (1988) Kinetic analysis of the interactions between plasminogen activator inhibitor 1 and both urokinase and tissue plasminogen activator. Arch Biochem Biophys 262:199–210

Henry M, Tregouet DA, Alessi MC, Aillaud MF, Visvikis S, Siest G, Tiret L, Juhan-Vague I (1998) Metabolic determinants are much more important than genetic polymorphisms in determining the PAI-1 activity and antigen plasma concentrations: a family study with part of the Stanislas Cohort. Arterioscler Thromb Vasc Biol 18:84–91

Hesselvik JF, Blomback M, Brodin B, Maller R (1989) Coagulation, fibrinolysis, and kallikrein systems in sepsis: relation to outcome. Crit Care Med 17:724–733

Holmes WE, Nelles L, Lijnen HR, Collen D (1987) Primary structure of human α_2-antiplasmin, a serine protease inhibitor (serpin). J Biol Chem 262:1659–1664

Holvoet P, Lijnen HR, Collen D (1986) A monoclonal antibody directed against the high-affinity lysine-binding site (LBS) of human plasminogen. Role of LBS in the regulation of fibrinolysis. Eur J Biochem 157:65–69

Ichinose A, Fujikawa K, Suyama T (1986) The activation of pro-urokinase by plasma kallikrein and its inactivation by thrombin. J Biol Chem 261:3486–3489

Iwai N, Shimoike H, Nakamura Y, Tamaki S, Kinoshita M (1998) The 4G/5G polymorphism of the plasminogen activator inhibitor gene is associated with the time course of progression to acute coronary syndromes. Atherosclerosis 136:109–114

Jansen JW, Haverkate F, Koopman J, Nieuwenhuis HK, Kluft C, Boschman TA (1987) Influence of factor XIIIa activity on human whole blood clot lysis in vitro. Thromb Haemost 57:171–175

Jansson JH, Norberg B, Nilsson TK (1989) Impact of acute phase on concentrations of tissue plasminogen activator and plasminogen activator inhibitor in plasma after deep-vein thrombosis or open-heart surgery. Clin Chem 35:1544–1545

Jensen PH, Schüler E, Woodrow G, Richardson M, Goss N, Hojrup P, Petersen TE, Rasmussen LK (1994a) A unique interhelical insertion in plasminogen activator inhibitor-2 contains three glutamines, Gln[83], Gln[84], Gln[86], essential for transglutaminase-mediated cross-linking. J Biol Chem 269:15394–15398

Jensen PH, Cressey LI, Gjertsen BT, Madsen P, Mellgren G, Hokland P, Gliemann J, Doskeland SO, Lanotte M, Vintermyr OK (1994b) Cleaved intracellular plasminogen activator inhibitor-2 in human myeloleukemia cells is a marker of apoptosis. Br J Cancer 70:834–840

Jensen PJ, Wu QY, Janowitz P, Ando Y, Schechter NM (1995) Plasminogen activator inhibitor type 2: An intracellular keratinocyte differentiation product that is incorporated into the cornified envelope. Exp Cell Res 217:65–71

Juhan-Vague I, Alessi MC (1993) Plasminogen activator inhibitor 1 and atherothrombosis. Thromb Haemost 70:138–143

Juhan-Vague I, Thompson SG, Jespersen J (1993) Involvement of the hemostatic system in the insulin resistance syndrome. A study of 1500 patients with angina pectoris. The ECAT Angina Pectoris Study Group. Arterioscler Thromb 13:1865–1873

Juhan-Vague I, Alessi MC (1997) PAI-1, obesity, insulin resistance and risk of cardiovascular events. Thromb Haemost 78:656–660

Kassis J, Hirsh J, Podor TJ (1992) Evidence that postoperative fibrinolytic shutdown is mediated by plasma factors that stimulate endothelial cell type I plasminogen activator inhibitor biosynthesis. Blood 80:1758–1764

Kato A, Hirosawa S, Toyota S, Nakamura Y, Nishi H, Kimura A, Sasazuki T, Aoki N (1993) Localization of the human α_2-plasmin inhibitor gene (PLI) to 17p13. Cytogenet Cell Genet 62:190–191

Keeton MR, Curriden SA, van Zonneveld AJ, Loskutoff DJ (1991) Identification of regulatory sequences in the type 1 plasminogen activator inhibitor gene responsive to transforming growth factor β. J Biol Chem 266:23048–23052

Keeton M, Eguchi Y, Sawdey M, Ahn C, Loskutoff DJ (1993) Cellular localization of type 1 plasminogen activator inhibitor messenger RNA and protein in murine renal tissue. Am J Pathol 142:59–70

Kjøller L, Kanse SM, Kirkegaard T, Rodenburg KW, Rønne E, Goodman SL, Preissner KT, Ossowski L, Andreasen PA (1997) Plasminogen activator inhibitor-1 represses integrin- and vitronectin-mediated cell migration independently of its function as an inhibitor of plasminogen activation. Exp Cell Res 232:420–429

Klinger KW, Winqvist R, Riccio A, Andreasen PA, Sartorio R, Nielsen LS, Stuart N, Stanislovitis P, Watkins P, Douglas R, et al (1987) Plasminogen activator inhibitor type 1 gene is located at region q21.3-q22 of chromosome 7 and genetically linked with cystic fibrosis. Proc Natl Acad Sci USA 84:8548–8552

Kluft C, Nieuwenhuis HK, Rijken DC, Groeneveld E, Wijngaards G, van Berkel W, Dooijewaard G, Sixma JJ (1987) alpha 2-Antiplasmin Enschede: dysfunctional alpha 2-antiplasmin molecule associated with an autosomal recessive hemorrhagic disorder. J Clin Invest 80:1391–1400

Knudsen H, Olesen T, Riccio A, Ungaro P, Christensen L, Andreasen PA (1994) A common response element mediates differential effects of phorbol esters and forskolin on type-1 plasminogen activator inhibitor gene expression in human breast carcinoma cells. Eur J Biochem 220:63–74

Koh SC, Anandakumar C, Montan S, Ratnam SS (1993) Plasminogen activators, plasminogen activator inhibitors and markers of intravascular coagulation in preeclampsia. Gynecol Obstet Invest 35:214–221

Koide T, Foster D, Yoshitake S, Davie EW (1986) Amino acid sequence of human histidine-rich glycoprotein derived from the nucleotide sequence of its cDNA. Biochemistry 25:2220–2225

Kornelisse RF, Hazelzet JA, Savelkoul HF, Hop WC, Suur MH, Borsboom AN, Risseeuw-Appel IM, van der Voort E, de Groot R (1996) The relationship between plasminogen activator inhibitor-1 and proinflammatory and counterinflammatory mediators in children with meningococcal septic shock. J Infect Dis 173:1148–1156

Koyama T, Koike Y, Toyota S, Miyagi F, Suzuki N, Aoki N (1994) Different NH2-terminal form with 12 additional residues of α_2-plasmin inhibitor from human plasma and culture media of Hep G2 cells. Biochem Biophys Res Commun 200:417–422

Kruithof EKO, Tran-Thang C, Ransijn A, Bachmann F (1984) Demonstration of a fast-acting inhibitor of plasminogen activators in human plasma. Blood 64:907–913

Kruithof EKO, Vassalli JD, Schleuning WD, Mattaliano RJ, Bachmann F (1986) Purification and characterization of a plasminogen activator inhibitor from the histiocytic lymphoma cell line U-937. J Biol Chem 261:11207–11213

Kruithof EKO, Tran-Thang C, Gudinchet A, Hauert J, Nicoloso G, Genton C, Welti H, Bachmann F (1987a) Fibrinolysis in pregnancy: a study of plasminogen activator inhibitors. Blood 69:460–466

Kruithof EKO, Nicolosa G, Bachmann F (1987b) Plasminogen activator inhibitor 1: development of a radioimmunoassay and observations on its plasma concentration during venous occlusion and after platelet aggregation. Blood 70:1645–1653

Kruithof EKO, Baker MS, Bunn CL (1995) Biological and clinical aspects of plasminogen activator inhibitor type 2 (PAI-2). 86:4007–4024

Kumar S, Baglioni C (1991) Protection from tumor necrosis factor-mediated cytolysis by overexpression of plasminogen activator inhibitor type-2. J Biol Chem 266:20960–20964

Laug WE, Cao XR, Yu YB, Shimada H, Kruithof EKO (1993) Inhibition of invasion of HT1080 sarcoma cells expressing recombinant plasminogen activator inhibitor 2. Cancer Res 53:6051–6057

Lawn RM (1994) The apolipoprotein(a) gene: characterization of 5' flanking regions and expression in transgenic mice. Chem Phys Lipids 67–68:19–23

Lawrence DA (1997) The role of reactive-center loop mobility in the serpin inhibitory mechanism. Adv Exp Med Biol 425:99–108

Lecander I, Åstedt B (1987) Specific plasminogen activator inhibitor of placental type PAI-2 occurring in amniotic fluid and cord blood. J Lab Clin Med 110:602–605

Lee MH, Vosburgh E, Anderson K, McDonagh J (1993) Deficiency of plasma plasminogen activator inhibitor 1 results in hyperfibrinolytic bleeding. Blood 81:2357–2362

Levi M, Biemond BJ, van Zonneveld AJ, ten Cate JW, Pannekoek H (1992) Inhibition of plasminogen activator inhibitor-1 activity results in promotion of endogenous thrombolysis and inhibition of thrombus extension in models of experimental thrombosis. Circulation 85:305–312

Levin EG, Santell L (1987) Conversion of the active to latent plasminogen activator inhibitor from human endothelial cells. Blood 70:1090–1098

Lijnen HR, Hoylaerts M, Collen D (1980) Isolation and characterization of a human plasma protein with affinity for the lysine binding sites in plasminogen. Role in the regulation of fibrinolysis and identification as histidine-rich glycoprotein. J Biol Chem 255:10214–10222

Lindahl TL, Sigurdardottir O, Wiman B (1989a) Stability of plasminogen activator inhibitor 1 (PAI-1). Thromb Haemost 62:748–751

Lindahl G, Gersdorf E, Menzel HJ, Duba C, Cleve H, Humphries S, Utermann G (1989b) The gene for the Lp(a)-specific glycoprotein is closely linked to the gene for plasminogen on chromosome 6. Hum Genet 81:149–152

Lindoff C, Lecander I, Åstedt B (1993) Fibrinolytic components in individual consecutive plasma samples during normal pregnancy. Fibrinolysis 7:190–194

Lindoff C, Åstedt B (1994) Plasminogen activator of urokinase type and its inhibitor of placental type in hypertensive pregnancies and in intrauterine growth retardation: possible markers of placental function. Am J Obstet Gynecol 171:60–64

Lorente JA, García-Frade LJ, Landin L, de Pablo R, Torrado C, Renes E, García-Avello A (1993) Time course of hemostatic abnormalities in sepsis and its relation to outcome. Chest 103:1536–1542

Loscalzo J, Weinfeld M, Fless GM, Scanu AM (1990) Lipoprotein(a), fibrin binding, and plasminogen activation. Arteriosclerosis 10:240–245

Loskutoff DJ (1991) Regulation of PAI-1 gene expression. Fibrinolysis 5:197–206
Loskutoff DJ, Curriden SA, Hu G, Deng G (1999) Regulation of cell adhesion by PAI-1. Acta Pathol Microbiol Immunol Scand 107:54–61
Loskutoff DJ, Samad F (1998) The adipocyte and hemostatic balance in obesity: studies of PAI-1. Arterioscler Thromb Vasc Biol 18:1–6
Lu H, Bruckert J, Soria J, Li H, de Gennes JL, Legrand A, Peynet J, Soria C (1990) Absence of inhibition by lipoprotein (a) inhibition of tPA induced thrombolysis in a patient's plasma milieu. Blood Coagul Fibrinolysis 1:513–516
Lupu F, Bergonzelli GE, Heim DA, Cousin E, Genton CY, Bachmann F, Kruithof EKO (1993) Localization and production of plasminogen activator inhibitor-1 in human healthy and atherosclerotic arteries. 13:1090–1100
Lyons-Giordano B, Loskutoff D, Chen CS, Lazarus G, Keeton M, Jensen PJ (1994) Expression of plasminogen activator inhibitor type 2 in normal and psoriatic epidermis. Histochem 101:105–112
Mahony D, Stringer BW, Dickinson JL, Antalis TM (1998) DNase I hypersensitive sites in the 5' flanking region of the human plasminogen activator inhibitor type 2 (PAI-2) gene are associated with basal and tumor necrosis factor-alpha-induced transcription in monocytes. Eur J Biochem 256:550–559
Mannucci PM, Mari D, Merati G, Peyvandi F, Tagliabue L, Sacchi E, Taioli E, Sansoni P, Bertolini S, Franceschi C (1997) Gene polymorphisms predicting high plasma levels of coagulation and fibrinolysis proteins. A study in centenarians. Arterioscler Thromb Vasc Biol 17:755–759
Mao SJ, Tucci MA (1990) Lipoprotein(a) enhances plasma clot lysis in vitro. FEBS Lett 267:131–134
Margaglione M, Cappucci G, Colaizzo D, Giuliani N, Vecchione G, Grandone E, Pennelli O, Di Minno G (1998a) The PAI-1 gene locus 4G/5G polymorphism is associated with a family history of coronary artery disease. Arterioscler Thromb Vasc Biol 18:152–156
Margaglione M, Cappucci G, d'Addedda M, Colaizzo D, Giuliani N, Vecchione G, Mascolo G, Grandone E, Di Minno G (1998b) PAI-1 plasma levels in a general population without clinical evidence of atherosclerosis: relation to environmental and genetic determinants. Arterioscler Thromb Vasc Biol 18:562–567
McLean JW, Tomlinson JE, Kuang WJ, Eaton DL, Chen EY, Fless GM, Scanu AM, Lawn RM (1987) cDNA sequence of human apolipoprotein(a) is homologous to plasminogen. Nature 330:132–137
Medcalf RL, Kruithof EKO, Schleuning WD (1988a) Plasminogen activator inhibitor 1 and 2 are tumor necrosis factor/cachectin-responsive genes. J Exp Med 168:751–759
Medcalf RL, Van den Berg E, Schleuning WD (1988b) Glucocorticoid-modulated gene expression of tissue- and urinary-type plasminogen activator and plasminogen activator inhibitor 1 and 2. J Cell Biol 106:971–978
Medina R, Socher SH, Han JH, Friedman PA (1989) Interleukin-1, endotoxin or tumor necrosis factor/cachectin enhance the level of plasminogen activator inhibitor messenger RNA in bovine aortic endothelial cells. Thromb Res 54:41–52
Menoud PA, Sappino N, Boudal–Khoshbeen M, Vassalli JD, Sappino AP (1996) The kidney is a major site of alpha(2)–antiplasmin production. J Clin Invest 97:2478–2484
Mesters RM, Florke N, Ostermann H, Kienast J (1996) Increase of plasminogen activator inhibitor levels predicts outcome of leukocytopenic patients with sepsis. Thromb Haemost 75:902–907
Mestries JC, Kruithof EKO, Gascon MP, Herodin F, Agay D, Ythier A (1994) In vivo modulation of coagulation and fibrinolysis by recombinant glycosylated human interleukin-6 in baboons. Eur Cytokine Netw 5:275–281
Mikus P, Urano T, Liljeström P, Ny T (1993) Plasminogen-activator inhibitor type 2 (PAI-2) is a spontaneously polymerising SERPIN. Biochemical characterisation of

the recombinant intracellular and extracellular forms. Eur J Biochem 218:1071–1082

Miles LA, Fless GM, Levin EG, Scanu AM, Plow EF (1989) A potential basis for the thrombotic risks associated with lipoprotein(a). Nature 339:301–303

Miura O, Hirosawa S, Kato A, Aoki N (1989a) Molecular basis for congenital deficiency of α_2-plasmin inhibitor. A frameshift mutation leading to elongation of the deduced amino acid sequence. J Clin Invest 83:1598–1604

Miura O, Sugahara Y, Aoki N (1989b) Hereditary α_2-plasmin inhibitor deficiency caused by a transport-deficient mutation (α_2-PI-Okinawa). Deletion of Glu137 by a trinucleotide deletion blocks intracellular transport. J Biol Chem 264:18213–18219

Moons L, Shi C, Ploplis V, Plow E, Haber E, Collen D, Carmeliet P (1998) Reduced transplant arteriosclerosis in plasminogen-deficient mice. J Clin Invest 102:1788–1797

Mottonen J, Strand A, Symersky J, Sweet RM, Danley DE, Geoghegan KF, Gerard RD, Goldsmith EJ (1992) Structural basis of latency in plasminogen activator inhibitor-1. Nature 355:270–273

Mueller BM, Yu YB, Laug WE (1995) Overexpression of plasminogen activator inhibitor 2 in human melanoma cells inhibits spontaneous metastasis in scid/scid mice. Proc Natl Acad Sci USA 92:205–209

Munkvad S, Gram J, Jespersen J (1990) A depression of active tissue plasminogen activator in plasma characterizes patients with unstable angina pectoris who develop myocardial infarction. Eur Heart J 11:525–528

Murakami J, Ohtani A, Murata S (1997) Protective effect of T-686, an inhibitor of plasminogen activator inhibitor-1 production, against the lethal effect of lipopolysaccharide in mice. Jpn J Pharmacol 75:291–294

Nagi DK, McCormack LJ, Mohamed-Ali V, Yudkin JS, Knowler WC, Grant PJ (1997) Diabetic retinopathy, promoter (4G/5G) polymorphism of PAI-1 gene, and PAI-1 activity in Pima Indians with type 2 diabetes. Diabetes Care 20:1304–1309

Nilsson T, Back O (1985) Elevated plasmin-alpha 2-antiplasmin complex levels in hereditary angioedema: evidence for the in vivo efficiency of the intrinsic fibrinolytic system. Thromb Res 40:817–821

Ny T, Hansson L, Lawrence D, Leonardsson G, Åstedt B (1989) Plasminogen activator inhibitor type 2 cDNA transfected into Chinese hamster ovary cells is stably expressed but not secreted. Fibrinolysis 3:189–196

Planus E, Barlovatz-Meimon G, Rogers RA, Bonavaud S, Ingber DE, Wang N (1997) Binding of urokinase to plasminogen activator inhibitor type-1 mediates cell adhesion and spreading. J Cell Sci 110:1091–1098

Ploplis VA, Carmeliet P, Vazirzadeh S, Van Vlaenderen I, Moons L, Plow EF, Collen D (1995) Effects of disruption of the plasminogen gene on thrombosis, growth, and health in mice. Circulation 92:2585–2593

Plow EF, Herren T, Redlitz A, Miles LA, Hoover-Plow JL (1995) The cell biology of the plasminogen system. FASEB J 9:939–945

Potempa J, Korzus E, Travis J (1994) The serpin superfamily of proteinase inhibitors: structure, function and regulation. J Biol Chem 269:15957–15960

Pralong G, Calandra T, Glauser MP, Schellekens J, Verhoef J, Bachmann F, Kruithof EKO (1989) Plasminogen activator inhibitor 1: a new prognostic marker in septic shock. Thromb Haemost 61:459–462

Presta M, Ennas MG, Torelli S, Ragnotti G, Gremo F (1990) Synthesis of urokinase-type plasminogen activator and of type-1 plasminogen activator inhibitor in neuronal cultures of human fetal brain: stimulation by phorbol ester. J Neurochem 55:1647–1654

Racanelli AL, Diemer MJ, Dobies AC, Dubin JR, Reilly TM (1992) Comparison of recombinant plasminogen activator inhibitor-1 and epsilon amino caproic acid in a hemorrhagic rabbit model. Thromb Haemost 67:692–696

Rakoczi I, Wiman B, Collen D (1978) On the biological significance of the specific interaction between fibrin, plasminogen and antiplasmin. Biochim Biophys Acta 540:295–300

Redlitz A, Tan AK, Eaton DL, Plow EF (1995) Plasma carboxypeptidases as regulators of the fibrinolytic system. J Clin Invest 96:2534–2538

Redlitz A, Nicolini FA, Malycky JL, Topol EJ, Plow EF (1996) Inducible carboxypeptidase activity. A role in clot lysis in vivo. Circulation 93:1328–1330

Reilly TM, Mousa SA, Seetharam R, Racanelli AL (1994) Recombinant plasminogen activator inhibitor type 1: a review of structural, functional, and biological aspects. Blood Coagul Fibrinolysis 5:73–81

Reith A, Booth NA, Moore NR, Cruickshank DJ, Bennett B (1993) Plasminogen activator inhibitors (PAI-1 and PAI-2) in normal pregnancies, pre-eclampsia and hydatidiform mole. Br J Obstet Gynaecol 100:370–374

Ridker PM, Hennekens CH, Lindpaintner K, Stampfer MJ, Miletich JP (1997) Arterial and venous thrombosis is not associated with the 4G/5G polymorphism in the promoter of the plasminogen activator inhibitor gene in a large cohort of US men. Circulation 95:59–62

Rijken DC, Juhan-Vague I, Collen D (1983) Complexes between tissue-type plasminogen activator and proteinase inhibitors in human plasma, identified with an immunoradiometric assay. J Lab Clin Med 101:285–294

Risse BC, Chung NM, Baker MS, Jensen PJ (2000) Evidence for intracellular cleavage of plasminogen activator inhibitor type 2 (PAI-2) in normal epidermal keratinocytes. J Cell Physiol 182:281–289

Ritchie H, Booth NA (1998) The distribution of the secreted and intracellular forms of plasminogen activator inhibitor 2 (PAI-2) in human peripheral blood monocytes is modulated by serum. Thromb Haemost 79:813–817

Saito H (1988) Alpha 2-plasmin inhibitor and its deficiency states. J Lab Clin Med 112:671–678

Sakharov DV, Plow EF, Rijken DC (1997) On the mechanism of the antifibrinolytic activity of plasma carboxypeptidase B. J Biol Chem 272:14477–14482

Sancho E, Declerck PJ, Price NC, Kelly SM, Booth NA (1995) Conformational studies on plasminogen activator inhibitor (PAI-1) in active, latent, substrate, and cleaved forms. Biochemistry 34:1064–1069

Santell L, Levin EG (1988) Cyclic AMP potentiates phorbol ester stimulation of tissue plasminogen activator release and inhibits secretion of plasminogen activator inhibitor-1 from human endothelial cells. J Biol Chem 263:16802–16808

Sawdey M, Podor TJ, Loskutoff DJ (1989) Regulation of type 1 plasminogen activator inhibitor gene expression in cultured bovine aortic endothelial cells. Induction by transforming growth factor-β, lipopolysaccharide, and tumor necrosis factor-α. J Biol Chem 264:10396–10401

Scanu AM (1992) Genetic basis and pathophysiological implications of high plasma Lp(a) levels. J Intern Med 231:679–683

Scanu AM, Pfaffinger D, Lee JC, Hinman J (1994) A single point mutation (Trp72-Arg) in human apo(a) kringle 4–37 associated with a lysine binding defect in Lp(a). Biochim Biophys Acta 1227:41–45

Schjetlein R, Haugen G, Wisloff F (1997) Markers of intravascular coagulation and fibrinolysis in preeclampsia: association with intrauterine growth retardation. Acta Obstet Gynecol Scand 76:541–546

Schatteman KA, Goossens FJ, Scharpé SS, Hendriks DF (2000) Proteolytic activation of purified human procarboxypeptidase U. Clin Chim Acta 292:25–40

Scherrer A, Kruithof EKO, Grob JP (1991) Plasminogen activator inhibitor-2 in patients with monocytic leukemia. Leukemia 5:479–486

Schneider SS, Schick C; Fish KE; Miller E; Pena JC; Treter SD; Hui SM; Silverman GA (1995) A serine proteinase inhibitor locus at 18q21.3 contains a tandem duplication of the human squamous cell carcinoma antigen gene. Proc Natl Acad Sci USA 92:3147–3151

Schuster WA, Medcalf RL, Kruithof EKO (1994) Localization and characterization of a retinoic acid response-like element in the plasminogen activator inhibitor-2 gene promoter. Fibrinolysis 8:113–119

Sharp AM, Stein PE, Pannu NS, Carrell RW, Berkenpas MB, Ginsburg D, Lawrence
DA, Read RJ (1999) The active conformation of plasminogen activator inhibitor
1, a target for drugs to control fibrinolysis and cell adhesion. Structure 7: 111–118
Silverman GA, Bartuski AJ, Cataltepe S, Gornstein ER, Kamachi Y, Schick C, Uemura
Y (1998) SCCA1 and SCCA2 are proteinase inhibitors that map to the serpin
cluster at 18q21.3. Tumour Biol 19:480–487
Simon DI, Fless GM, Scanu AM, Loscalzo J (1991) Tissue-type plasminogen activator
binds to and is inhibited by surface-bound lipoprotein(a) and low-density lipoprotein. Biochemistry 30:6671–6677
Smith EB, Crosbie L (1991) Does lipoprotein(a) [Lp(a)] compete with plasminogen in
human atherosclerotic lesions and thrombi? Atherosclerosis 89:127–136
Spencer-Green G (1994) Retinoic acid effects on endothelial cell function: interaction
with interleukin 1. Clin Immunol Immunopathol 72:53–61
Stahl A, Mueller BM (1997) Melanoma cell migration on vitronectin: regulation by
components of the plasminogen activation system. Int J Cancer 71:116–122
Stefansson S, Lawrence DA (1996) The serpin PAI-1 inhibits cell migration by blocking integrin $\alpha_v\beta_3$ binding to vitronectin. Nature 383:441–443
Steven J, Cottingham IR, Berry SJ, Chinery SA, Goodey AR, Courtney M, Ballance
DJ (1991) Purification and characterisation of plasminogen activator inhibitor 2
produced in Saccharomyces cerevisiae. Eur J Biochem 196:431–438
Stringer HA, van Swieten P, Heijnen HF, Sixma JJ, Pannekoek H (1994) Plasminogen
activator inhibitor-1 released from activated platelets plays a key role in thrombolysis resistance. Studies with thrombi generated in the Chandler loop. Arterioscler Thromb 14:1452–1458
Suffredini AF, Harpel PC, Parrillo JE (1989) Promotion and subsequent inhibition of
plasminogen activation after administration of intravenous endotoxin to normal
subjects. N Engl J Med 320:1165–1172
Sugiyama N, Sasaki T, Iwamoto M, Abiko Y (1988) Binding site of alpha 2-plasmin
inhibitor to plasminogen. Biochim Biophys Acta 952:1–7
Thalacker FW, Nilsen-Hamilton M (1992) Opposite and independent actions of cyclic
AMP and transforming growth factor β in the regulation of type 1 plasminogen
activator inhibitor expression. Biochem J 287:855–862
Thompson SG, Kienast J, Pyke SD, Haverkate F, van de Loo JC (1995) Hemostatic
factors and the risk of myocardial infarction or sudden death in patients with
angina pectoris. European Concerted Action on Thrombosis and Disabilities
Angina Pectoris Study Group. N Engl J Med 332:635–641
Thorsen S, Philips M, Selmer J, Lecander I, Åstedt B (1988) Kinetics of inhibition of
tissue-type and urokinase-type plasminogen activator by plasminogen-activator
inhibitor type 1 and type 2. Eur J Biochem 175:33–39
Tone M, Kikuno R, Kume-Iwaki A, Hashimoto-Gotoh T (1987) Structure of human
alpha 2-plasmin inhibitor deduced from the cDNA sequence. J Biochem (Tokyo)
102:1033–1041
Urano T, Kojima Y, Takahashi M, Serizawa K, Sakakibara K, Takada Y, Takada A (1993)
Impaired fibrinolysis in hypertension and obesity due to high plasminogen activator inhibitor-1 level in plasma. Jpn J Physiol 43:221–228
Urano T, Ihara H, Takada Y, Nagai N, Takada A (1996) The inhibition of human factor
Xa by plasminogen activator inhibitor type 1 in the presence of calcium ion, and
its enhancement by heparin and vitronectin. Biochim Biophys Acta 1298:199–208
Van Hinsbergh VW, Kooistra T, van den Berg EA, Princen HM, Fiers W, Emeis JJ
(1988) Tumor necrosis factor increases the production of plasminogen activator
inhibitor in human endothelial cells in vitro and in rats in vivo. Blood 72:1467–1473
Van Meijer M, Pannekoek H (1995) Structure of plasminogen activator Inhibitor 1
(PAI-1) and its function in fibrinolysis: an update. Fibrinolysis 9:263–276
Van Meijer M, Smilde A, Tans G, Nesheim ME, Pannekoek H, Horrevoets AJ (1997)
The suicide substrate reaction between plasminogen activator inhibitor 1 and
thrombin is regulated by the cofactors vitronectin and heparin. Blood 90:
1874–1882

Van Zonneveld AJ, Curriden SA, Loskutoff DJ (1988) Type 1 plasminogen activator inhibitor gene: functional analysis and glucocorticoid regulation of its promoter. Proc Natl Acad Sci USA 85:5525–5529

Vassalli JD, Dayer JM, Wohlwend A, Belin D (1984) Concomitant secretion of prourokinase and of a plasminogen activator-specific inhibitor by cultured human monocytes- macrophages. J Exp Med 159:1653–1668

Vaughan DE, De Clerck PJ, De Mol M, Collen D (1989) Recombinant plasminogen activator inhibitor-1 reverses the bleeding tendency associated with the combined administration of tissue-type plasminogen activator and aspirin in rabbits. J Clin Invest 84:586–591

Vervloet MG, Thijs LG, Hack CE (1998) Derangements of coagulation and fibrinolysis in critically ill patients with sepsis and septic shock. Semin Thromb Hemost 24:33–44

Von Heijne G, Liljeström P, Mikus P, Andersson H, Ny T (1991) The efficiency of the uncleaved secretion signal in the plasminogen activator inhibitor type 2 protein can be enhanced by point mutations that increase its hydrophobicity. J Biol Chem 266:15240–15243

Waltz DA, Natkin LR, Fujita RM, Wei Y, Chapman HA (1997) Plasmin and plasminogen activator inhibitor type 1 promote cellular motility by regulating the interaction between the urokinase receptor and vitronectin. J Clin Invest 100: 58–67

Westerhausen DR Jr, Hopkins WE, Billadello JJ (1991) Multiple transforming growth factor-β-inducible elements regulate expression of the plasminogen activator inhibitor type-1 gene in Hep G2 cells. J Biol Chem 266:1092–1100

Whisstock J, Skinner R, Lesk AM (1998) An atlas of serpin conformations. Trends Biochem Sci 23:63–67

Wohlwend A, Belin D, Vassalli JD (1987) Plasminogen activator-specific inhibitors produced by human monocytes/macrophages. J Exp Med 165:320–339

Xue Y, Björquist P, Inghardt T, Linschoten M, Musil D, Sjölin L, Deinum J (1998) Interfering with the inhibitory mechanism of serpins: crystal structure of a complex formed between cleaved plasminogen activator inhibitor type 1 and a reactive-centre loop peptide. Structure 6:627–636

Ye RD, Wun TC, Sadler JE (1988) Mammalian protein secretion without signal peptide removal. Biosynthesis of plasminogen activator inhibitor-2 in U-937 cells. J Biol Chem 263:4869–4875

Ye S, Green FR, Scarabin PY, Nicaud V, Bara L, Dawson SJ, Humphries SE, Evans A, Luc G, Cambou JP, Arveiler D, Henney AM, Cambien F (1995) The 4G/5G genetic polymorphism in the promoter of the plasminogen activator inhibitor-1 (PAI-1) gene is associated with differences in plasma PAI-1 activity but not with risk of myocardial infarction in the ECTIM study. Etude CasTemoins de I'nfarctus du Mycocarde. Thromb Haemost 74:837–841

Zelezna B, Rydzewski B, Lu D, Olson JA, Shiverick KT, Tang W, Sumners C, Raizada MK (1992) Angiotensin-II induction of plasminogen activator inhibitor-1 gene expression in astroglial cells of normotensive and spontaneously hypertensive rat brain. Mol Endocrinol 6:2009–2017

Zhu Y, Carmeliet P, Fay WP (1999) Plasminogen activator inhibitor-1 is a major determinant of arterial thrombolysis resistance. Circulation 99:3050–3055

Zöller B, García de Frutos P, Dahlbäck B (1998) A common 4G allele in the promoter of the plasminogen activator inhibitor-1 (PAI-1) gene as a risk factor for pulmonary embolism and arterial thrombosis in hereditary protein S deficiency. Thromb Haemost 79:802–807

CHAPTER 6
Assembly of the Plasminogen System on Cell Surfaces

E.F. PLOW, A. REDLITZ, S.B. HAWLEY, S. XUE, T. HERREN, J.L. HOOVER-PLOW, and L.A. MILES

A. Introduction

The contribution of the plasminogen system to fibrinolysis is subject to sophisticated and multifaceted regulatory mechanisms. Plasminogen activators, inhibitors of these activators and of plasmin, play critical roles in the initiation, propagation, and control of fibrinolysis. Additionally, the fibrin surface, per se, is an important regulator of its own degradation by providing specific binding sites for several components of the plasminogen system (plasminogen, tPA, α_2-antiplasmin). Aberrations in any of these regulatory mechanisms can upset the delicate balance between thrombus formation and dissolution, resulting in thrombotic or bleeding tendencies.

In addition to its central role in fibrinolysis, the plasminogen system has also been implicated in a broad range of physiologic and pathophysiologic events. As selected examples, physiological roles for the plasminogen system have been proposed in reproduction and development (STRICKLAND et al. 1976; JIA et al. 1990; REICH et al. 1985; MENOUD et al. 1989; SAPPINO et al. 1989) and in host defense mechanisms (KIRCHHEIMER and REMOLD 1989b; Lo et al. 1989); pathophysiological roles have been demonstrated in tumor cell invasion and metastasis (GOLDFARB et al. 1986; HOLLAS et al. 1992; OSSOWSKI et al. 1991b; STAHL and MUELLER 1994; KIM et al. 1998), bacterial infections (PARKKINEN and RAUVALA 1991; BERGE and SJÖBRING 1993; KUUSELA et al. 1992; COLEMAN et al. 1995; LOTTENBERG et al. 1992; LOTTENBERG 1997; SHAPIRO et al. 1997) and intimal hyperplasia (JACKSON and REIDY 1992; HERBERT and CARMELIET 1997; CARMELIET et al. 1997, 1998). These diverse functions can all be ascribed to the capacity of the plasminogen system to facilitate cell migration by assisting in cellular penetration of protein barriers, including fibrin barriers.

Studies reported in the 1980s began to establish a molecular mechanism for the role of the plasminogen system in cell migration; namely, the existence of specific binding sites for components of the system on cell surfaces (MILES and PLOW 1985; VASSALLI et al. 1985; BEEBE 1987; HAJJAR et al. 1987; PLOW et al. 1995; HAJJAR 1995). We now know that many of the principles ascribed to the interaction of the plasminogen system with the fibrin surface apply to cell surfaces. Specifically, plasmin(ogen) and plasminogen activators bind to both surfaces; recognition of both surfaces is mediated by similar structural motifs in plasmin(ogen) and its activators; plasminogen activation is enhanced as a consequence of surface binding; surface binding leads to directed substrate recognition and degradation; and the proteolytic activity of surface-bound plasmin is sustained in an inhibitor-rich environment.

This chapter considers cellular recognition and assembly of components of the plasminogen system. Current information regarding the interactions of the individual components of the system with cells is summarized and then the functional consequences of assembly of these components on cell-surfaces is discussed.

B. Plasminogen Receptors

I. General Characteristics

The defining features of plasminogen receptors are their low affinity, their high density, and their ubiquitous distribution. The binding of Glu-plasminogen, the native form of the molecule, to cells is a saturable process, and plasminogen binding isotherms provide evidence for a single class of sites with respect to affinity (e.g., MILES and PLOW 1985, 1987; HAJJAR et al. 1986; GONIAS et al. 1989; GONZALEZ-GRONOW et al. 1989) with dissociation constants generally falling within the 100–2000 nmol/l range (reviewed by MILES and PLOW 1987; PLOW et al. 1995). Since blood concentrations of plasminogen are 2 μmol/l, significant receptor occupancy is predicted and has been experimentally verified on blood cells (MILES et al. 1988a; ADELMAN et al. 1988) and tissue-fixed cells (BURTIN et al. 1987; CORREC et al. 1992). Moreover, certain naturally occurring derivatives of plasminogen may exhibit higher affinities for cells although their affinities still remain modest (BAUER et al. 1984; GONZALEZ-GRONOW et al. 1989; SILVERSTEIN et al. 1988). Since cell-bound plasmin is protected from inhibitors, its low affinity for cells may be essential to limit excessive proteolysis – as plasmin dissociates from the cell, it becomes rapidly neutralized by inhibitors.

The plasminogen binding capacity of cells can be extraordinarily high. As examples, platelets may bind ~40 000 plasminogen molecules per cell, equal to the number of their most abundant membrane glycoprotein; and endothelial cells may bind several million plasminogen molecules (HAJJAR et al. 1986; MILES et al. 1988b; GANZ et al. 1991). Indeed, many cells have plasminogen binding capacities in the 10^6–10^7 range (Table 1).

A survey of cells with demonstrated plasminogen binding potential is provided in Table 1. With the exception of erythrocytes, all blood cells bind plasminogen (MILES and PLOW 1987). Many tumor cell lines have high plasminogen binding capacities (BURTIN and FONDANÈCHE 1988; CORREC et al. 1990, 1992). A growing body of evidence links the plasminogen system with brain function (CHEN et al. 1997; HUANG et al. 1996; WANG et al. 1998), and neurons (NAKAJIMA et al. 1993) and microglial cells (NAKAJIMA et al. 1994) bind

Table 1. The cellular distribution and number of plasminogen receptors

Cell Type	Plasminogen Receptors Sites $(\times 10^5)$/Cell
Granulocytes	1.6
Platelets	0.37
Monocytes	4.4
Monocytoid cell lines (U937, THP-1)	160–4200
Lymphocytes	4.9
Leukemic cells (WEHI-3B, RPMI-1788)	4–190
Fibroblasts (GM1380)	310
Fibrosarcoma cells (HT-1080)	6
Epithelial (Glomerular) Cells	60
Osteosarcoma cells (MG-63)	75
Colonic carcinoma (SW1116)	0.15
Glioma cells (C6)	36
Endothelial cells	1.4–210

plasminogen. Thus, a permissive affinity, a high density of binding sites, and an extremely broad cellular distribution are consistent with diverse roles of plasminogen receptors in cellular responses.

II. Nature of Plasminogen Receptors

The high density and broad distribution implies that plasminogen receptors are heterogeneous in nature. Nevertheless, these receptors behave as a single class of binding sites with respect to affinity because the lysine binding sites (LBS) of plasminogen mediate the interactions (reviewed in MILES and PLOW 1987; REDLITZ and PLOW 1995). The LBS, resident within the kringles of the heavy chain region of plasminogen, preferentially recognize carboxy-terminal lysines. Of the five kringles within plasminogen, the LBS associated with kringle 1 and kringle 5 have been implicated in cellular recognition (MILES et al. 1988a). Thus, cell-surface proteins with carboxy-terminal lysines may function as plasminogen receptors. Accordingly, plasminogen will ligand blot several membrane proteins (MILES et al. 1991); each interaction can be blocked by carboxy-terminal lysine analogs (MILES et al. 1988a), which react with the LBS, and many can be eliminated by treatment of the cells with carboxypeptidase B (FELEZ et al. 1993; REDLITZ et al. 1994, 1995), which removes carboxy-terminal lysines. In addition, other proteins with LBS function, such as apoprotein(a) of lipoprotein(a), can recognize these binding sites and interfere with plasminogen binding to cells (MILES et al. 1989b, 1995a,c; HAJJAR et al. 1989; GONZALEZ-GRONOW et al. 1989; DUDANI et al. 1996). Such competition or shared recognition specificity may contribute to the pathogenesis of elevated levels of lipoprotein(a) (MILES and PLOW 1990; PLOW et al. 1995).

Table 2 lists some of the identified plasminogen receptors. Four representatives of the category of plasminogen receptors with carboxy-terminal lysines

Table 2. Candidate plasminogen receptors

Receptor	Sources[a]
Enolase-related proteins	Leukocytes, tumor cells, neurons
Annexin II	Endothelial cells
Amphoterin	Brain, transformed cells
GPIIb-IIIa	Platelets
GPIIb-IIIa-related protein	Fibroblast (arthritis)
Cytokeratin-8	Hepatocytes
Gangliosides	Monocytoid cells
PAM	Bacteria
Glutaraldehyde-6-phosphase dehydrogenase (related)	Bacteria
Actin	Endothelial cells
Complement C9	Endothelial cells

[a] First or independent identification but may be found in multiple cell types.

are α-enolase (Miles et al. 1991; Redlitz et al. 1994; Andronicos et al. 1997; Pancholi et al. 1998), annexin II (Cesarman et al. 1994), a cyto-keratin 8-like protein (Hembrough et al. 1995), and complement C9 (Christiansen et al. 1997). Enolase, cytokeratin 8, and complement C9 possess a naturally occurring carboxy-terminal lysine, whereas proteolysis of the Lys^{307}-Arg^{308} bond is required for the tetramer annexin II (Kassam et al. 1998) to function as a plasminogen receptor (Schwartz-Albiez et al. 1992; Cesarman et al. 1994). Glutaraldehyde 6-phosphate dehydrogenase, which was isolated as a high affinity plasmin receptor from bacteria (Lottenberg et al. 1992), also has a carboxy-terminal lysine. The binding of plasminogen to actin (Dudani et al. 1996) and to the surface of many bacteria (reviewed by Boyle and Lottenberg 1997) is also inhibited by lysine, suggesting an interaction with C-terminal lysine. Amphoterin, which was isolated as a high affinity plasminogen receptor from brain, appears to represent a receptor without a carboxy-terminal lysine (Parkkinen et al. 1993; Parkkinen and Rauvala 1991). The plasminogen binding protein of the group A streptococci, PAM (Berge and Sjöbring 1993; Ringdahl et al. 1998), also falls into this category. It is uncertain whether carboxy-terminal lysines are involved in the binding of plasminogen to GPIIb-IIIa from platelets (Miles et al. 1986), a GPIIb-IIIa-related molecule from fibroblasts (Gonzalez-Gronow et al. 1994), or cytok-eratin-8 from hepatocytes (Hembrough et al. 1995). In addition to these proteins, gangliosides (Miles et al. 1989a) and glycosaminoglycans (Andrade-Gordon and Strickland 1986) bind plasminogen and may contribute to the plasminogen binding capacity of cells. Receptor heterogeneity raises the possibility of functional differences among plasminogen receptors. For example, specific subsets of receptors may be particularly effective in supporting plasminogen activation (Felez et al. 1996) while others may elicit specific cellular responses (Gonzalez-Gronow et al. 1994; Syrovets et al. 1997).

III. Regulation of Plasminogen Receptors

The capacity of cells to modulate their plasminogen binding provides an important mechanism for regulation of the plasminogen system. Indeed, plasminogen receptor expression is highly malleable, and the responses are cell-type specific. Thrombin can cause either an up- or down-regulation of plasminogen receptor expression depending upon the cell type (MILES and PLOW 1985; MILES et al. 1988b). Glucocorticoid hormones, such as dexamethasone, down-regulate the binding of plasminogen to human fibrosarcoma cells by decreasing the affinity of the cells for ligand (PÖLLÄNEN 1989). On the other hand, glucocorticoids do not influence the binding of plasminogen to U937 monocytoid cells (FELEZ et al. 1990).

Upon treatment of U937 cells with PMA, adherent and nonadherent subpopulations of cells can be separated. Plasminogen receptor expression is down-regulated in the adherent and up-regulated in the nonadherent cells (FELEZ et al. 1990). These changes are dependent upon differences in the number but not the affinity of their plasminogen binding sites. In human peripheral blood monocytes, PMA-stimulation also up-regulates the number of plasminogen receptors, but does so in both the adherent and nonadherent cell populations. As with PMA, stimulation of U937 cells with 1,25-dihydroxyvitamin D also supports development of an adherent and a nonadherent cell population, the nonadherent cells expressing four times more plasminogen receptors (FELEZ et al. 1990). An additional increment in the plasminogen receptor expression is observed if 1,25-dihydroxyvitamin D is combined with interferon-γ (LU et al. 1993). In addition, increased plasminogen binding is associated with metastatic cancer cells (RANSON et al. 1998).

The observations described above indicate a close relationship between plasminogen receptor expression and cell adhesion. Indeed, cell adhesion can directly influence plasminogen binding capacity (KIM et al. 1996). Upon adhesion of monocytoid cells to fibronectin, vitronectin, or laminin substrata, the plasminogen binding capacity of the nonadherent cells increases, and binding to the adherent population decreases. If the nonadherent cells are separated and presented with additional substratum, they will adhere and will decrease their plasminogen receptor expression. Thus, when cells are in a migratory state (nonadherent), their plasminogen binding capacity is increased to accelerate plasmin generation and facilitate movement. When the cells become adherent, their plasminogen binding and plasmin generating capacities decrease to minimize injurious proteolysis. Since adherence and apoptosis are linked for many cell types, the relationship between plasminogen receptor expression and apoptosis is particularly interesting.

As up-regulation of plasminogen binding capacity can occur rapidly and does not require new protein synthesis (FELEZ et al. 1990), post-translational modifications, such as proteolysis, may be involved in receptor modulation. Plasmin has the capacity to up-regulate plasminogen binding (Fig. 1A) (CAMACHO et al. 1989; GONZALEZ-GRONOW et al. 1991). This increase depends

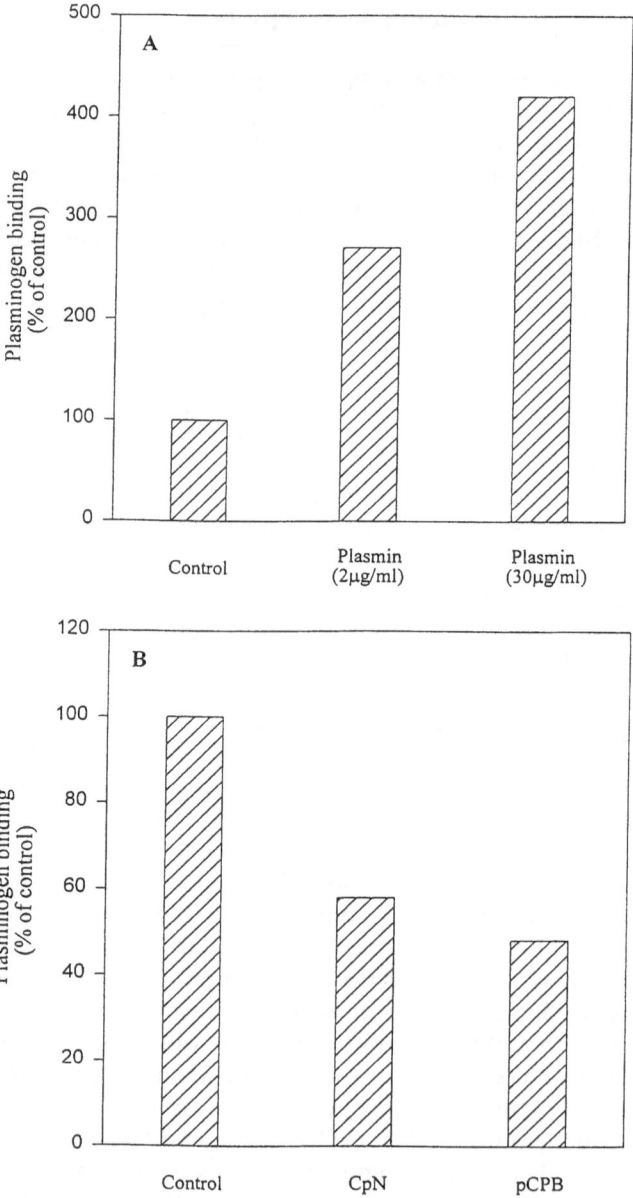

Fig. 1A–C. Proteolytic regulation of plasminogen receptor expression. **A** Data for an amplification loop in which plasmin treatment of U937 monocytoid cells is shown to proteolytically generate additional carboxy-terminal lysines which can serve as additional plasminogen binding sites. **B** The two major plasma carboxypeptidases, CpN and pCPB, are shown to suppress plasminogen binding to U937 cells by removing carboxy-terminal lysines from cell-surface proteins. Data are adapted from REDLITZ et al. (1995). **C** Model depicting the proteolytic mechanisms for up- and down-regulation of plasminogen binding to cells by plasmin and the carboxypeptidases, respectively

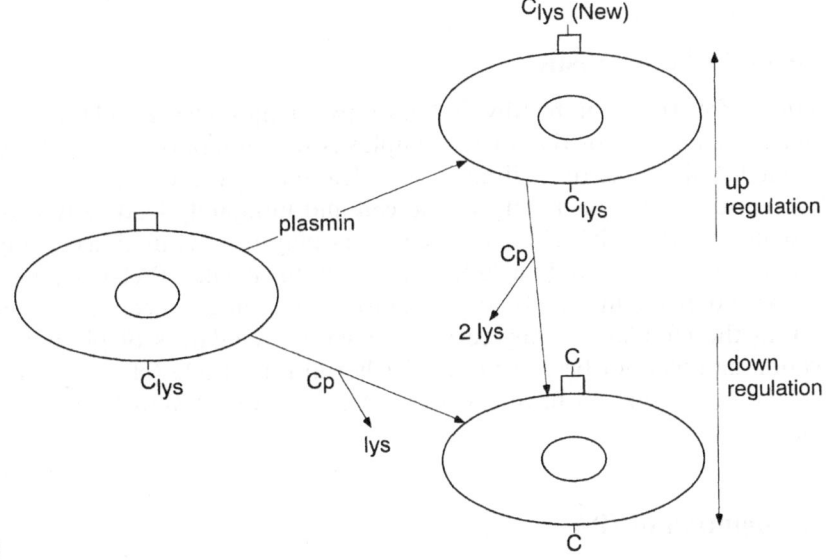

Fig. 1A–C. *Continued*

upon the generation of new carboxy-terminal lysine residues which can bind plasminogen. At the same time, proteolysis can suppress plasminogen binding to cells. Basic carboxypeptidases, which remove carboxy-terminal lysines, are present in plasma (SAKHAROV et al. 1997; SKIDGEL 1988; EATON et al. 1991). The two major carboxypeptidases of plasma are CpN and pCPB [which is likely to be the same as CpU (WANG et al. 1994a), and more recently has been designated as thrombin-activable fibrinolysis inhibitor (TAFI) (BOFFA et al. 1998)]. CpN is constitutively active (SKIDGEL 1995), whereas pCPB circulates as a proenzyme in complex with plasminogen (EATON et al. 1991). Both CpN and pCPB can suppress plasminogen binding to cells (REDLITZ et al. 1995) (Fig. 1B). TAFI is activated in vivo in dogs in whom an experimental thrombosis of the coronary artery had been produced it and reduced the lysis of this thrombus by tPA (REDLITZ et al. 1996). Taken together, these proteolytic events establish a mechanism for physiological up- and down-regulation of plasminogen (and tPA) receptor expression (Fig. 1C). When peripheral blood monocytes are cultured, they can up-regulate their binding sites for plasminogen by a factor of 100 (FELEZ et al. 1990). This observation suggests that plasminogen receptor expression may be maintained at a low level on blood cells by plasma carboxypeptidases. Explosive up-regulation may occur either when the dampening effects of carboxypeptidases are overcome and/or when the cell surface is subjected to proteolysis by plasmin or other proteinases.

C. tPA Receptors

I. General Characteristics

Receptors for tPA can be divided into two major classes: (1) receptors involved in clearance of tPA or its complexes with inhibitors; and (2) receptors that localize tPA on cell surfaces. Clearance receptors are associated with the endocytotic machinery of the cell and ultimately lead to lysosomal degradation of tPA. tPA clearance receptors may be classified according to recognition specificity and include carbohydrate-specific, PAI-1-dependent, and PAI-independent clearance receptors. As these receptors remove tPA from the circulation, they serve as negative regulators of plasminogen activation and will not be considered further in this article (cf. Chap. 3). The tPA clearance receptors have recently been reviewed (Redlitz and Plow 1995).

II. Recognition of tPA

In addition to the clearance receptors, binding sites have been identified which localize tPA to the cell surface and promote pericellular proteolysis. The amino-terminal domains of tPA, which are not required for catalytic activity, contain recognition sites for such tPA receptors. The amino-terminal finger domain, homologous to the fingers of fibronectin, has been implicated in the recognition of tPA by cells (Beebe et al. 1989). As both plasminogen and tPA have kringles with LBS function, these ligands will compete with each other for binding to certain cells (Felez et al. 1993).

III. Cellular Distribution of tPA Receptors

Like plasminogen receptors, tPA receptors are heterogeneous; unlike plasminogen receptors, the heterogeneity of tPA receptors leads to distinctive properties on different cell types. Thus, it is easier to consider the various tPA receptors from the standpoint of specific cell types.

1. Endothelial Cells

Numerous studies have shown that endothelial cells express binding sites for tPA. PAI-1, synthesized by endothelial cells, can associate with the cell surface (Sakata et al. 1988) or subcellular matrix (Schleef et al. 1990) and can mediate tPA binding at these sites. Several studies also have documented PAI-1 independent mechanisms of tPA binding (Beebe et al. 1989; Hajjar and Hamel 1990; Sanzo et al. 1990). Annexin II was isolated as a tPA binding protein from endothelial cells (Hajjar 1991; Hajjar et al. 1994). Overexpression of annexin II in renal epithelial cells increases their ability to bind tPA, while antisense suppression of annexin II-expression in endothelial cells reduced tPA binding (HAJJAR et al. 1994). As noted above, annexin II also binds plasminogen but the sites of plasminogen and tPA binding are different. Whilst binding of plasminogen to annexin II requires a posttranslational hydrolysis, generating the C-terminal Lys^{307}, tPA binding is direct. It has been localized to the N-terminal region (amino acids 8–13). The C9G mutant of annexin II lost its affinity for tPA, and the addition of homocysteine to native annexin resulted in an annexin C9-homocysteine adduct and complete loss of binding to tPA (HAJJAR et al. 1998). These results may explain, at least in part, the thrombogenic effect of homocysteinemia. Other candidate tPA receptors from endothelial cells are less well defined. Tubulin (BEEBE et al. 1990) and a 20 kDa (FUKAO et al. 1992) protein have also been described as tPA binding proteins.

2. Circulating Blood Cells

Specific binding of tPA to platelets (VAUGHAN et al. 1989) and leukocytes (FELEZ et al. 1991) has been reported. As these cells also bind plasminogen, they can assemble a functional plasminogen activation system on their surfaces. The receptor for tPA on platelets is unknown. tPA appears to share binding sites with plasminogen on monocytes, including proteins with carboxy-terminal lysine residues (FELEZ et al. 1993).

3. Cells of Neuronal Origin

Studies on granule neurons from mouse cerebellum (VERRALL and SEEDS 1989) and human sympathetic neurons (PITTMAN et al. 1989) suggest that tPA plays a role in neurite outgrowth. tPA binds to these cells but with different affinities: 50 pmol/l for granule (VERRALL and SEEDS 1989) and 23 nmol/l for sympathetic neurons (PITTMAN et al. 1989). Amphoterin was isolated as a tPA binding protein from developing neurites in rat brain (PARKKINEN and RAUVALA 1991), and anti-amphoterin antibodies inhibit neurite outgrowth (MERENMIES et al. 1991). Amphoterin also is present in many transformed cells and appears to be concentrated at the leading edge of motile cells, consistent with a role in cell migration (PARKKINEN et al. 1993).

4. Vascular Smooth Muscle Cells

Using a functional assay, ELLIS and WHAWELL (1997) demonstrated two classes of binding sites for tPA with Kds of 25 nmol/l and 3000 nmol/l on vascular smooth muscle cells. tPA binding to the higher affinity site resulted in a 90-fold enhancement of the activation of cell bound plasminogen.

5. Tumor Cells

Binding of tPA to melanoma cells was shown to be mediated via the kringle-2 domain of tPA (BIZIK et al. 1997).

D. Urokinase Receptors

I. General Characteristics

While the receptors for plasminogen and t-PA are characterized by their het-erogeneity, the binding of urokinase-type plasminogen activator (uPA)to cells is mediated predominantly by a single entity, uPAR. This section is organized to consider the structure and function of uPAR (CD87) and then briefly con-siders other uPA receptors.

II. The Urokinase Receptor, uPAR

1. Cellular Distribution of uPAR

uPAR is widely distributed. It is present on normal circulating blood cells (Miles and Plow 1987; reviewed by Plesner et al. 1997), megacaryoblasts (Wohn et al. 1997), endothelial cells (Miles et al. 1988b; Barnathan et al. 1990) and tissue resident cells including alveolar macrophages (Chapman et al. 1990), mast cells (Sillaber et al. 1997), smooth muscle cells (Reuning and Bang 1992), myogenic cells (Quax et al. 1992) and keratinocytes (Del Rosso et al. 1990). Many tumors and tumor cell lines also express uPAR (Needham et al. 1988; Cohen et al. 1991; Hollas et al. 1991; Schlechte et al. 1989; Ossowski et al. 1991a). The number of receptors typically ranges from 50,000 to 200000 sites per cell. The interaction of uPA with cells is of high affinity, with the Kd values generally falling within the 0.1 to 2 nmol/l range (Vassalli et al. 1985). To date, all uPAR-expressing cells that have been examined can also bind plasminogen but not all plasminogen binding cells express uPAR (Jardi et al. 1993). For example, among circulating blood cells, monocytes and neutrophils bind uPA (Miles and Plow 1987) while unstimulated lymphocyte populations (Nykjaer et al. 1994; Jardi et al. 1996) do not.

2. Recognition Specificity of uPAR

uPA is produced by a single proteolytic cleavage of its zymogen, single chain uPA (sc-uPA) (55kDa), resulting in a two-chain disulfide bonded molecule. This conversion can be mediated by a variety of enzymes (see Chap. 4). Both sc-uPA and its amino terminal fragment (ATF), consisting of amino acid residues 1–135, interact with the receptor with similar affinity as uPA (Cubellis et al. 1986; Stopelli et al. 1985). The low molecular weight form of uPA (33kDa), which is composed of the B chain and a small carboxyl termi-nal region of the A chain and still exhibits proteolytic activity, does not inter-act with uPAR (Vassali et al. 1985; Plow et al. 1986; Bajpai and Baker 1985). Therefore, the proteolytic activity of uPA is not required for the interaction with uPAR and the conversion of high molecular weight to low molecular weight uPA may regulate receptor occupancy. The region within uPA that interacts with cells has been further localized to the growth factor domain within the A chain. A linear peptide corresponding to amino acid residues

12–32, contained within this growth factor region, competes for binding of the uPA ligand with its receptor (APPELLA et al. 1987) with relatively low affinity, whereas a cyclic peptide comprising amino acids 19–31 competed with higher affinity the binding of uPA to uPAR (BÜRGLE et al. 1997). The ligand-receptor interaction exhibits species specificity (APPELLA et al. 1987; ESTRE-ICHER et al. 1989) which are due to intraspecific differences in both uPA and uPAR (QUAX et al. 1998).

3. The uPAR Gene

The cDNA sequence for human uPAR encodes a signal peptide of 21 amino acid residues followed by an open reading frame of 313 amino acids (ROLDAN et al. 1990). The human uPAR gene has been localized on chromosome 19q13 and extends over 23kb of genomic DNA (BØRGLUM et al. 1992). It contains seven exons, each protein domain being coded for by a pair of exons. uPAR from murine (KRISTENSEN et al. 1991; SUH et al. 1994), bovine (KRÄTZSCHMAR et al. 1993), and rat sources (RABBANI et al. 1994) have an overall homology of ~60% at the amino acid level. The 5′ flanking region of the human uPAR gene lacks conventional TATA and CAAT boxes but contains a CpG rich element and sequences related to consensus cis-acting elements for the activator protein-1, nuclear factor NF-κB, and specific promoter-1 transcription factors (SORAVIA et al. 1995).

4. The uPAR Protein

uPAR is not a transmembrane receptor, but is linked to the membrane via a glycosylphosphatidylinositol (GPI) anchor (PLOUG et al. 1991) (see Fig. 2). The signal for attachment of the anchor is present in the highly hydrophobic car-boxyl terminal 31 residues of the uPAR. By site directed mutagenesis, the preferred attachment site for GPI attachment is Gly^{283}, but Ser^{282} and Ala^{284} may function as alternate sites (MØLLER et al. 1992). As a result of this GPI attachment, phosphatidylinositol-specific phospholipase C treatment causes release of the uPAR from cells (PLOUG et al. 1992). Some forms of paroxysmal nocturnal hemoglobinuria are associated with a deficiency of GPI-linked proteins on cell surfaces including uPAR (PLOUG et al. 1992). It has been speculated that the deficiency of uPAR may contribute to the thrombotic tendencies in this disease (KLEIN and HARTMANN 1989).

Human uPAR has five potential N-linked glycosylation sites, and at least some of these sites are utilized in the mature protein (ESTREICHER et al. 1989; NIELSEN et al. 1988). Differential glycosylation occurs in different cell types and may account for differences in the affinity of various cell types for uPA (PICONE et al. 1989; PLOUG et al. 1998a). By site directed mutagenesis, replacement of Asn^{52} with Gln to preclude glycosylation at this site decreases affinity of the receptor (MØLLER et al. 1993; PLOUG et al. 1993). Mutation of either four or five of the glycosylation sites prevents efficient cell surface expression of uPAR (MØLLER et al. 1993).

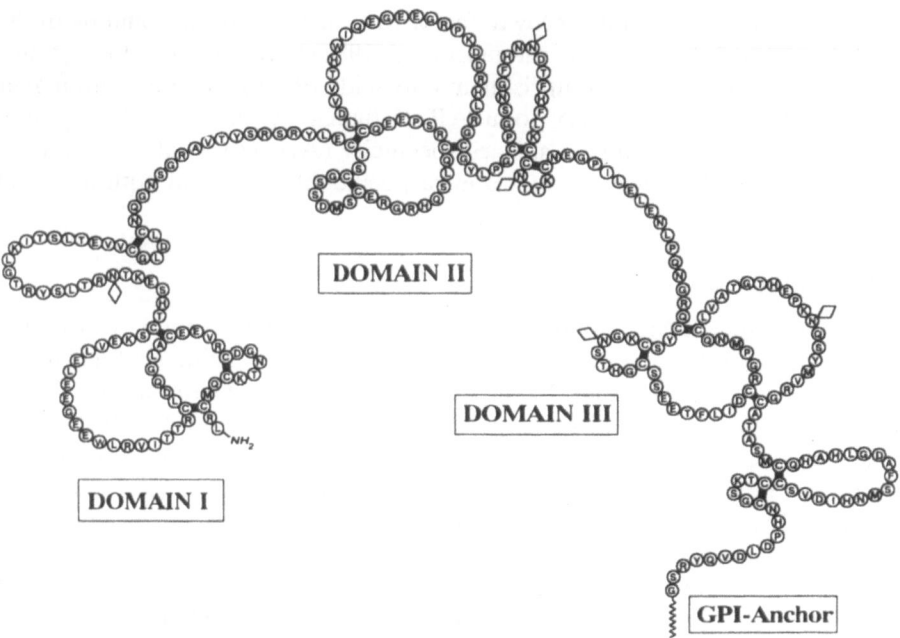

Fig. 2. Depiction of the human urokinase receptor, uPAR. The three domains are linked by inter-domain connectors. The N-linked carbohydrate attachment sites are specified by diamonds. Adapted from Danø et al. (1994)

uPAR is rich in cysteine with 28 present in mature human uPAR. The disulfide bonding pattern has been determined (Ploug et al. 1993) and organizes the molecule into three homologous domains of ~90 amino acids each. Based primarily on the pattern of cysteine residues, the internal sequence homology in each domain is also present in several other proteins. These include the murine Ly-6 antigens, the murine B-cell, and thymocyte antigen ThB, a herpes virus saimiri protein HVS-15, CD59, and MIRL (membrane inhibitor of reactive lysis) (Ploug et al. 1993; Behrendt et al. 1991; Palfree 1991; Albrecht et al. 1992). All of these otherwise unrelated proteins are GPI-linked molecules. In addition, the snake venom α-neurotoxins have a similar disulfide bond structure as uPAR (Ploug and Ellis 1994). Moreover, the gene structures of murine uPAR (Suh et al. 1994) and snake neurotoxin erabutoxin-c (Fuse et al. 1990) are very similar, and the predicted secondary structures of the two proteins contain extensive β-sheet structure (Moestrup et al. 1993; Møller et al. 1992).

The main ligand recognition site within the uPAR has been localized to domain 1 since monoclonal antibodies directed against this region block the interaction with uPA (Rønne et al. 1991). In addition, the isolated domain 1 fragment exhibits ligand binding activity (Møller et al. 1993) although the affinity is lower than that exhibited by intact uPAR (Ploug et al. 1994). The interaction of uPAR with its ligand includes an hydrophobic component (PLOUG et al. 1994). Domains 2 and 3 also appear to play a role in the ligand binding interaction (BEHRENDT et al. 1996; PLOUG et al. 1998b; HIGAZI et al. 1997).

A chemotactic epitope resides in the proteinase-sensitive linker region connecting domains 1 and 2, which is cleaved by u-PA and other proteases (FAZIOLI et al. 1997; HØYER-HANSEN et al. 1997).

5. Regulation of uPAR Expression

The functions of uPAR are subject to modulation by changes in the cell surface expression of this receptor. This can be effected at several levels. First, uPAR is present in intracellular granules and can be translocated to the cell surface following stimulation (BUSSO et al. 1994). Second, changes in uPAR expression are also observed during cellular ontogeny: immature leukemic cells do not express uPAR while uPAR is expressed on more mature leukemic cells (JARDI et al. 1996). Conversely, megakaryocytes express uPAR, but platelets do not (JIANG et al. 1996; WOHN et al. 1997). Third, uPAR expression can be modulated by growth factors, hormones and cytokines (MILES and PLOW 1985; TODD et al. 1985; NIELSEN et al. 1988; KIRCHHEIMER et al. 1988; PICONE et al. 1989; LUND et al. 1991a,b; MØLLER et al. 1992; WANG et al. 1994b; MANDRIOTA et al. 1995). Interestingly, the Mo3 lymphocyte activation antigen was found to be identical to uPAR (SUH et al. 1994). The recently cloned proximal promoter within 180 bp from the transcription initiation site which drives basal and some induced transcription of this molecule (SORAVIA et al. 1995) now provides a tool to examine the molecular basis for modulation of uPAR expression at the transcriptional level.

6. Non-proteolytic Functions of uPAR

In addition to its role in controlling plasminogen activation (see below), several distinct functions have been ascribed to uPAR (BLASI 1997). uPAR can also serve as a receptor for vitronectin (WALTZ and CHAPMAN 1994; WEI et al. 1994) and for integrins (WEI et al. 1996). The interaction of vitronectin with the uPAR is of high affinity and is blocked by antibodies recognizing domains 2/3 of the uPAR (WEI et al. 1994). Recombinant uPAR binds vitronectin in the absence of uPA, but vitronectin binding is promoted by concurrent binding of either uPA or uPA fragments containing the uPAR binding domain. PAI-1 can inhibit the interaction of uPAR with vitronectin (WALTZ et al. 1997). The formation of a stable uPAR/β-integrin complex inhibits native integrin adhesive function and promotes the binding of uPAR to vitronectin (WEI et al. 1996).

Several intracellular signaling events have been associated with uPAR occupancy by uPA and many of these functions appear to be independent of uPA proteolytic activity. A discussion of the extensive literature on uPAR-mediated signal transduction goes beyond the scope of this book. The reader is referred to two recent reviews by BESSER et al. (1997) and by DEAR and

MEDCALF (1998). Signaling events include, among others, monocyte migration during inflammation (KIRCHHEIMER and REMOLD 1989b; KIRCHHEIMER et al. 1991; ESTREICHER et al. 1990; GYETKO et al. 1994), promotion of myeloid differentiation (NUSRAT and CHAPMAN 1991), stimulation of endothelial cell migration (FIBBI et al. 1988; ODEKON et al. 1992), growth stimulating activity (RABBANI et al. 1992), stimulation of the expression of matrix metalloprotease 9 and cathepsin B (KIRCHHEIMER et al. 1988), activation of the Jak/Stat pathway (DUMLER et al. 1997), stimulation of diacylglycerol formation (DEL ROSSO et al. 1993), increases in *c-fos*, *jun* and *myc* mRNA levels (DUMLER et al. 1994; RABBANI et al. 1997) and activation of the extracellular signal-regulated kinases ERK1 and ERK2 (KONAKOVA et al. 1998; NGUYEN et al. 1998).

Since the uPAR is a GPI-anchored protein, the mechanisms by which such signal transduction occurs is presumably similar to that of other GPI-anchored proteins (STEFANOVÀ et al. 1991). Such signal probably requires an adapter protein to provide a link into the cytoplasm. A 38kDa protein which undergoes tyrosine phosphorylation following stimulation of cells with uPA (DUMLER et al. 1993) may be such an adapter protein.

Transmembrane functions of the uPAR are also implicated in the localization of uPAR on the leading edge of migrating cells (GYETKO et al. 1994). uPAR is colocalized with cytoskeletal proteins at focal adhesions (TAKAHASHI et al. 1990). Studies have demonstrated a close physical association between uPAR and the $\alpha_M\beta_2$ integrin (CR3, MAC-1), using co-capping and resonance energy transfer (XUE et al. 1994; CAO et al. 1995). Co-precipitation studies of monocyte lysates have shown that uPAR exists in large complexes with both $\alpha_M\beta_2$ and $\alpha_L\beta_2$ (LFA-1) (BOHUSLAV et al. 1995). Co-localization of uPAR with $\alpha_M\beta_2$ also plays a role in leukocyte chemotaxis (GYETKO et al. 1994).

III. Other uPA Receptors

A 70kDA uPA binding protein which did not interact with antibodies directed against uPAR was demonstrated in human platelets (JIANG et al. 1996). In contrast to the protection of cell-bound plasmin from α_2-antiplasmin, the uPA inhibitors, PAI-1, PAI-2, and protease nexin I (PN-1) can interact with cell bound uPA (ESTREICHER et al. 1990; CUBELLIS et al. 1989) [although some resistance of cell-bound uPA to PAI-1 on certain cell types has been reported (KIRCHHEIMER and REMOLD 1989a)]. uPAR alone is unable to internalize uPA or its complexes with these inhibitors (MILES et al. 1988b; VASSALLI et al. 1985). Nevertheless, rapid internalization does occur (ESTREICHER et al. 1990; JENSEN et al. 1990) via an initial interaction of the complexes with uPAR, followed by internalization by the LDL Receptor-Related Protein (LRP) (PLOUG and ELLIS 1994; CONESE et al. 1994; MOESTRUP et al. 1993). LRP is a member of the LDL receptor gene family (KRIEGER and HERZ 1994) and is the same as the α_2-macroglobulin receptor (KRISTENSEN et al. 1990; RABBANI et al. 1994). Internalization of uPA:PAI-1 complexes can also occur via two receptors related to LRP. These are glycoprotein 330 (MOESTRUP et al. 1993) and the very low density lipoprotein receptor (HEEGAARD et al. 1995). The internalization is thought to require the formation of a quaternary complex between uPA, the inhibitor, uPAR, and LRP (CONESE and BLASI 1995). It has been documented that uPAR is recycled to the cell surface after dissociation of the uPA:inhibitor complex (CONESE and BLASI 1995; NYKJAER et al. 1997). LRP is also involved in clearance of t-PA complexes.

E. Functional Consequences of Receptor Occupancy

I. Kinetic Consequences

1. Plasminogen Receptors

Plasminogen receptors, in concert with plasminogen activator receptors, function to attain efficient generation and expression of plasmin activity at the cell surface. The contribution of plasminogen receptors to these ends is fourfold (see Fig. 3).

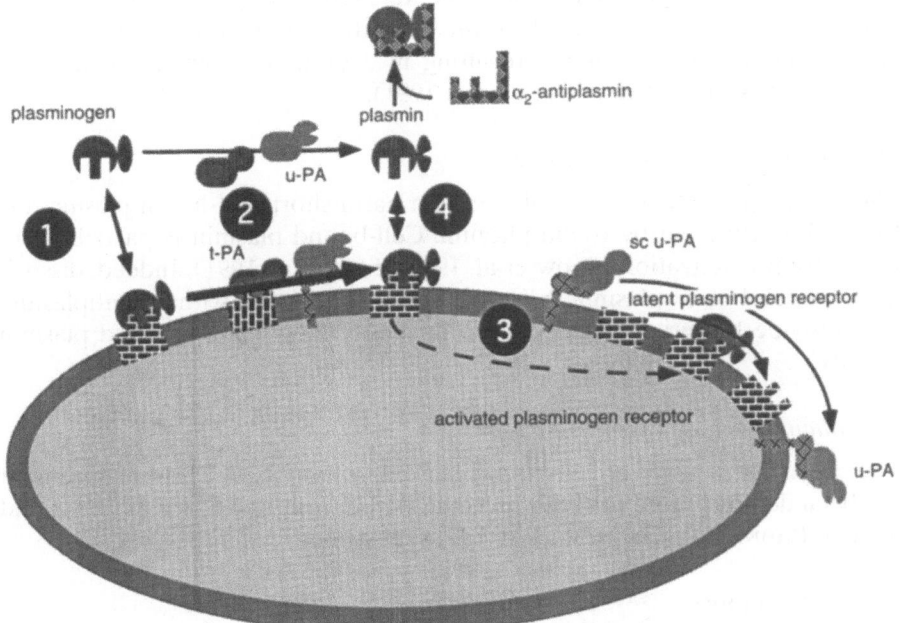

Fig. 3. Mechanisms for assembly and activation of the plasminogen system on cell surfaces. Reaction 1 is the equilibrium of plasminogen between the soluble phase and the cell surface. Reaction 2 describes the activation of plasminogen to plasmin by cell-bound or free plasminogen activators, tPA or uPA. Activation on the cell-surface is favored. Reaction 3 entails mechanism by which plasmin can enhance plasminogen activation, either by sc-uPA to uPA or by proteolytically generating additional carboxy-terminal lysines which can bind additional plasminogen. Reaction 4 indicates the neutralization of plasmin by α_2-antiplasmin with cell-bound plasmin being protected from inactivation. Reprinted with permission from PLOW et al. (1995)

a) Enhanced Plasminogen Activation

Studies with a variety of plasminogen-binding cells – platelets (Miles and Plow 1985), monocytoid cells (Plow et al. 1986), endothelial cells (Hajjar et al. 1986), and isolated plasminogen-receptor proteins such as annexin II (Cesarman et al. 1994) and enolase (Redlitz et al. 1994) – have clearly shown that receptor-bound plasminogen is more readily activated to plasmin than free plasminogen. This effect is attributable mainly to a decrease in K_m (Ellis et al. 1991). While this effect is independent of the plasminogen activator, it may be most pronounced with t-PA (Hajjar et al. 1986; Deguchi et al. 1985; Stricker et al. 1986).

b) Amplification Loops

Once activated, plasmin can participate in mechanisms that further enhance plasminogen activation. One such amplification mechanism involves the proteolysis of membrane proteins by cell-bound plasmin. The new carboxy-terminal lysine residues generated can then serve as additional plasminogen binding sites (Camacho et al. 1989; Gonzalez-Gronow et al. 1991). A second amplification mechanism involves the activation by cell-bound plasmin of receptor-bound pro-urokinase, resulting in a dramatic increase in its plasminogen activator activity (Ellis et al. 1991).

c) Protection of Plasmin Activity

The enzymatic activity of soluble plasmin has a short half-life in plasma due to rapid inactivation by α_2-antiplasmin. Cell-bound plasmin is partially protected from inactivation (Plow et al. 1986; Hall et al. 1991). Indeed, dissociation of cell-bound plasmin followed by rapid reaction with α_2-antiplasmin may provide the primary mechanism for controlling cell-associated plasmin activity.

d) Enhancement of Plasmin Activity

The enzymatic activity of cell-bound plasmin is increased. This enhancement has been demonstrated for both protein (Duval-Jobe and Parmely 1994) and peptide (Gonzalez-Gronow et al. 1991) substrates.

2. uPA Receptors

uPA bound to uPAR activates plasminogen more efficiently than uPA in solution (Ellis et al. 1989). The binding of uPA to uPAR decreases the K_m for plasminogen activation by 37-fold (from $25\,\mu mol/l$ to $0.67\,\mu mol/l$) and increases the overall catalytic efficiency of plasminogen activation about 6-fold (Ellis et al. 1991; Ploug et al. 1991; Stephens et al. 1989). Several mechanisms contribute to the enhancement in plasminogen activation. First, uPAR places uPA in a favorable position for plasminogen activation. Treatment with either the ATF (Ploug et al. 1991) or a monoclonal antibody against uPAR (Rønne et al. 1991) to disrupt the interaction of uPA with uPAR abolishes the cell-dependent enhancement in plasminogen activation. The importance of cell-surface localization of uPA in enhancing plasminogen activation can be mimicked by transfection of cells with a mutant uPA anchored to the cell membrane by a GPI linkage (LEE et al. 1992, 1994). Second, plasminogen activation is increased approximately 1000-fold by the conversion of sc-uPA to uPA (LIJNEN et al. 1989), and cleavage of sc-uPA by plasmin is enhanced on the cell surface (STEPHENS et al. 1989; RØNNE et al. 1991). Third, the interaction of sc-uPA with either cell-bound or soluble uPAR may expose the catalytic site of the proenzyme without cleavage (MANCHANDA and SCHWARTZ 1991; HIGAZI et al. 1995).

3. t-PA Receptors

Plasminogen activation by t-PA in the presence of endothelial cells is enhanced more than tenfold compared to the rate in fluid phase (HAJJAR et al. 1987). Because both plasminogen and t-PA bind to the cells, the acceleration may depend upon effects on either ligand or upon their co-localization on the cell surface. Condensation of both t-PA and plasminogen on the fibrin surface is required for enhancement of plasmin generation, and it appears that a similar mechanism occurs on cell surfaces (ELLIS and WHAWELL 1997).

II. Functional Consequences

The broad spectrum proteinase activity of plasmin can directly mediate breakdown of a variety of extracellular matrix constituents (reviewed by SAKSELA and RIFKIN 1988), including fibronectin, laminin, and vitronectin. Cell-bound plasmin also activates other matrix-degrading proteinases (WONG et al. 1992). Together, these proteolytic functions may facilitate the migration of cells through extracellular matrices and basement membrane barriers. As examples, the degradation of extracellular matrices by endothelial cells (NIEDBALA and PICARELLA 1992), mesangial cells (WONG et al. 1992), and squamous carcinoma cells (NIEDBALA and SARTORELLI 1990) is not only plasmin-dependent but is also inhibited by EACA, indicative of a requirement for cell-surface binding. The migration of endothelial cells (PEPPER et al. 1993), corneal epithelial cells (MORIMOTO et al. 1993), and a variety of tumor cells (melanoma cells (MEISSAUER et al. 1992), neuroblastoma cells (SHEA and BEERMANN 1992), ovarian cancer cells (KOBAYASHI et al. 1994), colon adenocarcinoma cells (COHEN et al. 1991), osteosarcoma cells (KARIKÓ et al. 1993)) is facilitated by the cell surface-associated plasminogen system.

Occupancy of uPAR has also been implicated in tumor cell biology. Blockade of the interaction between uPA, and blockade of uPAR decreases tumor cell invasiveness both in vivo and in vitro (PICONE et al. 1989; HOLLAS et al. 1991; SCHLECHTE et al. 1989; Ossowski et al. 1991a; CROWLEY et al. 1993). Also, malignant tissues exhibit increased uPA and uPAR expression compared to benign tissues (e.g., JANKUN et al. 1993). Therefore, the development of uPAR antagonists is being pursued as anti-cancer drugs. In vivo evidence also supports the importance of the plasminogen system in cell migration. Transexamic acid, an antagonist of plasminogen binding to cells, significantly inhibits smooth muscle cell migration into the intima after experimental vascular injury in rats (JACKSON and REIDY 1992).

Additional functional activities of cell-bound plasminogen and plasmin have also been described. Plasmin may modulate migration and proliferation of cells indirectly by activating growth factors which are synthesized and secreted as inactive precursors. Plasmin has been shown to activate basic FGF (FALCONE et al. 1993) and TGF-β (TAIPALE et al. 1992). Activation of these growth factors is inhibited by lysine analogs, indicative of involvement of cell-bound plasmin(ogen). These activated growth factors then, in turn, modify cellular behavior. In addition, and importantly, several functional consequences of plasminogen receptor occupancy do not rely on plasmin's proteolytic activity. These include calcium mobilization in fibroblasts (GONZALEZ-GRONOW et al. 1994), endothelial cell retraction (CONFORTI et al. 1994), and neutrophil aggregation (RYAN et al. 1992). Finally, while many of the functions ascribed to receptors for components of the plasminogen system have been deduced from in vitro systems, the recent generation of mice deficient in plasminogen supports a role of this system in growth and development (PLOPLIS et al. 1995), cell migration (CARMELIET et al. 1997), inflammation (PLOPLIS et al. 1998) as well as in thrombosis and fibrinolysis (BUGGE et al. 1995; PLOPLIS et al. 1995).

List of Abbreviations

ATF	amino terminal fragment
GPI	glycosylphosphatidylinositol
LBS	lysine binding site
LRP	LDL receptor-related protein
PAI-1(2)	plasminogen activator inhibitor type-1(2)
PAM	plasminogen binding antiphagocytic M-protein
PMA	phorbol 12-myristate 13-acetate
PN	protease nexin
sc-uPA	single-chain uPA
TAFI	thrombin-activable fibrinolysis inhibitor
tPA	tissue-type plasminogen activator
uPA	urinary-type (or urokinase-type) plasminogen activator
uPAR	uPA receptor

References

Adelman B, Rizk A, Hanners E (1988) Plasminogen interactions with platelets in plasma. Blood 72:1530–1535

Albrecht JC, Nicholas J, Cameron KR, Newman C, Fleckenstein B, Honess RW (1992) Herpesvirus saimiri has a gene specifying a homologue of the cellular membrane glycoprotein CD59. Virology 190:527–530

Andrade-Gordon P, Strickland S (1986) Interaction of heparin with plasminogen activators and plasminogen: Effects on the activation of plasminogen. Biochemistry 25:4033–4040

Andronicos NM, Ranson M, Bognacki J, Baber MS (1997) The human ENO1 gene product (recombinant human α-enolase) displays characteristics required for a plasminogen binding protein. Biochim Biophys Acta 1337:27–39

Appella E, Robinson EA, Ullrich SJ, Stoppelli MP, Corti A, Cassani G, Blasi F (1987) The receptor-binding sequence of urokinase. A biological function for the growth-factor module of proteases. J Biol Chem 262:4437–4440

Bajpai A, Baker JB (1985) Urokinase binding sites on human foreskin cells. Evidence for occupancy with endogenous urokinase. Biochem Biophys Res Commun 133: 994–1000

Barnathan ES, Kuo A, Rosenfeld L, Karikó K, Leski M, Robbiati F, Nolli ML, Henkin J, Cines DB (1990) Interaction of single-chain urokinase-type plasminogen activator with human endothelial cells. J Biol Chem 265:2865–2872

Bauer PI, Machovich R, Buki K, Csonka E, Koch S, Horvath I (1984) Interaction of plasmin with endothelial cells. Biochem J 218:119–124

Beebe DP (1987) Binding of tissue plasminogen activator to human umbilical vein endothelial cells. Thromb Res 46:241–254

Beebe DP, Miles LA, Plow EF (1989) A linear amino acid sequence involved in the interaction of t-PA with its endothelial cell receptor. Blood 74:2034–2037

Beebe DP, Wood LL, Moos M (1990) Characterization of tissue plasminogen activator binding proteins isolated from endothelial cells and other cell types. Thromb Res 59:339–350

Behrendt N, Ploug M, Patthy L, Houen G, Blasi F, Danø K (1991) The ligand-binding domain of the cell surface receptor for urokinase-type plasminogen activator. J Biol Chem 266:7842–7847

Behrendt N, Rønne E, Danø K (1996) Domain interplay in the urokinase receptor. Requirement for the third domain in high affinity ligand binding and demonstration of ligand contact sites in distinct receptor domains. J Biol Chem 271: 22885–22894

Berge A, Sjöbring U (1993) PAM, a novel plasminogen-binding protein from *Streptococcus pyogenes*. J Biol Chem 268:25417–25424

Besser D, Verde P, Nagamine Y, Blasi F (1997) Signal transduction and the u-PA/u-PAR system. Fibrinolysis 10:215–237

Bizik J, Trancikova D, Felnerova D, Verheijen JH, Vaheri A (1997) Spatial orientation of tissue-type plasminogen activator bound at the melanoma cell surface. Biochem Biophys Res Commun 239:322–328

Blasi F (1997) uPA, uPAR, PAI-1: key intersection of proteolytic, adhesive and chemotactic highways? Immunol Today 18:415–417

Boffa MB, Wang W, Bajzar L, Nesheim ME (1998) Plasma and recombinant thrombin-activable fibrinolysis inhibitor (TAFI) and activated TAFI compared with respect to glycosylation, thrombin/thrombomodulin-dependent activation, thermal stability, and enzymatic properties. J Biol Chem 273:2127–2135

Bohuslav J, Horejsí V, Hansmann C, Stöckl J, Weidle UH, Majdic O, Bartke I, Knapp W, Stockinger H (1995) Urokinase plasminogen activator receptor, β2-integrins, and Src-kinases within a single receptor complex of human monocytes. J Exp Med 181:1381–1390

Børglum AD, Byskov A, Ragno P, Roldan AL, Tripputi P, Cassani G, Danø K, Blasi F, Kruse TA (1992) Assignment of the urokinase-type plasminogen activator receptor gene (PLAUR) to chromosome 19q13.1–q13.2. Am J Hum Genet 50:492–497

Boyle MDP, Lottenberg R (1997) Plasminogen activation by invasive human pathogens. Thromb Haemost 77:1–10

Bugge TH, Flick MJ, Daugherty CC, Degen JL (1995) Plasminogen deficiency causes severe thrombosis but is compatible with development and reproduction. Genes Dev 9:794–807

Bürgle M, Koppitz M, Riemer C, Kessler H, König B, Weidle UH, Kellermann J, Lottspeich F, Graeff H, Schmitt M, Goretzki L, Reuning U, Wilhelm O, Magdolen V (1997) Inhibition of the interaction of urokinase-type plasminogen activator (uPA) with its receptor (uPAR) by synthetic peptides. Biol Chem 378:231–237

Burtin P, Chavanel G, Andre-Bougaran J, Gentile A (1987) The plasmin system in human adenocarcinomas and their metastases. A comparative immunofluorescence study. Int J Cancer 39:170–178

Burtin P, Fondanèche M-C (1988) Receptor for plasmin on human carcinoma cells. J Natl Cancer Inst 80:762–765

Busso N, Masur SK, Lazega D, Waxman S, Ossowski L (1994) Induction of cell migration by pro-urokinase binding to its receptor: Possible mechanism for signal transduction in human epithelial cells. J Cell Biol 126:259–270

Camacho M, Fondanèche M-C, Burtin P (1989) Limited proteolysis of tumor cells increases their plasmin-binding ability. FEBS Lett 245:21–24

Cao D, Mizukami IF, Garni-Wagner BA, Kindzelskii AL, Todd RF, III, Boxer LA, Petty HR (1995) Human urokinase-type plasminogen activator primes neutrophils for superoxide anion releases. Possible roles of complement receptor type III and calcium. J Immunol 154:1817–1829

Carmeliet P, Moons L, Dewerchin M, Rosenberg S, Herbert JM, Lupu F, Collen D (1998) Receptor-independent role of urokinase-type plasminogen activator in pericellular plasmin and matrix metalloproteinase proteolysis during vascular wound healing in mice. J Cell Biol 140:233–245

Carmeliet P, Moons L, Ploplis V, Plow E, Collen D (1997) Impaired arterial neointima formation in mice with disruption of the plasminogen gene. J Clin Invest 99: 200–208

Cesarman GM, Guevara CA, Hajjar KA (1994) An endothelial cell receptor for plasminogen/tissue plasminogen activator (t-PA). II. Annexin II-mediated enhancement of t-PA-dependent plasminogen activation. J Biol Chem 269:21198–21203

Chapman HA, Bertozzi P, Sailor LZ, Nusrat AR (1990) Alveolar macrophage urokinase receptors localize enzyme activity to the cell surface. Am J Physiol 259: L432–L438

Chen ZL, Strickland S (1997) Neuronal death in the hippocampus is promoted by plasmin-catalyzed degradation of laminin. Cell 91:917–925

Christiansen VJ, Sims PJ, Hamilton KK (1997) Complement C5b-9 increases plasminogen binding and activation on human endothelial cells. Arterioscler Thromb Vasc Biol 17:164–171

Cohen RL, Xi X-P, Crowley CW, Lucas BK, Levinson AD, Shuman MA (1991) Effects of urokinase receptor occupancy on plasmin generation and proteolysis of basement membrane by human tumor cells. Blood 78:479–487

Coleman JL, Sellati TJ, Testa JE, Kew RR, Furie MB, Benach JL (1995) *Borrelia burgdorferi* binds plasminogen, resulting in enhanced penetration of endothelial monolayers. Infect Immun 63:2478–2484

Conese M, Blasi F (1995) Urokinase/urokinase receptor system: internalization/degradation of urokinase-serpin complexes: mechanism and regulation. Biol Chem Hoppe-Seyler 376:143–155

Conese M, Olson D, Blasi F (1994) Protease nexin-1-urokinase complexes are internalized and degraded through a mechanism that requires both urokinase receptor and α_2-macroglobulin receptor. J Biol Chem 269:17886–17892

Conforti G, Dominguez-Jimenez C, Rønne Z, Høyer-Hansen G, Dejana E (1994) Cell-surface plasminogen activation causes a retraction of in vitro cultured human umbilical vein endothelial cell monolayer. Blood 83:994–1005

Correc P, Fondanèche M-C, Bracke M, Burtin P (1990) The presence of plasmin receptors on three mammary carcinoma MCF-7 sublines. Int J Cancer 46:745–750

Correc P, Zhang S, Komano O, Laurent M, Burtin P (1992) Visualization of the plasmin receptor on carcinoma cells. Int J Cancer 50:767–771

Crowley CW, Cohen RL, Lucas BK, Liu G, Shuman MA, Levinson AD (1993) Prevention of metastasis by inhibition of the urokinase receptor. Proc Natl Acad Sci USA 90:5021–5025

Cubellis MV, Andreasen P, Ragno P, Mayer M, Danø K, Blasi F (1989) Accessibility of receptor-bound urokinase to type-1 plasminogen activator inhibitor. Proc Natl Acad Sci USA 86:4828–4832

Cubellis MV, Nolli ML, Cassani G, Blasi F (1986) Binding of single-chain prourokinase to the urokinase receptor of human U937 cells. J Biol Chem 261:15819–15822

Danø K, Behrendt N, Brünner N, Ellis V, Ploug M, Dyke C (1994) The urokinase receptor: protein structure and role in plasminogen activation and cancer invasion. Fibrinolysis 8:189–203

Dear AE, Medcalf RL (1998) The urokinase-type-plasminogen-activator receptor (CD87) is a pleiotropic molecule. Eur J Biochem 252:185–193

Deguchi K, Murashima S, Shirakawa S, Soria C, Soria J, Dunn F, Tobelem G (1985) The potentiating effect of platelet on plasminogen activation by tissue plasminogen activator. Thromb Res 40:853–861

Del Rosso M, Anichini E, Pedersen N, Blasi F, Fibbi G, Pucci M, Ruggiero M (1993) Urokinase-urokinase receptor interaction: non-mitogenic signal transduction in human epidermal cells. Biochem Biophys Res Commun 190:347–352

Del Rosso M, Fibbi G, Dini G, Grappone C, Pucci M, Caldini R, Magnelli L, Fimiani M, Lotti T, Panconesi E (1990) Role of specific membrane receptors in urokinase-dependent migration of human keratinocytes. J Invest Dermatol 94:310–316

Dudani AK, Ganz PR (1996) Endothelial cell surface actin serves as a binding site for plasminogen, tissue plasminogen activator and lipoprotein(a). Br J Haematol 95: 168–178

Dumler I, Petri T, Schleuning W-D (1993) Interaction of urokinase-type plasminogen activator (u-PA) with its cellular receptor (u-PAR) induces phosphorylation on tyrosine of a 38kDa protein. FEBS Lett 322:37–40

Dumler I, Petri T, Schleuning W-D (1994) Induction of c-fos gene expression by urokinase-type plasminogen activator in human ovarian cancer cells. FEBS Lett 343:103–106

Dumler I, Weis A, Mayboroda OA, Maasch C, Jerke U, Haller H, Gulba DC (1998) The Jak/Stat pathway and urokinase receptor signaling in human aortic vascular smooth muscle cells. J Biol Chem 273:315–321

Duval-Jobe C, Parmely MJ (1994) Regulation of plasminogen activation by human U937 promonocytic cells, J Biol Chem 269:21353–21357

Eaton DL, Malloy BE, Tsai SP, Henzel W, Drayna D (1991) Isolation, molecular cloning, and partial characterization of a novel carboxypeptidase B from human plasma, J Biol Chem 266:21833–21838

Ellis V, Behrendt N, Danø K (1991) Plasminogen activation by receptor-bound urokinase. A kinetic study with both cell-associated and isolated receptor. J Biol Chem 266:12752–12758

Ellis V, Scully MF, Kakkar VV (1989) Plasminogen activation initiated by single-chain urokinase-type plasminogen activator. J Biol Chem 264:2185–2188

Ellis V, Whawell SA (1997) Vascular smooth muscle cells potentiate plasmin generation by both urokinase and tissue plasminogen activator-dependent mechanisms: evidence for a specific tissue-type plasminogen activator receptor on these cells. Blood 90:2312–2322

Estreicher A, Mühlhauser J, Carpentier J-L, Orci L, Vassalli J-D (1990) The receptor for urokinase type plasminogen activator polarizes expression of the protease to the leading edge of migrating monocytes and promotes degradation of enzyme inhibitor complexes. J Cell Biol 111:783–792

Estreicher A, Wohlwend A, Belin D, Schleuning W-D, Vassalli J-D (1989) Characterization of the cellular binding site for the urokinase-type plasminogen activator. J Biol Chem 264:1180–1189

Falcone DJ, McCaffrey TA, Haimovitz-Friedman A, Vergilio J-A, Nicholson AC (1993) Macrophage and foam cell release of matrix-bound growth factors. Role of plasminogen activation. J Biol Chem 268:11951–11958

Fazioli F, Resnati M, Sidenius N, Higashimoto Y, Appella E, Blasi F (1997) A urokinase-sensitive region of the human urokinase receptor is responsible for its chemotactic activity. EMBO J 16:7279–7286

Felez J, Chanquia CJ, Fabregas P, Plow EF, Miles LA (1993) Competition between plasminogen and t-PA for cellular binding sites. Blood 82:2433–2441

Felez J, Chanquia CJ, Levin EG, Miles LA, Plow EF (1991) Binding of tissue plasminogen activator to human monocytes and monocytoid cells. Blood 78:2318–2327

Felez J, Miles LA, Fabregas P, Jardi M, Plow EF, Lijnen RH (1996) Characterization of cellular binding sites and interactive regions within reactants required for enhancement of plasminogen activation by tPA on the surface of leukocytic cells. Thromb Haemost 76:577–584

Felez J, Miles LA, Plescia J, Plow EF (1990) Regulation of plasminogen receptor expression on human monocytes and monocytoid cell lines. J Cell Biol 111:1673–1683

Fibbi G, Ziche M, Morbidelli L, Magnelli L, Del Rosso M (1988) Interaction of urokinase with specific receptors stimulates mobilization of bovine adrenal capillary endothelial cells. Exp Cell Res 179:385–395

Fukao H, Hagiya Y, Nonaka T, Okada K, Matsuo O (1992) Analysis of binding protein for tissue-type plasminogen activator in human endothelial cells. Biochem Biophys Res Commun 187:956–962

Fuse N, Tsuchiya T, Nomomura Y, Menez A, Tamiya T (1990) Structure of the snake short-chain neurotoxin, erabutoxin c, precursor gene. Eur J Biochem 193:629–633

Ganz PR, Dupuis D, Dudani AK, Hashemi S (1991) Characterization of plasminogen binding to human capillary and arterial endothelial cells. Biochem Cell Biol 69:442–448

Goldfarb RH, Ziche M, Murano G, Liotta LA (1986) Plasminogen activators (urokinase) mediate neovascularization: Possible role in tumor angiogenesis. Semin Thromb Hemost 12:294–307

Gonias SL, Geary WA, VandenBerg SR (1989) Plasminogen binding to rat hepatocytes in primary culture and to thin slices of rat liver. Blood 75:729–736

Gonzalez-Gronow M, Edelberg JM, Pizzo SV (1989) Further characterization of the cellular plasminogen binding site: Evidence that plasminogen 2 and lipoprotein(a) compete for the same site. Biochemistry 28:2374–2377

Gonzalez-Gronow M, Gawdi G, Pizzo SV (1994) Characterization of the plasminogen receptors of normal and rheumatoid arthritis human synovial fibroblasts. J Biol Chem 269:4360–4366

Gonzalez-Gronow M, Stack S, Pizzo SV (1991) Plasmin binding to the plasminogen receptor enhances catalytic efficiency and activates the receptor for subsequent ligand binding. Arch Biochem Biophys 286:625–628

Gyetko MR, Todd RF, III, Wilkinson CC, Sitrin RG (1994) The urokinase receptor is required for human monocyte chemotaxis in vitro, J Clin Invest 93:1380–1387

Hajjar KA (1991) The endothelial cell tissue plasminogen activator receptor. Specific interaction with plasminogen. J Biol Chem 266:21962–21970

Hajjar KA (1995) Cellular receptors in the regulation of plasmin generation. Thromb Haemost 74:294–301

Hajjar KA, Gavish D, Breslow JL, Nachman RL (1989) Lipoprotein(a) modulation of endothelial cell surface fibrinolysis and its potential role in atherosclerosis. Nature 339:303–305

Hajjar KA, Hamel NM (1990) Identification and characterization of human endothelial cell membrane binding sites for tissue plasminogen activator and urokinase. J Biol Chem 265:2908–2916

Hajjar KA, Hamel NM, Harpel PC, Nachman RL (1987) Binding of tissue plasminogen activator to cultured human endothelial cells. J Clin Invest 80:1712–1719

Hajjar KA, Harpel PC, Jaffe EA, Nachman RL (1986) Binding of plasminogen to cultured human endothelial cells. J Biol Chem 261:11656–11662

Hajjar KA, Jacovina AT, Chacko J (1994) An endothelial cell receptor for plasminogen/tissue plasminogen activator. I. Identity with annexin II. J Biol Chem 269: 21191–21197

Hajjar KA, Mauri L, Jacovina AT, Zhong F, Mirza UA, Padovan JC, Chait BT (1998) Tissue plasminogen activator binding to the annexin II tail domain. Direct modulation by homocysteine. J Biol Chem 273:9987–9993

Hall SW, Humphries JE, Gonias SL (1991) Inhibition of cell surface receptor-bound plasmin by α_2-antiplasmin and α_2-macroglobulin. J Biol Chem 266:12329–12336

Heegaard CW, Simonsen AC, Oka K, Kjoller L, Christensen A, Madsen B, Ellgaard L, Chan L, Andreasen PA (1995) Very low density lipoprotein receptor binds and mediates endocytosis of urokinase-type plasminogen activator-type-1 plasminogen activator inhibitor complex. J Biol Chem 270:20855–20861

Hembrough TA, Vasudevan J, Allietta MM, Glass WF, II, Gonias SL (1995) A cytokeratin 8-like protein with plasminogen-binding activity is present on the external surfaces of hepatocytes, HepG2 cells and breast carcinoma cell lines. J Cell Sci 108:1071–1082

Herbert JM, Carmeliet P (1997) Involvement of u-PA in the anti-apoptotic activity of TGFb for vascular smooth muscle cells. FEBS Lett 413:401–404

Higazi AA-R, Cohen RL, Henkin J, Kniss D, Schwartz BS, Cines DB (1995) Enhancement of the enzymatic activity of single-chain urokinase plasminogen activator by soluble urokinase receptor. J Biol Chem 270:17375–17380

Higazi AA-R, Mazar A, Wang J, Quan N, Griffin R, Reilly R, Henkin J, Cines DB (1997) Soluble human urokinase receptor is composed of two active units. J Biol Chem 272:5348–5353

Hollas W, Blasi F, Boyd D (1991) Role of the urokinase receptor in facilitating extracellular matrix invasion by cultured colon cancer. Cancer Res 51:3690–3695

Hollas W, Hoosein N, Chung LWK, Mazar A, Henkin J, Karikó K, Barnathan ES, Boyd D (1992) Expression of urokinase and its receptor in invasive and non-invasive prostate cancer cell lines. Thromb Haemost 68:662–666

Høyer-Hansen G, Ploug M, Behrendt N, Rønne E, Danø K (1997) Cell-surface acceleration of urokinase-catalyzed receptor cleavage. Eur J Biochem 243:21–26

Hu LT, Pratt SD, Perides G, Katz L, Rogers RA, Klempner MS (1997) Isolation, cloning, and expression of a 70-kilodalton plasminogen binding protein of Borrelia burgdorferi. Infect Immun 65:4989–4995

Huang Y-Y, Bach ME, Lipp HP, Zhuo M, Wolfer DP, Hawkins RD, Schoonjans L, Kandel ER, Godfraind J-M, Mulligan R, Collen D, Carmeliet P (1996) Mice lacking the gene encoding tissue-type plasminogen activator show a selective interference with late-phase long-term potentiation in both Schaffer collateral and mossy fiber pathways. Proc Natl Acad Sci USA 93:8699–8704

Jackson CL, Reidy MA (1992) The role of plasminogen activation in smooth muscle cell migration after arterial injury. Ann NY Acad Sci 667:141–150

Jankun J, Merrick HW, Goldblatt PJ (1993) Expression and localization of elements of the plasminogen activation system in benign breast disease and breast cancers. J Cell Biochem 53:135–144

Jardí M, Inglés-Esteve J, Burgal M, Azqueta C, Velasco F, López-Pedrera C, Miles LA, Félez J (1996) Distinct patterns of urokinase receptor (uPAR) expression by leukemic cells and peripheral blood cells. Thromb Haemost 76:1009–1019

Jensen PH, Christensen EI, Ebbesen P, Gliemann J, Andreasen PA (1990) Lysosomal degradation of receptor-bound urokinase-type plasminogen activator is enhanced by its inhibitors in human trophoblastic choriocarcinoma cells. Cell Regulation 1:1043–1056

Jia X-C, Ny T, Hsueh AJW (1990) Synergistic effect of glucocorticoids and androgens on the hormonal induction of tissue plasminogen activator activity and messenger ribonucleic acid levels in granulosa cells. Mol Cell Endocrinol 68:143–151

Jiang YP, Pannell R, Liu JN, Gurewich V (1996) Evidence for a novel binding protein to urokinase-type plasminogen activator in platelet membranes. Blood 87:2775–2781

Karikó K, Kuo A, Boyd D, Okada SS, Cines DB, Barnathan ES (1993) Overexpression of urokinase receptor increases matrix invasion without altering cell migration in a human osteosarcoma cell line. Cancer Res 53:3109–3117

Kassam G, Choi KS, Ghuman J, Kang HM, Fitzpatrick SL, Zackson T, Zackson S, Toba M, Shinomiya A, Waisman DM (1998) The role of annexin II tetramer in the activation of plasminogen. J Biol Chem 273:4790–4799

Kim J, Yu W, Kovalski K, Ossowski L (1998) Requirement for specific proteases in cancer cell intravasation as revealed by a novel semiquantitative PCR-based assay. Cell 94:353–362

Kim S-O, Plow EF, Miles LA (1996) Regulation of plasminogen receptor expression on monocytoid cells by β_1-integrin-dependent cellular adherence to extracellular matrix proteins. J Biol Chem 271:23761–23767

Kirchheimer JC, Nong Y-H, Remold HG (1988) IFN-gamma, tumor necrosis factor-α and urokinase regulate the expression of urokinase receptors on human monocytes. J Immunol 141:4229–4234

Kirchheimer JC, Remold HG (1989a) Functional characteristics of receptor-bound urokinase on human monocytes: catalytic efficiency and susceptibility to inactivation by plasminogen activator inhibitors. Blood 74:1396–1402

Kirchheimer JC, Remold HG (1989b) Endogenous receptor-bound urokinase mediates tissue invasion of human monocytes. J Immunol 143:2634–2639

Kirchheimer JC, Remold HG, Wanivenhaus A, Binder BR (1991) Increased proteolytic activity on the surface of monocytes from patients with rheumatoid arthritis. Arthritis Rheum 34:1430–1433

Klein KL, Hartmann RC (1989) Acute coronary artery thrombosis in paroxysmal nocturnal hemoglobinuria. South Med J 82:1169–1171

Kobayashi H, Shinohara H, Takeuchi K, Itoh M, Fujie M, Saitoh M, Terao T (1994) Inhibition of the soluble and the tumor cell receptor-bound plasmin by urinary trypsin inhibitor and subsequent effects on tumor cell invasion and metastasis. Cancer Res 54:844–849

Konakova M, Hucho F, Schleuning WD (1998) Downstream targets of urokinase-type plasminogen-activator-mediated signal transduction. Eur J Biochem 253:421–429

Krätzschmar J, Haendler B, Kojima S, Rifkin DB, Schleuning W-D (1993) Bovine urokinase-type plasminogen activator and its receptor: Cloning and induction by retinoic acid. Gene 125:177–183

Krieger M, Herz J (1994) Structures and functions of multiligand lipoprotein receptors macrophage scavenger receptors and LDL receptor-related protein (LRP). Annu Rev Biochem 63:601–637

Kristensen P, Eriksen J, Blasi F, Danø K (1991) Two alternatively spliced mouse urokinase receptor mRNAs with different histological localization in the gastrointestinal tract. J Cell Biol 115:1763–1771

Kristensen T, Moestrup SK, Gliemann J, Bendtsen L, Sand O, Sottrup-Jensen L (1990) Evidence that the newly cloned low-density-lipoprotein receptor related protein (LRP) is the alpha-2-macroglobulin receptor. FEBS Lett 276:151–155

Kuusela P, Ullberg M, Kronvall G, Tervo T, Tarkkanen A, Saksela O (1992) Surface-associated activation of plasminogen on gram-positive bacteria. Effect of plasmin on the adherence of staphylococcus aureus. Acta Ophthalmol [Suppl 202]:42–46

Lee SW, Ellis V, Dichek D (1994) Characterization of plasminogen activation by glycosyl-phosphatidylinositol-anchored urokinase. J Biol Chem 269:2411–2418

Lee SW, Khan ML, Dichek D (1992) Expression of an anchored urokinase in the apical endothelial cell membrane. J Biol Chem 267:13020–13029

Lijnen HR, Van Hoef B, De Cock F, Collen D (1989) The mechanism of plasminogen activation and fibrin dissolution by single chain urokinase-type plasminogen activator in a plasma milieu in vitro. Blood 73:1864–1872

Lo SK, Ryan TJ, Gilboa N, Lai L, Malik AB (1989) Role of catalytic and lysine-binding sites in plasmin-induced neutrophil adherence to endothelium. J Clin Invest 84: 793–801

Lottenberg R (1997) A novel approach to explore the role of plasminogen in bacterial pathogenesis. Trends Microbiol 5:466–47

Lottenberg R, Broder CC, Boyle MDP, Kain SJ, Schroeder BL, Curtiss I (1992) Cloning, sequence analysis, and expression in Escherichia coli of a streptococcal plasmin receptor. J Bacteriol 174:5204–5210

Lu H, Li H, Mirshahi SS, Soria C, Soria J, Menashi S (1993) Comparative study of fibrinolytic activity on U937 line after stimulation by interferon gamma, 1,25 dihydroxyvitamin D_3 and their combination. Thromb Res 69:353–359

Lund LR, Rønne E, Roldan AL, Behrendt N, Rømer J, Blasi F, Danø K (1991a) Urokinase receptor mRNA level and gene transcription are strongly and rapidly increased by phorbol myristate acetate in human monocyte-like U937 cells. J Biol Chem 266:5177–5181

Lund LR, Rømer J, Rønne E, Ellis V, Blasi F, Danø K (1991b) Urokinase-receptor biosynthesis, mRNA level and gene transcription are increased by transforming growth factor $\beta1$ in human A549 lung carcinoma cells. EMBO J 10:3399–3407

Manchanda N, Schwartz BS (1991) Single chain urokinase. Augmentation of enzymatic activity upon binding to monocytes. J Biol Chem 266:14580–14584

Mandriota SJ, Seghezzi G, Vassalli JD, Ferrara N, Wasi S, Mazzieri R, Mignatti P, Pepper MS (1995) Vascular endothelial growth factor increases urokinase receptor expression in vascular endothelial cells. J Biol Chem 270:9709–9716

Meissauer A, Kramer MD, Schirrmacher V, Brunner G (1992) Generation of cell surface-bound plasmin by cell-associated urokinase-type or secreted tissue-type plasminogen activator: A key event in melanoma cell invasiveness in vitro. Exp Cell Res 199:179–190

Menoud PA, Debrot S, Schowing J (1989) Mouse neural crest cells secrete both urokinase-type and tissue-type plasminogen activators in vitro. Development 106:685–690

Merenmies J, Pihlaskari R, Laitinen J, Wartiovaara J, Rauvala H (1991) 30-kDa heparin-binding protein of brain (amphoterin) involved in neurite outgrowth. Amino acid sequence and localization in the filopodia of the advancing plasma membrane. J Biol Chem 266:16722–16729

Miles LA, Dahlberg CM, Levin EG, Plow EF (1989a) Gangliosides interact directly with plasminogen and urokinase and may mediate binding of these fibrinolytic components to cells. Biochemistry 28:9337–9343

Miles LA, Dahlberg CM, Plescia J, Felez J, Kate K, Plow EF (1991) Role of cell-surface lysines in plasminogen binding to cells: identification of alpha-enolase as a candidate plasminogen receptor. Biochemistry 30:1682–1691

Miles LA, Dahlberg CM, Plow EF (1988a) The cell-binding domains of plasminogen and their function in plasma. J Biol Chem 263:11928–11934

Miles LA, Fless GM, Levin EG, Scanu AM, Plow EF (1989b) A potential basis for the thrombotic risks associated with lipoprotein(a). Nature 339:301–303

Miles LA, Fless GM, Scanu AM, Baynham P, Sebald MT, Skocir P, Curtiss LK, Levin EG, Hoover-Plow JL, Plow EF (1995a) Interaction of Lp(a) with plasminogen binding sites on cells. Thromb Haemost 73:458–465

Miles LA, Ginsberg MH, White JG, Plow EF (1986) Plasminogen interacts with human platelets through two distinct mechanisms. J Clin Invest 77:2001–2009

Miles LA, Levin EG, Plescia J, Collen D, Plow EF (1988b) Plasminogen receptors, urokinase receptors, and their modulation on human endothelial cells. Blood 72:628–635

Miles LA, Plow EF (1985) Binding and activation of plasminogen on the platelet surface. J Biol Chem 260:4303–4311

Miles LA, Plow EF (1987) Receptor mediated binding of the fibrinolytic components, plasminogen and urokinase, to peripheral blood cells. Thromb Haemost 58: 936–942

Miles LA, Plow EF (1990) Lp(a): an interloper into the fibrinolytic system? Thromb Haemost 63:331–335

Moestrup SK, Nielsen S, Andreasen P, Jorgensen KE, Nykjaer A, Rolgaard H, Gliemann J, Christensen EJ (1993) Epithelial glycoprotein-330 mediates endocytosis of plasminogen activator-plasminogen activator inhibitor type-1 complexes. J Biol Chem 268:16564–16570

Møller LB, Ploug M, Blasi F (1992) Structural requirements for glycosyl-phosphatidylinositol anchor attachment in the cellular receptor for urokinase plasminogen activator. Eur J Biochem 208:493–500

Møller LB, Pöllänen J, Rønne E, Pedersen N, Blasi F (1993) N-linked glycosylation of the ligand-binding domain of the human urokinase receptor contributes to the affinity for its ligand. J Biol Chem 268:11152–11159

Morimoto K, Mishima H, Nishida T, Otori T (1993) Role of urokinase type plasminogen activator (u-PA) in corneal epithelial migration. Thromb Haemost 69:387–391

Nakajima K, Nagata K, Hamanoue M, Takemoto N, Shimada A, Kohsaka S (1993) Plasminogen-binding protein associated with the plasma membrane of cultured embryonic rat neocortical neurons. FEBS Lett 333:223–228

Nakajima K, Nagata K, Kohsaka S (1994) Plasminogen mediates an interaction between microglia and dopaminergic neurons. Eur Neurol 34 [Suppl 3]:10–16

Needham GK, Nicholson S, Angus B, Farndon JR, Harris AL (1988) Relationship of membrane-bound-tissue type and urokinase type plasminogen activators in human breast cancers to estrogen and epidermal growth factor receptors. Cancer Res 48:6603–6607

Nguyen DH, Hussaini IM, Gonias SL (1998) Binding of urokinase-type plasminogen activator to its receptor in MCF-7 cells activates extracellular signal-regulated kinase 1 and 2 which is required for increased cellular motility. J Biol Chem 273:8502–8507

Niedbala MJ, Picarella MS (1992) Tumor necrosis factor induction of endothelial cell urokinase-type plasminogen activator mediated proteolysis of extracellular matrix and its antagonism by gamma-interferon. Blood 79:678–687

Niedbala MJ, Sartorelli AC (1990) Plasminogen activator mediated degradation of subendothelial extracellular matrix by human squamous carcinoma cell lines. Cancer Communications 2:189–199

Nielsen LS, Kellerman GM, Behrendt N, Picone R, Danø K, Blasi F (1988) A 55000–60000 Mr receptor protein for urokinase-type plasminogen activator. Identification in human tumor cell lines and partial purification. J Biol Chem 263:2358–2363

Nusrat AR, Chapman HA (1991) An autocrine role for urokinase in phorbol ester-mediate differentiation of myeloid cell lines. J Clin Invest 87:1091–1097

Nykjaer A, Conese M, Christensen EI, Olson D, Cremona O, Gliemann J, Blasi F (1997) Recycling of the urokinase receptor upon internalization of the uPA:serpin complexes. EMBO J 16:2610–2620

Nykjaer A, Møller B, Todd RF, III, Christensen T, Andreasen PA, Gliemann J, Petersen CM (1994) Urokinase receptor: an activation antigen in human T lymphocytes. J Immunol 152:505–516

Odekon LE, Sato Y, Rifkin DB (1992) Urokinase-type plasminogen activator mediates basic fibroblast growth factor-induced bovine endothelial cell migration independent of its proteolytic activity. J Cell Physiol 150:258–263

Ossowski L, Clunie G, Masucci M-T, Blasi F (1991a) In vivo paracrine interaction between urokinase and its receptor: effect on tumor cell invasion. J Cell Biol 115:1107–1112

Ossowski L, Russo-Payne H, Wilson EL (1991b) Inhibition of urokinase-type plasminogen activator by antibodies: The effect on dissemination of a human tumor in the nude mouse. Cancer Res 51:274–281

Palfree RGE (1991) The urokinase-type plasminogen activator receptor is a member of the Ly-6 superfamily. Immunol Today 12:170–171

Pancholi V, Fischetti VA (1998) α-enolase, a novel strong plasmin(ogen) binding protein on the surface of pathogenic streptococci. J Biol Chem 273:14503–14515

Parkkinen J, Raulo E, Merenmies J, Nolo R, Kajander EO, Baumann M, Rauvala H (1993) Amphoterin, the 30-kDa protein in a family of HMG1-type polypeptides. Enhanced expression in transformed cells, leading edge localization, and interactions with plasminogen activation. J Biol Chem 268:19726–19738

Parkkinen J, Rauvala H (1991) Interactions of plasminogen and tissue plasminogen activated (t-PA) with amphoterin. Enhancement of t-PA-catalyzed plasminogen activation by amphoterin. J Biol Chem 266:16730–16735

Pepper MS, Sappino A-P, Stöcklin R, Montesano R, Orci L, Vassalli J-D (1993) Upregulation of urokinase receptor expression on migrating endothelial cells. J Cell Biol 122:673–684

Picone R, Kajitaniak EL, Nielsen LS, Behrendt N, Mastronicola MR, Cubellis MV, Stoppelli MP, Pedersen S, Danø K, Blasi F (1989) Regulation of urokinase receptors in monocyte-like U937 cells by phorbol ester phorbol myristate acetate. J Cell Biol 108:693–702

Pittman RN, Ivins JK, Buettner HM (1989) Neuronal plasminogen activators: Cell surface binding sites and involvement in neurite outgrowth. J Neurosci 9: 4269–4286

Plesner T, Behrendt N, Ploug M (1997) Structure, function and expression on blood and bone marrow cells of the urokinase-type plasminogen activator receptor, uPAR. Stem Cells 15:398–408

Ploplis VA, Carmeliet P, Vazirzadeh S, Van Vlaenderen I, Moons L, Plow EF, Collen D (1995) Effects of disruption of the *plasminogen* gene in mice on thrombosis, growth and health. Circulation 92:2585–2593

Ploplis VA, French EL, Carmeliet P, Collen D, Plow EF (1998) Plasminogen deficiency differentially affects recruitment of inflammatory cell populations in mice. Blood 91:2005–2009

Ploug M, Ellis V (1994) Structure-function relationships in the receptor for urokinase-type plasminogen activator. Comparison to other members of the Ly-6 family and snake venom α-neurotoxins. FEBS Lett 349:163–168

Ploug M, Ellis V, Danø K (1994) Ligand interaction between urokinase-type plasminogen activator and its receptor probed with 8-anilino-1-naphthalenesulfonate. Evidence for a hydrophobic binding site exposed only on the intact receptor. Biochemistry 33:8991–8997

Ploug M, Kjalke M, Rønne E, Weidle U, Høyer-Hansen G, Danø K (1993) Localization of the disulfide bonds in the NH_2-terminal domain of the cellular receptor for human urokinase-type plasminogen activator. A domain structure belonging to a novel superfamily of glycolipid-anchored membrane proteins. J Biol Chem 268: 17539–17546

Ploug M, Ostergaard S, Hansen LB, Holm A, Danø K (1998) Photoaffinity labeling of the human receptor for urokinase-type plasminogen activator using a decapeptide antagonist. Evidence for a composite ligand-binding site and a short interdomain separation. Biochemistry 37:3612–3622

Ploug M, Plesner T, Rønne E, Ellis V, Høyer-Hansen G, Hansen NE, Danø K (1992) The receptor for urokinase-type plasminogen activator is deficient on peripheral blood leukocytes in patients with paroxysmal nocturnal hemoglobinuria. Blood 79:1447–1455

Ploug M, Rahbek-Nielsen H, Nielsen PF, Roepstorff P, Danø K (1998) Glycosylation profile of a recombinant urokinase-type plasminogen activator receptor expressed in Chinese hamster ovary cells. J Biol Chem 273:13933–13943

Ploug M, Rønne E, Behrendt N, Jensen AL, Blasi F, Danø K (1991) Cellular receptor for urokinase plasminogen activator. Carboxyl-terminal processing and membrane anchoring by glycosyl-phosphatidylinositol, J Biol Chem 266:1926–1933

Plow EF, Freaney DE, Plescia J, Miles LA (1986) The plasminogen system and cell surfaces: Evidence for plasminogen and urokinase receptors on the same cell type. J Cell Biol 103:2411–2420

Plow EF, Herren T, Redlitz A, Miles LA, Hoover-Plow JL (1995) The cell biology of the plasminogen system. FASEB J 9:939–945

Pöllänen J (1989) Down-regulation of plasmin receptors on human sarcoma cells by glucocorticoids. J Biol Chem 264:5628–5632

Quax P, Frisdal E, Pedersen N, Bonavaud S, Thibert P, Martelly I, Verheijen JH, Blasi F (1992) Modulation of activities and RNA level of the components of the plasminogen activation system during fusion of human myogenic satellite cells in vitro. Dev Biol 151:166–175

Quax PHA, Grimbergen JM, Lansink M, Bakker AHF, Blatter MC, Belin D, Van Hinsbergh VWM, Verheijen JH (1998) Binding of human urokinase-type plasminogen activator to its receptor: residues involved in species specificity and binding. Arterioscler Thromb Biol 18:693–701

Rabbani SA, Mazar AP, Bernier SM, Haq M, Bolivar I, Henkin J, Goltzman D (1992) Structural requirements for the growth factor activity of the amino-terminal domain of urokinase. J Biol Chem 267:14151–14156

Rabbani SA, Gladu J, Mazar AP, Henkin J, Goltzman D (1997) Induction in human osteoblastic cells (SaOS2) of the early response genes fos, jun, and myc by the amino terminal fragment (ATF) of urokinase. J Cell Physiol 172:137–145

Rabbani SA, Rajwans N, Achbarou A, Murthy KK, Goltzman D (1994) Isolation and characterization of multiple isoforms of the rat urokinase receptor in osteoblasts. FEBS Lett 338:69–74

Ranson M, Andronicos NM, O'Mullane MJ, Baker MS (1998) Increased plasminogen binding is associated with metastatic breast cancer cells: differential expression of plasminogen binding proteins. Br J Cancer 77:1586–1597

Redlitz A, Fowler BJ, Plow EF, Miles LA (1994) The role of an enolase-related molecule in plasminogen binding to cells. Eur J Biochem 227:407–415

Redlitz A, Nicolini FA, Malycky JL, Topol EJ, Plow EF (1996) Inducible carboxypeptidase activity. A role in clot lysis in vivo. Circulation 93:1328–1330

Redlitz A, Plow EF (1995) Receptors for plasminogen and t-PA: an update. Baillieres Clin Haematol 8:313–327

Redlitz A, Tan AK, Eaton DL, Plow EF (1995) Plasma carboxypeptidases as regulators of the plasminogen system. J Clin Invest 96:2534–2538

Reich R, Miskin R, Tsafriri A (1985) Follicular plasminogen activator: involvement in ovulation. Endocrinology 116:516–521

Reuning U, Bang NU (1992) Regulation of the urokinase-type plasminogen activator receptor on vascular smooth muscle cells is under the control of thrombin and other mitogens. Arterioscler Thromb 12:1161–1170

Ringdahl U, Svensson M, Wistedt AC, Renné T, Kellner R, Müller-Esterl W, Sjöbring U (1998) Molecular co-operation between protein PAM and streptokinase for plasmin acquisition by Streptococcus pyogenes. J Biol Chem 273:6424–6430

Roldan AL, Cubellis MV, Masucci M-T, Behrendt N, Lund LR, Danø K, Appella E, Blasi F (1990) Cloning and expression of the receptor for human urokinase plasminogen activator, a central molecule in cell surface, plasmin dependent proteolysis. EMBO J 9:467–474

Rønne E, Behrendt N, Ellis V, Ploug M, Danø K, Høyer-Hansen G (1991) Cell-induced potentiation of the plasminogen activation system is abolished by a monoclonal antibody that recognizes the NH$_2$-terminal domain of the urokinase receptor. FEBS Lett 288:233–236

Ryan TJ, Lai L, Malik AB (1992) Plasmin generation induces neutrophil aggregation: Dependence on the catalytic and lysine binding sites. J Cell Physiol 151:255–261

Sakata Y, Okada M, Noro A, Matsuda M (1988) Interaction of tissue-type plasminogen activator and plasminogen activator inhibitor 1 on the surface of endothelial cells. J Biol Chem 263:1960–1969

Sakharov DV, Plow EF, Rijken DC (1997) On the mechanism of the antifibrinolytic activity of plasma carboxypeptidase B. J Biol Chem 272:14477–14482

Saksela O, Rifkin DB (1988) Cell-associated plasminogen activation: Regulation and physiological functions. Annu Rev Cell Biol 4:93–126

Sanzo MA, Howard SC, Wittwer AJ, Cochrane HM (1990) Binding of tissue plasminogen activator to human aortic endothelial cells. Biochem J 269:475–482

Sappino A-P, Huarte J, Belin D, Vassalli J-D (1989) Plasminogen activators in tissue remodeling and invasion: mRNA localization in mouse ovaries and implanting embryos. J Cell Biol 109:2471–2479

Schlechte W, Murano G, Boyd D (1989) Examination of the role of the urokinase receptor in human colon cancer mediated laminin degradation. Cancer Res 49: 6064–6069

Schleef RR, Podor TJ, Dunne E, Mimuro J, Loskutoff DJ (1990) The majority of type 1 plasminogen activator inhibitor associated with cultured human endothelial cells is located under the cells and is accessible to solution-phase tissue-type plasminogen activator. J Cell Biol 110:155–163

Schwartz-Albiez R, Heidtmann H-H, Wolf D, Schirrmacher V, Moldenhauer G (1992) Three types of human lung tumour cell lines can be distinguished according to surface expression of endogenous urokinase and their capacity to bind exogenous urokinase. Br J Cancer 65:51–57

Shapiro RL, Duquette JG, Nunes I, Roses DF, Harris MN, Wilson EL, Rifkin DB (1997) Urokinase-type plasminogen activator-deficient mice are predisposed to staphylococcal botryomycosis, pleuritis, and effacement of lymphoid follicles. Am J Pathol 150:359–369

Shea TB, Beermann ML (1992) Regulation of neuronal migration and neuritogenesis by distinct surface proteases. Relative contribution of plasmin and a thrombin-like protease. FEBS Lett 307:190–194

Sillaber C, Baghestanian M, Hofbauer R, Virgolini I, Bankl HC, Fureder W, Agis H, Willheim M, Leimer M, Scheiner O, Binder BR, Kiener HP, Bevec D, Fritsch G, Majdic O, Kress HG, Gadner H, Lechner K, Valent P (1997) Molecular and functional characterization of the urokinase receptor on human mast cells. J Biol Chem 272:7824–7832

Silverstein RL, Friedlander RJ, Jr, Nicholas RL, Nachman RL (1988) Binding of lysplasminogen to monocytes/macrophages. J Clin Invest 82:1948–1955

Skidgel RA (1988) Basic carboxypeptidases: regulators of peptide hormone activity. Trends Pharmacol Sci 9:299–304

Skidgel RA (1995) Human carboxypeptidase N (lysine carboxypeptidase). Methods Enzymol 248:653–663

Soravia E, Grebe A, De Luca P, Helin K, Suh TT, Degan JL, Blasi F (1995) A conserved TATA-less proximal promoter drives basal transcription from the urokinase-type plasminogen activator receptor gene. Blood 86:624–635

Stahl A, Mueller BM (1994) Binding of urokinase to its receptor promotes migration and invasion of human melanoma cells in vitro. Cancer Res 54:3066–3071

Stefanovà I, Horejsi V, Ansotegui IJ, Knapp W, Stockinger H (1991) GPI-anchored cell-surface molecules complexed to protein tyrosine kinases. Science 254:1016–1019

Stephens RW, Pöllänen J, Tapiovaara H, Leung K-C, Sim P-S, Salonen E-M, Rønne E, Behrendt N, Danø K, Vaheri A (1989) Activation of pro-urokinase and plasminogen on human sarcoma cells: A proteolytic system with surface-bound reactants. J Cell Biol 108:1987–1995

Stopelli MP, Corti A, Soffientini A, Cassani G, Blasi F, Assolan RK (1985) Differentiation-enhanced binding of the amino-terminal fragment of human urokinase plasminogen activator to a specific receptor on U937 monocytes. Proc Natl Acad Sci USA 82:4939–4943

Stricker RB, Wong D, Shiv DT, Reyes PT, Shuman MA (1986) Activation of plasminogen by tissue plasminogen activator on normal and thrombasthenic platelets: effects on surface proteins and platelet aggregation. Blood 68:275–280

Strickland S, Reich E, Sherman MI (1976) Plasminogen activator in early embryogenesis: Enzyme production by trophoblast and parietal endoderm. Cell 9:231–240

Suh TT, Nerlov C, Danø K, Degen JL (1994) The murine urokinase-type plasminogen activator receptor gene. J Biol Chem 269:25992

Syrovets T, Tippler B, Rieks M, Simmet T (1997) Plasmin is a potent and specific chemoattractant for human peripheral monocytes acting via a cyclic guanosine monophosphate-dependent pathway. Blood 89:4574–4583

Taipale J, Koli K, Keski-Oja J (1992) Release of transforming growth factor-beta 1 from the pericellular matrix of cultured fibroblasts and fibrosarcoma cells by plasmin and thrombin. J Biol Chem 267:25378–25384

Takahashi K, Ikeo K, Gojobori T, Tanifuji M (1990) Local function of urokinase receptor at the adhesion contact sites of a metastatic tumor cell, Thromb Res 57 [Suppl 10]:55–61

Todd RF III, Alvarez PA, Brott DA, Liu DY (1985) Bacterial lipopolysaccharide, phorbol myristate acetate and muramyl dipeptide stimulate the expression of a human monocyte surface antigen Mo3e. J Immunol 135:3869–3877

Vassalli J-D, Baccino D, Belin D (1985) A cellular binding site for the M_r 55,000 form of the human plasminogen activator, urokinase. J Cell Biol 100:86–92

Vaughan DE, Mendelsohn ME, Declerck PJ, Van Houtte E, Collen D, Loscalzo J (1989) Characterization of the binding of human tissue-type plasminogen activator to platelets. J Biol Chem 264:15869–15874

Verrall S, Seeds NW (1989) Characterization of ^{125}I-tissue plasminogen activator binding to cerebellar granule neurons. J Cell Biol 109:265–271

Waltz DA, Chapman HA (1994) Reversible cellular adhesion to vitronectin linked to urokinase receptor occupancy. J Biol Chem 269:14746–14750

Waltz DA, Natkin LR, Fujita RM, Wei Y, Chapman HA (1997) Plasmin and plasminogen activator inhibitor type 1 promote cellular motility by regulating the interaction between the urokinase receptor and vitronectin. J Clin Invest 100: 58–67

Wang W, Hendriks DF, Scharpé SS (1994) Carboxypeptidase U, a plasma carboxypeptidase with high affinity for plasminogen, J Biol Chem 269:15937–15944

Wang Y, Jones CJ, Dang J, Liang X, Olsen JE, Doe WF (1994) Human urokinase receptor expression is inhibited by amiloride and induced by tumor necrosis factor and phorbol ester in colon cancer cells. FEBS Lett 353:138–142

Wang YF, Tsirka SE, Strickland S, Stieg PE, Soriano SG, Lipton SA (1998) Tissue plasminogen activator (tPA) increases neuronal damage after focal cerebral ischemia in wild-type and tPA-deficient mice. Nat Med 4:228–231

Wei Y, Lukashev M, Simon DI, Bodary SC, Rosenberg S, Doyle MV, Chapman HA (1996) Regulation of integrin function by the urokinase receptor. Science 273: 1551–1555

Wei Y, Waltz DA, Rao N, Drummond RJ, Rosenberg S, Chapman HA (1994) Identification of the urokinase receptor as an adhesion receptor for vitronectin. J Biol Chem 269:32380–32388

Wohn KD, Kanse SM, Deutsch V, Schmidt T, Eldor A, Preissner KT (1997) The urokinase-receptor (CD87) is expressed in cells of the megakaryoblastic lineage. Thromb Haemost 77:540–547

Wong AP, Cortez SL, Baricos WH (1992) Role of plasmin and gelatinase in extracellular matrix degradation by cultured rat mesangial cells. Am J Physiol 263:F1112–F1118

Xue W, Kindzelskii A, Todd R III, Petty H (1994) Physical association of complement receptor type 3 and urokinase-type plasminogen activator receptor in neutrophil membranes. J Immunol 152:4630–4640

Section II
Clinical Use of Thrombolytic Agents

In Acute Myocardial Infarction

CHAPTER 7

Streptokinase and Anisoylated Lys-Plasminogen Streptokinase Activator Complex

V.J. MARDER and F. BACHMANN

A. Introduction

In 1933, the observation was made that an extract of hemolytic strepto-cocci lyses fibrin (TILLETT and GARNER 1933), and it took 16 years for a ther-apeutic preparation, streptococcal fibrinolysin (streptokinase), to be used clinically to dissolve fibrinous pleural effusions (TILLETT and SHERRY 1949). By the 1950s, investigators had explored the use of streptokinase (SK) in the dissolution of experimental thrombi and in acute myocardial infarction (AMI) (JOHNSON and TILLETT 1952; FLETCHER et al. 1959a,b), and in the late 1960s and 1970s it was widely applied in Europe for the treatment of deep venous thrombosis and pulmonary embolism (reviewed by FRANCIS and MARDER 1991). SK is derived from a non-pathogenic, group C (Lancefield) strain of streptococci (see Chap. 2), and was the first thrombolytic agent devel-oped for clinical usage, providing the basis for comparison with other throm-bolytic agents.

I. Pharmacology

1. Streptokinase (SK)

SK forms a complex with plasminogen which in turn converts plasminogen to plasmin (see Chap. 2). In man, the SK-plasmin(ogen) complex has a half-life of about 23 min (MENTZER et al. 1986). Plasmin degrades not only fibrin and fibrinogen but also other proteins, such as factor V (OMAR and MANN 1987), factor IX (SAMIS et al. 2000) von Willebrand factor (vWf) (FEDERICI et al. 1993), vitronectin (CHAIN et al. 1991), and complement (AGOSTONI et al. 1994). When plasmin forms on a fibrin surface (i.e., through activation of fibrin-bound plas-minogen), fibrin lysis occurs; when it forms in plasma (i.e., through activation of plasma plasminogen), fibrinogen degradation occurs. SK can form a complex with plasminogen bound to fibrin, but fibrin contains no specific binding sites for SK. Consequently, when SK is administered intravenously in patients, two independent actions take place, namely, activation of fibrin-bound plasminogen and activation of plasma plasminogen.

2. Anisoylated Lys-Plasmin(ogen) Streptokinase Activator Complex (APSAC, anistreplase)

Plasmin generated in whole blood or plasma is immediately inactivated, but plasmin bound via its high-affinity lysine binding site to fibrin is inactivated at a much lower rate by α_2-antiplasmin. In the search for more efficient thrombolytic agents, some researchers arrived at the original idea to acylate the active site center of the SK-plasmin(ogen) activator complex in which a configurational change exposes the active plasmin(ogen) site. Coupling of p-anisoyl to the active site serine of the plasmin(ogen) portion of activator complex led to a compound originally designated as BRL 26921, subsequently named APSAC (anisoylated plasminogen-SK activator complex) for the formal generic name anistreplase (Eminase, Smith-Kline Beecham). Thus, the catalytic site of anistreplase is blocked but its affinity to fibrin via the lysine binding site is preserved. Anistreplase deacylates with a half-life of about 40 min at 37 °C (HIBBS et al. 1989), and since it binds to fibrin, it is feasible to administer this drug as a bolus injection. In experimental animal models anistreplase had better clot specificity, i.e., caused less fibrinogenolysis and consumption of plasminogen and of α_2-antiplasmin, and a lesser hypotensive effect in vivo than SK (reviewed by FEARS 1989). Clinical studies revealed that anistreplase is indeed an effective thrombolytic agent, but the systemic proteolytic effects are hardly different from those observed with SK (HOFFMANN et al. 1985; SAMAMA et al. 1987; ANDERSON 1989; BASSAND et al. 1991; reviewed in MARDER 1989; VERSTRAETE 1989).

II. The Proteolytic State

During therapy with sufficient SK or anistreplase to overcome anti-SK antibody (the only inhibitor to the action of SK or its activator complex), prompt and extensive activation of plasma plasminogen takes place. The hyperplasminaemic state that results produces a hypocoagulable state in which fibrinogen, vWf, and factor V and VIII levels fall precipitously, the former often to levels of less than 1 g/l. The breakdown products of fibrinogen proteolysis (FDPs) that are formed impede normal clotting and contribute to the hypocoagulable state. These effects significantly prolong the thrombin clotting time, and to a lesser extent the activated partial thromboplastin time (APTT) and the prothrombin time. Hemostasis is impaired, primarily because of the degradation of fibrinogen (MENTZER et al. 1986) and the fact that no native fibrinogen remains in the circulation, the circulating clottable protein being fragment X (Fig. 1), a proteolytic fragment of fibrinogen (MARDER et al. 1969; MENTZER et al. 1986). The hemostatic defect not only increases the risk and severity of bleeding, but at the same time diminishes the likelihood of arterial rethrombosis. The anticoagulant state produced during and for some time after thrombolytic therapy with SK or anistreplase is a useful feature because the thrombogenic stimulus of a ruptured plaque, or an inadequate flow rate, may persist. The presence of an anticoagulant state avoids the necessity for

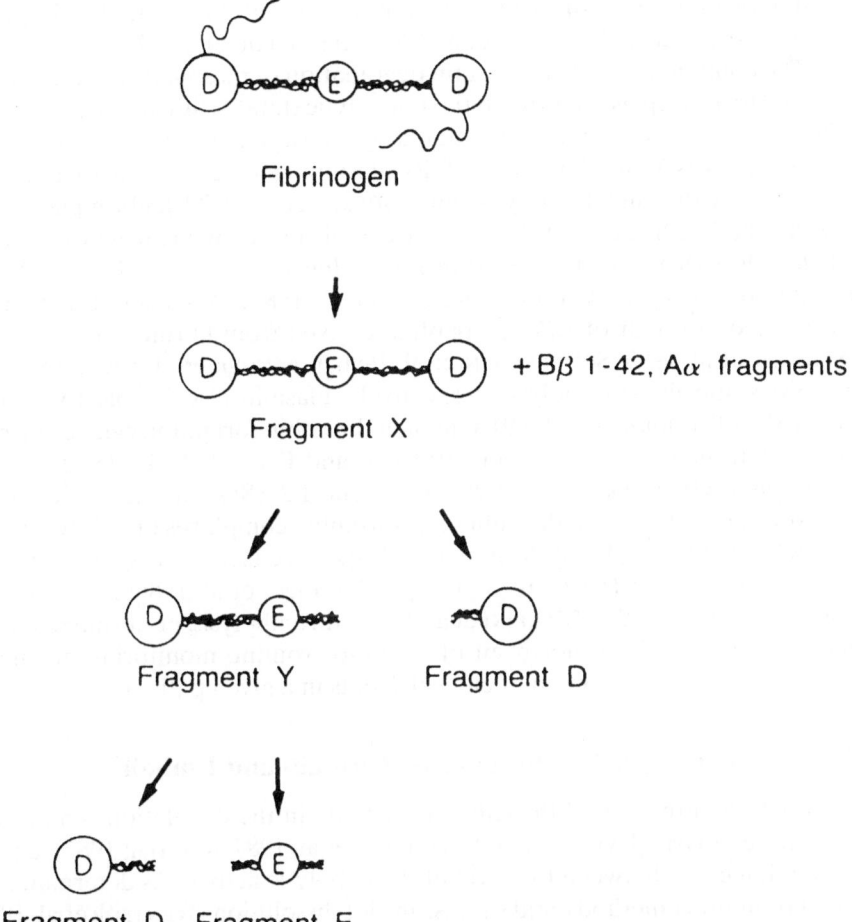

Fig. 1. Schema of degradation of fibrinogen by plasmin with the sequential formation of fragment X (slowly coagulable by thrombin), fragment Y and D (inhibitors of the conversion of fibrinogen to fibrin), and the end-products fragments D and E. Reproduced with permission from FRANCIS and MARDER (1994)

the simultaneous use of heparin, an agent that can add significantly to the bleeding risk (TIMMIS et al. 1986) (see also Sect. B.I.f).

Fibrinogen is the major determinant of plasma viscosity (MERRILL et al. 1966; REPLOGLE et al. 1967; reviewed by BECKER 1991). Thus, the breakdown of fibrinogen, associated with a reduction of blood and plasma viscosity (MORIARTY et al. 1988), may improve flow rate and reduce afterload and could be of benefit in patients with AMI, given that an increase in blood viscosity can be associated with coronary heart disease (JAN et al. 1975; MORIARTY et al. 1988; YARNELL et al. 1991; LOWE et al. 1991). After SK or anistreplase therapy restoration of a normocoagulable state depends on the synthesis of new fibrinogen and clearance of FDPs. Fibrinogen levels usually reach normal values within 36–48 h, and the half-life of FDPs is 6–8h (MENTZER et al. 1986), with higher molecular forms cleared more slowly (FLETCHER et al. 1962; ARDAILLOU et al. 1977).

The combined effects of plasmin degradation of plasma proteins are often referred to as the plasma proteolytic state ("lytic state") (SHERRY et al. 1959a,b). While the lytic state can be quantitated by measurement of plasminogen, fibrinogen, factors V and VIII, and FDPs, plasminogen activator activity can be measured by the euglobulin lysis time (SHERRY et al. 1959b), fibrin plate assay (ASTRUP and MÜLLERTZ 1952), or release of fibrin fragments from isotopically labeled clots (SAWYER et al. 1960) or fibrin films (MOROZ and GILMORE 1975). In addition, assay of D-dimer (RYLATT et al. 1983), a specific fragment of crosslinked fibrin, or of $B\beta_{1-42}$, a peptide cleaved from fibrinogen by plasmin (SCHMIDT et al. 1993; SCHARFSTEIN et al. 1996; GRANGER et al. 1998), reflect fibrinolysis and fibrinogenolysis, respectively. Plasmin also is able to activate factor VII (TSUJIOKA et al. 1999). Ongoing thrombin formation can be assessed by the determination of fibrinopeptides A and B (FPA, FPB) (SCHARFSTEIN et al. 1996; GRANGER et al. 1998), fragment 1.2 (SCHARFSTEIN et al. 1996; GRANGER et al. 1998), and thrombin-antithrombin complexes (TAT) (GULBA et al. 1988; TRIPODI et al. 1990). Although such assays reflect processes of increased thrombin formation (EISENBERG et al. 1987; GULBA et al. 1988; RAPOLD et al. 1989; TRIPODI et al. 1990; MERLINI et al. 1995), thrombolysis, or fibrinogen/fibrin degradation, they have not been of value for routine monitoring of thrombolytic therapy or prediction of clinical events in a given patient.

III. Factors Influencing the Lysis of Thrombi and Emboli

Although the presence of thrombolytic activity in the circulation is a prerequisite for clot lysis (RYLATT et al. 1983; JAFFE et al. 1987; MARDER 1987), a poor correlation exists between the level of thrombolytic activity, as determined by direct or indirect methods, and success in clot dissolution (RENTROP et al. 1981; ROTHBARD et al. 1985; MARDER 1987; RAO et al. 1988; BRENNER et al. 1990; FRANCIS and MARDER 1994; EL GAYLANI et al. 1996); this is because local factors in and around a thrombus are critical in determining the success of thrombolysis and the absence of vascular injury is the major determinant for avoiding fibrinolytic hemorrhage.

Effective thrombolysis depends on the penetration of activator into the clot with activation of fibrin-bound plasminogen. The rate of lysis in vitro is a direct function of the proximate concentration of the plasminogen activator and of fibrin-bound plasminogen, with the solubilizing process beginning at the surface and progressing inward (FRANCIS et al. 1984). In vivo this relationship is influenced greatly by anatomical and biological factors that are not controllable. These factors include (1) the surface area of clot exposed to blood, (2) the concentration of fibrin-bound plasminogen, (3) the affinity of the plasminogen activator to fibrin and plasminogen, and (4) the age of the thrombus.

Total occlusions of coronary arteries, defined as grade 0 in the TIMI 1 study (THE TIMI STUDY GROUP 1985), which expose a small surface area to the clot, are not as readily dissolved as when there is penetration of blood into and especially around the clot (grade 1) or when the vessel is only partially occluded (grade 2). That old thrombi are more resistant to lysis by plasminogen activators is well-recognized, partly because such thrombi are often inaccessible to activator, but also due to chemical changes in fibrin (MARDER et al. 1994). Factors that lead to increased resistance include (1) rapid chemical cross-linking of the fibrin polymers (GORMSEN et al. 1967), (2) formation of ultralarge fibrin α-chain polymers as occurs in the presence of high concentrations of factor XIII or large numbers of platelets (FRANCIS and MARDER 1988), (3) cross-linking of α_2-antiplasmin to fibrin by factor XIII (REED et al. 1992; REED and HOUNG 1999), (4) a high content of platelets containing plasminogen activater inhibitor 1 (PAI-1) (ZHU et al. 1999), (5) clot retraction, during which serum is squeezed out and fibrin becomes more compacted, and (6) progressive impaction of the thrombus so as to occlude the vessel totally. Resistance to lysis by SK in coronary arteries is not as evident with tissue-type plasminogen activator (tPA) (THE TIMI STUDY GROUP 1985; STEG et al. 1998) since the latter binds to fibrin and works longer and more effectively on older, more lysis-resistant fibrin.

B. Therapeutic Results in Acute Myocardial Infarction

I. Streptokinase

1. Early Studies with Intravenous Streptokinase

Paul Dudley White, a leading cardiologist in the first half of this century and a well known clinician and investigator among hemostasis and thrombosis experts, described in the first edition of his book on heart disease (WHITE 1931) that AMI could be due to the ulceration of atheromatous plaques and the formation of a coronary thrombus. This knowledge appears to have been ignored by subsequent cardiologists and pathologists, as exemplified by opinions such as "The infrequency of coronary thrombi in patients who died of acute cardiovascular collapse ... suggests that coronary thrombi are consequences rather than causes of acute myocardial infarction" (ROBERTS and BUJA 1972). This approach was certainly not conducive to the study of thrombolytic agents in patients with AMI, and in the 1960s and 1970s nearly all of the early thrombolytic trials in AMI were organized by hemostasis experts.

The first clinical report on a series of 17 men and 5 women with AMI, treated with an initial titrated dose of SK, followed by an intravenous infusion of 35000 to 150000 units/h over a period of 8h was reported by FLETCHER et al. (1958). Subsequently, from 1963 to 1979, there were 19 additional trials with intravenous SK, usually involving a 24-h period of treatment. The results of all of these trials have been reviewed (YUSUF et al. 1985; GRÜNEWALD and SEIFRIED 1994) and eight trials that met more rigid inclusion criteria (DEWAR et al. 1963; AMERY et al. 1969; HEIKINHEIMO et al. 1971; DIOGUARDI et al. 1971; BETT et al. 1973; BREDDIN et al. 1973; ABER et al. 1976; EUROPEAN CO-OPERATIVE STUDY GROUP FOR STREPTOKINASE TREATMENT IN ACUTE MYOCARdial Infraction 1979) were analyzed by Stampfer et al. (1982). A meta-analysis of the results, a basis for analysis when individual sample sizes are inadequate for mortality studies (L'abbe et al. 1987), suggested a 20%–22% reduction in the odds of death.

During this same period several pathologists clarified the pathogenesis of AMI by demonstrating that coronary thrombi were indeed frequently found in patients who died from AMI (Davies et al. 1976; reviewed by Davies 1994) and that the triggering event often was a ruptured plaque (Falk 1983). The seminal report by Dewood et al. (1980) demonstrated by means of coronary angiography a thrombus in 110 of 126 patients (87%) who were evaluated within 4h of the onset of symptoms. This report and the early work of Rentrop et al. (1979) represent a turning point in the management of AMI and led, in the early 1980s, to a dramatic increase in interest by the cardiology community.

2. Intracoronary Administration

Intracoronary administration of preparations containing SK is not a recent innovation. Nearly 30 years ago, Boucek et al. (1960) developed a novel method for the use of their SK-containing preparation for local perfusion, and Chazov et al. (1976) were the first to use coronary catheterization to instill SK directly into the coronary artery. Nevertheless, it remained for Rentrop et al. (1979, 1981) to demonstrate that intracoronary administration of SK into the obstructed coronary artery early after the onset of symptoms frequently resulted in rapid reperfusion. The latter observation soon was confirmed by others (Mason 1981) who infused 2000 U/min for periods of 60–90 min. When therapy was initiated within 3h of onset of symptoms, on average 75% of occluded vessels were reperfused in 30 min; serious bleeding complications were low (4.8%) and originated primarily at the catheter insertion site. A significant incidence of reocclusion (18%) was related to underlying residual stenosis (Serruys et al. 1983; Gold et al. 1984) with unlysed thrombus a prominent factor in the reduction of adequate reflow (Gash et al. 1986; Brown et al. 1986). However, ventriculography showed a 55% return of function of the initially stunned myocardium, which, in the absence of reperfusion, would have been expected to undergo necrosis and irreversible loss of function (Spann and Sherry 1984). Studies with [201]thallium scintigraphy demonstrated that coronary thrombolysis was associated with improved regional perfusion (Markis et al. 1981).

However, consistent with the observations by Reimer et al. (1979) on the progression of irreversible necrosis as a function of time in experimental coronary occlusion in dogs, Bergmann et al. (1982) demonstrated by positron emission tomography that there was little evidence of myocardial salvage when therapy was begun more than 4h after the onset of coronary occlusion. Along with timely reperfusion, there is a rapid disappearance of chest pain, early peaking of the MB fraction of creatine phosphokinase, appearance of reperfusion arrhythmias, and rapid reduction of the injury current on the electro-cardiogram (MASON 1981). Registries of intracoronary SK treatment (WEINSTEIN 1982; KENNEDY et al. 1985) strongly suggested improved mortality, and provided the basis for several randomized trials that produced short-term (KENNEDY et al. 1983; SIMOONS et al. 1986) and long-term (KENNEDY et al. 1985; SIMOONS et al. 1986) evidence of a reduction in mortality, especially among patients with anterior infarcts.

3. High-Dose Intravenous Administration of Streptokinase

a) Dose-Ranging Studies and Anti-SK Antibodies

The technical and logistic problems associated with emergency coronary catheterization, as well as its high cost led to the adoption of high-dose intravenous administration of SK. Dose-ranging studies of intravenous SK have not been exhaustive (COL et al. 1989; SIX et al. 1990), but the dose that has been studied most extensively and which has been generally accepted for routine use is 1,500,000 units (1.5 MU) infused over 1 h. This SK dose floods the circulation with high levels of SK, and neutralizes anti-SK antibodies in over 95% of all patients. SK and anistreplase are highly antigenic and most humans exhibit anti-SK antibodies (BACHMANN 1968), probably due to previous infection with α-hemolytic streptococci (SHAILA et al. 1994). Very high levels of anti-SK antibodies may prevent the lytic state and (LEW et al. 1984; SIGWART et al. 1985; HOFFMANN et al. 1988; BUCHALTER et al. 1992; JUHLIN et al. 1999) result in non-reperfusion. In most cases, the neutralization of anti-SK antibodies typically results in an activity equivalent of 65 U/ml of SK at the end of the SK infusion, and measurable SK activity remains in the circulation for about 1–2 h (MENTZER et al. 1986). Recovery from the lytic state begins 2 h later and is complete within 36–48 h (MARDER 1987). Anti-SK antibody titers rise on the fifth day after SK or anistreplase administration and rapidly reach 1000 U/ml of plasma at 2–4 weeks before slowly receding (SHAILA et al. 1994). In some patients high anti-SK titers may be found up to 7.5 years after the intravenous administration of SK or anistreplase (ELLIOTT et al. 1993; FEARS et al. 1992; SQUIRE et al. 1999). For this reason, it is recommended to not treat a patient who previously received SK or anistreplase a second time with an SK-containing agent. A mutant lacking the C-terminal 42 amino acids has been shown to be less immunogenic in baboons, but has not yet been used clinically in man (TORRÈNS et al. 1999).

Variations of the above dosing schedule have been utilized, for example, 3 MU (SIX et al. 1990; THEISS et al. 1996), or combinations with tPA and either 1.5 MU or reduced doses of SK (GRINES et al. 1991; reviewed by MARDER et al. 1994). Totally unexploited is the potential of genetically-engineered mutants of SK (MALKE and FERRETTI 1984; MÜLLER et al. 1989; WU et al. 1998; SHI et al. 1998). For example, replacement of plasmin-cleavable peptide bonds by non-cleavable bonds (WU et al. 1998; SHI et al. 1998) would likely result in mutants which remain active in the circulation for longer times. Deletion of the

N-terminal 59 amino acids conferred fibrin specificity to SK, i.e. resulted in markedly diminished fibrinogen degradation in vitro (REED et al. 1999), and a hirulog-SK fusion protein was found to bind to clot-bound thrombin, enhancing clot lysis (WANG et al. 1999). To date there have been no reports which have evaluated the clinical efficacy of such genetically engineered forms of SK.

b) Patency vs Recanalization Rate

In most trials where coronary angiography was performed in patients with suspected AMI, about 20% of patients exhibited open (patent) coronary arteries prior to treatment (GRANGER et al. 1994). A review of the literature shows that achievement of patency after intracoronary SK treatment was clearly associated with a lowered mortality, 2.5% of the 76% of patients with patent vessels vs 14% in the 24% of patients with vessels that remained occluded (KENNEDY et al. 1983).

Most recent studies using coronary angiography as one of the endpoints of thrombolytic therapy have evaluated reperfusion grade criteria established by the TIMI-1 investigators (THE TIMI STUDY GROUP 1985), with grades 0 (no distal flow) and 1 (minimal flow) perfusion being designated as thrombolysis failures and grades 2 (partial perfusion) and 3 (complete perfusion) as thrombolysis successes. Several studies have lent strong support to the "open artery" hypothesis which states that coronary reperfusion is associated with improved cardiac function and reduced morbidity and mortality (THE GUSTO ANGIO-GRAPHIC INVESTIGATORS 1993; SIMES et al. 1995; reviewed by FIBRINOLYTIC THERAPY TRIALISTS' (FTT) COLLABORATIVE GROUP 1994). Furthermore, although achievement of TIMI grade 2 flow may still benefit patients (REINER et al. 1996), it has become increasingly clear that only early and complete grade 3 flow is associated with optimal survival and clinical outcomes (DALEN et al. 1988; KARAGOUNIS et al. 1992; ANDERSON et al. 1993; STADIUS 1993; VOGT et al. 1993a; THE GUSTO ANGIOGRAPHIC INVESTIGATORS 1993; LINCOFF et al. 1995; GIBSON et al. 1995, 2000; reviewed by GRANGER et al. 1994). In the TEAM-2 trial, comparing anistreplase with SK, patients with TIMI grade 2 flow at 90–240 min after thrombolysis had indexes of infarct size comparable with those of patients with TIMI grade 0 or 1 flow (KARAGOUNIS et al. 1992). The retrospective review of four German multicenter studies revealed that in-hospital mortality was higher for patients with TIMI grade 0 or 1 flow (7.1%) and grade 2 flow (6.6%) than in those with grade 3 flow (2.7%) (VOGT et al. 1993a). In the GUSTO-I angiographic substudy, the mortality rate at 30 days among patients with grade 3 flow was 4.4% but 7.4% among those with grade 2 flow (THE GUSTO ANGIOGRAPHIC INVESTIGATORS 1993). This study also showed that all indexes of left ventricular function, including ejection fraction, end-systolic volume index, and global and regional wall motion, were clearly better at 90 min and at 7 days in the group who had early TIMI grade 3 flow.

A recent review analyses TIMI grade 2 and 3 flows and reocclusion rates obtained with three different thrombolytic agents (BARBAGELATA et al. 1997)

Table 1. TIMI grade 2 and 3 flow rates with different thrombolytic agents. Pooled analysis

Time[a]	Treatment Standard tPA	Accelerated tPA	Streptokinase	Anistreplase
TIMI grade 3 flow (%)				
60 min	39 (29–50)	58 (52–62)	–	40 (33–47)
90 min	50 (46–54)	63 (60–67)	31 (28–35)	50 (45–55)
180 min	–	43 (33–53)	46 (41–50)	60 (53–67)
1 day	73 (66–80)	71 (66–75)	51 (45–57)	78 (74–82)
3–21 days	77 (63–92)	75 (70–80)	54 (47–62)	88 (83–93)
TIMI grade 2 flow (%)				
60 min	23 (14–32)	17 (13–21)	–	20 (15–25)
90 min	22 (19–26)	20 (17–23)	25 (22–29)	18 (14–22)
180 min	–	33 (24–43)	27 (22–30)	12 (7–17)
1 day	13 (7–18)	15 (12–19)	30 (24–36)	12 (8–15)
3–21 days	3 (0–9)	12 (8–16)	24 (18–30)	2 (0–4)

[a] Time from start of thrombolytic therapy to angiography. Values are point estimates, with 95% confidence intervals in parenthesis.

(Table 1). The pooled data of 5170 patients who had coronary angiography after thrombolytic treatment show that the rate of early TIMI 3 flow was lowest with SK, highest with accelerated tPA, and intermediate and similar with standard dose t-PA and anistreplase at 90 min. At 180 min, there was a decrease with accelerated tPA to about the same level as with SK (about 44%, so-called "catch-up" phenomenon for SK) and then an increase to about 70% by 24 h. The rate for anistreplase increased to 60% by 180 min and to 90% between 3 and 21 days. SK was associated with stable TIMI 3 flow rates, remaining about 55% at the latest angiogram. The overall reocclusion rate was 12% with standard-dose and 6% with accelerated tPA. The non-fibrin specific agents anistreplase and SK had overall reocclusion rates of 4.2% and 3% respectively. It is possible that the higher reocclusion rate in patients treated with standard-dose tPA, compared with those given SK, offsets a potential benefit of earlier TIMI grade 3 perfusion in the tPA group, as evidenced by the equivalent functional parameters between tPA and SK groups in the TIMI-1 study (SHEEHAN et al. 1987) and mortality rates in the GISSI-2 trial (GRUPPO ITALIANO PER LO STUDIO DELLA SOPRAVVIVENZA NELL'INFARTO MIOCARDICO (GISSI) 1990). With the introduction of core angiographic laboratories the reproducibility of TIMI flow grades has been questioned because of interobserver variability, its categorical nature, its limited statistical power and the fact that the nonculprit flow (used to gauge TIMI 3 flow) is abnormally slow early in the course of AMI. Recently the corrected TIMI frame count (CTFC) has been developed as a more reproducible method of quantifying infarct artery blood flow. In this method the number of frames required for dye to first reach standardized distal landmarks is counted. Low CTFC of 10–50 therefore corresponds to free flow and are associated with lower mortality (GIBSON et al. 1996, 1999a,b; FRENCH et al. 1998, 1999a,b).

Table 2. Short-term mortality reduction using SK (vs placebo) administered in the first 3 h after acute myocardial infarction

Study	Year	No. of Patients	Mortality %		Survival benefit
			SK	Placebo	
ISAM	1986 1987	940	5.2	6.5	20% (n.s.)
GISSI	1986 1987	6094	9.2	12.0	23% ($p = 0.0005$)
WHITE et al. (1987)	1987	219	3.7	12.5	70% ($p = 0.012$)
Western Washington	1988	194	5.2	11.3	54% (n.s.)
ISIS-2	1987 1988	5108	8.1	12.2	34% ($p < 0.0001$)

Follow-up at approximately 3–4 weeks after treatment. Overall reduction of approximately 30% in favor of treatment with SK.
References: ISAM: THE I.S.A.M. STUDY GROUP (1986); SCHRÖDER et al. (1987); GISSI: GRUPPO ITALIANO PER LO STUDIO DELLA STREPTOCHINASI NELL'INFARTO MIOCARDICO (GISSI) (1986, 1987); Western Washington: KENNEDY et al. (1988); ISIS-2: ISIS STEERING COMMITTEE (1987); THE ISIS-2 (SECOND INTERNATIONAL STUDY OF INFARCT SURVIVAL) COLLABORATIVE GROUP (1988)

c) Mortality Reduction

There have been six randomized trials comparing mortality rates in patients with an acute myocardial infarction assigned either to a placebo control group or to a group receiving 1.5 MU of SK over a 30–60 min period (THE I.S.A.M. STUDY GROUP 1986; SCHRÖDER et al. 1987; VOTH et al. 1991; GRUPPO ITALIANO PER LO STUDIO DELLA STREPTOCHINASI NELL'INFARTO MIOCARDICO (GISSI) 1986, 1987; WHITE et al. 1987; KENNEDY et al. 1988; RITCHIE et al. 1988; CERQUEIRA et al. 1992; ISIS STEERING COMMITTEE 1987; THE ISIS-2 (SECOND INTERNATIONAL STUDY OF INFARCT SURVIVAL) COLLABORATIVE GROUP 1988; BAIGENT et al. 1998; EMERAS (ESTUDIO MULTICÉNTRICO ESTREPTOQUINASA REPÚBLICAS DE AMÉRICA DEL SUR) COLLABORATIVE GROUP 1993). Table 2 summarizes the results on those patients who received thrombolytic therapy within 3 h of symptom onset and who were evaluated for short term survival at 2–4 weeks after treatment. Two of the studies showed a trend that was not statistically significant towards a survival benefit using SK. The 20% advantage (5.2% vs 6.6%) in survival in the ISAM trial (THE I.S.A.M. STUDY GROUP 1986) was however associated with significantly better ventricular function in patients with anterior MI; the 20% improvement persisted up to three years after treatment (VOTH et al. 1991). The Western Washington trial (KENNEDY et al. 1988) showed a marked 54% improved survival, but the number of patients was insufficient to reach significance at 14 days after treatment; follow-up at two years in patients who had had anterior AMI did show a significant 20% advantage in the SK group. A study reported by WHITE et al. (1987) was also quite small, but the dramatic difference in mortality (3.7% vs 12.5%) among only 219 patients in the trial reached statistical significance for the 70% survival advantage with SK.

However, most weight of evidence is provided by two much larger studies. The GISSI trial (GRUPPO ITALIANO PER LO STUDIO DELLA STREPTOCHINASI NELL'INFARTO MIOCARDICO 1986, 1987), which compared more than 5000 patients treated in this time frame, showed a 23% survival advantage with SK (9.2% vs 12% mortality), which was highly significant ($p = 0.0005$), and likewise, the ISIS-2 experience showed an increase of 34% in survival (8.1% vs 12.2%; $p < 0.0001$) (THE ISIS-2 (SECOND INTERNATIONAL STUDY OF INFARCT SURVIVAL) COLLABORATIVE GROUP 1988). The overall experience of the five trials summarized in Table 2 is a reduction of 29% for patients treated within 3 h of symptom onset, as evaluated at approximately 2–4 weeks after such therapy. The sixth trial is the South American EMERAS study which enrolled only patients presenting more than 6 h after the onset of symptoms [EMERAS (ESTUDIO MULTICÉNTRICO ESTREPTOQUINASA REPÚBLICAS DE AMÉRICA DEL SUR) COLLABORATIVE GROUP 1993]. Among the 2080 patients in whom the time to treatment was 7–12 h from symptom onset there was a nonsignificant trend towards fewer deaths with SK (11.7% vs 13.2%), whereas there was little difference among the 1791 patients presenting after 13–24 h (11.4% vs 10.7%).

d) Delay to Treatment

Clearly, the earlier that patients are treated with thrombolytic therapy after symptom onset, the better the results [GRUPPO ITALIANO PER LO STUDIO DELLA STREPTOCHINASI NELL'INFARTO MIOCARDICO (GISSI) 1987; HACKWORTHY et al. 1988; THE ISIS-2 (SECOND INTERNATIONAL STUDY OF INFARCT SURVIVAL) COLLABORATIVE GROUP 1988; FRENCH and WHITE 1995; RAITT et al. 1996; NEWBY et al. 1996; LANGER et al. 1996; KELLY et al. 1997; STEG et al. 1998; ZEYMER et al. 1999; reviewed by BOERSMA et al. 1996]. The GISSI study showed that treatment provided during the first hour reduced mortality significantly by 47% as compared with 23% when treatment was delayed for up to 3 h and only 17%, still a significant reduction ($p = 0.03$), when treatment was provided at 3–6 h after symptom onset. Similar results were obtained in the ISIS-2 trial, which showed a 35% reduced mortality with treatment in the first 4 h, as compared with a 17% reduction with treatment provided at 5–24 h. The data on survival vs delay in the administration of thrombolytic therapy have been summarized for all of the plasminogen activator studies comprising more than 1000 patients [FIBRINOLYTIC THERAPY TRIALISTS' (FTT) COLLABORATIVE GROUP 1994]. These investigators arrive at a straight line relationship of lives saved with delay after symptom onset, from a high value of 35 lives/1000 in the first hours to about 20 for those treated between 7 h and 12 h and a non-statistically significant trend in favor of thrombolytic treatment (approximately 6 lives/1000 treatments) between 12 h and 24 h after symptom onset. Using the same trials and in addition trials specifically designed to study the impact of the delay to treatment, BOERSMA et al. (1996) did not find a straight line relationship between time to treatment and number of lives saved. They arrive at an even higher number of lives saved if treatment is given within the first hour after onset of symptoms (65/1000; see Fig. 2). Follow-up at one or two years [GRUPPO

A

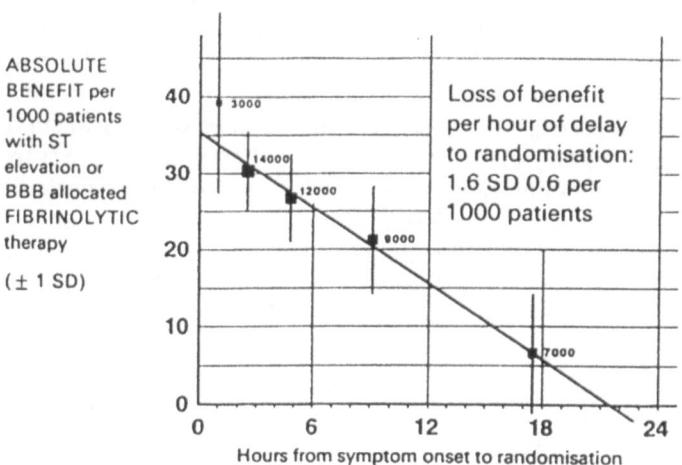

B

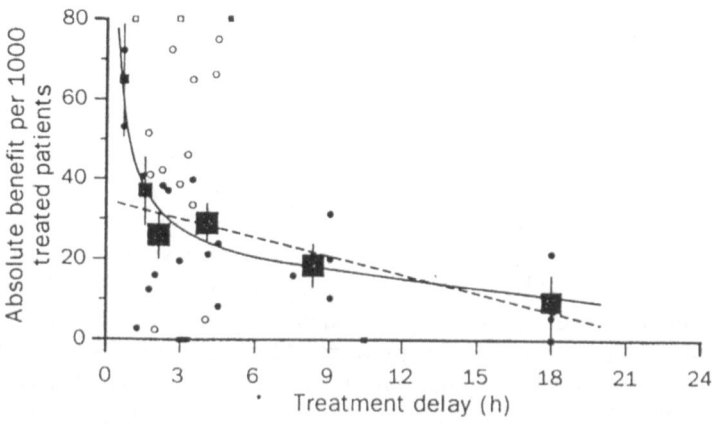

Fig. 2. A Absolute reduction in 35-day mortality vs delay from symptom onset to randomization among 45 000 patients with ST elevation or bundle-branch block. Reproduced with permission from FIBRINOLYTIC TRIALISTS' (FTT) COLLABORATIVE GROUP (1994). **B** Absolute 35-day mortality reduction vs treatment delay. *Small closed circles*: information from trials included in FTT analysis (above); *open circles*: information from additional trials; *small squares*: data beyond x/y cross. The linear and the non-linear regression lines are fitted within these data, weighted by inverse of the variance of the absolute benefit in each datapoint. *Black large squares*: average effects in six time-to-treatment groups (areas of squares inversely proportional to variance of absolute benefit described). Reproduced with permission from BOERSMA et al. (1996)

ITALIANO PER LO STUDIO DELLA STREPTOCHINASI NELL'INFARTO MIOCARDICO (GISSI) 1987; SCHRÖDER et al. 1987; THE ISIS-2 (SECOND INTERNATIONAL STUDY OF INFARCT SURVIVAL) COLLABORATIVE GROUP 1988], or even up to 14 years (BAIGENT et al. 1998; FRANZOSI et al. 1998; FRENCH et al. 1999; MAAS et al. 1999) after treatment clearly show that the survival advantage achieved at about 1 month is maintained long-term. While individual reports suggested that intracoronary SK administration can reduce mortality (KENNEDY et al. 1983; ANDERSON et al. 1983; KHAJA et al. 1983; KENNEDY et al. 1985; SIMOONS et al. 1986), a retrospective analysis of eight randomized studies of the intracoronary route (vs placebo) by FURBERG (1984) showed no advantage in mortality (11% vs 12.4%, $p = 0.64$) at 2–54 weeks after the infarction. It is likely that the greater delay required to provide intracoronary therapy accounts for this suboptimal response.

In distinction to the marginal [FIBRINOLYTIC THERAPY TRIALISTS' (FTT) COLLABORATIVE GROUP 1994] or frank absence of benefit if SK [EMERAS (ESTUDIO MULTICÉNTRICO ESTREPTOQUINASA REPÚBLICAS DE AMÉRICA DEL SUR) COLLABORATIVE GROUP 1993] or tPA (LATE STUDY GROUP 1993) is administered more than 12 h following symptom onset, efforts to minimize the time delay before treatment appear to translate into a survival advantage. The administration of plasminogen activator in the ambulance or even at home can cut an average of about one hour from the usual time delay before treatment with an overall decrease in mortality from 15% to 25% (LE FEUVRE et al. 1993; ROZENMAN et al. 1994; BROUWER et al. 1996; RAWLES 1997; reviewed by CARLSSON et al. 1997) (see also below). Factors which contribute to delays in symptom to needle time are advanced age, female gender, diabetes (MAYNARD et al. 1995; NEWBY et al. 1996; LAMBREW et al. 1997; LEIZOROVICZ et al. 1997; BERGLIN BLOHM et al. 1998) and the availability of primary PTCA. Indeed, in university hospitals offering 24 h a day PTCA, physicians who are confronted with the difficult choice of selecting a particular therapy for a patient may contribute to delay of instituting thrombolytic therapy for patients not suited for primary PTCA (DOOREY et al. 1998). Increased delays in instituting thrombolytic therapy are particularly disadvantageous for patients to be treated with SK (STEG et al. 1998). Guidelines on how to reduce symptom to needle time have been published (WILLIAMS 1998).

e) Bleeding Complications

All thrombolytic agents produce a risk for bleeding complications (RAO et al. 1988; SANE et al. 1989; MARDER 1990; LAUER et al. 1995; LEVINE et al. 1995; BERKOWITZ et al. 1997). It has become increasingly evident that the major cause of bleeding is lysis of hemostatic plugs at sites of previous invasive procedures or internal vascular injury (MARDER 1979; MARDER and SHERRY 1988). Thrombolytic therapy can induce platelet disaggregation (LOSCALZO and VAUGHAN 1987; RUDD et al. 1990) and dissolve fibrin in a hemostatic plug. Another contributing factor to bleeding is the use of heparin in conjunction with or immediately after thrombolytic therapy (National Institutes of Health Consensus Delvelopment Conference 1980; Ganz et al. 1984; Hillis et al. 1985; The Timi Study Group 1985; De Bono et al. 1992; Timmis et al. 1986; Verstraete et al. 1985; Levine et al. 1995; Bovill et al. 1997). For example, delayed subcutaneous heparin added to thrombolytic therapy of AMI is associated with a higher incidence of bleeding, including an increase of intracranial hemorrhage from 0.4% to 0.6% [Gruppo Italiano Per Lo Studio Della Sopravvivenza Nell'infarto Miocardico (GISSI) 1990; The International Study Group 1990; ISIS-3 (Third International Study Of Infarct Survival) Collaborative Group 1992]. Furthermore, intravenous heparin causes more bleeding than does subcutaneous heparin when administered with aspirin plus thrombolytic agents (SK, anistreplase or tPA) [ISIS-3 (Third International Study Of Infarct Survival) Collaborative Group 1992; Ridker et al. 1993; Bovill et al. 1997]. Higher doses of intravenous heparin and hirudin increase the bleeding risk even more, for example in the GUSTO-IIa [The Global Use of Strategies to Open Occluded Coronary Arteries (GUSTO) IIA Investigators 1994] and the TIMI-9 A (Antman 1994) trials which showed intracranial hemorrhage rates with SK and tPA of about 2%, resulting in early termination of the studies and reassessment of anticoagulant dosages. Heparin therapy apparently adds significantly to the bleeding risk after a hemostatic plug is dissolved by thrombolytic agents. Thus the incidence of intracranial hemorrhage using heparin plus aspirin was 0.3% in GUSTO-IIa, vs 1.8% for heparin plus aspirin use in addition to a thrombolytic agent [The Global Use of Strategies to Open Occluded Coronary Arteries (GUSTO) IIA Investigators 1994]. The use of effective glycoprotein IIb/IIIa receptor blockage with abciximab also increases the bleeding risk of thrombolytic therapy (Yaryura et al. 1998).

Bleeding, including severe bleeding, is most frequent when invasive procedures such as arterial catheterization are part of the treatment protocol (The Timi Study Group 1985; Verstraete et al. 1985). In the trials with high-dose, brief-duration intravenous SK for AMI, when invasive procedures were not performed as part of the protocol, the incidence of severe bleeding in the absence of compulsory anticoagulation has been about 0.3%–0.5%; with compulsory anticoagulation it has averaged 0.5%–1.0% [The I.S.A.M. Study Group 1986; Schreiber et al. 1986; Gruppo Italiano Per Lo Studio Della Streptochinasi Nell'infarto Miocardico (GISSI) 1986; Isis (International Studies Of Infarct Survival) Pilot Study Investigators 1987; White et al. 1987; Kennedy et al. 1988; Gruppo Italiano Per Lo Studio Della Sopravvivenza Nell'infarto Miocardico (GISSI) 1990; The International Study Group 1990; ISIS-3 (Third International Study Of Infarct Survival) Collaborative Group 1992].

If major bleeding occurs, thrombolytic therapy should be discontinued, blood replaced, and cryoprecipitate and platelets administered. The use of antifibrinolytic agents such as ε-amino caproic acid or aprotinin (Efstratiadis et al. 1991; Marder 1990) may be required in some instances.

Table 3. Contraindications and cautions for thrombolytic use in myocardial infarction[a]

Previous hemorrhagic stroke at any time, other strokes or cerebrovascular events within 1 year
Known intracranial neoplasm
Active internal bleeding (does not include menses)
Suspected aortic dissection
Cautions/relative contraindications
Severe uncontrolled hypertension on presentation (blood pressure >180/110 mm Hg)[b]
History or prior cerebrovascular accident or known intracerebral pathology not covered in contraindications
Current use of oral anticoagulation in therapeutic doses (INR ≥2–3); known bleeding diathesis
Recent trauma (within 2–4 weeks), including head trauma or traumatic or prolonged (>10 min) cardiopulmonary resuscitation or major surgery (<3 weeks)
Noncompressible vascular punctures
Recent (within 2–4 weeks) internal bleeding
For SK/anistreplase: prior exposure (especially within 5 days–2 years) or prior allergic reaction
Pregnancy
Active peptic ulcer
History of chronic severe hypertension

Reproduced from RYAN et al. (1996)
[a] Viewed as advisory for clinical decision making and may not be all-inclusive or definitive.
[b] Could be an absolute contraindication in low-risk patients with myocardial infarction.

f) Patient Selection, Contraindications

Many bleeding complications can be avoided by patient selection, elimination of invasive procedures, limited duration of therapy, and judicious use of anticoagulation (FLETCHER et al. 1965; MARDER 1979; SHARMA et al. 1982; LEVINE et al. 1995; CAIRNS et al. 1995; SIMOONS, FOR THE REPERFUSION THERAPY CONSENSUS GROUP 1997). Table 3 lists the contraindications and cautions recommended by the ACC/AHA guidelines for the management of patients with acute myocardial infarction (RYAN et al. 1996). Although patients over 70 years of age have an increased risk of developing cerebral hemorrhage (VERSTRAETE and COLLEN 1996; WHITE et al. 1996; RIDKER and HENNEKENS 1996) and have therefore often been excluded from trials with thrombolytic agents, age per se is not a contraindication for thrombolytic therapy since the benefit of thrombolytic treatment outweighs the risk. For every 1000 elderly patients treated with thrombolytic drugs, approximately 36 are saved compared with those receiving control treatment (VERSTRAETE and COLLEN 1996, SHERRY and MARDER 1991). However, a prior history of a clinical event involving the cerebral arteries, such as stroke, transient ischemic attack, or intracranial neoplasm represent a significant risk for an intracranial hemorrhage with administration of a thrombolytic agent. A category of patients that has often been excluded from some trials are those with diabetic retinopathy (see above), although MAHAFFEY et al. (1997), in analyzing the GUSTO-I data base, found that the incidence of intraocular hemorrhage in patients with diabetes was only 0.05%. Menstruation should not be considered an important contraindication (Lanter et al. 1994; Karnash et al. 1995).

g) Other Side Effects

Allergic reactions are not uncommon after the use of SK or anistreplase and include fever, rash, rigor, and bronchospasm. Anaphylactic shock appears to be rare and might have been overreported in a setting where hypotension is common. In the GISSI trial the incidence was 0.1% (Gruppo Italiano Per Lo Studio Della Streptochinasi Nell'infarto Miocardico 1987) and in the ISIS-2 study 0.2% [ISIS-2 (Second International Study Of Infarct Survival) Collaborative Group 1988]. In the ISIS-3 trial [ISIS-3 (Third International Study Of Infarct Survival) Collaborative Group 1992] possible allergic reactions with persistent shock occurred in 0.3% of the patients treated with SK, in 0.5% of the anistreplase arm, and in 0.1% of patients treated with t-PA. Hypotension is often observed during the administration of SK (Herlitz et al. 1993) and has been attributed to the proteolytic state which leads to bradykinin release from kallikrein (Hoffmeister et al. 1998a,b). In the ISIS-3 trial, profound hypotension was found with SK in 5.8%, with anistreplase in 5.9%, with tPA in 2.8%, and in open controls in 0.96% of the patients. Thrombocytopenia (Harrington et al. 1994) and an acquired thrombocytopathy (Serebruany et al. 1998) are uncommon. Interactions between SK and anistreplase with other drugs are likewise uncommon (De Boer and Van Griensven 1995).

h) Laboratory Monitoring

Repetitive assays of blood coagulation and fibrinolysis parameters are of little value in SK or anistreplase therapy because the information is predictive of neither a successful outcome nor of the risk of bleeding (Marder 1987; Rao et al. 1988; Bell 1995). Prior to thrombolytic therapy with SK or anistreplase for an AMI we recommend drawing blood for the following screening tests: platelet count, activated partial thromboplastin time (aPTT), prothrombin time, and fibrinogen concentration, which can serve as baseline to determine whether a lytic state will have been achieved or as retrospective information in cases of bleeding complications.

i) Rethrombosis

Some 10%–20% of arteries that are successfully reperfused with SK are subject to early rethrombosis despite anticoagulant therapy. The most important factors that predispose to rethrombosis are essentially the same as those that led to the initial occlusive event, the presence of a thrombogenic stimulus and altered flow dynamics due to residual thrombus or an underlying atherostenotic lesion (Gibson et al. 1995; Anderson 1997; Davies and Ormerod 1998). Ruptured atheromatous plaques are rich in tissue factor

(WILCOX et al. 1989; MARMUR et al. 1996; TOSCHI et al. 1997), a powerful trigger for blood coagulation, and are often covered by undissolved thrombus (GASH et al. 1986; BROWN et al. 1986; GULBA et al. 1990; VAN BELLE et al. 1998) possessing fibrin-bound thrombin activity (FRANCIS et al. 1983) which is poorly inactivated by heparin/antithrombin III (HOGG and JACKSON 1989; VON DEM BORNE et al. 1996). VAN BELLE and collaborators (1998) performed angioscopic examinations of the infarct-related artery 1–30 days after AMI and found that 45% of patients had evidence for a non-healed ruptured plaque and 77% had undissolved thrombus. Similar findings were reported by ARAKAWA et al. (1997). Platelets are activated during thrombolytic therapy (GURBEL et al. 1998; McREDMOND et al. 2000) and platelet-rich thrombi as they occur in the arterial circulation are more resistant to lysis by thrombolytic agents (ANDERSON 1997; ZHU et al. 1999). The thrombogenic stimulus and flow dynamics are interrelated, e.g., in the presence of a fully patent vessel with normal flow, the effects of a thrombogenic stimulus may be diluted and washed away. Conversely, a ruptured plaque or a clot that possesses only a mild thrombogenic stimulus but also causes luminal encroachment and altered hemodynamics (HARRISON et al. 1984) may lead to dyskinetic or even static zones of reduced flow, thereby predisposing to recurrent thrombosis. Several approaches have been used to reduce the incidence of rethrombosis after thrombolytic therapy. These are discussed in Chap. 11.

II. Anisoylated Lys-Plasmin(ogen) Streptokinase Activator Complex (APSAC, Anistreplase)

1. Randomized Clinical Trials

Up to 1990 about a dozen randomized clinical trials of anistreplase had been conducted in patients with AMI. Nine smaller trials, not including those of SEABRA-GOMES et al. (1987); LÓPEZ-SENDÓN et al. (1988); CHARBONNIER et al. (1989); HOGG et al. (1990) have been reviewed by HELD et al. (1990). Randomized trials that evaluated mortality as an endpoint and compared anistreplase with placebo, streptokinase, or tPA are listed in Table 4. The composite results of these early trials and of the AIMS trial (AIMS TRIAL STUDY GROUP 1988) looked promising and yielded a 44% short-term survival gain of anistreplase over placebo. In the very large ISIS-3 trial, however, anistreplase was found to be equivalent to SK and to tPA (ISIS-3 (THIRD INTERNATIONAL STUDY OF INFARCT SURVIVAL COLLABORATIVE GROUP 1992). The major advantage of anistreplase appears to lie in the convenience of administering the drug as an i.v. bolus which permits easier pre-hospital treatment. Side effects with anistreplase were as common as and similar to those encountered with SK.

2. Pre-hospital Treatment

Various measures have been undertaken and proposed to reduce the time delay between symptoms and the start of thrombolytic treatment (WILLIAMS

Table 4. Randomized trials comparing anistreplase (Anistr.) with placebo, SK, or tPA

No.	Trial	Year	Delay	No. of patients		Patency (%)[a]		Mortality (%)			Early reocclusion/ reinfarction	
				Anistr.	Other	Anistr.	Other	Time	Anistr.	Other	Anistr.	Other
1	AIMS	1988 1990	≤6	624	634			30 d 1 year	6.4 11.0	12.1 18.0		
2	9 early trials	1990		533	504				5.8	9.5		
3	TEAM-2	1991	≤4	188	182	73	73	ih	5.9	7.1	1	2
4	ISIS-3	1992	<24[b]	13773	13780			35 days 6 months	10.5 13.7	10.6 14.0	3.5	3.5
5	Vogt	1993	≤3	89	86	72	68	ih	6.7	3.5	6.7	5.8
6	Bassand	1991	≤4	90	93	73	84	ih	5.5	7.5	7.7	4.3
7	ISIS-3	1992	<24[b]	13773	13746			35 days 6 months	10.5 13.7	10.3 14.1	3.5	2.9
8	TAPS	1992	≤6	211	210	70	84	ih	8.1	2.4	2.5	10.3
9	TEAM-3	1992	≤4	161	164	89	86	30 days	6.2	7.9	1.9	3.7
10	TIMI-4	1994	≤6	147	138	74	85	6 weeks 1 year	8.8 11.0	2.2 5.3	6.8	4.3
11	DUCCS-II	1996	≤12	83	79	76	76	ih	4.8	1.3		

Other: in trials 1 and 2 placebo; trials 3-5 streptokinase, 1.5 Mio units infused over 90 min; trials 6-11 tPA, trials 6,8,10, and 11 used front-loaded tPA (alteplase); trial 7 used duteplase: initial bolus of 0.04 Mio units/kg, then 0.36 Mio units/kg during the first hour, followed by 0.067 Mio units/kg per h for the next 3h; trial 9 used a 3-h infusion of 100mg of alteplase. ih: in hospital mortality.

References: 1. AIMS TRIAL STUDY GROUP (1988, 1990); 2. HELD et al. (1990); 3. ANDERSON et al. (1991); 4. ISIS-3 (THIRD INTERNATIONAL STUDY OF INFARCT SURVIVAL) COLLABORATIVE GROUP (1992); 5. VOGT et al. (1993b); 6. BASSAND et al. (1991); 7. ISIS-3 (THIRD INTERNATIONAL STUDY OF INFARCT SURVIVAL) COLLABORATIVE GROUP (1992); 8. NEUHAUS et al. (1992); 9. ANDERSON et al. (1992); 10. CANNON et al. (1994); 11. O'CONNOR et al. (1994).

[a] Patency rates, in all trials TIMI flow grade 2 and 3 after 90min, in trial 9 after 1-2 days.
[b] Delay to treatment, in the ISIS-3 trial the median delay was 4h.

Table 5. Randomized clinical trials comparing pre-hospital and in-hospital thrombolytic therapy in suspected acute myocardial infarction

Study	Treatment	Gain (min)	No. of patients		Mortality			
			Pre	H	Time	Pre	H	p
Castaigne et al. (1989)	Anistreplase 30 U	60	50	50	ih	6.0	4.0	0.6
Schofer et al. (1990)	Urokinase 2 × 2 MU	43	40	38	ih	2.5	5.3	0.5
Great Group (1993)	Anistreplase 30 U	130	163	148	3 months	8.0	15.5	0.04
Rawles (1994)					1 year	10.4	21.6	0.007
Rawles (1997)					5 years	25.1	35.8	0.025
Weaver et al. 1993	Aspirin, Alteplase	33	175	185	ih	5.7	8.1	0.49
Brouwer et al. (1996)					2 years	11	9	0.46
European Myocardial Infarction Project Group (EMIP) (1993)	Anistreplase 30 U	55	2750	2719	1 month	9.7	11.1	0.08

Total adjusted proportional reduction in risk: 17% (95% CI 2%–29%); $p = 0.03$.
In the MITI study (WEAVER et al. 1993; BROUWER et al. 1996) alteplase was given i.v 100 mg/3 h.
Time, time of mortality assessment; ih, in-hospital; Pre, pre-hospital thrombolysis; H, treatment administered in hospital.

1998). Media campaigns, addressed particularly to subjects over 65 years old and to women, have had variable success. While in the USA the reduction of the symptoms to door delay was negligible (Ho et al. 1989), better results were obtained in Germany (Rustige et al. 1990) and in Sweden (Herlitz et al. 1989). In Göteborg, the time delay between symptoms and arrival at the hospital could be reduced from 3h to 2h. The feasibility to cut down on the door to needle time has been generally acknowledged (Kereiakes et al. 1990; MacCallum et al. 1990; Birkhead 1992; Pell et al. 1992) and includes start of the treatment in the emergency room and development of special protocols to shorten door to needle time (Maynard et al. 1995; Boisjolie et al. 1995).

As an alternative, the treatment in the patient's home or in the ambulance by general practitioners, nurses, ambulance personnel, or special cardiac teams has been investigated (reviewed by Carlsson et al. 1997). Most of the studies investigating pre-hospital treatment have used anistreplase. Table 5 lists randomized studies that have also evaluated mortality as an endpoint. The only trials exhibiting a trend or significant advantage in favor of pre-hospital thrombolytic therapy were those utilizing anistreplase. The GREAT trial was conducted among general practitioners whose practices were on average about 50km away from the next coronary care unit in Aberdeen. The administration of anistreplase at the patients' home resulted in a gain of 130min and in a significant survival benefit at 3 month, 1 year, and even 5 years after the administration of anistreplase (Great Group 1993; Rawles 1994, 1997). The largest of these trials, the EMIP study, demonstrated a slight benefit in favor of home treatment (The European Myocardial Infarction Project Group 1993). The MITI trial is interesting because the institution of pre-hospital treatment motivated the hospital staff and resulted in a considerably shorter door to needle time delay for in-hospital treatment. This could have contributed to the lack of a statistical survival advantage for the group receiving alteplase at home or in the ambulance (Weaver et al. 1993; Brouwer et al. 1996). Several other, non-randomized, trials arrived in general at the conclusion that a small benefit could be derived from starting treatment at home or in the ambulance (Schofer et al. 1990; Rozenman et al. 1995; Grijseels et al. 1995). It is thus very likely that in places where local transport conditions are difficult and more than 1h can be gained by starting a pre-hospital treatment, a survival benefit will be obtained for patients in whom treatment is started before undertaking the trip to the hospital. In metropolitan areas pre-hospital treatment has not produced enough benefit to warrant wider application at the present time. However, the development of tPA mutants which can be administered as a bolus, such as TNK-tPA, reteplase, and lanoteplase, may be an occasion to restudy the value of pre-hospital treatment with these improved new thrombolytic agents.

This chapter does not deal systematically with the comparison of SK or anistreplase and other thrombolytic agents. Such information can be found in Chaps. 8–11.

List of Abbreviations

AIMS	APSAC Intervention Mortality Study
AMI	acute myocardial infarction
APSAC	Anisoylated Lys-Plasminogen Streptokinase Activator Complex
aPTT	activated partial thromboplastin time
CTFC	corrected TIMI frame count
DUCCS	Duke University Clinical Cardiology group Study
EMERAS	Estudio Multicéntrico Estreptoquinasa Repúblicas de América del Sur
EMIP	European Myocardial Infarction Project
FDPs	fibrin(ogen) degradation products
FPA	FPB, fibrinopeptide A or B
GISSI	Gruppo Italiano per lo Studio della Streptochinasi nell'Infarto miocardico
GREAT	Grampian Region Early Anistreplase Trial
GUSTO	Global Utilization of Streptokinase and Tissue plasminogen activator for Occluded coronary arteries
ISAM	Intravenous Streptokinase in Acute Myocardial infarction
ISIS	International Study of Infarct Survival
LATE	Late Assessment of Thrombolytic Efficacy
MITI	Myocardial Infarction Triage and Intervention
MU	1 million (mega) units
PTCA	Percutaneous Transluminal Coronary Angioplasty
SK	streptokinase
TAPS	TPA-APSAC Patency Study
TAT	thrombin-antithrombin complex
TEAM	Thrombolytic Trial of Eminase in acute Myocardial infarction
TIMI	Thrombolysis In Myocardial Infarction
tPA	tissue-type plasminogen activator
vWF	von Willebrand factor

References

Aber CP, Bass NM, Berry CL, Carson PHM, Dobbs RJ, Fox KM, Hamblin JJ, Haydu SP, Howitt G, MacIver JE, Portal RW, Raftery EB, Rousell RH, Stock JPP (1976) Streptokinase in acute myocardial infarction: a controlled multicentre study in the United Kingdom. Br Med J 2:1100–1104

Agostoni A, Gardinali M, Frangi D, Cafaro C, Conciato L, Sponzilli C, Salvioni A, Cugno M, Cicardi M (1994) Activation of complement and kinin systems after thrombolytic therapy in patients with acute myocardial infarction. A comparison between streptokinase and recombinant tissue-type plasminogen activator. Circulation 90:2666–2670

AIMS Trial Study Group (1988) Effect of intravenous APSAC on mortality after acute myocardial infarction: preliminary report of a placebo-controlled clinical trial. Lancet 1:545–549

AIMS Trial Study Group (1990) Long-term effects of intravenous anistreplase in acute myocardial infarction. Lancet 335:427–431

Amery A, Roeber G, Vermeulen HJ, Verstraete M (1969) Single-blind randomized multicentre trial comparing heparin and streptokinase treatment in recent myocardial infarction. Acta Med Scand 505 [Suppl]:1–35

Anderson JL (1989) Reperfusion, patency and reocclusion with anistreplase (APSAC) in acute myocardial infarction. Am J Cardiol 64:12A–17A

Anderson JL (1997) Why does thrombolysis fail? Breaking through the reperfusion ceiling. Am J Cardiol 80:1588–1590

Anderson JL, Becker LC, Sorensen SG, Karagounis LA, Browne KF, Shah PK, Morris DC, Fintel DJ, Mueller HS, Ross AM, Hall SM, Askins JC, Doorey AJ, Grines CL, Moreno FL, Marder VJ, for the TEAM-3 Investigators (1992) Anistreplase versus alteplase in acute myocardial infarction: comparative effects on left ventricular function, morbidity and 1-day coronary artery patency. J Am Coll Cardiol 20: 753–766

Anderson JL, Karagounis LA, Becker LC, Sorensen SG, Menlove RL, for the TEAM-3 Investigators (1993) TIMI perfusion grade 3 but not grade 2 results in improved outcome after thrombolysis for myocardial infarction. Ventriculographic, enzymatic, and electrocardiographic evidence from the TEAM-3 Study. Circulation 87:1829–1839

Anderson JL, Marshall HW, Bray BE, Lutz JR, Frederick PR, Yanowitz FG, Datz FL, Klausner SC, Hagan AD (1983) A randomized trial of intracoronary streptokinase in the treatment of acute myocardial infarction. N Engl J Med 308:1312–1318

Anderson JL, Sorensen SG, Moreno FL, Hackworthy RA, Browne KF, Dale HT, Leya F, Dangoisse V, Eckerson HW, Marder VJ, and the TEAM-2 Study Investigators (1991) Multicenter patency trial of intravenous anistreplase compared with streptokinase in acute myocardial infarction. Circulation 83:126–140

Antman EM for the TIMI 9 A Investigators (1994) Hirudin in acute myocardial infarction. Safety report from the Thrombolysis and Thrombin Inhibition in Myocardial Infarction (TIMI) 9 A trial. Circulation 90:1624–1630

Arakawa K, Mizuno K, Shibuya T, Etsuda H, Tabata H, Nagayoshi H, Satomura K, Isojima K, Kurita A, Nakamura H (1997) Angioscopic coronary macromorphology after thrombolysis in acute myocardial infarction. Am J Cardiol 79:197–202

Ardaillou N, Dray L, Budzynski AZ, Marder VJ, Larrieu MJ (1977) The half-life of plasmic degradation products of human fibrinogen in rabbits. Thromb Haemost 37:201–209

Astrup T, Müllertz S (1952) The fibrin plate method for estimating fibrinolytic activity. Arch Biochem Biophys 40:346–356

Bachmann F (1968) Development of antibodies against perorally and rectally administered streptokinase in man. J Lab Clin Med 72:228–238

Baigent C, Collins R, Appleby P, Parish S, Sleight P, Peto R, on behalf of the ISIS-2 (Second International Study of Infarct Survival) Collaborative Group (1998) ISIS-2: 10 year survival among patients with suspected acute myocardial infarction in randomized comparison of intravenous streptokinase, oral aspirin, both, or neither. BMJ 316:1337–1343

Barbagelata NA, Granger CB, Oqueli E, Suárez LD, Borruel M, Topol EJ, Califf RM (1997) TIMI grade 3 flow and reocclusion after intravenous thrombolytic therapy: a pooled analysis. Am Heart J 133:273–282

Bassand J-P, Cassagnes J, Machecourt J, Lusson J-R, Anguenot T, Wolf J-E, Maublant J, Bertrand B, Schiele F (1991) Comparative effects of APSAC and rt-PA on infarct size and left ventricular function in acute myocardial infarction. A multicenter randomized study. Circulation 84:1107–1117

Becker RC (1991) Seminars in thrombosis, thrombolysis, and vascular biology. Part 5: Cellular rheology and plasma viscosity. Cardiology 79:265–270

Bell WR (1995) Laboratory monitoring of thrombolytic therapy. Clin Lab Med 15:165–178

Berglin Blohm M, Hartford M, Karlsson T, Herlitz J (1998) Factors associated with prehospital and in-hospital delay time in acute myocardial infarction: a 6-year experience. J Intern Med 243:243–250

Bergmann SR, Lerch RA, Fox KA, Ludbrook PA, Welch MJ, Ter-Pogossian MM, Sobel BE (1982) Temporal dependence of beneficial effects of coronary thrombolysis characterized by position tomography. Am J Med 73:573–581

Berkowitz SD, Granger CB, Pieper KS, Lee KL, Gore JM, Simoons M, Topol EJ, Califf RM, for the Global Utilization of Streptokinase and Tissue Plasminogen activator for Occluded Coronary Arteries (GUSTO) I Investigators (1997) Incidence and predictors of bleeding after contemporary thrombolytic therapy for myocardial infarction. Circulation 95:2508–2516

Bett JH, Castaldi PA, Hale GS, Isbister JP, McLean KH, O'Sullivan EF, Biggs JC, Chesterman CN, Hirsh J, McDonald IG, Morgan JJ, Rosenbaum M (1973) Australian multicentre trial of streptokinase in acute myocardial infarction. Lancet 1:57–60

Birkhead JS (1992) Time delays in provision of thrombolytic treatment in six district hospitals. BMJ 305:445–448

Boersma E, Maas ACP, Deckers JW, Simoons ML (1996) Early thrombolytic treatment in acute myocardial infarction: reappraisal of the golden hour. Lancet 348:771–775

Boisjolie CR, Sharkey SW, Cannon CP, Brunette D, Haugland JM, Thatcher JL, Henry TD (1995) Impact of a thrombolysis research trial on time to treatment for acute myocardial infarction in the emergency department. Am J Cardiol 76:396–398

Boucek RJ, Murphy WP Jr, Sommer L, Voudoukis I (1960) Segmental perfusion of the coronary arteries with fibrinolysin in man following a myocardial infarction. Am J Cardiol 6:525–533

Bovill EG, Tracy RP, Knatterud GL, Stone PH, Nasmith J, Gore JM, Thompson BW, Tofler GH, Kleiman NS, Cannon C, Braunwald E (1997) Hemorrhagic events during therapy with recombinant tissue plasminogen activator, heparin, and aspirin for unstable angina (Thrombolysis in Myocardial Ischemia, Phase IIIB Trial). Am J Cardiol 79:391–396

Breddin K, Ehrly AM, Fechler L, Frick D, König H, Kraft H, Krause H, Krzywanek HJ, Kutschera J, Losch HW, Ludwig O, Mikat B, Rausch F, Rosenthal P, Sartory S, Voigt G, Wylicil P (1973) Die Kurzzeitfibrinolyse beim akuten Myokardinfarkt. Dtsch med Wochenschr 98:861–873

Brenner B, Francis CW, Totterman S, Kessler CM, Rao AK, Rubin R, Kwaan HC, Gabriel KR, Marder VJ (1990) Quantitation of venous clot lysis with the D-Dimer immunoassay during fibrinolytic therapy requires correction for soluble fibrin degradation. Circulation 81:1818–1825

Brouwer MA, Martin JS, Maynard C, Wirkus M, Litwin PE, Verheugt FWA, Weaver WD, for the MITI project investigators (1996) Influence of early prehospital thrombolysis on mortality and event-free survival (The Myocardial Infarction Triage and Intervention [MITI] randomized trial). Am J Cardiol 78:497–502

Brown BG, Gallery CA, Badger RS, Kennedy JW, Mathey D, Bolson EL, Dodge HT (1986) Incomplete lysis of thrombus in the moderate underlying atherosclerotic lesion during intracoronary infusion of streptokinase for acute myocardial infarction: quantitative angiographic observations. Circulation 73:653–661

Buchalter MB, Suntharalingam G, Jennings I, Hart C, Luddington RJ, Chakraverty R, Jacobson SK, Weissberg PL, Baglin TP (1992) Streptokinase resistance: When might streptokinase administration be ineffective? Br Heart J 68:449–453

Cairns JA, Fuster V, Gore J, Kennedy JW (1995) Coronary thrombolysis. Chest 108 [Suppl]:401S–423S

Cannon CP, McCabe CH, Diver DJ, Herson S, Greene RM, Shah PK, Sequeira RF, Leya F, Kirshenbaum JM, Magorien RD, Palmeri ST, Davis V, Gibson CM, Poole WK, Braunwald E, TIMI 4 Investigators (1994) Comparison of front-loaded recombinant tissue-type plasminogen activator, anistreplase and combination thrombolytic therapy for acute myocardial infarction: Results of the thrombolysis in myocardial infarction (TIMI) 4 trial. J Am Coll Cardiol 24:1602–1610

Carlsson J, Schuster HP, Tebbe U (1997) Prähospitale thrombolytische Therapie bei akutem Myokardinfarkt. Anaesthesist 46:829–839

Castaigne AD, Hervé C, Duval-Moulin AM, Gaillard M, Dubois-Rande JL, Boesch C, Wolf M, Lellouche D, Jan F, Vernant P, Huguenard P (1989) Prehospital use of APSAC: results of a placebo-controlled study. Am J Cardiol 64:30A–33A

Cerqueira MD, Maynard C, Ritchie JL, Davis KB, Kennedy JW (1992) Long-term survival in 618 patients from the Western Washington Streptokinase in Myocardial Infarction Trials. J Am Coll Cardiol 20:1452–1459

Chain D, Kreizman T, Shapira H, Shaltiel S (1991) Plasmin cleavage of vitronectin: identification of the site and consequent attenuation in binding plasminogen activator inhibitor-. FEBS Lett 285:251–256

Charbonnier B, Cribier A, Monassier JP, Favier JP, Materne P, Brochier ML, Letac B, Hanssen M, Sacrez A, Kulbertus H (1989) Etude européenne multicentrique et randomisée de l'APSAC versus streptokinase dans l'infarctus du myocarde. Arch Mal Coeur Vaiss 9:1565–1571

Chazov EI, Mateeva LS, Mazaev AV, Sargin KE, Sadovskaia GV, Ruda MI (1976) Intracoronary administration of fibrinolysin in acute myocardial infarction. Ter Arkh 48:3–14

Col JJ, Col-De-Berys CM, Renkin JP, Lavenne-Pardonge EM, Bachy JL, Moriau MH (1989) Pharmacokinetics, thrombolytic efficacy and hemorrhagic risk of different streptokinase regimens in heparin-treated acute myocardial infarction. Am J Cardiol 63:1185–1192

Dalen JE, Gore JM, Braunwald E, Borer J, Goldberg RJ, Passamani ER, Forman S, Knatterud G, the T (1988) Six- and twelve-month follow-up of the phase 1 thrombolysis in myocardial infarction (TIMI) trial. Am J Cardiol 62:179–185

Davies CH, Ormerod OJ (1998) Failed coronary thrombolysis. Lancet 351:1191–1196

Davies M, Woolf N, Robertson W (1976) Pathology of acute myocardial infarction with particular reference to occlusive coronary thrombi. Br Heart J 38:669–664

Davies MJ (1994) Pathology of arterial thrombosis. Br Med Bull 50:789–802

de Boer A, van Griensven JMT (1995) Drug interactions with thrombolytic agents. Current perspectives. Clin Pharmacokinet 28:315–326

de Bono DP, Simoons ML, Tijssen J, Arnold AER, Betriu A, Burgersdijk C, Bescos LL, Mueller E, Pfisterer M, Van de Werf F, Zijlastra F, Verstraete M (1992) Effect of early intravenous heparin on coronary patency, infarct size, and bleeding complications after alteplase thrombolysis: results of a randomized double blind European Cooperative Study Group trial. Br Heart J 67:122–128

Dewar HA, Stephenson P, Horler AR, et al. (1963) Fibrinolytic therapy of coronary thrombosis: controlled trial of 75 cases. Br Med J 1:915–920

DeWood MA, Spores J, Notske R, Mouser LT, Burroughs R, Golden MS, Lang HT (1980) Prevalence of total coronary occlusion during the early hours of transmural myocardial infarction. N Engl J Med 303:897–902

Dioguardi N, Lotto A, Levi GF, Rota M, Proto C, Mannucci PM, Rossi P, Lomanto B, Mattei G, Fiorelli G, Agostini A (1971) Controlled trial of streptokinase and heparin in acute myocardial infarction. Lancet 2:891–895

Doorey A, Patel S, Reese C, O'Connor R, Geloo N, Sutherland S, Price N, Gleasner E, Rodrigue R (1998) Dangers of delay of initiation of either thrombolysis or primary angioplasty in acute myocardial infarction with increasing use of primary angioplasty. Am J Cardiol 81:1173–1177

Efstratiadis T, Munsch C, Crossman D, Taylor K (1991) Aprotinin used in emergency coronary operation after streptokinase treatment. Ann Thorac Surg 52:1320–1321

Eisenberg PR, Sherman LA, Jaffe AS (1987) Paradoxic elevation of fibrinopeptide A after streptokinase: evidence for continued thrombosis despite intense fibrinolysis. J Am Coll Cardiol 10:527–529

El Gaylani N, Davies S, Tovey J, Kinnarid T, Duly E, Buchalter MB (1996) Systemic lytic state is not a predictor of coronary reperfusion in acute myocardial infarction. Int J Cardiol 57:45–50

Elliott JM, Cross DB, Cederholm-Williams SA, White HD (1993) Neutralizing antibodies to streptokinase four years after intravenous thrombolytic therapy. Am J Cardiol 71:640–645

EMERAS (Estudio Multicéntrico Estreptoquinasa Repúblicas de América del Sur) Collaborative Group (1993) Randomized trial of late thrombolysis in patients with suspected acute myocardial infarction. Lancet 342:767–772

European Co-operative Study Group for Streptokinase Treatment in Acute Myocardial Infraction (1979) Streptokinase in acute myocardial infarction. N Engl J Med 301:797–802

Falk E (1983) Plaque rupture with severe pre-existing stenosis precipitating coronary thrombosis: characteristics of coronary atherosclerotic plaques underlying fatal occlusive thrombi. Br Heart J 50:127–134

Fears R (1989) Development of anisoylated plasminogen-streptokinase activator complex from the acyl enzyme concept. Semin Thromb Hemost 15:129–139

Fears R, Ferres H, Glasgow E, Standring R, Hogg KJ, Gemmill JD, Burns JMA, Rae AP, Dunn FG, Hillis WS (1992) Monitoring of streptokinase resistance titre in acute myocardial infarction patients up to 30 months after giving streptokinase or anistreplase and related studies to measure specific antistreptokinase IgG. Br Heart J 68:167–170

Federici AB, Berkowitz SD, Lattuada A, Mannucci PM (1993) Degradation of von Willebrand factor in patients with acquired clinical conditions in which there is heightened proteolysis. Blood 81:720–725

Fibrinolytic Therapy Trialists' (FTT) Collaborative Group (1994) Indications for fibrinolytic therapy in suspected acute myocardial infarction: Collaborative overview of early mortality and major morbidity results from all randomized trials of more than 1000 patients. Lancet 343:311–322

Fletcher AP, Alkjaersig N, Sherry S (1959a) The maintenance of a sustained thrombolytic state in man. I. Induction and effects. J Clin Invest 38:1096–1110

Fletcher AP, Alkjaersig N, Sherry S (1962) Pathogenesis of the coagulation defect developing during pathological plasma proteolytic (fibrinolytic) states. I. The significance of fibrinogen proteolysis and circulating fibrinogen breakdown products. J Clin Invest 41:896–916

Fletcher AP, Alkjaersig N, Sherry S, Genton E, Hirsh J, Bachmann F (1965) The development of urokinase as a thrombolytic agent: maintenance of a sustained thrombolytic state in man by its intravenous infusion. J Lab Clin Med 65:713–731

Fletcher AP, Alkjaersig N, Smyrniotis FE, Sherry S (1958) Treatment of patients suffering from early myocardial infarction with massive and prolonged streptokinase therapy. Trans Assoc Am Physicians 71:287–296

Fletcher AP, Sherry S, Alkjaersig N, Smyrniotis FE, Jick S (1959b) The maintenance of a sustained thrombolytic state in man. II. Clinical observations on patients with myocardial infarction and other thromboembolic disorders. J Clin Invest 38:1111–1119

Francis CW, Marder VJ (1988) Increased resistance to plasmic degradation of fibrin with highly crosslinked alpha-polymer chains formed at high factor XIII concentrations. Blood 71:1361–1365

Francis CW, Marder VJ (1991) Fibrinolytic therapy for venous thrombosis. Prog Cardiovasc Dis 34:193–204

Francis CW, Marder VJ (1994) Physiologic regulation and pathologic disorders of fibrinolysis. In: Colman RW, Hirsh J, Marder VJ, Salzman EW (eds) Hemostasis and Thrombosis. Basic Principles and Clinical Practice, 3rd Ed., Lippincott, Philadelphia, pp 1076–1103

Francis CW, Markham RE, Marder VJ (1984) Demonstration of in situ fibrin degradation in pathologic thrombi. Blood 63:1216–1224

Francis CW, Markham RE, Jr, Barlow GH, Florack TM, Dobrzynski DM, Marder VJ (1983) Thrombin activity of fibrin thrombi and soluble plasmic derivatives. J Lab Clin Med 102:220–230

Franzosi MG, Santoro E, De Vita C, Geraci E, Lotto A, Maggioni AP, Mauri F, Rovelli
    F, Santoro L, Tavazzi L, Tognoni G, on behalf of the GISSI Investigators (1998)
    Ten-year follow-up of the first megatrial testing thrombolytic therapy in patients
    with acute myocardial infarction. Results of the Gruppo Italiano per lo Studio
    della Sopravvivenza nell'Infarto-1 study. Circulation 98:2659–2665
French JK, Ellis CJ, White HD (1998) The corrected TIMI frame count. The new gold
    standard? Aust N Z J Med 28:569–573
French JK, Hyde TA, Patel H, Amos DJ, McLaughlin SC, Webber BJ, White HD (1999a)
    Survival 12 years after randomization to streptokinase: the influence of throm-
    bolysis in myocardial infarction flow at three to four weeks. J Am Coll Cardiol
    34:62–69
French JK, Straznicky IT, Webber BJ, Aylward PE, Frey MJ, Adgey AA, Williams BF,
    McLaughlin SC, White HD, for the HERO-1 Investigators (1999b) Angiographic
    frame counts 90 minutes after streptokinase predict left ventricular function at 48
    hours following myocardial infarction. Heart 81:128–133
French JK, White HD (1995) Left ventricular function following thrombolytic therapy
    for myocardial infarction. Clin Exp Pharmacol Physiol 22:173–179
Furberg CD (1984) Clinical value of intracoronary streptokinase. Am J Cardiol
    53:626–627
Ganz W, Geft I, Shah PK, Lew AS, Rodriguez L, Weiss T, Maddahi J, Berman DS,
    Charuzi Y, Swan HJ (1984) Intravenous streptokinase in evolving acute myocar-
    dial infarction. Am J Cardiol 53:1209–1216
Gash AK, Spann JF, Sherry S, Belber AD, Carabello BA, McDonough MT, Mann RH,
    McCann WD, Gault JH, Gentzler RD, et al (1986) Factors influencing reocclusion
    after coronary thrombolysis for acute myocardial infarction. Am J Cardiol
    57:175–177
Gibson CM, Cannon CP, Daley WL, Dodge JT, Jr., Alexander B, Jr., Marble SJ,
    McCabe CH, Raymond L, Fortin T, Poole WK, Braunwald E, for the TIMI 4 Study
    Group (1996) TIMI frame count: a quantitative method of assessing coronary
    artery flow. Circulation 93:879–888
Gibson CM, Cannon CP, Murphy SA, Ryan KA, Mesley R, Marble SJ, McCabe CH,
    Van de Werf F, Braunwald E, for the TIMI (Thrombolysis in Myyocardial Infarc-
    tion) Study Group (2000) Relationship of TIMI myocardial perfusion grade
    to mortality after administration of thrombolytic drugs. Circulation 101:125–
    130
Gibson CM, Cannon CP, Piana RN, Breall JA, Sharaf B, Flatley M, Alexander B, Diver
    DJ, McCabe CH, Flaker GC, Baim DS, Braunwald E (1995) Angiographic pre-
    dictors of reocclusion after thrombolysis: results from the thrombolysis in myocar-
    dial infarction (TIMI) 4 trial. J Am Coll Cardiol 25:582–589
Gibson CM, Murphy S, Menown IBA, Sequeira RF, Greene R, Van de Werf F,
    Schweiger MJ, Ghali M, Frey MJ, Ryan KA, Marble SJ, Giugliano RP, Antman
    EM, Cannon CP, Braunwald E, for the TIMI Study Group (1999a) Determinants
    of coronary blood flow after thrombolytic administration. J Am Coll Cardiol
    34:1403–1412
Gibson CM, Ryan KA, Kelley M, Rizzo MJ, Mesley R, Murphy S, Swanson J, Marble
    SJ, Dodge JT, Giugliano RP, Cannon CP, Antman EM, for the TIMI Study Group
    (1999b) Methodologic drift in the assessment of TIMI grade 3 flow and its impli-
    cations with respect to the reporting of angiographic trial results. Am Heart J
    137:1179–1184
Gold HK, Cowley MJ, Palacios IF, Vetrovec GW, Akins CW, Block PC, Leinbach RC
    (1984) Combined intracoronary streptokinase infusion and coronary angioplasty
    during acute myocardial infarction. Am J Cardiol 53:122C–125C
Gormsen J, Fletcher AP, Alkjaersig N, Sherry S (1967) Enzymatic lysis of clots: the influ-
    ence of fibrin stabilization on lysis rates. Arch Biochem Biophys 120:654–665
Granger CB, Becker R, Tracy RP, Califf RM, Topol EJ, Pieper KS, Ross AM, Roth S,
    Lambrew C, Bovill EG, for the GUSTO-I Hemostasis Substudy Group (1998)
    Thrombin generation, inhibition and clinical outcomes in patients with acute

myocardial infarction treated with thrombolytic therapy and heparin: results from the GUSTO-I Trial. J Am Coll Cardiol 31:497–505

Granger CB, White HD, Bates ER, Ohman EM, Califf RM (1994) A pooled analysis of coronary arterial patency and left ventricular function after intravenous thrombolysis for acute myocardial infarction. Am J Cardiol 74:1220–1228

GREAT Group (1993) Feasibility, safety, and efficacy of domiciliary thrombolysis by general practitioners: Grampian region early anistreplase trial. BMJ 305:548–553

Grijseels EWM, Bouten MJM, Lenderink T, Deckers JW, Hoes AW, Hartman JAM, van der Does E, Simoons ML (1995) Pre-hospital thrombolytic therapy with either alteplase or streptokinase. Practical applications, complications and long-term results in 529 patients. Eur Heart J 16:1833–1838

Grines CL, Nissen SE, Booth DC, Gurley JC, Chelliah N, Wolf R, Blankenship J, Branco MC, Bennett K, DeMaria AN, the Kentucky Acute Myocardial Infarction Trial (KAMIT) Group (1991) A prospective, randomized trial comparing combination half-dose tissue-type plasminogen activator and streptokinase with full- dose tissue-type plasminogen activator. Circulation 84:540–549

Gruppo Italiano per lo Studio della Sopravvivenza nell'Infarto Miocardico (GISSI) (1990) GISSI-2: A factorial randomised trial of alteplase versus streptokinase and heparin versus no heparin among 12 490 patients with acute myocardial infarction. Lancet 336:65–71

Gruppo Italiano per lo Studio della Streptochinasi nell'Infarto Miocardico (GISSI) (1986) Effectiveness of intravenous thrombolytic treatment in acute myocardial infarction. Lancet 1:397–402

Gruppo Italiano per lo Studio della Streptochinasi nell'Infarto Miocardico (GISSI) (1987) Long-term effects of intravenous thrombolysis in acute myocardial infarction: a final report of the GISSI study. Lancet 2:871–874

Grünewald M, Seifried E (1994) Meta-analysis of all available published clinical trials (1958–1990) on thrombolytic therapy for AMI: relative efficacy of different therapeutic strategies. Fibrinolysis 8:67–86

Gulba DC, Barthels M, Reil G, Lichtlen PR (1988) Thrombin/antithrombin-III complex level as early predictor of reocclusion after successful thrombolysis. Lancet 2: 97

Gulba DC, Bode C, Topp J, Höpp HW, Westhoff-Bleck M, Rafflenbeul W, Lichtlen PR (1990) High incidence of residual coronary artery thrombi after successful thrombolysis in myocardial infarction. Relevance for the rate of early reocclusions. A report from the multicenter dose-finding study for thrombolysis with urokinase preactivated natural pro-urokinase (TCL 598). Z Kardiol 79:279–285

Gurbel PA, Serebruany VL, Shustov AR, Bahr RD, Carpo C, Ohman EM, Topol EJ, for the GUSTO-III Investigators (1998) Effects of reteplase and alteplase on platelet aggregation and major receptor expression during the first 24 hours of acute myocardial infarction treatment. J Am Coll Cardiol 31:1466–1473

Hackworthy RA, Sorensen SG, Fitzpatrick PG, Barry WH, Menlove RL, Rothbard RL, Anderson JL, for the ASPSAC Investigators (1988) Effect of reperfusion on electrocardiographic and enzymatic infarct size: results of a randomized multicenter study on intravenous anisoylated plasminogen streptokinase activator complex (APSAC) versus intracoronary streptokinase in acute myocardial infarction. Am Heart J 116:903–914

Harrington RA, Sane DC, Califf RM, Sigmon KN, Abbottsmith CW, Candela RJ, Lee KL, Topol EJ (1994) Clinical importance of thrombocytopenia occurring in the hospital phase after administration of thrombolytic therapy for acute myocardial infarction. J Am Coll Cardiol 23:891–898

Harrison DG, Ferguson DW, Collins SM, Skorton DJ, Ericksen EE, Kioschos JM, Marcus ML, White CW (1984) Rethrombosis after reperfusion with streptokinase: importance of geometry of residual lesions. Circulation 69:991–999

Heikinheimo R, Ahrenberg P, Honkapohja H, Iisalo E, Kallio V, Konttinen Y, Leskinen O, Mustaniemi H, Reinikainen M, Siitonen L (1971) Fibrinolytic treatment in acute myocardial infarction. Acta Med Scand 189:7–13

Held PH, Teo KK, Yusuf S (1990) Effects of tissue-type plasminogen activator and anisoylated plasminogen streptokinase activator complex on mortality in acute myocardial infarction. Circulation 82:1668–1674

Herlitz J, Hartford M, Aune S, Karlsson T (1993) Occurrence of hypotension during streptokinase infusion in suspected acute myocardial infarction, and its relation to prognosis and metoprolol therapy. Am J Cardiol 71:1021–1024

Herlitz J, Hartford M, Blohm M, Karlson BW, Ekström L, Risenfors M, Wennerblom B, Luepker RV, Holmberg S (1989) Effect of media campaign on delay times and ambulance use in suspected myocardial infarction. Am J Cardiol 64:90–93

Hibbs MJ, Fears R, Ferres H, Standring R, Smith RAG (1989) Determination of the deacylation rate of p-anisoyl plasminogen streptokinase activator complex (APSAC, EMINASE) in human plasma, blood and plasma clots. Fibrinolysis 3:235–240

Hillis LD, Borer J, Braunwald E, Chesebro JH, Cohen LS, Dalen J, Dodge HT, Francis CK, Knatterud G, Ludbrook P, et al (1985) High dose intravenous streptokinase for acute myocardial infarction: preliminary results of a multicenter trial. J Am Coll Cardiol 6:957–962

Ho MT, Eisenberg MS, Litwin PE, Schaeffer SM, Damon SK (1989) Delay between onset of chest pain and seeking medical care: the effect of public education. Ann Emerg Med 18:727–731

Hoffmann JJML, Fears R, Bonnier JJRM, Standring R, Ferres H, De Swart JBRM (1988) Significance of antibodies to streptokinase in coronary thrombolytic therapy with streptokinase or APSAC. Fibrinolysis 2:203–210

Hoffmann JJML, van Rey FJW, Bonnier JJRM (1985) Systemic effects of BRL 26921 during thrombolytic treatment of acute myocardial infarction. Thromb Res 37:567–572

Hoffmeister HM, Ruf M, Wendel HP, Heller W, Seipel L (1998a) Streptokinase-induced activation of the kallikrein-kinin system and of the contact phase in patients with acute myocardial infarction. J Cardiovasc Pharmacol 31:764–772

Hoffmeister HM, Szabo S, Kastner C, Beyer ME, Helber U, Kazmaier S, Wendel HP, Heller W, Seipel L (1998b) Thrombolytic therapy in acute myocardial infarction: comparison of procoagulant effects of streptokinase and alteplase regimens with focus on the kallikrein system and plasmin. Circulation 98:2527–2533

Hogg KJ, Gemmill JD, Burns JMA, Lifson WK, Rae AP, Dunn FG, Hillis WS (1990) Angiographic patency study of anistreplase versus streptokinase in acute myocardial infarction. Lancet 335:254–258

Hogg PJ, Jackson CM (1989) Fibrin monomer protects thrombin from inactivation by heparin- antithrombin III: Implications for heparin efficacy. Biochemistry 86:3619–3623

ISIS (International Studies of Infarct Survival) Pilot Study Investigators (1987) Randomized factorial trial of high-dose intravenous streptokinase, of oral aspirin and of intravenous heparin in acute myocardial infarction. Eur Heart J 8:634–642

ISIS Steering Committee (1987) Intravenous streptokinase given within 0–4 hours of onset of myocardial infarction reduced mortality in ISIS-2. Lancet 1:502

ISIS-3 (Third International Study of Infarct Survival) Collaborative Group (1992) ISIS-3: a randomised comparison of streptokinase vs tissue plasminogen activator vs anistreplase and of aspirin plus heparin vs aspirin alone among 41 299 cases of suspected acute myocardial infarction. Lancet 339:753–770

Jaffe AS, Eisenberg PR, Wilner GD (1987) In vivo assessment of thrombosis and fibrinolysis during acute myocardial infarction. Prog Hematol 15:71–89

Jan K-M, Chien S, Bigger JT Jr (1975) Observations on blood viscosity changes after acute myocardial infarction. Circulation 51:1079–1084

Johnson AJ, Tillett WS (1952) Lysis in rabbits of intravascular blood clots by the streptococcal fibrinolytic system (streptokinase). J Exp Med 95:449–440

Juhlin P, Boström PA, Torp A, Bredberg A (1999) Streptokinase antibodies inhibit reperfusion during thrombolytic therapy with streptokinase in acute myocardial infarction. J Intern Med 245:483–488

Karagounis L, Sorensen SG, Menlove RL, Moreno F, Anderson JL, for the TEAM-2 Investigators (1992) Does thrombolysis in myocardial infarction (TIMI) perfusion grade 2 represent a mostly patent artery or a mostly occluded artery? Enzymatic and electrocardiographic evidence from the TEAM-2 study. J Am Coll Cardiol 19:1–10

Karnash SL, Granger CB, White HD, Woodlief LH, Topol EJ, Califf RM, for the GUSTO-I Investigators (1995) Treating menstruating women with thrombolytic therapy: insights from the global utilization of streptokinase and tissue plasminogen activator for occluded coronary arteries (GUSTO-I) trial. J Am Coll Cardiol 26:1651–1656

Kelly PA, Nolan J, Wilson I, Perrins EJ (1997) Preservation of autonomic function following successful reperfusion with streptokinase within 12 hours of the onset of acute myocardial infarction. Am J Cardiol 79:203–205

Kennedy JW, Gensini GG, Timmis GC, Maynard C (1985) Acute myocardial infarction treated with intracoronary streptokinase: a report of the Society for Cardiac Angiography. Am J Cardiol 55:871–877

Kennedy JW, Martin GV, Davis KB, Maynard C, Stadius M, Sheehan FH, Ritchie JL (1988) The Western Washington intravenous streptokinase in acute myocardial infarction randomized trial. Circulation 77:345–352

Kennedy JW, Ritchie JL, Davis KB, Fritz JK (1983) Western Washington randomized trial of intracoronary streptokinase in acute myocardial infarction. N Engl J Med 309:1477–1482

Kennedy JW, Ritchie JL, Davis KB, Stadius ML, Maynard C, Fritz JK (1985) The Western Washington randomized trial of intracoronary streptokinase in acute myocardial infarction: a 12 month follow-up report. N Engl J Med 212:1073–1078

Kereiakes DJ, Weaver WD, Anderson JL, Feldman T, Gibler B, Aufderheide T, Williams DO, Martin LH, Anderson LC, Martin JS, McKendall G, Sherrid M, Greenberg H, Teichman SL (1990) Time delays in the diagnosis and treatment of acute myocardial infarction: a tale of eight cities. Report from the Pre-hospital Study Group and the Cincinnati Heart Project. Am Heart J 120:773–780

Khaja F, Walton JA Jr, Brymer JF, Lo E, Osterberger L, O'Neill WW, Colfer HT, Weiss R, Lee T, Kurian T, Goldberg AD, Pitt B, Goldstein S (1983) Intracoronary fibrinolytic therapy in acute myocardial infarction. Report of a prospective randomized trial. N Engl J Med 308:1305–1311

L'Abbe KA, Detsky AS, O'Rourke K (1987) Meta-analysis in clinical research. Ann Intern Med 107:224–233

Lambrew CT, Bowlby LJ, Rogers WJ, Chandra NC, Weaver WD, for the Time to Thrombolysis Substudy of the National Registry of Myocardial Infarction (1997) Factors influencing the time to thrombolysis in acute myocardial infarction. Time to Thrombolysis Substudy of the National Registry of Myocardial Infarction-1. Arch Intern Med 157:2577–2582

Langer A, Goodman SG, Topol EJ, Charlesworth A, Skene AM, Wilcox RG, Armstrong PW, for the LATE Study Investigators (1996) Late assessment of thrombolytic efficacy (LATE) study: prognosis in patients with non-Q wave myocardial infarction. J Am Coll Cardiol 27:1327–1332

Lanter PL, Jennings CF, Roberts CS, Jesse RL (1994) Safety of thrombolytic therapy in normally menstruating women with acute myocardial infarction. Am J Cardiol 74:179–181

LATE Study Group (1993) Late assessment of thrombolytic efficacy (LATE) study with alteplase 6–24 hours after onset of acute myocardial infarction. Lancet 342: 759–766

Lauer JE, Heger JJ, Mirro MJ (1995) Hemorrhagic complications of thrombolytic therapy. Chest 108:1520–1523

Le Feuvre C, Yusuf S, Flather M, Farkouh M (1993) Maximizing benefits of therapies in acute myocardial infarction. Am J Cardiol 72:145G–155G

Leizorovicz A, Haugh MC, Mercier C, Boissel JP, on behalf of the EMIP Group (1997) Pre-hospital and hospital time delays in thrombolytic treatment in patients with suspected acute myocardial infarction. Analysis of data from the EMIP study. Eur Heart J 18:248–253

Levine MN, Goldhaber SZ, Gore JM, Hirsh J, Califf RM (1995) Hemorrhagic complications of thrombolytic therapy in the treatment of myocardial infarction and venous thromboembolism. Chest 108 [Suppl]:291S–301S

Lew AS, Neer T, Rodriguez L, Geft IL, Shah PK, Ganz W (1984) Clinical failure of streptokinase due to an unsuspected high titer of antistreptokinase antibody. J Am Coll Cardiol 4:183–185

Lincoff AM, Topol EJ, Califf RM, Sigmon KN, Lee KL, Ohman EM, Rosenschein U, Ellis SG (1995) Significance of a coronary artery with thrombolysis in myocardial infarction grade 2 flow "patency" (outcome in the thrombolysis and angioplasty in myocardial infarction trials). Am J Cardiol 75:871–876

Loscalzo J, Vaughan DE (1987) Tissue plasminogen activator promotes platelet disaggregation in plasma. J Clin Invest 79:1749–1755

Lowe GDO, Wood DA, Douglas JT, Riemersma RA, Macintyre CCA, Takase T, Tuddenham EGD, Forbes CD, Elton RA, Oliver MF (1991) Relationships of plasma viscosity, coagulation and fibrinolysis to coronary risk factors and angina. Thromb Haemost 65:339–343

López-Sendón J, Seabra-Gomes R, Macaya C, Santons FM, Muñoz J, Sobrino N, Calvo L, Silva J, Miguel J, López de Sá E, Jadrague LM, La Paz H (1988) Intravenous anisoylated plasminogen streptokinase activator complex versus intravenous streptokinase in myocardial infarction. A randomized multicenter study. Circulation 78 [Suppl 2]:II-277

Maas ACP, van Domburg RT, Deckers JW, Vermeer F, Remme WJ, Kamp O, Manger Cats V, Simoons ML (1999) Sustained benefit at 10–14 years follow-up after thrombolytic therapy in myocardial infarction. Eur Heart J 20:819–826

MacCallum AG, Stafford PJ, Jones C, Vincent R, Perez-Avila C, Chamberlain DA (1990) Reduction in hospital time to thrombolytic therapy by audit of policy guidelines. Eur Heart J 11 [Suppl F]:48–52

Mahaffey KW, Granger CB, Toth CA, White HD, Stebbins AL, Barbash GI, Vahanian A, Topol EJ, Califf RM, for the GUSTO-1 investigators (1997) Diabetic retinopathy should not be a contraindication to thrombolytic therapy for acute myocardial infarction: review of ocular hemorrhage incidence and location in the GUSTO-I trial. J Am Coll Cardiol 30:1606–1610

Malke H, Ferretti JJ (1984) Streptokinase: Cloning, expression, and excretion by Escherichia coli. Proc Natl Acad Sci USA 81:3557

Marder VJ (1979) The use of thrombolytic agents: choice of patient, drug administration, laboratory monitoring. Ann Intern Med 90:802–808

Marder VJ (1987) Relevance of changes in blood fibrinolytic and coagulation parameters during thrombolytic therapy. Am J Med 83:15–19

Marder VJ (1989) Comparison of thrombolytic agents: selected hematologic, vascular and clinical events. Am J Cardiol 64:2A–7A

Marder VJ (1990) Bleeding complications of thrombolytic treatment. Am J Hosp Pharm 47:S15–S19

Marder VJ, Hirsh J, Bell WR (1994) Rationale and practical basis of thrombolytic therapy. In: Colman RW, Hirsh J, Marder VJ, Salzman EW (eds) Hemostasis and Thrombosis. Basic Principles and Clinical Practice, 3rd edn. Lippincott, Philadelphia, pp 1514–1542

Marder VJ, Sherry S (1988) Thrombolytic therapy: current status. N Engl J Med 318:1512–1520, 1585–1595

Marder VJ, Shulman NR, Carroll W (1969) High molecular weight derivatives of human fibrinogen produced by plasmin. I. Physicochemical and immunological characterization. J Biol Chem 244:2111–2119

Markis JE, Malagold M, Parker JA, Silverman KJ, Barry WH, Als AV, Paulin S, Grossman W, Braunwald E (1981) Myocardial salvage after intracoronary thrombolysis with streptokinase in acute myocardial infarction. N Engl J Med 305:777–782

Marmur JD, Thiruvikraman SV, Fyfe BS, Guha A, Sharma SK, Ambrose JA, Fallon JT, Nemerson Y, Taubman MB (1996) Identification of active tissue factor in human coronary atheroma. Circulation 94:1226–1232

Mason DT (1981) International experience with percutaneous transluminal coronary recanalization by streptokinase-thrombolysis reperfusion in acute myocardial infarction: new, safe, landmark therapeutic approach salvaging ischemic muscle and improving ventricular function. Am Heart J 102 [Pt 2]:1126–1133

McRedmond JP, Harriott P, Walker B, Fitzgerald DJ (2000) Streptokinase-induced platelet activation involves antistreptokinase antibodies and cleavage of protease-activated receptor-1. Blood 95:1301–1308

Maynard C, Weaver WD, Lambrew C, Bowlby LJ, Rogers WJ, Rubison RM, for the Participants in the National Registry of Myocardial Infarction (1995) Factors influencing the time to administration of thrombolytic therapy with recombinant tissue plasminogen activator (data from the National Registry of Myocardial Infarction). Am J Cardiol 76:548–552

Mentzer RL, Budzynski AZ, Sherry S (1986) High-dose, brief-duration intravenous infusion of streptokinase in acute myocardial infarction: description of effects in the circulation. Am J Cardiol 57:1220–1226

Merlini PA, Bauer KA, Oltrona L, Ardissino D, Spinola A, Cattaneo M, Broccolino M, Mannucci PM, Rosenberg RD (1995) Thrombin generation and activity during thrombolysis and concomitant heparin therapy in patients with acute myocardial infarction. J Am Coll Cardiol 25:203–209

Merrill EW, Gilliland ER, Lee TS, Salzman EW (1966) Blood rheology: effect of fibrinogen deduced by addition. Circ Res 18:437–446

Moriarty AJ, Hughes R, Nelson SD, Balnave K (1988) Streptokinase and reduced plasma viscosity: a second benefit. Eur J Haematol 41:25–36

Moroz LA, Gilmore NJ (1975) A rapid and sensitive ^{125}I-fibrin solid phase fibrinolytic assay for plasmin. Blood 46:543–553

Müller J, Reinert H, Malke H (1989) Streptokinase mutations relieving Escherichia coli K-12 (prlA4) of detriments caused by the wild-type SKC gene. J Bacteriol 171:2202–2208

National Institute of Health Consensus Development Conference (1980) Thrombolytic Therapy in Thrombosis. Ann Intern Med 93:141–144

Neuhaus KL, von Essen R, Tebbe U, Vogt A, Roth M, Riess M, Niederer W, Forycki F, Wirtzfeld A, Maeurer W, Limbourg P, Merx W, Haerten K (1992) Improved thrombolysis in acute myocardial infarction with front- loaded administration of alteplase: results of the rt-PA-APSAC patency study (TAPS). J Am Coll Cardiol 19:885–891

Newby LK, Rutsch WR, Califf RM, Simoons ML, Aylward PE, Armstrong PW, Woodlief LH, Lee KL, Topol EJ, Van de Werf F, for the GUSTO-1 investigators (1996) Time from symptom onset to treatment and outcomes after thrombolytic therapy. J Am Coll Cardiol 27:1646–1655

O'Connor CM, Meese R, Carney R, Smith J, Conn E, Burks J, Hartman C, Roark S, Shadoff N, Heard M,III, Mittler B, Collins G, Navetta F, Leimberger J, Lee K, Califf RM, for the DUCCS Study Group (1994) A randomized trial of intravenous heparin in conjunction with anistreplase (anisoylated plasminogen streptokinase activator complex) in acute myocardial infarction: The Duke University Clinical Cardiology Study (DUCCS) 1. J Am Coll Cardiol 23:11–18

Omar MN, Mann KG (1987) Inactivation of Factor Va by plasmin. J Biol Chem 262:9750–9755

Pell ACH, Miller HC, Robertson CE, Fox KAA (1992) Effect of "fast track" admission for acute myocardial infarction on delay to thrombolysis. BMJ 304:83–87

Raitt MH, Maynard C, Wagner GS, Cerqueira MD, Selvester RH, Weaver WD (1996) Relation between symptom duration before thrombolytic therapy and final myocardial infarct size. Circulation 93:48–53

Rao AK, Pratt C, Berke A, Jaffe A, Ockene I, Schreiber TL, Bell WR, Knatterud G, Robertson TL, Terrin ML (1988) Thrombolysis in myocardial infarction (TIMI) trial. Phase I: hemorrhagic manifestations and changes in plasma fibrinogen and the fibrinolytic system in patients treated with recombinant tissue plasminogen activator and streptokinase. J Am Coll Cardiol 11:1–11

Rapold HJ, Kuemmerli H, Weiss M, Baur H, Haeberli A (1989) Monitoring of fibrin generation during thrombolytic therapy of acute myocardial infarction with recombinant tissue-type plasminogen activator. Circulation 79:980–989

Rawles J (1994) Halving of mortality at 1 year by domiciliary thrombolysis in the Grampian Region Early Anistreplase Trial (GREAT). J Am Coll Cardiol 23:1–5

Rawles JM (1997) Quantification of the benefit of earlier thrombolytic therapy: five-year results of the Grampian Region Early Anistreplase Trial (GREAT). J Am Coll Cardiol 30:1181–1186

Reed GL, Houng AK (1999) The contribution of activated factor XIII to fibrinolytic resistance in experimental pulmonary embolism. Circulation 99:299–304

Reed GL, Houng AK, Liu L, Parhami-Seren B, Matsueda LH, Wang S, Hedstrom L (1999) A catalytic switch and the conversion of streptokinase to a fibrin-targeted plasminogen activator. Proc Natl Acad Sci USA 96:8879–8883

Reed GL, Matsueda GR, Haber E (1992) Platelet factor XIII increases the fibrinolytic resistance of platelet-rich clots by accelerating the crosslinking of α_2-antiplasmin to fibrin. Thromb Haemost 68:315–320

Reimer KA, Jennings RB (1979) "The wavefront phenomenon" of myocardial ischemic death: transmural progression of necroses within the framework of ischemic bed size (myocardium at risk) and collateral flow. Lab Invest 40:633–644

Reiner JS, Lundergan CF, Fung A, Coyne K, Cho S, Israel N, Kazmierski J, Pilcher G, Smith J, Rohrbeck S, Thompson M, Van de Werf F, Ross AM, for the GUSTO-1 Angiographic Investigators (1996) Evolution of early TIMI 2 flow after thrombolysis for acute myocardial infarction. Circulation 94:2441–2446

Rentrop KP, Blanke H, Karsch KR, Wiegand V, Köstering H, Oster H, Leitz K (1979) Acute myocardial infarction: intracoronary application of nitroglycerin and streptokinase. Clin Cardiol 2:354–363

Rentrop P, Blanke H, Karsch KR, Kaiser H, Köstering H, Leitz K (1981) Selective intracoronary thrombolysis in acute myocardial infarction and unstable angina pectoris. Circulation 63:307–317

Replogle RL, Meiselman HJ, Merrill EW (1967) Clinical implications of blood rheology studies. Circulation 36:148–160

Ridker PM, Hebert PR, Fuster V, Hennekens CH (1993) Are both aspirin and heparin justified as adjuncts to thrombolytic therapy for acute myocardial infarction? Lancet 341:1574–1577

Ridker PM, Hennekens CH (1996) Age and thrombolytic therapy. Circulation 94:1807–1808

Ritchie JL, Cerqueira M, Maynard C, Davis K, Kennedy JW (1988) Ventricular function and infarct size: the Western Washington Intravenous Streptokinase in Acute Myocardial Infarction Trial. J Am Coll Cardiol 11:689–697

Roberts WC, Buja LM (1972) The frequency and significance of coronary arterial thrombi and other observations in fatal myocardial infarction. Am J Med 52:425–443

Rothbard RL, Fitzpatrick PG, Francis CW, Caton DM, Hood WB, Marder VJ (1985) Relationship of the lytic state to successful reperfusion with standard low-dose intracoronary streptokinase. Circulation 71:562–570

Rozenman Y, Gotsman M, Weiss T, Lotan C, Mosseri M, Sapoznikov D, Welber S, Nassar H, Hasin Y, Gilon D (1994) Very early thrombolysis in acute myocardial infarction – a light at the end of the tunnel. Isr J Med Sci 30:99–107

Rozenman Y, Gotsman MS, Weiss AT, Lotan C, Mosseri M, Sapoznikov D, Welber S, Hasin Y, Gilon D (1995) Early intravenous thrombolysis in acute myocardial infarction: the Jerusalem experience. Int J Cardiol 49 [Suppl]:S21–S28

Rudd MA, George D, Amarante P, Vaughan DE, Loscalzo J (1990) Temporal effects of thrombolytic agents on platelet function in vivo and their modulation by prostaglandins. Circ Res 67:1175–1181

Rustige J, Burczyk U, Werner AS, Senges J (1990) Akuter Herzinfarkt. Verkürzung der Prähospitalphase durch Massenaufklärung möglich. Dtsch Aerztebl 18:1450–1454

Ryan TJ, Anderson JL, Antman EM, Braniff BA, Brooks NH, Califf RM, Hillis LD, Hiratzka LF, Rapaport E, Riegel BJ, Russell RO, Smith EE III., Weaver WD (1996) ACC/AHA guidelines for the management of patients with acute myocardial infarction: a report of the American College of Cardiology/American Heart Association Task Force on Practice Guidelines (Committee on Management of Acute Myocardial Infarction). J Am Coll Cardiol 28:1328–1419

Rylatt DB, Blake AS, Cottis LE, Massingham DA, Fletcher WA, Masci PP, Whitaker AN, Elms M, Bunce I, Webber AJ, et al (1983) An immunoassay for human D dimer using monoclonal antibodies. Thromb Res 31:767–778

Samama M, Conard J, Verdy E, et al (1987) Biological study of intravenous anisoylated plasminogen streptokinase activator complex in acute myocardial infarction. Drugs 33 [Suppl 3]:268–274

Samis JA, Ramsey GD, Walker JB, Nesheim ME, Giles AR (2000) Proteolytic processing of human coagulation factor IX by plasmin. Blood 95:943–951

Sane DC, Califf RM, Topol EJ, Stump DC, Mark DB, Greenberg CS (1989) Bleeding during thrombolytic therapy for acute myocardial infarction: mechanisms and management. Ann Intern Med 111:1010–1022

Sawyer WD, Fletcher AP, Alkjaersig N, Sherry S (1960) Studies on the thrombolytic activity of human plasma. J Clin Invest 39:426–434

Scharfstein JS, Abendschein DR, Eisenberg PR, George D, Cannon CP, Becker RC, Sobel B, Cupples LA, Braunwald E, Loscalzo J, for the TIMI-5 Investigators (1996) Usefulness of fibrinogenolytic and procoagulant markers during thrombolytic therapy in predicting clinical outcomes in acute myocardial infarction. Am J Cardiol 78:503–510

Schmidt B, Vegh P, Johnston M, Andrew M, Weitz J (1993) Do coagulation screening tests detect increased generation of thrombin and plasmin in sick newborn infants? Thromb Haemost 69:418–421

Schofer J, Büttner J, Geng G, Gutschmidt K, Herden HN, Mathey DG, Moecke HP, Polster P, Raftopoulo A, Sheehan FH, Voelz P (1990) Prehospital thrombolysis in acute myocardial infarction. Am J Cardiol 66:1429–1433

Schreiber TL, Miller DH, Silvasi DA, Moses JW, Borer JS (1986) Randomized double-blind trial of intravenous streptokinase for acute myocardial infarction. Am J Cardiol 58:47–52

Schröder R, Neuhaus K-L, Leizorovicz A, Linderer T, Tebbe U (1987) A prospective placebo-controlled double-blind multicenter trial of intravenous streptokinase in acute myocardial infarction (ISAM): long-term mortality and morbitdity. J Am Coll Cardiol 9:197–203

Seabra-Gomes R, Silva A, Alceixo A, Real T, Freire I, Torres I, Freitas R, Crespo F (1987) Evaluation of 2 intravenous thrombolytic agents (anisoylated plasminogen streptokinase activator complex versus streptokinase) in patients with acute myocardial infarction. Drugs 33 [Suppl 3]:169–174

Serebruany VL, Gurbel PA, Shustov AR, Dalesandro MR, Gumbs CI, Grabletz LB, Bahr RD, Ohman EM, Topol EJ, for the GUSTO-III Investigators (1998) Depressed platelet status in an elderly patient with hemorrhagic stroke after thrombolysis for acute myocardial infarction. Stroke 29:235–238

Serruys PW, Wijns W, van den Brand M, Ribeiro V, Fioretti P, Simoons ML, Kooijman CJ, Reiber JH, Hugenholtz PG (1983) Is transluminal coronary angioplasty mandatory after successful thrombolysis? Quantitative coronary angiographic study. Br Heart J 50:257–265

Shaila G, Chandrashekhar YS, Kumar N, Ganguly NK, Anand IS (1994) Antistreptokinase antibodies before and after streptokinase therapy in patients with acute myocardial infarction from areas endemic for streptococcal infection and influence on reperfusion rates. Am J Cardiol 74:187–189

Sharma GV, Cella G, Parisi AF, Sasahara AA (1982) Thrombolytic therapy. N Engl J Med 306:1268–1276

Sheehan FH, Braunwald E, Canner P, Dodge HT, Gore J, Van Natta P, Passamani ER,
    Williams DO, Zaret B (1987) The effect of intravenous thrombolytic therapy on
    left ventricular function: a report on tissue-type plasminogen activator and strep-
    tokinase from the Thrombolysis in Myocardial Infarction (TIMI Phase I) Trial.
    Circulation 75:817–829
Sherry S, Fletcher AP, Alkjaersig N (1959a) Fibrinolysis and fibrinolytic activity in man.
    Physiol Rev 39:343–382
Sherry S, Lindemeyer RI, Fletcher AP, Alkjaersig N (1959b) Studies on enhanced fib-
    rinolytic activity in man. J Clin Invest 38:810–822
Sherry S, Marder VJ (1991) Mistaken guidelines for thrombolytic therapy of acute
    myocardial infarction in the elderly. J Am Coll Cardiol 17:1237–1238
Shi GY, Chang BI, Su SW, Young KC, Wu DH, Chang LC, Tsai YS, Wu HL (1998)
    Preparation of a novel streptokinase mutant with improved stability. Thromb
    Haemost 79:992–997
Sigwart U, Grbic M, Bachmann F (1985) Measurement of streptokinase antibodies. J
    Am Coll Cardiol 5:1500
Simes RJ, Topol EJ, Holmes DR Jr, White HD, Rutsch WR, Vahanian A, Simoons ML,
    Morris D, Betriu A, Califf RM, Ross AM, for the GUSTO-I Investigators (1995)
    Link between the angiographic substudy and mortality outcomes in a large ran-
    domized trial of myocardial reperfusion. Importance of early and complete infarct
    artery reperfusion. Circulation 91:1923–1928
Simoons MI, for the Reperfusion Therapy Consensus Group (1997) Selection of reper-
    fusion therapy for individual patients with evolving myocardial infarction. Eur
    Heart J 18:1371–1381
Simoons ML, Serruys PW, van den Brand M, Res J, Verheugt FWA, Krauss XH, Remme
    WJ, Bär F, de Zwaan C, van der Laarse A, Vermeer F, Lubsen J, for the Working
    Group on Thrombolytic Therapy in Acute Myocardial Infarction of the Nether-
    lands Interuniversity Cardiology Institute (1986) Early thrombolysis in acute
    myocardial infarction: limitations of infarct size and improved survival. J Am Coll
    Cardiol 7:717–728
Six AJ, Louwerenburg HW, Braams R, Mechelse K, Mosterd WL, Bredero AC, Dun-
    selman PHJM, van Hemel NM (1990) A double-blind randomized multicenter
    dose-ranging trial of intravenous streptokinase in acute myocardial infarction. Am
    J Cardiol 65:119–123
Spann JF, Sherry S (1984) Coronary thrombolysis for evolving myocardial infarction.
    Drugs 28:465–483
Squire IB, Lawley W, Fletcher S, Holme E, Hillis WS, Hewitt C, Woods KL (1999)
    Humoral and cellular immune responses up to 7.5 years after administration of
    streptokinase for acute myocardial infarction. Eur Heart J 20:1245–1252
Stadius ML (1993) Angiographic monitoring of reperfusion therapy for acute myocar-
    dial infarction. TIMI grade 3 perfusion is the goal. Circulation 87:2055–2057
Stampfer MJ, Goldhaber SZ, Yusuf S, Peto R, Hennekens CH (1982) Effect of intra-
    venous streptokinase on acute myocardial infarction: pooled results from ran-
    domized trials. N Engl J Med 307:1180–1182
Steg PG, Laperche T, Golmard J-L, Juliard J-M, Benamer H, Himbert D, Aubry P, for
    the PERM Study Group (1998) Efficacy of streptokinase, but not tissue-type plas-
    minogen activator, in achieving 90-minute patency after thrombolysis for acute
    myocardial infarction decreases with time to treatment. J Am Coll Cardiol 31:
    776–779
The European Myocardial Infarction Project Group (1993) Prehospital thrombolytic
    therapy in patients with suspected acute myocardial infarction. N Engl J Med
    329:383–389
The Global Use of Strategies to Open Occluded Coronary Arteries (GUSTO) IIa
    Investigators (1994) Randomized trial of intravenous heparin versus recombinant
    hirudin for acute coronary syndromes. Circulation 90:1631–1637
The GUSTO Angiographic Investigators (1993) The effects of tissue plasminogen
    activator, streptokinase, or both on coronary-artery patency, ventricular func-

tion, and survival after acute myocardial infarction. N Engl J Med 329:1615–1622

The I.S.A.M. Study Group (1986) A prospective trial of intravenous streptokinase in acute myocardial infarction (I.S.A.M.). Mortality, morbidity, and infarct size at 21 days. N Engl J Med 314:1465–1471

The International Study Group (1990) In-hospital mortality and clinical course of 20,891 patients with suspected acute myocardial infarction randomised between alteplase and streptokinase with or without heparin. Lancet 336:71–75

The ISIS-2 (Second International Study of Infarct Survival) Collaborative Group (1988) Randomised trial of intravenous streptokinase, oral aspirin, both, or neither among 17 187 cases of suspected acute myocardial infarction: ISIS-2. Lancet 2:349–360

The TIMI Study Group (1985) The thrombolysis in myocardial infarction (TIMI) trial: phase 1 findings. N Engl J Med 312:932–936

Theiss W, Busch U, Renner U, Blömer H (1996) High-dose streptokinase therapy with three million units in one hour significantly improves early coronary patency rates in acute myocardial infarction. Am J Cardiol 78:1167–1169

Tillett WS, Garner RL (1933) The fibrinolytic activity of hemolytic streptococci. J Exp Med 68:485–502

Tillett WS, Sherry S (1949) The effect in patients of streptococcal fibrinolysin (streptokinase) and streptococcal deoxyribonuclease on fibrinous, purulent and sanguineous pleural exudations. J Clin Invest 28:173–190

Timmis GC, Mammen EF, Ramos RG, Gordon S, Gangadharan V, Hauser AM, Westveer DC, Stewart JR (1986) Hemorrhage vs rethrombosis after thrombolysis for acute myocardial infarction. Arch Intern Med 146:667–672

Torrèns I, Ojalvo AG, Seralena A, Hayes O, De la Fuente J (1999) A mutant streptokinase lacking the C-terminal 42 amino acids is less immunogenic. Immunol Lett 70:213–218

Toschi V, Gallo R, Lettino M, Fallon JT, Gertz SD, Fernandez-Ortiz A, Chesebro JH, Badimon L, Nemerson Y, Fuster V, Badimon JJ (1997) Tissue factor modulates the thrombogenicity of human atherosclerotic plaques. Circulation 95:594–599

Tripodi A, Bottasso B, Mannucci PM (1990) Elevation of thrombin-antithrombin complexes during thrombolytic therapy in patients with myocardial infarction. Res Clin Lab 20:197–202

Tsujioka H, Suehiro A, Kakishita E (1999) Activation of coagulation factor VII by tissue-type plasminogen activator. Am J Hematol 61:34–39

Van Belle E, Lablanche JM, Bauters C, Renaud N, McFadden EP, Bertrand ME (1998) Coronary angioscopic findings in the infarct-related vessel within 1 month of acute myocardial infarction: natural history and the effect of thrombolysis. Circulation 97:26–33

Verstraete M (1989) Effects of thrombolytic agents on coronary recanalization and patency, cardiac enzyme release, left ventricular function, and mortality in patients with acute myocardial infarction. J Cardiovasc Pharmacol 14:S38–S48

Verstraete M, Bernard R, Bory M, Brower RW, Collen D, de Bono DP, Erbel R, Huhmann W, Lennane RJ, Lubsen J, et al (1985) Randomised trial of intravenous recombinant tissue-type plasminogen activator versus intravenous streptokinase in acute myocardial infarction. Report from the European Cooperative Study Group for Recombinant Tissue-type Plasminogen Activator. Lancet 1:842–847

Verstraete M, Collen D (1996) Optimising thrombolytic therapy in elderly patients with acute myocardial infarction. Drugs Aging 8:17–22

Vogt A, von Essen R, Tebbe U, Feuerer W, Appel K-F, Neuhaus K-L (1993a) Impact of early perfusion status in the infarct-related artery on short-term mortality after thrombolysis for acute myocardial infarction: retrospective analysis of four German multicenter studies. J Am Coll Cardiol 21:1391–1395

Vogt P, Monnier P, Schaller M-D, Goy J-J, Beuret P, Essinger A, Bachmann F, Hauert J, Perret C, Sigwart U, Kappenberger L (1993b) Comparison of results of intra-

venous infusion of anistreplase versus streptokinase in acute myocardial infarction. Am J Cardiol 71:274–280

von dem Borne PAK, Meijers JCM, Bouma BN (1996) Effect of heparin on the activation of factor XI by fibrin-bound thrombin. Thromb Haemost 76:347–353

Voth E, Tebbe U, Schicha H, Neuhaus K-L, Schröder R, The I.S.A.M. Study Group
(1991) Intravenous streptokinase in acute myocardial infarction (I.S.A.M.) trial:
serial evaluation of left ventricular function up to 3 years after infarction estimated
by radionuclide ventriculography. J Am Coll Cardiol 18:1610–1616

Wang H, Song H, Yang VC (1999) A recombinant prodrug type approach for triggered
delivery of streptokinase. J Contr Release 59:119–122

Weaver WD, Cerqueira M, Hallstrom AP, Litwin PE, Martin JS, Kudenchuk PJ,
Eisenberg M (1993) Prehospital-initiated vs hospital-initiated thrombolytic
therapy. The Myocardial Infarction Triage and Intervention Trial. JAMA 270:
1211–1216

Weinstein J (1982) Treatment of myocardial infarction with intracoronary streptokinase: efficacy and safety data from 209 United States cases in the Hoechst-Roussel
registry. Am Heart J 104 [Pt 2]:894–898

White HD, Barbash GI, Califf RM, Simes RJ, Granger CB, Weaver WD, Kleiman NS,
Aylward PE, Gore JM, Vahanian A, Lee KL, Ross AM, Topol EJ, for the GUSTO-
1 investigators (1996) Age and outcome with contemporary thrombolytic therapy.
Results from the GUSTO-1 trial. Circulation 94:1826–1833

White HD, Norris RM, Brown MA, Takayama M, Maslowski A, Bass NM, Ormiston
JA, Whitlock T (1987) Effect of intravenous streptokinase on left ventricular function and early survival after acute myocardial infarction. N Engl J Med 317:
850–855

White PD (1931) Coronary heart disease. In: White PD (ed) Heart Disease,
MacMillan, New York

Wilcox JN, Smith KM, Schwartz SM, Gordon D (1989) Localization of tissue factor in
the normal vessel wall and in the atherosclerotic plaque. Proc Natl Acad Sci USA
86:2839–2843

Williams WL (1998) Guidelines to reducing delays in administration of thrombolytic
therapy in acute myocardial infarction. Drugs 55:689–698

Wu XC, Ye R, Duan Y, Wong SL (1998) Engineering of plasmin-resistant forms of streptokinase and their production in Bacillus subtilis: streptokinase with longer functional half-life. Appl Environ Microbiol 64:824–829

Yarnell JWG, Baker IA, Sweetnam PM, Bainton D, O'Brien JR, Whitehead PJ, Elwood
PC (1991) Fibrinogen, viscosity, and white blood cell count are major risk factors
for ischemic heart disease. The Caerphilly and Speedwell collaborative heart
disease studies. Circulation 83:836–844

Yaryura RA, Zaqqa M, Ferguson JJ (1998) Complications associated with combined
use of abciximab and an intracoronary thrombolytic agent (urokinase or tissue-
type plasminogen activator). Am J Cardiol 82:518–519

Yusuf S, Collins R, Peto R, Furberg C, Stampfer MJ, Goldhaber SZ, Hennekens CH
(1985) Intravenous and intracoronary fibrinolytic therapy in acute myocardial
infarction: overview of results on mortality, reinfarction and side-effects from 33
randomized controlled trials. Eur Heart J 6:556–585

Zeymer U, Tebbe U, Essen Rv , Haarmann W, Neuhaus K-L, for the ALKK-Study
Group (1999) Influence of time to treatment on early infarct-related artery
patency after different thrombolytic regimens. Am Heart J 137:34–38

Zhu Y, Carmeliet P, Fay WP (1999) Plasminogen activator inhibitor-1 is a major determinant of arterial thrombolysis resistance. Circulation 99:3050–3055

CHAPTER 8

Tissue-Type Plasminogen Activator (tPA)

D. COLLEN

A. Introduction

Cardiovascular diseases, comprising acute myocardial infarction (AMI), stroke, and venous thromboembolism have, as their immediate underlying cause, thrombosis of critically situated blood vessels with loss of blood flow to vital organs. One approach to the treatment of thrombosis consists of the pharmacologic dissolution of the blood clot via the intravenous infusion of plasminogen activators, which activate the blood fibrinolytic system.

Several plasminogen activator have been used for the thrombolytic treatment of AMI (Fig. 1). Streptokinase, anistreplase, and two-chain urokinase induce extensive systemic activation of the fibrinolytic system, deplete plasminogen, and degrade circulating fibrinogen, Factor V, and Factor VIII (so-called proteolytic state). In contrast, the physiologic plasminogen activators, tPA and single-chain uPA (sc-uPA), activate plasminogen preferentially at the fibrin surface. The same holds true for staphylokinase (see Chap. 16) and for the vampire bat plasminogen activator (see Chap. 17). Plasmin, associated with the fibrin surface, is protected from rapid inhibition by α_2-antiplasmin and may thus efficiently degrade the fibrin of a thrombus (WIMAN and COLLEN 1978) (Fig. 2).

Thrombolytic therapy of AMI is based on the premise that coronary artery thrombosis is its proximate cause. Rupture of an atheromatous plaque leads to occlusive thrombosis that produces myocardial ischemia and cell necrosis, leading to loss of ventricular function and possibly death. The hypothesis underlying thrombolytic therapy in AMI is that early and sustained recanalization prevents cell death, reduces infarct size, preserves myocardial function, and reduces early and late mortality. The beneficial effects of thrombolytic therapy in AMI are now well established, and it has become routine treatment, given to more than 500000 patients per year world-wide.

The current indications to thrombolytic therapy in patients with acute myocardial infarction are summarized in Table 1. Contraindications are listed on Table 3, Chap. 7. It has been estimated that only 20%–25% of the 1.5–2 million patients hospitalized world-wide each year with acute myocardial infarction receive thrombolytic therapy but that at least twice as many patients

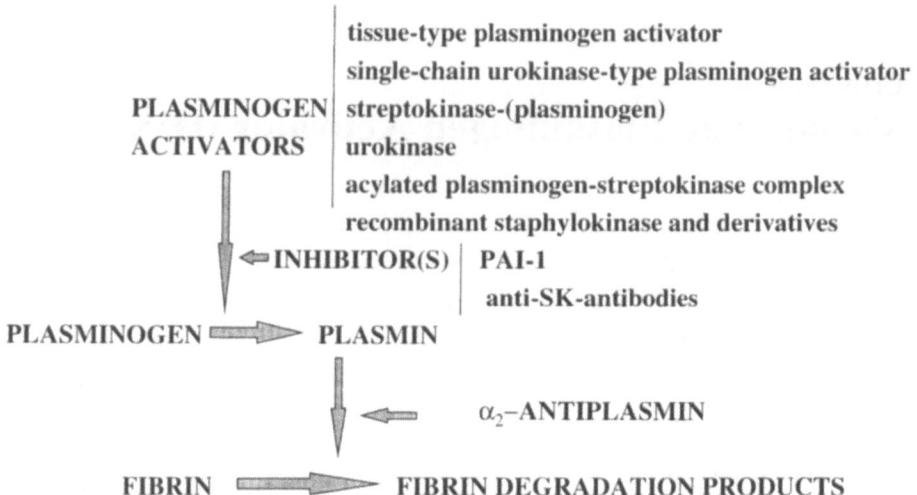

Fig. 1. Schematic representation of the fibrinolytic system. The zymogen plasminogen is converted to the active serine proteinase plasmin, which degrades fibrin into soluble degradation products, by the physiological plasminogen activators (PA), tissue-type PA (tPA) or urokinase-type PA (uPA) or by exogenously administered PAs. Inhibition may occur at the level of plasminogen activators by plasminogen activator inhibitor-1 (PAI-1) or plasminogen activator inhibitor-2 (PAI-2), or at the level of plasmin, mainly by α_2-antiplasmin

THROMBOLYSIS AND FIBRIN-SPECIFICITY

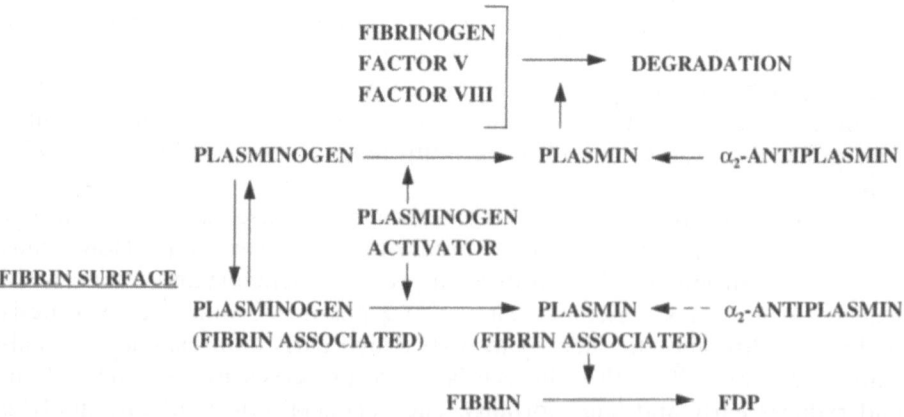

Fig. 2. Schematic representation of the concept of fibrin-selectivity. Non-fibrin-selective plasminogen activators (e.g., streptokinase, urokinase, anistreplase = APSAC) activate both plasminogen in the fluid phase and fibrin-associated plasminogen relatively indiscriminately. Systemic plasminogen activation leads to exhaustion of the substrate (reduced thrombolytic efficacy) and depletion of α_2-antiplasmin (systemic fibrinolytic state with fibrinogen breakdown). Fibrin-specific plasminogen activators (tPA, sc-uPA, staphylokinase) preferentially activate fibrin-associated plasminogen. Fibrin-selective agents preserve plasminogen and spare fibrinogen

Table 1. ACC/AHA guidelines for thrombolytic therapy in acute myocardial infarction (AMI)

Class I: (Evidence and/or general agreement that a given treatment is beneficial, useful and effective)
1. ST elevation (>0.1 mV, two or more contiguous leads)[a], time to therapy 12 h or less[b], age less than 75 years
2. Bundle branch block (obscuring ST-segment analysis) and history suggesting acute AMI

Comment: Treatment benefit is present regardless of gender, presence of diabetes, blood pressure (if less than 180 mm Hg systolic), heart rate, or history of previous AMI (FIBRINOLYTIC THERAPY TRIALISTS' (FTT) COLLABORATIVE GROUP 1994). Benefit is greater in the setting of anterior AMI, diabetes, low blood pressure (less than 100 mm Hg systolic), or high heart rate (greater than 100 bpm). The earlier therapy begins, the better the outcome, with the greatest benefit decidedly occurring when therapy is given within the first 3 h; proven benefit occurs, however, up to at least 12 h of the onset of symptoms. Benefit is less with inferior acute AMI, except for the subgroup with associated right ventricular infarction (ST elevation RV-4) or anterior-segment depression

Class IIa: (Weight of evidence/opinion is in favor of usefulness/efficacy)
1. ST elevation[a], age 75 or older

Comment: In persons older than 75 years, the overall risk of mortality from infarction is high without and with therapy. Although the proportionate reduction in mortality is less than in patients younger than 75, the absolute reduction results in 10 lives saved per 1000 patients treated in those over 75. The relative benefit of therapy is reduced (FIBRINOLYTIC THERAPY TRIALISTS' (FTT) COLLABORATIVE GROUP 1994).

Class IIb: (Usefulness/efficacy is less well established by evidence/opinion)
1. ST elevation[a], time to therapy greater than 12–24 h[b].
2. Blood pressure on presentation >180 mm Hg systolic and/or >110 mm Hg diastolic associated with high risk MI.

Comment: Generally there is only a small trend for benefit of therapy after a delay of more than 12–24 h, but thrombolysis may be considered for selected patients with ongoing ischemic pain and extensive ST elevation. Risk of intracerebral hemorrhage is greater when presenting blood pressure is greater than 180/110 mm Hg, and in this situation the potential benefit of therapy must be weighed carefully against the risk of hemorrhagic stroke. Risk of cardiac rupture appeared to increase with prolonged time to therapy in an earlier meta-analysis (HONAN et al. 1990) but was not associated with increased risk of rupture in later, larger study (BECKER et al. 1995). Generally patients presenting more than 12 h after symptom onset were excluded from some but not all trials. An attempt to lower blood pressure first (with nitrates, β-adrenoceptor blocker, etc.) is recommended, but is not of proven benefit in lowering the risk of intracerebral hemorrhage. Primary PTCA or CABG may be considered if available

Class III: (Evidence and/or general agreement that a treatment is not useful/effective and in some cases may be harmful)
1. ST elevation[a], time to therapy greater than 24 h[b], ischemic pain resolved
2. ST-segment depression only

Comment: In the absence of ST elevation, there is no evidence of benefit for patients with normal electrocardiographic or non-specific changes, and, using current thrombolytic regimens, there is some suggestion of harm (including increased bleeding risk) for patients with ST-segment depression only (FIBRINOLYTIC THERAPY TRIALISTS' (FTT) COLLABORATIVE GROUP 1994; TIMI IIIB INVESTIGATORS 1994). When marked ST-segment depression is confined to leads V1 through V4, there is a likelihood that his reflects a posterior current of injury and suggests a circumflex

Table 1. *Continued*

---

artery occlusion for which thrombolytic therapy would be considered inappropriate. Very recent retrospective analysis of the Late Assessment of Thrombolytic Efficacy (LATE) Trial (LATE STUDY GROUP 1993; LANGER et al. 1996; BRAUNWALD and CANNON 1996) also casts some uncertainties about withholding thrombolytic therapy from this heterogeneous group of patients

---

Taken from RYAN et al. (1996, 1999).
[a] Repeat ECGs recommended during medical observation in suggestive clinical settings when initial ECG is nondiagnostic of ST elevation.
[b] Time of symptoms onset is defined as the beginning of continuous, persistent discomfort that brought the patient to the hospital.

might be eligible for such treatment (EUROPEAN SECONDARY PREVENTION STUDY GROUP 1996; FRENCH et al. 1996; JHA et al. 1996; JULIARD et al. 1997; KRUMHOLZ et al. 1997). The two most widely used thrombolytic agents for intravenous administration to patients with AMI are non-fibrin-selective streptokinase and fibrin-selective alteplase (tPA) (Fig. 2). Recommendations for the choice of a particular thrombolytic agent and dosage regimens have been established by the Task Force of the International Society and Federation of Cardiology (now called World Heart Federation) and the World Health Organization (Table 2).

B. Development of tPA as a Thrombolytic Agent

I. Isolation and Purification of tPA

Astrup and collaborators were first in describing that many human and animal tissues contain a plasminogen activator (ASTRUP and PERMIN 1947; ASTRUP and STAGE 1952; reviewed in BACHMANN and KRUITHOF 1984). THORSEN et al. (1972) made the original observation that fibrin binds tPA, but not urokinase. Many investigators have since confirmed this phenomenon (BINDER et al. 1979; AASTED 1980; ALLEN and PEPPER 1981; RADCLIFFE and HEINZE 1981; KRUITHOF and BACHMANN 1982; RÅNBY 1982; reviewed in FEARS 1989). tPA is a very inefficient activator of plasminogen in the absence of fibrin but in its presence the activation of plasminogen is greatly potentiated (CAMIOLO et al. 1971; HOYLAERTS et al. 1982). This is explicable by the assembly of plasminogen and of tPA on the fibrin surface. Most agree that binding of sc-tPA and of tc-tPA is roughly comparable, although the single-chain form may bind slightly better (HIGGINS and VEHAR 1987; RIJKEN et al. 1982).

These observations greatly stimulated research directed at purifying tPA. However, the isolation and purification of tPA from tissues has been difficult, because this enzyme is present in very small quantities and is bound firmly to particulate cell constituents. Purified or enriched tPA preparations have been obtained from pig heart (COLE and BACHMANN 1977; WALLÉN et al. 1982),

Table 2. Currently used regimens of coronary thrombolysis

Streptokinase and aspirin	Streptokinase (SK) 1.5 million U IV over 30–60 min, combined with acetylsalicylic acid (ASA) 160–325 mg daily started as soon as possible and continued indefinitely. The safety and efficacy of this regimen in terms of mortality reduction was established in ISIS-2, GISSI-2, and ISIS-3. It is associated with moderate efficacy for early coronary artery recanalization; about 55% patency at 90 min with a catch-up to about 80% at 3 h. It is less efficient for mortality reduction in patients treated within the first 6 h than accelerated alteplase and intravenous heparin as demonstrated by GUSTO.
Alteplase and intravenous heparin	Alteplase (tPA) accelerated regimen: 100 mg IV over 90 min (15 mg bolus, 0.75 mg/kg not exceeding 50 mg over 30 min, and 0.5 mg/kg not exceeding 35 mg over the next hour) combined with 160–325 mg ASA and immediate intravenous heparin (5000 U bolus and 1000 U per hour, preferably monitored with activated partial thromboplasin time). In GUSTO, the accelerated regimen was associated with a statistically significant lower mortality than SK (6.3% vs 7.3%, $p = 0.001$) but with a slightly higher incidence (0.1%) of survival with disabling stroke.
Anistreplase and aspirin	Anistreplase (APSAC) 30 U IV given as a slow bolus over 3–5 min, in combination with aspirin 160 mg/day. The comparative mortality reduction was similar to that with SK/ASA in ISIS-3. The efficacy for early coronary artery recanalization is probably between that of SK and front-loaded alteplase.
Selection of regimen	The GUSTO trial has demonstrated a significant overall survival benefit of accelerated tPA given with intravenous heparin over previous regimens, particularly SK with subcutaneous or with intravenous heparin (14% mortality reduction with 95% confidence intervals of 6–21%) and a consistent pattern of fewer complications, including allergic reactions, clinical indicators of left ventricular dysfunction, and arrhythmias. The survival benefit is largest in patients <75 years old, with anterior infarction, and <4 h from onset of symptoms. No subgroups were identified in which tPA was significantly worse than SK but, possibly because of a lack of statistical power, no statistically significant benefit of tPA compared with SK was documented in patients >75 years old, in patients with small inferior infarcts, and in patients presenting >4 h after the onset of symptoms. Therefore, if cost considerations become a limiting factor, tPA should be reserved primarily for the former subgroups and SK for the latter. If financial constraints do not permit the use of any thrombolytic agents, ASA should be administered as soon as possible.

Taken from SCHLANT (1994).

human uterus (RIJKEN et al. 1979), and from vascular perfusion fluids (BINDER et al. 1979; AASTED 1980; ALLEN and PEPPER 1981). Real progress in the characterization of tPA was made when it was discovered that larger quantities of this enzyme could be obtained from certain cell cultures, such as the human Bowes melanoma cell line (RIJKEN and COLLEN 1981). The Bowes melanoma cell line was provided to us by Dr. D.B. RIFKIN, New York University Medical School, toward the end of 1978. It had been obtained originally from metastatic melanoma cells from a patient named Bowes. The Bowes cell line is unique because it secretes large amounts of a plasminogen activator which does not crossreact with antibodies directed against uPA (WILSON et al. 1980). When mixtures of fibrinogen and culture supernatant of the Bowes melanoma cell line were mixed and clotted, all plasminogen activator activity remained fixed to the clot. Thus, the Bowes plasminogen activator resembled tPA. RIJKEN et al. (1979), who had developed a purification method for tPA, yielding approximately 1 mg from 5 kg of human uterine tissue joined our laboratory in 1979. Using a slightly modified procedure, we were able to purify a plasminogen activator from the supernatant of the Bowes melanoma cell line and to demonstrate that this material was identical to the human uterine tPA (RIJKEN and COLLEN 1981). Subsequently, the purification procedure was scaled upward to produce a total amount of approximately 2 g of tPA (COLLEN et al. 1982) permitting the systematic characterization of its biochemical (RIJKEN et al. 1982; HOYLAERTS et al. 1982), biological, and physiologic properties and to investigate its thrombolytic potential in animal models of thrombosis (COLLEN et al. 1983).

The first two renal transplant patients suffering from an ileofemoral/renal vein thrombosis were treated with purified Bowes melanoma cell tPA in 1981 (WEIMAR et al. 1981). Very low doses (5–7.5 mg given over 24 h) resulted in complete lysis of the venous thrombi in these uremic patients.

II. Cloning and Expression of the tPA cDNA

In 1980, at the 5th Congress on Fibrinolysis in Malmö, Sweden, Dr. DIANE PENNICA, a scientist from the Department of Molecular Biology at the Genentech Corporation, explored the possibility of developing tPA as a thrombolytic agent and sought collaborations. We began a very fruitful exchange of know-how, reagents, cell lines, and antibodies directed against human tPA, and just two years later Dr. PENNICA reported on the successful cloning of tPA at the 6th Congress on Fibrinolysis in Lausanne, Switzerland. The seminal report was published in Nature (PENNICA et al. 1983). With the approval of the FDA, recombinant tPA (rtPA) was first administered to a patient on February 11, 1994. The rapid progress was possible only because of the concerted efforts of many scientists at Genentech and academic institutions in the United States and Belgium. We were able to demonstrate that the kinetics of activation of plasminogen by melanoma cell and by recombinant tPA (ZAMARRON et al.

1984) and the turnover in rabbits were similar and that the thrombolytic potential of rtPA in a rabbit jugular vein thrombosis preparation was indistinguishable from that of the plasminogen activator of melanoma origin (COLLEN et al. 1984).

III. Experimental Coronary Thrombosis Models for AMI

Around 1980, several cardiology groups had begun to administer intracoronary streptokinase (SK) to patients with AMI (see Chap. 7). Late in 1981, at an NIH workshop on coronary thrombolysis, we met with Dr. B.E. SOBEL from Washington University, St. LOUIS, and agreed to initiate a collaboration exploring the use of tPA for thrombolysis in AMI. Coronary thrombosis in dogs was produced by advancing a copper coil into the left anterior descending coronary artery (LAD). Intravenous infusion of human melanoma cell tPA resulted in prompt coronary recanalization without systemic activation of the fibrinolytic system. Furthermore, it restored myocardial blood flow and intermediary metabolism in the region at risk, as demonstrated tomographically with ^{11}C-labeled palmitate and $H_2^{15}O$ given intravenously (BERGMANN et al. 1983). Subsequently these studies were extended to rtPA (VAN DE WERF et al. 1984). In a concurrent collaborative study with Dr. H. K. GOLD, Massachusetts General Hospital, Boston, coronary thrombosis was produced between two ligatures of the LAD in open-chest dogs. Infusion of tPA elicited clot lysis and myocardial salvage (GOLD et al. 1984). Finally, in a collaborative study with Dr. W. Flameng at the University of Leuven, the coronary thrombolytic properties, clot-specificity, and myocardial protection achievable with rtPA were confirmed in baboons (FLAMENG et al. 1985).

IV. First Administration of tPA to Patients with AMI

The first study in which melanoma cell tPA was administered to patients with AMI was performed in 1983. Participants included Dr. F. VAN DE WERF and collaborators at the University of Leuven and Dr. SOBEL and co-workers at Washington University. Intravenously administered tPA in doses of 0.2–0.4 mg/min recanalized occluded coronary arteries within 30–60 min in six of seven patients without producing a proteolytic state. Assays of three parameters, sensitive for the detection of a proteolytic state, namely fibrinogen, plasminogen, and α_2-antiplasmin, demonstrated that none of these markers of proteolysis was significantly reduced and that the level of fibrinogen degradation products (FDPs) had not increased after tPA (VAN DE WERF et al. 1984).

These observations stimulated initiation of a tricenter, dose-finding randomized trial with rtPA provided by Genentech. Fifty patients were enrolled in the first half of 1984 at Washington University Medical Center, the Massachusetts General Hospital, and Johns Hopkins University Medical Center. The intravenous infusion of 0.5 mg/kg rtPA over 30–60 min in 18 patients produced recanalization in 12 (67%). If the same dose was followed by 0.25 mg/kg over an additional 60 min recanalization of the culprit artery resulted in 13 of 15 patients (87%). However, of the 25 patients with restoration of vascular patency, 20% exhibited a reocclusion during the 30 min after discontinuation of the infusion (COLLEN et al. 1984). This observation lead to the design of the TIMI I study.

C. Clinical Trials with tPA in AMI

I. Early Clinical Studies in AMI

The Thrombolysis In Myocardial Infarction (TIMI) Study Group was established in 1983 by the National Heart, Lung and Blood Institute (Bethesda, Md.) in order to assess the efficacy of intravenous SK and other thrombolytic agents in the treatment of AMI. The study group initially included 13 clinical sites, a data co-ordinating center, a drug-distributing center, and core laboratories for radiographic, radionuclear, electrocardiographic, coagulation, and pathological studies. The TIMI phase I study assessed 90 min recanalization rates of totally occluded infarct-related coronary arteries. A preliminary report was published in 1985 (THE TIMI STUDY GROUP 1985) and the full report in 1987 (CHESEBRO et al. 1987). In this double-blind study patients were randomly assigned to either 1.5 Mio U SK, infused over 1 h and a 3-h infusion of tPA placebo, or to tPA administered by continuous intravenous infusion over 3 h (40 mg, 20 mg, 20 mg, in the first, second, and third hour respectively) and a 1-h infusion of SK placebo. All patients received 5000 U of heparin prior to coronarography, followed by the administration of the respective thrombolytic agent. Intravenous heparin, 1000 U/h, was given to all patients 3 h after the initial bolus dose of heparin. Treatment with aspirin and dipyridamole was started prior to discharge. Coronarographies were evaluated centrally without knowledge of treatment assignment. On February 5, 1985 the NHLBI stopped phase I trial on the recommendation of the TIMI Policy Advisory and Data Monitoring Board, because of substantial, statistically significant differences in recanalization rates between the patients given tPA and those given SK. Indeed, improvement from TIMI grade 0 flow (for definition see Chap. 7) to grade 2 or 3 flow, 90 min after start of thrombolytic treatment, occurred in 26% of patients receiving SK but in 56% of those treated with tPA. Improvement from TIMI grade 0 or 1 to grade 2 or 3 occurred in 31% of patients with SK and in 62% of those given tPA (CHESEBRO et al. 1987) (Table 3). Twenty-one-day mortality, occurrence of clinical signs of reinfarction, bleeding at the catheter site, and number of units of blood transfused were similar in the two treatment arms. The 6- and 12-month follow-up demonstrated that the mor-

Table 3. Early comparative trials with tPA and placebo or SK in AMI

	ECSG-II		Topol		TIMI-1		ECSG-I	
	tPA	Plac	tPA	Plac	tPA	SK	tPA	SK
No. of patients treated	64	65	75	25	147	143	64	65
Percentage of reperfusion 90 min after start (%)								
From TIMI grade 0 to grades 2 or 3					56	26		
From TIMI grades 0, 1 to grades 2, 3					62	31		
Patency rates 90 min after start (%)	61	21	69	24			70	55
21-day or in-hospital mortality (%)	1	4	4	8	5	4	5	5
Reocclusion by coronarography (%)			16	12	24[a]	14		
Clinical reinfarction (%)	5	0	4	4	13	12	3	5
Bleeding at catheter site (%)	5	0	26	8	66	67	3	8

[a] If episodes of reinfarction or death are attributed to reocclusion the rates were 29% for tPA and 30% for SK.
References: ECSG-II: VERSTRAETE et al. (1985b); TOPOL et al. (1987); TIMI-1: CHESEBRO et al. (1987), ECSG-I: VERSTRAETE et al. (1985a).

tality rates were lower in the tPA group (7.7% and 10.5%) than in the SK group (9.5% and 11.6%; n.s.) (DALEN et al. 1988).

The same year a similar study was performed in Europe. The treatment scheme was simpler. Initially all patients received an i.v. bolus of 5000 U of heparin. To reduce the symptom to needle time as much as possible, coronary angiography was performed only 90 min after the start of treatment and thus assessed only patency rates. tPA was given in a body weight adjusted dosage of 0.75 mg/kg over a period of 90 min. The patients in the SK group received in addition 0.5 g of aspirin and 1.5 Mio U of SK infused over 60 min (VERSTRAETE et al. 1985a). Patency rates 90 min after the start of thrombolytic treatment were 70% in the tPA and 55% in the SK group ($p = 0.054$). Two further trials that compared tPA with placebo are listed in Table 3. In both, the patency rates with tPA were two to three times higher than those in the placebo group (VERSTRAETE et al. 1985b; TOPOL et al. 1987). There were no strokes in these small trials, bleeding occurred mainly at the catheter insertion sites, but reocclusion remained a major problem and it was not clear whether and at what intensity heparin should be given as an adjunct treatment. Furthermore, cardiologists became increasingly aware of the fact that the prognosis of patients exhibiting TIMI grade 2 flow was significantly worse than of those demonstrating TIMI grade 3 flow (DALEN et al. 1988; KARAGOUNIS et al. 1992; ANDERSON et al. 1993; STADIUS 1993; VOGT et al. 1993; THE GUSTO ANGIOGRAPHIC

Investigators 1993; Lincoff et al. 1995; Gibson et al. 1995; reviewed by Granger et al. 1994 and by Barbagelata et al. 1997; Gibson et al. 2000).

II. tPA Trials, 1986–1989

In the following years tPA thrombolysis trials attempted to resolve these problems by (1) administering aspirin to all patients prior to treatment with a thrombolytic drug based on the findings of the ISIS-2 trial (THE ISIS-2 (SECOND INTERNATIONAL STUDY OF INFARCT SURVIVAL) COLLABORATIVE GROUP 1988), (2) increasing the dose of tPA to 100 mg or even to 150 mg (Topol et al. 1987; Passamani et al. 1987), (3) prolonging the period of the tPA infusion to several hours (Verstraete et al. 1987; Topol et al. 1988), (4) adopting a front-loading schema of tPA administration (Neuhaus et al. 1989), and (5) administering heparin early and in higher doses (reviewed by Mahaffey et al. 1996; Sobel 1997; White 1997).

Table 4 lists most of the trials conducted in the years 1986 to 1989. It does not contain comparative trials of tPA vs SK or anistreplase which were begun during this period and which are discussed in Chap. 10. In the earliest trials Genentech provided a preparation that was composed mostly of two-chain tPA. In later trials primarily single-chain material was used. In the trials with short term mortality as the primary endpoint there was a 27% (Wilcox et al. 1988) and 45% (Van de Werf et al. 1988) reduction of the 21–30 day mortality in the tPA group compared to placebo. Three trials investigated early left ventricular function with absolute LVEF in the average 6%–7% higher in the tPA groups (Guerci et al. 1987; National Heart Foundation of Australia Coronary Thrombolysis Group 1988; O'Rourke et al. 1988). Two trials were designed to explore the effect of continued infusion of tPA on the reocclusion rate. In the ECSG-III study two groups of patients received 40 mg of tPA during the first 90 min, at which time point angiographies revealed a 66% patency rate. In the first group this was followed by further infusion of 30 mg of tPA over 6 h; in the second by the administration of a 6 h placebo infusion. The reocclusion rates were not statistically different between the two groups (Verstraete et al. 1987). In the TAMI-II study 150 mg of tPA were administered in two different dosage schedules. One group received 60 mg in the first hour and the remaining 90 mg over 7 h. To investigate also the importance of weight adjustment, the second group received 1 mg/kg over 60 min and the remaining dose over the next 5 h. Patency at 90 min was borderline significantly higher in the body weight adjusted group (Topol et al. 1988). Two studies attempted to enforce early reperfusion/patency by raising the total dose of tPA to 150 mg (Topol et al. 1987; Passamani et al. 1987). However this scheme was soon abandoned because of an unacceptably high rate of intracerebral hemorrhage (Braunwald et al. 1987a,b).

Several studies addressed the problem of nonperfusion or early reocclusion after thrombolysis by means of emergency or rescue PTCA (Topol et al. 1987; Simoons et al. 1988; The Timi Research Group 1988). In none of these

Table 4. Main features of trials with tPA conducted in the years 1986–89

Study	Dose of tPA (mg)	Adjunct therapy	Patients (no.)	Primary endpoint	Main results
ECSG-III, VERSTRAETE et al. (1987)	40 (tc) 40 (tc)	30 mg tPA/6h Placebo	42 40	Reocclusion	No difference
TIMI-II Pilot, PASSAMANI et al. (1987)	150 (sc)	H	317	Patency	at 90 min 82% open
TAMI-II, TOPOL et al. (1988)	60 1 mg/kg	90 mg/7h; H, A, D rest/5h; H, A, D	178 206	90 min patency	64% 77% (p = 0.06)
Hopkins, GUERCI et al. (1987)	80–100[a] Placebo	H, A, D, N H, A, D, N	72 66	LV function	LVEF 53% LVEF 46%
NATIONAL HEART FOUNDATION AUSTRALIA CORONARY THROMBOLYSIS GROUP (1988)	100 (sc) Placebo	H H	73 71	LV function	LVEF 58% LVEF 52%
TICO, O'ROURKE et al. (1988)	100 (sc) Placebo	H, D H, D	74 71	LV function	LVEF 61% LVEF 54%
ASSET, WILCOX et al. (1988)	100 (sc) Placebo	H H	2516 2495	30 day mortality	7.2% (27% reduction) 9.8%
ECSG-V, VAN DE WERF et al. (1988)	100 (sc) Placebo	H, A H, A	355 366	21 day mortality	3.7% (45% reduction) 6.8%
NEUHAUS et al. (1989)	100 front-loaded	H	74	early patency	60 min patency 74% 90 min patency 91%

(tc), early, mainly two-chain tPA preparation; (sc), mainly single-chain tPA; H, heparin; A, aspirin; D, dipyridamole; N, nitroglycerin; LVEF, left ventricular ejection fraction.

All patency rates refer to TIMI grade 2 and 3 flow.

[a]The first patients received tc-tPA; at later stages sc-tPA was given.

trials was there a beneficial effect seen of early emergency PTCA after thrombolysis.

Finally, still in an attempt to improve further early reperfusion, Neuhaus and collaborators developed the front-loaded (also called accelerated) tPA regimen and obtained in a non-randomized trial the astonishingly high TIMI grade 2 and 3 flow rates of 83% (TIMI grade 3, 70%) after 60 min and 96% (TIMI grade 3, 87%) after 90 min (NEUHAUS et al. 1989). Most of the trials prior to 1990 had used the standard dosage regimen which consists of a 10 mg bolus, then 50 mg during the first hour and the remaining 40 mg given over the second and third hour. In the accelerated scheme, patients receive a bolus of 15 mg initially, followed by 50 mg given over the first 30 min and the remaining 35 mg during the last 60 min (see Table 2). A pharmacokinetic study revealed that, using this scheme, tPA concentrations during the first 30 min were approximately 50% higher than those during a standard infusion schedule. Plasma half-lives were similar (3.5 min) with both schemes. Markers of a proteolytic state (fibrinogen, plasminogen, α_2-antiplasmin, and FDPs were not significantly different (TANSWELL et al. 1992).

III. tPA Trials 1990–1995

This is the period of the three megatrials GISSI-2, ISIS-3, and GUSTO (for a detailed discussion consult Chap. 10).

A similar accelerated scheme as the one described by NEUHAUS was used in one treatment arm in the GUSTO study, consisting of a 15 mg bolus, followed by 0.75 mg/kg (not to exceed 50 mg) over 30 min and 0.5 mg/kg (not to exceed 35 mg) over the next 60 min (THE GUSTO INVESTIGATORS 1993). The 30-day mortality and the rate of disabling stroke were as follows: (1) streptokinase with delayed subcutaneous heparin, 7.2% and 0.5%; (2) streptokinase with concurrent intravenous heparin, 7.4% and 0.5%; (3) accelerated tPA with concurrent intravenous heparin, 6.3% and 0.6%; and (4) combination of streptokinase and tPA with intravenous heparin, 7.0% and 0.6%. Compared to the 30-day mortality of 6.3% in patients receiving accelerated tPA and intravenous heparin, that of the combined groups of patients receiving streptokinase with either subcutaneous or intravenous heparin was 7.3% ($p = 0.001$). A combined end point of death or disabling stroke was also significantly lower in the accelerated tPA group than in the streptokinase-only groups (6.9% vs 7.8%, $p = 0.006$). In the GUSTO trial the accelerated alteplase with intravenous heparin produced somewhat over 50% complete recanalization (TIMI grade 3 flow) at 90 min (which is the main predictor of clinical benefit) compared to around 30% with streptokinase and aspirin (THE GUSTO ANGIOGRAPHIC INVESTIGATORS 1993).

The GISSI-2 mortality study (20 891 patients) (GRUPPO ITALIANO PER LO STUDIO DELLA SOPRAVVIVENZA NELL'INFARTO MIOCARDICO (GISSI) 1990; THE INTERNATIONAL STUDY GROUP 1990) compared SK and single-chain alteplase (100 mg intravenously over 3 h). The ISIS-3 trial (41 299 patients) [ISIS-3

(THIRD INTERNATIONAL STUDY OF INFARCT SURVIVAL) COLLABORATIVE GROUP 1992] compared SK and double-chain duteplase (0.6MU/kg representing approximately 2mg/kg intravenously over 4h). Delayed subcutaneous heparin or no heparin was used in most patients in both studies. Both trials showed no difference in survival between streptokinase and tPA. Several explanations have been proposed for the lack of apparent correlation between mortality in these studies and initial (90-min) patency of the infarct-related artery in other interventional studies. The prevailing view at present is that these trials did not routinely use adjunctive intravenous heparin with tPA to protect against reocclusion, resulting in similar patency rates and consequently similar mortality benefits with streptokinase and tPA.

IV. Further Attempts to Improve the Efficacy of Thrombolysis and to Reduce the Incidence of Reocclusion

Several important developments have taken place in the 1990s. To improve the efficacy of thrombolysis in AMI, new thrombolytic agents, such as staphylokinase and vampire bat plasminogen activator as well as mutants of tPA (TNK-tPA, reteplase, lanoteplase, E6010) were developed and tested in clinical trials. Large cardiology centers began using primary PTCA as the preferred treatment of AMI.

Ruptured atherosclerotic plaques are very thrombogenic, in large measure due to their content of tissue factor (WILCOX et al. 1989; TOSCHI et al. 1997; MARMUR et al. 1996). Furthermore, even after successful thrombolysis, small thrombi may remain in the culprit coronary artery (VAN BELLE et al. 1998; ARAKAWA et al. 1997). Fibrin binds thrombin (FRANCIS et al. 1983) which is not effectively inhibited by heparin/antithrombin (HOGG and JACKSON 1989; VON DEM BORNE et al. 1996). All thrombolytic agents can activate platelets (GURBEL et al. 1998) and it is well known that platelet-rich thrombi as they occur in the coronary circulation are more resistant to thrombolytic agents than red thrombi (ANDERSON 1997; ZHU et al. 1999). For these reasons the effectiveness of adjunctive therapy with hirudin and other direct thrombin inhibitors, and with Gp IIb/IIIa receptor antagonists, was widely explored. These developments are discussed in Chaps. 11, 16, 17, and 19.

In this section two further developments will be described. The first concerns the appropriateness of administering thrombolytic therapy to patients who arrive late at the hospital, the second concerns further dosing schedules with tPA.

In the LATE study 5711 patients with symptoms and ECG criteria consistent with AMI were randomized in double-blind fashion to tPA (100mg over 3h) or matching placebo, between 6h and 24h from symptom onset. Both groups received immediate oral aspirin and at later stages of the trial intravenous heparin (LATE STUDY GROUP 1993; LANGER et al. 1996). The 35-day mortality in patients treated within 12h of the onset of symptoms was 8.9% with tPA and 12.0% in the placebo group ($p = 0.023$). In patients in whom the symptom to needle interval exceeded 12h there was no benefit (tPA: 8.7%; placebo 9.2%; $p = 0.6$). Based on the results of this important study the ACC/AHA guidelines recommend to administer thrombolytic therapy up to 12h after onset of symptoms (Table 1, class 1). The accelerated dosage scheme used in the GUSTO trial was further modified by GULBA et al. (1997) by increasing the front-loading dose to 20mg and giving the remaining 80mg over 60min, resulting in over 80% TIMI grade 3 flow at the 90min angiogram. Pilot angiographic studies suggested that also a double bolus administration of 50mg of tPA 30min apart would further increase the TIMI grade 3 patency rate (GEMMILL et al. 1991; PURVIS et al. 1994). In the COBALT trial this scheme was investigated in a double-blind fashion in 7169 patients [VAN DE WERF FOR THE CONTINUOUS INFUSION VERSUS DOUBLE-BOLUS ADMINISTRATION OF ALTEPLASE (COBALT) INVESTIGATORS 1997]. Thirty-day mortality was higher in the double-bolus group (8.0%) compared to that of the accelerated tPA group (7.5%). Stroke rates were also higher in the double-bolus group (any stroke 1.9% vs 1.1%; hemorrhagic stroke 1.5% vs 0.8%). The authors concluded that accelerated tPA remains the preferred regimen for the treatment of AMI.

List of Abbreviations

AMI	Acute Myocardial Infarction
APSAC	Anisoylated Plasminogen Streptokinase Activator Complex
ASA	Acetyl Salicylic Acid
APTT	Activated Partial Thromboplastin Time
COBALT	COntinuous infusion versus double-Bolus Administration of ALTeplase
ECSG	European Collaborative Study Group
FDA	Food and Drug Administration (U.S.A.)
FDPs	Fibrin(ogen) Degradation Products
GISSI	Gruppo Italiano per lo Studio della Streptochinasi nell'Infarto miocardico
GUSTO	Global Utilization of Streptokinase and Tissue plasminogen activator for Occluded coronary arteries
ISIS	International Study of Infarct Survival
LAD	Left Anterior Descending coronary artery
LVEF	Left Ventricular Ejection Fraction
LATE	Late Assessment of Thrombolytic Efficacy
NHLBI	National Heart, Lung and Blood Institute (U.S.A.)
NIH	National Institute of Health (U.S.A.)
PTCA	Percutaneous Transluminal Coronary Angioplasty
sc	single-chain
SK	StreptoKinase
TAMI	Thrombolysis and Angioplasty in Myocardial Infarction

tc   two-chain
TICO   Thrombolysis In acute Coronary Occlusion
TIMI   Thrombolysis In Myocardial Infarction
tPA   tissue-type Plasminogen Activator
uPA   urinary-type Plasminogen Activator (urokinase)

References

Aasted B (1980) Purification and characterization of human vascular plasminogen activator. Biochim Biophys Acta 621:241–254

Allen RA, Pepper DS (1981) Isolation and properties of human vascular plasminogen activator. Thromb Haemost 45:43–50

Anderson JL (1997) Why does thrombolysis fail? Breaking through the reperfusion ceiling. Am J Cardiol 80:1588–1590

Anderson JL, Karagounis LA, Becker LC, Sorensen SG, Menlove RL, for the TEAM-3 Investigators (1993) TIMI perfusion grade 3 but not grade 2 results in improved outcome after thrombolysis for myocardial infarction. Ventriculographic, enzymatic, and electrocardiographic evidence from the TEAM-3 Study. Circulation 87:1829–1839

Arakawa K, Mizuno K, Shibuya T, Etsuda H, Tabata H, Nagayoshi H, Satomura K, Isojima K, Kurita A, Nakamura H (1997) Angioscopic coronary macromorphology after thrombolysis in acute myocardial infarction. Am J Cardiol 79:197–202

Astrup T, Permin PM (1947) Fibrinolysis in the animal organism. Nature 159:681–683

Astrup T, Stage A (1952) Isolation of a soluble fibrinolytic activator from animal tissue. Nature 170:929–930

Bachmann F, Kruithof EKO (1984) Tissue plasminogen activator: chemical and physiological aspects. Sem Thromb Hemost 10:6–17

Barbagelata NA, Granger CB, Oqueli E, Suárez LD, Borruel M, Topol EJ, Califf RM (1997) TIMI grade 3 flow and reocclusion after intravenous thrombolytic therapy: a pooled analysis. Am Heart J 133:273–282

Becker RC, Charlesworth A, Wilcox RG, Hampton J, Skene A, Gore JM, Topol EJ (1995) Cardiac rupture associated with thrombolytic therapy: impact of time to treatment in the late assessment of thrombolytic efficacy (LATE) study. J Am Coll Cardiol 25:1063–1068

Bergmann SR, Fox KAA, Ter-Pogossian MM, Sobel BE, Collen D (1983) Clot-selective coronary thrombolysis with tissue-type plasminogen activator. Science 220:1181–1183

Binder BR, Spragg J, Austen KF (1979) Purification and characterization of human vascular plasminogen activator derived from blood vessel perfusates. J Biol Chem 254:1998–2003

Braunwald E, Cannon CP (1996) Non-Q wave and ST segment depression myocardial infarction: is there a role for thrombolytic therapy? J Am Coll Cardiol 27:1333–1334

Braunwald E, Knatterud GL, Passamani E, Robertson TL, Solomon R (1987a) Update from the Thrombolysis in Myocardial Infarction Trial. J Am Coll Cardiol 10:970

Braunwald E, Knatterud GL, Passamani ER, Robertson TL (1987b) Announcement of protocol change in thrombolysis in myocardial infarction trial. J Am Coll Cardiol 9:467

Camiolo SM, Thorsen S, Astrup T (1971) Fibrinogenolysis and fibrinolysis with tissue plasminogen activator, urokinase, streptokinase-activated human globulin, and plasmin. Proc Soc Exp Biol Med 138:277–280

Chesebro JH, Knatterud G, Roberts R, Borer J, Cohen LS, Dalen J, Dodge HT, Francis CK, Hillis D, Ludbrook P, Markis JE, Mueller H, Passamani ER, Powers ER, Rao AK, Robertson T, Ross A, Ryan TJ, Sobel BE, Willerson J, Williams DO, Zaret BL,

Braunwald E (1987) Thrombolysis in myocardial infarction (TIMI) trial, phase I: a comparison between intravenous tissue plasminogen activator and intravenous streptokinase. Clinical findings through hospital discharge. Circulation 76:142–154

Cole ER, Bachmann F (1977) Purification and properties of a plasminogen activator from pig heart. J Biol Chem 252:3729–3737

Collen D, Rijken DC, Van Damme J, Billiau A (1982) Purification of human tissue-type plasminogen activator in centigram quantities from human melanoma cell culture fluid and its conditioning for use in vivo. Thromb Haemost 48:294–296

Collen D, Stassen JM, Marafino BJ Jr, Builder S, De Cock F, Ogez J, Tajiri D, Pennica D, Bennett WF, Salwa J, Hoyng CF (1984) Biological properties of human tissue-type plasminogen activator obtained by expression of recombinant DNA in mammalian cells. J Pharmacol Exp Ther 231:146–152

Collen D, Stassen JM, Verstraete M (1983) Thrombolysis with human extrinsic (tissue-type) plasminogen activator in rabbits with experimental jugular vein thrombosis. J Clin Invest 71:368–376

Collen D, Topol EJ, Tiefenbrunn AJ, Gold HK, Weisfeldt ML, Sobel BE, Leinbach RC, Ludbrook PA, Yasuda T, Bulkley BH, Robison AK, Hutter AM, Bell WR, Spadaro JJ, Khaw BA, Grossbard EB (1984) Coronary thrombolysis with recombinant human tissue-type plasminogen activator: a prospective, randomized, placebo-controlled trial. Circulation 70: 1012–1017

Dalen JE, Gore JM, Braunwald E, Borer J, Goldberg RJ, Passamani ER, Forman S, Knatterud G, and the TIMI Investigators (1988) Six- and twelve-month follow-up of the phase I Thrombolysis in Myocardial Infarction (TIMI) trial . Am J Cardiol 62:179–185

European Secondary Prevention Study Group (1996) Translation of clinical trials into practice: a European population-based study of the use of thrombolysis for acute myocardial infarction . Lancet 347:1203–1207

Fears R (1989) Binding of plasminogen activators to fibrin: characterization and pharmacological consequences. Biochem J 261:313–324

Fibrinolytic Therapy Trialists' (FTT) Collaborative Group (1994) Indications for fibrinolytic therapy in suspected acute myocardial infarction: Collaborative overview of early mortality and major morbidity results from all randomized trials of more than 1000 patients. Lancet 343:311–322

Flameng W, Van de Werf F, Vanhaecke J, Verstraete M, Collen D (1985) Coronary thrombolysis and infarct size reduction after intravenous infusion of recombinant tissue-type plasminogen activator in nonhuman primates. J Clin Invest 75:84–90

Francis CW, Markham RE Jr, Barlow GH, Florack TM, Dobrzynski DM, Marder VJ (1983) Thrombin activity of fibrin thrombi and soluble plasmic derivatives. J Lab Clin Med 102:220–230

French JK, Williams BF, Hart HH, Wyatt S, Poole JE, Ingram C, Ellis CJ, Williams MG, White HD (1996) Prospective evaluation of eligibility for thrombolytic therapy in acute myocardial infarction. BMJ 312:1637–1641

Gemmill JD, Hogg KJ, MacIntyre PD, Booth N, Rae AP, Dunn FG, Hillis WS (1991) A pilot study of the efficacy and safety of bolus administration of alteplase in acute myocardial infarction. Br Heart J 66:134–138

Gibson CM, Cannon CP, Murphy SA, Ryan KA, Mesley R, Marble SJ, McCabe CH, Van de Werf F, Braunwald E, for the TIMI (Thrombolysis in Myyocardial Infarction) Study Group (2000) Relationship of TIMI myocardial perfusion grade to mortality after administration of thrombolytic drugs. Circulation 101:125–130

Gibson CM, Cannon CP, Piana RN, Breall JA, Sharaf B, Flatley M, Alexander B, Diver DJ, McCabe CH, Flaker GC, Baim DS, Braunwald E (1995) Angiographic predictors of reocclusion after thrombolysis: results from the thrombolysis in myocardial infarction (TIMI) 4 trial. J Am Coll Cardiol 25:582–589

Gold HK, Fallon JT, Yasuda T, Leinbach RC, Khaw BA, Newell JB, Guerrero JL, Vislosky FM, Hoyng CF, Grossbard E, Collen D (1984) Coronary thrombolysis with recombinant human tissue-type plasminogen activator. Circulation 70: 700–707

Granger CB, White HD, Bates ER, Ohman EM, Califf RM (1994) A pooled analysis of coronary arterial patency and left ventricular function after intravenous thrombolysis for acute myocardial infarction. Am J Cardiol 74:1220–1228

Gruppo Italiano per lo Studio della Sopravvivenza nell'Infarto Miocardico (GISSI) (1990) GISSI-2: A factorial randomised trial of alteplase versus streptokinase and heparin versus no heparin among 12490 patients with acute myocardial infarction. Lancet 336:65–71

Guerci AD, Gerstenblith G, Brinker JA, Chandra NC, Gottlieb SO, Bahr RD, Weiss JL, Shapiro EP, Flaherty JT, Bush DE, Chew PH, Gottlieb SH, Halperin HR, Ouyang P, Walford GD, Bell WR, Fatterpaker AK, Llewellyn M, Topol EJ, Healy B, Siu CO, Becker LC, Weisfeldt ML (1987) A randomized trial of intravenous tissue plasminogen activator for acute myocardial infarction with subsequent randomization to elective coronary angioplasty. N Engl J Med 317:1613–1618

Gulba DC, Tanswell P, Dechend R, Sosada M, Weis A, Waigand J, Uhlich F, Hauck S, Jost S, Rafflenbeul W, Lichtlen PR, Dietz R (1997) Sixty-minute alteplase protocol: a new accelerated recombinant tissue-type plasminogen activator regimen for thrombolysis in acute myocardial infarction. J Am Coll Cardiol 30:1611–1617

Gurbel PA, Serebruany VL, Shustov AR, Bahr RD, Carpo C, Ohman EM, Topol EJ, for the GUSTO-III Investigators (1998) Effects of reteplase and alteplase on platelet aggregation and major receptor expression during the first 24 hours of acute myocardial infarction treatment. J Am Coll Cardiol 31:1466–1473

Higgins DL, Vehar GA (1987) Interaction of one-chain and two-chain tissue plasminogen activator with intact and plasmin-degraded fibrin. Biochemistry 26:7786–7791

Hogg PJ, Jackson CM (1989) Fibrin monomer protects thrombin from inactivation by heparin- antithrombin III: Implications for heparin efficacy. Biochemistry 86:3619–3623

Honan MB, Harrell FE Jr, Reimer KA, Califf RM, Mark DB, Pryor DB, Hlatky MA (1990) Cardiac rupture, mortality and a timing of thrombolytic therapy: a meta-analysis. J Am Coll Cardiol 16:359–367

Hoylaerts M, Rijken DC, Lijnen HR, Collen D (1982) Kinetics of the activation of plasminogen by human tissue plasminogen activator. Role of fibrin. J Biol Chem 257:2912–2919

ISIS-3 (Third International Study of Infarct Survival) Collaborative Group (1992) ISIS-3: a randomised comparison of streptokinase vs tissue plasminogen activator vs anistreplase and of aspirin plus heparin vs aspirin alone among 41299 cases of suspected acute myocardial infarction. Lancet 339:753–770

Jha P, Deboer D, Sykora K, Naylor CD (1996) Characteristics and mortality outcomes of thrombolysis trial participants and nonparticipants: a population-based comparison. J Am Coll Cardiol 27:1335–1342

Juliard J-M, Himbert D, Golmard J-L, Aubry P, Karrillon GJ, Boccara A, Steg PG (1997) Can we provide reperfusion therapy to all unselected patients admitted with acute myocardial infarction? J Am Coll Cardiol 30:157–164

Karagounis L, Sorensen SG, Menlove RL, Moreno F, Anderson JL, for the TEAM-2 Investigators (1992) Does thrombolysis in myocardial infarction (TIMI) perfusion grade 2 represent a mostly patent artery or a mostly occluded artery? Enzymatic and electrocardiographic evidence from the TEAM-2 study. J Am Coll Cardiol 19:1–10

Kruithof EKO, Bachmann F (1982) Studies on the binding of tissue plasminogen activator to fibrinogen and fibrin. In: Henschen A, Graeff H, Lottspeich A (eds) Fibrinogen – recent biomedical and medical aspects. W. de Gruyter, Berlin, pp 377–387

Krumholz HM, Murillo JE, Chen J, Vaccarino V, Radford MJ, Ellerbeck EF, Wang Y (1997) Thrombolytic therapy for eligible elderly patients with acute myocardial infarction. JAMA 277:1683–1688

Langer A, Goodman SG, Topol EJ, Charlesworth A, Skene AM, Wilcox RG, Armstrong PW, for the LATE Study Investigators (1996) Late assessment of thrombolytic efficacy (LATE) study: prognosis in patients with non-Q wave myocardial infarction. J Am Coll Cardiol 27:1327–1332

LATE Study Group (1993) Late assessment of thrombolytic efficacy (LATE) study with alteplase 6–24 hours after onset of acute myocardial infarction. Lancet 342:759–766

Lincoff AM, Topol EJ, Califf RM, Sigmon KN, Lee KL, Ohman EM, Rosenschein U, Ellis SG (1995) Significance of a coronary artery with thrombolysis in myocardial infarction grade 2 flow "patency" (outcome in the thrombolysis and angioplasty in myocardial infarction trials). Am J Cardiol 75:871–876

Mahaffey KW, Granger CB, Collins R, O'Connor CM, Ohman EM, Bleich SD, Col JJ, Califf RM (1996) Overview of randomized trials of intravenous heparin in patients with acute myocardial infarction treated with thrombolytic therapy. Am J Cardiol 77:551–556

Marmur JD, Thiruvikraman SV, Fyfe BS, Guha A, Sharma SK, Ambrose JA, Fallon JT, Nemerson Y, Taubman MB (1996) Identification of active tissue factor in human coronary atheroma. Circulation 94:1226–1232

National Heart Foundation of Australia Coronary Thrombolysis Group (1988) Coronary thrombolysis and myocardial salvage by tissue plasminogen activator given up to 4 hours after onset of myocardial infarction. Lancet 1:203–208

Neuhaus KL, Feuerer W, Jeep-Tebbe S, Niederer W, Vogt A, Tebbe U (1989) Improved thrombolysis with a modified dose regimen of recombinant tissue-type plasminogen activator. J Am Coll Cardiol 14:1566–1569

O'Rourke M, Baron D, Keogh A, Kelly R, Nelson G, Barnes C, Raftos J, Graham K, Hillman K, Newman H, Healey J, Woolridge J, Rivers J, White H, Whitlock R, Norris R (1988) Limitation of myocardial infarction by early infusion of recombinant tissue-type plasminogen activator. Circulation 77:1311–1315

Passamani E, Hodges M, Herman M, Grose R, Chaitman B, Rogers W, Forman S, Terrin M, Knatterud G, Robertson T, Braunwald E (1987) The thrombolysis in myocardial infarction (TIMI) phase II pilot study: tissue plasminogen activator followed by percutaneous transluminal coronary angioplasty. J Am Coll Cardiol 10:51B–564

Pennica D, Holmes WE, Kohr WJ, Harkins RN, Vehar GA, Ward CA, Bennett WF, Yelverton E, Seeburg PH, Heynecker HL, Goeddel DV, Collen D (1983) Cloning and expression of human tissue-type plasminogen activator cDNA in *E. coli*. Nature 301:214–221

Purvis JA, McNeill AJ, Siddiqui RA, Roberts MJD, McClements BM, McEneaney D, Campbell NPS, Khan MM, Webb SW, Wilson CM, Adgey AAJ (1994) Efficacy of 100 mg of double-bolus alteplase in achieving complete perfusion in the treatment of acute myocardial infarction. J Am Coll Cardiol 23:6–10

Radcliffe R, Heinze T (1981) Stimulation of tissue plasminogen activator by denatured proteins and fibrin clots: A possible additional role for plasminogen activator? Arch Biochem Biophys 211:750–761

Rijken DC, Collen D (1981) Purification and characterization of the plasminogen activator secreted by human melanoma cells in culture. J Biol Chem 256:7035–7041

Rijken DC, Hoylaerts M, Collen D (1982) Fibrinolytic properties of one-chain and two-chain human extrinsic (tissue-type) plasminogen activator. J Biol Chem 257:2920–2925

Rijken DC, Wijngaards G, Zaal De Jong M, Welbergen J (1979) Purification and partial characterization of plasminogen activator from human uterine tissue. Biochim Biophys Acta 580:140–153

Ryan TJ, Anderson JL, Antman EM, Braniff BA, Brooks NH, Califf RM, Hillis LD, Hiratzka LF, Rapaport E, Riegel BJ, Russell RO, Smith EE III, Weaver WD (1996) ACC/AHA guidelines for the management of patients with acute myocardial infarction: a report of the American College of Cardiology/American Heart Association Task Force on Practice Guidelines (Committee on Management of Acute Myocardial Infarction). J Am Coll Cardiol 28:1328–1419

Ryan TJ, Antman EM, Brooks NH, Califf RM, Hillis LD, Hiratzka LF, Rapaport E, Riegel B, Russell RO, Smith EE, Weaver WD, Gibbons RJ, Alpert JS, Eagle KA,

Gardner TJ, Garson A, Jr., Gregoratos G, Smith SC, Jr. (1999) 1999 update: ACC/AHA Guidelines for the Management of Patients With Acute Myocardial Infarction: Executive Summary and Recommendations: A report of the American College of Cardiology/American Heart Association Task Force on Practice Guidelines (Committee on Management of Acute Myocardial Infarction). Circulation 100:1016–1030

Rånby M (1982) Tissue Plasminogen Activator: Isolation, Enzymatic Properties and Assay Procedure, Thesis. Umea, Sweden, Umea University Medical Dissertations. New Series No. 90

Schlant RC (1994) Reperfusion in acute myocardial infarction. Circulation 90: 2091–2102

Simoons ML, Arnold AER, Betriu A, de Bono DP, Col J, Dougherty FC, von Essen R, Lambertz H, Lubsen J, Meier B, Michel PL, Raynaud P, Rutsch W, Sanz GA, Schmidt W, Serruys PW, Thery C, Uebis R, Vahanian A, Van de Werf F, Willems GM, Wood D, Verstraete M, for the European Cooperative Study Group for Recombinant Tissue-Type Plasminogen Activator (rTPA) (1988) Thrombolysis with tissue plasminogen activator in acute myocardial infarction: no additional benefit from immediate percutaneous coronary angioplasty. Lancet i:197–203

Sobel BE (1997) Heparin and streptokinase. Cardiovasc Drugs Ther 11:97–100

Stadius ML (1993) Angiographic monitoring of reperfusion therapy for acute myocardial infarction. TIMI grade 3 perfusion is the goal. Circulation 87:2055–2057

Tanswell P, Tebbe U, Neuhaus KL, Gläsle-Schwarz L, Wojcik J, Seifried E (1992) Pharmacokinetics and fibrin specificity of alteplase during accelerated infusions in acute myocardial infarction. J Am Coll Cardiol 19:1071–1075

The GUSTO Angiographic Investigators (1993) The effects of tissue plasminogen activator, streptokinase, or both on coronary-artery patency, ventricular function, and survival after acute myocardial infarction. N Engl J Med 329:1615–1622

The GUSTO Investigators (1993) An international randomized trial comparing four thrombolytic strategies for acute myocardial infarction. N Engl J Med 329:673–682

The International Study Group (1990) In-hospital mortality and clinical course of 20 891 patients with suspected acute myocardial infarction randomised between alteplase and streptokinase with or without heparin. Lancet 336:71–75

The ISIS-2 (Second International Study of Infarct Survival) Collaborative Group (1988) Randomised trial of intravenous streptokinase, oral aspirin, both, or neither among 17 187 cases of suspected acute myocardial infarction: ISIS-2. Lancet 2:349–360

The TIMI Research Group (1988) Immediate vs delayed catheterization and angioplasty following thrombolytic therapy for acute myocardial infarction. TIMI II A results. JAMA 260:2849–2858

The TIMI Study Group (1985) The thrombolysis in myocardial infarction (TIMI) trial: phase 1 findings. N Engl J Med 312:932–936

Thorsen S, Glas-Greenwalt P, Astrup T (1972) Differences in the binding to fibrin of urokinase and tissue plasminogen activator. Thromb Diath Haemorrh 28:65–74

TIMI IIIB Investigators (1994) Effects of tissue plasminogen activator and a comparison of early invasive and conservative strategies in unstable angina and non-Q-wave myocardial infarction: Results of the TIMI IIIB trial. Circulation 89:1545–1556

Topol EJ, Califf RM, George BS, Kereiakes DJ, Abbottsmith CW, Candela RJ, Lee KL, Pitt B, Stack RS, O'Neill WW, for the Thrombolysis and Angioplasty in Myocardial Infarction Study Group (1987) A randomized trial of immediate versus delayed elective angioplasty after intravenous tissue plasminogen activator in acute myocardial infarction. N Engl J Med 317:581–588

Topol EJ, George BS, Kereiakes DJ, Candela RJ, Abbottsmith CW, Stump DC, Boswick JM, Stack RS, Califf RM, and the TAMI Study Group (1988) Comparison of two dose regimens of intravenous tissue plasminogen activator for acute myocardial infarction. Am J Cardiol 61:723–728

Topol EJ, Morris DC, Smalling RW, Schumacher RR, Taylor CR, Nishikawa A, Liberman HA, Collen D, Tufte ME, Grossbard EB, O'Neill WW (1987) A multicenter, randomized, placebo-controlled trial of a new form of intravenous recombinant tissue-type plasminogen activator (activase) in acute myocardial infarction. J Am Coll Cardiol 9:1205–1213

Toschi V, Gallo R, Lettino M, Fallon JT, Gertz SD, Fernandez-Ortiz A, Chesebro JH, Badimon L, Nemerson Y, Fuster V, Badimon JJ (1997) Tissue factor modulates the thrombogenicity of human atherosclerotic plaques. Circulation 95:594–599

Van Belle E, Lablanche JM, Bauters C, Renaud N, McFadden EP, Bertrand ME (1998) Coronary angioscopic findings in the infarct-related vessel within 1 month of acute myocardial infarction: natural history and the effect of thrombolysis. Circulation 97:26–33

Van de Werf F, Arnold AER, for the European Cooperative Study Group for recombinant tissue type plasminogen activator (1988) Intravenous tissue plasminogen activator and size of infarct, left ventricular function, and survival in acute myocardial infarction. BMJ 297:1374–1379

Van de Werf F, Bergmann SR, Fox KAA, De Geest H, Hoyng CF, Sobel BE, Collen D (1984) Coronary thrombolysis with intravenously administered human tissue-type plasminogen activator produced by recombinant DNA technology. Circulation 69:605–610

Van de Werf F, for the Continuous Infusion versus Double-Bolus Administration of Alteplase (COBALT) Investigators (1997) A comparison of continuous infusion of alteplase with double-bolus administration for acute myocardial infarction. N Engl J Med 337:1124–1130

Van de Werf F, Ludbrook PA, Bergmann SR, Tiefenbrunn AJ, Fox KAA, De Geest H, Verstraete M, Collen D, Sobel BE (1984) Coronary thrombolysis with tissue-type plasminogen activator in patients with evolving myocardial infarction. N Engl J Med 310:609–613

Verstraete M, Arnold AER, Brower RW, Collen D, de Bono DP, de Zwaan C, Erbel R, Hillis WS, Lennane RJ, Lubsen J, Mathey D, Reid DS, Rutsch W, Schartl M, Schofer J, Serruys PW, Simoons ML, Uebis R, Vahanian A, Verheugt F, von Essen R (1987) Acute coronary thrombolysis with recombinant human tissue-type plasminogen activator: initial patency and influence of maintained infusion on reocclusion rate. Am J Cardiol 60:231–237

Verstraete M, Bernard R, Bory M, Brower RW, Collen D, de Bono DP, Erbel R, Huhmann W, Lennane RJ, Lubsen J et al. (1985a) Randomised trial of intravenous recombinant tissue-type plasminogen activator versus intravenous streptokinase in acute myocardial infarction. Report from the European Cooperative Study Group for Recombinant Tissue-type Plasminogen Activator. Lancet 1:842–847

Verstraete M, Bleifeld W, Brower RW, Charbonnier B, Collen D, de Bono DP, Dunning AJ, Lennane RJ, Lubsen J, Mathey DG, Michel PL, Raynaud P, Schofer J, Vahanian A, Vanhaecke J, Van de Kley GA, Van de Werf F, von Essen R (1985b) Double-blind randomised trial of intravenous tissue-type plasminogen activator versus placebo in acute myocardial infarction. Lancet 2:965–969

Vogt A, von Essen R, Tebbe U, Feuerer W, Appel K-F, Neuhaus K-L (1993) Impact of early perfusion status in the infarct-related artery on short-term mortality after thrombolysis for acute myocardial infarction: retrospective analysis of four German multicenter studies. J Am Coll Cardiol 21:1391–1395

von dem Borne PAK, Meijers JCM, Bouma BN (1996) Effect of heparin on the activation of factor XI by fibrin-bound thrombin. Thromb Haemost 76:347–353

Wallén P, Bergsdorf N, Rånby M (1982) Purification and identification of two structural variants of porcine tissue plasminogen activator by affinity adsorption on fibrin. Biochim Biophys Acta 719:318–328

Weimar W, van Seyen AJ, Stibbe J, Billiau A, de Somer D, Collen D (1981) Specific lysis of an iliofemoral thrombus by administration of extrinsic (tissue-type) plasminogen activator. Lancet ii:1018–1020

White HD (1997) Is heparin of value in the management of acute myocardial infarction? Cardiovasc Drugs Ther 11:111–119

Wilcox JN, Smith KM, Schwartz SM, Gordon D (1989) Localization of tissue factor in the normal vessel wall and in the atherosclerotic plaque. Proc Natl Acad Sci USA 86:2839–2843

Wilcox RG, von der Lippe G, Olsson CG, Jensen G, Skene AM, Hampton JR, for the ASSET Study Group (1988) Trial of tissue plasminogen activator for mortality reduction in acute myocardial infarction. Anglo-Scandinavian Study of Early Thrombolysis (ASSET). Lancet 2:525–530

Wilson EL, Becker MLB, Hoal EG, Dowdle EG (1980) Molecular species of plasminogen activators secreted by normal and neoplastic human cells. Cancer Res 40:933–938

Wiman B, Collen D (1978) Molecular mechanism of physiological fibrinolysis. Nature 272:549–550

Zamarron C, Lijnen HR, Collen D (1984) Kinetics of the activation of plasminogen by natural and recombinant tissue-type plasminogen activator. J Biol Chem 259: 2080–283

Zhu Y, Carmeliet P, Fay WP (1999) Plasminogen activator inhibitor-1 is a major determinant of arterial thrombolysis resistance. Circulation 99:3050–3055

CHAPTER 9

Urokinase and Single-Chain Urokinase-Type Plasminogen Activator (Pro-urokinase)

V. Gurewich

A. History

The discovery of urokinase (UK) is described in chap. 4. The UK in urine is not a waste product as shown by studies of the fate of intravenous injections of UK, which was found not to be excreted by the kidney (Fletcher et al. 1965). Instead, it is secreted by renal tubular cells (Sappino et al. 1991) where it helps prevent obstruction of the tubules by fibrin and other proteinaceous deposits, as first proposed by Astrup and Sterndorff (1952). Kidney cells in culture are another current source of UK and Barlow and Lazer (1972) showed that UK from this source was identical to that purified from urine.

Husain et al. (1979) purified a novel form of UK from urine which was resistant to reducing conditions on electrophoresis. It represented, therefore, a single-chain form of UK, which had not previously been described. It was shown that 10%–25% of the UK in normal urine was, in fact, single-chain UK. The explanation for why this had eluded detection for 28 years was believed to be related to the multi-step purification procedure used to isolate UK from urine, which permitted proteolytic cleavage of single-chain UK to take place. By contrast, the method which resulted in this chance discovery was a novel, single-step isolation procedure. Partial characterization of the single-chain UK showed that it had the same molecular weight as two-chain HMW-UK, but had a significantly lower specific activity. This suggested that it was a pro-enzyme, as also indicated by the finding that it was resistant to inactivation by diisopropylflurophosphate. It was believed to represent the native form of UK (Husain et al. 1979, 1983).

Single-chain UK was later, but independently, purified from human plasma (Wun et al. 1982b; Tissot et al. 1982), from glioblastoma cell culture medium (Nielsen et al. 1982), and from epidermoid carcinoma by Wun et al. (1982a), who also showed that it was a proenzyme which could be activated by plasmin. Previous evidence of a precursor form of UK in certain cell culture media was published by Bernik and Oller (1973, 1976), who found evidence of a latent UK activity which could be activated by plasmin, thrombin, or trypsin. Later, Nolan et al. (1977) showed that the trypsin activatable activity in embryonic kidney cell culture fluid failed to bind to benzamidine Sepharose in contrast to UK, consistent with it being a proactivator. However, purification of this pro-enzyme and demonstration that it was a single-chain form of UK was not achieved at that time.

B. Nomenclature

The two-chain plasminogen activator of urine was named urokinase (UK) by SOBEL et al. (1952) and therefore the single-chain precursor was originally named single-chain UK (HUSAIN et al. 1979) and later pro-urokinase (pro-UK) (WUN et al. 1982a) when its proenzyme status was further characterized by demonstrating activation by plasmin. Subsequently, this property was put into question by a kinetic study of recombinant pro-UK (rpro-UK) which suggested that its catalytic efficiency against plasminogen was comparable to that of UK (COLLEN et al. 1986b). On the basis of this finding, the designation pro-UK seemed to be a misnomer. It was recommended and accepted by the Subcommittee on Fibrinolysis that the names sc-uPA and tc-uPA, for single-chain and two-chain urokinase-type plasminogen activator, be used (COLLEN 1985). The designation uPA, for all forms of UK, was a logical parallel to the previously accepted designation tPA for all forms of tissue-type plasminogen activator. However, subsequent studies showed that sc-uPA was a pro-enzyme form of uPA (see below), and therefore the term pro-UK is appropriate and has persisted in the literature as a more descriptive and historically more coherent name.

C. Biochemistry of Pro-UK/UK (sc-uPA/tc-uPA)

The primary sequence of the A and B chains of uPA is shown in chap. 4, Fig. 1.

The two principal enzymes that cleave pro-UK at its activation site are plasmin (WUN et al. 1982a) and kallikrein (ICHINOSE et al. 1986; HAUERT et al. 1989) (for more details see chap. 4). The catalytic efficiency of this activation reaction is relatively low (PANNELL and GUREWICH 1986) which may explain the difficulty encountered converting the small amount of single-chain UK initially isolated from urine to its two-chain derivative (HUSAIN et al. 1983; STUMP et al. 1986).

It is likely that the biological function of pro-UK/UK is better served by this activation reaction being relatively slow at physiological concentrations of the reactants. As a result, there is more opportunity for pro-UK itself to activate plasminogen, which it can do efficiently under certain conditions, such as in the presence of fibrin fragment E (LIU and GUREWICH 1991, 1992) or on cell surfaces (MANCHANDA and SCHWARTZ 1991). Moreover, on cells, rapid UK generation may actually be counter-productive since the more rapid the activation, the more rapidly the reaction will be subject to inhibition by plasminogen activator inhibitor-1 (PAI-1) and the complex internalized by the cell.

A hyperactive transitional state between pro-UK and UK has also been identified whose expression is also favored by a less efficient conversion of pro-UK to UK. This transitional state was shown to have a catalytic efficiency against plasminogen at least threefold greater than that of UK. The phenomenon was postulated to be a consequence of an unusual property of pro-UK, i.e., a K_M against plasminogen lower than that of UK. During plasminogen activation by pro-UK/UK, a transitional hyperactivity develops corresponding to the K_M of pro-UK and the K_{cat} of UK (Liu et al. 1992). Therefore, pro-UK/UK appears to be a somewhat more potent plasminogen activator than UK.

Several cleavage forms of pro-UK exist which are discussed in chap. 4. Thrombin also cleaves pro-UK at Arg^{156}-Phe^{157} (chap. 4, Fig. 1), just two residues from the activation site (Ichinose et al. 1986), yielding a two-chain form which is indistinguishable from tc-uPA on gel electrophoresis under reducing conditions. This thrombin cleavage product has little catalytic activity (Gurewich and Pannell 1987; Braat et al. 1999) and is resistant to activation by plasmin. Thrombomodulin promotes the reaction between thrombin and pro-UK (De Munk et al. 1991; Braat et al. 1998), suggesting a biological role for this reaction or its product on the endothelial surface.

The catalytic constants of thrombin-cleaved pro-UK (thromb-UK) against both synthetic substrate and plasminogen are comparable to those of the intrinsic activity of pro-UK. Moreover, like pro-UK, the K_M of thromb-UK against plasminogen is lower than that of UK and its catalytic activity against plasminogen is enhanced several hundred-fold by fibrin-fragment E. The role of thromb-UK in plasminogen activation in the presence of fibrin fragment E or on cell surfaces, where its activity may also be promoted, remains to be established. It has been previously reported that thromb-UK is an effective and highly fibrin-specific thrombolytic in vivo in experimental animals (Abercrombie et al. 1990). Therefore, thromb-UK represents a two-chain form of uPA which closely resembles single-chain pro-UK, except for its resistance to activation, which diminishes its efficacy but increases its specificity. The reduced activity indicates that the N-terminal isoleucine on the B-chain of UK (present in tc-uPA but missing in thromb-UK) is important for stabilization of the catalytic site and full expression of its activity (Liu and Gurewich 1993).

I. The Intrinsic Activity of Pro-UK (sc-uPA)

Measurement of the intrinsic catalytic activity of pro-UK against plasminogen is complicated by a positive feedback resulting from the activation of pro-UK by formed plasmin. The generation of even trace amounts of UK will substantially affect the kinetics of the reaction. This probably explains the wide range of catalytic activities for pro-UK which have been reported, from an activity equivalent to that of UK (Collen et al. 1986b) to no activity at all (Husain 1991). However, there is now a consensus that pro-UK is a zymogen, but one with a significant intrinsic activity, which under conditions where pro-

UK activation is adequately inhibited corresponds to about 0.4% that of UK (PANNELL and GUREWICH 1987; PETERSEN et al. 1988). The existence of a true intrinsic activity is also supported by findings with a plasmin-resistant Lys[158] mutant pro-UK which had a comparable catalytic activity against plasmino-gen to that estimated for pro-UK (GUREWICH et al. 1988b). Recently, studies of the structural determinants of the intrinsic catalytic activity of pro-UK have shown that a site directed mutation at Lys[300] essentially eliminates the intrin-sic activity of pro-UK (LIU et al. 1996; SUN et al. 1997), further excluding the possibility that trace UK contaminants were responsible for it.

The relatively high intrinsic activity of pro-UK compared with other serine protease zymogens is unusual. Pro-UK has a lower K_M against plasminogen than its enzymatic form, UK (COLLEN et al. 1986b; LIU et al. 1992) In addition, the inactivation of pro-UK by diisopropylfluorophosphate is reversible (LIU and GUREWICH 1995) whereas that of all other serine proteases is irreversible. Finally, the intrinsic activity of pro-UK is subject to considerable promotion without activation to UK. In the presence of fibrin fragment E, the catalytic activity of pro-UK against native Glu-plasminogen is promoted 500-fold, giving it an activity equivalent to that of UK. This promotion is not due to ternary complex formation but, instead, has been related to a particular con-formational change in Glu-plasminogen induced specifically by fragment E (LIU and GUREWICH 1992). A more modest promotion of plasminogen activa-tion by pro-UK by cell surfaces has also been reported (MANCHANDA and SCHWARTZ 1991). It is therefore likely that the intrinsic activity of pro-UK is physiologically important, and that activation to UK is not a pre-requisite of pro-UK-mediated catalysis.

II. Fibrinolysis by Pro-UK/UK (sc-uPA/tc-uPA)

Fibrinolysis mediated by plasminogen activators is dependent on the activa-tion of fibrin-bound, rather than free, plasminogen. Plasmin which is not bound to fibrin, by one or more of its lysine binding kringles, does not induce lysis (SUENSON et al. 1990; PÂQUES et al. 1992). Therefore, non-specific plasminogen activation is not useful in fibrinolysis.

Plasminogen binds to fibrin via an internal lysine residue on the D-domain of intact fibrin (NIEUWENHUIZEN et al. 1983). Early fibrin degradation exposes new plasminogen binding sites (HARPEL et al. 1985; TRAN-THANG et al. 1986) which are carboxy-terminal lysines, in particular the three carboxy-terminal lysines on the E-domain of fibrin (VÁRADI and PATTHY 1983). Urokinase has little preference for fibrin-bound vs free plasminogen and, therefore, induces fibrinolysis accompanied by non-specific plasminogen activation. Since plasmin is a relatively non-specific protease, its systemic generation causes degradation of numerous substrates other than fibrin, including fibrinogen, clotting factors V and VIII, certain platelet membrane glycopro-teins, constituents of vascular basement membrane, and certain complement zymogens, causing generation of anaphylotoxins (for review see GUREWICH 1989a).

By contrast, pro-UK induces relatively fibrin specific clot lysis (GUREWICH et al. 1984), since it activates fibrin-bound plasminogen preferentially. The mechanism responsible for this has remained surprisingly elusive. This fibrin-selectivity of plasminogen activation by pro-UK appears to be related to a conformational change in Glu-plasminogen when it binds to fibrin, since fibrin-specificity in a plasma milieu was lost when Lys-plasminogen was substituted (PANNELL and GUREWICH 1986). A dependence on carboxy-terminal lysine binding sites on fibrin for plasminogen binding was implicated when it was observed that treatment of clots with carboxypeptidase-B, which cleaves carboxy-terminal lysines, inhibited lysis by pro-UK (but not tPA) (PANNELL et al. 1988).

This finding led to studies with fibrin fragment E, which, as noted above, contains three carboxy-terminal lysines. The specific and selective, several hundred-fold, promotion of plasminogen activation by pro-UK in the presence of fragment E was not related to conversion of pro-UK to UK, since it also occurred with a plasmin resistant Lys^{158} pro-UK mutant. Its extent appears sufficient to explain the fibrin selectivity of pro-UK by this mechanism (LIU and GUREWICH 1991, 1992). Since promotion is dependent on the fibrin E-domain, on which plasminogen binding sites are available only after fibrin degradation has been initiated, this mechanism of action is consistent with the lag phase, characteristic of clot lysis by pro-UK (in contrast to tPA) in a plasma milieu. It also explains why this lag phase is attenuated by pretreatment of the clot with small amounts of UK or tPA added to the plasma (GUREWICH 1987).

Another mechanism to explain the fibrin specificity of pro-UK has been proposed by HUSAIN (1993) who found that pro-UK, but not UK, binds to fibrin in the presence of Zn^{2+}. However, since a concentration considerably higher than the physiological one for Zn^{2+} was used, the biological or pharmacological implications of this observation are unclear. Moreover, when clot lysis is performed in citrate plasma, in which Zn^{2+} has been largely taken out of solution, the fibrin specificity of pro-UK is nevertheless retained.

Therefore, the preferential activation by pro-UK of Glu-plasminogen bound to the carboxy-terminal lysines on the E-domain of fibrin remains the explanation for its fibrin specificity. At the same time local activation of pro-UK to UK on the fibrin surface is an additional element which greatly enhances the fibrinolytic effect of pro-UK. Clot lysis studies with pro-UK were invariably accompanied by a small (<10%) conversion of pro-UK to UK (PANNELL and GUREWICH 1986; DECLERCK et al. 1990). When clot lysis is not accompanied by local UK generation, as with a plasmin resistant mutant pro-UK, lysis is quite inefficient, requiring high doses (GUREWICH et al. 1988b). The confinement of UK to the clot surface during clot lysis mediated by pro-UK is probably related to the action of plasma inhibitors.

III. Contrasting Properties of Pro-UK and tPA

The complementary mechanisms of fibrin-dependent plasminogen activation by tPA and pro-UK have helped to explain their synergy in clot lysis (PANNELL et al. 1988). However, the complementarity of tPA and pro-UK is a function of their restricted and different plasminogen substrates, and will be lost after significant conversion of pro-UK to UK has taken place during clot lysis. Accordingly, when a Lys^{158} mutant of pro-UK, unconvertible to UK, was used in clot lysis together with tPA, more extensive synergy was found (GUREWICH et al. 1988b). Therefore, the synergy of tPA and pro-UK in fibrinolysis is limited by pro-UK activation, since UK and tPA are not synergistic, as evidenced in the clinical TAMI II trial (TOPOL et al. 1988). This limitation complicates its experimental demonstration, and probably explains why this subject has been controversial, synergy having been confirmed by some investigators under certain conditions (GUREWICH and PANNELL 1986; COLLEN et al. 1987; FRY et al. 1989; SABOVIC and KEBER 1995) but not by others under different experimental conditions (COLLEN et al. 1986a; NGUYEN et al. 1989).

It is evident that the fibrinolytic properties of tPA and pro-UK are different and remarkably complementary, as summarized in Table 1. Their principal functional differences are the following:

1. Both tPA and pro-UK induce equivalently fibrin-specific clot lysis, but by distinctly different mechanisms. The fibrin-specificity of tPA is related to its high fibrin-affinity, whereas that of pro-UK is unrelated to fibrin-binding.
2. Pro-UK is a zymogen in blood and remains inert, at fibrin-specific concentrations, until it reaches a blood clot where it selectively activates fibrin-bound plasminogen. By contrast, tPA is an enzyme which is rapidly inactivated by plasma inhibitors.
3. Plasminogen activation by tPA is specifically and selectively promoted by fibrin fragment D, which contains an internal lysine plasminogen binding site, whereas that by pro-UK is promoted by fibrin fragment E, which contains a carboxy-terminal lysine plasminogen binding site.
4. Pro-UK is activated to UK on the fibrin surface, changing it from a restricted to an unrestricted plasminogen activator. By contrast, the conversion of single-chain to two-chain tPA is not accompanied by any significant change in its properties.

Table 1. The contrasting properties of pro-UK and tPA

	Pro-UK	tPA
Fibrin-specificity:	+	+
Zymogen (inert in plasma):	+	0
Fibrin-clot binding:	0	+
Plasminogen activation promoted by:	Fibrin fragment E	Fibrin fragment D
Fibrinolytic importance of local activation to two-chain form:	+	0
Uptake and promotion of fibrinolysis by platelets	+	0
Upregulation by its own EGF domain	+	0

5. Pro-UK, but not tPA, is specifically taken up from whole blood by platelets (GUREWICH et al. 1993), which contain a novel membrane receptor which binds to a domain on the A-chain of pro-UK (JIANG et al. 1996). Platelets in a clot have also been shown to promote lysis by pro-UK, whereas they inhibit lysis by tPA.

6. The EGF domain of uPA up-regulates pro-UK expression by endothelial cells (see below). These latter two properties of pro-UK may help explain the extended duration of its fibrinolytic effect reported in experimental animals after bolus administration (BADYLAK et al. 1988) and the low rate of reocclusion reported following coronary thrombolysis with pro-UK (see Table 2).

It has not yet been tested in comparative clinical trials whether these distinguishing properties of pro-UK, given in doses at which systemic conversion to UK is minimized, provide it with clinical advantages in therapeutic thrombolysis compared with higher doses of pro-UK or with other plasminogen activators.

IV. Pro-UK/UK and Platelets

A number of experimental studies have shown that the fibrinolytic effect of pro-UK is promoted by the presence of platelets in a fibrin clot (GUREWICH et al. 1988a; LOZA et al. 1994) whereas that of tPA is inhibited (GUREWICH and PANNELL 1986; FAY et al. 1994). The latter observation is readily explained by the fact that platelets are rich in plasminogen activator inhibitor-1 (PAI-1), which rapidly inactivates tPA (ZHU et al. 1999). In fact, the inhibition of tPA by PAI-1 has been postulated to explain the clinical finding that patients with acute myocardial infarction (AMI) treated with tPA in the morning hours, when PAI-1 levels are higher, are more resistant to thrombolysis by tPA (KURNIK 1995). This morning resistance to tPA may be of special significance since the morning is also the time of day associated with the highest incidence of AMI (MULLER et al. 1985). However, resistance to UK in the early morning and in the evening has also been described (KONO et al. 1996).

Pro-UK is not inactivated by PAI-1, which explains its lack of inhibition by platelets but does not explain its promotion by platelets. An explanation for this promotion came from studies which showed that prekallikrein is associated with platelets which is relevant since kallikrein activates pro-UK to UK (ICHINOSE et al. 1986; HAUERT et al. 1989). The promotion of pro-UK-induced clot lysis by platelets was shown to be related to this enzyme (LOZA et al. 1994). Platelets provide a surface which can facilitate plasminogen activation since plasminogen also binds to the platelets, and this binding is enhanced three- to nine-fold by thrombin stimulation of platelets (MILES and PLOW 1985). In addition, pro-UK is also tightly associated with platelets, having been identified in the outer leaflet of the platelet membrane (PARK et al. 1989). About 20% of the endogenous pro-UK in blood was found to be associated with platelets

(GUREWICH et al. 1993) and uptake by platelets of pro-UK from whole blood, at physiological concentrations of pro-UK, was demonstrated (GUREWICH et al. 1995). A similar uptake of HMW-UK, but not LMW-UK, by platelets from a buffer milieu was also demonstrated, implicating the A-chain of uPA. A novel ≈70kDa, high affinity receptor for uPA was identified in the platelet membrane (JIANG et al. 1996). Therefore, platelets have a novel, high affinity surface receptor for uPA distinct from the uPAR (WOHN et al. 1997) found on many other cells.

The importance of platelets in pro-UK-mediated fibrinolysis has not been established in man but has been demonstrated in dogs, a species in which the fibrinolytic system is known to be remarkably efficient and related primarily to a high uPA activity, contributing to the rapid lysis of experimental pulmonary emboli in the dogs (LANG et al. 1993). Platelets, therefore, provide a carrier for some of the pro-UK administered therapeutically. It remains to be determined if these pro-UK enriched platelets extend the thrombolytic effect of pro-UK.

D. Pharmacokinetics

The turnover of pro-UK in blood plasma is short, having an initial half-life of approximately 7min in man (COLLEN and VAN DE WERF 1987; KÖHLER et al. 1991; DE BOER et al. 1993; MICHELS et al. 1999) and shorter in other mammalian species. The clearance of pro-UK is similar to that of UK (SCHNEIDER et al. 1982; KÖHLER et al. 1991) indicating that clearance is not mediated by complexation with inhibitors. uPA is cleared principally by the liver (VAN GRIENSVEN et al. 1997), and hepatectomy has been shown to be associated with considerable prolongation of the T/2 (STUMP et al. 1987).

The recognition site on the uPA molecule responsible for its rapid clearance has not been identified. It is evident that it is not on the A-chain, where the uPAR binding site is located, since LMW forms of uPA, missing the A-chain (BARLOW et al. 1981), are cleared at essentially the same rate as HMW forms in man. By contrast, in the rat, about a threefold prolongation of clearance was reported for a deletion mutant missing the EGF domain (HIRAMATSU et al. 1989). Since unglycosylated rpro-UK from *E. coli* is cleared at the same rate as natural pro-UK (DE BOER et al. 1993), carbohydrate receptors cannot be significantly involved in the clearance of uPA either. The clearance from the platelet compartment has not been studied, but since the uPA associated with platelets resists dissociation, clearance is probably dependent on the half-life of platelets.

Finally, it has been reported that when UK is administered into the duodenum of dogs (SUMI et al. 1980; SASAKI et al. 1985) or rats (SUMI et al. 1985) or given in enteric capsules by mouth to volunteers (TOKI et al. 1985) some absorption takes place, but more importantly, endogenous synthesis and intravascular release of uPA is stimulated. In these studies, the amount of activity generated was sufficient to lyse an experimentally induced venous thrombus (SUMI et al. 1980). When oral administrations of UK were given daily for seven days, no evidence of attenuation of the fibrinolytic effect was seen. A measurable increase in fibrinolytic activity started within 1 h of the oral administration of 120,000 IU of UK, and had a duration of effect, measured by amidolytic and plasminogen activator activity of more than 6 h (TOKI et al. 1985).

Little further is known about these surprising findings. However, since they suggested that uPA up-regulated its own synthesis, this was tested in endothelial cells and monocytes in culture. Consistent with the above findings, it was shown that the EGF domain of uPA induced about a fivefold stimulation of uPA synthesis (LI et al., 1996). This phenomenon may help explain the unanticipated finding of BADYLAK et al. (1988) that pro-UK "had a longer than expected thrombolytic effect based on the known half-life" when given by bolus injection to dogs.

E. Clinical Studies in Acute Myocardial Infarction

Due to the important differences in properties between pro-UK and UK, their clinical use will be taken up separately.

I. Urokinase

1. Controlled Trials of UK vs Controls

The first controlled trial with UK-activated plasmin (thrombolytic effect probably mainly due to UK) was reported by LIPPSCHUTZ et al. (1965). In-hospital mortality of 43 patients with acute myocardial infarction (AMI), randomized to receive UK-activated plasmin, was 14%, and that of the 41 control patients was 17%. Two other small trials similarly showed a survival advantage for patients treated with UK compared to controls. In a very small Scandinavian trial mortality in the group treated with UK was 14%, in the control group 21% (n.s.) (GORMSEN et al. 1973). In a larger French study 2 out of 60 patients in the UK group died (3%), but 8 out of 60 patients in the control group (13%; $p < 0.05$) (BROCHIER et al. 1975).

A bolus dose of 7200 IU/kg UK followed by an 18-h infusion of 7200 IU/kg was given to 172 patients in the EUROPEAN COLLABORATIVE STUDY (1975). Compared with 169 controls, the ST segment returned to normal quicker in the UK-treated patients. No differences in mortality were found, which in retrospect is hardly surprising considering the size of this study. A bolus of UK (2 million IU) was used in another study and patency was determined by angiography at 1.1 ± 0.6 h (MATHEY et al. 1985). Patency, defined as prompt and complete filling, occurred in 30 out of 50 patients (60%). Repeat angiography after three weeks revealed patency in 23 out of 24 patients studied whose arteries were patent initially. No bleeding complications were reported. A similar patency rate and low incidence of reocclusions were reported in 10 patients given 3 million IU of UK administered by infusion over 45–60 min (Wall et al. 1990).

In the large USIM study 2531 patients with AMI were randomized to receive an intravenous bolus of 1 Mio units of UK, repeated after 60 min plus heparin (bolus of 10,000 U, followed by 1000 U/h for 48 h) or heparin alone. At 16 days, overall hospital mortality was 8% in the UK and 8.3% in the heparin group (n.s.). The incidence of major bleeding and stroke was similar in the two groups (Rossi and Bolognese 1991).

2. Prehospital vs In-hospital Thrombolysis with UK

In an attempt to reduce the symptom to needle time in patients with AMI several studies have investigated the feasibility and effect of administering thrombolytic agents in the home of the patients or in the ambulance (reviewed by Carlsson et al. 1997). The potential benefit and the risk of administering prehospital UK for AMI was evaluated by Schofer et al. (1990) in a double blind study. Patients presenting <4 h after symptom onset received 2 Mio units of UK as an intravenous bolus either before (Group A, $n = 40$) or after (Group B, $n = 38$) hospital admission. The mean time interval from symptom to needle was 85 min in Group A and 137 min in group B ($p < 0.0005$). Coronary angiography before discharge, mortality, left ventricular function, and complication rates did not differ between the two group and the authors concluded that "the saving of 45 min in the early stage of an AMI through prehospital thrombolysis did not appear to be important for salvage of myocardial function."

3. Comparative Trials of UK vs SK or tPA

In a small French recanalization study 42 patients were randomized into 4 groups to receive either different schedules of intracoronary SK (maximal dose 350000 U) or UK (maximal dose 250000 U, supplemented with a small amount of 75 mg of Lys-plasminogen). Recanalization was better in the SK (87%) than in the UK group (64%). Systemic fibrinolysis occurred to a lesser degree with UK (De Prost et al. 1983).

In three studies UK was compared with tPA. A Japanese double-blind study randomized 198 patients into 3 groups: 33 mg of tPA, 50 mg of tPA, or 960,000 units of UK. All patients underwent baseline and 60 min coronary angiography. Recanalization rates were 76%, 75%, and 42%, respectively. There was no difference in the incidence of major bleeding or hospital mortality between the three groups (Kanemoto et al. 1991).

The German GAUS study investigated the effects of rtPA (alteplase) and UK on patency and early reocclusion in a single-blind, randomized multicenter trial of 246 patients with AMI of <6 h duration. rtPA was administered as an initial 10 mg bolus followed by 60 mg of rtPA given intravenously over 90 min. UK was administered as an initial bolus of 1.5 Mio units, followed by another 1.5 Mio units infused over 90 min (Neuhaus et al. 1988). Coronary angiography performed at the end of the infusion revealed a TIMI grade 2 or 3 patency rate of 69% in the patients given tPA and 66% on those given UK (n.s.). Among patients treated within 3h from symptom onset, patency rates were 64% in the tPA and 70% in the UK group (n.s.). Reocclusions after 24 h were more frequent in the tPA group (Table 2).

The TIMIKO study compared a double bolus UK vs a front-loaded tPA regimen for AMI. In this prospective, multicenter, randomized study 618 patients presenting within 6h after onset of chest pain were assigned to receive 1.5Mio units of UK, followed by a 20000U/kg (max. 1.5Mio units) bolus 30min after the first dose, or the front-loaded regimen of alteplase used in the GUSTO-1 trial (THE GUSTO INVESTIGATORS 1993). All patients received 200mg of aspirin immediately and continued to take it daily. All patients also received intravenous heparin for 3days, adjusted to prolong the APTT to 1.5–2 times normal. Coronary angiography was performed in some centers after 60min and 90min. TIMI grade 3 flow was achieved at 90min in 70% of the patients receiving UK and in 68% of those treated with tPA (n.s.). There was no difference in the incidence of death, reinfarction, or major bleedings. Hemorrhagic stroke was more common in the tPA group (1.1%) than in the UK group (0.3%; n.s.). The authors concluded that the double bolus UK regimen is an easy, safe, and effective thrombolytic treatment, equivalent to front-loaded alteplase [PARK, FOR THE THROMBOLYSIS IN MYOCARDIAL INFARCTION IN KOREA (TIMIKO) STUDY GROUP 1998].

Taken all trials with UK together, one arrives at the conclusion that UK is an effective thrombolytic agents, but offers no convincing advantages over other thrombolytic agents.

II. Pro-urokinase

A comparison of the fibrinolytic properties of pro-UK and UK (GUREWICH et al. 1984; ZAMARRON et al. 1984; COLLEN et al. 1984, 1985; FLAMENG et al. 1986) showed that pro-UK is far more specific than UK in vitro and also more effective at inducing clot lysis in vivo. These promising experimental results formed the basis for clinical trials of pro-UK in AMI, which is the first clinical indication for which it has been evaluated.

1. Early Dose-finding Studies

The first reported study consisted of six patients infused with 40mg of natural, glycosylated pro-UK infused over 60min. Complete reperfusion was achieved in four patients with little associated fibrinogen degradation, consistent with the experience in experimental animals (VAN DE WERF et al. 1986a). In a subsequent study, 17 patients with AMI were given non-glycosylated rpro-UK from *E. coli*. In eight patients, 40mg was given over 1h and in nine patients, 70mg was administered. The lower dose achieved fibrin specific recanalization in 75%, but recanalization was incomplete in 50%. The higher dose achieved recanalization with normal distal run-off in 78% of the patients, but

Table 2. Randomized trials comparing UK and tPA

No.	Trial	Year	Delay (h)	No. of patients		Patency (%)		Mortality			Reocclusion/reinfarction (%)	
				UK	tPA	UK	tPA	Time	UK	tPA	UK	tPA
1	GAUS	1988	≤6	117	121	66 70	69[b] 64[c]	30 days	4.3	5.0	1.6 6.5	10.5[d] 14.8[e]
2	TIMIKO	1998	≤6	350	268	81 80	83[a] 83[b]	30 days	4.6	4.4	3.1	3.4[e]

References: 1. Neuhaus et al. (1988); 2. Park, for the Thrombolysis In Myocardial Infarction In Korea (TIMIKO) Study Group (1998). Patency rates after start of treatment:
[a] 60 min.
[b] 90 min.
[c] Patients with ≤3h delay from symptoms to treatment.
[d] Reocclusion without additional intervention.
[e] Reinfarction.

fibrinogen degradation ranged from 6% to 96% (VAN DE WERF et al. 1986b). A similar dose-finding study with rpro-UK was performed by DIEFENBACH et al. (1988). Twelve patients with angiographically proven AMI of less than 4 h duration were given 20 mg of pro-UK as a bolus, followed by 60 mg infused over 1 h, and another 12 subjects received a 10 mg bolus and 30 mg as infusion. Time to reperfusion was 43 min in the first and 67 min in the second group (p < 0.005). TIMI flow grade 2 and 3 perfusion rates 90 min after the start of the treatment were 91% in the higher dose group, but only 50% in the 40 mg group (p < 0.001). A systemic lytic state occurred in 33% of patients in the first group, but only in 9% of those given the low dose of pro-UK. Taken together these two studies suggest that doses of approximately 40 mg of pro-UK, which do not cause a significant proteolytic state, are not as efficient as doses of approximately 80 mg (corresponding to about 8 Mio units of UK).

The broad variability in the fibrin specificity of pro-UK at higher infusion rates observed in these studies is quite typical. Non-specificity is related to the systemic conversion of pro-UK to UK by plasmin. Since non-glycosylated rpro-UK is more sensitive to activation by plasmin (LENICH et al. 1992) a non-specific effect is apt to occur more readily with this form of pro-UK, and this appears to be consistent with clinical experience.

During the same time period Japanese investigators evaluated the effect of small doses of intracoronary administration of natural pro-UK and of UK in a randomized multicenter study (KAMBARA et al. 1988). Fifty patients (Group H) received 6000 U of pro-UK i.c., 44 subjects (Group L) were given 3000 U of pro-UK i.c., and 54 patients (Group U) received 3000 U of UK. Coronary recanalization rates (TIMI 2 or 3) determined angiographically after 45 min of i.c. infusion were 90% in group H, 59% in group L, and 61% in group U. Bleeding complications were more common in group U.

2. Randomized Trials of Pro-UK vs SK

The first randomized double blind trial compared 80 mg non-glycosylated rpro-UK (Saruplase), given as a 20 mg bolus followed by an infusion of 60 mg in 1 h, with standard dose SK (1.5 million IU in 1 h) (PRIMI TRIAL STUDY GROUP 1989). A total of 401 patients with AMI of 4 h or less in duration were randomized into the two treatment groups. Angiographic patency (TIMI 2 or 3) at 60 min was 72% for rpro-UK and 48% for SK (p < 0.001), but at 90 min the differences became insignificant (71% and 64% for pro-UK and SK, respectively). A major drop in fibrinogen occurred in both groups, being 86% and 94% for pro-UK and SK, respectively, but the transfusion requirements were significantly (p < 0.01) less in the pro-UK treated patients (4.0% vs 11.3% for SK). There were no significant differences between the two treatment groups with respect to hospital mortality, recurrent ischemic events, arrhythmias, or heart failure (Table 3). A one-year follow-up in 387 patients enrolled in the PRIMI trial revealed a low and similar overall mortality and functional status in the two treatment groups (DIEFENBACH 1992). The five-

Table 3. Randomized trials comparing pro-UK with SK or tPA

No.	Trial	Year	Delay (h)	No. of patients		Patency (%)		Mortality			Reocclusion/ reinfarction (%)	
				pro-UK	SK	pro-UK	SK	Time	pro-UK	SK	pro-UK	SK
1	PRIMI	1989 1992 1999	≤6	198	203	72 71	48[b] 64[c]	ih 1 year 5 years	3.5 6.3 20.1	4.9 5.6 16.9	5.0 4.0 19.0	4.4[d] 2.5[e] 10.8
2	COMPASS	1998	≤6	1542	1547			30 days 1 year	5.7 8.2	6.7 9.6	5.4	4.5[e]
				pro-UK	tPA	pro-UK	tPA		pro-UK	tPA	pro-UK	tPA
3	SESAM	1997	≤6	236	237	75 80 80	69[a] 75[b] 81[c]	ih	4.7	3.8	1.2 4.2	2.4[d] 4.2[e]

References: 1. PRIMI TRIAL STUDY GROUP et al. (1989); DIEFENBACH (1992); SPIEKER et al. (1999); 2. TEBBE et al. (1998); 3. BÄR et al. (1997).
ih, in-hospital mortality.
Patency rates after start of treatment:
[a] 45 min.
[b] 60 min.
[c] 90 min.
[d] Reocclusion.
[e] Reinfarction.

year follow-up in 255 patients revealed comparable mortality rates in the two treatment groups; 20.8% in the saruplase and 16.9% in the SK group (n.s.) (SPIECKER et al. 1999).

The major systemic fibrinolytic effect observed with rpro-UK in this study reflected the systemic conversion of most of the rpro-UK to UK (KOSTER et al. 1994). In light of this, the PRIMI study represents more of a high-dose UK trial than a pro-UK trial. Not surprisingly, therefore, when rpro-UK (Saruplase) at this dose was compared with UK in the SUTAMI study of 543 patients randomly allocated into the two treatment groups, no differences in patency rates (75.4% vs 74.2%) or bleeding complications were found between pro-UK and UK (MICHELS et al. 1995).

The large double blind, multicenter COMPASS trial compared pro-UK (saruplase) with SK in a total of 3089 patients with AMI of <6h duration. SK was given at the dose of 1.5 Mio U, infused over 60 min, and pro-UK as a 20mg bolus, followed by 60mg infused over 60min (TEBBE et al. 1998). At the beginning of the treatment the pro-UK group received 5000 U of heparin i.v., the SK group a placebo injection of heparin and a dummy bolus injection of placebo in lieu of the pro-UK bolus. All patients received i.v. heparin for ≥24h starting 30min after the end of the thrombolytic infusion; heparin was adjusted to maintain an APTT at 1.5–2.5 times that of normal. The main results are listed in table 3. Mortality up to day 30 was 5.7% in the pro-UK group and 6.7% in the SK group ($p < 0.01$). Hemorrhagic strokes occurred more often in patients receiving pro-UK (0.9% vs 0.3%), whereas thromboembolic strokes were more often seen in the SK-treated patients (0.5% vs 1.0%). The rate of bleeding was similar in the two treatment groups (10.4% vs 10.9%). Hypotension and cardiogenic shock occurred less frequently in the pro-UK group.

3. Randomized Trial of Pro-UK vs tPA

The medium-size SESAM Study randomized 473 patients with AMI of <6h duration into a pro-UK (saruplase) group or a tPA (alteplase) group (BÄR et al.). Co-medication included heparin and aspirin. Angiographies were performed at 45, 60, and 90min after the start of the thrombolytic treatment. Saruplase was given as a 20mg bolus, followed by a 60min infusion of 60mg. Alteplase was administered as a 10mg initial bolus, an infusion of 50mg over the next hour, and 40mg over the second and third hour. The main results are listed in Table 3. Just 45min after the start of the treatment TIMI 2 or 3 perfusion rates were 75% in the pro-UK group and 69% in the tPA group. The 90min patency rates were very similar. Reocclusion rates, including patients who underwent revascularization interventions were slightly lower in the pro-UK group (6.7% vs 10.3%). Complication rates were similar in the two treatment groups.

III. Non-specific Plasminogen Activation

Fibrin specific coronary thrombolysis by pro-UK was demonstrated in a study of 40 patients in whom 36–69mg of natural pro-UK was infused over 90min

(Loscalzo et al. 1989). A TIMI grade 2 or 3 patency at 65 ± 22 min occurred in 51% of the patients and fibrinogen levels fell by only 10%. However, most cardiologists aim at achieving much higher TIMI grade 2 or 3 patency rates and have historically been inclined to push doses of thrombolytic agents higher and higher (for tPA up to 150 mg) until an unacceptably high rate of cerebral hemorrhages is encountered. It is unfortunate that larger dose finding studies have not been performed with pro-UK.

When therapeutic thrombolysis by any plasminogen activator is associated with a major systemic activation of plasminogen, there are adverse consequences which may undermine the clinical benefits of restoring coronary patency. These side effects are related not only to bleeding into the myocardial injury zone and elsewhere, but paradoxically also to hypercoagulability due to activation of certain clotting factors and platelet activation. In addition, there is activation of certain mediators of tissue necrosis. These consequences of non-specific plasminogen activation are related to the many substrates which are hydrolyzed by plasmin and have recently been reviewed (Gurewich and Muller 1996). It is noteworthy that when fibrin selectivity was retained during coronary thrombolysis with pro-UK, hypercoagulability, evidenced by intravascular thrombin generation, was not found (Weaver et al. 1994).

Another consequence of systemic plasminogen activation is the depletion of the plasma plasminogen reservoir ("plasminogen steal") which also compromises thrombolytic efficacy (Torr et al. 1992). This "plasminogen steal" concept is consistent with a dose finding trial of glycosylated rpro-UK in which the efficacy of 60 mg and 80 mg doses in 1 h were compared. At 60 min, the 60 mg dose gave a higher TIMI 3 coronary artery patency (65% vs 30%) and higher TIMI 2 or 3 patency (73% vs 61%) than the 80 mg dose (Torr et al. 1992). The 60 mg dose induced only 24% reduction in fibrinogen, compared with 41% with the 80 mg dose. Although the number of patients studied was too small for the efficacy differences to reach statistical significance, a similar finding that 60 mg induced comparable coronary artery patency to 80 mg, while being significantly more fibrin-specific, was obtained in a dose-finding study of unglycosylated rpro-UK by Farmitalia.

IV. Combinations of Pro-UK and UK or tPA

Pro-urokinase has been shown to be complementary to tPA in its fibrinolytic effect (Pannell et al. 1988; Gurewich 1989b) and the two activators are synergistic in clot lysis under certain conditions (Gurewich and Pannell 1986). Therefore, trials evaluating various combinations of these two agents were undertaken by a number of investigators.

1. Early Dose-finding Studies of Pro-UK and UK

Bode et al. (1988) investigated three different low-dose schedules of pro-UK. In one group a combination of a bolus of only 3.7 mg of pro-UK and of

250000 U of UK, followed by 44.3 mg of pro-UK, yielded a higher TIMI 2 or 3 patency rate (65%) than that observed in the two groups receiving the same or a higher dose of pro-UK only. In another, open-label, prospective study thrombolysis was started with 250000 U of UK, followed by either 33 mg or 48 mg of natural pro-UK (TCL 598, Sandoz) (GULBA et al. 1989b). With 33 mg of pro-UK TIMI 2 or 3 patency was observed in only 33% of patients, whereas the intermediate dose of 48 mg produced TIMI 2 or 3 flow in 75% of patients ($p < 0.01$). A second angiogram performed 24–36 h later revealed reocclusion in 60% of the patients given the lower dose of pro-UK and in 8% in the higher dose group ($p < 0.05$). In both groups fibrinogen decreased <15%. Bleeding complications were more frequently observed in the higher dose group, but were mostly related to puncture sites.

2. Early Dose-finding Studies of Pro-UK and tPA

The first of these studies tested a combined infusion of 10 mg rtPA and 10 mg non-glycosylated rpro-UK given over 1 h. Nine patients with documented AMI were treated, in three of whom a prior infusion of 10 mg tPA had failed to achieve reperfusion. Complete recanalization was achieved in seven out of nine patients and transient recanalization occurred in an eighth patient. No change in the fibrinogen level occurred in any of the patients (COLLEN and VAN DE WERF 1987). In a second study the simultaneous infusion of 12 mg tPA over 30 min and 48 mg non-glycosylated rpro-UK infused over 40 min in 38 patients with AMI was evaluated. Patency (TIMI grade 2 or 3) was observed in 61% at 60 min and in 82% in 90 min. At 90 min TIMI 3 patency was seen in 58% of the patients. The mean decline in fibrinogen was less than 20% and no significant bleeding complications occurred (BODE et al. 1990c). In a third study, tPA was given as a 10 mg bolus followed by a 10 mg infusion over 1 h. The pro-UK dose was 16.3 mg given as a 4 mg bolus followed by an infusion of 12.3 mg over 1 h. The doses were chosen to represent approximately 20% of the monotherapy dose of each agent. Twenty-three patients with AMI were treated and TIMI grade 2 or 3 patency at 90 min was found in 70%. Two additional patients were observed to lyse at 113 min and 150 min. These results were considered to be consistent with a synergistic effect and showed that effective lysis could be obtained without fibrinogen degradation, which was less than 15% (KIRSHENBAUM et al. 1991).

3. Larger Clinical Trial with Pro-UK and tPA

A study to evaluate a sequential combination of tPA and pro-UK was published by ZARICH et al. (1995). The design of the PATENT study was based on the experimental findings that tPA and pro-UK are sequential as well as synergistic in their effects. It has been shown that tPA followed by pro-UK, but not the reverse order, induces synergistic thrombolysis in experimental animals (COLLEN et al. 1987). According to this understanding of fibrinolysis, only a small bolus of tPA should be required to initiate lysis and optimize the conditions for pro-UK/UK. In the first 10 patients in this study, a 10 mg bolus of tPA was given and in the remaining 91 patients, only 5 mg of tPA was administered. The tPA bolus was followed in all patients by an infusion of unglycosylated rpro-UK of 40 mg/h for 90 min. In the entire group of 101 patients, a TIMI 2 or 3 patency at 90 min of 77% and a TIMI 3 patency of 60% was observed. No reocclusions were seen in 28 patients who were patent and reexamined at 24 h. The fibrinogen decline was highly variable, as noted in other studies with an intermediate dose, and averaged 50%. Bleeding complications were essentially limited to puncture sites and only 1 out of 101 patients in the study died in hospital (Zarich et al. 1995), similar to the low death rate (1/128 patients) reported by Weaver et al. (1994) with a comparable dose of pro-UK.

V. Bolus administration of Pro-UK

To study the safety and efficacy of saruplase as a bolus, the angiographic and clinical outcomes of three bolus regimens were investigated in the BASE pilot study in 192 patients with an acute myocardial infarction and compared with the standard regimen (Bär et al. 1999). Fifty-two patients received a double bolus of 40 mg and 40 mg after 30 minutes, 51 patients a bolus of 80 mg, and 36 patients a bolus of 60 mg. Fifty-three patients received the standard regimen (a bolus of 20 mg and 60 mg IV infusion over 1 hour). At 60 and 90 min TIMI 3 flow rates were, respectively, 61% and 73% with the 40/40-mg bolus, 51% and 57% with the 80-mg bolus, 31% and 36% with the 60-mg bolus, and 55% and 72% with the standard 20/60-mg infusion.

The primary endpoint, persistent patency (TIMI 2 + 3) at 24–45 hours, was seen in 69%, 65%, 44%, and 68% of patients. Inclusion in the 60-mg bolus group was prematurely stopped because of the low patency rates. The 40/40-mg bolus group had the highest mortality rate (13.5%). Other adverse event rates were similar in the four groups. The 80-mg single was further tested in the BIRD study (Original report not yet published, summarized in Ferguson 1999). A total of 2410 patients presenting within 6 h of symptoms were randomized to standard therapy with saruplase (20 mg bolus plus 60 mg infusion over 1 h (n = 1214) or a single bolus of 80 mg (n = 1196). The primary endpoint of the study, 30-day mortality, was 6.0% with standard saruplase therapy and 5.9% with single-bolus administration. Other 30-day endpoints included reinfarction (5.0% with standard therapy versus 6.5% with single bolus treatment), recurrent angina (26% versus 25%), hemorrhagic stroke (0.7% versus 0.8%) and major bleeding events (2.4% versus 2.8%). The authors concluded that the single bolus therapy is clinically equivalent to standard bolus plus infusion treatment. This would allow to administer saruplase at home or in the ambulance to shorten the symptom to needle time.

VI. Heparin and Pro-UK

The use of heparin with pro-UK has also been studied. In the treatment of AMI heparin is used as an adjunctive agent after thrombolytic therapy with all agents in order to prevent reocclusions. When tPA is administered a bolus of heparin is routinely given at the start of thrombolytic therapy to increase patency rates. In the LIMITS study the effect of heparin on patency has been evaluated. In randomized allocation, 56 patients were given a bolus of 5000 IU heparin and 62 patients were given a placebo prior to coronary thrombolysis with unglycosylated rpro-UK (saruplase) given as a 20 mg bolus followed by an infusion of 60 mg over 1 h. TIMI 2 or 3 patency at 90 min occurred in 78.6% of the heparin group compared with 56.5% in the placebo ($p = 0.01$). It was concluded that heparin promotes coronary thrombolysis by pro-UK (TEBBE et al. 1995).

The mechanism responsible for this promotion is unclear since heparin has no effect on clot lysis by pro-UK in a plasma milieu in vitro. It has been proposed that the effect is related to heparin inactivation of thrombin, thereby preventing pro-UK inactivation by thrombin associated with the clot (GULBA et al. 1989a). Another additional possibility relates to the observation that heparin induces release of tPA from the vessel wall (KLEIN 1989). Therefore, heparin may mimic the effect of a small bolus of tPA in its promotion of fibrinolysis by pro-UK.

F. Second Generation Pro-UK

As noted above, the sensitivity of pro-UK to systemic conversion to UK limits its specificity in therapeutic thrombolysis, especially in individuals with a higher plasminogen concentration or lower PAI-1 level. This property is related, in part, to the intrinsic activity of pro-UK which is relatively high for a serine protease zymogen. As a result, plasma concentrations in excess of $\approx 1-2\,\mu g/ml$ for 60 min will induce non-specific plasminogen activation in most patients and result in the systemic conversion of pro-UK to UK by plasmin.

Many attempts have been made to engineer mutants of pro-UK or UK with better fibrin specificity or longer half-life in the circulation using mainly molecular biotechnology (LIJNEN et al. 1988b; EGUCHI et al. 1990; LIJNEN et al. 1990; LIU and GUREWICH 1992; LI et al. 1992; LIEBER et al. 1995; MA et al. 1996). Chimeras have been produced in an attempt to combine the best features of tPA and of pro-UK (LIJNEN et al. 1988a; NELLES et al. 1990). Constructs of u-PA and antibodies specific for fibrin have been engineered to confer high affinity of uPA to thrombi (BODE et al. 1987, 1990a,b) and the Gly-Pro-Arg sequence attached to portions of the UK molecule to confer affinity to platelet-rich thrombi (HUA et al. 1996). Many of these constructs have only been tested in vitro and only a few have shown any promise when tested in thrombosis models in experimental animals (RUNGE et al. 1996). To date no clinical trials have been conducted with these mutants, chimeras, or bispecific agents.

We have undertaken a study of the catalytic domain of pro-UK to determine the cause of its relatively high intrinsic activity in an effort to attenuate it. Computer modeling studies suggested that Lys[300] in a flexible loop region of the molecule could, by virtue of this flexibility, interact with Asp[355] to induce a transient active site conformation. This hypothesis was tested by constructing a site directed mutant of Lys[300]-Ala. Characterization of the mutant showed that its intrinsic activity was reduced about 100-fold (to 0.005% of UK rather than 0.4%) and its plasma stability was greatly enhanced, confirming the findings from the modeling studies. When incubated in plasma, pro-UK at concentrations of $\geq 1\,\mu g/ml$ induced 100% degradation of fibrinogen after 1h, whereas the Ala[300]-pro-UK mutant at a concentration of $50\,\mu g/ml$ induced no fibrinogen degradation under the same conditions (Liu et al. 1996). These studies have led to the development of a number of site-directed pro-UK mutants (Sun et al. 1997) which induce thrombolysis with far greater fibrin-specificity than does natural pro-UK.

Attempts to prolong the plasma clearance of pro-UK have, so far, been restricted to the coupling of uPA to albumin (Breton et al. 1995) or to polyethyleneglycol (Kajihara et al. 1994; Sakuragawa et al. 1986). Mutations of the pro-UK molecule itself have not been successful to prolong the T/2, due to the fact that clearance appears to be independent of the A-chain of pro-UK and of the carbohydrate side chains. However, since a portion of pro-UK has an extended half-life by virtue of binding to platelets, and since bolus pro-UK has been shown to have a surprisingly protracted fibrinolytic effect in animals (Badylak et al. 1988) altering this property of pro-UK may be of lesser clinical importance.

G. Conclusions

Pro-urokinase, the single-chain precursor (sc-uPA) of two-chain UK (tc-uPA), is a fibrin-selective plasminogen activator, which is inert in plasma in contrast to other plasminogen activators. When administered at infusion rates (40–60mg/h) at which its systemic conversion to UK is minimized, pro-UK induces a high rate of TIMI 3 patency, accompanied by low rates of both reocclusion and in-hospital mortality (Weaver et al. 1994; Zarich et al. 1995; Vermeer et al. 1999). The low reocclusion rate appears to be a property of pro-UK itself, suggesting that it will not require any of the new adjunctive therapies (antithrombins or IIb/IIIa platelet receptor antagonists) which have been developed to prevent reocclusions after coronary thrombolysis with other plasminogen activators. A comparative trial of pro-UK given at these intermediate infusion rates and SK or tPA has not yet been undertaken. At higher infusion rates (80mg/h), extensive systemic conversion of pro-UK to UK takes place, and the advantages of pro-UK over UK are lost (Michels et al. 1995). The sensitivity of unglycosylated rpro-UK to conversion to UK is slightly greater than that of glycosylated pro-UK (Lenich et al. 1992), a finding which appears consistent with the clinical experience.

An unusual property of pro-UK is that it is taken up from whole blood by platelets (Gurewich et al. 1993, 1995) where it binds to their outer membrane via a specific receptor distinct from uPAR (Jiang et al. 1996). This property may extend the intravascular half-life of a portion of therapeutically administered pro-UK considerably.

List of Abbreviations

APTT   activated partial thromboplastin time
BASE   Bolus Administration of Saruplase in Europe
COMPASS   COMPArison trial of Saruplase and Streptokinase
EGF   epidermal growth factor
GAUS   German Activator Urokinase Study
HMW   high molecular weight
LIMITS   Liquemin In Myocardial Infarction during Thrombolysis with
              Saruplase
LMW   low molecular weight
PAI-1   plasminogen activator inhibitor type
PATENT   Pro-urokinase And TPA Enhancement of Thrombolysis
PRIMI   Pro-urokinase In Myocardial Infarction
pro-UK   zymogen form of urokinase
rpro-UK   recombinant pro-UK
rtPA   recombinant tissue-type plasminogen activator
sc-uPA   single-chain uPA, also called pro-urokinase
SESAM   Study in Europe with Saruplase and Alteplase in Myocardial
              infarction
SUTAMI   Saruplase and Urokinase in the Treatment of Acute Myocardial
              Infarction
TAMI   Thrombolysis in Acute Myocardial Infarction
tc-uPA   two-chain uPA, also called urokinase
TIMIKO   Thrombolysis in Myocardial Infarction In Korea
tPA   tissue-type plasminogen activator
UK   urokinase
uPA   urinary-type (or urokinase-type) plasminogen activator, also
              called urokinase
USIM   Urochinasi per via Sisternica nell'Infarto Miocardico

References

Abercrombie DM, Buchinski B, Salvato KA, Vovis GF, Stump DC, Broeze RJ (1990)
    Fibrin specific thrombolysis by two-chain urokinase-type plasminogen activator
    cleaved after arginine 156 by thrombin. Thromb Haemost 64:426–432
Astrup T, Sterndorff I (1952) An activator of plasminogen in normal urine. Proc Soc
    Exp Biol Med 81:675–678
Badylak SF, Voytik S, Klabunde RE, Henkin J, Leski M (1988) Bolus dose response
    characteristics of single chain urokinase plasminogen activator and tissue plas-
    minogen activator in a dog model of arterial thrombosis. Thromb Res 52:295–312
Bär FW, Meyer J, Boland J, Betriu A, Artmeyer B, Charbonnier B, Michels HR, Tebbe
    U, Spiecker M, Vermeer F, von Fisenne MJ, Hopkins GR, Barth H (1998) Bolus
    Administration of Saruplase in Europe (BASE), a Pilot Study in Patients with
    Acute Myocardial Infarction. J Thromb Thrombolysis 6:147–153
Bär FW, Meyer J, Vermeer F, Michels R, Charbonnier B, Haerten K, Spiecker M,
    Macaya C, Hanssen M, Heras M, Boland JP, Morice MC, Dunn FG, Uebis R,
    Hamm C, Ayzenberg O, Strupp G, Withagen AJ, Klein W, Windeler J, Hopkins G,

Barth H, von Fisenne MJ, for the SESAM Study Group (1997) Comparison of saruplase and alteplase in acute myocardial infarction. Am J Cardiol 79:727–732

Barlow GH, Francis CW, Marder VJ (1981) On the conversion of high molecular weight urokinase to the low molecular weight form by plasmin. Thromb Res 23:541–547

Barlow GH, Lazer LA (1972) Characterization of the plasminogen activator isolated from human embryo kidney cells: comparison with urokinase. Thromb Haemost 1:201–208

Bernik MB, Oller EP (1973) Increased plasminogen activator (urokinase) in tissue culture after fibrin deposition. J Clin Invest 42:823–834

Bernik MB, Oller EP (1976) Plasminogen activator and proactivator (urokinase precursor) in lung cultures. J Am Med Wom Assoc 31:465–472

Bode C, Matsueda GR, Hui KY, Haber E (1990a) Antibody-directed urokinase: a specific fibrinolytic agent. Science 229:765–767

Bode C, Runge MS, Newell JB, Matsueda GR, Haber E (1987) Characterization of an antibody-urokinase conjugate. A plasminogen activator targeted to fibrin. J Biol Chem 262:10819–10823

Bode C, Runge MS, Schönermark S, Eberle T, Newell JB, Küber W, Haber E (1990b) Conjugation to antifibrin Fab' enhances fibrinolytic potency of single-chain urokinase plasminogen activator. Circulation 81:1974–1980

Bode C, Schoenermark S, Schuler G, Zimmermann R, Schwarz F, Kuebler W (1988) Efficacy of intravenous prourokinase and a combination of prourokinase and urokinase in acute myocardial infarction. Am J Cardiol 61:971–974

Bode C, Schuler G, Nordt T, Schönermark S, Baumann H, Richardt G, Dietz R, Gurewich V, Kübler W (1990c) Intravenous thrombolytic therapy with a combination of single-chain urokinase-type plasminogen activator and recombinant tissue-type plasminogen activator in acute myocardial infarction. Circulation 81:907–913

Braat EAM, Levi M, Bos R, Haverkate F, Lassen MR, de Maat MPM, Rijken DC (1999) Inactivation of single-chain urokinase-type plasminogen activator by thrombin in human subjects. J Lab Clin Med 134:161–167

Braat EAM, Los P, Rijken DC (1998) The inactivation of single-chain urokinase-type plasminogen activator by thrombin in a plasma milieu: effect of thrombomodulin. Blood Coagul Fibrinolysis 9:419–427

Breton J, Pezzi N, Molinari A, Bonomini L, Lansen J, Gonzalez De Buitrago G, Prieto I (1995) Prolonged half-life in the circulation of a chemical conjugate between a pro-urokinase derivative and human serum albumin. Eur J Biochem 231:563–569

Brochier M, Raynaud R, Planiol T, Fauchier JP, Griguer P, Archambaud D, Pellois A, Clisson M (1975) Le traitement par l'urokinase des infarctus du myocarde et syndromes de menace. Arch Mal Coeur Vaiss 68:563–569

Carlsson J, Schuster HP, Tebbe U (1997) Prähospitale thrombolytische Therapie bei akutem Myokardinfarkt. Anaesthesist 46:829–839

Collen D (1985) Report of the Meeting of the Subcommittee on Fibrinolysis, San Diego, CA. Thromb Haemost 54:893

Collen D, De Cock F, Demarsin E, Lijnen HR, Stump DC (1986a) Absence of synergism between tissue-type plasminogen activator (t-PA), single-chain urokinase-type plasminogen activator (scu-PA) and urokinase on clot lysis in a plasma milieu in vitro. Thromb Haemost 56:35–39

Collen D, Stassen J-M, De Cock F (1987) Synergistic effect of thrombolysis of sequential infusion of tissue-type plasminogen activator (t-PA), single chain urokinase-type plasminogen activator (scu-PA) and urokinase in the rabbit jugular vein thrombosis model. Thromb Haemost 58:943–946

Collen D, Stassen JM, Blaber M, Winkler M, Verstraete M (1984) Biological and thrombolytic properties of proenzyme and active forms of human urokinase. 3. Thrombolytic properties of natural and recombinant urokinase in rabbits with experimental jugular vein thrombosis. Thromb Haemost 52:27–30

Collen D, Stump D, Van de Werf F (1985) Coronary thrombolysis in dogs with intravenously administered pro-urokinase. Circulation 72:384–388

Collen D, Van de Werf F (1987) Coronary arterial thrombolysis with low-dose syner-
    gistic combinations of recombinant tissue-type plasminogen activator (rt-PA) and
    recombinant single-chain urokinase-type plasminogen activator (rscu-PA) for
    acute myocardial infarction. Am J Cardiol 60:431–434
Collen D, Zamarron C, Lijnen HR, Hoylaerts M (1986b) Activation of plasminogen by
    pro-urokinase. 2. Kinetics. J Biol Chem 261:1259–1266
de Boer A, Kluft C, Gerloff J, Dooijewaard G, Günzler WA, Beier H, van der Meer
    FJM, Cohen AF (1993) Pharmacokinetics of saruplase, a recombinant unglycosy-
    lated human single-chain urokinase-type plasminogen activator and its effects on
    fibrinolytic and haemostatic parameters in healthy male subjects. Thromb
    Haemost 70:320–325
de Munk GAW, Groeneveld E, Rijken DC (1991) Acceleration of the thrombin inac-
    tivation of single chain urokinase-type plasminogen activator (pro-urokinase) by
    thrombomodulin. J Clin Invest 88:1680–1684
de Prost D, Guerot C, Laffay N, Horellou MH, Samama M (1983) Intra-coronary
    thrombolysis with streptokinase or lys-plasminogen/urokinase in acute myocardial
    infarction: effects on recanalization and blood fibrinolysis. Thromb Haemost
    50:792–796
Declerck PJ, Lijnen HR, Verstreken M, Moreau H, Collen D (1990) A monoclonal anti-
    body specific for two-chain urokinase-type plasminogen activator. Application to
    the study of the mechanism of clot lysis with single-chain urokinase-type plas-
    minogen activator in plasma. Blood 75:1794–1800
Diefenbach C (1992) One-year follow-up in a randomized double-blind trial of saru-
    plase versus streptokinase in acute myocardial infarction: Long-term results of the
    PRIMI trial. Coron Artery Dis 3:925–931
Diefenbach C, Erbel R, Pop T, Mathey D, Schofer J, Hamm C, Ostermann H, Schmitz
    Hubner U, Bleifeld W, Meyer J (1988) Recombinant single-chain urokinase-
    type plasminogen activator during acute myocardial infarction. Am J Cardiol
    61:966–970
Eguchi Y, Sakata Y, Matsuda M, Osada H, Numao N, Ohmori M, Kondo K (1990) Char-
    acterization of thrombin- and plasmin-resistant mutants of recombinant human
    single chain urokinase-type plasminogen activator. J Biochem (Tokyo) 108:72–
    79
European Collaborative Study (1975) Controlled trial of urokinase in myocardial
    infarction. Lancet 2:624–627
Fay WP, Eitzman DT, Shapiro AD, Madison EL, Ginsburg D (1994) Platelets inhibit
    fibrinolysis in vitro by both plasminogen activator inhibitor-1-dependent and
    –independent mechanisms. Blood 83:351–356
Ferguson JJ (1999) Highlights of the 20th Congress of the European Society of Car-
    diology. Circulation 99:1127–1131
Flameng W, Vanhaecke J, Stump DC, Van de Werf F, Holmes W, Guenzler WA, Flohe
    L, Collen D (1986) Coronary thrombolysis by intravenous infusion of recombi-
    nant single chain urokinase-type plasminogen activator or recombinant urokinase
    in baboons: effect on regional blood flow, infarct size and hemostasis. J Am Coll
    Cardiol 8:118–124
Fletcher AP, Alkjaersig N, Sherry S, Genton E, Hirsh J, Bachmann F (1965) The devel-
    opment of urokinase as a thrombolytic agent: maintenance of a sustained throm-
    bolytic state in man by its intravenous infusion. J Lab Clin Med 65:713–731
Fry ET, Mack DL, Sobel BE (1989) The nature of synergy between tissue-type and
    single chain urokinase-type plasminogen activators. Thromb Haemost 62:909–916
Gormsen J, Tidstrom B, Feddersen C, Ploug J (1973) Biochemical evaluation of low
    dose of urokinase in acute myocardial infarction. A double-blind study. Acta Med
    Scand 194:191–198
Gulba DC, Fischer K, Barthels M, Jost S, Möller W, Frombach R, Reil G-H, Daniel WG,
    Lichtlen PR (1989a) Potentiative effect of heparin in thrombolytic therapy
    of evolving myocardial infarction with natural pro-urokinase. Fibrinolysis
    3:165–173

Gulba DCL, Fischer K, Barthels M, Polensky U, Reil G-H, Daniel WG, Welzel D, Lichtlen PR (1989b) Low dose urokinase preactivated natural prourokinase for thrombolysis in acute myocardial infarction. Am J Cardiol 63:1025–1031

Gurewich V (1987) Experiences with pro-urokinase and potentiation of its fibrinolytic effect by urokinase and by tissue plasminogen activator. J Am Coll Cardiol 10:16B–121

Gurewich V (1989a) Importance of fibrin specificity in therapeutic thrombolysis and the rationale of using sequential and synergistic combinations of tissue plasminogen activator and pro-urokinase. Semin Thromb Hemost 15:123–128

Gurewich V (1989b) The sequential, complementary and synergistic activation of fibrin-bound plasminogen by tissue plasminogen activator and pro-urokinase. Fibrinolysis 3:59–66

Gurewich V, Emmons F, Pannell R (1988a) Spontaneous clot lysis in whole human plasma by endogenous tissue type and urokinase type plasminogen activators: demonstration of a promoting effect by t-PA and by platelets on urokinase. Fibrinolysis 2:143–149

Gurewich V, Johnstone M, Loza J-P, Pannell R (1993) Pro-urokinase and prekallikrein are both associated with platelets: Implications for the intrinsic pathway of fibrinolysis and for therapeutic thrombolysis. FEBS Lett 318:317–321

Gurewich V, Johnstone MT, Pannell R (1995) The selective uptake of high molecular weight urokinase-type plasminogen activator by human platelets. Fibrinolysis 9:188–195

Gurewich V, Muller J (1996) Is coronary thrombolysis associated with side effects that significantly compromise clinical benefits. Am J Cardiol 77:756–758

Gurewich V, Pannell R (1986) A comparative study of the efficacy and specificity of tissue plasminogen activator and pro-urokinase: demonstration of synergism and of different thresholds of non-selectivity. Thromb Res 44:217–228

Gurewich V, Pannell R (1987) Inactivation of single-chain urokinase (pro-urokinase) by thrombin and thrombin-like enzymes: relevance of the findings to the interpretation of fibrin-binding experiments. Blood 69:769–772

Gurewich V, Pannell R, Broeze RJ, Mao J (1988b) Characterization of the intrinsic fibrinolytic properties of pro- urokinase through a study of plasmin-resistant mutant forms produced by site-specific mutagenesis of lysine158. J Clin Invest 82:1956–1962

Gurewich V, Pannell R, Louie S, Kelley P, Suddith L, Greenlee R (1984) Effective and fibrin-specific clot lysis by a zymogen precursor form of urokinase (pro-urokinase): A study in vitro and in two animal species. J Clin Invest 73:1731–1739

Harpel PC, Chang T, Verderber E (1985) Tissue plasminogen activator and urokinase mediate the binding of Glu-plasminogen to plasma fibrin I. Evidence for new binding sites in plasmin-degraded fibrin I. J Biol Chem 260:4432–4440

Hauert J, Nicoloso G, Schleuning WD, Bachmann F, Schapira M (1989) Plasminogen activators in dextran sulfate-activated euglobulin fractions: a molecular analysis of factor XII- and prekallikrein- dependent fibrinolysis. Blood 73:994–999

Hiramatsu R, Kasai S, Amatsuji Y, Kawai T, Hirose M, Morita M, Tanabe T, Kawabe H, Arimura H, Yokoyama K (1989) Effect of deletion of epidermal growth factor-like domain on plasma clearance of pro-urokinase. Fibrinolysis 3:147–151

Hua Z-C, Chen X-C, Dong C, Zhu D-X (1996) Characterization of a recombinant chimeric plasminogen activator composed of Gly-Pro-Arg-Pro tetrapeptide and truncated urokinase-type plasminogen activator expressed in *Escherichia coli*. Biochem Biophys Res Commun 222:576–583

Husain SS (1991) Single-chain urokinase-type plasminogen activator does not possess measurable intrinsic amidolytic or plasminogen activator activities. Biochemistry 30:5797–5805

Husain SS (1993) Fibrin affinity of urokinase-type plasminogen activator. Evidence that Zn2+ mediates strong and specific interaction of single-chain urokinase with fibrin. J Biol Chem 268:8574–8579

Husain SS, Gurewich V, Lipinski B (1983) Purification and partial characterization of a single-chain molecular weight form of urokinase from human urine. Arch Biochem Biophys 220:31–38

Husain SS, Lipinski B, Gurewich V (1979) Isolation of plasminogen activators useful as therapeutic and diagnostic agents (single-chain, high-affinity urokinase). US Patent 4:381–346

Ichinose A, Fujikawa K, Suyama T (1986) The activation of pro-urokinase by plasma kallikrein and its inactivation by thrombin. J Biol Chem 261:3486–3489

Jiang YP, Pannell R, Liu JN, Gurewich V (1996) Evidence for a novel binding protein to urokinase-type plasminogen activator in platelet membranes. Blood 87:2775–2781

Kajihara J, Shibata K, Nakano Y, Nishimuro S, Kato K (1994) Physicochemical characterization of PEG-PPG conjugated human urokinase. Biochim Biophys Acta Gen Subj 1199:202–208

Kambara H, Kawai C, Kajiwara N, Niitani H, Sasayama S, Kanmatsuse K, Kodama K, Sato H, Nobuyoshi M, Nakashima M, Matsuo O, Matsuda T (1988) Randomized, double-blinded multicenter study. Comparison of intracoronary single-chain urokinase-type plasminogen activator, pro-urokinase (GE-0943), and intracoronary urokinase in patients with acute myocardial infarction. Circulation 78:899–905

Kanemoto N, Goto Y, Hirosawa K, Kawai C, Kimata S, Yui Y, Yamamoto Y (1991) Intravenous recombinant tissue-type plasminogen activator (rt- PA) and urokinase (UK) in patients with evolving myocardial infarction – A multicenter double-blind, randomized trial in Japan. Jpn Circ J 55(3):250–261

Kirshenbaum JM, Bahr RD, Flaherty JT, Gurewich V, Levine HJ, Loscalzo J, Schumacher RR, Topol EJ, Wahr DW, Braunwald E, for the Pro-Urokinase for Myocardial Infarction Study Group (1991) Clot-selective coronary thrombolysis with low-dose synergistic combinations of single-chain urokinase-type plasminogen activator and recombinant tissue-type plasminogen activator. Am J Cardiol 68:1564–1569

Klein P (1989) Plasma levels of t-PA:AG under standard heparin therapy. Thromb Res 56:649–653

Kono T, Morita H, Nishina T, Fujita M, Hirota Y, Kawamura K, Fujiwara A (1996) Circadian variations of onset of acute myocardial infarction and efficacy of thrombolytic therapy. J Am Coll Cardiol 27:774–778

Koster RW, Cohen AF, Hopkins GR, Beier H, Günzler WA, van der Wouw PA (1994) Pharmacokinetics and pharmacodynamics of Saruplase, an unglycosylated single-chain urokinase-type plasminogen activator, in patients with acute myocardial infarction. Thromb Haemost 72:740–744

Kurnik PB (1995) Circadian variation in the efficacy of tissue-type plasminogen activator. Circulation 91:1341–1346

Köhler M, Sen S, Miyashita C, Hermes R, Pindur G, Heiden M, Berg G, Mörsdorf S, Leipnitz G, Zeppezauer M, Schieffer H, Wenzel E, Schönberger A, Hollemeyer K (1991) Half-life of single-chain urokinase-type plasminogen activator (scu-PA) and two-chain urokinase-type plasminogen activator (tcu-PA) in patients with acute myocardial infarction. Thromb Res 62:75–81

Lang IM, Marsh JJ, Konopka RG, Olman MA, Binder BR, Moser KM, Schleef RR (1993) Factors contributing to increased vascular fibrinolytic activity in mongrel dogs. Circulation 87:1990–2000

Lenich C, Pannell R, Henkin J, Gurewich V (1992) The influence of glycosylation on the catalytic and fibrinolytic properties of pro-urokinase. Thromb Haemost 68:539–544

Li C, Jiang Y, Gurewich V, Liu JN (1996) Urokinase-type plasminogen activator (u-PA) stimulates its own expression in endothelial cells and monocytes. Fibrinolysis 10:257 (abstr)

Li X-K, Lijnen HR, Nelles L, Hu M-H, Collen D (1992) Biochemical properties of recombinant mutants of nonglycosylated single chain urokinase-type plasminogen activator. Biochim Biophys Acta Protein Struct Mol Enzymol 1159:37–43

Lieber A, Peeters MJ, Gown A, Perkins J, Kay MA (1995) A modified urokinase plasminogen activator induces liver regeneration without bleeding. Hum Gene Ther 6:1029–1037

Lijnen HR, Nelles L, Van Houtte E, Collen D (1990) Pharmacokinetic properties of mutants of recombinant single chain urokinase-type plasminogen activator obtained by site- specific mutagenesis of Lys158, Ile159 and Ile160. Fibrinolysis 4:211–214

Lijnen HR, Pierard L, Reff ME, Gheysen D (1988a) Characterization of a chimaeric plasminogen activator obtained by insertion of the second kringle structure of tissue-type plasminogen activator (amino acids 173 through 262) between residues Asp 130 and Ser 139 of urokinase-type plasminogen activator. Thromb Res 52:431–441

Lijnen HR, Van Hoef B, Nelles L, Holmes WE, Collen D (1988b) Enzymatic properties of single-chain and two-chain forms of a Lys^{158}®Glu^{158} mutant of urokinase-type plasminogen activator. Eur J Biochem 172:185–188

Lippschutz EJ, Ambrus JL, Ambrus CM, et al (1965) Controlled study on the treatment of coronary occlusion with urokinase-activated human plasmin. Am J Cardiol 16:93–98

Liu JN, Gurewich V (1991) A comparative study of the promotion of tissue plasminogen activator and pro-urokinase-induced plasminogen activation by fragments D and E-2 of fibrin. J Clin Invest 88:2012–2017

Liu JN, Gurewich V (1992) Fragment E-2 from fibrin substantially enhances pro-urokinase-induced Glu-plasminogen activation. A kinetic study using the plasmin-resistant mutant pro-urokinase Ala-158-rpro-UK. Biochemistry 31:6311–6317

Liu JN, Gurewich V (1993) The kinetics of plasminogen activation by thrombin-cleaved pro-urokinase and promotion of its activity by fibrin fragment E-2 and by tissue plasminogen activator. Blood 81:980–987

Liu JN, Gurewich V (1995) Inactivation of the intrinsic activity of pro-urokinase by diisopropyl fluorophosphate is reversible. J Biol Chem 270:8408–8410

Liu JN, Pannell R, Gurewich V (1992) A transitional state of pro-urokinase that has a higher catalytic efficiency against glu-plasminogen than urokinase. J Biol Chem 267:15289–15292

Liu JN, Tang W, Sun ZY, Kung W, Pannell R, Sarmientos P, Gurewich V (1996) A site-directed mutagenesis of pro-urokinase which substantially reduces its intrinsic activity. Biochemistry 35:14070–14076

Loscalzo J, Wharton TP, Kirshenbaum JM, Levine HJ, Flaherty JT, Topol EJ, Ramaswamy K, Kosowsky BD, Salem DN, Ganz P, et al (1989) Clot-selective coronary thrombolysis with pro-urokinase. Circulation 79:776–782

Loza J-P, Gurewich V, Johnstone M, Pannell R (1994) Platelet-bound prekallikrein promotes pro-urokinase-induced clot lysis: a mechanism for targeting the factor XII dependent intrinsic pathway of fibrinolysis. Thromb Haemost 71:347–352

Ma Z, Yu RR, Hua ZC, Zhu DX (1996) Urokinase mutant with better fibrin-specificity. Science in China 39:523–533

Manchanda N, Schwartz BS (1991) Single chain urokinase. Augmentation of enzymatic activity upon binding to monocytes. J Biol Chem 266:14580–14584

Mathey DG, Schofer J, Sheehan FH, Becher H, Tilsner V, Dodge HT (1985) Intravenous urokinase in acute myocardial infarction. Am J Cardiol 55:878–882

Michels HR, Hoffman JJ, Bär FW (1999) Pharmacokinetics and hemostatic effects of saruplase in patients with acute myocardial infarction: comparison of infusion, single-bolus, and split-bolus administration. J Thromb Thrombolysis 8:213–221

Michels R, Hoffmann H, Windeler J, et al. (1995) A double-blind multicenter comparison of the efficacy and safety of Saruplase and urokinase in the treatment of acute myocardial infarction: report of the SUTAMI Study Group. J Thrombos Thrombolys 2:117–124

Miles LA, Plow EF (1985) Binding and activation of plasminogen on the platelet surface. J Biol Chem 260:4303–4311

Muller JE, Stone PH, Turi ZG, Rutherford JD, Czeisler CA, Parker C, Poole WK, Passamini E, Roberts R, Robertson T, Sobel BE, Willerson JT, Braunwald E, and the MILIS Study Group (1985) Circadian variation in the frequency of onset of acute myocardial infarction. N Engl J Med 313:1315–1322

Nelles L, Lijnen HR, Van Nuffelen A, Demarsin E, Collen D (1990) Characterization of domain deletion and/or duplication mutants of a recombinant chimera of tissue-type plasminogen activator and urokinase-type plasminogen activator (rt-PA/u-PA). Thromb Haemost 64:53–60

Neuhaus KL, Tebbe U, Gottwik M, Weber MAJ, Feuerer W, Niederer W, Haerer W, Praetorius F, Grosser KD, Huhmann W, Hoepp HW, Alber G, Sheikhzadeh A, Schneider B (1988) Intravenous recombinant tissue plasminogen activator (rt-PA) and urokinase in acute myocardial infarction: results of the German Activator Urokinase Study (GAUS). J Am Coll Cardiol 12:581–587

Nguyen G, Fretault J, Samama M, Coursaget J (1989) Thrombolytic properties of tissue plasminogen activator (t-PA) and single chain urokinase plasminogen activator (scu-PA) in vitro: absence of synergy as demonstrated by two quantitative analytical methods. Fibrinolysis 3:23–26

Nielsen LS, Hansen JG, Skriver L, Wilson EL, Kaltaft K, Zeuthen J, Danø K (1982) Purification of zymogen to plasminogen activator from human glioblastoma cells by affinity chromatography with monoclonal antibody. Biochemistry 21:6410–6415

Nieuwenhuizen W, Vermond A, Voskuilen M, Traas DW, Verheijen JH (1983) Identification of a site in fibrin(ogen) which is involved in the acceleration of plasminogen activation by tissue-type plasminogen activator. Biochim Biophys Acta 748: 86–92

Nolan C, Hall LS, Barlow GH, Tribby II (1977) Plasminogen activator from human embryonic kidney cell cultures. Evidence for a proactivator. Biochim Biophys Acta 496:384–400

Pannell R, Black J, Gurewich V (1988) Complementary modes of action of tissue-type plasminogen activator and pro-urokinase by which their synergistic effect on clot lysis may be explained. J Clin Invest 81:853–859

Pannell R, Gurewich V (1986) Pro-urokinase: a study of its stability in plasma and of a mechanism for its selective fibrinolytic effect. Blood 67:1215–1223

Pannell R, Gurewich V (1987) Activation of plasminogen by single-chain urokinase or by two-chain urokinase – A demonstration that single-chain urokinase has a low catalytic activity (pro-urokinase). Blood 69:22–26

Park S, Harker LA, Marzec UM, Levin EG (1989) Demonstration of single chain urokinase-type plasminogen activator on human platelet membrane. Blood 73:1421–1425

Park SJ, for the Thrombolysis in Myocardial Infarction in Korea (TIMIKO) Study Group (1998) Comparison of double bolus urokinase versus front-loaded alteplase regimen for acute myocardial infarction. Am J Cardiol 82:811–813

Petersen LC, Lund LR, Nielsen LS, Danø K, Skriver L (1988) One-chain urokinase-type plasminogen activator from human sarcoma cells is a proenzyme with little or no intrinsic activity. J Biol Chem 263:11189–11195

PRIMI Trial Study Group, Meyer J, Bär F, Barth H, Charbonnier B, El Deeb MF, Erbel R, Flohé L, Gülker H, Heikkilä J, Massberg I (1989) Randomised double-blind trial of recombinant pro-urokinase against streptokinase acute myocardial infarction. Lancet i:863–867

Pâques E-P, Karges H-E, Römisch J (1992) Prevention of fibrinogenolysis without impairment of thrombolysis: combination of a plasminogen activator with a_2-antiplasmin. Fibrinolysis 6:111–117

Rossi P, Bolognese L (1991) Comparison of intravenous urokinase plus heparin versus heparin alone in acute myocardial infarction. Am J Cardiol 68:585–592

Runge MS, Harker LA, Bode C, Ruef J, Kelly AB, Marzec UM, Allen E, Caban R, Shaw S-Y, Haber E, Hanson SR (1996) Enhanced thrombolytic and antithrombotic potency of a fibrin-targeted plasminogen activator in baboons. Circulation 94:1412–1422

Sabovic M, Keber D (1995) In-vitro synergism between t-PA and scu-PA depends on clot retraction. Fibrinolysis 9:101–105

Sakuragawa N, Shimizu K, Kondo K, Kondo S, Niwa M (1986) Studies on the effect of PEG-modified urokinase on coagulation-fibrinolysis using beagles. Thromb Res 41:627–635

Sappino A-P, Huarte J, Vassalli J-D, Belin D (1991) Sites of synthesis of urokinase and tissue-type plasminogen activators in the murine kidney. J Clin Invest 87:962–970

Sasaki K, Moriyama S, Tanaka Y, Sumi H, Robbins KC (1985) The transport of ^{125}I-labeled human high molecular weight urokinase across the intestinal tract in a dog model with stimulation of synthesis and/or release of plasminogen activators. Blood 66:69–75

Schneider P, Bachmann F, Sauser D (1982) Urokinase. A short review of its properties and of its metabolism. In: Mannucci PM, D'Angelo A (eds) Urokinase: basic and clinical aspects. Academic Press, London, pp 1–15

Schofer J, Büttner J, Geng G, Gutschmidt K, Herden HN, Mathey DG, Moecke HP, Polster P, Raftopoulo A, Sheehan FH, Voelz P (1990) Prehospital thrombolysis in acute myocardial infarction. Am J Cardiol 66:1429–1433

Sobel GW, Mohler SR, Jones NW, Dowdy ABC, Guest MM (1952) Urokinase: an activator of plasma profibrinolysin extracted from human urine. Am J Physiol 171:768–769

Spiecker M, Windeler J, Vermeer F, Michels R, Seabra-Gomes R, vom Dahl J, Kerber S, Verheugt FWA, Westerhof PW, Bär FW, Nixdorff U, Barth H, Hopkins GR, von Fisenne MJM, Meyer J, for the PRIMI Investigators (1999) Thrombolysis with saruplase versus streptokinase in acute myocardial infarction: five-year results of the PRIMI trial. Am Heart J 138:518–524

Stump DC, Kieckens L, De Cock F, Collen D (1987) Pharmacokinetics of single chain forms of urokinase-type plasminogen activator. J Pharmacol Exp Ther 242:245–250

Stump DC, Thienpont M, Collen D (1986) Urokinase-related proteins in human urine. Isolation and characterization of single-chain urokinase (pro-urokinase) and urokinase-inhibitor complex. J Biol Chem 261:1267–1273

Suenson E, Bjerrum P, Holm A, Lind B, Meldal M, Selmer J, Petersen LC (1990) The role of fragment X polymers in the fibrin enhancement of tissue plasminogen activator-catalyzed plasmin formation. J Biol Chem 265:22228–22237

Sumi H, Sasaki K, Toki N, Robbins KC (1980) Oral administration of urokinase. Thromb Res 20:711–714

Sumi H, Seiki M, Morimoto N, Tsushima H, Maruyama M, Mihara H (1985) Plasma fibrinolysis after intraduodenal administration of urokinase in rats. Enzyme 33:122–127

Sun Z, Jiang Y, Ma Z, Wu H, Liu BF, Xue Y, Tang W, Chen Y, Li C, Zhu D, Gurewich V, Liu JN (1997) Identification of a flexible loop region (297–313) of urokinase-type plasminogen activator, which helps determine its catalytic activity. J Biol Chem 272:23818–23823

Tebbe U, Michels R, Adgey J, Boland J, Caspi A, Charbonnier B, Windeler J, Barth H, Groves R, Hopkins GR, Fennell W, Betriu A, Ruda M, Mlczoch J, for the Comparison Trial of Saruplase and Streptokinase (COMPASS) Investigators (1998) Randomized, double-blind study comparing saruplase with streptokinase therapy in acute myocardial infarction: the COMPASS Equivalence Trial. J Am Coll Cardiol 31:487–493

Tebbe U, Windeler J, Boesl I, Hoffmann H, Wojcik J, Ashmawy M, Schwarz ER, von Loewis of Menar P, Rosemeyer P, Hopkins G, Barth H, on behalf of the LIMITS Study Group (1995) Thrombolysis with recombinant unglycosylated single-chain urokinase-type plasminogen activator (saruplase) in acute myocardial infarction: influence of heparin on early patency rate (LIMITS study). J Am Coll Cardiol 26:365–373

The GUSTO Investigators (1993) An international randomized trial comparing four thrombolytic strategies for acute myocardial infarction. N Engl J Med 329:673–682

Tissot J-D, Schneider P, Hauert J, Ruegg M, Kruithof EKO, Bachmann F (1982) Isolation from human plasma of plasminogen activator identical to urinary high molecular weight urokinase. J Clin Invest 70:1320–1323

Toki N, Sumi H, Sasaki K, Boreisha I, Robbins KC (1985) Transport of urokinase across the intestinal tract of normal human subjects with stimulation of synthesis and/or release of urokinase-type proteins. J Clin Invest 75:1212–1222

Topol EJ, Califf RM, George BS, Kereiakes DJ, Rothbaum D, Candela RJ, Abbotsmith CW, Pinkerton CA, Stump DC, Collen D, Lee KL, Pitt B, Kline EM, Boswick JM, O'Neill WW, Stack RS, and the TAMI Study Group (1988) Coronary arterial thrombolysis with combined infusion of recombinant tissue-type plasminogen activator and urokinase in patients with acute myocardial infarction. Circulation 77:1100–1107

Torr SR, Nachowiak DA, Fujii S, Sobel BE (1992) "Plasminogen steal" and clot lysis. J Am Coll Cardiol 19:1085–1090

Tran-Thang Ch, Kruithof EKO, Atkinson J, Bachmann F (1986) High-affinity binding sites for human Glu-plasminogen unveiled by limited plasmic degradation of human fibrin. Eur J Biochem 160:599–604

Van de Werf F, Nobuhara M, Collen D (1986a) Coronary thrombolysis with human single-chain, urokinase-type plasminogen activator (pro-urokinase) in patients with acute myocardial infarction. Ann Intern Med 104:345–348

Van de Werf F, Vanhaecke J, De Geest H, Verstraete M, Collen D (1986b) Coronary thrombolysis with recombinant single-chain urokinase-type plasminogen activator in patients with acute myocardial infarction. Circulation 74:1066–1070

Van Griensven JMT, Koster RW, Hopkins GR, Beier H, Günzler WA, Kroon R, Schoemaker RC, Cohen AF (1997) Effect of changes in liver blood flow on the pharmacokinetics of saruplase in patients with acute myocardial infarction. Thromb Haemost 78:1015–1020

Váradi A, Patthy L (1983) Location of plasminogen-binding sites in human fibrin(ogen). Biochemistry 22:2440–2446

Vermeer F, Bosl I, Meyer J, Bär F, Charbonnier B, Windeler J, Barth H (1999) Saruplase is a safe and effective thrombolytic agent; observations in 1,698 patients. Results of the PASS study. Practical Applications of Saruplase Study. J Thrombos Thrombolys 8:143–150

Wall TC, Phillips HR,III, Stack RS, Mantell S, Aronson L, Boswick J, Sigmon K, DiMeo M, Chaplin D, Whitcomb D, Pasi D, Zawodniak M, Hajisheik M, Hegde S, Barker W, Tenney R, Califf RM (1990) Results of high dose intravenous urokinase for acute myocardial infarction. Am J Cardiol 65:124–131

Weaver WD, Hartmann JR, Anderson JL, Reddy PS, Sobolski JC, Sasahara AA (1994) New recombinant glycosylated prourokinase for treatment of patients with acute myocardial infarction. J Am Coll Cardiol 24:1242–1248

Wohn KD, Kanse SM, Deutsch V, Schmidt T, Eldor A, Preissner KT (1997) The urokinase-receptor (CD87) is expressed in cells of the megakaryoblastic lineage. Thromb Haemost 77:540–547

Wun TC, Ossowski L, Reich E (1982a) A proenzyme of human urokinase. J Biol Chem 257:7262–7268

Wun TC, Schleuning D, Reich E (1982b) Isolation and characterization of urokinase from human plasma. J Biol Chem 257:3276–3283

Zamarron C, Lijnen HR, Van Hoef B, Collen D (1984) Biological and thrombolytic properties of proenzyme and active forms of human urokinase. 1. Fibrinolytic and fibrinogenolytic properties in human plasma in vitro of urokinase obtained from human urine or by recombinant DNA technology. Thromb Haemost 52:19–23

Zarich SW, Kowalchuk GJ, Weaver WD, Loscalzo J, Sassower M, Manzo K, Byrnes C, Muller JE, Gurewich V, for the PATENT Study Group (1995) Sequential combination thrombolytic therapy for acute myocardial infarction: results of the Pro-Urokinase and t-PA Enhancement of Thrombolysis (PATENT) Trial. J Am Coll Cardiol 26:374–379

Zhu Y, Carmeliet P, Fay WP (1999) Plasminogen activator inhibitor-1 is a major determinant of arterial thrombolysis resistance. Circulation 99:3050–3055

Comparative Evaluations of tPA vs SK with Particular Reference to the GUSTO-I Trial

A.C. CHIU and E.J. TOPOL

A. Introduction

The benefits of intravenous (i.v.) thrombolytic therapy in acute myocardial infarction (AMI) are now well established. This chapter will briefly review the evidence demonstrating therapeutic benefit of thrombolytic therapy and then focus on the recent major comparative randomized clinical trials of tissue-type plasminogen activator (tPA) vs streptokinase (SK) and, in particular, the Global Utilization of SK and tPA for Occluded Coronary Arteries (GUSTO-I) randomized trial of strategies employing tissue plasminogen activator or SK or both.

B. Evolution of Thrombolytic Therapies

I. Rationale for Thrombolysis

The elucidation of the primary pathophysiologic mechanism leading to acute myocardial infarction, that of coronary artery thrombotic occlusion triggered by atheromatous plaque rupture, prompted investigators to develop strategies aimed at restoring infarct related artery patency. Early interventions utilizing intracoronary administration of SK were reported by RENTROP et al. (1981). Subsequent interventions aimed at restoring patency and, by inference, myocardial perfusion, included percutaneous coronary artery angioplasty (PTCA) and coronary artery bypass grafting (CABG). The delay and logistical difficulties inherent in relying on emergency angioplasty or intracoronary administration of thrombolytic agents for timely restoration of coronary patency soon forced investigators to consider other alternatives. The application of i.v. thrombolytic agents provided an attractive practical strategy for intensive clinical investigation.

II. Early Angiographic Patency Trials, tPA vs SK

The first medium-size, randomized trial comparing i.v. tPA with SK, the Thrombolysis in Myocardial Infarction (TIMI) trial, began in 1984 enrolling patients presenting within 6 h of symptom onset and exhibiting acute injury patterns on electrocardiogram (ECG). Chesebro et al. (1987) compared infarct related artery recanalization rates at 90 min in 143 patients treated with i.v. administered SK vs that obtained in 147 patients given i.v. tPA and heparin (The TIMI Study Group 1985). Though this trial illustrated efficacy of both regimens, tPA clearly demonstrated a higher TIMI grade 2 or 3 recanalization rate (62%) at the 90 min mark than SK (31%; $p < 0.001$). TIMI-1 demonstrated a survival advantage at 6- and 12-month follow-up of patients treated with tPA (mortality rates of 7.7% and 10.5% for tPA, and 9.5% and 11.6% for SK; n.s.). Early patency at 90 min, accompanied by sustained patency at the time of hospital discharge, resulted in a very low mortality of 1.9% after 6 months and 3.8% after 12 months, irrespective of the agent used (Dalen et al. 1988). It was as yet unclear how or if the differential patency rates between the two agents would affect clinical outcome as measured by long term mortality.

During the same period a similar randomized study was performed by the European Collaborative Study Group (ECSG). However, in order to reduce the time delay from hospital admission to start of thrombolytic therapy, no baseline coronary angiography was performed. Initially all patients received an i.v. bolus of 5000 U of heparin. tPA was given in a body weight adjusted dosage of 0.75 mg/kg over a period of 90 min. The patients in the SK group received in addition 0.5 g of aspirin and 1.5 MU of SK infused over 60 min (Verstraete et al. 1985). Patency rates were assessed 90 min after the start of thrombolytic treatment and were 70% in the tPA and 55% in the SK group ($p = 0.054$).

The third trial was conducted in New Zealand in double-blind fashion using a double-dummy (placebo) technique (White et al. 1989). One hundred and thirty five patients were randomized to receive 1.5 MU of SK, infused over 30 min, and 135 patients received tPA, infused over 3 h. An initial bolus of 10 mg of tPA was followed by an infusion of 50 mg during the first hour and 20 mg each for the second and third hour. I.v. heparin was started 30 min after the beginning of thrombolytic therapy and continued for 48 h. Routine coronary angiography was only performed 3 weeks later; patency rates were 76% in the tPA and 75% in the SK group. The effects of the two agents on left ventricular function and on reinfarction rates were similar. Mortality at 30 days was 3.7% in the tPA and 7.4% in the SK group ($p = 0.2$).

The smaller Italian PAIMS study assessed reperfusion in AMI patients by non-invasive signs 4 h after the start of thrombolytic therapy (Magnani et al. 1989). Eighty six patients randomized to the tPA group received a 10 mg bolus of tPA, followed by 90 mg over 3 h, at doses of 50 mg, 20 mg, and 20 mg at each successive hour. The 85 patients allocated to receive SK were given 1.5 MU of SK, infused over 60 min. Reperfusion was considered to have taken place when the following three events occurred in a close temporal relation:

1. An abrupt abatement or a progressive reduction of chest pain
2. A reduction of the magnitude of ST segment shift that continued in sequential ECGs until it was <50% of the basal value

3. A rapid increase in serum levels of MBCK with the peak occurring within 13 h.

Reperfusion at 4 h occurred in 79% of patients in both groups. Patency of the infarct-related vessel at coronary angiography, approximately 4 days after thrombolytic therapy was 81% in the tPA and 74% in the SK group.

As pointed out by LINCOFF and TOPOL (1993), therapeutic efficacy as measured by 90 min infarct related artery patency is problematic on several counts. It is often taken for granted that infarct related artery patency is equivalent to myocardial reperfusion. However, this assumption is unwarranted and lies at the crux of the apparent dissociation between angiographic patency, left ventricular function and mortality noted in most of the early trials. Angiographic "patency" is inclusive of not only arteries reopened by thrombolytic therapy but also includes arteries never completely occluded and arteries spontaneously reopened. "Recanalization" would be a more discriminating term to describe only those arteries reopened by thrombolytic therapy. "Reflow" referring to the quality and briskness of contrast or blood flow through a previously occluded epicardial coronary artery has been categorized by the TIMI grade system. Myocardial "reperfusion" at the tissue level is ultimately the primary purpose of thrombolytic therapy and underlies the open artery hypothesis but epicardial coronary artery patency assessed only at one given point in time does not necessarily equate to tissue reperfusion as has been demonstrated by advances in contrast perfusion echocardiography and continuous 12-lead digital electrocardiography. What has been typically referred to as "patency" in the literature actually represents combined TIMI grade 2 and 3 flow assessed at only a few points in time and hence fails to appreciate the dynamic state of blood flow in proximity of an unstable atheromatous plaque. Variables, other than TIMI flow rates, were shown to influence final infarct size and mortality, such as myocardium at risk and collateral blood flow, as demonstrated by sequential tomographic myocardial perfusion imaging using technetium-99 sestamibi (KLARICH et al. 1999; BRUCE et al. 1999). Definition of the exact relationship between TIMI grade 2 or 3 flow to preservation of left ventricular function and reduction in mortality would have to await the findings of the Angiographic Substudy of the GUSTO-I comparative thrombolytic trial.

III. Adjunctive Therapies; Heparin and Aspirin

Chapter 11 deals with adjunctive therapies currently in use or under development. This section discusses the use of heparin and of aspirin only. The mortality benefit of aspirin therapy in AMI (23% decrease) was convincingly demonstrated by the ISIS-2 trial of i.v. SK [THE ISIS-2 (SECOND INTERNATIONAL STUDY OF INFARCT SURVIVAL) COLLABORATIVE GROUP 1988]. While specific data are lacking for the role and dosage of aspirin with other thrombolytic agents, ISIS-2 results have been extrapolated to justify the current clinical practice of at least 160 mg of aspirin per day in AMI treated with any thrombolytic agent.

This strategy was incorporated in the subsequent major comparative thrombolytic trials including GUSTO-I.

While it was elucidated by Topol et al. (1989) in the third report of the Thrombolysis and Angioplasty in Myocardial Infarction Study (TAMI-3) group that heparin conferred no early patency advantage over tPA alone, the discovery of the role of heparin as a corequisite to maintaining coronary artery patency following therapy with any of the relatively fibrin-specific plasminogen activators, can be credited to a host of investigators. Among the investigators, the work of Bleich et al. (1990), De Bono et al. (1992) for the European Cooperative Study Group sixth report, Hsia et al. (1990, 1992) for the Heparin-Aspirin Reperfusion Trial Investigators, and Arnout et al. (1992) clearly illustrated the effect of heparin in sustaining patency after tPA administration. The latter two trials were particularly notable for the observations that sustained patency required a given level of anticoagulation with heparin. Failure to achieve or maintain a therapeutic level of the activated partial thromboplastin time (APTT) resulted in a marked decrement in rates of sustained patency. The work of these investigators were to have a major impact on the design of the subsequent GUSTO-I trial as well as on the interpretation of the GISSI-2 and the ISIS-3 trials.

IV. Major Comparative Thrombolytic Trials

The large scale "megatrial" in search for more optimal thrombolytic regimens was initiated in Italy [Gruppo Italiano per lo Studio Della Sopravvivenza nell'Infarcto Miocardico (GISSI-2) 1990] and combined with the International Study Group (1990) trial for a total of 20,891 patients presenting within 6h of onset of ischemic symptoms and acute myocardial injury pattern (ST-segment elevation) on ECG. GISSI-2 compared SK 1.5MU i.v. over 60min against the then accepted "conventional" tPA regimen of alteplase 100mg i.v. over 3h. The trial utilized a 2×2 factorial design assigning patients also to either heparin 12500U s.c. twice daily commencing 12h after thrombolytic therapy or to no heparin. i.v. heparin was not recommended in the protocol and was rarely used in the centers participating in the study. The major clinical events are given in Table 1. There was no difference in mortality or reinfarction rate between the tPA and the SK groups nor between those who did or did not receive s.c. heparin. The incidence of total stroke was higher in the tPA group (1.3% vs 0.9%, a relative risk (RR) for tPA of 1.41; 95% confidence interval (CI) 1.09–1.83). However, the difference in definite hemorrhagic stroke, 0.4% and 0.3% respectively, was negligible. On the other hand major bleeds were less frequent with tPA (0.6% vs 0.9% in the SK group; RR 0.67, CI 0.49–0.91). The administration of heparin also was associated with a higher incidence of major bleeds (1.0% with and 0.5% without heparin, RR 1.79, CI 1.31–2.45).

GISSI-2 was soon followed by the ISIS-3 (Third International Study of Infarct Survival) Collaborative Group 1992) which enrolled 41,299

Table 1. In-hospital mortality and major clinical events in the four treatment groups of the GISSI-2 and International Study Group trial

Event	tPA + heparin $n = 5170$	tPA $n = 5202$	SK + heparin $n = 5191$	SK $n = 5205$
Death (%)	9.2	8.7	7.9	9.2
Reinfarction (%)	2.5	2.8	3.0	3.0
Ventricular fibrillation (%)	6.4	7.3	6.4	6.6
Cardiogenic shock (%)	5.8	5.4	5.6	6.2
Stroke total (%)	1.2	1.4	1.0	0.9
Hemorrhagic (%)	0.4	0.4	0.2	0.3
Ischemic (%)	0.5	0.6	0.5	0.4
Undefined (%)	0.3	0.4	0.3	0.1
Major bleeds (%)	0.8	0.5	1.2	0.6

Taken from: THE INTERNATIONAL STUDY GROUP (1990).

patients. ISIS-3 differed from its predecessor in several respects. Chief among these was the inclusion of a trial arm for anistreplase administered conventionally as a 30 U i.v. bolus over 3 min. The form of tPA utilized in ISIS-3 was duteplase dosed at 0.6 MU/kg i.v. over 4 h which was validated by the work of GRINES and DEMARIA (1990) who reported a 90 min patency rate of 67%, roughly comparable to the results of conventional alteplase angiographic patency trials. ISIS-3 utilized a 4×2 factorial design randomizing patients to one of three thrombolytic agents and to either s.c. heparin commencing 4 h after thrombolytic therapy or no heparin. All patients received 162 mg of aspirin per day. ISIS-3 also had wider entry criteria for patients including those without electrocardiographic acute injury pattern, those presenting up to 24 h from the onset of infarction and those who were deemed to have an "uncertain" indication for thrombolytic therapy by the attending physician. The major results of the ISIS-3 trial are listed in Table 2.

Thus, neither the ISIS-3 nor the GISSI-2 mortality trials were able to demonstrate superiority of one regimen over another. The results of these two trials have been interpreted to show no advantage in clinical benefit of tPA over SK despite earlier evidence in the ability of tPA to achieve higher 90 min patency. In ISIS-3, the tPA group exhibited a 1.4% total stroke rate of which 0.7% was hemorrhagic as opposed to the SK group which sustained a 1.0% total stroke rate of which 0.2% was hemorrhagic. While tPA as well as anistreplase were associated with an increased total stroke rate and stroke death compared with the SK group, the ISIS-3 investigators pointed out that the lack of mortality advantage with tPA was not accounted for by the difference in stroke deaths. Furthermore, despite an apparent advantage in favor of SK with regards to the rate of stroke, the actual numbers of stroke and, in particular, hemorrhagic stroke in all groups were low. As for adjunctive therapy, there was no sustained benefit seen in those patients randomized to receive s.c. heparin compared to no heparin in either trial.

Table 2. 35-day mortality and major clinical events in the ISIS-3 trial

Event	Adjunctive agents		Thrombolytic agent		
	Asp + H $n = 20400$	Asp $n = 20375$	tPA $n = 13569$	SK $n = 13607$	Anistrepl. $n = 13599$
Death (%)	10.3	10.3	10.3	10.6	10.5
3-month survival (%)	87.9	87.3	87.6	87.6	87.5
6-month survival (%)	86.1	86.0	85.9	86.0	86.3
Reinfarction (%)	3.2	3.5	2.9	3.5[c]	3.55
Ventricular fibrillation (%)	5.4	5.6	5.5	5.7	5.3
Cardiogenic shock	6.9	7.1	6.8	7.1	7.1
Any stroke (%)	1.28	1.18	1.39	1.04[d]	1.26
Hemorrhagic (%)*	0.56	0.40[a]	0.66[e]	0.23	0.55[f]
Transfused or other major bleeds	1.0	0.8[b]	0.8	0.9	1.0

Taken from: ISIS-3 (Third International Study of Infarct Survival) Collaborative Group (1992).
* Definitely or probably hemorrhagic.
2p (2-sided probability): Aspirin + heparin vs aspirin: [a]<0.02; [b]<0.01.
tPA vs SK: [c]<0.05; [d]<0.01; [e]<0.00001.
SK vs anistreplase: [f]<0.0001.

Several theories were advanced to account for the apparent lack of increased benefit from earlier reperfusion with tPA. Postulated mechanisms include insufficient heparinization of the group receiving tPA, differences in the pharmocodynamic effect of alteplase and duteplase, less reocclusion with SK, SK-mediated reduction of blood viscosity, the putative cerebral protective effect of hypotension induced mainly by SK, enhancement of tPA-induced but not of SK- or urokinase-mediated plasminogen activation by amyloid β-peptide deposits in the brain which might explain the higher incidence of cerebral hemorrhage during thrombolytic therapy with tPA (Kingston et al. 1995; Wnendt et al. 1997), or the lack of critical importance of re-establishing early and complete infarct related vessel patency. Certainly, some of these hypotheses appear to have little support from the totality of evidence compiled in the angiographic trials.

The experience of 9 randomized trials comprising each more than 1000 patients with AMI were evaluated in a meta-analysis by the Fibrinolytic Therapy Trialists' (FTT) Collaborative Group (1994). The major finding was that the mortality benefit from thrombolytic therapy was also seen in patients presenting with AMI and bundle branch block, in patients presenting beyond 12 h since symptom onset, in elderly patients, and in patients presenting with hypotension, tachycardia, or both. Clinically, these findings support the broadening of indications for thrombolytic therapy in the setting of AMI. The FTT investigators were also able to evaluate critically the excess early mortality and stroke, known as the "early hazard" associated with thrombolytic therapy, as well as to show a large net clinical benefit despite any excess early mortality or stroke associated with thrombolytic therapy.

Citing the correlation between efficacy and infusion rate of tPA demonstrated in earlier studies by Topol et al. (1988) and others, Neuhaus et al. (1989) described the improved thrombolytic efficacy of an "accelerated" or "front-loaded" tPA regimen achieving "a high early patency rate of the infarct-related artery without an increase in reocclusion rate and adverse reaction." This pivotal accomplishment was achieved by a 100-mg dose given within 90 min divided into an initial 15 mg bolus, a 50 mg infusion over 30 min, and a remaining 35 mg infused over 60 min. The rtPA-APSAC Patency Study (TAPS) reported by Neuhaus et al. (1992) demonstrated the superiority of "front-loaded" or "accelerated" tPA when administered with immediate and continued i.v. heparin over anistreplase given with similar adjunctive therapy. Marked improvements were found in 90-min patency and in in-hospital mortality in the tPA treatment arm. Of the 199 patients treated with the accelerated tPA regimen, 84% had achieved TIMI grade 2 or 3 flow by 90 min compared with 70% in the conventional anistreplase group ($p = 0.0007$). Indeed, by 60 min the accelerated regimen had already achieved 73% patency, a rate found only at 90 min when utilizing the conventional tPA regimen. In-hospital mortality in the accelerated tPA group was 2.4% compared with that of 8.1% for the anistreplase group ($p = 0.0095$). These findings are consistent with and support the open artery hypothesis of limiting infarct size by rapid restoration of perfusion and hence limiting mortality.

The superiority of the accelerated regimen over any other tPA regimen was confirmed that same year by the report of Wall et al. (1992) for the seventh Thrombolysis and Angioplasty in Myocardial Infarction (TAMI-7) Study Group and by the report of Carney et al. (1992) for the RAAMI Study Investigators who conducted a randomized, angiographic study directly comparing the accelerated against the conventional tPA regimen. The accelerated tPA regimen clearly represented a more aggressive and more successful protocol than either of the conventional tPA strategies employed in ISIS-3 or GISSI-2.

The s.c. heparin regimens in ISIS-3 and GISSI-2 were similar. The preference for delayed s.c. heparin reflected theoretical concerns the ISIS-3 and GISSI-2 investigators had with regards to potential bleeding complications during the thrombolytic state. The relative logistical advantages of s.c. over i.v. heparin, with its requirement for close monitoring of the APTT are also obvious. The expected APTT with the s.c. heparin regimen utilized in these trials has been stated by the ISIS-3 authors as being approximately 50 s and was believed to represent a safe but effective degree of anticoagulation.

However, many investigators became increasingly aware of the fact that sustained patency after tPA thrombolysis is best maintained when immediate and concurrent i.v. heparin is administered to achieve an APTT value ≥70 s or 1.5–2.5 times the mean laboratory standard value. This level of anticoagulation was associated with a known but small increase in the risk of bleeding complications. It is fair to conclude that tPA regimens in the ISIS-3 and GISSI-2 trials were disadvantaged by not having either timely or sufficient adjunctive anticoagulation with i.v. heparin. This represents a serious handicap for tPA strategies in these trials since tPA is known to have a far shorter half-life than SK and produces considerably less proteolysis (which has an anticoagulant effect) than SK (TRACY et al. 1997).

It was apparent at this juncture that thrombolytic therapy and aspirin both provide additive therapeutic benefit by substantially lowering the mortality following AMI. What remained undetermined was which of the available thrombolytic strategies would provide the greatest incremental benefit in a head-to-head comparison. The role of heparin among the strategies under consideration would also require definitive study.

C. The GUSTO-I Trial

I. Design of the Trial

The Global Utilization of SK and Tissue Plasminogen Activator for Occluded Coronary Arteries (GUSTO-I) trial was conceived to test the open artery hypothesis (THE GUSTO INVESTIGATORS 1993). Specifically, it was theorized that earlier, more complete and sustained infarct related artery patency and, hence, myocardial reperfusion, would result in reduced mortality. The hypothesis would be supported if the agent which accomplished the earliest sustained reperfusion, thought to be accelerated tPA, yielded the lowest mortality relative to an active treatment control arm consisting of the then accepted standard thrombolytic agent. The trial would thus compare the best tPA regimen, using a weight adjusted "Neuhaus" accelerated regimen together with immediate and sustained concurrent heparin against SK with identical heparin adjunctive therapy against a combination of tPA and SK identified as a theoretically promising alternative for lowering reocclusion by CALIFF et al. (1991) in the TAMI-5 randomized trial.

Following the announcement of the ISIS-3 trial preliminary results in March 1991 with GUSTO enrolment at 1160 patients, an SK arm with s.c. heparin was added to the trial since the ISIS-3 preliminary results suggested that s.c. heparin was superior to no adjunctive therapy with SK. This finding was later found to be incorrect by the final ISIS-3 report published more than one year later [ISIS-3 (THIRD INTERNATIONAL STUDY OF INFARCT SURVIVAL) COLLABORATIVE GROUP 1992]. Nevertheless, at the time the fourth treatment arm was added it was decided that, should there be no significant mortality difference between SK treatment arms ($p > 0.10$), the results of the two arms would be combined for purposes of analysis. However, the data for the two SK treatment groups would also be analyzed individually.

The regimens used in the four arms were:

1. SK (Kabikinase, Kabi Vitrum, Sweden) 1.5 MU i.v. over 60 min followed by sodium heparin (Sanofi, Paris) 12 500 U s.c. twice daily commencing 4 h after start of thrombolytic therapy (ISIS-3 regimen)
2. SK 1.5 MU i.v. over 60 min with heparin 5000 U i.v. bolus followed by 1000 U/hour i.v. infusion (1200 U/hour for patients >80 kg) with subsequent adjustments to maintain the APTT between 60 s and 85 s
3. Accelerated alteplase tPA (Genentech, San Francisco, CA) 15 mg i.v. bolus, 0.75 mg/kg not to exceed 50 mg infused over 30 min, 0.50 mg/kg not to exceed 35 mg infused over next 60 min with the same i.v. heparin regimen
4. The combination of i.v. tPA (1 mg/kg not to exceed 90 mg over 60 min with 10% given as initial i.v. bolus) and concurrent SK (1.0 MU i.v. over 60 min) through separate i.v. catheters with the same i.v. heparin regimen

For s.c. heparin, treatment was continued for 7 days or until discharge from hospital; i.v. heparin was given for at least 48 h or longer at the investigators' discretion and the APTT was monitored at 6 h, 12 h, 24 h for heparin dose titration. Subjects in all treatment arms received chewable aspirin (Bayer, New York, NY) ≥160 mg administered orally as soon as possible followed by 160–325 mg once daily and, if not contraindicated, atenolol (ICI Pharmaceuticals, Wilmington, DE) 5 mg i.v. divided into two doses followed by 50 to 100 mg oral once daily. All other medications and procedures including coronary angiography, PTCA and CABG were left to the discretion of the investigator.

GUSTO-I was designed as a randomized open label trial with 30-day mortality as the primary endpoint. There has been previous validation of carefully conducted open label trial designs by blinded trial designs (GISSI-1 validated by ISIS-2). Furthermore, logistical considerations including the manufacturer's refusal to supply SK and the requirement, to maintain blinding, an additional second i.v. line would have been required in 30,000 patients (the combination tPA/SK treatment arm required two separate i.v. lines) making a blinded approach untenable.

In view of the small but significant risk of stroke associated with thrombolytic therapy recognized in prior studies, GUSTO-I established a Stroke Review Committee. Blinded to treatment assignment, the Stroke Review Committee was to investigate, classify according to etiology and functional disability and follow up any and all strokes occurring in trial patients. In addition, unlike prior megatrials, all patients suffering a neurologic deficit were required to undergo computed tomography scan or magnetic resonance imaging of the brain. Accordingly, the secondary endpoints were defined as death plus nonfatal stroke or nonfatal disabling stroke or nonfatal hemorrhagic stroke. The degree of disability resulting from stroke was prospectively defined. The disability definitions were validated by quality of life interviews with disabled stroke survivors. Bleeding complications were also prospectively defined and monitored.

Adequate statistical power to detect a 15% reduction in mortality rate or an absolute decrease of 1% for either of the tPA containing treatment arms vs the active control arms employing SK strategies necessitated a sample size of 41,000 patients. This figure, calculated on the basis of an assumed alpha of 0.05 and two-tailed testing, would yield 90% power to detect differences in treatment effect if the active control arm mortality was 8% or higher. Such a sample size would still yield 80% power in the unlikely event that the active control arm mortality was lower than the 8% figure which is well below mortality rates reported in previous randomized trials of SK. Of note, both ISIS-3 and GISSI-2 were also designed to detect a 15% reduction in mortality rates.

Three subgroups were prespecified for subsequent analysis and these were subgroups stratified by age <75 or ≥75 years, by anterior or other location of infarct, and subgroups categorized by time to treatment. Most importantly, an angiographic substudy compiling mechanistic data not obtained in the previous megatrials was conducted prospectively studying the relationships between infarct related artery patency, reocclusion, ventricular function, and survival. This substudy serves as the only attempt by a large scale thrombolytic trial to provide angiographic evidence to support or refute the open artery hypothesis.

Eligibility criteria were: (1) at least 20 min of chest discomfort; (2) presentation less than 6 h after symptom onset; and (3) accompanying electrocardiographic signs of acute injury (≥0.1 mV ST elevation in at least two limb leads or ≥0.2 mV ST elevation in at least two contiguous precordial leads). Exclusion criteria were previous stroke, active bleeding, prior treatment with SK or anistreplase, recent trauma or major surgery, previous participation in GUSTO-I, and noncompressible vascular punctures. Refractory uncontrolled severe hypertension was considered a strong relative contraindication.

Standard blinded interim analysis, data management, and quality assurance procedures were employed. All patients who suffered major adverse events and a randomly selected ten percent of all data forms were audited by a contracted independent research organization. Topol et al. (1992) for the GUSTO-I Steering Committee published the most stringent efforts to date to ensure patient safety. An independent Data and Safety Monitoring Committee was charged with the responsibility of adjudicating safety data with particular attention paid to hemorrhagic stroke data interpreted within the concept of net clinical benefit and against the expected rates of hemorrhagic stroke. The committee would also monitor the incidence of stroke between treatment arms as it was foreseen that the mandated extensive imaging evaluation of any neurologic event might yield higher rates of stroke due to greater reporting. The Data Co-ordinating Centre, cataloguing adverse events via mandatory facsimile reports transmitted within 24 h of either death or discharge of every enrolled patient, reported on the rate of hemorrhagic stroke on a biweekly basis.

The GUSTO-I Steering Committee also took the unprecedented step to require written statements from participating investigators at all levels attesting to the absence of any financial conflict of interest. By unanimous vote the Steering Committee also extended the then accepted requirements of investigators to include banning the acceptance of any honoraria or reimbursement for travel expenses. It also extended the period of potential conflict of interest to one year after formal publication of the trial's primary results. With the exception of two biostatisticians at the Data Co-ordinating Centre, all other investigators and sponsors of GUSTO-I remained blinded to study data and were prohibited from access to data until the prescribed analysis was complete.

II. Primary Results of GUSTO-I

Between December 27, 1990 and February 22, 1993, 41,021 patients were enrolled in the main trial. Of those, 2431 were enrolled in the Angiographic Substudy. A total of 1081 hospitals in 15 countries participated. The four treatment groups differed slightly by the actual number enrolled due to the delayed addition of the SK/s.c. heparin arm. As might be expected in a randomized study of the size of GUSTO-I, there were no significant differences in baseline characteristics between the four treatment groups. In general, patients were on average 61–62 years of age (52–70), 25% were female, time to treatment was 164–165 min (average time to treatment was 5 min longer for the combination therapy arm due to the additional time required for preparation of the two thrombolytic agents), and 40% had anterior location of myocardial infarction. In all four arms 17% had experienced a prior infarct, 5% had prior CABG, 2% were Killip class III, and 1% were Killip class IV. There were similarly no significant differences with respect to history of either hypertension, diabetes, or tobacco use.

A high degree of medication compliance to the protocol was achieved in all areas including the use of s.c. heparin. Though not previously utilized in the United States, the overall rate of compliance with s.c. heparin was 89%; lack of compliance was 14% in the United States and 7% in other countries. Of patients in the SK plus s.c. heparin arm, 36% received i.v. heparin during the hospitalization, the principal reasons advocated being for the treatment of recurrent ischemia or for cardiac catheterization. It is noteworthy that s.c. heparin yielded no benefit in terms of 35-day mortality as reported in the ISIS-3 final report. Moreover, had the final results of ISIS-3 been known earlier, no treatment arm allowing for s.c. heparin would have been added to the GUSTO-I design.

Analysis of the GUSTO-I primary end point (see Table 3) demonstrated that accelerated tPA with i.v. heparin resulted in a statistically significant reduction of the 30-day mortality of 15% (95% confidence interval 5.9–21.3, $p = 0.001$). In practical terms, this translates into 10 lives saved per 1000 patients treated or the prevention of 1 out of 7 deaths that would be expected from the active control thrombolytic therapies. The mortality advantage of accelerated tPA is now known to be sustained at 1 year follow-up (CALIFF et al. 1996).

Table 3. GUSTO-I primary endpoint and outcome

Outcome	SK + SQH n = 9796	SK + IVH n = 10377	Accelerated tPA + IVH n = 10344	Combined Therapy n = 10328	p Value (accelerated tPA vs SK)
24-h mortality (%)	2.8	2.9	2.3	2.8	0.005
30-day mortality (%)	7.2	7.4	6.3	7.0	0.001
Or nonfatal stroke (%)	7.9	8.2	7.2	7.9	0.006
Or nonfatal disabling stroke (%)	7.7	7.9	6.9	7.6	0.006
Or nonfatal hemorrhagic stroke (%)	7.4	7.6	6.6	7.4	0.004

IVH, i.v. heparin; SQH, s.c. heparin; tPA, tissue plasminogen activator.
Adapted from The Gusto Investigators (1993).

Analysis of the secondary endpoints of net clinical benefit also yielded highly significant differences favoring accelerated tPA with i.v. heparin over combination therapy or SK strategies. Furthermore, these benefits were sustained when comparisons were made to each of the SK arms separately or to the SK treatment arms taken together. In terms of 24-h mortality rates, accelerated tPA yielded a 19.2% ($p = 0.005$) reduction over the SK treatment arms (Table 3). With respect to the combined secondary endpoints of 30-day mortality or nonfatal stroke, nonfatal hemorrhagic stroke, or nonfatal disabling stroke rates, accelerated tPA yielded a 11.2% ($p = 0.006$), 12% ($p = 0.004$), and 11.5% ($p = 0.006$) reduction respectively over the SK treatment arms. Aylward et al. (1996) have performed a thorough analysis on the effect of hypertension on mortality due to hemorrhagic stroke in the GUSTO-I patients. Systolic blood pressure >175 mm Hg was found in 1229 patients; in these the rate of intracranial hemorrhage was 1.71% compared to the overall rate of 0.7% ($p < 0.0001$).

Analysis of prespecified subgroups deserves special mention as much confusion has arisen over the meaning of subgroup analysis. It is important to remember that subgroups are often underpowered to draw any definitive conclusions. The value in subgroup analysis lies in the assessment of directionality and consistency of results when compared to the primary analysis of the full, adequately powered cohort. The isolation and piecemeal consideration of a specific subgroup from the main body of the trial for the purpose of finding statistically discrepant results is at best a limited "ice-pick" or univariate view. As such, it is prone to spurious results with the attendant loss of statistical power in small subgroups. Any lack of consistency and direction of effect across subgroups may well justify concern. The subgroup analysis is otherwise confirmatory or neutral.

The age subgroup analysis was consistent between both subgroups and the total GUSTO-I population (Table 4). The advantage in mortality reduction in those younger than 65, in those 65–74 and in those 75–85 years of age who were treated with the accelerated tPA regimen was sustained at one year

Table 4. GUSTO-I 30-day mortality rates with odds ratios and 95% confidence intervals (CI) in prespecified subgroups by age, infarct location and time to treatment

Subgroups	Patients (%)	Mortality rate (%)		Odds Ratio and 95% CI
		SK	tPA	
Age (years)				
<75	88	5.5	4.4	
≥75	12	20.6	19.3	
Infarct Location				
Anterior	39	10.5	8.6	
Other	61	5.3	4.7	
Time to Treatment (h)				
0–2	27	5.5	4.6	
>2–4	51	6.5	5.6	
>4–6	19	9.6	8.7	
>6	4	8.9	9.1	
				0.5 1.0 1.5

Abbreviations as in Table 1. Adapted from: THE GUSTO INVESTIGATORS (1993) with corrections from TOPOL et al. (1994).

follow-up as reported by HOLMES et al. (1995b). Only in the small underpowered subgroup of patients over 85 years of age was the relative mortality reduction between thrombolytic agents unresolved. Clearly, the advantage of early, complete and sustained reperfusion in ameliorating the high mortality risk of myocardial infarction in the elderly proportionally outweighs any risk associated with thrombolytic therapy in the aged. Those at the greatest risk stood to benefit the most. CALIFF et al. (1997) developed a logistic regression model from the GUSTO-I patient data and arrived at similar conclusions. Patients with AMI who had more high risk characteristics, such as advanced age, anterior infarction, higher classification (except Killip class IV), lower blood pressure, and an increased heart rate had the greatest absolute benefit with accelerated tPA vs SK.

Similar findings pertain to the subgroup classified by infarct location. Anterior infarction patients clearly benefited from accelerated tPA with mortality rates of 8.6% compared with the SK group mortality of 10.5%. Those with nonanterior infarct locations also benefited more from tPA than from SK though, due to its lower mortality rate, this subgroup had insufficient statistical power with the odds ratio confidence interval (CI) crossing the line of unity. Nonetheless, the direction of benefit for this subgroup is consistent with that of the total GUSTO-I population.

With respect to time to treatment, the direction of effect was consistent for all subgroups. In the small underpowered subgroup with treatment initiated after 6h, there was a small late patency benefit which still justifies thrombolytic therapy up to 12h. The time to treatment data are also noteworthy with respect to the marked difference in mortality rates between the subgroups. It was lowest for those treated at 0–2 h (tPA: 4.6%, SK: 5.5%), intermediate for those treated at 2–4 h (tPA: 5.6%, SK: 6.5%), and highest for those treated at 4–6 h (tPA: 8.7%, SK: 9.6%). The observation that earlier restoration of perfusion results in lower mortality should be of no great surprise since it was predicted on the physiologic principles enunciated in the open artery hypothesis (see Chap. 7). The link between earlier reperfusion, probably resulting in preserved left ventricular function, and lower mortality, would be subsequently demonstrated by the Angiographic Substudy.

III. The GUSTO-I Angiographic Substudy

The Angiographic Substudy (THE GUSTO-I ANGIOGRAPHIC INVESTIGATORS 1993) provides a unique mechanistic view accounting for the observed relative mortality benefits of accelerated tPA over other thrombolytic regimens. Among the 2431 patients randomized between the 75 angiographic centers, 1210 were randomized to 90 min angiography with the remainder nearly evenly divided between 3 h, 24 h, and 7 day angiography. The cineangiograms were then reviewed and classified by TIMI grade at a blinded core angiography laboratory.

The Angiographic Substudy demonstrated a clear superiority of accelerated tPA over SK in achieving 90 min TIMI grade 3 flow. TIMI grade 3 flow was accomplished in 54% in the tPA group, 38% in the combined treatment group, 32% in the SK and i.v. heparin group, and 29% in the SK and s.c. heparin group (comparisons between the tPA group and each of the other three regimens $p < 0.001$). Rates of reocclusion were low and not significantly different between treatment groups. Left ventricular function at 90 min as measured by preserved regional wall motion also demonstrated benefit in the accelerated tPA treatment group (29% of the group) when compared to the SK treatment groups (18.5%; $p < 0.001$) and to the combination therapy group (21%; $p = 0.035$). A similar benefit was found in the accelerated tPA and the SK treatment groups with respect to the number of abnormal segment chords. Differences in indices of left ventricular function at 90 min and at 5–7 days, when stratified by TIMI grade flow were even more striking and significant. Statistically significant benefit as measured by ejection fraction, end-systolic volume index, wall motion score, number of abnormal chords, and preserved regional wall motion, were found for the TIMI grade 3 patients compared to TIMI grade 0, 1, 2 patients at 90 min and at 5–7 days (Table 5). Preservation of regional wall motion was found in 31% of the TIMI grade 3 group at 90 min compared to 11%, 17%, and 19% at 90 min for the TIMI grade 0, 1, 2 groups respectively ($p < 0.001$ for all comparisons). The benefit in preserved regional wall motion was sustained at the 5–7 day evaluation in 18%, 22%, 27%, and 39% of the TIMI grade 0, 1, 2, 3 groups respectively ($p < 0.001$ for all comparisons). REINER et al. (1996) using multivariate regression analysis found TIMI flow to be the only significant determinants of global and regional left

Table 5. The effect of TIMI grade patency on ventricular function at 90 min and 5–7 day follow-up

LV Functional Indices	TIMI 0	TIMI 1	TIMI 2	TIMI 3[c]
At 90 min	$n = 233$	$n = 84$	$n = 275$	$n = 370$
Ejection Fraction (%)	55 ± 15	55 ± 15	56 ± 15	62 ± 14
End Systolic Volume Index (ml/m^2)	31 ± 17	33 ± 21	29 ± 14	26 ± 14
Wall motion (SD/chord)	-2.8 ± 1.3	-2.7 ± 1.4	-2.6 ± 1.4	-2.2 ± 1.5
Number abnormal chords	26 ± 17	26 ± 19	27 ± 19	18 ± 17
Preserved regional wall motion (% of group)	11	17	19[a]	31
At 5–7 Days	$n = 171$	$n = 63$	$n = 212$	$n = 284$
Ejection Fraction (%)	56 ± 14	54 ± 12	56 ± 14	61 ± 14
End Systolic Volume Index (ml/m^2)	32 ± 16	34 ± 13	30 ± 13	26 ± 14
Wall motion (SD/chord)	-2.5 ± 1.2	-2.7 ± 1.2	-2.3 ± 1.4	-1.8 ± 1.7
Number abnormal chords	23 ± 18	25 ± 19	22 ± 18	15 ± 16
Preserved regional wall motion (% of group)	18	22	27[b]	39[d]

± values give one standard deviation from the respective mean values. Wall motion is expressed as the mean magnitude of depressed infarct zone chords. Wall motion was considered preserved if infarct zone chords were normal. Chords in the infarct zone were considered abnormal if more than two standard deviations below normal. Adapted from THE GUSTO ANGIOGRAPHIC INVESTIGATORS (1993).
[a] $p = 0.026$ for the comparison of this group with the groups with TIMI 0, 1.
[b] $p = 0.034$ for the comparison of this group with the groups with TIMI 0, 1.
[c] $p < 0.001$ for the comparison of this group with the groups with TIMI 0, 1, 2 except where otherwise noted.
[d] $p = 0.007$ for the comparison of this group with the group with TIMI 2.

ventricular function assessed at 5–7 days after thrombolytic therapy. These findings relating left ventricular function to TIMI grade flow dramatically underscore the link between early TIMI grade 3 patency and mortality reduction as predicted by the open artery hypothesis. The observations of early TIMI 3 flow obtained by the accelerated tPA regimen has been confirmed in the pooled analysis of 5474 angiographic observations from 15 studies (BARBAGELATA et al. 1997; see also Chap. 7, Table 1).

Significantly, KLEIMAN et al. (1994) in studying early 24-h mortality had noted differences based on TIMI grade flow with TIMI 0 or 1 associated mortality of 2.35%, TIMI 2 with 2.92% and TIMI 3 with 0.89%. This observation helped to account for the 19% reduction in 24-h mortality rate observed in the accelerated tPA treatment group compared to the SK treatment groups. Also, as can be seen from the above TIMI grade stratified 24-h mortality, TIMI grade 2 flow, unlike TIMI grade 3 flow, did not confer any mortality advantage when compared against TIMI grade 0 and 1. In light of the concerns of LINCOFF and TOPOL (1993) regarding the apparent dissociation between angiographic "patency," left ventricular function, and subsequent mortality, these findings strongly suggest that TIMI grade 2 flow, heretofore considered to represent "patency," is inadequate reperfusion failing to limit infarct size or reduce mortality and hence accounts for the apparent dissociation between 90 min angiographic patency and clinical outcome.

SIMES et al. (1995) reported that TIMI grade 3 flow, regardless of the treatment received, was associated with the lowest 30-day mortality rates (4.0% for TIMI grade 3 vs 7.9% for TIMI grade 2 vs 9.2% for TIMI grade 1 and vs 8.4% for TIMI grade 0, $p < 0.01$). Moreover, the ability of TIMI grade 3 flow to serve as a significant independent predictor of 30-day mortality was demonstrated after logistic regression analysis and multivariate regression analysis adjusting for baseline clinical variables ($p = 0.015$, odds ratio 0.46, 95% CI, 0.25–0.86). Most importantly, in order to test whether the observed mortality rates in the four treatment arms of the total GUSTO-I population could be reliably explained by differences in 90 min TIMI grade patency observed in the Angiographic Substudy population, these investigators undertook a study utilizing mathematical modeling methods. Predictions of mortality rates in the main GUSTO-I population based on models constructed from 90 min TIMI stratified patency data in the four treatment arms of the Angiographic Substudy were closely correlated with the actual observed results in the main study population. The degree of correlation between predicted and observed mortality was high ($r = 0.97$) with the proportion of squared error explained (R^2) equal to 0.92. This suggests that 92% of the variation in mortality rates between the treatment arms could be explained on the basis of differences in TIMI grade flow at 90 min. There were no significant differences in baseline clinical characteristics between the main trial population and the Angiographic Substudy patients.

KLEIMAN et al. (1994) also prospectively examined the 24-h mortality rates in GUSTO-I in response to the previously observed apparent paradoxically increased risk of death within the first 24 h in patients treated with thrombolytic agents. Utilizing multiple regression analysis it was determined that the most significant predictors of early death were hypotension, tachycardia, shorter height, and absence of prior smoking. By 6 h no differences in early mortality were evident between thrombolytic regimens; however, by 24 h mortality for the SK group was 2.89%, for the combination group 2.84%, and for the tPA group 2.36% ($p = 0.005$). Thus, the observed increased 24-h mortality after thrombolysis is more a function of initial poor hemodynamic presentation due to acute left ventricular failure, and a thrombolytic strategy aimed at early, complete, and sustained restoration of coronary patency rather than paradoxically increasing mortality, actually protects against early death.

In concert with the findings of factors predicting 24-h mortality, LEE et al. (1995) identified five factors influencing 30-day mortality. Age, together with lower systolic blood pressure, higher Killip class, elevated heart rate, and anterior infarction accounted for 90% of the prognostic data found in the baseline characteristics. Taken together, these findings, indicative of mortality associated with the hemodynamic consequences of acute left ventricular failure, support strategies aimed at the preservation of left ventricular function through earlier, complete and sustained restoration of patency.

The long-term follow-up data presented recently (Ross et al. 1998) clearly demonstrate a marked difference in survival rates between TIMI grades 0, 1, 2, and TIMI grade 3 which become even more pronounced over time (Fig. 1). TIMI grade 3 survival rates, compared against the survival rates of all other TIMI grade subgroups, demonstrate an initial 3 lives saved per 100 treated at 30 days with an additional 5 lives saved per 100 from 30 days to 2 years. These data reflect the reperfusion advantage achieved in all patients with TIMI grade 3 flow and the uniformly worse mortality rates seen over time in all other TIMI grade subgroups. It would be reasonable to expect that the Kaplan-Meier survival curves of TIMI 3 vs TIMI 2 and TIMI 0 and 1 will continue to diverge for some time beyond the present length of follow-up indicating sustained mortality benefit though evidence to support this must await the 5 year follow up data. Two-year survival curves for the patients stratified by last in-hospital left ventricular ejection fraction demonstrated that 30-day deaths occurred in 3.1% of those whose ejection fraction was >40% and in 12% of those with more severely depressed function. The overall 2-year mortality in patients with ejection fractions >40% was 7.2% vs 27% in those with ejection fractions ≤40%. Between 30 days and 2 years after AMI the corresponding mortality rates were 4.3% and 16% respectively. Thus the survival advantage for the preserved ventricular function group was approximately 20%. This lends support to what has long been clinically suspected; that not only is there a

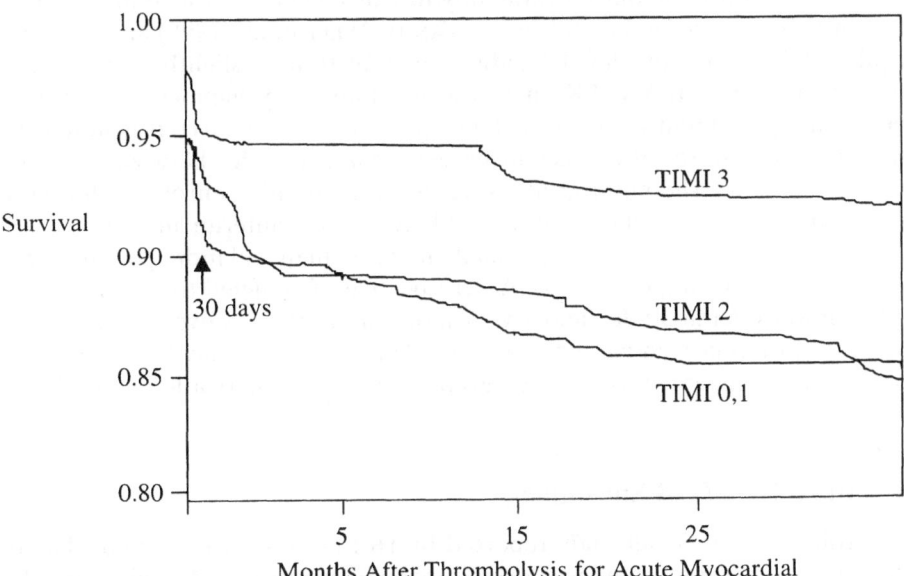

Fig. 1. GUSTO-I survival rates stratified by TIMI grade patency. Adapted from Ross et al. (1998)

benefit in mortality reduction but also a benefit in functional status related to preservation of left ventricular function.

IV. Further Analysis of GUSTO-I Results

Several non-prespecified subgroup analyses have been conducted in response to concerns regarding possible confounding effects of nonprotocol interventions such as PTCA or CABG surgery. In patients with cardiogenic shock, Berger et al. (1997) found that an aggressive invasive catheterization and revascularization strategy (mainly PTCA) was associated with a reduced 30-day mortality (odds ratio 0.43, $p < 0.0001$). Likewise Anderson et al. (1997) demonstrated that, in patients with cardiogenic shock, the early use of intraaortic balloon counterpulsation was associated with a trend toward lower 30-day and 1-year mortality, albeit at the risk of bleeding and other adverse events. On the other hand, Tardiff et al. (1997) reported that CABG surgery in patients enrolled in GUSTO-I after thrombolytic therapy entailed a higher rather than lower mortality risk. Nine percent of patients treated with tPA and 8.3% of those treated with SK underwent CABG during the initial hospital period. Multivariate analysis revealed that CABG surgery was an independent predictor of 30-day mortality (RR 1.87) and a weaker predictor of 1-year mortality (RR 1.21). The authors concluded that the survival benefit of accelerated tPA was not related to CABG and that the short-term mortality associated with CABG may be balanced by anticipated long-term benefit in specific groups of patients.

Analysis for intercountry variability has also been performed in response to controversy surrounding this issue. Van de Werf et al. (1995) and Holmes et al. (1997) found that while US patients demonstrated a slightly greater mortality benefit from tPA vs SK, there was no statistically significant difference between US and non-US subjects (chi-square 1.06, $p = 0.304$). It is noteworthy that there were differences in baseline characteristics between US and non-US patients. The latter group were on average older, enrolled later with more advanced Killip class, and more likely to have anterior infarctions. US centers tended to include proportionally more women and more patients who had prior CABG surgery or PTCA. Irrespective of assigned treatment, there was a lower overall 30-day and 1-year mortality in the US even after adjustment for differences in baseline variables. This difference may be a reflection of the more aggressive use of angiographic and revascularization procedures in the US.

V. Other GUSTO-I Substudies

A cardiogenic shock substudy reported by Holmes et al. (1995a) concluded that patients who presented in shock as opposed to patients who subsequently developed shock did equally poorly with 57% mortality and 55% mortality respectively. However, accelerated tPA treated patients were significantly less likely to develop shock than either SK or combination therapy treated patients. The rate of shock developing in accelerated tPA treated patients was 5.5% compared to 7.4% in the SK/s.c. heparin arm, 6.9% in the SK/i.v. heparin arm, and 6.3% in the combined therapy arm with $p < 0.001$ for all comparisons. While patients presenting in shock exhibited lower mortality with SK than tPA (probably due to the perfusion independent manner in which SK restores patency), the only significant factor improving outcome in shock patients, regardless of thrombolytic treatment, was PTCA.

HASDAI et al. (1999) analyzed demographic, clinical, and hemodynamic characteristics in the 2968 patients with cardiogenic shock. Using logistic regression modeling techniques they found that the odds of dying were 1.49 times higher for patients 10 years older and 1.7 times higher for patients with prior infarction. Findings derived from physical examination, such as altered sensorium and cold, clammy skin were independent predictors of death (odds 1.68 times higher for each). In patients with oliguria the odds were 2.25 times higher. Mortality rate was lowest for cardiac output of 5.1 l/min and pulmonary capillary wedge pressure of 20 mm Hg and increased with either higher or lower values. Patients with shock on admission had better outcomes than those in whom shock developed later. In a follow-up study HASDAI et al. (2000) refined their analysis of clinical factors predictive of cardiogenic shock. The four major factors, namely age (hazard ratio 1.47), systolic blood pressure, heart rate, and Killip class (1.7 for Killip class II versus I, and 2.95 for Killip class III versus I) accounted for >85% of the predictive information. On the basis of these analyses a prognostic algorithm was developed permitting prediction of outcome in patients with cardiogenic shock undergoing thrombolytic therapy.

The subject of diabetes and thrombolytic therapy of AMI in the GUSTO-I trial has been addressed by several investigators (MAK and TOPOL 2000). Mortality in patients with diabetes and AMI is approximately twice that of nondiabetic patients. In the GUSTO Angiographic Trial 12.8% of the 2431 patients had diabetes (WOODFIELD et al. 1996). This cohort had a significantly higher proportion of female and elderly patients and they were more often hypertensive, came to the hospital later, had more congestive heart failure, and a higher number of previous AMI and CABG. There were no differences in TIMI flow grade 3 and of ejection fractions 90 min after starting thrombolytic therapy between diabetics and nondiabetics, but reocclusion rates were higher in diabetics (n.s.). Diabetic patients had less compensatory hyperkinesia in the noninfarct zone ($p < 0.01$). The 30-day mortality was 11.3% in diabetic vs 5.9% in nondiabetic patients ($p < 0.0001$). After adjustment for clinical and angiographic variables, diabetes remained an independent determinant of 30-day mortality ($p = 0.02$). A similar analysis was conducted in the 41,021 patients enrolled in the GUSTO-I trial by MAK et al. (1997). In the 5044 patients with diabetes mortality at 1-year follow-up was 14.5% vs 8.9% in the nondiabetic patients. Diabetic retinopathy had previously been identified as a contraindication to thrombolytic therapy without clear evidence that these patients have an increased risk of ocular hemorrhage. MAHAFFEY et al. (1997) in reviewing all cases in the GUSTO-I trial found only 12 patients with intraocular hemorrhage, but only one nondiabetic patient with intraocular hemorrhage. The authors conclude that diabetic retinopathy should not be considered a contraindication to thrombolysis.

The stroke substudy reported by GORE et al. (1995) revealed a 1.4% incidence of stroke in GUSTO-I patients, the majority of which were non-hemorrhagic in all treatment arms. As expected, the total stroke rate was lowest in the SK arms, next lowest in the accelerated tPA arm, and the highest in the combined treatment arm. Unfortunately, but again not unexpectedly, 41% of all strokes were fatal, 31% were disabling. and 60% of hemorrhagic strokes proved fatal during the hospitalization. Multiple regression analysis identified advanced age, lower body weight, prior cerebrovascular disease, hypertension, treatment with tPA, and an interaction between hypertension and age as the significant predictors of intracranial hemorrhage. Nonetheless, it was clear that the net clinical benefit of accelerated tPA vs SK was maintained across all subgroups reflecting the fact that the small total risk of stroke and the even smaller risk of intracranial hemorrhage was far outweighed by the reduction in mortality achieved by accelerated tPA. MAHAFFEY et al. (1998) performed an in-depth analysis of the risk factors for nonhemorrhagic stroke in the GUSTO-I patients. Of the 247 patients who developed this adverse event, 17% died and another 40% were disabled by the 30-day follow-up. The most important predictors were older age, followed by higher heart rate, history of stroke or transient ischemic attacks, diabetes, previous angina, and history of hypertension. The authors developed a simple nomogram that can predict the risk of nonhemorrhagic stroke on the basis of baseline characteristics.

Parenthetically, THE NATIONAL INSTITUTE OF NEUROLOGICAL DISORDERS AND STROKE rTPA STROKE STUDY GROUP (1995) reported on the successful use of tPA for the treatment of ischemic stroke in a randomized, double blind, placebo controlled trial of 624 patients. Alteplase tPA was dosed at 0.9 mg per kg not to exceed 90 mg with 10% given as initial bolus and the remaining 90% infused over 60 min. When administered within 3 h of ischemic stroke onset, tPA resulted in a net clinical benefit despite an increased incidence of intracranial hemorrhage. The benefit, measured in terms of neurological function, was not at the expense of any additional mortality.

In response to the heightened concerns related to health care expenditure and the implications for cost in treating AMI (CALIFF et al. 1999) with an agent as expensive as tPA, MARK et al. (1995) reported a cost effectiveness study. If it is assumed that tPA is substituted for SK in the approximately 250,000 eligible myocardial infarction patients seen each year, it has been estimated that the cost would be approximately $500 million per year yielding 3.5 million patient years saved. Cost effectiveness ratios when calculated across the spectrum of patient variables show tPA to be cost effective when compared to a host of customarily accepted expensive medical therapies. tPA therapy cost effectiveness ratios were least effective when tPA was employed in the treatment of patients under the age of 40 years and in patients under 60 years treated for inferior infarction. In the US, as of 1996, nearly 80% of patients with AMI receiving thrombolytic therapy receive accelerated tPA.

VI. Epilogue

CANNON et al. (1994) reported the double blind TIMI-4 trial which compared accelerated tPA against anistreplase and a combination treatment arm. As in GUSTO-I, the TIMI-4 patients also received aspirin and immediate i.v. heparin. Both angiographic data at 90 min and 1 year mortality mirror the findings of GUSTO-I thus providing confirmatory evidence of the superior performance of accelerated tPA and of its consequent mortality benefit.

The impact of GUSTO-I and its Angiographic Substudy is clear. The treatment of AMI with accelerated tPA and i.v. heparin results in complete (TIMI grade 3), sustained reperfusion resulting in preservation of left ventricular function, and consequent reduction in mortality. The crucial importance of the rapid restoration of TIMI grade 3 blood flow is powerfully demonstrated in the GUSTO-I Angiographic Substudy. These conclusions, in confirmation of the postulated mechanism of benefit stated in the open artery hypothesis, is perhaps the most compelling, important, and enduring contribution of GUSTO-I. Finally, by any measure, the utilization of accelerated tPA, in preventing or avoiding one of every seven deaths due to AMI, is a major therapeutic advance among all efforts heretofore directed at the treatment of the leading cause of death in Western society.

List of Abbreviations and Acronyms

AMI	Acute myocardial infarction
APSAC	Anisoylated Plasminogen SK Activator Complex
APTT	Activated Partial Thromboplastin Time
CABG	Coronary Artery Bypass Graft
CI	Confidence Interval
ECG	Electrocardiogram
ECSG	European Collaborative Study Group
FDPs	Fibrin(ogen) Degradation Products
GISSI	Gruppo Italiano per lo Studio della Streptochinasi nell'Infarto miocardico
GUSTO	Global Utilization of SK and Tissue plasminogen activator for Occluded coronary arteries
ISIS	International Study of Infarct Survival
i.v.	intravenous
MBCK	MB isoform of creatine kinase
MU	Million Units
NIH	National Institute of Health (U.S.A.)
PAIMS	Plasminogen Activator Italian Multicenter Study
PTCA	Percutaneous Transluminal Coronary Angioplasty
RAAMI	Rapid Administration of Alteplase in Myocardial Infarction
RR	relative risk

s.c.	subcutaneous
SK	Streptokinase
TAMI	Thrombolysis and Angioplasty in Myocardial Infarction
TAPS	TPA-APSAC Patency Study
TICO	Thrombolysis In acute Coronary Occlusion
TIMI	Thrombolysis In Myocardial Infarction
tPA	tissue-type Plasminogen Activator

References

Anderson RD, Ohman EM, Holmes DR, Jr, Col I, Stebbins AL, Bates ER, Stomel RJ, Granger CB, Topol EJ, Califf RM, for the GUSTO-1 investigators (1997) Use of intraaortic balloon counterpulsation in patients presenting with cardiogenic shock: observations from the GUSTO-I Study. Global Utilization of Streptokinase and TPA for Occluded Coronary Arteries. J Am Coll Cardiol 30:708–715

Arnout J, Simoons M, de Bono D, Rapold HJ, Collen D, Verstraete M (1992) Correlation between level of heparinization and patency of the infarct-related coronary artery after treatment of acute myocardial infarction with alteplase (rt-PA). J Am Coll Cardiol 20:513–519

Aylward PE, Wilcox RG, Horgan JH, White HD, Granger CB, Califf RM, Topol EJ, for the GUSTO-I Investigators (1996) Relation of increased arterial blood pressure to mortality and stroke in the context of contemporary thrombolytic therapy for acute myocardial infarction. A randomized trial. Ann Intern Med 125:891–900

Barbagelata NA, Granger CB, Oqueli E, Suárez LD, Borruel M, Topol EJ, Califf RM (1997) TIMI grade 3 flow and reocclusion after intravenous thrombolytic therapy: a pooled analysis. Am Heart J 133:273–282

Berger PB, Holmes DR, Jr, Stebbins AL, Bates ER, Califf RM, Topol EJ, for the GUSTO-1 investigators (1997) Impact of an aggressive invasive catheterization and revascularization strategy on mortality in patients with cardiogenic shock in the Global Utilization of Streptokinase and Tissue Plasminogen Activator for Occluded Coronary Arteries (GUSTO-I) trial. An observational study. Circulation 96:122–127

Bleich SD, Nichols TC, Schumacher RR, Cook DH, Tate DA, Teichman SL (1990) Effect of heparin on coronary arterial patency after thrombolysis with tissue

Bruce CJ, Christian TF, Schaer GL, Spaccavento LJ, Jolly MK, O'Connor MK, Gibbons RJ (1999) Determinants of infarct size after thrombolytic treatment in acute myocardial infarction. Am J Cardiol 83:1600–1605

Califf RM, Stump D, Topol EJ, Mark DB (1999) Economic and cost-effectiveness in evaluating the value of cardiovascular therapies. The impact of the cost-effectiveness study of GUSTO-1 on decision making with regard to fibrinolytic therapy. Am Heart J 137 [Suppl S]:S90–S93

Califf RM, Topol EJ, Stack RS, Ellis SG, George BS, Kereiakes DJ, Samaha JK, Worley SJ, Anderson JL, Harrelson-Woodlief L, Wall TC, Phillips III HR, Abbottsmith CW, Candela RJ, Flanagan WH, Sasahara AA, Mantell SJ, Lee KL, for the TAMI Study Group (1991) Evaluation of combination thrombolytic therapy and timing of cardiac catheterization in acute myocardial infarction. Results of thrombolysis and angioplasty in myocardial infarction – phase 5 randomized trial. Circulation 83:1543–1556

Califf RM, White HD, Van de Werf EV, Sadowski Z, Armstrong PW, Vahanian A, Simoons ML, Simes RJ, Lee KL, Topol EJ, for the GUSTO-I Investigators (1996) One-year results from the global utilization of streptokinase and TPA for occluded coronary arteries (GUSTO-I) trial. Circulation 94:1233–1238

Califf RM, Woodlief LH, Harrell FE, Jr, Lee KL, White HD, Guerci A, Barbash GI, Simes RJ, Weaver WD, Simoons ML, Topol EJ, for the GUSTO-I Investigators (1997) Selection of thrombolytic therapy for individual patients: development of a clinical model. Am Heart J 133:630–639

Cannon CP, McCabe CH, Diver DJ, Herson S, Greene RM, Shah PK, Sequeira RF, Leya F, Kirshenbaum JM, Magorien RD, Palmeri ST, Davis V, Gibson CM, Poole WK, Braunwald E, for the TIMI 4 Investigators (1994) Comparison of front-loaded recombinant tissue-type plasminogen activator, anistreplase and combination thrombolytic therapy for acute myocardial infarction: Results of the thrombolysis in myocardial infarction (TIMI) 4 trial. J Am Coll Cardiol 24:1602–1610

Carney RJ, Murphy GA, Brandt TR, Daley PJ, Pickering E, White HJ, McDonough TJ, Vermilya SK, Teichman SL, for the RAAMI Study Investigator (1992) Randomized angiographic trial of recombinant tissue-type plasminogen activator (alteplase) in myocardial infarction. J Am Coll Cardiol 20:17–23

Chesebro JH, Knatterud G, Roberts R, Borer J, Cohen LS, Dalen J, Dodge HT, Francis CK, Hillis D, Ludbrook P, Markis JE, Mueller H, Passamani ER, Powers ER, Rao AK, Robertson T, Ross A, Ryan TJ, Sobel BE, Willerson J, Williams DO, Zaret BL, Braunwald E (1987) Thrombolysis in myocardial infarction (TIMI) trial, phase I: a comparison between intravenous tissue plasminogen activator and intravenous streptokinase. Clinical findings through hospital discharge. Circulation 76:142–154

Dalen JE, Gore JM, Braunwald E, Borer J, Goldberg RJ, Passamani ER, Forman S, Knatterud G, and the TIMI Investigators (1988) Six- and twelve-month follow-up of the phase I Thrombolysis in Myocardial Infarction (TIMI) trial. Am J Cardiol 62:179–185

de Bono DP, Simoons ML, Tijssen J, Arnold AER, Betriu A, Burgersdijk C, Bescos LL, Mueller E, Pfisterer M, Van de Werf F, Zijlastra F, Verstraete M (1992) Effect of early intravenous heparin on coronary patency, infarct size, and bleeding complications after alteplase thrombolysis: results of a randomised double blind European Cooperative Study Group trial (ECSG-6). Br Heart J 67:122–128

Fibrinolytic Therapy Trialists' (FTT) Collaborative Group (1994) Indications for fibrinolytic therapy in suspected acute myocardial infarction: Collaborative overview of early mortality and major morbidity results from all randomised trials of more than 1000 patients. Lancet 343:311–322

Gore JM, Granger CB, Simoons ML, Sloan MA, Weaver WD, White HD, Barbash GI, Van de Werf F, Aylward PE, Topol EJ, Califf RM, for the GUSTO Investigators (1995) Stroke after thrombolysis. Mortality and functional outcomes in the GUSTO-trial. Circulation 92:2811–2818

Grines CL, DeMaria AN (1990) Optimal utilization of thrombolytic therapy for acute myocardial infarction: concepts and controversies. J Am Coll Cardiol 16:223–231

Gruppo Italiano per lo Studio della Sopravvivenza nell'Infarto Miocardico (GISSI) (1990) GISSI-2: A factorial randomised trial of alteplase versus streptokinase and heparin versus no heparin among 12 490 patients with acute myocardial infarction. Lancet 336:65–71

Hasdai D, Califf RM, Thompson TD, Hochman JS, Ohman EM, Pfisterer M, Bates ER, Vahanian A, Armstrong PW, Criger DA, Topol EJ, Holmes DR Jr. (2000) Predictors of cardiogenic shock after thrombolytic therapy for acute myocardial infarction. J Am Coll Cardiol 35:136–143

Hasdai D, Holmes DR Jr, Califf RM, Thompson TD, Hochman JS, Pfisterer M, Topol EJ, for the GUSTO-I Investigators (1999) Cardiogenic shock complicating acute myocardial infarction: predictors of death. Am Heart J 138:21–31

Holmes DR Jr, Bates ER, Kleiman NS, Sadowski Z, Horgan JH, Morris DC, Berger PB, Topol EJ, for the GUSTO-I Investigators (1995a) Contemporary reperfusion therapy for cardiogenic shock: the GUSTO-I trial experience. J Am Coll Cardiol 26:668–674

Holmes DR Jr, Califf RM, Topol EJ (1995b) Lessons we have learned from the GUSTO trial. J Am Coll Cardiol 25 [Suppl]:10S–17S

Holmes DR Jr, Califf RM, Van de Werf F, Berger PB, Bates ER, Simoons MLW, White HD, Thompson TD, Topol EJ, for the GUSTO-I Investigators (1997) Difference in countries' use of resources and clinical outcome for patients with cardiogenic shock after myocardial infarction: results from the GUSTO trial. Lancet 349:75–78

Hsia J, Hamilton WP, Kleiman N, Roberts R, Chaitman BR, Ross AM, for the Heparin-Aspirin Reperfusion Trial (HART) Investigators (1990) A comparison between heparin and low-dose aspirin as adjunctive therapy with tissue plasminogen activator for acute myocardial infarction. N Engl J Med 323:1433–1437

Hsia J, Kleiman N, Aguirre F, Chaitman BR, Roberts R, Ross AM, for the Heparin-Aspirin Reperfusion Trial (HART) Investigators (1992) Heparin-induced prolongation of partial thromboplastin time after thrombolysis: relation to coronary artery patency. J Am Coll Cardiol 20:31–35

ISIS-3 (Third International Study of Infarct Survival) Collaborative Group (1992) ISIS-3: a randomised comparison of streptokinase vs tissue plasminogen activator vs anistreplase and of aspirin plus heparin vs aspirin alone among 41 299 cases of suspected acute myocardial infarction. Lancet 339:753–770

Kingston IB, Castro MJM, Anderson S (1995) In vitro stimulation of tissue-type plasminogen activator by Alzheimer amyloid β-peptide analogues. Nat Med 1:138–142

Klarich KW, Christian TF, Higano ST, Gibbons RJ (1999) Variability of myocardium at risk for acute myocardial infarction. Am J Cardiol 83:1191–1195

Kleiman NS, White HD, Ohman EM, Ross AM, Woodlief LH, Califf RM, Holmes DR, Bates E, Pfisterer M, Vahanian A, Topol EJ, for the GUSTO Investigators (1994) Mortality within 24 hours of thrombolysis for myocardial infarction. The importance of early reperfusion. Circulation 90:2658–2665

Lee KL, Woodlief LH, Topol EJ, Weaver WD, Betriu A, Col J, Simoons M, Aylward P, Van de Werf F, Califf RM, for the GUSTO-I Investigators (1995) Predictors of 30-day mortality in the era of reperfusion for acute myocardial infarction. Results from an international trial of 41 021 patients. Circulation 91:1659–1668

Lincoff AM, Topol EJ (1993) Illusion of reperfusion. Does anyone achieve optimal reperfusion during acute myocardial infarction? . Circulation 88:1361–1374

Magnani B, for the PAIMS Investigators (1989) Plasminogen Activator Italian Multicenter Study (PAIMS): Comparison of intravenous recombinant single-chain human tissue- type plasminogen activator (rt-PA) with intravenous streptokinase in acute myocardial infarction. J Am Coll Cardiol 13:19–26

Mahaffey KW, Granger CB, Sloan MA, Thompson TD, Gore JM, Weaver WD, White HD, Simoons ML, Barbash GI, Topol EJ, Califf RM, for the GUSTO-I Investigators (1998) Risk factors for in-hospital nonhemorrhagic stroke in patients with acute myocardial infarction treated with thrombolysis: results from GUSTO-I. Circulation 97:757–764

Mahaffey KW, Granger CB, Toth CA, White HD, Stebbins AL, Barbash GI, Vahanian A, Topol EJ, Califf RM, for the GUSTO-1 investigators (1997) Diabetic retinopathy should not be a contraindication to thrombolytic therapy for acute myocardial infarction: review of ocular hemorrhage incidence and location in the GUSTO-I trial. J Am Coll Cardiol 30:1606–1610

Mak K-H, Moliterno DJ, Granger CB, Miller DP, White HD, Wilcox RG, Califf RM, Topol EJ, for the GUSTO-1 investigators (1997) Influence of diabetes mellitus on clinical outcome in the thrombolytic era of acute myocardial infarction. J Am Coll Cardiol 30:171–179

Mak K-H, Topol EJ (2000) Emerging concepts in the management of acute myocardial infarction in patients with diabetes mellitus. J Am Coll Cardiol 35:563–568

Mark DB, Hlatky MA, Califf RM, Naylor CD, Lee KL, Armstrong PW, Barbash G, White H, Simoons ML, Nelson CL, Clapp-Channing N, Knight JD, Harrell FE Jr, Simes J, Topol EJ (1995) Cost effectiveness of thrombolytic therapy with tissue plasminogen activator as compared with streptokinase for acute myocardial infarction. N Engl J Med 332:1418–1424

Neuhaus KL, Feuerer W, Jeep-Tebbe S, Niederer W, Vogt A, Tebbe U (1989) Improved thrombolysis with a modified dose regimen of recombinant tissue-type plasminogen activator. J Am Coll Cardiol 14:1566–1569

Neuhaus KL, von Essen R, Tebbe U, Vogt A, Roth M, Riess M, Niederer W, Forycki F, Wirtzfeld A, Maeurer W, Limbourg P, Merx W, Haerten K (1992) Improved thrombolysis in acute myocardial infarction with front- loaded administration of alteplase: results of the rt-PA-APSAC patency study (TAPS). J Am Coll Cardiol 19:885–891

Reiner JS, Lundergan CF, Fung A, Coyne K, Cho S, Israel N, Kazmierski J, Pilcher G, Smith J, Rohrbeck S, Thompson M, Van de Werf F, Ross AM, for the GUSTO-1 Angiographic Investigators (1996) Evolution of early TIMI 2 flow after thrombolysis for acute myocardial infarction. Circulation 94:2441–2446

Rentrop P, Blanke H, Karsch KR, Kaiser H, Köstering H, Leitz K (1981) Selective intracoronary thrombolysis in acute myocardial infarction and unstable angina pectoris. Circulation 63:307–317

Ross AM, Coyne KS, Moreyra E, Reiner JS, Greenhouse SW, Walker PL, Simoons ML, Draoui YC, Califf RM, Topol EJ, Van de Werf F, Lundergan CF, for the GUSTO-I Angiographic Investigators (1998) Extended mortality benefit of early postinfarction reperfusion. GUSTO-I Angiographic Investigators. Circulation 97:1549–1556

Simes RJ, Topol EJ, Holmes DR, Jr., White HD, Rutsch WR, Vahanian A, Simoons ML, Morris D, Betriu A, Califf RM, Ross AM, for the GUSTO-I Investigators (1995) Link between the angiographic substudy and mortality outcomes in a large randomized trial of myocardial reperfusion. Importance of early and complete infarct artery reperfusion. Circulation 91:1923–1928

Tardiff BE, Califf RM, Morris D, Bates E, Woodlief LH, Lee KL, Green C, Rutsch W, Betriu A, Aylward PE, Topol EJ, for the GUSTO Investigators (1997) Coronary revascularization surgery after myocardial infarction: impact of bypass surgery on survival after thrombolysis. J Am Coll Cardiol 29:240–249

The GUSTO Angiographic Investigators (1993) The effects of tissue plasminogen activator, streptokinase, or both on coronary-artery patency, ventricular function, and survival after acute myocardial infarction. N Engl J Med 329:1615–1622

The GUSTO Investigators (1993) An international randomized trial comparing four thrombolytic strategies for acute myocardial infarction. N Engl J Med 329:673–682

The International Study Group (1990) In-hospital mortality and clinical course of 20 891 patients with suspected acute myocardial infarction randomised between alteplase and streptokinase with or without heparin. Lancet 336:71–75

The ISIS-2 (Second International Study of Infarct Survival) Collaborative Group (1988) Randomised trial of intravenous streptokinase, oral aspirin, both, or neither among 17 187 cases of suspected acute myocardial infarction: ISIS-2. Lancet 2:349–360

The National Institute of Neurological Disorders and Stroke rt-PA Stroke Study Group (1995) Tissue plasminogen activator for acute ischemic stroke. N Engl J Med 333:1581–1587

The TIMI Study Group (1985) The thrombolysis in myocardial infarction (TIMI) trial: phase 1 findings. N Engl J Med 312:932–936

Topol EJ, Armstrong P, Van de Werf F, Kleiman N, Lee K, Morris D, Simoons M, Barbash G, White H, Califf RM, on behalf of the GUSTO Steering Committee (1992) Confronting the issues of patient safety and investigator conflict of interest in an international clinical trial of myocardial reperfusion. J Am Coll Cardiol 19:1123–1128

Topol EJ, Califf RM, Lee KL (1994) More on the GUSTO trial. N Engl J Med 331:277–278

Topol EJ, George BS, Kereiakes DJ, Candela RJ, Abbottsmith CW, Stump DC, Boswick JM, Stack RS, Califf RM, and the TAMI Study Group (1988) Comparison of two dose regimens of intravenous tissue plasminogen activator for acute myocardial infarction. Am J Cardiol 61:723–728

Topol EJ, George BS, Kereiakes DJ, Stump DC, Candela RJ, Abbottsmith CW, Aronson L, Pickel A, Boswick JM, Lee KL, Ellis SG, Califf RM, and the TAMI Study Group (1989) A randomized controlled trial of intravenous tissue plasminogen activator and early intravenous heparin in acute myocardial infarction (TAMI-3). Circulation 79:281–286

Tracy RP, Rubin DZ, Mann KG, Bovill EG, Rand M, Geffken D, Tracy PB (1997) Thrombolytic therapy and proteolysis of factor V. J Am Coll Cardiol 30:716–724

Van de Werf F, Topol EJ, Lee KL, Woodlief LH, Granger CB, Armstrong PW, Barbash GI, Hampton JR, Guerci A, Simes RJ, Ross AM, Califf RM, for the GUSTO Investigators (1995) Variations in patient management and outcomes for acute myocardial infarction in the United States and other countries: results from the GUSTO trial. JAMA 273:1586–1591

Verstraete M, Bernard R, Bory M, Brower RW, Collen D, de Bono DP, Erbel R, Huhmann W, Lennane RJ, Lubsen J, Mathey D, Meyer J, Michels HR, Rutsch W, Schartl M, Schmidt W, Uebis R, von Essen R (1985) Randomised trial of intravenous recombinant tissue-type plasminogen activator versus intravenous streptokinase in acute myocardial infarction. Report from the European Cooperative Study Group for Recombinant Tissue-type Plasminogen Activator. Lancet 1:842–847

Wall TC, Califf RM, George BS, Ellis SG, Samaha JK, Kereiakes DJ, Worley SJ, Sigmon K, Topol EJ, for the TAMI-7 Study Group (1992) Accelerated plasminogen activator dose regimens for coronary thrombolysis: the TAMI-7 Study Group. J Am Coll Cardiol 19:482–489

White HD, Rivers JT, Maslowski AH, Ormiston JA, Takayama M, Hart HH, Sharpe DN, Whitlock RML, Norris RM (1989) Effect of intravenous streptokinase as compared with that of tissue plasminogen activator on left ventricular function after first myocardial infarction. N Engl J Med 320:817–821

Wnendt S, Wetzels I, Günzler WA (1997) Amyloid-β peptides stimulate tissue-type plasminogen activator but not recombinant prourokinase. Thromb Res 85:217–224

Woodfield SL, Lundergan CF, Reiner JS, Greenhouse SW, Thompson MA, Rohrbeck SC, Deychak Y, Simoons ML, Califf RM, Topol EJ, Ross AM, for the GUSTO-I Angiographic Investigators (1996) Angiographic findings and outcome in diabetic patients treated with thrombolytic therapy for acute myocardial infarction: the GUSTO-I experience. J Am Coll Cardiol 28:1661–1669

Conjunctive Therapy to Reduce the Occurrence of Coronary Reocclusion After Thrombolytic Treatment of AMI

R. Hayes, R. Gallo, V. Fuster and J. Chesebro

A. Introduction

Thrombolytic therapy aims to dissolve the occlusive thrombus present in over 90% of acute myocardial infractions (AMI) (Dewood et al. 1980; Falk 1983; Davies and Thomas 1989; Fuster et al. 1992; reviewed by Rentrop 2000). Prompt reperfusion reduces infarct size, preserves ventricular function, and reduces mortality.

However several obstacles remain: (1) resistance to therapy by 90 min angiographic patency remains at 15%–40%; (2) acute coronary reocclusion occurs in 5%–25% of patients; (3) an average of 45 min is needed for reperfusion of the infarct-related artery; (4) the efficacy of currently available therapy has a cost, increased bleeding in approximately 2%–5% of cases, including a 0.2%–0.7% rate of intracerebral hemorrhage (The ISIS-2 [Second International Study of Infarct Survival] Collaborative Group 1988; ISIS-3 [Third International Study of Infarct Survival] Collaborative Group 1992; Collen 1993). Conjunctive antithrombotic therapy aims to achieve optimal and lasting reperfusion.

B. The Role of Thrombin and Platelets in Thrombus Formation

The occlusive thrombus which leads to myocardial infarction results from disruption of a lipid-rich atherosclerotic plaque in three quarters of patients (Fuster et al. 1992; Falk et al. 1995). The disrupted plaque is composed of various thrombogenic substances including (1) a lipid-rich core surrounded by macrophages and containing tissue factor; and (2) smooth muscle cells and types I and III collagen (Drake et al. 1989; Fuster et al. 1992; Falk et al. 1995; Toschi et al. 1997). Tissue factor produced by macrophages binds circulating factor VIIa; the complex directly induces thrombin generation through the extrinsic pathway via activation of factors IX and X (Fuster et al. 1992; Marmur et al. 1996; Toschi et al. 1997; Taubman et al. 1999). Minute amounts of thrombin activate platelets and cause aggregation and upregulation, and binding of von Willebrand Factor (vWF) to the glycoprotein IIb/IIIa receptor

PLATELETS AND COAGULATION

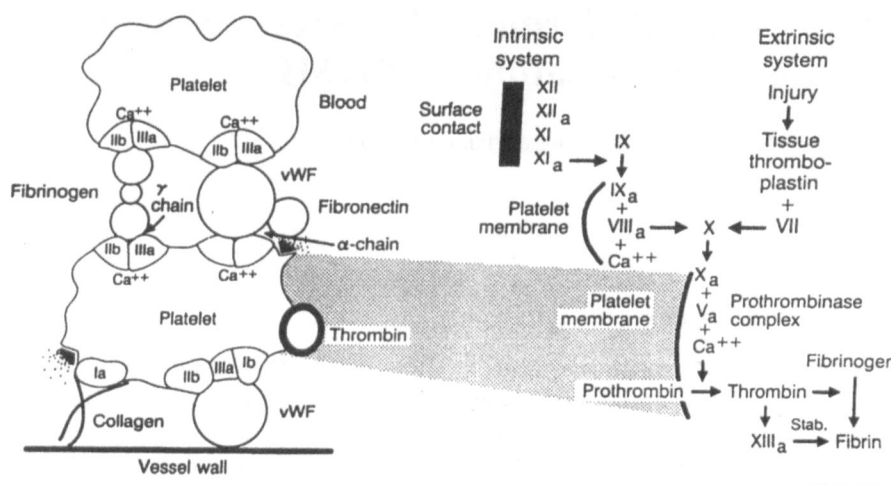

Fig. 1. Platelets and coagulation: The platelet plays two roles in thrombosis. As shown on the *left side of the figure* a platelet plug is formed by adhesion to the vessel wall through von Willebrand Factor (vWF) or collagen, and aggregation to other platelets through fibrinogen, fibronectin, and vWF. Such binding is mediated by integrin receptors (i.e., GP Ia/IIa, IIb/IIIa). The intrinsic and extrinsic systems of the coagulation cascade are shown to the *right of the figure*. The platelet membrane acts as a surface which greatly amplifies the assembly and activity of the prothrombinase complex (modified with permission from STEIN et at. 1989)

at high shear stress and of fibrinogen at low shear rates (COLLER 1995). Medial smooth muscle cells initiate mural thrombosis (HERAS et al. 1989, 1990). Collagen is able to activate platelets directly via the GP Ia/IIa receptor (Fig. 1) (BADIMON et al. 1988).

Thrombin has many actions (Fig. 2) both pro- and anti-thrombotic, is a strong platelet agonist at minute concentrations, converts fibrinogen to fibrin, and activates factor XIII which crosslinks and stabilizes the fibrin clot. Thrombin activates factors V and VIII which increase thrombin generation at a 278 000 times greater rate than factor Xa alone (MANN et al. 1985; CHESEBRO and FUSTER 1991). Thrombin also binds to thrombomodulin and activates protein C and protein S which inactivate factors Va and VIIIa and lead to an important negative feedback of the coagulation cascade. Fibrin-bound thrombin in thrombi is a potent agonist of further clot propagation via platelet activation and more thrombin generation (CHESEBRO and FUSTER 1991; MEYER et al. 1994). Thrombin activity is necessary for generation and maintenance of in vivo platelet-rich thrombus (HERAS et al. 1989, 1990; WYSOKINSKI et al. 1996).

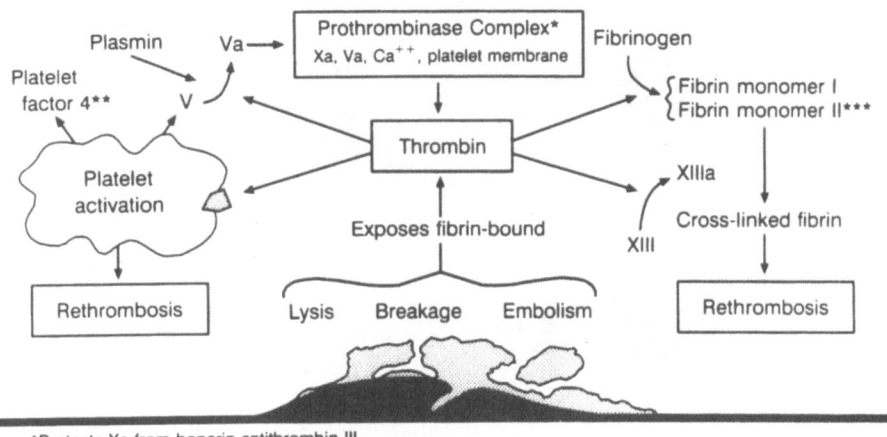

*Protects Xa from heparin-antithrombin III
**Neutralizes heparin
***Inhibits heparin-antithrombin III interaction with thrombin

CG-127526X-1A

Fig. 2. Biologic activities of thrombin: Thrombin, formed by factor Xa from the pro-thrombin complex, has many actions such as (1) platelet activation, (2) activation of factor XIII which stabilizes fibrin, (3) activation of factors V and VIII which further generates thrombin, and (not shown here), (4) activation of protein C and S via binding to thrombomodulin which acts as a negative feedback by inactivating factors Va and VIIIA (WEBSTER et al. 1991)

Platelets are also activated by other strong agonists such as thromboxane (TXA$_2$) and platelet activating factor, and weak agonists such as epinephrine, collagen, and serotonin. Thrombin appears to be the primary activator since minute amounts initiate platelet aggregation, and specific thrombin inhibition with hirudin limits platelet deposition in vivo to a single layer and causes dissolution of arterial thrombus (SCHMID et al. 1962; HERAS et al. 1990; WYSOKINSKI et al. 1996). Strong agonists activate platelets independent of cyclooxygenase through the phospholipase C second messenger system. Thus, such agonists are not inhibited by aspirin. Weak agonists activate platelets via the cyclooxygenase pathway to produce TXA$_2$ and are inhibited by aspirin (Fig. 3). Platelets enhance thrombin generation by providing an effective surface for formation of the prothrombinase complex (factor Va, Xa, pro-thrombin, and Ca^{2+}, Fig. 1) (COLLER 1995).

C. Evidence for the Efficacy of Thrombolytic Therapy

Thrombolytic therapy has decreased mortality approximately 30% in randomized controlled trials (GRUPPO ITALIANO PER LO STUDIO DELLA STREPTOCHINASI NELL'INFARTO MIOCARDICO [GISSI] 1986; THE I.S.A.M. STUDY GROUP 1986; THE ISIS-2 [SECOND INTERNATIONAL STUDY OF INFARCT SURVIVAL] COLLABORATIVE GROUP 1988; AIMS TRIAL STUDY GROUP 1988; WILCOX et al.

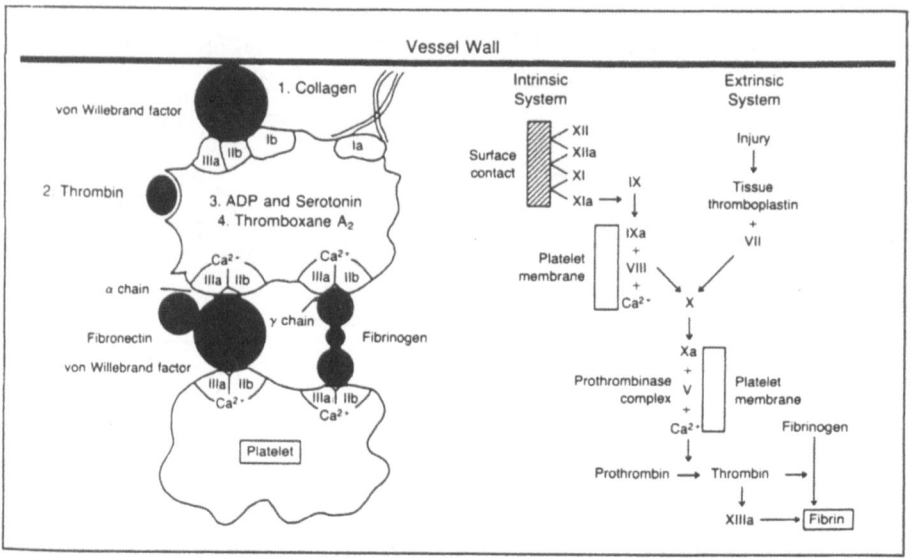

Fig. 3. Activation of platelets. Platelets are activated by various in vivo agonists and by adhesion to the vessel wall (binding to collagen) (1,2). Arachidonic acid (AA) is converted to thromboxane (TXA2) (4) by the action of two enzymes, cyclooxygenase and thromboxane synthase. By binding to a receptor various platelet agonists (i.e., ADP, serotonin) are released from platelet granules (3). Thrombin (via the intrinsic and extrinsic cascade, *right*) and collagen can still activate the platelet even if AA formation is blocked. Tissue factor from the plaque fatty gruel or arterial mural thrombus, initiates the extrinsic pathway. In the activated state, the platelet expresses the glycoprotein IIb/IIIa receptor on its surface membrane, which leads to platelet aggregation (FUSTER et al. 1992)

1988; reviewed by MARDER and SHERRY 1988). The GUSTO trial reported a 6.3% mortality with front loaded tissue-type plasminogen activator (tPA) plus intravenous heparin (THE GUSTO INVESTIGATORS 1993; THE GUSTO ANGIOGRAPHIC INVESTIGATORS 1993). Early recanalisation without reocclusion salvages myocardium, prevents reinfarction, and decreases congestive heart failure (CHF) and death.

I. Importance of Achieving Patency of the Infarct Related Artery

The efficacy of thrombolytic therapy is "time-dependent"; early treatment especially increases the incidence of reperfusion with streptokinase (SK) (CHESEBRO et al. 1987; KIM and BRAUNWALD 1993). The earlier thrombolytic therapy is given, the greater the increase in lives saved (GRUPPO ITALIANO PER LO STUDIO DELLA STREPTOCHINASI NELL'INFARTO MIOCARDICO [GISSI] 1986; THE GUSTO INVESTIGATORS 1993; GREAT GROUP 1993; WEAVER et al. 1993; THE EUROPEAN MYOCARDIAL INFARCTION PROJECT GROUP 1993). GISSI-1 showed a 50% reduction in mortality for patients treated within 1 h. For treatment given within 70 min, the mortality was 1.2% as compared to 10% for patients treated after 70 min. Of those patients treated within 1 h, 40% had no evidence of infarct on thallium scintigraphy. By meta-analysis at least 1.6/1000 lives are saved for each hour earlier treatment is initiated (FIBRINOLYTIC THERAPY TRIALISTS' [FTT] COLLABORATIVE GROUP 1994; BOERSMA et al. 1996).

A patent infarct-related artery (IRA) reduces coronary events. Many studies have clearly demonstrated that an open IRA, particularly TIMI grade 3 flow patency, 60–90 min after start of thrombolytic therapy is correlated with improved cardiac function and survival (the open artery hypothesis). Thus, patients with a patent IRA had a 2.5% 1-year mortality as compared to 15% for those with an occluded IRA (KENNEDY et al. 1985). In TIMI-1, the one-year mortality was 8.1% in patients with a patent IRA, as compared to 14.8% in those with an occluded IRA (CHESEBRO et al. 1987). In the TAMI study in-hospital mortality was 5.2% in patients with open and 10.4% in those with occluded IRA (TOPOL et al. 1987). In the GUSTO angiographic study the 30-day mortality, regardless of treatment assignment, was 8.9% in patients with TIMI grade 0 and 1 flow, 7.4% in those exhibiting TIMI grade 2 flow, and 4.4% in those with TIMI grade 3 flow ($p < 0.01$) (THE GUSTO ANGIOGRAPHIC INVESTIGATORS 1993). After 6 and, perhaps, 12 h the benefits of a patent IRA seem to persist at a time when salvage of viable myocardium is unlikely; treatment with tPA within 6–12 h resulted in a significant 27% reduction in 35-day mortality (LATE STUDY GROUP 1993). Patients receiving SK within 6–12 h had a 14% reduction in mortality (n.s.) (EMERAS [ESTUDIO MULTICÉNTRICO ESTREPTOQUINASA REPÚBLICAS DE AMÉRICA DEL SUR] COLLABORATIVE GROUP 1993). The benefit of a patent IRA involves improved left ventricular function due to a decrease of infarct expansion and remodelling (documented by decreased left ventricular end systolic volumes), increased electrical stability (fewer late potentials on signal averaged ECGs), and salvage of hibernating myocardium (KIM and BRAUNWALD 1993).

Early randomized trials have reported 90 min patency as a primary end-point. The patency rate (TIMI grade 2 or 3) with SK was reported as approximately 43%–64%, with standard dose tPA 63%–79%, and with front-loaded tPA 81%–91% (with a mean of 85%); within 2–3 h, patency with tPA is approximately equal to that with SK (I.S.A.M. STUDY GROUP 1986; CHESEBRO et al. 1987; THE GUSTO ANGIOGRAPHIC INVESTIGATORS 1993; LINCOFF and TOPOL 1993). However, the mean incidence in 15 studies of complete reperfusion (TIMI grade 3 or normal coronary flow) at 90 min (63% for accelerated tPA, 50% for standard dose tPA and 31.5% for SK, respectively) best predicted mortality reduction and an improved clinical outcome (BARBAGELATA et al. 1997) and is presumably what accounted for the 14% reduction in mortality achieved in GUSTO-I (OHMAN et al. 1990; KARAGOUNIS et al. 1992; THE GUSTO INVESTIGATORS 1993; THE GUSTO ANGIOGRAPHIC INVESTIGATORS 1993; REINER et al. 1994; SIMES et al. 1995). In these studies TIMI grade 2 flow (delayed flow of contrast) was shown to be no better than TIMI grade 0 or 1 flow. In the GUSTO angiographic study, TIMI grade 3 flow was associated with a 30-day mortality of 4.4% and TIMI grade 2 flow with a mortality of 7.4% (The GUSTO Investigators 1993; Simes et al. 1995).

II. Limitations of Thrombolytic Therapy

Even with the success of front-loaded tPA in the GUSTO trial, no reperfusion occurred in 19% of patients, and only 54% achieved TIMI grade 3 flow. The cost of increased reperfusion was a risk of bleeding in 0.8%–1.5% of patients with intracranial hemorrhage in 0.5%–0.9%. In addition 4.9%–6.4% of arteries initially patent at 90 min reoccluded at 5–7 days (The GUSTO Investigators 1993). Reocclusion is associated with a 2.5-times increased rate of in-hospital mortality (11% vs 4.5%, $p = 0.001$), an increased rate of pulmonary edema, hypotension and atrioventricular block, and less recovery of global or infarct-zone left ventricular function (Ohman et al. 1990). The ability of angiography to predict reocclusion using such factors as degree of residual stenosis, minimal luminal diameter, presence of thrombus, or eccentricity has met with mixed results (Reiner et al. 1994; Gibson et al. 1995). The largest study to date reported by the GUSTO investigators demonstrated that none of these parameters adequately predicted reocclusion (Reiner et al. 1994). The use of techniques such as ST-segment monitoring may be more predictive (Krucoff et al. 1986; Dellborg et al. 1991).

III. The Role of Platelets and Thrombin in Resistance to Therapy

Thrombosis and thrombolysis are dynamic, simultaneous, and opposing processes. The key goal of thrombolytic therapy is to activate the plasminogen system to form plasmin, the enzyme which lyses the fibrin clot. However, fibrinolysis activates local and systemic factors, and leads to the generation of thrombin which may result in thrombosis and reocclusion of the IRA. Thus, blocking thrombosis with antithrombotic therapy enhances lysis (Chesebro and Fuster 1991).

1. Thrombin

Numerous studies have shown that thrombolytic therapy leads to increased generation and activity of thrombin (Eisenberg et al. 1987, 1992; Owen et al. 1988; Rao et al. 1988; Rapold et al. 1989, 1992; Stump et al. 1989; Hoffmeister et al. 1999). Thrombin is generated from prothrombin with release of prothrombin fragment 1.2 (F1.2) via the assembly of the prothrombinase complex on phospholipid membranes (Figs. 1, 4). Free thrombin forms a 1:1 stoichiometric complex (so-called thrombin-antithrombin complexes [TAT]) with the physiologic inhibitor of thrombin, antithrombin III [AT-III]. Thrombin bound to fibrin and/or to damaged artery is very thrombogenic since it is poorly inactivated by AT-III, even in the presence of heparin (Weitz et al. 1990). Meyer

BIOCHEMICAL MARKERS OF FIBRIN AND THROMBIN FORMATION AND FIBRINOLYSIS

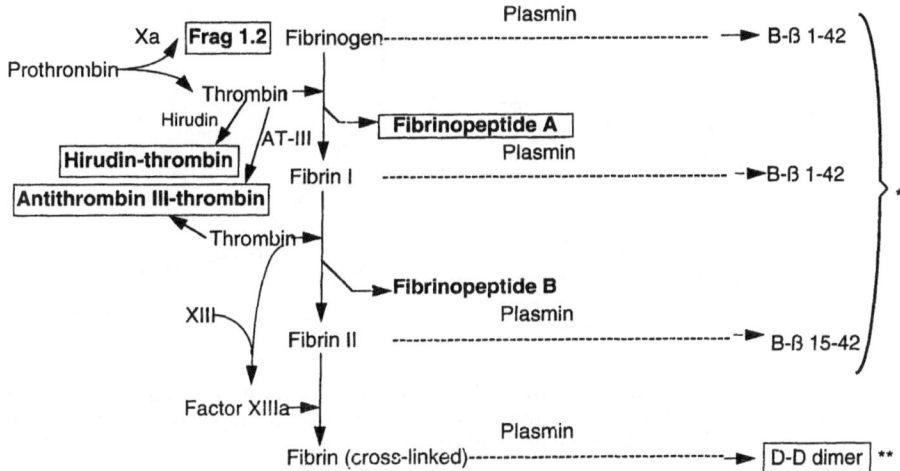

* Plus other soluble, uncross-linked fibrinogen degradation products
**Plus other soluble, cross-linked fibrin degradation products

Fig. 4. Biochemical markers of thrombin formation and fibrinolysis: Prothrombin fragment 1.2 is formed when prothrombin is converted to thrombin by factor Xa. Therefore, it is a marker of thrombin generation. Fibrinopeptide A (FPA) is formed when fibrinogen is converted to fibrin by thrombin. Therefore it is a marker of thrombin activity (CHESEBRO et al. 1995)

et al. (1994) have demonstrated that a thrombus forms rapidly on an injured arterial wall. Thrombus growth, as measured by platelet and fibrin deposition, was effectively inhibited by hirudin and to a much lesser extent by heparin, suggesting that thrombin was the principal agent responsible for thrombus growth. The best antithrombotic therapy will maximally accelerate thrombolysis and result in the least residual mural thrombus (MRUK et al. 1996). This study showed the critical role of thrombin inhibition in enhancing thrombolysis and reducing residual mural thrombus.

Thrombin converts fibrinogen to fibrin during which fibrinopeptide A (FPA) is formed. Thus, both thrombin activity and generation of fibrin can be assayed. Fibrinolysis leads to the formation of fibrin degradation products (FDPs) which also bind active thrombin (STUMP et al. 1989; MERLINI et al. 1994; WEITZ et al. 1998).

EISENBERG et al. (1987, 1992) and RAPOLD et al. (1989, 1992) showed in separate studies that patients with AMI treated with thrombolytic agents had an immediate increase in FPA levels. FPA levels rose more with SK therapy than with tPA and could be abolished or significantly attenuated by an infusion of heparin (Fig. 5). Patients with high plasma FPA levels (>50 ng/ml) 24 h

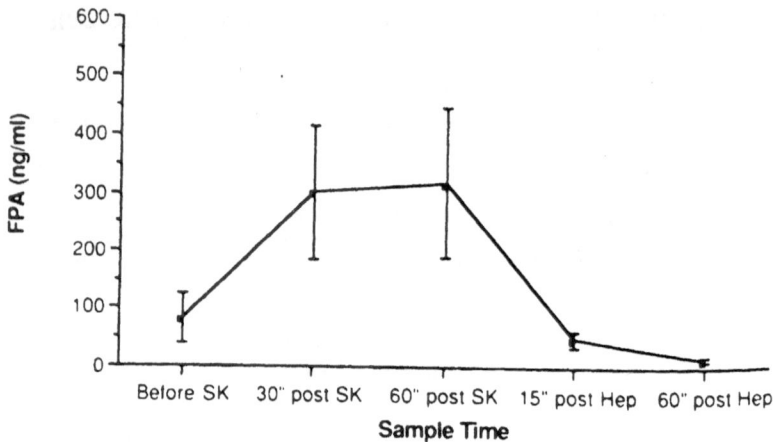

Fig. 5. FPA levels in 19 patients with AMI treated with SK. Administration of SK leads to an elevation of FPA levels. Conjunctive administration of heparin leads to prompt A decline in FPA levels (Eisenberg et al. 1987)

after tPA therapy had a significantly increased incidence of reocclusion by angiography (88% vs 55%) with a specificity of 86% but a sensitivity of only 48%. FPA levels did not correlate with recurrent ischemia or left ventricular mural thrombus (Rapold et al. 1989) and were not attenuated by heparin in all patients (Owen et al. 1988). Patients whose FPA levels remained elevated despite adequate anticoagulation with heparin had reocclusion of the IRA (Rapold et al. 1992). Granger et al. (1998) determined markers of thrombin generation in 292 patients enrolled in the GUSTO-I trial. Baseline FPA levels were elevated to ten times normal and increased further during thrombolytic therapy. F1.2 and TAT levels also increased during thrombolysis. The former was a predictive marker of clinical outcome. Patients who died or experienced reinfarction during the first 30 days had higher F1.2 levels (0.42 nmol/l) than those without such an event (0.34 nmol/l; $p = 0.008$). Although hirudin better inhibits clot-bound thrombin than does heparin, persistent thrombin generation was still found in 18 subjects with coronary heart disease given 3 different doses of hirudin IV for 6h (Zoldhelyi et al. 1994).

The TIMI-1 and TAMI-1 studies showed that markers of fibrinolysis predicted patency of the IRA (Rao et al. 1988; Stump et al. 1989). Stump et al. (1989) showed that tPA induced a rapid fibrinogenolysis with lowering of fibrinogen and increase in FDPs. The risk of reocclusion of the IRA at 7–10 days correlated inversely with the magnitude of its drop to nadir levels, and to peak levels of FDPs. Although fibrinogen levels decreased and FDP levels increased with infusion of SK (to a lesser extent with tPA), by 5h the level of fibrinogen began to rise and was still rising above normal at hospital discharge, suggesting a hypercoagulable state after infarction (Rao et al. 1988; Merlini et al. 1994).

2. Platelets

Numerous studies have shown that, during thrombolysis, platelets are activated either directly by the thrombolytic agent, or indirectly by thrombin and,
perhaps, plasmin (NIEWIAROWSKI et al. 1973; SCHAFER and ADELMAN 1985;
FITZGERALD et al. 1988; LUCORE and SOBEL 1988; GOLINO et al. 1988; KERINS et
al. 1989; LEE and MANN 1989; JANG et al. 1989; POTTER VAN LOON et al. 1990;
GERTZ et al. 1990; VAUGHAN et al. 1991; KUNITADA et al. 1992; ARONSON et al.
1992; KAWANO et al. 1998). Platelet-rich compared to platelet-poor thrombi are
much less sensitive to thrombolysis in vitro and in vivo (JANG et al. 1989).

The composition of residual thrombi depends both on the extent of lysis
and the influence of shear force. Shear force is a measure of the difference in
velocity between different areas of laminar flow, and is related directly to
velocity and inversely to the third power of the luminal diameter. In a blood
vessel, the velocity of flow is greatest in the center of the vessel, and the highest
shear force is at the blood-vessel wall interface where the greatest difference
in velocities exists (BACK et al. 1977; ALEVRIADOU et al. 1993). In stenotic
lesions, shear force is increased at the apex of the stenosis where platelet deposition is greatest (BADIMON et al. 1986; BADIMON and BADIMON 1989; MAILHAC
et al. 1994). Fibrin deposition also increases with increasing shear force but
more so proximally and distally to the stenosis where flow separation and eddy
currents also lead to red cell deposition (MAILHAC et al. 1994). The result is a
thrombus composed of a "white-head" and a "red-tail" (AMBROSE 1992).

FITZGERALD et al. (1988) and KERINS et al. (1989) showed that thrombolysis with either SK or tPA induced synthesis of thromboxane as measured by
the urinary metabolites 2,3-dinor-TXB_2 and prostacyclin. Synthesis of these
compounds was less with successful reperfusion, and was abolished by pretreatment with aspirin. GOLINO et al. (1988), using the Folts model in dogs,
showed that cyclic flow variations were abolished when both thromboxane and
serotonin receptor antagonists were administered.

On balance, plasmin appears to increase platelet activation in vivo during
thrombolysis. Early studies proposed that plasmin causes platelet activation
(NIEWIAROWSKI et al. 1973) by activating prothrombin and factor V which act
to increase thrombin generation and platelet activation (LEE and MANN 1989;
EISENBERG et al. 1992). ARONSON et al. (1992) also showed that platelets treated
with tPA or SK increased thrombin generation. Others reports showed that
plasmin, generated by SK, lead to inhibition of platelet aggregation (ADELMAN
et al. 1985; SCHAFER and ADELMAN 1985; GOUIN et al. 1992).

Platelets attenuate thrombolysis by several mechanisms. ZHU et al. (1999)
showed that platelet derived plasminogen activator inhibitor-1 (PAI-1) is a
major determinant of the resistance of platelet-rich arterial thrombi to lysis
by pharmacological concentrations of tPA. PAI-1 reduced tPA activity by 47%
in rabbits (LUCORE and SOBEL 1988). Efficacy of thrombolysis correlated inversely with PAI-1 levels (BARBASH et al. 1989; POTTER VAN LOON et al. 1992).
However, KUNITADA et al. (1992) showed that platelets inhibit thrombolysis

with tPA not only via release of PAI-1, but also via clot retraction, with decreased access of the fibrinolytic proteins. Antibodies induced by SK cause platelet aggregation by binding to the SK/plasminogen complex on the platelet surface and counteract SK-mediated thrombolysis (Vaughan et al. 1991; McRedmond et al. 2000).

D. Antithrombotic Therapy

The high risk of rethrombosis after thrombolysis underscores the need for conjunctive antithrombotic therapy. Such therapy may target thrombin (heparin, low-molecular weight heparin, hirudin and its derivatives, hirulog, etc.) or other serine proteases involved in the generation of thrombin such as factor VIIa (i.e., tissue factor pathway inhibitor) or factor Xa (tick anticoagulant peptide, etc.), or platelets (aspirin, ticlopidine, GP IIb/IIIa antagonists, prostacyclin, inhibitors of thromboxane or serotonin synthesis, receptor antagonists, etc.).

I. Thrombin Inhibition

1. Heparin

Heparin is a glycosaminoglycan composed of a heterogeneous mixture of molecules of different molecular weights. The anticoagulant action of heparin against thrombin is due to the simultaneous binding of one-third of these heparin molecules (those with ≥18 saccharides) to AT-III and thrombin, forming a ternary complex. Heparin/AT-III binding accelerates inactivation of thrombin, factors XIa, IXa, and Xa, and to a lesser extent also of factor VIIa. Binding of heparin to heparin cofactor II (at much higher doses) can also inactivate thrombin (Hirsh et al. 1995). The efficacy and safety of heparin administration by either continuous IV or subcutaneous routes are comparable provided the dosages are adequate. Subcutaneous administration requires higher doses (averaging 17000 U/12 per hour) to counteract the reduced bioavailability. An initial simultaneous IV bolus is needed if an immediate effect is required because heparin binding sites in the circulation must be saturated to maintain plasma levels, and subcutaneous absorption is delayed by 1–2 h.

Heparin has several disadvantages for treatment of platelet-rich arterial thrombosis (Hirsh et al. 1995; Chesebro et al. 1995):

1. It has a weak antiplatelet effect which is directly related to dosage (plasma levels). Both, increased dosage and aspirin increase the bleeding risk.
2. It is inactivated by platelet factor 4 which is released from activated platelets and by fibrin monomer II formed as fibrinogen is converted to fibrin.
3. Heparin also binds to endothelial cells throughout the vasculature and to macrophages which internalize, depolymerize, and metabolize heparin. It

also binds to a variety of plasma proteins such as histidine-rich glycoprotein, vitronectin, fibronectin, and vWF and to as yet not clearly defined
acute-phase reactant proteins. This protein binding contributes to reduced
bioavailability at low doses, variable anticoagulant effect at fixed doses, and
the laboratory phenomenon of heparin resistance (YOUNG et al. 1997).
4. Heparin dose/response curves are non-linear, and increase disproportion-
   ately with intensity and duration of increasing dose.
5. Heparin acts indirectly through the activation of AT-III and at high doses
   via heparin cofactor II.
6. AT-III binds poorly to thrombin which is bound to fibrin in a thrombus, to
   circulating fibrin-split products, or to subendothelial matrix (WEITZ et al.
   1990, 1998; HIRSH 1995).

Since heparin at usual doses mainly inactivates free thrombin and poorly
blocks growth of thrombus, arterial thrombus may grow to occlusion
(WYSOKINSKI et al. 1996).

Despite these problems, the lowest reported incidence of reocclusion at
57 days by angiography (4.9%–6.4%) was achieved with heparin (continuous IV or subcutaneous administration) plus aspirin therapy (THE GUSTO
INVESTIGATORS 1993). Heparin at high doses potentiates lysis in animal models.
It reduced the time to reperfusion in canine arterial segments injured with a
copper coil, and treated with thrombolysis and heparin (200 U/kg or 300 U/kg)
(CERCEK et al. 1986; TOMARU et al. 1989), and in the occluded porcine carotid
artery after percutaneous transluminal angioplasty (PTA) and catheter
endarterectomy, and reduced residual mural thrombus (MRUK et al. 1996). In
the HEAP pilot study, an acute bolus of heparin (300 U/kg) resulted in a 90-
min coronary TIMI 3 patency in 37% of patients (VERHEUGT et al. 1998).

Early randomized trials included heparin with thrombolytic agents
(GRUPPO ITALIANO PER LO STUDIO DELLA STREPTOCHINASI NELL'INFARTO
MIOCARDICO [GISSI] 1986; THE I.S.A.M. STUDY GROUP 1986; THE ISIS-2
[SECOND INTERNATIONAL STUDY OF INFARCT SURVIVAL] COLLABORATIVE GROUP
1988; AIMS TRIAL STUDY GROUP 1988; WILCOX et al. 1988). Several studies have
documented the therapeutic benefit of conjunctive heparin therapy with
fibrin-specific thrombolytic agents (TOPOL et al. 1989a; HSIA et al. 1990; BLEICH
et al. 1990; DE BONO et al. 1992) (Fig. 6).

Heparin in the usual therapeutic doses safely given to humans does not
enhance thrombolysis (TOPOL et al. 1989a). The National Heart Foundation of
Australia Study Group showed that 24h of treatment with heparin followed
by treatment with aspirin (300 mg) and dipyridamole (300 mg) achieved the
same, relatively poor, patency rate at 7–10 days as 7 days of additional intravenous heparin (81% vs 80%) (THOMPSON et al. 1991).

Achieving an activated partial thromboplastin time (aPTT) in the therapeutic range is important in maintaining adequate perfusion of the IRA.
Patients whose aPTT at 8h and 12h was <45s had a patency rate of only 45%
at 18h, whereas those patients with aPTTs >45s or >60s had patency rates of

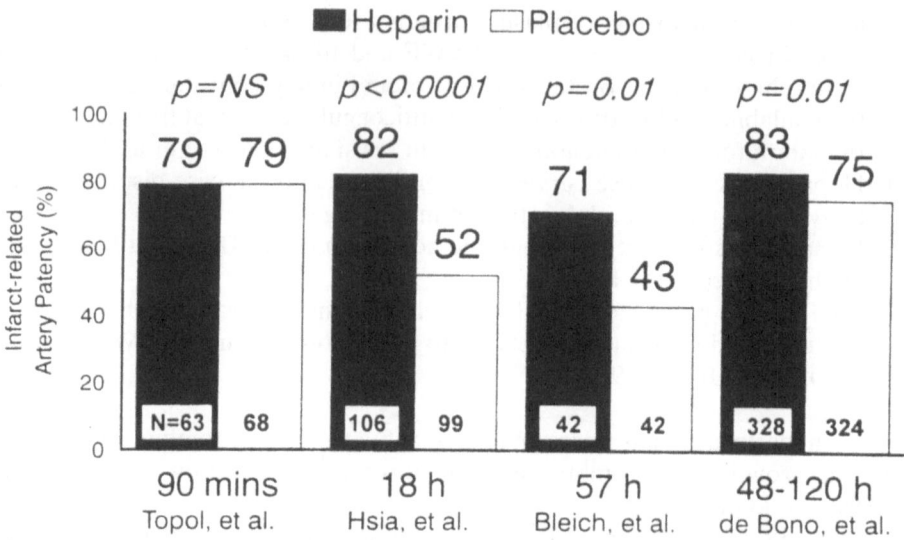

Fig. 6. Thrombolytic trials with heparin vs placebo: Data showing the effect of heparin vs placebo on patency of the IRA at later time points in patients receiving tPA from the four major trials (Cannon et al. 1995)

88% and 95% respectively (Hsia et al. 1992) (Fig. 7). Patients with an aPTT >2 times control at 3–36h had a patency rate at 81h (mean) of 90%, compared to 80% and 72% in those exhibiting aPTTs of 1.3–2.0 or <1.3 times control, respectively. Only 32% of patients achieved an aPTT in the therapeutic range (Arnout et al. 1992). Finally, Kaplan et al. (1987) showed that the aPTT level in patients treated with conjunctive heparin therapy was inversely correlated with recurrent ischemic events. Therefore, it is currently recommended to measure the aPTT at 6h, 12h, and 24h after infusion of tPA and other fibrin specific thrombolytic agents to attain an aPTT of 1.5–2.5 times control (The GUSTO Investigators 1993; Cannon et al. 1994). Granger et al. (1996) examined the aPTT in 29,656 patients in the GUSTO-I trial and found that, at 12h, the value associated with the lowest 30-day mortality, stroke, and bleeding rate was 50–70s. There was a clustering of reinfarction observed in the first 10h after discontinuation of IV heparin. However, in the COBALT trial (Van de Werf F, for the Continuous Infusion versus Double-Bolus Administration of Alteplase [COBALT] Investigators 1997), aPTT values at 12h, 24h, and 48h were not significantly different between patients who did or did not develop a hemorrhagic stroke, whereas aPTT values at 6h after thrombolysis correlated positively with the incidence of hemorrhagic stroke (Kruik et al. 1998). This strongly suggests that heparin adjustments should be performed as soon as possible after completing the tPA infusion.

The role of conjunctive therapy with heparin in addition to non-fibrin specific agents is much less clear. There has been no large scale randomized

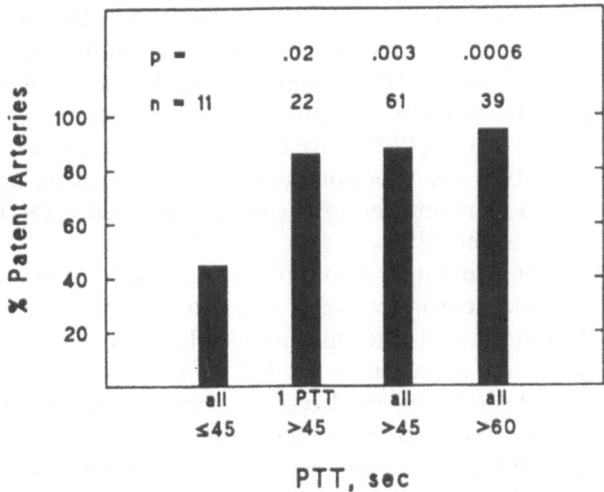

Fig. 7. Heparin aspirin reinfarction study: Hsia et al. demonstrated the advantage of a therapeutic APTT in achieving patency of the infarct related artery in those patients with AMI receiving tPA. Here 95% of patients with an aPTT > 60s had a patent IRA, as opposed to 88% and 45% in those patients with aPTTs of 45–60s and <45s respectively (Hsia et al. 1992)

trial to date comparing heparin vs no heparin in conjunction with SK therapy.

Although not designed as a mortality study, the SCATI study showed a mortality benefit when subcutaneous heparin 12500 U/every 12h was given in addition to SK alone (8.8% vs 4.5% $p = 0.05$) (The SCATI Group 1989). In an ISIS-2 pilot study, patients treated with intravenous heparin 12h after SK had lower reinfarction rates than those patients who did not receive heparin (2.2% vs 4.9%) (ISIS [International Studies of Infarct Survival] Pilot Study Investigators 1987). Furthermore, though the ISIS-2 study was not designed to test the effect of conjunctive heparin therapy, a sub-group analysis showed that mortality was 8.3% in those patients receiving IV heparin, and 10.1% in those who did not ($p < 0.05$), and vascular mortality was reduced by 19% (The ISIS-2 [Second International Study of Infarct Survival] Collaborative Group 1988). Finally, the reduction of FPA levels by heparin during SK therapy supports the efficacy of heparin with SK (Eisenberg et al. 1987; Owen et al. 1988).

Subcutaneous heparin after thrombolytic therapy with either tPA or SK (delayed 4h in ISIS-3 and 12h in GISSI-2) did not decrease mortality or reinfarction probably because of the delay in treatment with heparin, where aPTTs did not reach therapeutic levels for 24–48h after initiation when the risk of reocclusion is highest (Gruppo Italiano per lo Studio della Sopravvivenza nell'Infarto Miocardico 1990; Lincoff and Topol 1993). Although GUSTO was not designed to test the value of heparin therapy, the lowest reported rate of angiographic reocclusion with either continuous IV or subcutaneous heparin for 5–7 days plus aspirin, suggests that heparin therapy to an aPTT of 1.5–2.5 times control (target 2.0) starting 4 h after SK is beneficial for patients (Cairns 1995). In the majority of earlier studies heparin was administered in fixed doses. Hochman et al. (1999) have demonstrated that lower, body weight-adjusted heparin doses are superior than the traditional dosage schemes in achieving early aPTTs within the target range and reduce the need for dose changes over the ensuing 24 h.

The benefits of combining heparin and aspirin in acute coronary syndromes, the increased coronary event rate after stopping heparin, and the reduced post-hospital event-rate on anticoagulation plus aspirin all suggest that additional therapy to aspirin is needed for 6–12 weeks beyond the acute event (The GUSTO Angiographic Investigators 1993; Cohen et al. 1994; The Global Use of Strategies to Open Occluded Coronary Arteries [Gusto] IIb Investigators 1996; Wallentin for the Fragmin During Instability in Coronary Artery Disease [FRISC] Study Group 1996).

Support for this recommendation is provided by the angioscopic demonstration that healing of the IRA requires more than one month and that an unstable yellow plaque with adherent thrombus is common during that period (Van Belle et al. 1998). Glick et al. (1996) have examined whether extending the anticoagulant effect of heparin by low-molecular-weight heparin (LMWH) can prevent recurrent myocardial ischemia after AMI treated with SK. On the fifth day after AMI and after heparin therapy cessation, 103 patients were randomly assigned to either treatment with LMWH (enoxaparin 40 mg/d subcutaneously for 25 days) or control (no treatment). During the 6-months observation period 20% of the patients in the control group experienced reinfarction vs 4.6% in the LMWH group. Angina pectoris was diagnosed in 21.6% of the controls and in 9.3% of the LMWH patients.

2. Low-Molecular-Weight Heparin

Several studies have been published on the effect of low-molecular-weight heparin (LMWH) in unstable angina (reviewed by Antman and Cohen 1999), but only one study addressed the effect of LMWH as a conjunctive agent in the treatment of AMI with SK. In the randomized, double-blind BIOMACS II study 54 patients with AMI received subcutaneously 100 IU/kg of dalteparin prior to an infusion of SK and 47 patients received a placebo injection instead (Frostfeldt et al. 1999). A second subcutaneous injection of 120 IU/kg of dalteparin or of placebo was given 12 h later. All patients also received 300 mg aspirin as an oral loading dose and 75 mg on consecutive days. Immediately after the start of treatment monitoring with continuous vector-ECG was started and continued for 24–28 h, when coronary angiography was performed. In the dalteparin group TIMI grade 3 flow after 20–28 h was 68% and in the placebo group 51% ($p = 0.1$). Non-invasive investigation of early reperfusion by means of vector-ECG and myoglobin determinations also revealed a tendency for earlier reperfusion in the dalteparin group, and fewer post-thrombolytic ST vector magnitude increases, indicative of ischemic episodes ($p = 0.037$). In the upcoming HART II trial subcutaneous enoxaparin will be compared with IV unfractionated heparin as conjunctive agents in patients receiving front-loaded tPA for ST-segment elevation AMI (ANTMAN and COHEN 1999).

Vasoflux, a novel anticoagulant derived from LMWH, did not improve patency rates in patients treated with SK, compared to heparin and SK (PETERS et al. 2000).

3. Hirudin

Direct thrombin inhibitors such as hirudin are able to overcome many of the limitations of heparin. Hirudin is a 65-amino acid polypeptide originally extracted from the salivary gland of the leech *Hirudo medicinalis*, and now produced by recombinant technology. Hirudin forms a 1:1 complex directly with thrombin (more rapidly than AT-III) with its carboxy-terminus binding to the substrate recognition site (which recognizes fibrinogen or the platelet membrane thrombin receptor), and its amino-terminus binding to the catalytic site of thrombin (Fig. 8). Hirudin binds thrombin noncovalently but with high specificity and affinity. Furthermore, unlike heparin: (1) hirudin readily binds to thrombin bound to fibrin or arterial wall matrix, (2) there are no circulating inhibitors of hirudin, and (3) hirudin does not bind to endothelial cells or plasma proteins. The result is that hirudin achieves a more specific and con-

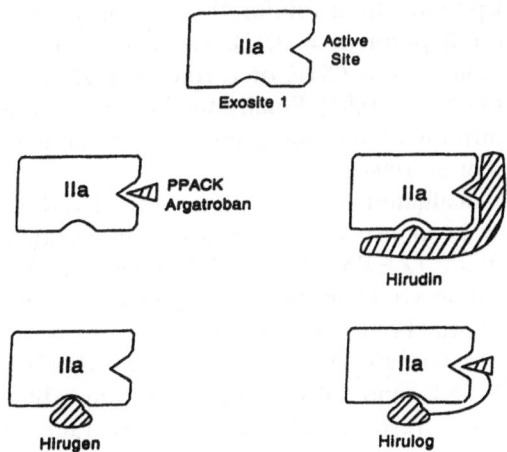

Fig. 8. Interaction of direct thrombin inhibitors with thrombin. The carboxy terminus of hirudin binds to exosite I (the substrate recognition site) while the amino terminus binds to the active (catalytic) site. Also depicted is the binding of hirugen (to the exosite), hirulog (to both sites), and PPACK and Argatroban (to the active site) (WEITZ et al. 1995)

sistent state of anticoagulation (Zoldhelyi et al. 1993; Topol et al. 1994; The Global Use of Strategies to Open Occluded Coronary Arteries [GUSTO] IIb Investigators 1996; Toschi et al. 1996). For example, in the TIMI-5 and 6 studies, the proportion of patients whose aPTT values varied <30s between the highest and the lowest aPTT was only 25% in patients treated with heparin as compared to 61%–74% of patients treated with hirudin (Cannon et al. 1994; Lee et al. 1995). The half-life of hirudin is 2–3h (Zoldhelyi et al. 1993). Hirudin may be antigenic. In patients with heparin-induced thrombocytopenia, treated with hirudin for more than 5 days, 56% developed antibodies against hirudin (Song et al. 1999).

Hirudin reduced platelet and fibrin deposition of deeply injured arterial wall segments to a single layer as compared to heparin which reduced deposition to a lesser extent but in a dose dependent manner (Heras et al. 1989). At an aPTT 2–3 times control, hirudin totally eliminated mural thrombosis and limited platelet deposition to a single layer (Heras et al. 1989, 1990). Hirudin enhances dissolution of platelet-rich arterial thrombi (Wysokinski et al. 1996). Salutary effects of hirudin in animal models (Heras et al. 1989, 1990; Mruk et al. 1996) led to the development of pilot studies with hirudin as conjunctive agents in thrombolysis (Cannon et al. 1994; Topol et al. 1994; Lee et al. 1995; Zeymer et al. 1995; Wysokinski et al. 1996).

The TIMI-5 trial randomized 246 patients to standard dose heparin vs one of four doses of hirudin in combination with front loaded tPA. The primary end point, TIMI grade 3 flow at 90min and 18–36h, was 61.8% in the hirudin group and 49.4% in the heparin group ($p = 0.07$), and at 18–36h patency was 98% in the hirudin and 89% in the heparin group ($p = 0.01$). Furthermore, reocclusion occurred in only 1.6% of hirudin patients vs 6.7% for heparin. The incidence of in-hospital death or reinfarction was only 6.8% for hirudin, compared to 16.7% for heparin ($p = 0.02$). The composite endpoint of death, myocardial infarction, severe CHF, or cardiogenic shock was reduced with hirudin (9.3% vs 19.0%, $p = 0.03$). Finally, the incidence of major spontaneous hemorrhage was infrequent, and more common in the heparin group (4.7% vs 1.2%) (Cannon et al. 1994).

The HIT-I trial evaluated hirudin with front loaded tPA in 40 patients and also showed a low incidence of reocclusion, reinfarction and bleeding for hirudin-treated patients (ZEYMER et al. 1995). The HIT-II trial was an open, sequential dose-finding study in 143 AMI patients treated with tPA (von Essen et al. 1998). Initial bolus and subsequent IV infusion doses/kg per hour during 48h were 0.1/0.06mg/kg, 0.2/0.1mg/kg, and 0.4/0.15mg/kg. TIMI 3 flow at 60min was 50%, 58%, and 63%, respectively. Major bleeding occurred in 0%, 4.8%, and 8.4%, respectively, but no intracranial hemorrhage was observed.

The TIMI-6 study, a non-angiographic trial in 193 patients using a 5-day infusion of antithrombotic drug, suggested that hirudin at a dose of 0.1mg/kg per hour was the optimal dose with SK to reduce the composite endpoint of death, CHF, or cardiogenic shock. At hospital discharge this was 17.6% in the heparin group compared to 21.6%, 9.7%, and 6% with 0.05, 0.1, and 0.2 mg/kg per hour dose of hirudin, respectively (LEE et al. 1995).

Three larger-scale trials comparing hirudin to heparin (TIMI 9A, GUSTO IIa, and HIT-III) were discontinued prematurely due to an increased incidence of major bleeding including intracerebral hemorrhage (ANTMAN et al. 1994; THE GLOBAL USE OF STRATEGIES TO OPEN OCCLUDED CORONARY ARTERIES [GUSTO] IIa INVESTIGATORS 1994; NEUHAUS et al. 1994). Major bleeding in the heparin treated patients occurred in those who were older, and had higher aPTTs (mean 115s within 12h), usually due to the higher heparin dosage given in heavier patients. In the hirudin group, there was an increased risk of bleeding in patients with elevated creatinine since hirudin is excreted by the kidney.

In GUSTO IIb and TIMI 9B trials the heparin dose was lowered to 1000 U/h (aPTT of 55–85s) and the dose of hirudin reduced to a bolus of 0.1 mg/kg plus an infusion at 0.1 mg/kg per hour (aPTTs 55–85s) (THE GLOBAL USE OF STRATEGIES TO OPEN OCCLUDED CORONARY ARTERIES [GUSTO] IIb INVESTIGATORS 1996; ANTMAN, FOR THE TIMI 9B INVESTIGATORS 1996). In the GUSTO IIb multicenter study a total of 12,141 patient with acute coronary syndrome were enrolled. The risk of death or AMI at 24h was significantly lower in the group assigned to hirudin therapy than in the group receiving heparin (1.3% vs 2.1%; $p = 0.001$). The primary endpoint of death or nonfatal AMI or reinfarction at 30 days was reached in 9.8% of the heparin group compared with 8.9% in the hirudin group ($p = 0.06$). All the benefit in the GUSTO IIb trial was in non-USA patients where therapy was administered 28 min after randomization compared to no benefit in USA patients receiving therapy 41 min after randomization. There was no significant difference in major bleeding complications. In the GUSTO IIb subgroup of patients with ST elevation 2274 patients received tPA and 1015 patients received SK. Among SK-treated patients, death or reinfarction at 30 days occurred more often in those treated with conjunctive heparin (14.4%) than in those receiving hirudin (8.6%; $p = 0.004$). Among tPA-treated patients the rates were 10.9% with heparin and 10.3% with hirudin ($p = 0.68$) (METZ et al. 1998). In the TIMI 9B study, where hirudin was given later after randomization (median 44 min), there was no difference between the hirudin and heparin outcome (ANTMAN, FOR THE TIMI 9B INVESTIGATORS 1996).

In the double blind HIT-4 trial 1208 patients with AMI of less than 6h duration were treated with aspirin and SK and randomized to receive recombinant hirudin (leptorudin, IV bolus of 0.2 mg/kg, followed by subcutaneous injections of 0.5 mg/kg b.i.d. for 5–7 days) or heparin (IV placebo bolus, followed by 12,500 IU b.i.d. for the same time period) (NEUHAUS et al. 1999). In the HIT-4 study hirudin was given before the infusion of SK (1.5 Mio U IV) was started. Although grade 3 flow tended to be higher in the hirudin (41%) than in the heparin group (34%) there were no significant differences in the two groups with respect to total stroke and reinfarction rate, death, rescue-PTCA, and refractory angina.

It is still not clear whether the aPTT is the most suitable test to monitor hirudin therapy (reviewed by Stürzebecher 1991). It may well be that the ecarin time (Pötzsch et al. 1997) is more suitable to measure free circulating hirudin concentrations in patients and that hirudin doses have to be adjusted, as is the case for heparin, to obtain optimal but not excessive anticoagulation.

4. Hirulog

As shown in Fig. 8, hirulog is composed of hirugen, a synthetic 12-amino acid peptide containing the same amino acids present at residues 53–64 in hirudin (the only exception being sulfation of tyrosine at position 63 to increase thrombin binding), linked to D-Phe-Pro-Arg-Pro-$(Gly)_4$. As depicted, the hirugen portion binds to the substrate recognition site on thrombin, and the linked peptide binds to the catalytic site. However hirulog is less effective than hirudin due to its higher K_D and because thrombin can cleave the Arg-Pro bond, thereby converting it to a weaker thrombin inhibitor. Hirulog has a half-life of only 35–40 min (Lidón et al. 1994; Théroux et al. 1995).

Two angiographic pilot studies evaluated hirulog versus standard dose heparin given during 5 days with SK (Lidón et al. 1994; Théroux et al. 1995). Patency of the IRA at 90 min was significantly better in the hirulog-treated patients (77% vs 47%, $p < 0.05$), and TIMI grade 3 flow in the IRA was 67% compared to 40% for heparin ($p = 0.08$).

In the HERO-1 trial, 412 patients with AMI, presenting within 12 h, were given aspirin and SK, and randomized in a double-blind manner to receive up to 60 h of either heparin (5000 U bolus, followed by 1000–1200 U/h), low-dose hirulog (bivalidin, 0.125 mg/kg bolus, followed by 0.25 mg/kg per hour for 12 h, then half this dose), high-dose hirulog (twice the dose of low-dose hirulog in all phases). TIMI grade 3 flow of the IRA at 90–120 min was 35% in the heparin, 46% in the low-dose hirulog, and 48% in the high-dose hirulog group (p: heparin vs hirulog in both groups <0.04). At 48 h reocclusion was more frequent in the heparin group (n.s.). By 35 days death, cardiogenic shock, or reinfarction had occurred in 18% of patients receiving heparin, in 14% of the low-dose, and in 12.5% of the high-dose hirulog group (n.s). (White et al. 1997). Major bleeding was less frequent in the two hirulog groups. Based on these promising results a larger HERO-2 study is now under way. A total of 17,000 patients will be enrolled. In this trial patients with AMI of less than 6 h duration will be randomized to hirulog or heparin and be treated with 1.5 Mio U of SK.

5. Argatroban

In the ARGAMI-2 study, 1200 patients with AMI, treated with SK ($n = 601$) or tPA ($n = 599$) were randomized to receive heparin or low or high doses of argatroban (Novastan), a synthetic direct thrombin inhibitor. An interim analysis revealed that the low dose argatroban treatment was ineffective. The final analysis of 1001 patients (high dose argatroban 494; heparin 507) showed no significant differences in mortality and in all primary endpoints between argatroban (20%) and heparin (19%). Major bleeding (0.4%), stroke (0.2%), and intracranial hemorrhage (0%) were less common in the argatroban group than in the heparin group (1.2%, 0.6%, and 0.4%, respectively). Target aPTT (55–85 s) was more frequently achieved in the argatroban than in the heparin group (BEHAR et al 1998).

In the relatively small MINT study, 125 patients with AMI within 6 h were randomized to heparin and low-dose or high-dose argatroban in addition to front-loaded tPA (JANG et al. 1999). TIMI grade 3 flow at 90 min was achieved in 42% of heparin, 57% of low-dose, and 59% of high-dose argatroban patients. In patients presenting after 3 h TIMI grade 3 flow was significantly more frequent in high-dose argatroban vs heparin patients (57% vs 20%; $p = 0.03$). Major bleeding was more common in the heparin group. The composite of death, recurrent AMI, cardiogenic shock or congestive heart failure, revascularisation, and recurrent ischemia at 30 days occurred in 37.5% of heparin, 32% of low-dose, and 25.5% of high-dose argatroban patients ($p = 0.23$). The results obtained in these two studies using a small-molecule direct thrombin inhibitor look promising and warrant further, larger trials.

6. Efegatran

In the randomized dose-finding ESCALAT study, comprising 245 patients with AMI of <12 h duration, 4 different doses of efegatran, a small-molecule direct thrombin inhibitor in combination with SK (1.5 Mio U) were compared with heparin and front-loaded tPA. The combination of efegatran plus SK was less effective than that of heparin plus tPA in achieving early patency. Death and major bleeding, although not significant, occurred more frequently in the efegatran group (FUNG et al. 1999).

II. Antiplatelet Agents

1. Aspirin

Aspirin is a partial inhibitor of platelet activation, able to block cyclooxygenase-dependent agonists such as ADP, collagen, epinephrine, and serotonin, though not fully able to block agonists which act independently of the cyclooxygenase pathway such as thrombin, thromboxane, and platelet activating factor (reviewed by AWTRY and LOSCALZO 2000). Some patients, however, are resistant to the action of aspirin due to a mutation in the cyclooxygenase gene (SCHMID et al. 1962; FUNK et al. 1991).

Although a dose as low as 30 mg/day may be all that is needed (THE DUTCH TIA TRIAL GROUP 1991), it is currently recommended to give 162.5–325 mg of aspirin immediately with thrombolysis, and daily thereafter (CAIRNS et al. 1995). This recommendation derives mainly from the ISIS-2 study. Patients randomized to SK or aspirin had similar 23% and 25% reductions in 5 week mortality ($p < 0.0001$). Combined therapy led to a 42% reduction in 5-week mortality and a 50% reduction in reinfarction. This benefit persisted at 15 months and 10 years follow-up [The ISIS-2 (Second International Study of Infarct Survival] Collaborative Group 1988; Baigent et al. 1998).

A meta-analysis by Roux et al. (1992) of 32 trials, including 4930 patients treated with either tPA or SK, showed that aspirin reduces angiographic reocclusion from 25% to 11%, and recurrent ischemia from 41% to 25%.

2. Glycoprotein IIb/IIIa Antagonists

GP IIb/IIIa platelet membrane receptor antagonists completely abolish platelet aggregation at a dosage resulting in more than 80% binding (Coller 1995). This has led to the development of a host of compounds which interfere with the binding of fibrinogen and of other ligands to the platelet membrane integrin receptor IIb/IIIa. The first compounds used in clinical trials is the humanized monoclonal mouse antibody c7E3 (abciximab, ReoPro) directed against the activated form of GP IIb/IIIa (Jang et al. 1994, reviewed by Faulds and Sorkin 1994). Other compounds inhibiting platelet aggregation are peptides comprising an RGD (Arg-Gly-Asp)-sequence or a KDG (Lys-Gly-Asp)-sequence which bind with high affinity to GP IIb/IIIa, such as eptifibatide (Integrilin, a cyclic heptapeptide). Another approach has been to mimic the charge and spatial conformation of the RGD sequence via engineered synthetic and semisynthetic compounds. Examples of such peptidomimetic intravenously used compounds are tirofiban (Aggrastat), lamifiban (Ro-44-9883) and fradafiban (BIBU 52ZW). Orally administered small molecules have also been developed such as xemilofiban (SC-54684A), sibrafiban (Ro-48-3657), lefradafiban (BIBU-104xx), lotrafiban (SB-214857), orbafiban (SC-57099B) and roxifiban (DMP-754) (reviewed by Madan et al. 1998 and Verstraete 2000). Most of these drugs have been initially used for the treatment of unstable angina and non-Q-wave myocardial infarction and in conjunction with PTCA or stent implantation where they have shown good efficacy in the prevention of later coronary events (recently reviewed by Ronner et al. 1998a; Kong et al. 1998; Adgey 1998; Harrington 1999; Kereiakes et al. 2000; Tcheng 2000; Kleiman et al. 2000). Investigators have long feared to use these agents in conjunction with thrombolytic therapy because of their potential to produce intracranial hemorrhage.

Only a few studies evaluated these agents with thrombolysis. The TAMI-8 dose-ranging pilot trial evaluated 60 patients treated with tPA, heparin, and aspirin and 50 patients with escalating bolus injections of c7E3 at 3h, 6h, and 15h after tPA; 10 patients received placebo. A patent IRA occurred in 92% of c7E3 treated patients, compared to 56% of control patients. Recurrent ischemia occurred in only 13% of c7E3 treated patients compared to 20% of controls. This benefit was achieved with a 25% risk of major bleeding compared to 50% in the control group predominantly during coronary artery bypass surgery (KLEIMAN et al. 1993). Two further studies have examined whether the adjunction of abciximab to thrombolysis in AMI patients allows reduction of the doses of the thrombolytic agent.

In the TIMI 14 trial, 888 patients with ST elevation were randomized to receive: (1) an accelerated full dose of alteplase; (2) abciximab without a thrombolytic agent in the form of a bolus of 0.25 mg/kg, followed by an IV infusion of 0.125 μg/kg/min for 12h; (3) reduced doses of alteplase (20–65 mg) plus abciximab; (4) reduced doses of SK plus abciximab. TIMI 3 flow at 90 min was achieved in 57% of patients in group 1, in 32% in group 2, in 47%–77% in group 3, and in 17%–33% in group 4 (ANTMAN et al. 1999; DE LEMOS et al. 2000). There was a fairly good consistency of results in the various subgroups analyzed (gender, age, diabetes, smokers).

The SPEED/GUSTO-4 dose-ranging trial investigated the combination of abciximab with reteplase. Patients ($n = 305$) with ST elevation AMI were randomized into six treatment groups: (1) abciximab alone given in the same dose as in the TIMI 14 trial; (2) abciximab plus a 5 U bolus of reteplase; (3) abciximab plus a 7.5 U bolus of reteplase; (4) abciximab plus a 10 U bolus of reteplase; (5) abciximab plus 2 boluses of 5 U and 2.5 U of reteplase (30 min apart); (6) abciximab plus double bolus of 5 U of reteplase. TIMI grade 3 flow at 60 min was 28% in group 1, 53% in group 2, 46% in group 3, 44% in group 4, 48% in group 5, and 63% in group 6. Respective corrected TIMI frame counts were 100, 40, 45, 44, 36, and 36. Hemorrhagic complications were not significantly increased with combination therapy (OHMAN et al. 1998; FERGUSON 1999). This trial continues and is the basis for a very large GUSTO-4 study which plans to randomize approximately 16,000 patients to treatment with either abciximab plus low-dose reteplase and low-dose weight-adjusted heparin or full-dose reteplase plus standard heparin. The primary endpoint of this study will be mortality at 30 days.

Three randomized pilot trials evaluated synthetic GP IIb/IIIa antagonists with tPA (integrilin in two trials, and lamifiban in the other). In the IMPACT-AMI trial, 180 patients with AMI were assigned to one of six integrilin doses or placebo. All patients received accelerated alteplase, aspirin, and IV heparin. TIMI grade 3 flow, 90 min after start of therapy, was achieved in 66% of the 49 patients who had received the highest integrilin doses, but in only 39% of the 52 patients receiving only alteplase and placebo. The incidence of bleeding was low in all groups (OHMAN et al. 1997). In the other integrilin study, 171 patients with AMI all received 1.5 Mio U of SK and were randomized to four conjunctive therapy modes: placebo or integrilin 0.18 mg/kg as a bolus followed by: (1) an IV infusion of 0.75 μg/kg per minute; (2) 1.33 μg/kg per minute; or (3) 2.0 μg/kg per minute. The highest dose was discontinued when an increased bleeding rate was observed. There were no intracranial hemorrhages and the vast majority of occurrences of bleeding were from arterial puncture sites. Coronary angiography was performed 90 min after the start of thrombolytic therapy. TIMI grade 3 flow was observed in 38% of patients in those receiving only SK, and in 53%, 44%, and 52% in the three integrilin groups respectively (RONNER et al. 1998b).

The PARADIGM trial was designed to assess the safety, pharmacody-namics, and effects on reperfusion of the GP IIb/IIIa inhibitor lamifiban when given with thrombolysis to 353 patients with AMI. Patients with ST segment elevation presenting within 12h of symptom onset who were treated with either tPA or SK were enrolled in this three-part phase II dose exploration study. In part A, all patients received lamifiban in an open-label, dose escala-tion scheme. Parts B and C were randomized, double-blind comparison of a bolus plus 24-h infusion of lamifiban vs placebo with patients randomized in a 2:1 ratio. The goal was to identify a dose or doses of lamifiban that provided >85% ADP-induced platelet aggregation inhibition. Platelet aggregation was inhibited by lamifiban in a dose-dependent manner with the highest doses exceeding 85% ADP-induced platelet aggregation inhibition. There was more bleeding associated with lamifiban; major bleeding occurred in 3.0% of patients receiving lamifiban and in 1.7% of controls; transfusions were given in 16.1% lamifiban-treated vs 10.3% placebo treated patients. Lamifiban induced more rapid reperfusion (75% vs 62% in controls as measured by con-tinuous ECG parameters (HARRINGTON 1998).

Heparin does not appear to be necessary when using abciximab for the reduction of arterial thrombosis assessed by platelet-deposition onto arterial media using the ex-vivo perfusion chamber in patients undergoing high-risk angioplasty with c7E3 (DANGAS et al. 1998). The use of heparin with synthetic GP IIb/IIIa inhibitors is currently being evaluated. Gp IIb/IIIa inhibitors may cause thrombocytopenia, probably through activation of platelets (PETER et al. 2000).

3. Other Drugs Inhibiting Platelet Function

In the AMISTAD trial, 236 patients with AMI were randomized to receive IV adenosine or placebo as conjunctive therapy in addition to tPA or SK. The primary endpoint was infarct size as assessed by SPECT-sestamibi at 6 days. Myocardial infarct size was significantly reduced in the adenosine group (33% relative reduction) (MAHAFFEY et al. 1999).

Clopidogrel, given at an initial dose of 75mg in addition to t-PA in an open, non-randomized study, was well tolerated in 116 patients with AMI (BASSAND et al. 1999). No efficacy data were given in this study and it is ques-tionable whether prompt platelet inhibition was achieved with this small dose; effective inhibition of platelet aggregation with clopidogrel within a few hours necessitates a loading dose of 300–375mg (BACHMANN et al. 1996).

Ridogrel resulted in no difference in patency at 7–14 days compared to aspirin in conjunction with SK, although recurrent ischemia was significantly reduced (13% vs 19%) (THE RAPT INVESTIGATORS 1994).

Therapy with prostacyclin either alone or in conjunction with tPA had limited success due to unacceptable side effects (TOPOL et al. 1989b). Prostaglandin E1 decreased the time to reperfusion with SK but not with tPA (SHARMA et al. 1986). Another prostacyclin analogue, taprostene was not effective with single chain urokinase (BAR et al. 1993).

Other conjunctive therapy with drugs which do not have an effect on the hemostasis system, such as β-blockers, ACE-inhibitors, nitrates, calcium channel blockers, antioxidants, and magnesium is not discussed here. The reader is referred to some recent publications dealing with these agents (KIZER et al. 1999; MICHAELS et al. 1999; DI PASQUALE et al. 1999; RAGHU et al. 1999; GERSH 1999).

4. Physical Conjunctive Therapy

Numerous in vitro and animal studies have demonstrated that low-frequency ultrasound augments thrombolysis (reviewed by ATAR et al. 1999). Ultrasound accelerates binding of tPA to crosslinked fibrin, increases binding affinity and the number of tPA-binding sites on crosslinked fibrin (SIDDIQI et al. 1998). In the ACUTE feasibility study intracoronary catheter-directed low-frequency ultrasound achieved TIMI grade 3 flow in 13 of 15 patients (87%) with AMI (ROSENSCHEIN et al. 1997).

In 24 dogs a thrombotic occlusion of the left anterior descending coronary artery was produced. After 60 min of occlusion tPA (1.4 mg/kg) was given IV over 90 min. In 12 dogs transcutaneous ultrasound (27 kHz) was applied over the chest. At 90 min after the start of thrombolytic therapy the TIMI 3 and TIMI 2+3 flow rates were, respectively, 67% and 83% in the ultrasound group and 25 and 33% in the group receiving only tPA ($p = 0.006$). Pathological examination did not reveal injury secondary to ultrasound in the skin, soft tissues, heart or lungs (SIEGEL et al. 2000). It thus appears that noninvasive transthoracic ultrasound has substantial potential as a conjunctive means to improve coronary thrombolysis.

E. Outlook

Compared to the presently optimal treatment of AMI, using the accelerated tPA regime and conjunctive therapy with aspirin and heparin, further improvements of conjunctive treatment schemes are to be expected. These include low-molecular-weight heparins, more individualized doses of hirudin and hirulog (possibly by means of the ecarin clotting time), administered over longer periods of time, and combination therapy of Gp IIb/IIIa receptor inhibitors with reduced doses of a newer thrombolytic agent.

List of Abbreviations

ACUTE	Analysis of Coronary Ultrasound Thrombolysis Eudpoints
AIMS	APSAC Intervention Mortality Study

AMI	Acute Myocardial Infarction
AMISTAD	Acute Myocardial Infarction Study of Adenosine
APSAC	Anisoylated Plasminogen Streptokinase Activator Complex
ARGAMI	ARGatroban vs heparin as adjuvant therapy to thrombolysis for Acute Myocardial Infarction
aPTT	Activated Partial Thromboplastin Time
AT-III	Antithrombin III
BIOMACS	BIOchemical Markers in Acute Coronary Syndromes
CHF	Congestive Heart Failure
COBALT	COntinuous infusion versus double-Bolus administration of ALTeplase
ECG	Electrocardiogram
EMERAS	Estudio Multicéntrico Estreptoquinasa Repúblicas de América del Sur
ESCALAT	Efegatran enhances Streptokinase patency of Coronary Arteries Like Accelerated TPA
F1.2	Prothrombin Fragment 1 + 2
FDPs	Fibrin(ogen) Degradation Products
FPA	Fibrinopeptide A
FTT	Fibrinolytic Therapy Trialists
GISSI	Gruppo Italiano per lo Studio della Streptochinasi nell'Infarto miocardico
GP	Glycoprotein
GREAT	Grampian Region Early Anistreplase Trial
GUSTO	Global Utilization of Streptokinase and Tissue plasminogen activator for Occluded coronary arteries
HEAP	Heparin in Early Patency
HERO	Hirulog Early Reperfusion/Occlusion
HIT	Hirudin for the Improvement of Thrombolysis
IMPACT	Integrilin to Minimize Platelet Aggregation and Coronary Thrombosis
IRA	Infarct-related Artery
I.S.A.M.	Intravenous Streptokinase in Acute Myocardial infarction
ISIS	International Study of Infarct Survival
IV	Intravenous
LATE	Late Assessment of Thrombolytic Efficacy
LMWH	Low-molecular-weight heparin(s)
MINT	Myocardial Infarction with Novostan and TPA
PAI-1	Plasminogen Activator Inhibitor type-1
PARADIGM	Platelet Aggregation Receptor Antagonist Dose Investigation and reperfusion Gain in Myocardial infarction
PTA	Percutaneous Transluminal Angioplasty
PTCA	Percutaneous Transluminal Coronary Angioplasty
RAPT	Ridogrel versus Aspirin Patency Trial

SCATI   Studio sulla Calciparina nell'Angina e nella Thrombosi Ven-
             tricolare nell'Infarto
SK   Streptokinase
SPEED   Strategies for Patency Enhancement in the Emergency
             Department
TAMI   Thrombolysis and Angioplasty in Myocardial Infarction
TAT   Thrombin-Antithrombin complex
TIMI   Thrombolysis In Myocardial Infarction
tPA   tissue-type Plasminogen Activator
TXA$_2$ Thromboxane

References

Adelman B, Michelson AD, Loscalzo J, Greenberg J, Handin RI (1985) Plasmin effect on platelet glycoprotein Ib–von Willebrand factor interactions. Blood 65:32–40

Adgey AA (1998) An overview of the results of clinical trials with glycoprotein IIb/IIIa inhibitors. Eur Heart J 19 [Suppl D]:D10–D21

AIMS Trial Study Group (1988) Effect of intravenous APSAC on mortality after acute myocardial infarction. Preliminary report of a placebo-controlled clinical trial. Lancet 1:545–549

Alevriadou BR, Moake JL, Turner NA, Ruggeri ZM, Folie BJ, Phillips MD, Schreiber AB, Hrinda ME, McIntire LV (1993) Real-time analysis of shear-dependent thrombus formation and its blockade by inhibitors of von Willebrand factor binding to platelets. Blood 81:1263–1276

Ambrose JA (1992) Plaque disruption and the acute coronary syndromes of unstable angina and myocardial infarction: if the substrate is similar, why is the clinical presentation different? J Am Coll Cardiol 19:1653–1658

Antman EM, for the TIMI 9B Investigators (1996) Hirudin in acute myocardial infarction. Thrombolysis and thrombin inhibition in myocardial infarction (TIMI) 9B trial. Circulation 94:911–921

Antman EM, for the TIMI 9A Investigators (1994) Hirudin in acute myocardial infarction. Safety report from the Thrombolysis and Thrombin Inhibition in Myocardial Infarction (TIMI) 9A trial. Circulation 90:1624–1630

Antman EM, Cohen M (1999) Newer antithrombin agents in acute coronary syndromes. Am Heart J 138 [Suppl]:S563–S569

Antman EM, Giugliano RP, Gibson CM, McCabe CH, Coussement P, Kleiman NS, Vahanian A, Adgey AA, Menown I, Rupprecht H-J, Van der Wieken R, Ducas J, Scherer J, Anderson K, Van de Werf F, Braunwald E, for the TIMI 14 Investigators (1999) Abciximab facilitates the rate and extent of thrombolysis. Results of the thrombolysis in myocardial infarction (TIMI) 14 trial. Circulation 99: 2720–2732

Arnout J, Simoons M, de Bono D, Rapold HJ, Collen D, Verstraete M (1992) Correlation between level of heparinization and patency of the infarct-related coronary artery after treatment of acute myocardial infarction with alteplase (rt-PA). J Am Coll Cardiol 20:513–519

Aronson DL, Chang P, Kessler CM (1992) Platelet-dependent thrombin generation after in vitro fibrinolytic treatment. Circulation 85:1706–1712

Awtry EH, Loscalzo J (2000) Aspirin. Circulation 101:1206–1218

Atar S, Luo H, Nagai T, Siegel RJ (1999) Ultrasonic thrombolysis: catheter-delivered and transcutaneous applications. Eur J Ultrasound 9:39–54

Bachmann F, Savcic M, Hauert J, Geudelin JB, Kieffer G, Cariou R (1996) Rapid onset of inhibition of ADP-induced platelet aggregation by a loading dose of clopidogrel. Eur Heart J 17 [Suppl]: 263 (Abstract)

Back LD, Radbill JR, Crawford DW (1977) Analysis of pulsatile, viscous blood flow through diseased coronary arteries of man. J Biomech 10:339–353

Badimon L, Badimon JJ (1989) Mechanisms of arterial thrombosis in nonparallel streamlines: platelet thrombi grow on the apex of stenotic severely injured vessel wall. Experimental study in the pig model. J Clin Invest 84:1134–1144

Badimon L, Badimon JJ, Galvez A, Chesebro JH, Fuster V (1986) Influence of arterial damage and wall shear rate on platelet deposition. Ex vivo study in a swine model. Arteriosclerosis 6:312–320

Badimon L, Badimon JJ, Turitto VT, Vallabhajosula S, Fuster V (1988) Platelet thrombus formation on collagen type I. A model of deep vessel injury. Influence of blood rheology, von Willebrand factor, and blood coagulation. Circulation 78:1431–1442

Baigent C, Collins R, Appleby P, Parish S, Sleight P, Peto R, on behalf of the ISIS-2 (Second International Study of Infarct Survival) Collaborative Group (1998) ISIS-2: 10 year survival among patients with suspected acute myocardial infarction in randomized comparison of intravenous streptokinase, oral aspirin, both, or neither. BMJ 316:1337–1343

Bär FW, Meyer J, Michels R, Uebis R, Lange S, Barth H, Groves R, Vermeer F (1993) The effect of taprostene in patients with acute myocardial infraction treated with thrombolytic therapy: results of the START study. Saruplase Taprostene Acute Reocclusion Trial. Eur Heart J 14:1118–1126

Barbagelata NA, Granger CB, Oqueli E, Suárez LD, Borruel M, Topol EJ, Califf RM (1997) TIMI grade 3 flow and reocclusion after intravenous thrombolytic therapy: a pooled analysis. Am Heart J 133:273–282

Barbash GI, Hod H, Roth A, Miller HI, Rath S, Zahav YH, Modan M, Zivelin A, Laniado S, Seligsohn U (1989) Correlation of baseline plasminogen activator inhibitor activity with patency of the infarct artery after thrombolytic therapy in acute myocardial infarction. Am J Cardiol 64:1231–1235

Bassand J-P, Cariou R, Grollier G, Kragten J, Wolf J-E, Heyndrickx GR (1999) Clopidogrel-rt-PA-heparin combination in the treatment of acute myocardial infarction. Semin Thromb Hemost 25 [Suppl 2]:69–75

Behar S, Hod H, Kaplinsky E, for the ARGAMI-2 Study Group (1998) Argatroban versus heparin as adjuvant therapy to thrombolysis for acute myocardial infarction: safety considerations – ARGAMI-2 Study. Circulation 98:453 (Abstract)

Bleich SD, Nichols TC, Schumacher RR, Cook DH, Tate DA, Teichman SL (1990) Effect of heparin on coronary arterial patency after thrombolysis with tissue plasminogen activator in acute myocardial infarction. Am J Cardiol 66:1412–1417

Boersma E, Maas ACP, Deckers JW, Simoons ML (1996) Early thrombolytic treatment in acute myocardial infarction: reappraisal of the golden hour. Lancet 348:771–775

Cairns JA, Fuster V, Gore J, Kennedy JW (1995) Coronary thrombolysis. Chest 108 [Suppl]:401S–423S

Cannon CP (1995) Thrombin inhibitors in acute myocardial infarction. Cardiol Clin 13:421–433

Cannon CP, McCabe CH, Henry TD, Schweiger MJ, Gibson RS, Mueller HS, Becker RC, Kleiman NS, Haugland JM, Anderson JL, Sharaf BL, Rogers WJ, Williams DO, Braunwald E (1994) A pilot trial of recombinant desulfatohirudin compared with heparin in conjunction with tissue-type plasminogen activator and aspirin for acute myocardial infarction: Results of the Thrombolysis in Myocardial Infarction (TIMI) 5 trial. J Am Coll Cardiol 23:993–1003

Cercek B, Lew AS, Hod H, Yano J, Reddy NK, Ganz W (1986) Enhancement of thrombolysis with tissue-type plasminogen activator by pretreatment with heparin. Circulation 74:583–587

Chesebro JH, Fuster V (1991) Dynamic thrombosis and thrombolysis. Role of antithrombins. Circulation 83:1815–1817

Chesebro JH, Knatterud G, Roberts R, Borer J, Cohen LS, Dalen J, Dodge HT, Francis
    CK, Hillis D, Ludbrook P, Markis JE, Mueller H, Passamani ER, Powers ER,
    Rao AK, Robertson T, Ross A, Ryan TJ, Sobel BE, Willerson J, Williams DO,
    Zaret BL, Braunwald E (1987) Thrombolysis in myocardial infarction (TIMI) trial,
    phase I: a comparison between intravenous tissue plasminogen activator and intra-
    venous streptokinase. Clinical findings through hospital discharge. Circulation
    76:142–154
Chesebro JH, Toschi V, Lettino M, Gallo R, Badimon JJ, Fallon JT, Fuster V (1995)
    Evolving concepts in the pathogenesis and treatment of arterial thrombosis. Mt
    Sinai J Med 62:275–286
Cohen M, Adams PC, Parry G, Xiong J, Chamberlain D, Wieczorek I, Fox KAA,
    Chesebro JH, Strain J, Keller C, Kelly A, Lancaster G, Ali J, Kronmal R, Fuster
    V, and the Antithrombotic Therapy in Acute Coronary Syndromes Research
    Group (1994) Combination antithrombotic therapy in unstable rest angina and
    non-Q-wave infarction in nonprior aspirin users. Primary end points analysis from
    the ATACS trial. Circulation 89:81–88
Collen D (1993) Towards improved thrombolytic therapy. Lancet 342:34–36
Coller BS (1995) Blockade of platelet GPIIb/IIIa receptors as an antithrombotic
    strategy. Circulation 92:2373–2380
Dangas G, Badimon JJ, Coller BS, Fallon JT, Sharma SK, Hayes RM, Meraj P, Ambrose
    JA, Marmur JD (1998) Administration of abciximab during percutaneous coro-
    nary intervention reduces both ex vivo platelet thrombus formation and fibrin
    deposition: implications for a potential anticoagulant effect of abciximab. Arte-
    rioscler Thromb Vasc Biol 18:1342–1349
Davies MJ, Thomas AC (1985) Plaque fissuring – the cause of acute myocardial
    infarction, sudden ischaemic death, and crescendo angina. Br Heart J 53:363–
    373
de Bono DP, Simoons ML, Tijssen J, Arnold AER, Betriu A, Burgersdijk C, Bescos LL,
    Mueller E, Pfisterer M, Van de Werf F, Zijlstra F, Verstraete M, for the European
    Cooperative Study Group (1992) Effect of early intravenous heparin on coronary
    patency, infarct size, and bleeding complications after alteplase thrombolysis:
    results of a randomized double blind European Cooperative Study Group trial
    (ECSG-6). Br Heart J 67:122–128
de Lemos JA, Antman EM, Gibson CM, McCabe CH, Giugliano RP, Murphy SA,
    Coulter SA, Anderson K, Scherer J, Frey MJ, Van der Wieken R, Van de Werf F,
    Braunwald E, for the TIMI 14 Investigators (2000) Abciximab improves both epi-
    cardial flow and myocardial reperfusion in ST-elevation myocardial infarction.
    Observations from the TIMI 14 trial. Circulation 101:239–243
Dellborg M, Topol EJ, Swedberg K (1991) Dynamic QRS complex and ST segment
    vectorcardiographic monitoring can identify vessel patency in patients with
    acute myocardial infarction treated with reperfusion therapy. Am Heart J 122:943–
    948
DeWood MA, Spores J, Notske R, Mouser LT, Burroughs R, Golden MS, Lang HT
    (1980) Prevalence of total coronary occlusion during the early hours of transmural
    myocardial infarction. N Engl J Med 303:897–902
Di Pasquale P, Lo Verso P, Bucca V, Cannizzaro S, Scalzo S, Maringhini G, Rizzo R,
    Paterna S (1999) Effects of trimetazidine administration before thrombolysis in
    patients with anterior myocardial infarction: short-term and long-term results.
    Cardiovasc Drugs Ther 13:423–428
Drake TA, Morrissey JH, Edgington TS (1989) Selective cellular expression of tissue
    factor in human tissues. Am J Pathol 134:1087–1097
Eisenberg PR, Sherman LA, Jaffe AS (1987) Paradoxic elevation of fibrinopeptide A
    after streptokinase: evidence for continued thrombosis despite intense fibrinoly-
    sis. J Am Coll Cardiol 10:527–529
Eisenberg PR, Sobel BE, Jaffe AS (1992) Activation of prothrombin accompanying
    thrombolysis with recombinant tissue-type plasminogen activator. J Am Coll
    Cardiol 19:1065–1069

EMERAS (Estudio Multicéntrico Estreptoquinasa Repúblicas de América del Sur) Collaborative Group (1993) Randomised trial of late thrombolysis in patients with suspected acute myocardial infarction. Lancet 342:767–772

Falk E (1983) Plaque rupture with severe pre-existing stenosis precipitating coronary thrombosis: characteristics of coronary atherosclerotic plaques underlying fatal occlusive thrombi. Br Heart J 50:127–134

Falk E, Shah PK, Fuster V (1995) Coronary plaque disruption. Circulation 92:657–671

Faulds D, Sorkin EM (1994) Abciximab (c7E3 Fab). A review of its pharmacology and therapeutic potential in ischaemic heart disease. Drugs 48:583–598

Ferguson JJ (1999) Meeting highlights. Highlights of the 71st scientific sessions of the American Heart Association. Circulation 99:2486–2491

Fibrinolytic Therapy Trialists' (FTT) Collaborative Group (1994) Indications for fibrinolytic therapy in suspected acute myocardial infarction: Collaborative overview of early mortality and major morbidity results from all randomised trials of more than 1000 patients. Lancet 343:311–322

Fitzgerald DJ, Catella F, Roy L, FitzGerald GA (1988) Marked platelet activation in vivo after intravenous streptokinase in patients with acute myocardial infarction. Circulation 77:142–150

Frostfeldt G, Ahlberg G, Gustafsson G, Helmius G, Lindahl B, Nygren A, Siegbahn A, Swahn E, Venge P, Wallentin L (1999) Low molecular weight heparin (dalteparin) as adjuvant treatment of thrombolysis in acute myocardial infarction – a pilot study: biochemical markers in acute coronary syndromes (BIOMACS II). J Am Coll Cardiol 33:627–633

Fung AY, Lorch G, Cambier PA, Hansen D, Titus BG, Martin JS, Lee JJ, Every NR, Hallstrom AP, Stock-Novack D, Scherer J, Weaver WD, and the ESCALAT Investigators (1999) Efegatran sulfate as an adjunct to streptokinase versus heparin as an adjunct to tissue plasminogen activator in patients with acute myocardial infarction. Am Heart J 138:696–704

Funk CD, Funk LB, Kennedy ME, Pong AS, FitzGerald GA (1991) Human platelet/-erythroleukemia cell prostaglandin G/H synthase: cDNA cloning, expression, and gene chromosomal assignment. FASEB J 5:2304–2312

Fuster V, Badimon L, Badimon JJ, Chesebro JH (1992) The pathogenesis of coronary artery disease and the acute coronary syndromes. N Engl J Med 326:242–250 &-310–318

Gersh BJ (1999) Optimal management of acute myocardial infarction at the dawn of the next millennium. Am Heart J 138 [Suppl 2]:S188–S202

Gertz DS, Kragel AH, Kalan JM, Braunwald E, Roberts WC, and the TIMI Investigators (1990) Comparison of coronary and myocardial morphologic findings in patients with and without thrombolytic therapy during fatal first acute myocardial infarction. Am J Cardiol 66:904–909

Gibson CM, Cannon CP, Piana RN, Breall JA, Sharaf B, Flatley M, Alexander B, Diver DJ, McCabe CH, Flaker GC, Baim DS, Braunwald E (1995) Angiographic predictors of reocclusion after thrombolysis: results from the Thrombolysis In Myocardial Infarction (TIMI) 4 Trial. J Am Coll Cardiol 25:582–589

Glick A, Kornowski R, Michowich Y, Koifman B, Roth A, Laniado S, Keren G (1996) Reduction of reinfarction and angina with use of low-molecular-weight heparin therapy after streptokinase (and heparin) in acute myocardial infarction. Am J Cardiol 77:1145–1148

Golino P, Ashton JH, Glas Greenwalt P, McNatt J, Buja LM, Willerson JT (1988) Mediation of reocclusion by thromboxane A_2 and serotonin after thrombolysis with tissue-type plasminogen activator in a canine preparation of coronary thrombosis. Circulation 77:678–684

Gouin I, Lecompte T, Morel M-C, Lebrazi J, Modderman PW, Kaplan C, Samama MM (1992) In vitro effect of plasmin on human platelet function in plasma. Inhibition of aggregation caused by fibrinogenolysis. Circulation 85:935–941

Granger CB, Becker R, Tracy RP, Califf RM, Topol EJ, Pieper KS, Ross AM, Roth S, Lambrew C, Bovill EG, for the GUSTO-I Hemostasis Substudy Group (1998)

Thrombin generation, inhibition and clinical outcomes in patients with acute myocardial infarction treated with thrombolytic therapy and heparin: results from the GUSTO-I Trial. J Am Coll Cardiol 31:497–505

Granger CB, Hirsch J, Califf RM, Col J, White HD, Betriu A, Woodlief LH, Lee KL, Bovill EG, Simes RJ, Topol EJ, for the GUSTO-I Investigators (1996) Activated partial thromboplastin time and outcome after thrombolytic therapy for acute myocardial infarction: results from the GUSTO-I trial. Circulation 93:870–878

GREAT Group (1993) Feasibility, safety, and efficacy of domiciliary thrombolysis by general practitioners: Grampian region early anistreplase trial. BMJ 305:548–553

Gruppo Italiano per lo Studio della Sopravvivenza nell'Infarto Miocardico (1990) GISSI-2: A factorial randomised trial of alteplase versus streptokinase and heparin versus no heparin among 12 490 patients with acute myocardial infarction. Lancet 336:65–71

Gruppo Italiano per lo Studio della Streptochinasi nell'Infarto Miocardico (GISSI) (1986) Effectiveness of intravenous thrombolytic treatment in acute myocardial infarction. Lancet 1:397–402

Harrington RA (1998) Combining thrombolysis with the platelet glycoprotein IIb/IIIa inhibitor lamifiban: Results of the platelet aggregation receptor antagonist dose investigation and reperfusion gain in myocardial infarction (PARADIGM) trial. J Am Coll Cardiol 32:2003–2010

Harrington RA (1999) Overview of clinical trials of glycoprotein IIb-IIIa inhibitors in acute coronary syndromes. Am Heart J 138 (Suppl.):S276–S286

Heras M, Chesebro JH, Penny WJ, Bailey KR, Badimon L, Fuster V (1989) Effects of thrombin inhibition on the development of acute platelet-thrombus deposition during angioplasty in pigs. Heparin versus recombinant hirudin, a specific thrombin inhibitor. Circulation 79:657–665

Heras M, Chesebro JH, Webster MWI, Mruk JS, Grill DE, Penny WJ, Bowie EJW, Badimon L, Fuster V (1990) Hirudin, heparin, and placebo during deep arterial injury in the pig. The in vivo role of thrombin in platelet-mediated thrombosis. Circulation 82:1476–1484

Hirsh J, Raschke R, Warkentin TE, Dalen JE, Deykin D, Poller L (1995) Heparin: mechanism of action, pharmacokinetics, dosing considerations, monitoring, efficacy, and safety. Chest 108 [Suppl]:258S–275S

Hochman JS, Wali AU, Gavrila D, Sim MJ, Malhotra S, Palazzo AM, De La Fuente B (1999) A new regimen for heparin use in acute coronary syndromes. Am Heart J 138:313–318

Hoffmeister HM, Szabo S, Kastner C, Beyer ME, Helber U, Kazmaier S, Wendel HP, Heller W, Seipel L (1998) Thrombolytic therapy in acute myocardial infarction: comparison of procoagulant effects of streptokinase and alteplase regimens with focus on the kallikrein system and plasmin. Circulation 98:2527–2533

Hsia J, Hamilton WP, Kleiman N, Roberts R, Chaitman BR, Ross AM, for the Heparin-Aspirin Reperfusion Trial (HART) Investigators (1990) A comparison between heparin and low-dose aspirin as adjunctive therapy with tissue plasminogen activator for acute myocardial infarction. N Engl J Med 323:1433–1437

Hsia J, Kleiman N, Aguirre F, Chaitman BR, Roberts R, Ross AM, for the Heparin-Aspirin Reperfusion Trial (HART) Investigators (1992) Heparin-induced prolongation of partial thromboplastin time after thrombolysis: relation to coronary artery patency. J Am Coll Cardiol 20:31–35

ISIS (International Studies of Infarct Survival) Pilot Study Investigators (1987) Randomized factorial trial of high-dose intravenous streptokinase, of oral aspirin and of intravenous heparin in acute myocardial infarction. Eur Heart J 8:634–642

ISIS-3 (Third International Study of Infarct Survival) Collaborative Group (1992) ISIS-3: a randomised comparison of streptokinase vs tissue plasminogen activator vs anistreplase and of aspirin plus heparin vs aspirin alone among 41 299 cases of suspected acute myocardial infarction. Lancet 339:753–770

Jang IK, Gold HK, Ziskind AA, Fallon JT, Holt RE, Leinbach RC, May JW, Collen D (1989) Differential sensitivity of erythrocyte-rich and platelet-rich arterial thrombi to lysis with recombinant tissue-type plasminogen activator. A possible explanation for resistance to coronary thrombolysis. Circulation 79:920–928

Jang I-K, Brown DF, Giugliano RP, Anderson HV, Losordo D, Nicolau JC, Dutra OP, Bazzino O, Viamonte VM, Norbady R, Liprandi AS, Massey TJ, Dinsmore R, Schwarz RP Jr, and the MINT Investigators (1999) A multicenter, randomized study of argatroban versus heparin as adjunct to tissue plasminogen activator (TPA) in acute myocardial infarction: myocardial infarction with novastan and TPA (MINT) study. J Am Coll Cardiol 33:1879–1885

Jang Y, Lincoff AM, Plow EF, Topol EJ (1994) Cell adhesion molecules in coronary artery disease. J Am Coll Cardiol 24:1591–1601

Kaplan K, Davison R, Parker M, Mayberry B, Feiereisel P, Salinger M (1987) Role of heparin after intravenous thrombolytic therapy for acute myocardial infarction. Am J Cardiol 59:241–244

Karagounis L, Sorensen SG, Menlove RL, Moreno F, Anderson JL, for the TEAM-2 Investigators (1992) Does thrombolysis in myocardial infarction (TIMI) perfusion grade 2 represent a mostly patent artery or a mostly occluded artery? Enzymatic and electrocardiographic evidence from the TEAM-2 study. J Am Coll Cardiol 19:1–10

Kawano K, Aoki I, Aoki N, Homori M, Maki A, Hioki Y, Hasumura Y, Terano A, Arai T, Mizuno H, Ishikawa K (1998) Human platelet activation by thrombolytic agents: effects of tissue-type plasminogen activator and urokinase on platelet surface P-selectin expression. Am Heart J 135:268–271

Kennedy JW, Ritchie JL, Davis KB, Stadius ML, Maynard C, Fritz JK (1985) The Western Washington randomized trial of intracoronary streptokinase in acute myocardial infarction: a 12 month follow-up report. N Engl J Med 212:1073–1078

Kereiakes DJ, McDonald M, Broderick T, Roth EM, Whang DD, Martin LH, Howard WL, Schneider J, Shimshak T, Abbottsmith CW (2000) Platelet glycoprotein IIb/IIIa receptor blockers: An appropriate-use model for expediting care in acute coronary syndromes. Am Heart J 139 [Suppl]:S53–S60

Kerins DM, Roy L, FitzGerald GA, Fitzgerald DJ (1989) Platelet and vascular function during coronary thrombolysis with tissue-type plasminogen activator. Circulation 80:1718–1725

Kim CB, Braunwald E (1993) Potential benefits of late reperfusion of infarcted myocardium. The open artery hypothesis. Circulation 88:2426–2436

Kizer JR, Cannon CP, McCabe CH, Mueller HS, Schweiger MJ, Davis VG, Perritt R, Antman EM, for the TIMI Investigators (1999) Trends in the use of pharmacotherapies for acute myocardial infarction among physicians who design and/or implement randomized trials versus physicians in routine clinical practice: the MILIS-TIMI experience. Am Heart J 137:79–92

Kleiman NS, Lincoff AM, Flaker GC, Pieper KS, Wilcox RG, Berdan LG, Lorenz TJ, Cokkinos DV, Simoons ML, Boersma E, Topol EJ, Califf RM, Harrington RA, for the PURSUIT Investigators (2000) Early percutaneous coronary intervention, platelet inhibition with eptifibatide, and clinical outcomes in patients with acute coronary syndromes. Circulation 101:751–757

Kleiman NS, Ohman EM, Califf RM, George BS, Kereiakes D, Aguirre FV, Weisman H, Schaible T, Topol EJ (1993) Profound inhibition of platelet aggregation with monoclonal antibody 7E3 Fab after thrombolytic therapy. Results of the Thrombolysis and Angioplasty in Myocardial Infarction (TAMI) 8 Pilot Study. J Am Coll Cardiol 22:381–389

Kong DF, Califf RM, Miller DP, Moliterno DJ, White HD, Harrington RA, Tcheng JE, Lincoff AM, Hasselblad V, Topol EJ (1998) Clinical outcomes of therapeutic agents that block the platelet glycoprotein IIb/IIIa integrin in ischemic heart disease. Circulation 98:2829–2835

Krucoff MW, Green CE, Satler LF, Miller FC, Pallas RS, Kent KM, Del Negro AA, Pearle DL, Fletcher RD, Rackley CE (1986) Noninvasive detection of coronary artery patency using continuous ST-segment monitoring. Am J Cardiol 57:916–922

Kruik JH, Verheugt FWA, Hacke W, von Kummer R, Van de Werf F, for the COBALT Investigators (1998) Correlation of activated partial thromboplastin time at 6 hours after thrombolysis for acute myocardial infarction and hemorrhagic stroke in the COBALT trial. Circulation 98 [Suppl I]:562–562 (Abstract)

Kunitada S, FitzGerald GA, Fitzgerald DJ (1992) Inhibition of clot lysis and decreased binding of tissue-type plasminogen activator as a consequence of clot retraction. Blood 79:1420–1427

LATE Study Group (1993) Late assessment of thrombolytic efficacy (LATE) study with alteplase 6–24 hours after onset of acute myocardial infarction. Lancet 342:759–766

Lee CD, Mann KG (1989) Activation/inactivation of human factor V by plasmin. Blood 73:185–190

Lee LV, for the TIMI 6 Investigators (1995) Initial experience with hirudin and streptokinase in acute myocardial infarction: results of the Thrombolysis In Myocardial Infarction (TIMI) 6 trial. Am J Cardiol 75:7–13

Lidón RM, Théroux P, Lespérance J, Adelman B, Bonan R, Duval D, Lévesque J (1994) A pilot, early angiographic patency study using a direct thrombin inhibitor as adjunctive therapy to streptokinase in acute myocardial infarction. Circulation 89:1567–1572

Lincoff AM, Topol EJ (1993) Illusion of reperfusion. Does anyone achieve optimal reperfusion during acute myocardial infarction? Circulation 88:1361–1374

Lucore CL, Sobel BE (1988) Interactions of tissue-type plasminogen activator with plasma inhibitors and their pharmacologic implications. Circulation 77:660–669

McRedmond JP, Harriott P, Walker B, Fitzgerald DJ (2000) Streptokinase-induced platelet activation involves antistreptokinase antibodies and cleavage of protease-activated receptor-1. Blood 95:1301–1308

Madan M, Berkowitz SD, Tcheng JE (1998) Glycoprotein IIb/IIIa integrin blockade. Circulation 98:2629–2635

Mahaffey KW, Puma JA, Barbagelata NA, DiCarli MF, Leesar MA, Browne KF, Eisenberg PR, Bolli R, Casas AC, Molina-Viamonte V, Orlandi C, Blevins R, Gibbons RJ, Califf RM, Granger CB, for the AMISTAD Investigators (1999) Adenosine as an adjunct to thrombolytic therapy for acute myocardial infarction: results of a multicenter, randomized, placebo-controlled trial: the Acute Myocardial Infarction STudy of ADenosine (AMISTAD) trial. J Am Coll Cardiol 34:1711–1720

Mailhac A, Badimon JJ, Fallon JT, Fernández-Ortiz A, Meyer B, Chesebro JH, Fuster V, Badimon L (1994) Effect of an eccentric severe stenosis on fibrin(ogen) deposition on severely damaged vessel wall in arterial thrombosis. Relative contribution of fibrin(ogen) and platelets. Circulation 90:988–996

Mann KG, Tracy PB, Nesheim ME (1985) Assembly and function of prothrombinase complex on synthetic and natural membranes. In: Oates JA, Harwiger J, Ross R (eds). Interaction of platelets with the vessel wall, American Physiologic Society, Washington D.C., pp 47–57

Marder VJ, Sherry S (1988) Thrombolytic therapy: current status. N Engl J Med 318:1512–1520, 1585–1595

Marmur JD, Thiruvikraman SV, Fyfe BS, Guha A, Sharma SK, Ambrose JA, Fallon JT, Nemerson Y, Taubman MB (1996) Identification of active tissue factor in human coronary atheroma. Circulation 94:1226–1232

Merlini PA, Bauer KA, Oltrona L, Ardissino D, Cattaneo M, Belli C, Mannucci PM, Rosenberg RD (1994) Persistent activation of coagulation mechanism in unstable angina and myocardial infarction. Circulation 90:61–68

Metz BK, White HD, Granger CB, Simes RJ, Armstrong PW, Hirsh J, Fuster V, MacAulay CM, Califf RM, Topol EJ, for the Global Use of Strategies to Open Occluded Coronary Arteries in Acute Coronary Syndromes (GUSTO-IIb) Investigators (1998) Randomized comparison of direct thrombin inhibition versus heparin in conjunction with fibrinolytic therapy for acute myocardial infarction: results from the GUSTO-IIb Trial. J Am Coll Cardiol 31:1493–1498

Meyer BJ, Badimon JJ, Mailhac A, Fernández-Ortiz A, Chesebro JH, Fuster V, Badimon L (1994) Inhibition of growth of thrombus on fresh mural thrombus. Targeting optimal therapy. Circulation 90:2432–2438

Michaels AD, Maynard C, Every NR, Barron HV, for the National Registry of Myocardial Infarction 2 Participants (1999) Early use of ACE inhibitors in the treatment of acute myocardial infarction in the United States: experience from the National Registry of Myocardial Infarction 2. Am J Cardiol 84:1176–1181

Mruk JS, Zoldhelyi P, Webster MWI, Heras M, Grill DE, Holmes DR, Fuster V, Chesebro JH (1996) Does antithrombotic therapy influence residual thrombus after thrombolysis of platelet-rich thrombus? Effects of recombinant hirudin, heparin, or aspirin. Circulation 93:792–799

Neuhaus K-L, Molhoek GP, Zeymer U, Tebbe U, Wegscheider K, Schröder R, Camez A, Laarman GJ, Grollier GM, Lok DJ, Kuckuck H, Lazarus P, for the HIT-4 Investigators (1999) Recombinant hirudin (lepirudin) for the improvement of thrombolysis with streptokinase in patients with acute myocardial infarction: results of the HIT-4 trial. J Am Coll Cardiol 34:966–973

Neuhaus K-L, von Essen R, Tebbe U, Jessel A, Heinrichs H, Mäurer W, Döring W, Harmjanz D, Kötter V, Kalhammer E, Simon H, Horacek T (1994) Safety observations from the pilot phase of the randomized r-hirudin for improvement of thrombolysis (HIT-III) study. A study of the Arbeitsgemeinschaft Leitender Kardiologischer Krankenhausärzte (ALKK). Circulation 90:1638–1642

Niewiarowski S, Senyi AF, Gillies P (1973) Plasmin-induced platelet aggregation and platelet release reaction. Effects on hemostasis. J Clin Invest 52:1647–1659

Ohman EM, Califf RM, Topol EJ, Candela R, Abbottsmith C, Ellis S, Sigmon KN, Kereiakes D, George B, Stack R, and the TAMI Study Group (1990) Consequences of reocclusion after successful reperfusion therapy in acute myocardial infarction. Circulation 82:781–791

Ohman EM, Kleiman NS, Gacioch G, Worley SJ, Navetta FI, Talley JD, Anderson HV, Ellis SG, Cohen MD, Spriggs D, Miller M, Kereiakes D, Yakubov S, Kitt MM, Sigmon KN, Califf RM, Krucoff MW, Topol EJ, for the IMPACT-AMI Investigators (1997) Combined accelerated tissue-plasminogen activator and platelet glycoprotein IIb/IIIa integrin receptor blockade with Integrilin in acute myocardial infarction. Results of a randomized, placebo-controlled, dose-ranging trial. Circulation 95:846–854

Ohman EM, Lincoff AM, Bode C, Bachinsky WB, Ardissino D, Betriu A, Schildcrout JS, Oliverio R, Barnathan E, Sherer J, Sketch MS, Topol EJ (1998) Enhanced early reperfusion at 60 minutes with low-dose reteplase combined with full-dose abciximab in acute myocardial infarction: preliminary results from the GUSTO-4 pilot (SPEED) dose-ranging trial. Circulation 98 [Suppl I]:504 (Abstract)

Owen J, Friedman KD, Grossman BA, Wilkins C, Berke AD, Powers ER (1988) Thrombolytic therapy with tissue plasminogen activator or streptokinase induces transient thrombin activity. Blood 72:616–620

Peter K, Straub A, Kohler B, Nordt T, Ruef J, Moser M, Schwarz M, Kübler W, Bode C (2000) Platelet activation as a potential mechanism of Gp IIb/IIIa inhibitor-induced thrombocytopenia. J Am Coll Cardiol 35 [Suppl]:371A (Abstract)

Peters RJG, Spickler W, Theroux P, White H, Gibson M, Molhoek PG, Hirsch J, Weitz JI, Weaver D (2000) Randomized comparison of vasoflux, a novel anticoagulant, and heparin as adjunctive therapy to streptkinase for acute myocardial infarction. J Am Coll Cardiol 35 [Suppl]:373A (Abstract)

Potter van Loon BJ, Rijken DC, Brommer EJ, van der Maas APC (1992) The amount of plasminogen, tissue-type plasminogen activator and plasminogen activator inhibitor type 1 in human thrombi and the relation to ex-vivo lysibility. Thromb Haemost 67:101–105

Pötzsch B, Hund S, Madlener K, Unkrig C, Müller-Berghaus G (1997) Monitoring of recombinant hirudin: assessment of a plasma-based ecarin clotting time assay. Thromb Res 86:373–383

Raghu C, Peddeswara Rao P, Seshagiri Rao D (1999) Protective effect of intravenous magnesium in acute myocardial infarction following thrombolytic therapy. Int J Cardiol 71:209–215

Rao AK, Pratt C, Berke A, Jaffe A, Ockene I, Schreiber TL, Bell WR, Knatterud G, Robertson TL, Terrin ML, for the TIMI Investigators (1988) Thrombolysis in myocardial infarction (TIMI) trial. Phase I: hemorrhagic manifestations and changes in plasma fibrinogen and the fibrinolytic system in patients treated with recombinant tissue plasminogen activator and streptokinase. J Am Coll Cardiol 11:1–11

Rapold HJ, de Bono D, Arnold AER, Arnout J, De Cock F, Collen D, Verstraete M, for the European Cooperative Study Group (1992) Plasma fibrinopeptide A levels in patients with acute myocardial infarction treated with alteplase. Correlation with concomitant heparin, coronary artery patency, and recurrent ischemia. Circulation 85:928–934

Rapold HJ, Kuemmerli H, Weiss M, Baur H, Haeberli A (1989) Monitoring of fibrin generation during thrombolytic therapy of acute myocardial infaction with recombinant tissue-type plasminogen activator. Circulation 79:980–989

Reiner JS, Lundergan CF, van den Brand M, Boland J, Thompson MA, Machecourt J, Py A, Pilcher GS, Fink CA, Burton JR, Simoons ML, Califf RM, Topol EJ, Ross AM, for the GUSTO Angiographic Investigators (1994) Early angiography cannot predict postthrombolytic coronary reocclusion: Observations from the GUSTO angiographic study. J Am Coll Cardiol 24:1439–1444

Rentrop KP (2000) Thrombi in acute coronary syndromes: revisited and revised. Circulation 101:1619–1626

Ronner E, Dykun Y, van den Brand MJBM, Van der Wieken LR, Simoons ML (1998a) Platelet glycoprotein IIB/IIIA receptor antagonists. An asset for treatment of unstable coronary syndromes and coronary intervention. Eur Heart J 19:1608–1616

Ronner E, van Kesteren HAM, Zijnen P, Tebbe U, Molhoek P, Cuffie C, Veltri E, Lorenz T, Neuhaus K, Simoons ML (1998b) Combined therapy with streptokinase and integrilin. Eur Heart J 19 [Suppl]:123 (Abstract)

Rosenschein U, Roth A, Rassin T, Basan S, Laniado S, Miller HI (1997) Analysis of coronary ultrasound thrombolysis endpoints in acute myocardial infarction (ACUTE trial). Results of the feasibility phase. Circulation 95:1411–1416

Roux S, Christeller S, Lüdin E (1992) Effects of aspirin on coronary reocclusion and recurrent ischemia after thrombolysis: a meta-analysis. J Am Coll Cardiol 19:671–677

Schafer AI, Adelman B (1985) Plasmin inhibition of platelet function and of arachidonic acid metabolism. J Clin Invest 75:456–461

Schmid JH, Jackson DP, Conley CL (1962) Mechanism of action of thrombin on platelets. J Clin Invest 41:543–553

Sharma B, Wyeth RP, Gimenez HJ, Franciosa JA (1986) Intracoronary prostaglandin E_1 plus streptokinase in acute myocardial infarction. Am J Cardiol 58:1161–1166

Siddiqi F, Odrljin TM, Fay PJ, Cox C, Francis CW (1998) Binding of tissue-plasminogen activator to fibrin: effect of ultrasound. Blood 91:2019–2025

Siegel RJ, Atar S, Fishbein MC, Brasch AV, Peterson TM, Nagai T, Pal D, Nishioka T, Chae J-S, Birnbaum Y, Zanelli C, Luo H (2000) Noninvasive, transthoracic, low-frequency ultrasound augments thrombolysis in a canine model of acute myocardial infarction. Circulation 101:2026–2029

Simes RJ, Topol EJ, Holmes DR Jr, White HD, Rutsch WR, Vahanian A, Simoons ML, Morris D, Betriu A, Califf RM, Ross AM, for the GUSTO-I Investigators (1995) Link between the angiographic substudy and mortality outcomes in a large randomized trial of myocardial reperfusion. Importance of early and complete infarct artery reperfusion. Circulation 91:1923–1928

Song X, Huhle G, Wang L, Hoffmann U, Harenberg J (1999) Generation of anti-hirudin antibodies in heparin-induced thrombocytopenic patients treated with r-hirudin. Circulation 100:1528–1532

Stein B, Fuster V, Halperin JL, Chesebro JH (1989) Antithrombotic therapy in cardiac disease. An emerging approach based on pathogenesis and risk. Circulation 80:1501–1513

Stump DC, Califf RM, Topol EJ, Sigmon K, Thornton D, Masek R, Anderson L, Collen D, and the TAMI Study Group (1989) Pharmacodynamics of thrombolysis with recombinant tissue-type plasminogen activator. Correlation with characteristics of and clinical outcomes in patients with acute myocardial infarction. Circulation 80:1222–1230

Stürzebecher J (1991) Methods for determination of hirudin. Semin Thromb Hemost 17:99–102

Taubman MB, Giesen PLA, Schecter AD, Nemerson Y (1999) Regulation of the procoagulant response to arterial injury. Thromb Haemost 82:801–805

Tcheng JE (2000) Clinical challenges of platelet glycoprotein IIb/IIIa receptor inhibitor therapy: bleeding, reversal, thrombocytopenia, and retreatment. Am Heart J 139 (Suppl.):S38–S45

The Dutch TIA Trial Study Group (1991) A comparison of two doses of aspirin (30 mg vs. 283 mg a day) in patients after a transient ischemic attack or minor ischemic stroke. N Engl J Med 325:1261–1266

The European Myocardial Infarction Project Group (1993) Prehospital thrombolytic therapy in patients with suspected acute myocardial infarction. N Engl J Med 329:383–389

The Global Use of Strategies to Open Occluded Coronary Arteries (GUSTO) IIa Investigators (1994) Randomized trial of intravenous heparin versus recombinant hirudin for acute coronary syndromes. Circulation 90:1631–1637

The Global Use of Strategies to Open Occluded Coronary Arteries (GUSTO) IIb Investigators (1996) A comparison of recombinant hirudin with heparin for the treatment of acute coronary syndromes. N Engl J Med 335:775–782

The GUSTO Angiographic Investigators (1993) The effects of tissue plasminogen activator, streptokinase, or both on coronary-artery patency, ventricular function, and survival after acute myocardial infarction. N Engl J Med 329:1615–1622

The GUSTO Investigators (1993) An international randomized trial comparing four thrombolytic strategies for acute myocardial infarction. N Engl J Med 329:673–682

The I.S.A.M. Study Group (1986) A prospective trial of intravenous streptokinase in acute myocardial infarction (I.S.A.M.). Mortality, morbidity, and infarct size at 21 days. N Engl J Med 314:1465–1471

The ISIS-2 (Second International Study of Infarct Survival) Collaborative Group (1988) Randomised trial of intravenous streptokinase, oral aspirin, both, or neither among 17 187 cases of suspected acute myocardial infarction: ISIS-2. Lancet 2:349–360

The RAPT Investigators (1994) Randomized trial of ridogrel, a combined thromboxane A_2 synthase inhibitor and thromboxane A_2/prostaglandin endoperoxide receptor antagonist, versus aspirin as adjunct to thrombolysis in patients with acute myocardial infarction. The Ridogrel Versus Aspirin Patency Trial (RAPT). Circulation 89:588–595

The SCATI Group (1989) Randomised controlled trial of subcutaneous calcium-heparin in acute myocardial infarction. Lancet 2:182–186

Thompson PL, Aylward PE, Federman J, Giles RW, Harris PJ, Hodge RL, Nelson GI, Thomson A, Tonkin AM, Walsh WF, for the National Heart Foundation of Australia Coronary Thrombolysis Group (1991) A randomized comparison of intravenous heparin with oral aspirin and dipyridamole 24 hours after recombinant tissue-type plasminogen activator for acute myocardial infarction. Circulation 83:1534–1542

Théroux P, Pérez-Villa F, Waters D, Lespérance J, Shabani F, Bonan R (1995) Randomized double-blind comparison of two doses of hirulog with heparin as adjunctive therapy to streptokinase to promote early patency of the infarct-related artery in acute myocardial infarction. Circulation 91:2132–2139

Tomaru T, Uchida Y, Nakamura F, Sonoki H, Tsukamoto M, Sugimoto T (1989) Enhancement of arterial thrombolysis with native tissue type plasminogen activator by pretreatment with heparin or batroxobin: an angioscopic study. Am Heart J 117:275–281

Topol EJ, Califf RM, George BS, Kereiakes DJ, Abbottsmith CW, Candela RJ, Lee KL, Pitt B, Stack RS, O'Neill WW, for the Thrombolysis and Angioplasty in Myocardial Infarction Study Group (1987) A randomized trial of immediate versus delayed elective angioplasty after intravenous tissue plasminogen activator in acute myocardial infarction. N Engl J Med 317:581–588

Topol EJ, Ellis SG, Califf RM, George BS, Stump DC, Bates ER, Nabel EG, Walton JA, Candela RJ, Lee KL, Kline EM, Pitt B, for the Thrombolysis and Angioplasty in Myocardial Infarction (TAMI) 4 Study Group (1989b) Combined tissue-type plasminogen activator and prostacyclin therapy for acute myocardial infarction. J Am Coll Cardiol 14:877–884

Topol EJ, Fuster V, Harrington RA, Califf RM, Kleiman NS, Kereiakes DJ, Cohen M, Chapekis A, Gold HK, Tannenbaum MA, Rao AK, Debowey D, Schwartz D, Henis M, Chesebro J (1994) Recombinant hirudin for unstable angina pectoris. A multicenter, randomized angiographic trial. Circulation 89:1557–1566

Topol EJ, George BS, Kereiakes DJ, Stump DC, Candela RJ, Abbottsmith CW, Aronson L, Pickel A, Boswick JM, Lee KL, Ellis SG, Califf RM, and the TAMI Study Group (1989a) A randomized controlled trial of intravenous tissue plasminogen activator and early intravenous heparin in acute myocardial infarction (TAMI-3). Circulation 79:281–286

Toschi V, Gallo R, Lettino M, Fallon JT, Gertz SD, Fernandez-Ortiz A, Chesebro JH, Badimon L, Nemerson Y, Fuster V, Badimon JJ (1997) Tissue factor modulates the thrombogenicity of human atherosclerotic plaques. Circulation 95:594–599

Toschi V, Lettino M, Gallo R, Badimon JJ, Chesebro JH (1996) Biochemistry and biology of hirudin. Coron Artery Dis 7:420–428

Van Belle E, Lablanche JM, Bauters C, Renaud N, McFadden EP, Bertrand ME (1998) Coronary angioscopic findings in the infarct-related vessel within 1 month of acute myocardial infarction: natural history and the effect of thrombolysis. Circulation 97:26–33

Van de Werf F, for the Continuous Infusion versus Double-Bolus Administration of Alteplase (COBALT) Investigators (1997) A comparison of continuous infusion of alteplase with double-bolus administration for acute myocardial infarction. N Engl J Med 337:1124–1130

Vaughan DE, Van Houtte E, Declerck PJ, Collen D (1991) Streptokinase-induced platelet aggregation. Prevalence and mechanism. Circulation 84:84–91

Verheugt FWA, Liem A, Zijlstra F, Marsh RC, Veen G, Bronzwaer JGF (1998) High dose bolus heparin as initial therapy before primary angioplasty for acute myocardial infarction: results of the Heparin in Early Patency (HEAP) pilot study. J Am Coll Cardiol 31:289–293

Verstraete M (2000) Synthetic inhibitors of platelet glycoprotein IIb/IIIa in clinical development. Circulation 101:e76–e80

von Essen R, Zeymer U, Tebbe U, Jessel A, Kwasny H, Mateblowski M, Niederer W, Wagner J, Mäurer W, von Leitner E-R, Haerten K, Roth M, Neuhaus K-L, for the Arbeitsgemeinschaft Leitender Kardiologischer Krankenhausärzte (1998) HBW 023 (recombinant hirudin) for the acceleration of thrombolysis and prevention of coronary reocclusion in acute myocardial infarction: results of a dose-finding study (HIT-II) by the Arbeitsgemeinschaft Leitender Kardiologischer Krankenhausärzte. Coron Artery Dis 9:265–272

Wallentin L, for the Fragmin during Instability in Coronary Artery Disease (FRISC) Study Group (1996) Low-molecular-weight heparin during instability in coronary artery disease. Lancet 347:561–568

Weaver WD, Cerqueira M, Hallstrom AP, Litwin PE, Martin JS, Kudenchuk PJ, Eisenberg M, for the Myocardial Infarction Triage and Intervention Project Group (1993) Prehospital-initiated vs hospital-initiated thrombolytic therapy: the Myocardial Infarction Triage and Intervention Trial. JAMA 270:1211–1216

Webster MWI, Chesebro JH, Fuster V (1991) Antithrombotic therapy in acute myocardial infarction: enhancement of reocclusion, and prevention of thromboembolism.

In: Gersh BJ, Rahimtoola SH (eds). Acute myocardial infarction, Elsevier, New York, pp 331–348

Weitz JI, Califf RM, Ginsberg JS, Hirsh J, Theroux P (1995) New antithrombotics. Chest 108 (Suppl):471S–485S

Weitz JI, Hudoba M, Massel D, Maraganore J, Hirsh J (1990) Clot-bound thrombin is protected from inhibition by heparin- antithrombin III but is susceptible to inactivation by antithrombin III-independent inhibitors. J Clin Invest 86:385–391

Weitz JI, Leslie B, Hudoba M (1998) Thrombin binds to soluble fibrin degradation products where it is protected from inhibition by heparin-antithrombin but susceptible to inactivation by antithrombin-independent inhibitors. Circulation 97:544–552

White HD, Aylward PE, Frey MJ, Adgey AA, Nair R, Hillis WS, Shalev Y, French JK, Collins R, Maraganore J, Adelman B, on behalf of the Hirulog Early Reperfusion/Occlusion (HERO) trial Investigators (1997) Randomized, double-blind comparison of hirulog versus heparin in patients receiving streptokinase and aspirin for acute myocardial infarction (HERO). Circulation 96:2155–2161

Wilcox RG, von der Lippe G, Olsson CG, Jensen G, Skene AM, Hampton JR, for the ASSET Study Group (1988) Trial of tissue plasminogen activator for mortality reduction in acute myocardial infarction. Anglo-Scandinavian Study of Early Thrombolysis (ASSET). Lancet 2:525–530

Wysokinski W, McBane R, Chesebro JH, Owen WG (1996) Reversibility of platelet thrombosis in vivo. Quantitative analysis in porcine carotid arteries. Thromb Haemost 76:1108–1113

Young E, Podor TJ, Venner T, Hirsh J (1997) Induction of the acute-phase reaction increases heparin-binding proteins in plasma. Arterioscler Thromb Vasc Biol 17:1568–1574

Zeymer U, von Essen R, Tebbe U, Michels H-R, Jessel A, Vogt A, Roth M, Appel K-F, Neuhaus K-L (1995) Recombinant hirudin and front-loaded alteplase in acute myocardial infarction: final results of a pilot study. HIT-I (hirudin for the improvement of thrombolysis). Eur Heart J 16 (Suppl D):22–27

Zhu Y, Carmeliet P, Fay WP (1999) Plasminogen activator inhibitor-1 is a major determinant of arterial throm321bolysis resistance. Circulation 99:3050–3055

Zoldhelyi P, Bichler J, Owen WG, Grill DE, Fuster V, Mruk JS, Chesebro JH (1994) Persistent thrombin generation in humans during specific thrombin inhibition with hirudin. Circulation 90:2671–2678

Zoldhelyi P, Webster MWI, Fuster V, Grill DE, Gaspar D, Edwards SJ, Cabot CF, Chesebro JH (1993) Recombinant hirudin in patients with chronic, stable coronary artery disease. Safety, half-life, and effect on coagulation parameters. Circulation 88 [Part 1]:2015–2022

*Thrombolytic Treatment of Other
Clinical Thromboembolic Conditions*

CHAPTER 12

New Concepts in Thrombolysis of Pulmonary Embolism

A.A. SASAHARA and G.V.R.K. SHARMA

A. Introduction

Venous thromboembolism continues to be a major health problem, arising mainly as a complication of the immobilized hospital patient, but which also occurs in ambulant, otherwise healthy individuals. The incidences of deep vein thrombosis (DVT) and pulmonary embolism (PE) have been estimated in a number of epidemiological surveys dealing with a variety of data sources, all of which are probably gross underestimates of the true occurrence rates. Some years ago it was estimated, on the basis of hospital statistics in the United States in 1966, that the total number of diagnosed cases of PE was about 106 000 and the total number of diagnosed cases of DVT was about 182 000 (HUME et al. 1970). Thus according to this estimate there were about a quarter of a million patients diagnosed with DVT and/or PE in the United States in 1966. A longitudinal study by COON et al. (1973), on the prevalence and incidence of venous thromboembolism in a Michigan community, provides additional epidemiological information. Data derived from this community was extrapolated to the 1970 U.S. census figures to arrive at an estimate of the annual incidence of DVT in the U.S. of over 250 000 cases. In addition, data from this study were used to estimate the prevalence of the postthrombotic sequellae in the U.S. population. The approximate frequency of stasis changes in the skin of the legs was 6–7 million persons, while about 500 000 have or have had leg ulcers. These figures for the prevalence of the post-thrombotic changes would represent a frequency of about 5% of the U.S. adult population. A more recent estimate of the magnitude of the PE problem was based upon the prevalence rate for fatal and non-fatal PE (SASAHARA and SHARMA 1988). It was estimated that approximately 3–4 patients/1000 inpatients and about 15 patients/1000 inpatients suffer fatal and non-fatal PE respectively, resulting in a total of about 150 000 patients suffering fatal PE and about 600 000 patients suffering non-fatal PE each year in the United States.

PE is indeed a modern paradox – as great advances are being made in medicine, the incidence of the disease appears to be rising. This rising incidence appears to be due to the increasing numbers of older and sicker patients who are hospitalized and subjected to longer periods of bed rest and more complex operations and procedures.

The aetiology of PE is usually due to deep vein thrombi embolizing to the lungs although substances such as fat, tumor, amniotic fluid, air, foreign particles, etc. may also embolize to the lungs in the appropriate clinical setting producing non-thrombotic PE. The majority of significant PE arise from thrombi in the popliteal, femoral, or iliac veins. Data from autopsies have shown that 80% or more of patients with PE have associated DVT (SEVITT and GALLAGHER 1961). Hence, the selection of treatment for PE should also be optimal for the treatment of its DVT source. Other possible sources of venous thrombi include the inferior vena cava, the subclavian vein, the internal jugular vein, the cavernous sinuses of the skull, and the right atrium and ventricle (WOLFE and SABISTON 1980).

Emboli, once released from the peripheral venous circulation, are distributed to both lungs in approximately 65% of cases, to the right lung alone in 25%, and to the left in 10% (SASAHARA et al. 1973). The lower lobes are involved four times more frequently than the upper lobes. The majority of thromboemboli lodge in the larger or intermediate pulmonary arteries with 35% or less reaching the smaller vessels.

The mortality of undiagnosed, and therefore untreated, PE is relatively high, ranging from 18% to 35% (SIDDIQUE et al. 1996). However, once diagnosis is made and the disease treated, there is a substantial reduction in mortality to about 8%–10% (SASAHARA et al. 1973; KASPER et al. 1997) Patients presenting with cardiogenic shock have a mortality of about 25% and those necessitating cardiopulmonary resuscitation of about 65% (KASPER et al. 1997). In the International Cooperative Pulmonary Embolism Registry (ICOPER) 2454 consecutive eligible patients with acute PE were registered from 62 hospitals in seven countries in Europe and North America. Overall crude mortality rate at 3 months was 17.4%. After exclusion of 61 patients in whom PE was first discovered at autopsy the 3-month mortality rate was 15.3%. Forty-five percent of the deaths were ascribed to PE and 18% to underlying cancer (GOLDHABER et al. 1999). On multiple regression analysis the following predictive factors for death were identified (hazard ratio given in parentheses):

- systolic arterial hypotension (2.9)
- congestive heart failure (2.4)
- cancer (2.3)
- tachypnea (2.0)
- right ventricular hypokinesis on echocardiography (2.0)
- chronic obstructive pulmonary disease (1.8) and
- age over 70 years (1.6)

This observation underscores the importance of early diagnosis and the institution of early and aggressive therapy in reducing mortality.

B. Treatment with Thrombolytic Therapy

In the long and continuing search for improved methods of treating pulmonary embolism, many modes of therapy have been developed. Thus, the clinician has several options available today and also bears the responsibility for selecting an approach which should satisfy both short- and long-term treatment objectives.

Short term objectives should include: (1) the prevention of thromboembolic propagation; (2) prevention of recurrent PE; and (3) the removal of thromboemboli as completely as possible from the pulmonary vasculature. Long term objectives should include: (1) the prevention of recurrent PE; (2) prevention of chronic venous insufficiency; and (3) the prevention of chronic pulmonary hypertension. We believe both short- and long-term objectives can only be achieved with the use of thrombolytic therapy (SHARMA et al. 1980, 1990).

Two thrombolytic agents, streptokinase (SK) and urokinase (UK), have been available to the professional community for the treatment of PE since the late 1970s, and prior to that time they were used in venous thromboembolism on an investigational basis. In 1990, tissue plasminogen activator (tPA) was approved by the Food and Drug Administration for the treatment of PE.

Streptokinase is a protein secreted by group C β-hemolytic streptococci with fibrinolytic properties. It combines with plasminogen to become an activator which then converts the remaining plasminogen into plasmin. UK is a natural fibrinolytic agent which was originally isolated from human urine, but is currently expressed by human kidney cells. Now it can also be made by genetic engineering, expressing it in a mouse hybridoma cell line (CREDO et al. 1997a), in E. Coli, yeast, and Chinese hamster ovary cells (see Chap. 4). The zymogen form of UK, pro-UK is also produced in these cells lines (CREDO et al. 1997b; see also Chap. 9). Tissue plasminogen activator is also made by genetic engineering, expressed in a Chinese hamster ovary cell line (see Chaps. 2 and 8).

The concept of thrombolytic agents actually dissolving existing clots, rather than merely preventing further clot development as with anticoagulants, has always been attractive to clinicians. As a result, two major cooperative clinical trials sponsored by the National Institutes of Health (NIH), comparing UK and SK with heparin therapy, were conducted. The Phase I Urokinase Pulmonary Embolism Trial (UPET) (SASAHARA et al. 1973) was designed to compare therapeutic efficacy between heparin and UK therapy and the Phase II trial (UROKINASE-STREPTOKINASE PULMONARY EMBOLISM TRIAL (USPET) 1974; SASAHARA et al. 1975) was designed to compare therapeutic efficacy among three thrombolytic regimens: 12h UK, 24h UK, and 24h SK. The results from these trials established the efficacy of SK and UK in the treatment of PE. From the data of these two large trials, we can separate out clinical effects and benefits which are established from those which are possible, but not yet proven.

C. Established Clinical Effects of Thrombolytic Therapy

I. Thrombolytic Agents Hasten Thromboembolic Resolution

The evidence that confirms this benefit was primarily derived from the two multicenter trials sponsored by the NIH. In the UPET, 160 patients received a 12-h infusion of either UK or heparin, followed by heparin in both groups. Comparison of control and posttherapy pulmonary angiograms showed that significantly more lysis of pulmonary emboli occurred in the UK-treated patients than in those given heparin. A particularly striking finding was the extent of clot lysis in patients with large pulmonary emboli that occluded over 35% of the pulmonary vasculature.

Moreover, comparison of baseline and posttherapy perfusion lung scans showed greater reperfusion in the UK-treated patients than in the heparin group. Although, on the average, significant differences in reperfusion did not persist beyond the first few days, they were evident for approximately a week in patients with massive emboli.

In a further objective assessment of drug efficacy, selected pressures in the right heart and pulmonary circulation were measured and compared. Of the eight evaluated measurements, six showed a greater improvement or more frequent return to normal in the UK-treated patients than in the heparin group. These improvements were in mean right atrial, right ventricular systolic and diastolic, and mean pulmonary artery (PA) pressures, as well as in total pulmonary vascular resistance (PVR) and systemic arterial oxygen tensions.

Because of favorable results obtained in other countries with purified SK, which had not been available for testing in Phase I, a second trial was deemed necessary to compare the two fibrinolytic agents. This study, the USPET, or Phase II, randomly assigned 167 patients to three thrombolytic treatment groups. Details of the protocol were very similar to the Phase I study and the same types of efficacy assessments were carried out.

II. Results of Pulmonary Angiography

Figure 1 shows the mean changes in thromboembolic obstruction as assessed by pulmonary angiograms obtained before and 24 h after initiating therapy. In the three fibrinolytic treatment groups, the mean improvement indicating embolic resolution was essentially the same and also equivalent to the mean improvement with 12 h UK therapy in the Phase I study. Equivalence of the 12 h UK effects in Phases I and II is important because it permits pooling of data from the two trials. In this combined analysis, all three groups treated with fibrinolytic agents showed significantly greater clot resolution than the group treated with heparin.

III. Results of Perfusion Lung Scans

In the three fibrinolytic therapy groups in Phase II, the mean lung scan perfusion defect before therapy was 33% of the total pulmonary perfusion. Fol-

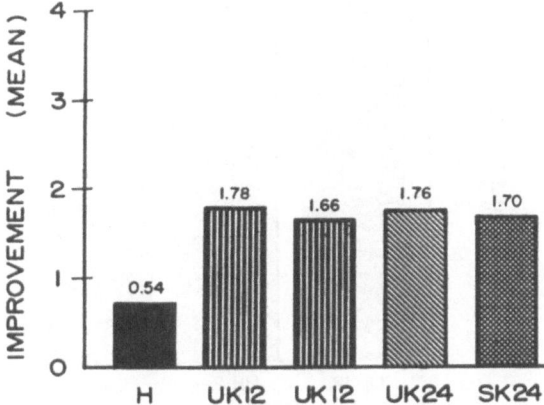

Fig. 1. Angiographic assessment of the mean changes following therapy in the five groups of patients in the UPET and USPET trials (see text). Note the virtual identical degrees of improvement in the lytic groups, all of which are superior to the heparin group. Reproduced from SASAHARA et al. (1975).

lowing the therapy period, the mean degrees of resolution revealed in lung scans made from 24 h to 30 h after the start of treatment were 9.0% with 12 h of UK, 11.6% with 24 h of UK, 7% with 24 h of SK, and 2.5% with 12 h of heparin. In terms of percentages of the mean pretreatment lesion extent in the three groups, the improvements were 20%, 29%, 18%, and 6% respectively.

The difference between the results with 24 h UK (29%) and 24 h SK (18%) almost achieved statistical significance (Fig. 2). All three fibrinolytic therapy groups showed significantly greater reperfusion than did the heparin group in the Phase I study.

IV. Hemodynamic Observations

Mean changes in four important pressure measurements 24 h after initiation of fibrinolytic therapy showed no consistent difference favoring one agent or regimen over another (Fig. 3). Patients treated for 24 h with SK showed less reduction in PA pressure but a greater increase in cardiac output than the two UK groups. These changes, however, were minor. All three lytic therapy groups showed significantly greater hemodynamic improvement or more frequent return to normal than did the heparin group in Phase I. In massive PE, the degree of lowering of the pulmonary hypertension was significantly greater with UK than with SK (Fig. 4).

Thus, from the two controlled trials it can be concluded that UK and SK conferred the following benefits within the first 24 h of therapy:

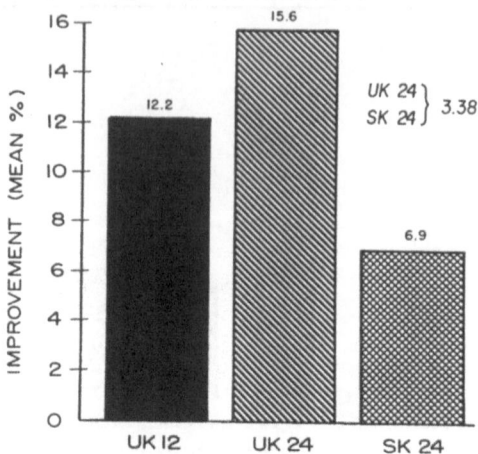

Fig. 2. Perfusion lung scan change in massive PE. The degree of reperfusion at 24 h in the 24 h UK group was significantly greater than the reperfusion in the 24 h SK group. The 12 h and 24 h UK groups were not significantly different, though there was a strong trend. Reproduced from Sasahara et al. (1975).

1. The induced fibrinolysis resulted in more rapid and extensive clot resolution within the pulmonary vasculature as assessed by pulmonary angiography. Spontaneous lysis during heparin therapy was minimal.
2. Clot dissolution resulted in greater reperfusion of the embolized lung as assessed by serial lung scans. The degree of reperfusion with heparin therapy was minimal.
3. Lysis of obstructing emboli resulted in greater reductions in hemodynamic abnormalities and more frequent return to normal cardiopulmonary function. Such changes were minimal with heparin therapy.
4. Equivalent improvements in resolution of pulmonary embolism can be achieved with 12 or 24 h of UK or 24 h of SK, except in massive PE.

V. Thrombolytic Therapy Removes Pulmonary Thromboemboli More Completely than Heparin Therapy

Many patients who recover from acute PE do so without apparent sequelae. Their lung scans return to normal as do their pulmonary angiograms. In the Phase I study, approximately 85% of patients had either normal lung scans or minimal (less than 10%) residual scan defects one year after embolization. However, neither conventional pulmonary angiography nor perfusion lung scanning is a sensitive indicator of changes in the pulmonary microcirculation, which principally influences changes in PA pressures.

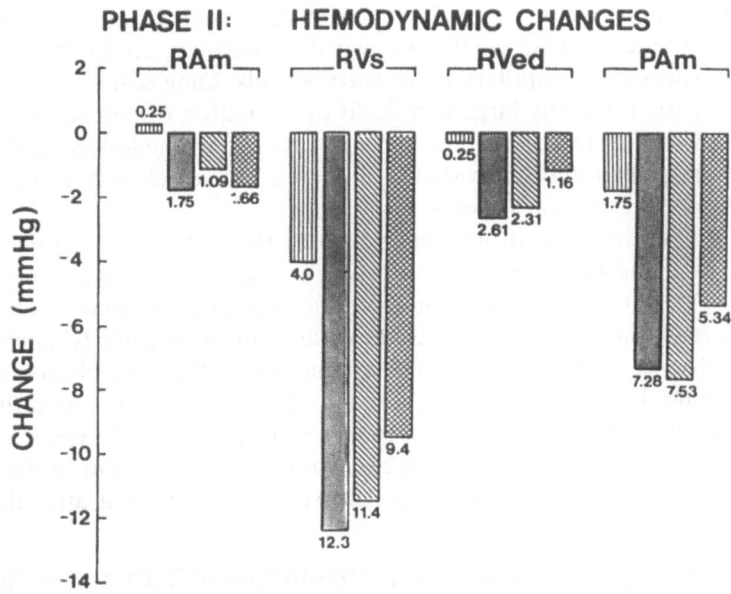

Fig. 3. Mean hemodynamic changes in the heparin and three thrombolytic groups (12 h UK, 24 h UK, 24 h SK). The grouping of the four treatment groups for each pressure measurement (right atrial mean, right ventricular systolic and enddiastolic, and pulmonary arterial mean pressures) are as follows, *left to right*: heparin, 12 h UK, 24 h UK and 24 h SK. Though minimal reductions in the abnormal pressures were noted in the heparin group, significant improvements occurred in the lytic groups. Reproduced from SASAHARA et al. (1975).

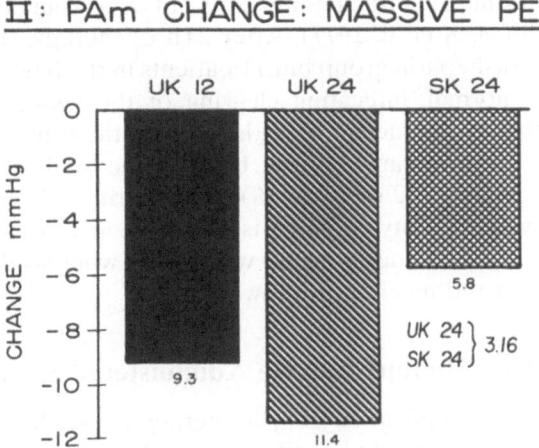

Fig. 4. Pulmonary arterial mean (PAm) pressure change in massive PE. The degree of lowering of PAm pressure was significantly greater in the 24 h UK group compared to the 24 h SK group. The two UK groups were not significantly different. Reproduced from SASAHARA et al. (1975).

Angiograms, when correctly performed, permit visualization of third-order PA branches, whereas it is known that there are about 25 successively smaller orders of precapillary pulmonary vessels. Lung scans are also greatly limited by the relatively large threshold of resolution of detector-collimation systems. Although radioisotope macroaggregates become lodged in the microcirculation, detection of individual aggregates and vessels is beyond the resolution limits of scanning systems.

As an indirect substitute, we measured the pulmonary capillary blood volume (V_c) and the pulmonary diffusing capacity (D_{co}), which are sensitive indices of the integrity of the pulmonary microcirculation (SHARMA et al. 1980). Patients from the Phase I and II trials, without prior or underlying cardiopulmonary disease, were selected for examination, so that any observed abnormality could be reasonably attributed to PE rather than other disorders. Studies of 40 such patients showed that the mean V_c and D_{co} were normal in patients treated with fibrinolytic agents, whereas they were low in the heparin group (Fig. 5). These differences, persisted for up to one year after therapy.

VI. Thrombolytic Agents Hasten Dissolution of Thrombi in the Venous System of the Legs

Rapid lysis of thrombi in the deep venous system tends to preserve the anatomy and function of the venous valve cusps (KAKKAR 1973) whereas slow resolution with heparin therapy appears to distort and destroy venous valves. We therefore employed impedance plethysmography (IPG) to assess the resolution of deep-vein thrombosis in most of the patients with PE treated at our center in the Phase I and II trials. IPG was performed daily for one week. Initially, 16 patients, randomized to receive heparin and 16 to receive fibrinolytic therapy, had abnormal IPG tracings indicative of deep-vein obstruction in one or both limbs (SHARMA et al. 1977). After 24 h of therapy, IPG showed that only 1 patient in the heparin group but 11 patients in the lytic group converted their tracings to normal, indicating clearing of the venous outflow block. During the ensuing week, tracings of 3 additional patients in the heparin group and 1 in the lytic group became normal, bringing the total to 4 of 16 (25%) in the heparin group and to 12 of 16 (75%) in the fibrinolytic group. It is worth noting that the vast majority of patients whose deep venous flow improved with lytic therapy showed improvement within 24 h, whereas the change in the heparin group occurred much more slowly.

VII. Thrombolytic Therapy Can Be Administered Safely

The only important complication of lytic therapy is bleeding. In the Phase I study, overt bleeding or unexplained drops in hematocrit were frequently noted. Bleeding complications occurred in 27% of the heparin group and 44% of the UK group. The majority of UK-treated patients who bled did so within 24 h of therapy. Thereafter, the frequency of bleeding appeared to be less than

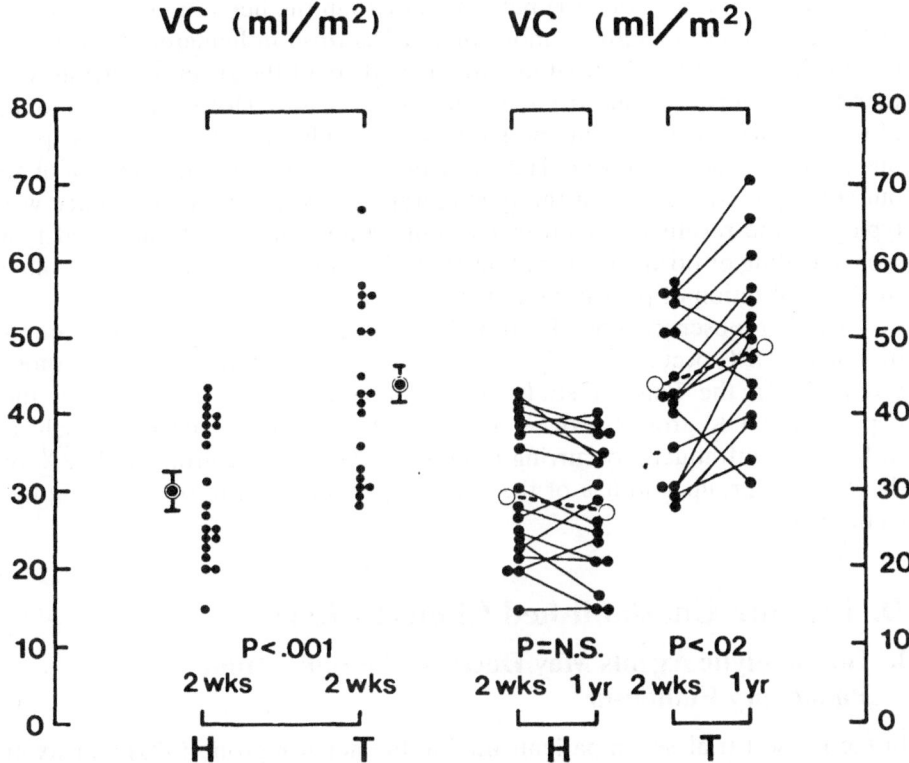

Fig. 5. Results of the pulmonary capillary blood volume (VC) measured in the heparin (*H*) and thrombolytic (*T*) treated groups, expressed as milliliter per square meter of body surface area. The *left panel* shows the results at 2 weeks and the *right panel* compares the 1-year results with the 2-week results in each group. The *open circles* represent mean values. There were significant differences in the normal pulmonary capillary blood volumes of the thrombolytic group compared with the abnormally reduced volumes in the heparin group at 2 weeks and at 1 year. Reproduced from SHARMA et al. (1980).

in the heparin group. Moreover, the increased bleeding in patients receiving UK was related to cutdowns and arterial punctures. Spontaneous hemorrhage occurred with equal frequency in the heparin and UK groups.

In the Phase II trial, bleeding occurred in 25% of the 12h UK group, 31% of the 24h UK group, and 22% of the 24h SK group. Careful analysis showed that from 39% to 72% of the patients who bled in the different groups did so from invaded sites such as arterial punctures, venous cutdowns in which angiographic catheters were left in place for at least 24h, and multiple venipunctures. Minimizing these invasive procedures sharply reduces the incidence of bleeding. This was demonstrated by BELL (1975) who reported that minimal bleeding occurred when invasive procedures were avoided.

In our own Phase I and II patients, 3 of 16 in the heparin group and 7 of 16 in the lytic group showed more than a 5% drop in hematocrit. Of these, two of the three heparin-treated patients and six of the seven receiving lytic therapy developed oozing from venous cutdown sites. However, the amount of oozing was not significant enough to warrant blood replacement except in one patient in the lytic group. Thus transfusion was required in only 3% of all our study patients or 6% of the lytic therapy recipients. Our experience was typical of the results obtained in the multicenter trial, which indicated that most bleeding or oozing occurs at invaded sites and can usually be controlled by the application of pressure bandages.

Another observed complication is fever. Mild temperature elevations of at least 0.8°C occurred in 15% of patients receiving UK and 24% of those given SK. In the Phase I study, the mean temperature elevations in the heparin and UK groups proved to be identical. Allergic reactions such as rashes were infrequent, occurring in none of the 12h UK group, only 2% of the 24h UK group, and 6% of the SK group. No patient developed anaphylactic shock.

D. Possible Unestablished Clinical Effects

I. Thrombolytic Agents May Decrease Mortality from Pulmonary Embolism

In the Phase I trial, seven patients died in the heparin group (9%) and six in the UK group (7%), resulting in an overall mortality of 8.1%. In the Phase II study, the overall mortality was 8.3% – respectively 7% in the 12h UK group, 9% in the 24h UK group, and 9% in the 24h SK group. None of these differences was significant. However, neither trial was designed to detect a mortality difference.

However, if one examines the causes of death in the trials, some of the observations suggest that lytic therapy might be lifesaving in appropriate clinical situations. For instance, in the Phase I study, four and possibly five of seven deaths (57% to 71%) in the heparin group resulted from shock, cardiac arrest, or heart failure attributable to PE compared with only one of six deaths (16%) in the fibrinolytic group. The majority of deaths in the lytic groups resulted from hemorrhages in clinical situations that we have now learned to avoid. It is possible that a trial limited to patients with massive PE might show a difference in mortality. The early lysis of massive emboli should lead to earlier lowering of pulmonary hypertension, restoration of right ventricular function, and stabilization of cardiopulmonary status. In contrast, the slow and incomplete lysis of thromboemboli with heparin therapy could unnecessarily prolong the hemodynamic abnormalities, leading to progressive deterioration and death. Analysis of mortality causes in the Phase I trial tends to support this hypothesis.

In a thoughtful analysis of Phase I results in patients with massive PE and shock, SAUTTER et al. (1972) cited the 80% mortality associated with pulmonary embolectomy in similar patients and the relatively low 18% mortality in those treated medically. They recommended institution of partial cardiopulmonary bypass to support such patients during administration of lytic therapy. It has been shown that pulmonary arterial pressure may be reduced as soon as 90 min after the start of fibrinolytic therapy. These observations and recommendations appear to support the concept that early institution of fibrinolytic therapy in patients with massive PE may be lifesaving. But a large-scale clinical trial, as recommended by Sautter, is required to secure convincing data.

II. Thrombolytic Therapy Improves Long-term Hemodynamic Status

In addition to the measurements of pulmonary capillary blood volume at 2 weeks and at 1 year, we also made long-term observations of the hemodynamic status in 23 patients (thrombolysis 12 patients; heparin 11 patients) after 5–10 years (SHARMA et al. 1990). The hemodynamic studies were carried out at rest and during exercise. After routine right heart catheterization with a thermodilution Swan-Ganz catheter and determination of PA pressures and cardiac output, the patients performed supine leg exercise on a bicycle ergonometer. A minimum duration of 3 min steady state exercise at a heart rate of at least 50% over the baseline was required before the hemodynamic measurements were made. All patients were male The two groups were matched for age, body surface area, and the severity of PE. The average period of follow-up was 7.3 years for the heparin group and 7.5 years for the thrombolytic group.

In the group randomized to heparin, these late posttherapy measurements of resting PA mean pressure and PVR did not change significantly from the abnormal, resting, immediate posttherapy values, measured 7.3 years earlier. In contrast, in the thrombolytic group, resting PA mean pressure and pulmonary vascular resistance were within normal limits in this late posttherapy measurement and not statistically significantly different from the immediate postthrombolysis values measured 7.5 years earlier. On exercise in the heparin group the PA mean pressure increased significantly more from its elevated resting value, while in the thrombolytic group there was no significant change from its normal baseline value. The PVR showed a similar pattern during exercise, increasing significantly from its abnormal elevated value in the heparin group, while showing little change within the normal limits in the thrombolytic group.

Our studies, therefore, demonstrate that the more complete clot resolution in the pulmonary vasculature brought about by thrombolytic agents has long-term implications in the improved hemodynamic status. However, the numbers were small in our series and confirmatory data from a larger group will be required before these important observations can be accepted as established.

E. New Thrombolytic Regimens

More recently, there has been renewed interest shown in thrombolysis of pulmonary embolism by employing new and novel regimens of high dose/short duration therapy of several plasminogen activators. Since one of the major factors involved in the bleeding from thrombolytic therapy is duration of therapy, investigators have sought to design regimens that employed high doses over a short period of time to maximize efficacy and safety (DICKIE et al. 1974; PETITPRETZ et al. 1984; GOLDHABER et al. 1992; DIEHL et al. 1992).

Another novel approach which seems to bridge the conventional and approved 24h regimen of UK and the abbreviated and bolus regimens of tPA and UK is the bolus +12h low-dose intrapulmonary infusion of UK (GONZÁLEZ-JUANATEY et al. 1992). In 16 patients with PE, UK therapy was administered with a bolus dose of 500000IU, followed by a right atrial catheter infusion of 1 million IU over 12h (83333IU/h). There was a mean improvement of the angiographic score of 57% when repeat angiography at 48h was compared with the control series, and significant (over 30%) improvement in all of the hemodynamic parameters measured. All patients improved and only one patient sustained a significant bleed at the catheter insertion site. The investigators concluded that this new and novel regimen was useful in the safe and efficacious administration of UK. Another trial of importance was the UKEP Study which compared two dosages of UK in patients with massive PE: the standard 4400IU/kg and a reduced dosage of 2000IU + heparin (THE UKEP STUDY RESEARCH GROUP 1987). Equal efficacy was achieved with improved safety: only 4.5% of the patients exhibited severe bleeding complications. UK also has been administered as intrapulmonary artery infusion in 26 patients with massive PE of whom 9 had contraindications to the use of thrombolytic therapy (recent surgical interventions). A loading dose of 4000U/kg given as a bolus, followed by 4000U/kg for 12 to 24h produced rapid and significant resolution of pulmonary emboli, as demonstrated by repeat pulmonary angiography. Intracerebral hemorrhage was not observed and none of the patients with bleeding needed blood transfusions (MCCOTTER et al. 1999).

Table 1 lists trials performed with tPA. All of these studies are relatively small and do not permit one to determine whether the treatment with thrombolytic agents results in a survival benefit. All trials have performed functional studies, most often pulmonary angiography, lung scans, and other functional tests before and after the administration of placebo, heparin, tPA, UK, or SK. The studies can be summarized as follows:

1. When comparing tPA with placebo or heparin alone, there was a much more rapid improvement of the Miller index by pulmonary angiography

Table 1. tPA thrombolysis trials in pulmonary embolism

Year	Author	No. of patients		Treatment		Improvement			Major bleeding[b]		ICH No.	
		tPA i.v.	Other	tPA dose (mg)[a]	Other	h/Test	tPA (%)	Other (%)	tPA (%)	Other (%)	tPA	Other
1986	Goldhaber et al.	36	–	50/2 h (+40/4 h)	–	2/A 6/A	21 49	– –	6	–	0	–
1987	Goldhaber et al.	47	–	50/2 h (+40/4 h)	–	6/A	impr.[c]	–	4	–	0	–
1988	Goldhaber et al.	22	UK 23	100/2 h	4400 U/kg/h for 24 h	2/A	23	8	18	48	0	0
1988	Verstraete et al.	15	tPA, ip. 19	50/2 h (+50/5 h)		2/A 38	15 36	12	12		0	0
1990	Proped Investigators	9	Plac 4	40–80/40–90 min	Plac	2/A	7	–8	11	0	0	0
1992	Levine et al.	33	Plac 25	0.33/kg/2 min	Plac	24/LS	37	19	0	0	0	0
1992	Goldhaber et al.	44	UK 46	100/2 h	3 Mio U/2 h	2/A	22	18	16	11	2	1
1992	Dalla-Volta et al.	20	Hep 20	100/2 h	1750 U/h	2/A	12	0.4	15	12	1	0
1992	Meyer et al.	34	UK 29	100/2 h	4400 U/kg/h for 24 h	12/A	20	30	21	28	0	1
1993	Goldhaber et al.	46	Hep 55	100/2 h	Bolus 5000 U + 1000 U/h	24LS	34	4	4	2	1	0
1994	Goldhaber et al.	27	tPA 60	100/2 h	0.6 mg/kg per 15 min	2/A 24/LS	23 31	9 19	7	3	2	1

Table 1. Continued

Year	Author	No. of patients		Treatment		Improvement			Major bleeding[b]		ICH No.	
		tPA i.v.	Other	tPA dose (mg)[a]	Other	h/Test	tPA (%)	Other (%)	tPA (%)	Other (%)	tPA	Other
1994	SORS et al.	17	tPA 36	100/2h	0.6 mg/kg per 15 min	1/TPR	36	29	6	8	0	0
1994	GULBA et al.	24	13	120/2h[d]	Embolectomy				28	15	1	0
1996	GISELBRECHT et al.	Y28	O27	1 mg/kg/10min		48/LS	10	9	18	12	1	1
1997	MENEVEAU et al.	25	SK 25	100/2h	100000 U/h for 12h	0.5 TPR / 1 / 2	18 / 30 / 42	5 / 14 / 13	16	12	0	0
1998	MENEVEAU et al.	23	SK 43	100/2h	1.5 Mio U/2h	0.5 TPR / 1 / 2	20 / 34 / 38	15 / 21 / 31	20	8	0	0
1998	KONSTANTINIDES et al.	27	HEP 13	100/2h	aPTT 2–3× normal	12/H	23–27	–5 to 7	n.r.	n.r.	n.r.	n.r.
1999	TEBBE et al.	13	rPA 27	100/2h	10 U×2	12/H	30–40	44–57	8	15	0	0

ICH, Intracerebral hemorrhage; Plac, placebo; Hep, heparin; ip, administration of thrombolytic agent via pulmonary artery; Y, patients <75 years of age; O, patients >75 years of age. Tests: A, angiography; LS, lung scan; TPR, total pulmonary resistance; aPTT, heparin was adjusted according to activated partial thromboplastin time; H, hemodynamics, improvements of mean pulmonary artery pressure, total pulmonary resistance and of cardiac index are given; n.r., not reported; rPA, reteplase.
[a]In most studies that administered 100 mg of tPA over 2h an initial bolus of 10 mg was given.
[b]Major bleeding or drop of hematocrit of >0.1 l/l.
[c]Improvement of angiographic perfusion in 83% of patients.
[d]Including an initial bolus of 20 mg.

during the first 1–4h. Differences in angiographic scores tended to disappear subsequently.

2. When comparing i.v. administration vs infusion of the tPA directly into the pulmonary artery there was no difference in efficacy or complications (VERSTRAETE et al. 1988).

3. Dosage regimens of 100mg of tPA over 2h resulted in a more rapid recanalization of the pulmonary vasculature than those administering the same dose over longer time periods, e.g., 6h.

4. There appears to be no statistically significant difference in efficacy between the three thrombolytic agents tPA, UK, and SK if these are given in high doses over a short time period.

5. Bleeding complications occur in a substantial number of patients and intracerebral hemorrhage was encountered in about 1.4–1.9% of patients treated with tPA (KANTER et al. 1997; DALEN et al. 1997). This incidence is higher than that encountered in the thrombolytic treatment of acute myocardial infarction.

6. In patients over 75 years of age the extent of improvement of pulmonary circulation and the incidence of bleeding complications was identical to that in patients less than 75 years old in the trial of GISSELBRECHT et al. (1996). However other investigators, analyzing the database of 5 studies comprising 312 patients arrived at the conclusion that patients over the age of 70 had a fourfold higher risk of bleeding complications than patients under the age of 50 (MIKKOLA et al. 1997). The study of GULBA et al. (1994) compared thrombolytic therapy and embolectomy in patients with massive pulmonary embolism and shock in an open trial. In general, patients who could not be promptly operated on got thrombolysis. Mortality in embolectomized patients in this study was remarkably low at 23%, but not significantly different from that of the thrombolyzed patients (33%).

Our belief currently is that thrombolytic therapy has significant benefits for patients with PE. The selection of an agent at this time is probably less important than making the decision to employ thrombolytic therapy in patients with "significant" PE. The recent studies offering different regimens are important in that the designs incorporate the principle of high dose/short duration therapy to achieve efficacy with maximal safety. Significant emboli previously indicated the involvement of several segmental defects on lung scan or three segmental pulmonary arteries on angiography, indicating that approximately 20%–25% of the lung vascular volume was compromised by the PE.

A more recent procedure to assess the overall hemodynamic impact of PE is echocardiography (COVARRUBIAS et al. 1977; STECKLEY et al. 1978; SHIINA et al. 1980; KASPER et al. 1986; COME et al. 1987). As the size of the embolic process increases, more alterations in normal right ventricular hemodynamics take place, e.g., right ventricular (RV) enlargement, left ventricular compression, RV wall hypokinesis in the absence of RV hypertrophy, and varying degrees of tricuspid regurgitation. Because of the correlation between RV hemodynamics and echocardiographic changes, the echocardiogram is capable of detecting the patient with hemodynamically significant PE. These patients, in the absence of any contraindication to thrombolytic therapy, may be the optimal candidates for thrombolytic therapy (Goldhaber 1998). Two recently published registries demonstrate that echocardiography has become a widely applied procedure in the decision making process regarding whether to use thrombolytic agents or not. In the ICOPER registry of 2454 patients with PE from 52 hospitals in 7 countries, 84% underwent perfusion lung scanning and 19% had pulmonary angiography. Of the 47% patients on whom echocardiography was performed, 40% had right ventricular hypokinesis. In this registry 13% of patients were given thrombolytic therapy (Goldhaber et al. 1997). In the German MAPPET registry 1001 patients with major PE from 204 centers were enrolled. Inclusion criteria included acute right heart failure and/or pulmonary hypertension. In this more severely affected patient population, two-dimensional and Doppler echocardiography was performed in 74% of patients, lung scans in 57%, right heart catheterization in 26%, and pulmonary angiography in 18%. Overall, the number of diagnostic studies decreased with increasing severity of the hemodynamic status of the patients (Kasper et al. 1997). In a substudy, including only 719 hemodynamically stable patients (i.e., without evidence of cardiogenic shock) primary thrombolytic treatment was given to 169 patients (23.5%), whereas the remaining 550 patients were initially treated with heparin alone. Overall 30-day mortality was significantly lowered in the patients who received thrombolytic treatment (4.7% vs 11.1%; $p = 0.0.016$). Recurrent PE was also much less frequent but bleeding was more common (see Table 2). Multiple logistic regression analysis revealed that the reduction of 30-day mortality associated with primary thrombolysis remained evident after adjustment for the influence of other relevant clinical characteristics at presentation (age, arterial hypotension, etc.) with an observed risk of 0.46 (confidence interval 0.21–1.00; $p = 0.05$) (Konstantinides et al. 1997).

Table 2. Adverse in-hospital events in the MAPPET registry

Event	Primary thrombolysis ($n = 169$)	Initial heparin alone ($n = 550$)	p
All deaths	8 (4.7)	61 (11.1)	0.016
From PE	7 (4.1)	58 (10.5)	
From underlying disease	0	1 (0.2)	
From complications of diagnostic or therapeutic procedures	1 (0.6)	1 (0.2)	
Recurrent PE	13 (7.7)	103 (18.7)	<0.001
Major bleeding	37 (21.9)	43 (7.8)	<0.001
Cerebral	2 (1.2)	2 (0.4)	
Other sites	35 (20.7)	41 (7.5)	

Reproduced from Konstantinides et al. (1997).

F. Surgical Therapy

Aside from pulmonary embolectomy, the surgical techniques employed in pulmonary embolism are for the purpose of preventing recurrent episodes. Because more than 90% of thrombi originate in the lower extremities, surgical maneuvers have been directed toward interruption of the inferior vena cava. As such, the following indications can be recommended for interruption:

1. Contraindications to anticoagulation
2. Recurrence during adequate anticoagulation
3. Septic pelvic thrombophlebitis with emboli
4. Recurrent pulmonary emboli
5. Pulmonary embolectomy

The majority of patients requiring surgical venous interruption are those in whom some contraindication to anticoagulation exists. These include severe systemic hypertension associated with grade III or IV hypertensive retinopathy, presence of an actively bleeding lesion in the gastrointestinal or genitourinary tract (symptomatic lesions without active bleeding are considered relative contraindications), craniotomy or cerebrovascular accident within the prior 4–6 months, or evidence of a lesion known or suspected to be associated with intracranial hemorrhage, including cerebral neoplasms, presence of an uncontrolled hypocoagulable state, including coagulation factor deficiencies, platelet abnormalities or other spontaneous hemorrhagic or purpuric phenomena, and severe renal or hepatic insufficiency. Relative contraindications to anticoagulation must be considered on an individual patient basis, weighing the risk and impact of bleeding against the morbidity of surgical venous interruptions.

The recognition of true recurrence during the acute phase of pulmonary embolism is a difficult clinical problem that cannot be resolved without the aid of lung scanning or pulmonary angiography. Only by angiography can the distinction between recurrent pulmonary embolism and fragmentation and distal migration of the original clot be answered with any assurance (SASAHARA and BARSAMIAN 1973; SASAHARA 1975). The importance of this distinction has considerable bearing on the therapy. Adequate anticoagulation implies the intravenous administration of heparin in adequate doses to achieve and maintain therapeutic levels. Should true recurrence occur during this period, the patient may be considered a "heparin-failure" and caval interruption can be subsequently performed. The fragmentation and distal migration process, on the other hand, requires only continuation of the heparin. In contrast, thrombolytic therapy can be continued for both recurrence and fragmentation with distal migration.

If recurrence develops during well-controlled orally administered anticoagulation therapy in the recovery period, or during long-term anticoagulation, reanticoagulation is recommended. Subsequently, several courses are available, depending on a number of factors: cardiopulmonary status of the patient, magnitude of reembolism, assessment of deep veins in the legs, and nature of the predisposing event. If the underlying cardiopulmonary status is satisfactory, the embolic episode submassive (less than 35%–40% of total pulmonary vasculature affected), the deep veins minimally abnormal, and the predisposing event temporary (e.g., leg trauma), retreatment with intravenously administered heparin, following the same time-course sequence as in the initial event, may be carried out. However, if the underlying cardiopulmonary status is unstable, the reembolic episode massive, the deep veins grossly abnormal, or the predisposing condition chronic (e.g., recurrent heart failure), caval interruption is recommended. Assessment of patency of deep veins in the legs is best performed by ascending phlebography or alternatively by one of the non-invasive methods, e.g., electrical impedance plethysmography (Sasahara et al. 1967; Wheeler et al. 1971) or duplex scanning, and is extremely helpful in this decision process. If a sizable clot in the deep vein is seen, with the proximal end "free floating", or if the non-invasive study indicates deep venous obstruction, it can be assumed that the patient is still at risk of sustaining recurrent pulmonary embolism.

Although interrupting the inferior vena cava will prevent further embolization in the acute period, recurrences can occur (Gurewich et al. 1966; Decousus et al. 1998). Decousus and his team studied 400 patients with proximal deep vein thrombosis, who were at risk for PE and randomized to receive a vena caval filter or no filter. There was an initial small clinical benefit of filters in preventing recurrent PE, but it was counterbalanced by an excess (20.8% vs 11.6%) of recurrent deep vein thrombosis without difference in mortality at two years. The use of temporary vena cava filters has been advocated in patients at high risk for re-embolization, particularly during thrombolysis, but whose clinical underlying condition is reversible (Zwaan et al. 1998). In a recent report of a multicenter registry of 188 patients the main indication for temporary filter insertion was indeed thrombolytic therapy (in 53% of patients). Average filter time was 5.4 days. Four patients (2.1%) died of PE despite filter protection. Major filter problems were filter thrombosis (16%) and filter dislocation (4.8%) (Lorch et al. 2000).

Pulmonary embolectomy, a dramatic and heroic procedure, is now being performed in fewer patients with pulmonary embolism. The procedure should be performed with cardiopulmonary bypass in patients with angiographically confirmed massive embolism with shock who do not respond to vigorous medical therapy and who, without mechanical removal of the embolism, would probably die (Sautter et al. 1975; Cross and Mowlem 1967). The mortality ranges between 30% and 100% in different series, but in the infrequent appropriate clinical situation it may be lifesaving (Sasahara and Barsamian 1973; Gulba et al. 1994).

More recently catheter thrombectomy has been introduced as an option for the treatment of high-risk PE (Brady et al. 1991; Greenfield et al. 1993; Koning et al. 1997). The current status of percutaneous mechanical thrombectomy has been recently reviewed by Sharafuddin and Hicks (1997, 1998a,b). In one study percutaneous catheter and guidewire fragmentation together with local infusion of tPA in 4 patients with massive PE produced excellent results (Murphy et al. 1999).

G. Adjunctive Therapy

Since most patients with clinically detectable pulmonary embolism will have some degree of hypoxemia, oxygen therapy is an important adjunct. Though cyanosis may be noted only in the very ill, arterial oxygen tension will invariably be reduced. Regardless of the mechanism of hypoxemia in the individual patient (venous admixture, ventilation-perfusion reduction, or diffusion defect), the administration of oxygen will relieve or diminish the symptoms of hypoxemia in many patients (SASAHARA et al. 1967). Since hypoventilation is rarely a cause of hypoxemia in pulmonary embolism, oxygen may be administered comfortably by nasal catheter without fear of suppressing ventilation.

In patients who sustain major to massive pulmonary embolism, cardiac failure is frequently observed, particularly when there is pre-existing cardiopulmonary disease. The cardiac index may fall below 2.0 l/min per m^2 (McINTYRE and SASAHARA 1972). In such circumstances, the administration of an isoproterenol hydrochloride drip (2–4 mg/500 ml of 5% dextrose in water) is helpful as a cardiotonic agent: it increases cardiac output and decreases pulmonary arterial pressure. In the event that hypotension is present and persists after isoproterenol administration, levarterenol bitartrate (2–6 ml of 0.2% in 500 ml of 5% dextrose in water) can be administered. Occasionally, when the central venous pressure is low, administration of intravenous fluids may be helpful in restoring near-normal hemodynamics. Digitalis glycosides, intravenously administered diuretics, and various antiarrhythmic agents should be used in the appropriate clinical situation in the usual dosages. In patients who complain of severe pleuritic pain or who exhibit apprehension, morphine sulfate, administered slowly and intravenously 1 mg at a time (up to 5–10 mg), may be useful. Apprehension is lessened and ventilation is frequently improved. Codeine sulfate (30–60 mg) may be given for lesser pain. Since the great majority of patients who suffer acute pulmonary embolism have their site of thromboemboli in the deep veins of the legs, appropriate measures, such as leg elevation for symptomatic treatment should be instituted.

List of Abbreviations

D$_{co}$	pulmonary diffusing capacity
DVT	deep venous thrombosis
ICOPER	International Cooperative Pulmonary Embolism Registry
IPG	impedance plethysmography
MAPPET	MAnagement Strategy and Prognosis of Pulmonary Embolism RegisTry
NIH	National Institutes Health (U.S.A.)
PA	pulmonary artery
PAIMS	Plasminogen Activator Italian Multicenter Study
PE	pulmonary embolism

PIOPED	Prospective Investigation of Pulmonary Embolism Diagnosis
PVR	pulmonary vascular resistance
RV	right ventricular
SK	streptokinase
tPA	tissue-type plasminogen activator
UK	urokinase
UKEP	Urokinase Embolie Pulmonaire
UPET	Urokinase Pulmonary Embolism Trial
USPET	Urokinase Streptokinase Pulmonary Embolism Trial
V_c	pulmonary capillary blood volume

References

Bell WR (1975) Thrombolytic therapy: a comparison between urokinase and strep-tokinase. Semin Thromb Hemost 2:1–5

Brady AJ, Crake T, Oakley CM (199 1) Percutaneous catheter fragmentation and distal dispersion of proximal pulmonary embolus. Lancet 338:1186–1189

Come PC, Kim D, Parker JA, Goldhaber SZ, Braunwald E, Markis JE (1987) Early reversal of right ventricular dysfunction in patients with acute pulmonary embolism after treatment with intravenous tissue plasminogen activator. J Am Coll Cardiol 10:971–978

Coon WW, Willis PW, Keller JB (1973) Venous thromboembolism and other venous disease in the Tecumseh Community Health Study. Circulation 48: 839–846

Covarrubias EA, Sheikh MU, Fox LM (1977) Echocardiography and pulmonary embolism. Ann Int Med 87:720–725

Credo RB, Burke SE, Barker WM, Schulz GA, Feibel JM, Henkin J, Sasahara AA (1997a) Recombinant urokinase (r UK): biochemistry, pharmacology, and clinical experience. In: Sasahara A, Loscalzo J (eds) New therapeutic agents in thrombosis and thrombolysis. Marcel Dekker, New York, pp 513–537

Credo RB, Burke SE, Barker WM, Villiard EM, Sweeny JM, Henkin J, Sasahara AA (1997b) Recombinant glycosylated pro-urokinase: biochemistry, pharmacology, and early clinical experience. In: Sasahara AA, Loscalzo J (eds) New therapeutic agents in thrombosis and thrombolysis. Marcel Dekker, New York, pp 561–589

Cross FS, Mowlem A (1967) A survey of the current status of pulmonary embolectomy for massive pulmonary embolism. Circulation 35 [Suppl 1]:186–191

Dalen JE, Alpert JS, Hirsh J (1997) Thrombolytic therapy for pulmonary embolism: is it effective? Is it safe? When is it indicated? Arch Intern Med 157:2550–2556

Dalla-Volta S, Palla A, Santolicandro A, Giuntini C, Pengo V, Visioli O, Zonzin P, Zanut-tini D, Barbaresi F, Agnelli G, Morpurgo M, Marini MG, Visani L (1992) PAIMS 2: alteplase combined with heparin versus heparin in the treatment of acute pulmonary embolism. Plasminogen Activator Italian Multicenter Study 2. J Am Coll Cardiol 20:520–526

Decousus H, Leizorovicz A, Parent F, Page Y, Tardy B, Girard P, Laporte S, Faivre R, Charbonnier B, Barral F, Huet Y, Simonneau G, for the Prevention du Risque d'Embolie Pulmonaire par Interruption Cave Study Group (1998) A clinical trial of vena caval filters in the prevention of pulmonary embolism in patients with proximal deep-vein thrombosis. N Engl J Med 338:409–415

Dickie KJ, Groot WJ, Cooley RN, Bond TP, Guest MM (1974) Hemodynamic effects of bolus infusion of urokinase in pulmonary thromboembolism. Am Rev Respir Dis 109:48–56

Diehl J-L, Meyer G, Igual J, Collignon MA, Giselbrecht M, Even P, Sors H (1992) Effec-tiveness and safety of bolus administration of alteplase in massive pulmonary embolism. Am J Cardiol 70:1477–1480

Gisselbrecht M, Diehl JL, Meyer G, Collignon MA, Sors H (1996) Clinical presentation and results of thrombolytic therapy in older patients with massive pulmonary embolism: a comparison with non-elderly patients. J Am Geriatr Soc 44:189–193

Goldhaber SZ (1998) Pulmonary embolism. N Engl J Med 339:93–104

Goldhaber SZ, Agnelli G, Levine NIN (1994) Reduced dose bolus alteplase vs conventional alteplase infusion for pulmonary embolism thrombolysis. An international multicenter randomized trial. Chest 106:718–724

Goldhaber SZ, de Rosa M, Visani L (1997) International cooperative pulmonary embolism registry detects high mortality rate. Circulation 96 [Suppl I]:1– 159 (Abstract)

Goldhaber SZ, Haire WD, Feldstein ML, Miller M, Toltzis R, Smith JL, Taveira da Silva AM, Come PC, Lee I, Parker JA, et al (1993) Alteplase versus heparin in acute pulmonary embolism: randomized trial assessing right-ventricular function and pulmonary perfusion. Lancet 341:507–511

Goldhaber SZ, Heit J, Sharma GVRK, Nagel JS, Kim D, Parker JA, Drum D, Reagan K, Anderson J, Kessler CN/1, Markis J, Dawley D, Meyerovitz M, Vaughan DE, Turneh SS, Loscalzo J, Selwyn AP, Braunwald E (1988) Randomized controlled trial of recombinant tissue plasminogen activator versus urokinase in the treatment of acute pulmonary embolism. Lancet 2:291–298

Goldhaber SZ, Kessler CM, Heit JA, Elliott CG, Friedenberg WR, Heiselman DE, Wilson DB, Parker JA, Bennett D, Feldstein ML, Selwyn AP, Kim D, Sharma GVRK, Nagel JS, Meyerovitz; MF (1992) Recombinant tissue-type plasminogen activator versus a novel dosing regimen of urokinase in acute pulmonary embolism: A randomized controlled multicenter trial. J Am Coll Cardiol 20:24–30

Goldhaber SZ, Meyerovitz MF, Markis JE, Kim D, Kessler CM, Sharma GVRK, Vaughan DE, Selwyn AP, Dawley DL, Loscalzo J, Sasahara AA, Grossbard EB, Braunwald E (1987) Thrombolytic therapy of acute pulmonary embolism: current status and future potential. J Am Coll Cardiol 10:9613–10413

Goldhaber SZ, Vaughan DE, Markis JE, Selwyn AP, Meyerowitz MF, Loscalzo J, Kim DS, Kessler CM, Dawley DL, Sharma GVRK, Sasahara A, Grossbard EB, Braunwald E (1986) Acute pulmonary embolism treated with tissue plasminogen activator. Lancet 2:886

Goldhaber SZ, Visani L, de Rosa M, for ICOPER (1999) Acute pulmonary embolism: clinical outcomes in the International Cooperative Pulmonary Embolism Registry (ICOPER). Lancet 353:1386–1389

González-Juanatey JR, Valdés L, Amaro A, Iglesias C, Alvarez D, Acuña JMG, De la Peña MG (1992) Treatment of massive pulmonary thromboembolism with low intrapulmonary dosages of urokinase. Short-term angiographic and hemodynamic evolution. Chest 102:341–346

Gulba DC, Schmid C, Borst H-G, Lichtlen P, Dietz H, Luft FC (1994) Medical compared with surgical treatment for massive pulmonary embolism. Lancet 343: 578–577

Gurewich V, Thomas DP, Rabinov KR (1966) Pulmonary embolism after ligation of the inferior vena cava. N Engl J Med 274:1350–1354

Hume M, Sevitt S, Thomas DP (1970) Venous Thrombosis and Pulmonary Embolism. Harvard University Press, Cambridge MA

Kakkar VV (1973) Results of streptokinase therapy in deep vein thrombosis. Postgrad Med J 49 [August Suppl]:60–67

Kanter DS, Mikkola KNI, Patel SR, Parker JA, Goldhaber SZ (1997) Thrombolytic therapy for pulmonary embolism. Frequency of intracranial hemorrhage and associated risk factors. Chest I 11:1241–1245

Kasper W, Konstantinides S, Geibel A, Olschewski M, Heinrich F, Grosser KD, Rauber K, Iversen S, Redecker M, Kienast J (1997) Management strategies and determinants of outcome in acute major pulmonary embolism: results of a multicenter registry. J Am Coll Cardiol 30:1165–1171

Kasper W, Meinertz T, Henkel B, Eissner D, Hahn K, Hofmann T, Zeiher A, Just H
   (1986) Echocardiographic findings in patients with proved pulmonary embolism.
   Am Heart J 112:1284–1290
Koning R, Cribier A, Gerber L, Eltchaninoff H, Tron C, Gupta V, Soyer R, Letac B
   (1997) A new treatment for severe pulmonary embolism: percutaneous rheolytic
   thrombectomy. Circulation 96:2498–2500
Konstantinides S, Geib IA, Olschewski M, Heinrich F, Grosser K, Rauber K, Iversen
   S, Redecker M, Kienast J, Just H, Kasper W (1997) Association between thrombolytic treatment and the prognosis of hemodynamically stable patients with
   major pulmonary embolism: results of a multicenter registry. Circulation 96:
   882–888
Konstantinides S, Tiede N, Geibel A, Olschewski M, Just H, Kasper W (1998) Comparison of alteplase versus heparin for resolution of major pulmonary embolism.
   Am J Cardiol 82:966–970
Levine M, Hirsh J, Weitz J, Cruickshank M, Neemeh J, Turpie AG, Gent M (1990) A
   randomized trial of a single bolus dosage regimen of recombinant tissue plasminogen activator in patients with acute pulmonary embolism. Chest 98:1473–1479
Lorch H, Welger D, Wagner V, Hillner B, Strecker EP, Herrmann H, Voshage G, Zur
   C, Schwarzbach C, Schroder J, Gullotta U, Pleissner J, Huttner S, Siering U, Mar-
   cklin C, Chavan A, Glaser F, Apitzsch DE, Moubayed K, Leonhardi J, Schuchard
   UM, Weiss HD, Zwaan M (2000) Current practice of temporary vena cava filter
   insertion: a multicenter registry. J Vasc Interv Radiol 11:83–88
McCotter CJ, Chiang KS, Fearrington EL (1999) Intrapulmonary artery infusion of
   urokinase for treatment of massive pulmonary embolism: a review of 26 patients
   with and without contraindications to systemic thrombolytic therapy. Clin Cardiol
   22:661–664
McIntyre KM, Sasahara AA (1972) Pulmonary angiography, scanning and hemody-
   namics in pulmonary embolism. Critical review and correlations. CRC Crit Rev
   Radiol Sci 3:489
Meneveau N, Schiele Metz D, Valette B, Attali P, Vuillemenot A, Grollier G, Elaerts J,
   Mossard J-M, Viel J-F, Bassand J-P (1998) Comparative efficacy of a two-hour
   regimen of streptokinase versus alteplase in acute massive pulmonary embolism:
   immediate clinical and hemodynamic outcome and one-year follow-up. J Am Coll
   Cardiol 31:1057–1063
Meneveau N, Schiele F, Willernenot A, Valette B, Grollier G, Bernard Y, Bassand JP
   (1997) Streptokinase vs alteplase in massive pulmonary embolism. A randomized
   trial assessing right heart hemodynamics and pulmonary vascular obstruction. Eur
   Heart J 18:1141–1148
Meyer G, Sors H, Charbonnier B, Kasper W, Bassand JP, Kerr IH, Lesaffre E, Vanhove
   P, Verstraete M (1992) Effects of intravenous urokinase versus alteplase on total
   pulmonary resistance in acute massive pulmonary embolism: a European multicenter double-blind trial. J Am Coll Cardiol 19:239–245
Mikkola KM, Patel S, Parker JA, Grodstein F, Goldhaber SZ (1997) Increasing age is
   a major risk factor for hemorrhagic complications after pulmonary embolism
   thrombolysis. Am Heart J 134:69–72
Murphy JM, Mulvihill N, Mulcahy D, Foley B, Smiddy P, Molloy MP (1999) Percutaneous catheter and guidewire fragmentation with local administration of recombinant tissue plasminogen activator as a treatment for massive pulmonary
   embolism. Eur Radiol 9:959–964
Petitpretz P, Simmoneau G, Cerrina J, Musset D, Dreyfus M, Vandenbroek MD, Duroux
   P (1984) Effects of a single bolus of urokinase in patients with life-threatening
   pulmonary emboli: a descriptive trial. Circulation 70:861–866
Sasahara AA (1975) Current problems in pulmonary embolism: Introduction. In:
   Sasahara AA, Sonnenblick EH, Lesch M (eds) Pulmonary embolism. Grune &
   Stratton, New York, pp 1–8
Sasahara AA, Barsamian EM (1973) Another look at pulmonary embolectomy. Ann
   Thorac Surg 16:317–320

Sasahara AA, Bell WR, Simon TL, Stengle JM, Sherry S (1975) The phase II urokinase streptokinase pulmonary embolism trial: a national cooperative study. Thromb Diath Haemorrh 33:464–476

Sasahara AA, Cannilla JE, Morse RL, Tremblay GM (1967) Clinical and physiologic studies in pulmonary thromboembolism. Am J Cardiol 20:10–20

Sasahara AA, Hyers TM, Cole CM, Wenger N, Ederer F, Stengle JM, Murray JA, Sherry S (1973) The urokinase pulmonary embolism trial: A national cooperative study. Circulation 47 (Suppl II):1–98

Sasahara AA, Sharma GVRK (1988) Thrombolytic therapy in the treatment of pulmonary embolism: short and long-term benefits. In: Comerota AJ (ed) Thrombolytic therapy. Grune and Stratton, Orlando, pp 51–63

Sautter RD, Myers WO, Ray JF, Wenzel FJ (1975) Pulmonary embolectomy: review and current status. In: Sasahara A, Sonnenblick EH, Lesch M (eds) Pulmonary embolism. Grune & Stratton, New York, pp 143–161

Sautter RD, Myers WO, Wenzel FJ (1972) Implications of the urokinase study concerning the surgical treatment of pulmonary embolism. J. Thorac Cardiovasc Surg 63:54–60

Sevitt S, Gallagher NG (1961) Venous thrombosis and pulmonary embolism. A clinico-pathological study in injured and burned patients. Brit J Surg 48:475–479

Sharafuddin MJ, Hicks ME (1997) Current status of percutaneous mechanical thrombectomy. 1. General principles. J Vasc Interv Radiol 8:911–921

Sharafuddin MJ, Hicks ME (1998a) Current status of percutaneous mechanical thrombectomy. 2. Devices and mechanisms of action. J Vasc Interv Radiol 9:15–31

Sharafuddin MJ, Hicks ME (1998b) Current status of percutaneous mechanical thrombectomy. 3. Present and future applications. J Vasc Interv Radiol 9:209–224

Sharma GVRK, Burleson VA, Sasahara AA (1980) Effect of thrombolytic therapy on pulmonary capillary blood volume in patients with pulmonary embolism. N Engl J Med 303:842–845

Sharma GVRK, Folland ED, McIntyre KM, Sasahara AA (1990) Long term hemodynamic benefit of thrombolytic therapy in pulmonary embolic disease. J Am Coll Cardiol 15:652–845

Sharma GVRK, O'Donnell DJ, Belko JS, Sasahara AA (1977) Thrombolytic therapy in deep vein thrombosis. In: Paoletti R, Sherry S (eds) Thrombosis and urokinase. Academic Press, London, pp 181–189

Shiina A, Kondo K, Kawai N, Hosada S (1980) Echocardiographic manifestations of acute pulmonary thromboembolism. A case report. Jpn Heart J 22:853

Siddique RM, Siddique MI, Connors AF, Rimm AA (1996) Thirty-day case-fatality rates for pulmonary embolism in the elderly. Arch Intern Med 156:2343–2347

Sors H, Pacouret G, Azarian R, Meyer G, Charbonnier B, Simonneau G (1994) Hemodynamic effects of bolus vs 2-h infusion of alteplase in acute massive pulmonary embolism. A randomized controlled multicenter trial. Chest 106:712–717

Steckley R, Smith CW, Robertson RM (1978) Acute right ventricular overload. An echocardiographic clue to pulmonary thromboembolism. Johns Hopkins Med J 143:122

Tebbe U, Graf A, Kamke W, Zahn R, Forycki F, Kratzsch G, Berg G (1999) Hemodynamic effects of double bolus reteplase versus alteplase infusion in massive pulmonary embolism. Am Heart J 138:39–44

The PIOPED Investigators (1990) Tissue plasminogen activator for the treatment of acute pulmonary embolism. Chest 97:528–533

The UKEP Study Research Group (1987) The UKEP study multicentre clinical trial on two local regimens of urokinase in massive pulmonary embolism. Europ Heart J 8:2–10

Urokinase-Streptokinase Pulmonary Embolism Trial (USPET) (1974) Phase 2 results. JAMA 229:1606–1612

Verstraete M, Miller GAH, Bounameaux H, Charbonnier B, Colle JP, Lecorf G, Marbet GA, Mornbaerts P, Olsson CG (1988) Intravenous and intrapulmonary recombinant tissue-type plasminogen activator in the treatment of acute massive pulmonary embolism. Circulation 77:353–360

Wheeler HB, Mullick C, Anderson JN, Pearson D (1971) Diagnosis of occult deep vein thrombosis by a noninvasive bedside technique. Surgery 70:20–28

Wolfe WG, Sabiston DC Jr (1980) In: Wolfe WG (ed) Pulmonary embolism. Saunders, Philadelphia, p 35

Zwaan M, Lorch H, Kulke C, Kagel C, Schweider G, Siemens HJ, Müller G, Eberhardt I, Wagner T, Weiss H-D (1998) Clinical experience with temporary vena caval filters. J Vasc Interv Radiol 9:594–601

CHAPTER 13

Deep Venous Thrombosis

H. BOUNAMEAUX

A. Introduction

Thrombosis of the deep veins (DVT) impedes the blood return from the affected limb to the heart and destroys the venous valvular system. Standard treatment with heparin prevents thrombus extension and embolization to the lungs but does not significantly reduce the formed thrombus. It has been proposed that rapid and complete lysis of the thrombus would result in preservation or restoration of the anatomy and function of the deep veins. Because venous thromboembolism accounts for more than 300 000 hospitalizations per year in the US (LANDEFELD and HANUS 1993) and because 400 000–500 000 Americans suffer from chronic leg ulcers (COON 1977), many of them being secondary to previous DVT, effective and safe thrombolysis might have important public health implications.

In 1980, a consensus report of the National Institutes of Health stated that "thrombolytic therapy represents a significant advance in the management of proximal acute deep-vein thrombosis, and . . . with further developments aimed at improving its efficacy and reducing the bleeding risk, this form of therapy could become the initial treatment for all forms of acute deep-vein thrombosis" (NATIONAL INSTITUTE OF HEALTH CONSENSUS DEVELOPMENT CONFERENCE 1980). Nonetheless, 20 years later, thrombolysis has not gained widespread acceptance in the treatment of acute DVT of the lower limbs. Moreover, therapeutic regimens are far from established in this indication in most countries and we still ignore whether the potential late benefit of obviating the disastrous postthrombotic syndrome (PTS) compensates the immediate risk of severe hemorrhagic complications.

This review summarizes the presently available data on immediate and late effects of thrombolytic treatment of DVT and tries to assess objectively its benefit-to-risk balance in this indication.

B. Treatment Regimens

Several treatment schemes have been used (DUCKERT 1984). In most trials, streptokinase (SK) has been given with an initial loading dose of 250000 U followed by an infusion of 100000 U/h for several days. In Germany, the so-called ultra-high SK (1.5–9 MU over 4–6 h daily for a few days) has also been used. In a few reports, urokinase (UK) has been administered with a loading dose of 250000 U to 1 MU over 20 min followed by an infusion of 1–3 MU per day for several days (GOLDHABER et al. 1994, 1996; MOIA et al. 1994). Treatment schemes using tissue-type plasminogen activator (tPA) have not been established yet. A dose ranging study revealed that treatment with an initial dose of approximately 0.5 mg/kg of tPA administered over 2 h, followed by a maintenance infusion of 0.06 mg/kg/h for an additional 4 h or 22 h was not very effective, whereas a maintenance infusion over 33 h resulted in significant lysis but also in unacceptable bleeding complications (MARDER et al. 1992). In the largest trial published so far a dose of 0.05 mg/kg/h for 24 h has been used (GOLDHABER et al. 1990) but alternative schemes have also been tested, including lower-dose infusions prolonged for 7 days (BOUNAMEAUX et al. 1992a) or repeated short infusions over 2 days (VERHAEGHE et al. 1989; TURPIE et al. 1990).

C. Immediate Effects of Thrombolytic Therapy of DVT: The Definite Risk

Acute symptoms and signs of DVT include leg swelling and pain which are diminished by bed rest and elastic bandages, and pulmonary embolism which can be prevented by anticoagulants. Thrombolytic treatment does not significantly modify these symptoms and signs. On the other hand, there is no evidence that lysis of leg thrombi would result in more pulmonary emboli, as some clinicians fear. Thus, the immediate effect of thrombolysis is restricted to more rapid and more extensive dissolution of the thrombi in the leg veins, as demonstrated in several randomized trials, most of them comparing heparin with SK or, in more recent reports, with recombinant tissue-type plasminogen activator (tPA) (Table 1). In a pooled analysis, GOLDHABER et al. (1984) calculated that significant thrombolysis on venogram was achieved 3.7 times (95% confidence interval 2.5–5.7) more often among patients treated with SK than among patients treated with heparin. However, definition of significant venographic improvement is poorly standardized and does not necessarily correlate with clinical outcome. Moreover, substantial differences exist between venographic assessments when they are performed by the investigator or by a central review panel, local assessment being significantly more optimistic than central judgment (BOUNAMEAUX et al. 1992b). Thus, more recent studies with the novel thrombolytic agent tPA, which have been performed according to better defined methodological standards, are more likely to reflect the reality than studies performed up to the early 1980s in which evaluation of venograms was often not performed in a centralized, blinded fashion.

In summary, despite their heterogeneity, regarding the dose regimen or the thrombolytic agent used (DUCKERT 1984), the vast majority of these trials

Table 1. Immediate results of thrombolysis in DVT: efficacy data[a]

First Author and Year	n	Agent	Duration	Assessment	Result
ROBERTSON (1967)	20	SK	12 h	Venogram	SK not better
ROBERTSON (1968)	16	SK	24 h	Venogram	SK better
KAKKAR (1969)	20	SK	>120 h	Venogram, clinical, FUT	SK better
ROBERTSON (1970)	16	SK	72 h	Venogram	SK better
TSAPOGAS (1973)	34	SK	48–72 h	Venogram, clinical	SK better
PORTER (1975)	49	SK	72 h	Venogram	SK better
BIEGER (1976)	10	SK	72 h	Venogram, venous pressure	SK better
SEAMAN (1976)	50	SK	72 h	Venogram	SK better
MARDER (1987)	24	SK	72 h	Venogram	SK better
ARNESEN (1978)	42	SK	4 days	Venogram	SK better
ELLIOT (1979b)	51	SK	72 h	Venogram, clinical	SK better
VERHAEGHE (1989)	21	tPA	8 h (2×)	Venogram	tPA not better
GOLDHABER (1990)	67	tPA	24 h	Venogram	tPA better
TURPIE (1990)	83	tPA	8 h (2×)	Venogram	tPA not better
GOLDHABER (1996)	17	UK	30 min (3×)	Venogram, MRI, Doppler	UK not better

SK streptokinase; tPA recombinant tissue-type plasminogen activator; UK urokinase; FUT fibrinogen uptake test; MRI magnetic resonance imaging.
[a] Restricted to randomized, controlled studies comparing heparin with the thrombolytic agent (with or without concomitant heparin).

suggests a superiority of the lytic agent over heparin with respect to diminution of venographic thrombus size.

This effect, however, is achieved at the cost of a serious risk of bleeding (Table 2). In his pooled analysis, GOLDHABER et al. (1984) observed a statistically significant 2.9-fold increase of the frequency of major bleeding (95% confidence interval 1.1–8.1) among SK-treated patients compared with those treated with heparin. Since a five-day course of heparin is associated with a major bleeding risk of about 5%, serious hemorrhagic complications can be anticipated in about one fifth of patients receiving SK. In the largest case series published, JACOBSEN (1989) reported a fatality rate of 7/1200 DVT patients treated with SK (0.6%), which differs probably not much from the mortality of DVT itself.

The hemorrhagic risk associated with the use of tPA does not seem to be clearly lower, with 8 individuals out of 25 tPA-treated patients experiencing major bleeding in the study of VERHAEGHE et al. (1989), 2/43 in the trial of GOLDHABER et al. (1990), 2/41 in that of TURPIE et al. (1990) and 5/32 in the European co-operative study that compared two tPA dose regimens, all hemorrhagic events occurring in the group allocated to the dose regimen of 0.5 mg/kg per 24 h over 7 days (BOUNAMEAUX et al. 1992a).

Although the type of plasminogen activator, the duration of infusion, the doses administered, and the use of concomitant heparin may play a role in the

Table 2. Immediate results of thrombolysis in DVT: safety data[a]

First Author and Year	n	Agent	Duration	Major bleeding	
				Thrombolysis n (%)	Heparin n (%)
Robertson (1968)	16	SK	24 h	3/8 (38)	1/8 (12)
Porter (1975)	49	SK	72 h	4/24 (17)	1/26 (4)
Arnesen (1978)	42	SK	4 days	3/21 (14)	3/21 (14)
Elliot (1979b)	51	SK	72 h	2/25 (8)	0/25 (0)
Verhaeghe (1989)	21	tPA	8 h (2×)	8/25 (32)	0/7 (0)
Goldhaber (1990)	67	tPA	24 h	1/53 (2)	0/12 (0)
Turpie (1990)	83	tPA	8 h (2×)	3/41 (7)	1/42 (2)
Bounameaux (1992a)[b]	32	tPA	7 days	5/17 (29)	0/15 (0)
Goldhaber (1996)	17	UK	30 min (3×)	0/8 (0)	1/9 (11)

[a] Restricted to randomized, controlled studies comparing heparin with the thrombolytic agent (with or without concomitant heparin).
[b] In that trial, very low doses of tPA were given in the comparative group.

occurrence of severe bleeding under thrombolytic therapy in DVT patients, it is obvious that all treatment schemes are associated with a definite increase of major hemorrhages compared with heparin. Interestingly, Levine et al. (1995) noted that there is a higher rate of bleeding associated with thrombolytic therapy in patients with DVT compared with myocardial infarction, probably due to the more prolonged treatment duration and presence of underlying diseases.

D. Late Effects of Thrombolytic Therapy of DVT: The Potential Benefit

Since immediate (and usually only partial) reduction of thrombus size cannot be a therapeutic goal in itself, long term follow-up studies are necessary to assess the efficacy of thrombolytic treatment on the occurrence of the late PTS and of its extreme manifestation, the chronic venous leg ulcer. Unfortunately, most of these studies lack a sufficient sample size or follow-up duration, lost a considerable proportion of patients, and rely often on venographic data rather than clinical evaluation (Table 3). In a comprehensive literature review, Rogers and Lutcher (1990) included results of non-randomized trials, which resulted in a more clear-cut advantage of thrombolytic treatment over heparin with respect to the presence of a normal venogram at follow-up (28/57 vs 2/56) and a normal valvular function (25/61 vs 8/53), but again without convincing clinical data.

Thus, the few available randomized controlled studies and several uncontrolled trials suggest that a normal venogram at follow-up is more likely to occur in thrombolysed patients than in heparin-treated patients. On the other hand, a normal venogram at long-term follow-up seems to correlate with the

Table 3. Long-term follow-up, randomized studies following thrombolytic treatment of DVT

First Author and Year	n	n (FU)	FU	Postthrombotic syndrome		Normal venogram	
				T	H	T	H
KAKKAR (1969)	20	15	6–12 months	ND	ND	4/7	1/7
ROBERTSON (1970)	16	7	9–12 years	2/4	2/3	0/1	0/3
BIEGER (1976)	10	10	3–4 months	2/5	3/5	5/5	1/5
COMMON (1976)[a]	50	27	7 months	ND	ND	6/15	1/12
ELLIOT (1979b)	51	47	3 months	10/26	ND	11/20	ND
JOHANSSON (1979)[b]	16	5	12–14 years	3/3	2/2	0/3	0/2
ARNESEN (1982)	42	35	6.5 years	7/17	12/18	7/17	0/18

All studies are with SK.
FU, follow-up; T, thrombolysis group; H, heparin group; ND, no data.
[a] Follow-up of patients initially reported by SEAMAN et al. (1976).
[b] Long-term follow-up of the patients initially reported by ROBERTSON et al. (1967).

absence of clinically overt PTS but there are too few patients to achieve statistical significance.

Additional interesting data are available from a non-randomized series of 223 patients with venographically proven DVT who were followed-up over 13 years (EICHLISBERGER et al. 1994). Among these patients, 144 were treated with lytic agents. Immediate substantial venographic improvement was observed in 100 patients vs none among the 79 heparin-treated patients. Thirteen years later, PTS and leg ulcers were present in 29% and 6% respectively of these patients. In those without initial venographic improvement, comprising 44 thrombolysed and 79 heparin-treated patients, the corresponding figures were 39% and 9.8% respectively (n.s.).

In summary, clinical trials so far failed to provide convincing evidence that thrombolytic therapy significantly reduces the frequency of the PTS several years after an acute proximal DVT.

E. The Benefit-to-Risk Balance

Thrombolysis in DVT is associated with a definite, immediate bleeding risk and a potential, long-term benefit in terms of risk reduction of developing PTS. Using basically the data presented in the previous section of this review, O'MEARA et al. (1994) constructed a decision-analysis model that assumed a threefold increase of major bleeding risk and a 50% reduction of PTS, the latter clearly favoring thrombolytic therapy. This model predicted that in a cohort of 10 000 patients treated for DVT, there would be 70 cerebral bleedings (39 being lethal) more in the SK-treated patients along with a reduction of 496 severe and 4226 mild PTSs.

Interestingly, these authors studied the patients' preferences regarding these outcomes. Although only 36 patients were interviewed, the results are far from equivocal: all patients were willing to accept only very small risks of immediate death or stroke to avoid late PTS and this was also true for the 16 individuals who suffered from this condition, thereby resulting in an advantage of heparin alone vs heparin plus SK treatment in proximal DVT. As compared with SK, heparin provided 29 days of additional life expectancy over the predicted life expectancy of 20 years. Although the difference between the two treatments was thus small, all meaningful sensitivity analyses did not modify the fact that heparin alone remained the preferred treatment. This kind of analysis underscores the importance of considering patients' values and preferences in making decisions about treatment. This is particularly true in a situation where risks and benefits of therapeutic options are not well established.

F. How to Reduce the Risk and Improve the Benefit?

I. Reducing Hemorrhagic Risk

The decision analysis model described above (O'Meara et al. 1994) that included patients' preferences and concluded as to the superiority of heparin alone over heparin plus SK for treating proximal DVT has an important implication if one aims at improving the benefit-to-risk ratio of thrombolytic therapy in that particular indication. Since patients are unwilling to accept risky treatments, it is probably more important to develop safer thrombolytic regimens or to select more carefully patients at lower risk of bleeding than to look primarily for more efficient treatments.

Usual contra-indications to thrombolytic therapy are listed in Table 4. They should all be considered absolute in the setting of DVT. It has been estab-

Table 4. Usual contra-indications to the use of thrombolytic agents

Recent major surgery (within 2 weeks)
Cerebrovascular accident (especially within 2 months)
Neurosurgical intervention (within 2 months)
Impairment of hemostasis
Oral anticoagulant treatment
Thrombocytopenia $<50\,000 \times 10^9/l$
Malignant hypertension
Traumatic reanimation
Recent gastro-intestinal bleeding
Pregnancy
Diabetic retinopathy
Age >75 years
Puncture of a non-compressible artery (within 2 weeks)
Streptokinase allergy[a]
Administration of streptokinase or anistreplase in the preceding year[a]

[a] Concerns only streptokinase and anistreplase.

lished by multivariate analysis in a series of 150 patients who had intracerebral hemorrhage under thrombolysis and 294 matched controls that, age over 65 years, body weight below 70 kg, hypertension at hospital admission, and administration of tPA, were independent risk predictors of intracranial bleeding (SIMOONS et al. 1993). Since these data were derived from a population of patients who received the thrombolytic agent because of acute myocardial infarction, they cannot be directly transferred to the setting of DVT but they provide some useful clues for evaluating a potential candidate to thrombolytic therapy, whatever the indication is.

II. Improving Thrombolytic Efficacy

Several dose regimens have been proposed that claim to be superior to others, like "ultrahigh dose SK" (1.5–3 MU/h or even more over 6 h every day for 2–5 days) which is largely used in Germany (THEISS et al. 1987) but no randomized, controlled study has ever been published to substantiate the claimed superiority. Regional perfusion with tPA in a prospective randomized trial of 137 patients did not turn out to be superior to systemic administration of tPA (SCHWIEDER et al. 1995).

Local, "catheter-directed" thrombolysis has also been said to produce greater venographic dissolution of acute and even chronic DVT but the data are restricted to non-randomized series lacking a sufficient follow-up (COMEROTA et al. 1994; BJARNASON et al. 1997; VERHAEGHE et al. 1997; HORNE et al. 2000). SEMBA and DAKE (1994) reported a 72% rate of complete lysis among 25 patients with iliofemoral DVT who were treated locally with urokinase and BJARNASON et al. (1997) stated that in their collective of 77 patients the primary patency rate (i.e., no further intervention needed), one year after catheter-directed thrombolysis, was 63% in femoral DVT and 40% in iliac DVT. A recent report of the National Multicenter Registry analyses the results obtained with catheter-directed urokinase infusions in 287 patients with proximal symptomatic lower limb DVT. After thrombolysis 99 iliac and five femoral vein lesions were treated with stents. Complete lysis was achieved in 31%, grade II lysis (50-99% clot dissolution) in 52% and <50 % lysis in 17% of cases. At 1 year primary patency rate was 60% (MEWISSEN et al. 1999). The use of Wallstents, followed by balloon angioplasty to relieve inferior vena cava obstruction in patients with femoral arterio-venous grafts, was reported to preserve patency of the afflicted vessels (CHANG et al. 1998).

Certainly, thrombus age is an important determinant of its sensitivity to lysis by plasminogen activators. Although thrombus age cannot be estimated accurately, there is some indication that it correlates with the duration of symptoms and most authors suggest that symptoms lasting more than 7 days are associated with an unacceptably low efficacy of thrombolytic agents.

Two independent groups recently suggested that initial venographic appearance would predict the short-term response to thrombolytic therapy in DVT. Thus, MEYEROVITZ et al. (1992) showed that significant lysis (more than 50%) of DVT by tPA was significantly more frequent in venous segments involved with nonobstructive thrombi than in those with obstructive thrombi (12 of 23 vs 5 of 51, $p < 0.005$). THÉRY et al. (1992) reached similar conclusions in a prospective study of 174 patients in which 27 of 45 (60%) of patients with nonocclusive clots were completely free of clots at the control venogram vs 17 of 116 (14%) with occlusive clots ($p < 0.001$). In addition, among nonocclusive thrombi, proximal (caval, iliac, or femoral) ones were more easily lysed than more distal (popliteal) clots. From these data, it appears that patients with occlusive DVT should not be submitted to the risk of thrombolytic therapy because the chance of reaching a complete dissolution of thrombi is only 10%–15%. However, nonocclusive thrombi represent the minority of clots in the deep veins of the leg and the risk of developing PTS in that particular subpopulation is unknown.

G. Miscellaneous Venous Indications

I. Phlegmasia Caerulea Dolens

This rare condition of extensive leg thrombosis is characterized by an impaired microcirculation with gangrene of the extremity and is associated with a 50% risk of amputation and a 25% lethality. It is often considered as a possible indication for thrombolytic therapy on the basis of case reports with favorable outcomes (ELLIOT et al. 1979a). However, patients with phlegmasia caerulea dolens suffer frequently from terminal cancer disease and the indication for aggressive therapy may not always be justified.

II. Superior or Inferior Vena Cava Syndrome

In a retrospective review of the experience of the Cleveland clinic, GRAY et al. (1991) reported complete clot lysis and relief of symptoms in 9/16 patients with superior vena cava syndrome under thrombolytic therapy. Three factors predicting success were the use of UK compared with SK, the presence of a central venous catheter, and a duration of symptoms of less than five days. The exact place of thrombolysis in inferior vena cava syndrome is not established.

III. Deep Venous Thrombosis of the Upper Limbs

Because subclavian or axillary vein thrombosis is rarely followed by late symptoms and the course of the disease appears to be usually benign, an aggressive therapy that provokes major hemorrhage in 20% of patients will hardly find justification in that indication (ELLIS et al. 2000).

IV. Venous Thromboses of Unusual Localization

Unusual thromboses like vena mesenterica, portal or hepatic vein thrombosis, or thromboses secondary to paroxysmal nocturnal hemoglobinuria have been proposed as potential targets for thrombolytic therapy (SHOLAR and BELL 1985). Since controlled, randomized trials will probably never be performed in such rare conditions, the decision to treat patients with thrombolytic agents will depend upon the severity and potential consequences attributed to the particular thrombotic event. In the particular setting of central retinal vein thrombosis, 40 patients could be randomized to receive SK (followed by heparin) or placebo (KOHNER et al. 1976). At follow-up, the visual acuity was significantly better in the treated group than in the untreated group with, however, three vitreous hemorrhages and permanent loss of vision in three patients under SK. In a retrospective analysis of 58 patients with central retinal vein occlusion visual acuity of 2 or more lines on the logarithmic visual acuity chart was found in 44% of patients treated with tPA within 11 days of onset of symptoms. Only 14% of patients treated with hemodilution showed a similar improvement (HATTENBACH et al. 1999). Anecdotal reports of successful local or systemic thrombolysis exist also in thrombosis of cerebral central or dural venous sinuses (FREY et al. 1999) but the superiority of this treatment over anticoagulant therapy is questionable.

V. Thrombosis of Central Venous Catheters

Thrombosis of central venous catheters can occur in up to 25% of cases. In order to re-open obstructed catheters, HAIRE et al. (1994) compared in a controlled randomized trial the administration of three successive boluses of 5000–10000 U of UK or one to two boluses of 2mg of tPA, the latter regimen being associated with a patency rate of 90%, which was significantly superior to that obtained with urokinase.

H. Laboratory Monitoring

Laboratory tests of blood coagulation and fibrinolytic processes have usually been considered as useless to adapt the dose of the plasminogen activator during thrombolytic therapy, especially in myocardial infarction and pulmonary embolism patients where the treatment lasts only 60–120min (MARDER 1987). On the other hand, early studies suggested that a plasma fibrinogen concentration above 0.8g/l under SK treatment for DVT was associated with a very low chance of complete clearance of the thrombi (DUCKERT et al. 1975). Recently, it was proposed that a patient with bleeding time prolongation during SK therapy for DVT was twice as likely to produce important clot lysis (HIRSCH et al. 1991). The series, however, consisted of only 16 patients with DVT of various localizations and deserves to be confirmed in a larger patients population.

In summary, at the present time there is no indication to monitor short-term thrombolysis or bleeding complications by laboratory controls unless a "lytic state" needs to be proven or its importance estimated, especially in prolonged infusions or where a patient needs to be evaluated for a possible surgical intervention. However, monitoring of concomitant or subsequent heparin therapy remains necessary.

I. Treatment of Bleeding

Major bleeding under thrombolytic therapy is an emergency, especially if the central nervous system is involved. Bleedings at catheter insertion sites can usually be controlled by manual compression and by discontinuing thrombolytic and adjuvant anticoagulant treatments. If necessary, heparin can be antagonized by protamine, and fibrinogen and coagulation factors infused by means of viro-inactivated plasma derivatives. In case of intracerebral event, CT or MRI imaging should be obtained without delay to assess the need and possibility of emergency neurosurgical intervention.

J. Conclusions and Perspectives

Over the past decade, thrombolytic therapy has become the most successful therapy in acute myocardial infarction and massive pulmonary embolism, two conditions that are associated with quite high lethality. DVT carries a definitely smaller risk. It is, therefore, questionable to apply a potentially dangerous treatment for that condition unless the long-term benefit would be undebatable and important (Verstraete 1995). In fact, the effects of thrombolytic treatment on the occurrence of late PTS is far from established, which leads us to consider this therapy experimental in DVT patients. This recommendation is further supported by the recent publication of a 12-year survey of 58 DVT patients initially treated with heparin followed by oral anticoagulants. In this series, marked trophic changes were found in only 5% of patients, and venous ulcer was observed in only one patient (Franzeck et al. 1996). If one were to consider treating a DVT patient with thrombolytic agents outside of a clinical trial, several measures should be considered which aim at maximizing treatment efficacy and minimizing hemorrhagic risk (Table 5). The individual benefit-to-risk ratio should be carefully assessed before ordering a treatment that specialists are reluctant to use in that particular indication

Table 5. Possible indications for thrombolytic therapy in DVT

Age under 50 (to minimize hemorrhagic risk and to maximize long-term benefit)
Proximal DVT with duration of symptoms of less than 7 days (to maximize immediate thrombolytic effect)
Nonocclusive character of the thrombi on venogram (to maximize immediate thrombolytic effect)
No history of previous DVT in the same leg (to maximize long-term benefit)
No contra-indication to thrombolytic therapy (to minimize hemorrhagic risk)

(Sɪᴅᴏʀᴏᴠ 1989; Wᴇɪɴᴍᴀɴɴ and Sᴀʟᴢᴍᴀɴ 1994; Vᴇʀsᴛʀᴀᴇᴛᴇ 1995) in most countries except Germany and Austria. It remains to be tested whether the transdermal application of ultrasound is able to enhance the thrombolytic efficacy in the treatment of DVT (without causing bleeding complications and local tissue damage), as was demonstrated in vitro (Sᴜᴄʜᴋᴏᴠᴀ et al. 1998) and in animal models (Bɪʀɴʙᴀᴜᴍ et al. 1998).

New thrombolytic agents are presently in development and/or in a clinical testing phase (staphylokinase, mutants of tPA, chimeric plasminogen activators; see Chaps. 16–19). They will probably, like the presently available agents, impair hemostatic mechanisms and be associated with a bleeding risk. Whether this risk will be diminished and outweighed by the therapeutic benefit or not should be studied in carefully designed trials before they can be recommended for the treatment of DVT.

List of Abbreviations and Acronyms

CT   computerized tomography
DVT   Deep venous thrombosis
FUT   fibrinogen uptake test
MRI   Magnetic resonance imaging
MU   Mega (1 million) units
PTS   postthrombotic syndrome
SK   Streptokinase
tPA   tissue-type plasminogen activator (alteplase)

References

Arnesen H, Heilo A, Jakobsen E, Ly B, Skaga E (1978) A prospective study of streptokinase and heparin in the treatment of deep vein thrombosis. Acta Med Scand 203:457–463

Arnesen H, Hoiseth A, Ly B (1982) Streptokinase or heparin in the treatment of deep vein thrombosis. Follow-up results of a prospective study. Acta Med Scand 211:65–68

Bieger R, Boekhout-Mussert RJ, Hohmann F, Loeliger EA (1976) Is streptokinase useful in the treatment of deep vein thrombosis? Acta Med Scand 199:81–88

Birnbaum Y, Luo H, Nagai T, Fishbein MC, Peterson TM, Li S, Kricsfeld D, Porter TR, Siegel RJ (1998) Noninvasive in vivo clot dissolution without a thrombolytic drug: recanalization of thrombosed iliofemoral arteries by transcutaneous ultrasound combined with intravenous infusion of microbubbles. Circulation 97:130–134

Bjarnason H, Kruse JR, Asinger DA, Nazarian GK, Dietz CA Jr, Caldwell MD, Key NS, Hirsch AT, Hunter DW (1997) Iliofemoral deep venous thrombosis: safety and efficacy outcome during 5 years of catheter-directed thrombolytic therapy. J Vasc Interv Radiol 8:405–418

Bounameaux H, Banga JD, Bluhmki E, Coccheri S, Fiessinger JN, Haarmann W, Lockner D, Mahler F, Ninet J, Schneider PA, de Torrente A, van der Meer J, Verhaeghe R (1992a) Double-blind, randomized comparison of systemic continuous infusion of 0.25 versus 0.50 mg/kg/24 h of alteplase over 3 to 7 days for treatment of deep venous thrombosis in heparinized patients: results of the European Thrombolysis with rt-PA in Venous Thrombosis (ETTT) Trial. Thromb Haemost 67:306–309

Bounameaux H, Prins TR, Schmitt HE, Schneider PA, and the ETTT Trial Investigators (1992b) Venography of the lower limbs. Pitfalls of the diagnostic standard. Invest Radiol 27:1009–1011

Chang TC, Zaleski GX, Lin BH, Funaki B, Leef J (1998) Treatment of inferior vena cava obstruction in hemodialysis patients using Wallstents: early and intermediate results. Am J Roentgenol 171:125–128

Comerota AJ, Aldridge SC, Cohen G, Ball DS, Pliskin M, White JV (1994) A strategy of aggressive regional therapy for acute iliofemoral venous thrombosis with contemporary venous thrombectomy or catheter-directed thrombolysis. J Vasc Surg 20:244–254

Common HH, Seaman AJ, Rösch J, Porter JM, Dotter CT (1976) Deep vein thrombosis treated with streptokinase or heparin. Follow-up of a randomized study. Angiology 27:645–654

Coon WW (1977) Epidemiology of venous thromboembolism. Ann Surg 186:149–164

Duckert F (1984) Thrombolytic therapy. Sem Thromb Hemost 10:87–103

Duckert F, Müller G, Nyman D, Benz A, Prisender S, Madar G, da Silva MA, Widmer LK, Schmitt HE (1975) Treatment of deep vein thrombosis with streptokinase. Br Med J 1:479–481

Eichlisberger R, Frauchiger B, Widmer MT, Widmer LK, Jäger K (1994) Spätfolgen der tiefen Venenthrombose; ein 13-Jahres Follow-up von 223 Patienten. VASA 23:234–243

Elliot MS, Immelman EJ, Jeffery P, Benatar SR, Funston MR, Smith JA, Jacobs P, Shepstone BJ, Ferguson AD, Louw JH (1979a) The role of thrombolytic therapy in the management of phlegmasia caerulea dolens. Br J Surg 66:422–424

Elliot MS, Immelman EJ, Jeffery P, Benatar SR, Funston MR, Smith JA, Shepstone BJ, Ferguson AG, Jacobs P, Walker W, Louw JH (1979b) A comparative randomized trial of heparin versus streptokinase in the treatment of acute proximal venous thrombosis: an interim report of a prospective trial. Br J Surg 66:838–843

Ellis MH, Manor Y, Witz M (2000) Risk factors and management of patients with upper limb deep vein thrombosis. Chest 117:43–46

Franzeck UK, Schalch I, Jäger KA, Schneider E, Grimm J, Bollinger A (1996) Prospective 12-year follow-up study of clinical and hemodynamic sequelae after deep vein thrombosis in low-risk patients (Zürich Study). Circulation 93:74–79

Frey JL, Muro GJ, McDougall CG, Dean BL, Jahnke HK (1999) Cerebral venous thrombosis: combined intrathrombus rtPA and intravenous heparin. Stroke 30:489–494

Goldhaber SZ, Buring JE, Lipnick RJ, Hennekens CH (1984) Pooled analyses of randomized trials of streptokinase and heparin in phlebographically documented acute deep venous thrombosis. Am J Med 76:393–397

Goldhaber SZ, Hirsch DR, Macdougall RC, Polak JF, Creager MA (1996) Bolus recombinant urokinase versus heparin in deep venous thrombosis: A randomized controlled trial. Am Heart J 132:314–318

Goldhaber SZ, Meyerovitz MF, Green D, Vogelzang RL, Citrin P, Heit J, Sobel M, Wheeler HB, Plante D, Kim H, Hopkins A, Tufte M, Stump D, Braunwald E (1990) Randomized controlled trial of tissue plasminogen activator in proximal deep venous thrombosis. Am J Med 88:235–240

Goldhaber SZ, Polak JF, Feldstein ML, Meyerovitz MF, Creager MA (1994) Efficacy and safety of repeated boluses of urokinase in the treatment of deep venous thrombosis. Am J Cardiol 73:75–79

Gray BH, Olin JW, Graor RA, Young JR, Bartholomew JR, Ruschhaupt WF (1991) Safety and efficacy of thrombolytic therapy for superior vena cava syndrome. Chest 99:54–59

Haire WD, Atkinson JB, Stephens LC, Kotulak GD (1994) Urokinase versus recombinant tissue plasminogen activator in thrombosed central venous catheters: a double-blinded, randomized trial. Thromb Haemost 72:543–547

Hattenbach L-O, Wellermann G, Steinkamp GWK, Scharrer I, Koch FHJ, Ohrloff C (1999) Visual outcome after treatment with low-dose recombinant tissue plasminogen activator or hemodilution in ischemic central retinal vein occlusion. Ophthalmologica 213:360–366

Hirsch DR, Reis SE, Polak JF, Donovan BC, Goldhaber SZ (1991) Prolonged bleeding time as a marker of venous clot lysis during streptokinase therapy. Am Heart J 122:965–971

Horne MK, Mayo DJ, Cannon RO, Chen CC, Shawker TH, Chang R (2000) Intraclot recombinant tissue plasminogen activator in the treatment of deep venous thrombosis of the lower and upper extremities. Am J Med 108:251–255

Jacobsen B (1989) Thrombolytic treatment in venous thromboembolism. In: Eklöf B, Gjöres JE, Thulesius O, Bergqvist D (eds) Controversies in the management of venous disorders. Butterworths, London, pp 116–126

Johansson L, Nylander G, Hedner U, Nilsson IM (1979) Comparison of streptokinase with heparin: late results in the treatment of deep venous thrombosis. Acta Med Scand 206:93–98

Kakkar VV, Flanc C, Howe CT, O'Shea M, Flute PT (1969) Treatment of deep vein thrombosis. A trial of heparin, streptokinase, and arvin. Br Med J 1:806–810

Kohner EM, Pettit JE, Hamilton AM, Bulpitt CJ, Dollery CT (1976) Streptokinase in central retinal vein occlusion: a controlled clinical trial. Br Med J 1:550–553

Landefeld CS, Hanus P (1993) Economic burden of venous thromboembolism. In: Goldhaber SZ (ed) Prevention of venous thromboembolism. Dekker, New York, pp 69–85

Levine MN, Goldhaber SZ, Gore JM, Hirsh J, Califf RM (1995) Hemorrhagic complications of thrombolytic therapy in the treatment of myocardial infarction and venous thromboembolism. Chest 108 [Suppl]:291S–301S

Marder VJ (1987) Relevance of changes in blood fibrinolytic and coagulation parameters during thrombolytic therapy. Am J Med 83 [Suppl 2A]:15–19

Marder VJ, Brenner B, Totterman S, Francis CW, Rubin R, Rao AK, Kessler CM, Kwaan HC, Sharma GVRK, Fong KLL (1992) Comparison of dosage schedules of rt-PA in the treatment of proximal deep vein thrombosis. J Lab Clin Med 119:485–495

Mewissen MW, Seabrook GR, Meissner MH, Cynamon J, Labropoulos N, Haughton SH (1999) Catheter-directed thrombolysis for lower extremity deep venous thrombosis: report of a National Multicenter Registry. Radiology 211:39–49

Meyerovitz MF, Polak JF, Goldhaber SZ (1992) Short-term response to thrombolytic therapy in deep venous thrombosis: predictive value of venographic appearance. Radiology 184:345–348

Moia M, Mannucci PM, Pini M, Prandoni P, Gurewich V (1994) A pilot study of pro-urokinase in the treatment of deep vein thrombosis. Thromb Haemost 72: 430–433

National Institute of Health Consensus Development Conference (1980) Thrombolytic therapy in thrombosis: a NIH development conference. Ann Intern Med 93:141–144

O'Meara JJ III, McNutt RA, Evans AT, Moore SW, Downs SM (1994) A decision analysis of streptokinase plus heparin as compared with heparin alone for deep-vein thrombosis. N Engl J Med 330:1864–1869

Porter JM, Seaman AJ, Common HH, Rosch J, Eidemiller LR, Calhoun AD (1975) Comparison of heparin and streptokinase in the treatment of venous thrombosis. Am Surg 41:511–519

Robertson BR, Nilsson IM, Nylander G (1968) Value of streptokinase and heparin in treatment of acute deep venous thrombosis. A coded investigation. Acta Chir Scand 134:203–208

Robertson BR, Nilsson IM, Nylander G (1970) Thrombolytic effect of streptokinase as evaluated by phlebography of deep venous thrombi of the leg. Acta Chir Scand 136:173–180

Robertson BR, Nilsson IM, Nylander G, Olow B (1967) Effect of streptokinase and heparin on patients with deep venous thrombosis. A coded examination. Acta Chir Scand 133:205–215

Rogers LQ, Lutcher CL (1990) Streptokinase therapy for deep vein thrombosis: a comprehensive review of the English literature. Am J Med 88:389–395

Schwieder G, Grimm W, Siemens HJ, Flor B, Hilden A, Gmelin E, Friedrich HJ, Wagner T (1995) Intermittent regional therapy with rt-PA is not superior to systemic thrombolysis in deep vein thrombosis (DVT) – a German multicenter trial. Thromb Haemost 74:1240–1243

Seaman AJ, Common HH, Rösch J, Dotter CT, Porter JM, Lindell TD, Lawler WL, Schlueter WJ (1976) Deep vein thrombosis treated with streptokinase or heparin. A randomized study. Angiology 27:549–556

Semba CP, Dake MD (1994) Iliofemoral deep venous thrombosis: aggressive therapy with catheter-directed thrombolysis. Radiology 191:487–494

Sholar PW, Bell WR (1985) Thrombolytic therapy for inferior vena cava thrombosis in paroxysmal nocturnal hemoglobinuria. Ann Intern Med 103:539–541

Sidorov J (1989) Streptokinase vs heparin for deep venous thrombosis. Can lytic therapy be justified? Arch Intern Med 149:1841–1845

Simoons ML, Maggioni AP, Knatterud G, Leimberger JD, de Jaegere P, van Domburg R, Boersma E, Franzosi MG, Califf R, Schroder R, et al (1993) Individual risk assessment for intracranial haemorrhage during thrombolytic therapy. Lancet 342:1523–1528

Suchkova V, Siddiqi FN, Carstensen EL, Dalecki D, Child S, Francis CW (1998) Enhancement of fibrinolysis with 40-kHz ultrasound. Circulation 98:1030–1035

Theiss W, Baumann G, Klein G (1987) Fibrinolytische Behandlung tiefer Venenthrombosen mit Streptokinase in ultrahoher Dosierung. Dtsch Med Wochenschr 112:668–674

Théry C, Bauchart JJ, Lesenne M, Asseman P, Flajollet J-G, Legghe R, Marache P (1992) Predictive factors of effectiveness of streptokinase in deep venous thrombosis. Am J Cardiol 69:117–122

Tsapogas MJ, Peabody RA, Wu KT, Karmody AM, Devaraj KT, Eckert C (1973) Controlled study of thrombolytic therapy in deep vein thrombosis. Surgery 74:973–984

Turpie AGG, Levine MN, Hirsh J, Ginsberg JS, Cruickshank M, Jay R, Gent M (1990) Tissue plasminogen activator (rt-PA) vs heparin in deep vein thrombosis. Results of a randomized trial. Chest 97 [Suppl]:172S–175S

Verhaeghe R, Besse P, Bounameaux H, Marbet GA (1989) Multicenter pilot study of the efficacy and safety of systemic rt-PA administration in the treatment of deep vein thrombosis of the lower extremities and/or pelvis. Thromb Res 55:5–11

Verhaeghe R, Stockx L, Lacroix H, Vermylen J, Baert AL (1997) Catheter-directed lysis of iliofemoral vein thrombosis with use of rt-PA. Eur Radiol 7:996–1001

Verstraete M (1995) Thrombolytic treatment. BMJ 311:582–583

Weinmann EE, Salzman EW (1994) Deep-vein thrombosis. N Engl J Med 331:1630–1641

CHAPTER 14

Peripheral Arterial Occlusions

D.C. BERRIDGE

A. Introduction

The management of peripheral arterial occlusions has seen many changes over the last 15 years. Traditional management received a great boost with the introduction of the Fogarty thromboembolectomy balloon in 1963 (FOGARTY et al. 1963). Until then management had largely been expectant using heparin, with surgical intervention in terms of reconstruction still in its comparative infancy. BLAISDELL et al. (1978) advocated a conservative approach using heparin. Whilst mortality was acceptably low (7%), 28% required a major amputation. In response to that publication, PLECHA compared the results of the Cleveland Vascular Registry in its annual report of 1977 (PLECHA et al. 1978). In this report the amputation rate was only 3%, but associated with a mortality of 24%. To quote Dr Plecha "I think it is obvious that Dr Blaisdell's conservative approach has traded lives for limbs – and I think that it is an excellent trade."

Thomas Fogarty pioneered the concept of the careful use of balloon catheters capable of removing the thrombo-embolic material from a remote entry site. Published results confirmed that this indeed resulted in a significant improvement in limb salvage (FOGARTY et al. 1963; TAWES et al. 1985). However, the presentation of patients with acute or acute on chronic ischemia has been changing. The classical white leg associated with embolus, with or without propagated thrombus, is a relatively rare presentation today. We are far more likely to see a patient with a preceding history of intermittent claudication, one in whom previous reconstructions or interventions have been performed, and/or patients who are heavy smokers. Whilst they may have previously been asymptomatic, most will undoubtedly have underlying arteriosclerotic disease. Associated with these anatomical alterations are changes of the hemostatic system in the direction of a hypercoagulable state. In the ADMIT trial fibrinogen and plasminogen activator inhibitor type 1 (PAI-1) were increased in patients with peripheral arterial disease compared to controls (PHILIPP et al. 1997). Furthermore, as was shown in patients with acute myocardial infarction and stroke, elevated CRP levels predicted future risk of developing symptomatic peripheral vascular disease and thus provide support for the hypothesis that chronic inflammation is a pathogenic factor of athero-genesis (RIDKER et al. 1998). Patients with emboli presenting later than 72h may have very adherent embolic and propagated thrombus material. Even careful use of an embolectomy catheter in these patients may result in intimal damage and in exacerbating the situation rather than assisting it. The advent of on-table arteriography including "digital subtraction" and "road-mapping" facilities have also allowed the initial "blind" embolectomy to proceed along more carefully controlled and monitored conditions. On-table angioplasty and thrombolysis add further to the arsenal of the vascular surgeon.

Initial interest in peripheral arterial thrombolysis involved the use of streptokinase (SK) administered systemically (AMERY et al. 1970). Whilst encouraging initial reports were published it never became adopted routinely due to the anxieties over side effects, particularly the risk of cerebro-vascular accident and/or major hemorrhage. With the advent of advanced percutaneous methods, and the local administration technique popularized by DOTTER et al. (1974), more acceptable efficacy and safety profiles were established, marking the basis of today's approach (HESS et al. 1982; KATZEN et al. 1984). We now also have the benefit of comparatively large randomized studies comparing thrombolysis with surgery, which have allowed us to be more precise with regard to the indications for the initial management of acute limb ischemia. The initial management clearly needs to be tailored to the individual patient, thereby allowing the optimal use of both minimal interventional radiological and open surgical techniques. There is no role for pure thrombolysis or pure surgery, and each patient should be managed in a team environment with the clinical skills of both the interventional vascular radiologist and the vascular surgeon.

Percutaneous aspiration techniques have allowed further modification of techniques especially for the more acutely ischemic leg, although distal embolization may occur in approximately 10% of cases (WAGNER and STARCK 1992). It should be noted that 88 patients (86%) of this particular series of 102 patients had cardiac disease as the apparent source of the embolus.

B. Current Indications

With the advent of percutaneous techniques, local administration using intra-thrombotic SK or urokinase (UK), and more recently recombinant tissue plasminogen activator (in the following abbreviated as tPA) has become estab-lished as a management option in the care of patients with recent peripheral arterial occlusions. The indications for its use have become better defined as experience has been gained.

Intrathrombotic thrombolysis is indicated in four main groups of patients:

1. Acute or acute on chronic ischemia of native vessels due to either in situ thrombosis or embolism, particularly if the latter is in distal, possibly inac-cessible vessels
2. Graft thrombosis, both prosthetic and autogenous venous

3. Thrombosed popliteal aneurysms
4. Post-procedural thrombosis

There has been considerable debate as to the length of duration of the history of presentation. In the United Kingdom it has been traditional to limit this technique to lesions of less than 30 days, with some centers allowing up to 6 weeks duration. On the other hand, in Europe and in the United States, lesions of up to 6 months duration have been considered for lytic treatment. Some authors advocate that if one can pass a guide wire through the thrombus which 'feels soft' then that patient could be considered for lysis regardless of the apparent clinical duration of history. There is some evidence to substantiate this view. In a study of 103 patients with limb-threatening symptoms lasting ≤14 days, wire traversal was accomplished in 84 (82%) of which 75 (89%) had successful lysis. On the other hand, of the remaining 19 patients (18%) in whom the guide wire could not be passed, lysis was achieved in just 3 (16%) (OURIEL et al. 1994a). This study failed to show any difference in relation to multiple or single end-hole catheters or between sheath and no-sheath systems. In an analysis of factors predicting the outcome of intra-arterial thrombolysis, good runoff and intrathrombotic infusion were the two principal predictors of a favorable outcome (SANDBAEK et al. 1996). LAMMER et al. (1986) reported a success rate of 78% recanalization within 1–7 h in lesions of up to 2 years duration (mean 4.5 months). However, published experience suggests that the longer the history, the more likely the lesion will be resistant to lysis. It will also require longer duration of infusions with higher doses of lytic agents and will be associated with a higher risk of hemorrhagic complications (VERSTRAETE et al. 1971). We do now have some information in the form of randomized studies that indicate that lysis may be the preferred option to surgery in the initial management of lesions of up to 7 days (OURIEL et al. 1994b) and 14 days (THE STILE INVESTIGATORS 1994; OURIEL et al. 1998), respectively. However, neither of these studies provided any hard information as to where the true cut-off between benefit and disadvantage lies. Is it really 14 days or could it be as long as 30 days or even 6 weeks? We cannot gain this information from these studies due to their recruitment protocols (in the STILE trial lesions of up to 6 months, in Ouriel's 1994 trial only lesions of ≤7 days and in Ouriel's 1998 trial duration of acute symptoms of ≤14 days were included). Caution should therefore be exercised before embarking on attempted lysis of lesions much in excess of these suggested time periods due to the lack of information regarding potential benefit.

Occluded femoro-popliteal grafts are particularly suited to an initial trial of percutaneous arterial thrombolysis. In a non-randomized comparison of UK and SK in thrombosed femoro-popliteal grafts, UK was superior in terms of success (77% vs 41%) and complications (23% vs 50%) (GARDINER et al. 1986). Native artery (78%) or prosthetic grafts (80%) are more likely to achieve success than venous bypass grafts (54%) (OURIEL et al. 1994a).

Thrombosed autogenous grafts are a separate entity. Many authors would suggest that a graft thrombosed for more than 24–48 h had probably caused such damage to the endothelium that, even if lysis was successful, there would be inevitable rethrombosis due to the highly thrombogenic endothelium (BELKIN et al. 1990). Patients with (patent) vein grafts also exhibit upregulation of PAI-1 which may represent a risk factor for the development of (re)thrombosis (PELTONEN et al. 1996). In a series of 22 thrombosed vein grafts in 21 patients, with a mean duration of symptoms before thrombolysis of 14 days, lysis only achieved a one-year patency rate in $37 \pm 12\%$ (BELKIN et al. 1990). In another study of 44 patients, primary graft patency at 1 year was 25%. Multivariate analysis revealed poorer outcome in patients with diabetes and those whose vein graft had been in place for less than 12 months (NACKMAN et al. 1997). Also, early thrombosis of a graft probably represents an error of selection of the patient, the inflow outflow site, the type of graft, or indeed a technical error itself. The risks of thrombolysis are unwarranted in these cases. Surgical exploration, revision with or without intra-operative lysis appears a safer proposition. However very acceptable results with lesions of up to 11 days duration (PATERSON et al. 1993) or even longer (SULLIVAN et al. 1991) have been reported. Clearly the best policy for the management of graft thrombosis is prevention. In this regard the concept of the "failing graft" picked up on graft surveillance will allow earlier intervention with improved salvage results (SANCHEZ et al. 1991).

Thrombosed popliteal aneurysms are a special case and deserve discussion in their own right. It is of paramount importance that the perfusion catheter is placed in the run-off vessel (i.e., distal popliteal) to attempt lysis and definition of the run-off vessels. It is unnecessary to attempt lysis of the aneurysm itself and to risk its associated much higher risk of distal embolization. Once definition of the run-off is established then a formal surgical proximal and distal ligation of the aneurysm and bypass can be performed. Thrombolysis of very distal occluded grafts is also possible and may yield favorable results (SPENCE et al. 1997).

C. Contra-Indications

For local administration, there are a few absolute contra-indications:

- Cerebrovascular accident within the last 2 months
- Bleeding diathesis (including serious recent gastrointestinal bleeding)
- A limb that is unsalvageable with fixed necrosis or dead muscle

In addition, relative but major contra-indications are:

- Recent major trauma or operation (less than 10 days)
- Pregnancy
- Inaccessible Dacron grafts due to the risk of potentially uncontrollable hemorrhage through the interstices of the graft
- Severe hypertension (>200 mm Hg systolic and/or >110 mm Hg diastolic)

Age over 75 years is not a contra-indication for thrombolysis but is associated with high complication rates. Surgical treatment appears to be superior for this group of patients (BRAITHWAITE et al. 1998).

D. Drugs Available

I. Streptokinase

SK was originally obtained in quite impure form, hence there were marked allergic side effects. Severe anaphylactic reactions, while rare, can prove fatal (KAKKAR et al. 1969). These undesirable effects have largely, though not completely, been eliminated with more modern production methods (COMEROTA 1988).

SK requires plasminogen as a cofactor, which then allows conversion of plasminogen to plasmin (see Chap. 2). Thus if thrombus reaccumulates in a patient receiving SK, there may be insufficient plasminogen within the thrombus to allow full lysis to occur.

It is known from studies related to coronary thrombolysis that, following the administration of 1.5 MU of SK for coronary thrombolysis, significant antibody levels can be detected for up to 4 years post treatment that would effectively neutralize about half of a further dose of the same magnitude (ELLIOTT et al. 1993). Whilst intra-thrombotic administration of SK may allow artificially high local levels of SK, a difference in outcome could be detected. Patients with high titers responded partially, with only 20% achieving lysis (VAN BREDA et al. 1987). It is suggested therefore that reuse of SK in a patient for peripheral arterial thrombolysis either following earlier coronary or indeed peripheral arterial thrombolysis may actually be condemning that patient to failure before one has even commenced the infusion. Despite these drawbacks very acceptable results have been reported with intra-arterial SK strictly adhering to a well defined protocol with appropriate radiological and/or surgical intervention (WALKER and GIDDINGS 1988).

II. APSAC, Anistreplase (Acylated-Plasminogen-Streptokinase-Activator Complex)

Although used quite extensively in the field of coronary thrombolysis, its use in peripheral arterial thrombolysis has been very restricted. In a comparative series reported by EARNSHAW et al. (1986), anistreplase was associated with poor lytic success and an unacceptably high hemorrhagic complication rate. Its theoretical advantage is the ability to administer it systemically because anistreplase binds to fibrin. There are no studies which have looked at its potential role delivered as an intra-thrombotic agent, although this is unlikely to show any major difference to results obtained with SK itself.

III. Urokinase and Pro-Urokinase

This has been the favored agent in a large number of both European and American series. UK has a number of theoretical advantages over SK. It does not excite an antibody response, and hence it can be reused. It has a higher lytic potency and incurs less hemorrhagic complications than the non-selective action of SK (VAN BREDA et al. 1987). Pro-urokinase, the single-chain zymogen form of UK (Scu-PA) has some fibrin specificity (see Chaps 4 and 9).

IV. Recombinant Tissue Plasminogen Activator (tPA)

tPA is a relatively fibrin specific agent that, like UK, does not incur an antibody response (see Chaps. 8 and 10) and hence is ideally suited to multiple reuse without alteration in treatment schedules. Early studies all used doses higher than those currently in use. Initial results appeared to suggest a higher lytic rate with higher doses, associated with increased limb salvage and with reduced hemorrhagic complications compared to historical control series using SK. A mutant of tPA, reteplase, has shown promise as an alternative to UK in a pilot study (DAVIDIAN et al. 2000).

V. Staphylokinase

Staphylokinase is still an investigative drug. It is produced using recombinant genetic techniques (see Chap. 16). Preliminary results in peripheral arterial occlusion have been published by VANDERSCHUEREN et al. (1995). In a recent report on 191 patients with native ($n = 122$) or bypass graft ($n = 69$) occlusions of <120 days duration staphylokinase was given as a 2mg bolus, followed by an infusion of 1mg/h. Revascularization was complete in 83%, partial in 13% and absent in only 4%. One-month and one-year mortality were 3.1% and 6.9% respectively. Four patients (2.1%), all over the age of 70 years developed fatal intracranial bleeding. Staphylokinase appears to be highly effective in peripheral arterial occlusion resulting in a low mortality and a high one-year amputation-free survival (84%) (HEYMANS et al. 2000).

E. Techniques Used

I. Systemic Administration

Original reports concerned with peripheral arterial thrombolysis used systemic administration of SK along the lines of management associated with coronary administration. However, whilst coronary lysis is involved with the very rapid lysis of comparatively recent (i.e., hours as opposed to days or even weeks) thrombosis of small volume and presumably little in the way of established fibrin cross-linking, peripheral arterial lysis is usually involved with much longer, mature, well-established thrombi of considerable volume. Hence it is not too surprising that we cannot adopt the large dose, short duration systemic regimes that our cardiology colleagues have developed and used so effectively [ISIS-3 (THIRD INTERNATIONAL STUDY OF INFARCT SURVIVAL) COLLABORATIVE GROUP 1992]. Nevertheless some reports demonstrated moderate benefit (EARNSHAW et al. 1990). Intravenous APSAC (5 mg) given 8-hourly was also tried, but was associated with a high hemorrhagic complication rate and a lysis rate of less than 30% (EARNSHAW et al. 1986). In a collected series of 440 cases reported by EARNSHAW (1994) the overall success rate was only 39% ($n = 170$), with 19% of cases experiencing major hemorrhage, of which 4% were fatal. In addition, a further 2% of patients had non-fatal strokes. Clearly these results are not as favorable as those achieved using intra-arterial delivery of a more limited dose of the thrombolytic agent.

The advent of tPA heralded a potential answer. We now had the opportunity of using a relatively fibrin-specific agent that could be administered systemically but would not cause significant systemic activation. However, in a randomized trial comparing intra-arterial tPA, intra-arterial SK, and intra-venous tPA, the systemic use of tPA resulted in a significantly reduced lytic success rate (45%) and an increased hemorrhagic complication rate (15% major; 45% minor), compared to either of the two intra-arterial regimens (BERRIDGE et al. 1991). It therefore seems that, for the foreseeable future, systemic administration is not to be recommended for peripheral arterial thrombolysis.

II. Local Low-dose Intra-thrombotic Administration

Following Dotter's publication (DOTTER et al. 1974), the percutaneous low-dose continuous infusion technique became the standard technique for peripheral arterial thrombolysis. A catheter usually with a single end-hole is placed within the thrombus. A contra-lateral femoral puncture may be preferable to an ipsilateral ante-grade puncture that may prove more difficult to compress and ultimately to achieve hemostasis. A variety of different regimens with different agents have been developed, The usual dose for SK is 5000–10000 IU/h intra-arterially, and for tPA 0.5–1 mg/h (low-dose), or 0.05 mg/kg per hour intra-arterially. UK has been used in both high-dose and low-dose formats, there being no obvious benefit when using the higher dose regime.

III. High-Dose Bolus

HESS et al. (1982) popularized the rapid advancement, multiple bolus technique using SK. SULLIVAN et al. (1989) showed a significantly reduced lysis time (10 h vs 34 h; $p < 0.001$), and reduced total dose of UK in the high-dose bolus group. Complications were also less in the high-dose bolus group (9%) compared to the standard regime (23%). WHOLEY et al. (1998) treated 235 patients in a multicenter retrospective study with a bolus of 250000 U of UK, followed by 120000–250000 U/h for 4h. Complete thrombolysis was achieved in 60 (86%) of the acute, 20 (77%) of the subacute and 105 (75%) of the chronic occlusions.

JUHAN et al. (1991) and later BUCKENHAM et al. (1992) reported faster recanalization times using a higher bolus dose regimen with tPA. Overall lytic times were reduced to less than half of those achieved with the local low-dose regimes. JUHAN et al. (1991) suggested 5 mg boluses of tPA repeated at 5–10 min intervals up to three times, then to continue with 0.05 mg/kg per hour (3.5 mg/h in a 70 kg man). In ten patients the duration of lysis varied from 5 min to 4 h 15 min (mean 2 h 30 min). Only two minor hemorrhages occurred, both at the puncture sites. However, WARD et al. (1994) also using a high-dose bolus technique reported 70% angiographic lysis ($n = 16$), with clinical success in only 48% ($n = 12$), with a shorter lysis time (14 ± 8 h), but with an apparent increased risk of hemorrhage (35%, $n = 8$). The dosage regime consisted of an initial bolus of 20 mg tPA followed by an infusion of 1 mg/h. Lesions of up to 90 days duration were included. However direct comparison with other studies is difficult as no data was given as to the median or range of duration of infusion or the length of lesion treated. Further we know from ODAVIC (1995) that there appears to be an increased risk of hemorrhage when more than 20 mg of tPA is used. Hence we may be placing all of these patients in the high-risk group automatically. More recently a randomized trial (BRAITH-WAITE et al. 1997) of high-dose vs low-dose regimens has been reported. Time to lysis was achieved in less than 4 h in half of the patients with the high-dose regime and 80% of patients achieved lysis by a mean period of 20 h. Further, there was a trend to an increased complication rate in the high-dose group which did not achieve statistical significance, possibly due to a type two error. Hence apart from the potential role of this technique in the more severely ischemic limb it does not appear to hold any other definite advantage over conventional low-dose regimes. Indeed the overall time within the radiology department was similar for both groups. The advantage of the low-dose group in this particular context is that it potentially allows a more efficient use of the Interventional Suite and does not occupy it for an indefinite time on each occasion lysis is attempted.

IV. Pulse-Spray Administration

Bookstein's group (VALJI et al. 1991; BOOKSTEIN et al. 1994a,b) has pioneered the use of this technique with its novel method of forced radial infusion of small aliquots of lytic agents through specially designed catheters with multiple side holes. The concern over the production of massive distal embolization was not fulfilled largely due to the utilization of a "distal plug" of thrombus which is retained until the later stages of lysis, hence acting as a safety net to allow further lysis before entry to run-off vessels is allowed. There has been considerable controversy in the literature as to the true benefits and mechanism of action (KANDARPA et al. 1993; BOOKSTEIN et al. 1994a,b). In the 1993 series from Bookstein's group (VALJI et al. 1993) a 33% distal embolization rate was encountered in those patients in whom early angioplasty was performed due to "lytic stagnation." This is clearly in excess of the 2% reported by GALLAND et al. (1993) using low-dose continuous infusions. KANDARPA et al. (1993) pointed out that after the initial bolus, subsequent bolusing may merely serve to achieve pericatheter flow of lytic agent rather than continued radial forced infusion. In part, efficiency of the pulse-spray (and of continuous perfusion) may depend on catheter design (Roy et al. 1998). Difference in duration of infusion may have been due, at least in part, to a difference in assessing the end-point. Valji et al. (1993) refer to "lytic stagnation" as an end-point "avoiding prolonged unsuccessful treatment." However this has not been validated in a randomized controlled study. In randomized studies of forced periodic infusion vs conventional infusion, no difference was found in overall time to lysis (Kandarpa et al. 1993). In an open non-randomized comparison of pulse-spray with tPA to a historical series of conventional low-dose lysis with tPA, a significantly shorter time to lysis was achieved with the pulse-spray regime (Yusuf et al. 1994). However the numbers involved were too small to make any valid judgment on any other parameter of efficacy, safety, or indeed long term outcome However there was no significant difference in overall success rate (pulse-spray 75% vs conventional lysis 60%). A follow-up of 100 cases with pulse-spray and an additional low dose infusion of tPA and/or catheter-directed intervention revealed a cumulative limb salvage rate at 30 months of 79% (Armon et al. 1997).

As with high-dose bolus administration, this technique does allow the application of thrombolysis to the more acutely ischemic limb (Yusuf et al. 1994). Experimental studies in rabbits suggest that greater lysis can be achieved by increasing the frequency from the previously used 1 pulse/2 min to 2 pulses/min, allowing concomitantly to decrease the dose of tPA administered (Bookstein and Bookstein 2000). Most acute or acute on chronic ischemic limbs presenting as emergency no longer fit the classical presentation of a "white leg." Hence these allow considered management usually performed on the next available list, or emergency list the following day. It also still allows the application of the conventional low-dose regime in the majority of patients without prejudicing their affected limb.

V. Enclosed Thrombolysis

The first description of enclosed thrombolysis in six patients with peripheral arterial occlusion was reported by Jørgensen at al. (1989). The recanalized segment was isolated with a 7-French double-balloon catheter and a bolus of 5 mg of tPA applied. At 10 days and 30 days all patients had evidence of recanalization. Tønnesen et al. (1991) reported 55 patients with chronic femoro-popliteal occlusions treated by percutaneous transluminal angioplasty and enclosed thrombolysis. Of 33 patients (60%) with 1–3 vessel run-off, 3-month patency was 100%, with 90% patency at 1 year. However of the 20 patients in whom there was no infra-popliteal run-off, 1-year patency, on an intention to treat basis, was only 62%. Hence, as the authors themselves comment, there is a need for a randomized trial of conventional angioplasty vs angioplasty plus enclosed thrombolysis to ascertain whether there is a true beneficial effect from the addition of enclosed thrombolysis.

VI. Intra-operative Lysis

This is a very useful adjunct to any surgical procedure that may be associated with residual thrombus or distal inaccessible emboli (COHEN et al. 1986; COMEROTA et al. 1989; BEARD et al. 1993). Up to 30% of thromboembolectomies may have residual thromboembolic material on an intra-operative angiogram (MAVOR et al. 1972). A dose of 100 000 IU of SK in 100 ml of normal saline infused into a closed system over 30 min followed by a repeat on-table angiogram has been proposed (BEARD et al. 1993). Other authors have used 5 mg aliquots of tPA, repeated after 10–15 min up to a maximum of three doses. Whilst there have been no randomized studies of different lytic agents in intra-operative lysis, Comerota's group found that 5/14 (36%) receiving SK achieved lysis and limb salvage compared to 13/24 (54%) of those receiving UK. There were five amputations in each group (SK 36%, UK 21%) (COMEROTA et al. 1989). In a subsequent study COMEROTA et al. (1993) randomized 134 patients to receive one of three bolus doses of UK (125,000 U, 250 000 U, or 500 000 U) or placebo infused intra-arterially before lower extremity bypass for chromic limb ischemia. The average duration of graft occlusion was 34 days. Catheter placement failed in 39% of patients randomized to thrombolysis. These patients were reverted to surgical revascularization. In those patients where catheter placement was successful, patency was restored in 84% and 42% had major reduction in their planned operation. In these patients, 1-year results compared favorably with the best surgical procedure which was new graft placement. Acutely ischemic patients (≤14 days) randomized to thrombolysis showed a trend toward a lower amputation rate at 30 days ($p = 0.07$) and at 1 year ($p = 0.026$) compared with surgical patients. MEYEROVITZ et al. (1995) randomized 20 patients with claudication or limb-threatening ischemia of at least 3 weeks duration due to iliac or femoro-popliteal occlusions either to thrombolysis with tPA followed by angioplasty, or to angioplasty alone. Life-table analysis revealed a significant improvement in the cumulative primary patency rate for patients treated initially with thrombolysis followed by angioplasty (86% at six months, 51% at 1 year) compared with angioplasty alone (11% at 6 months and 1 year).

Intra-operative lysis is particularly useful in the management of the thrombozed popliteal aneurysm. Here percutaneous thrombolysis is ideally performed of the run-off vessels only. If it is impossible to position the catheter in the run-off vessel, then lysis in any other position may cause an unacceptably high incidence of distal embolization of up to 12% (GALLAND et al. 1993). In these situations and also those where the limb is considered too ischemic to allow sufficient time to achieve patency with lysis, intra-operative lysis is the answer. The below knee popliteal artery is explored, a cannula inserted, and lysis commenced with one of the above regimes. During this period the operator can be employed dissecting out the above knee popliteal artery, harvesting the saphenous vein, and performing the proximal anastomosis.

F. Comparative Results

I. Streptokinase vs Recombinant Tissue Plasminogen Activator

Whilst there have been a number of series which show acceptable success and complication rates for each agent, there are relatively few trials that allow a direct randomized comparison of the safety and efficacy of the different thrombolytic agents and their regimens (Table 1). In a large retrospective review of over 400 patients, GRAOR et al. (1990) showed a large difference between the efficacy of UK or tPA with that seen following the use of SK. Similarly the safety profile of UK and tPA were superior to that seen with SK. There was a trend for a slight increase in patency with tPA compared to UK. Conversely there was a slight trend for more hemorrhagic complications with tPA compared to UK. Comparison of open or historical controls has the inevitable risk of bias including the influence of the learning curve for that particular institution. Nevertheless we can learn some useful information by their critical appraisal in the light of evidence from the randomized studies. In a randomized comparison of intra-arterial SK, intra-arterial tPA, and intra-venous tPA (BERRIDGE et al. 1991), a significant advantage of intra-arterial tPA over intra-arterial SK was seen. Limb salvage at 30 days was 20% higher with tPA, with a significantly lower major hemorrhagic complication rate and amputation rate than with intra-arterial SK. Intravenous SK certainly worked but its overall success rate (45%) and safety profile were inferior to that of intra-arterial SK (80% radiological success; 15% major hemorrhage; 15% minor hemorrhage).

Table 1. Dose regimes in peripheral arterial thrombolysis

Drug	Commonly used dosages	References
Streptokinase	5000 U/h+250 U heparin/h	WALKER and GIDDINGS (1985)
tPA	0.05 mg/h 0.5–1.0 mg/kg/h 10 mg/5 mg/3 mg/h 5 mg × 3, then 0.05 mg/kg 0.2 ml every 15 s for 15 min, then 0.2 ml every 30 s (0.33 mg/ml)	BERRIDGE et al. (1990) GRAOR et al. (1990) VERSTRAETE et al. (1988) JUHAN et al. (1991) YUSUF et al. (1994)
Urokinase	4000 U/min 250 000 bolus, then 4000 U/min for 4 h, followed by 2000 U/min for up to 36 h 4000 U/min for 2 h, followed by 2000 U/min, then 1000 U/min 250 000 U/h (4000 U/min), then 60 000–120 000 U/h 150 000 U over 20 min in 0.2-ml increments (25 000 U/ml) 250 000 U/h for 4 h, then 125 000 U/h or 50 000 U/h	GRAOR et al. (1990) THE STILE INVESTIGATORS (1994) OURIEL et al. (1994b) McNAMARA and FISCHER (1985) BOOKSTEIN et al. (1994b) CRAGG et al. (1991)

This order of magnitude was maintained during the three-month follow-up period included in the protocol. Hemorrhagic complications and the risk of amputation were all significantly less with tPA than with SK (BERRIDGE et al. 1991). In a follow-up study with more patients from the same institution, LONSDALE et al. (1992) reported that this advantage was clearly maintained in favor of tPA despite using rigorous criteria of success. In a large collective comparison, GRAOR et al. (1990) have shown that both tPA and UK have a clear advantage in terms of both efficacy and incidence of hemorrhagic complications.

II. UK vs Recombinant Tissue Plasminogen Activator

MEYEROVITZ et al. (1990) performed a randomized comparison of these two agents. Whilst tPA appeared to show earlier and more frequent lysis, there was no significant difference in any of the parameters measured at 30 days. However, due to the small numbers involved in this study, we may again be obscuring a real difference due to a type two effect.

SCHWEIZER et al. (1996) randomized 120 patients with acute and subacute peripheral arterial occlusions in tPA and UK treatment arms. Initial recanalization was achieved in 85% and 73% of patients in the tPA and UK groups respectively. At six-month follow-up, tPA patients had lower rates of Fontaine stage III and IV disease, amputation rates, and reocclusions than did UK patients. Graor's comparative series, involving several hundred patients, did not show any significant difference between the two agents either in terms of efficacy or in terms of safety. There was a trend to slightly less hemorrhagic complications using UK, with a slight advantage in lysis with tPA (GRAOR et al. 1990; reviewed in SEMBA et al. 2000).

III. Pro-UK vs. UK

In the PURPOSE trial three doses of intra-arterial pro-UK (2mg, 4mg, or 8mg/h for 8h, then 0.5mg/h) were compared to UK (4000U/min for 4h, then 2000U/min). Lysis was fastest with the 8mg/h dose of pro-UK but the rate of bleeding complications was also slightly higher (OURIEL et al. 1999).

IV. Surgery vs Thrombolysis

GRAOR et al. (1988) compared the outcome from surgery with that from peripheral arterial thrombolysis. Despite a relatively small series there did seem to be an advantage of lysis over surgery (median survival 195 days vs 30 days) for the management of thrombosed peripheral arterial grafts ($p = 0.01$).

The first randomized trial of surgery vs thrombolysis by NILSSON et al. (1992) involved lesions of less than 14 days duration, randomized to initial thrombolysis or to thromboembolectomy. The lytic regime used consisted of

30 mg of tPA over a 3 h period. Thrombectomy achieved patency in 65% compared to 40% in the lysis group. However care should be used in interpreting these data as there were only 20 patients in the trial.

A large randomized study in peripheral arterial thrombolysis is the STILE trial. Some 393 patients were randomized to either initial surgery or initial thrombolysis. In the latter group, patients either received tPA at 0.05 mg/kg per hour for up to 12 h, or a bolus of 250 000 U of UK, followed by 4000 U/min for up to 4 h and by 2000 U/min for up to 36 h. For lesions of less than 14 days duration, initial lysis treatment was associated with significantly lower amputation rates, and over half (56%) had a reduction in the magnitude of their surgical procedure. These changes remained at 6 months (THE STILE INVESTIGATORS 1994). There was no significant difference in efficacy or safety between UK and tPA.

In a substudy of the STILE trial, WEAVER et al. (1996) compared surgical revascularization vs thrombolysis in patients with native (non-embolic) lower-extremity occlusions. One hundred and fifty patients who had symptomatically deteriorated within the past 6 months were randomized to catheter-directed thrombolysis (84 to tPA, 0.05–0.1 mg/kg per hour; and 66 to UK, bolus of 250 000 U, followed by 4000 U/min for 4 h and then 2000 U/min up to 36 h). In 78% of patients randomized to thrombolysis the catheter could be positioned properly. This led to a reduction in the predetermined surgical procedure in 56% of the patients. Time to lysis was shorter with tPA (8 h) than with UK (24 h; $p < 0.05$) but no difference was observed in efficacy or safety. At one-year follow-up the incidence of recurrent ischemia and major amputations was higher in patients randomized to thrombolysis (Table 2). However, in diabetics one year mortality of thrombolysed patients was only 7.7% compared with that of 26.3% in surgical patients ($p < 0.05$). On the other hand, more ampu-

Table 2. Clinical outcome in a randomized trial comparing thrombolysis with surgical revascularization

	Surgery	Thrombolysis	p
Number of patients	87	150	
Overall death/Amputation (%)	15	18	
Death (%)	15	11	
Major Amputation (%)	0	10	0.0013
Duration of ischemia			
≤14 days, number of patients	16	32	
Death/Amputation (%)	19	12.5	
Death (%)	19	6	
Major Amputation (%)	0	6	
>14 days, number of patients	69	118	
Death/Amputation (%)	14.5	23	
Death (%)	14.5	14	
Major amputation (%)	0	13	

Reproduced from WEAVER et al. (1996).

tations had to be performed in thrombolysed diabetics (15.4%) than in surgical patients (0%; $p < 0.05$). Although in this trial thrombolysis led to a reduction in the surgical procedures for a majority of patients, long-term outcome was inferior, particularly in patients whose duration of ischemia had lasted for more than 14 days, or who had femoral-popliteal occlusions or critical ischemia (WEAVER et al. 1996).

Another substudy of the STILE trial evaluated safety and outcome of catheter-directed thrombolysis vs surgical vascularisation in 124 patients with lower limb bypass graft occlusion. Results were similar to those described above. Patients with ischemia of ≤14 days randomized to thrombolysis had a lower amputation rate at 30 days ($p = 0.07$) and at 1-year follow-up ($p = 0.03$) compared to surgical patients. In patients with >14 days ischemia, amputation rates were similar but thrombolysed patients had significantly more recurrent ischemia ($p < 0.001$) (COMEROTA et al. 1996).

In another randomized prospective study by OURIEL et al. (1994b), 144 patients with limb threatening ischemia of less than 7 days duration were randomized to either initial surgery or initial thrombolysis. Whilst there was no difference in limb salvage at 12 months (82%) between the two groups, there was a significantly higher survival rate in the thrombolysis group (84% vs 58% at 12 months, $p < 0.01$). This appeared to be primarily due to a difference of in-hospital cardiopulmonary complications in the operative treatment groups (49% vs 16%, $p = 0.001$).

In the phase I TOPAS trial 213 patients with acute lower-extremity ischemia for ≤14 days were randomized into four groups. The first three groups received catheter-directed UK of either 2000, 4000, or 6000 U/min for 4 h, then all groups were continued with 2000 U/min up to 48 h. The fourth group underwent surgery. The 4000 U/min dosage produced the most favorable efficacy/bleeding risk ratio (OURIEL et al. 1996). In this group 1-year mortality (14%) and amputation-free survival (75%) did not differ from the results achieved in the surgical group (16% and 65% respectively). In the subsequent phase II TOPAS trial, 544 patients with acute lower limb ischemia of ≤14 days duration from 113 North American and European institutions were randomized to receive catheter-directed thrombolysis with UK (4000 U/min for 4 h, then 2000 U/min up to 48 h) or surgical vascular recanalization. Angiograms available for 246 of the 272 patients treated with UK revealed recanalization in 80% and complete dissolution of thrombus in 68%. Amputation-free survival in the UK group was 72% at six months, 65% at one year, and 75% and 70% respectively in the 272 patients treated surgically (n.s.). At 6 months the surgery group had undergone 551 open operative procedures (excluding amputations) as compared with 315 in the thrombolysis group (Tables 3, 4). Major hemorrhage occurred in 12.5% of the UK patients and in 5.5% of the surgical group. There were four episodes of intracranial hemorrhage in the UK group (1.6%), one of which was fatal. The authors conclude that, despite the higher frequency of hemorrhagic complications, thrombolysis with UK compares favorably with surgical procedures because it reduced the need for

Table 3. Operative interventions in the TOPAS phase II trial comparing thrombolysis with surgical revascularisation

Operative Intervention	No. of interventions			
	UK group ($n = 272$)		Surgery group ($n = 272$)	
	6 months	1 year	6 months	1 year
Amputation	48	58	41	51
Above the knee	22	25	19	26
Below the knee	26	33	22	25
Open surgical procedures[a]	315	351	551	590
Major	102	116	177	193
Moderate	89	98	136	145
Minor	124	137	238	252
Percutaneous procedures	128	135	55	70

Reproduced from OURIEL et al. (1998).
[a] Major surgical procedures included the insertion of a new bypass graft, replacement of an existing graft, and excision or repair of aneurysm. Moderate procedures included graft revision, endarterectomy, profundaplasty, exploratory vascular surgery, and transmetatarsal amputation. Minor procedures included thromboembolectomy, amputations of digits, and fasciotomy. Endovascular procedures included percutaneous transluminal angioplasty, atherectomy, stent placement, suction thrombectomy, and thrombolytic therapy.

Table 4. Clinical outcomes in the TOPAS phase II trial comparing thrombolysis with surgical revascularisation

Worst outcome[a]	As a percentage of patients (%)			
	UK group ($n = 272$)		Surgery group ($n = 272$)	
	6 months	1 year	6 months	1 year
Death	16	20	12.3	17
Amputation	12.2	15	12.9	13.1
Above the knee	5.6	6.5	6.1	7.5
Below the knee	6.6	8.5	6.8	5.6
Open surgical procedures	40.3	39.3	69.0	65.4
Major	23.6	24.3	39.3	39.3
Moderate	10.3	8.7	16.3	13.4
Minor	6.4	6.3	13.4	12.7
Endovascular procedures	16.9	15.4	2.1	1.7
Medical treatment alone	14.6	10.3	3.7	2.8

Reproduced from OURIEL et al. (1998).
[a] The most severe clinical outcome for each patients is reported on the basis of Kaplan-Meier analysis.

surgical interventions and achieved similar results with respect to survival and incidence of amputation (OURIEL et al. 1998).

A separate report used the Cox proportional hazards multifactor analysis to determine the parameters predictive of successful therapy. Of 28 variables analyzed, 8 were predictive of amputation-free survival: white race, younger age, history of central nervous disease, history of malignancy, congestive heart failure, low body weight, presence of skin color changes, and pain at rest. The length of occlusion predicted whether the patient would fare better with thrombolysis or surgery. Occlusion lengths >30 cm fared better with thrombolysis, whereas the results were reversed for occlusion lengths <30 cm (OURIEL and VEITH 1998).

G. Long-Term Results

Once initial lysis is achieved, and particularly if the underlying aetiologic factor is corrected, it is apparent that long-term patency can be expected. In the Guildford series after four years, limb salvage was seen in 68%. Overall amputation was 22% with the majority required in the first 30 days (12%), with a further 10% up to 4 years later. Similarly mortality over the 4-year period was 36%, 14% occurring in the first 30 days, with no death or amputation apparently caused by the complications of treatment (GIDDINGS et al. 1993). HESS et al. (1987) reported a success rate of 59% after five years duration and LONSDALE et al. (1993b), reporting the Nottingham series, found a two-year cumulative patency of 81%.

H. Adjuvant Treatment

Concurrent heparin has been advocated as reducing the risk of pericatheter thrombosis, but there is no evidence to substantiate this. Adjuvant thromboxane receptor antagonists have also been tried but incurred a higher hemorrhagic complication rate (LONSDALE et al. 1993a). In chronic arterial occlusions which have not responded to conventional angioplasty and/or short-term thrombolysis, recanalization was achieved by prolonged and alternating intra-arterial perfusion with tPA and prostaglandin E1 (KRÖGER et al. 1998). Aspirin has been shown to be associated with an improved limb salvage and lower amputation and death rates than thrombolysis alone, without significantly affecting the risk of major hemorrhage or rethrombosis (BRAITHWAITE et al. 1995). Promising results also have been obtained in a pilot study using the glycoprotein IIb/IIIa receptor antagonist abciximab in conjunction with intra-arterial UK (TEPE et al. 1999). It remains to be evaluated whether newer techniques such as ultrasound-enhanced thrombolysis (HAMM et al. 1997; LUO et al. 1998; ROSENSCHEIN et al. 1999; reviewed in ATAR et al. 1999) which have shown promising results in the treatment of myocardial infarction will further improve the treatment of peripheral arterial occlusions. Mechanical thrombolysis using the Amplatz thrombectomy clot macerator may also prove to be of value in this condition (GÖRICH et al. 1998).

I. Complications

I. Major Hemorrhage

Major hemorrhage is defined as that causing hypotension, or requiring transfusion of blood or blood products, or causing a fall in hemoglobin in excess of 40 g/l. In this context the accepted risk of major hemorrhage was thought to be in the order of 5% (BERRIDGE et al. 1989). However, some later prospectively audited series have indicated that this risk may be closer to 10% (BRAITHWAITE et al. 1997). This may of course reflect the changing indications, experience, and increasing presence of graft rather than native vessel thrombosis. It is invariably associated with an increased risk in those patients in whom previous arterial punctures have been performed. Minimizing unnecessary patient movement may also contribute to a lesser incidence of hemorrhage, particularly that emanating from the catheter entry site. Unfortunately there are no useful predictors of impending hemorrhage, although some authors advocate the use of cryoprecipitate if the fibrinogen level falls below 1.5 g/l. The risk of major hemorrhage increases with increasing age. Treatment, including that of hemorrhagic stroke, involves correction of any derangement of the coagulation screen with fresh frozen plasma, correction of fibrinogen level with cryoprecipitate, and ensuring that the patient's platelet level is within the normal range. The author does not advocate the use of antifibrinolytics such as ε-amino caproic acid in view of the danger of causing a further complication as a result of thrombosis. It is well known that following cessation of thrombolytic treatment a rebound fibrinogenemia can result. Compounding this with an uncontrollable stimulus to favor coagulation does not appear logical.

II. Minor Hemorrhage

This is usually of little consequence and can be minimized by the strict avoidance of unnecessary intramuscular and subcutaneous administration of drugs. Invasive procedures should also be kept to an absolute minimum wherever possible. It is almost inevitable that a small hematoma will form at the site of catheter/sheath entry. With adequate fixation of this site, continued trauma to this area should be minimized and hence the risk of pericatheter hemorrhage reduced.

III. Cerebrovascular Accident

This is a devastating sequel to peripheral arterial thrombolysis. In an earlier review the risk was found to be approximately 1% (BERRIDGE et al. 1989). In a more recent presentation from the Thrombolysis Study group in the United Kingdom, this risk was reported as 2.2% (Dawson et al. 1996). The management of this condition is usually restricted. Apart from stopping the lytic agent and checking for the degree of systemic anticoagulation incurred, there is little else than can be done to benefit the patient. The use of anti-fibrinolytics has been advocated but one may cause a thrombotic catastrophe whilst trying to stop a fibrinolytic disaster. A computed tomographic study is essential to determine whether the event is embolic (in which case there is anecdotal evidence that one may be able to proceed (Nyamekye et al. 1993)) or hemorrhagic which obviously contra-indicates continuation of either thrombolysis or heparinization.

IV. Distal Embolization

Distal embolization occurs in approximately 2% of lower limb arterial thrombolysis cases. It is more likely to occur in cases of thrombosed popliteal aneurysms (12%) (Galland et al. 1993), particularly if the catheter is not situated in the run-off vessels alone. Whilst many cases can be resolved by continuation of the lytic regime, there is undoubtedly a risk of major morbidity.

V. Pericatheter Thrombosis

This is a variable entity that probably occurs more commonly than we appreciate. However, it usually resolves with continuation and/or repositioning of the catheter. Resistant cases of propagation of pericatheter thrombosis may be treated either by changing the lytic agent or abandoning lysis for a surgical approach. There is no convincing evidence for the simultaneous use of heparin when using either UK or tPA to reduce the incidence of pericatheter thrombosis. However post-lysis heparinization is essential in order to reduce the risk of rethrombosis, particularly in the presence of a rebound hyperfibrinogenemia.

VI. Allergic Reactions

As there is no evidence of antibody formation with either UK or with tPA, it is not surprising that allergic reactions are not recorded. The same cannot be said for SK. Despite modern manufacturing processes, it is inherently allergenic and is known to excite an antibody response. The level of response can be marked and prolonged as discussed earlier.

VII. Other Complications

Other complications are relatively rare or rather less well recognized. They include catheter related problems including vessel or graft perforation and the reperfusion syndrome itself.

J. Summary

In summary, peripheral arterial thrombolysis has recently seen many changes. We now have evidence from comparatively large randomized trials that initial thrombolytic therapy may allow significant improvement in survival of the affected limb but also of the patient. We have better information as to which patients are likely to benefit, which are likely to have an increased risk of complications, and the actual numeric risk of these sometimes devastating complications. The Working Party on Thrombolysis in the Management of Limb Ischemia in which this author had the privilege of participating came up with 33 specific recommendations dealing with diverse clinical situations (WORKING PARTY ON THROMBOLYSIS IN THE MANAGEMENT OF LIMB ISCHEMIA 1998). Thrombolytic therapy of the peripheral arteries should represent a team effort between vascular surgeons, interventional radiologists, and angiologists in order that the patient is afforded the best possible intervention, in the correct guise, at the correct time, and is not subjected to unnecessary risk of either thrombolysis, other radiological intervention, or surgery solely because the other options are not locally available.

List of Abbreviations and Acronyms

ADMIT	Arterial Disease Multiple Intervention Trial
APSAC	Anisoylated Plasminogen Streptokinase Activator Complex
CRP	C-reactive protein
ISIS	International Study of Infarct Survival
MU	Mega (1 million) units
PAI-1	Plasminogen activator inhibitor type 1
PURPOSE	Pro-urokinase vs Urokinase for Recanalization of Peripheral Occlusions Safety and Efficacy trial
rtPA	recombinant tissue-type plasminogen activator
SK	streptokinase
STILE	Surgery vs Thrombolysis for Ischemia of the Lower Extremity
TOPAS	Thrombolysis Or Peripheral Arterial Surgery trial
tPA	tissue-type plasminogen activator (abbreviation also used for rtPA)
UK	urokinase

References

Amery A, Deloof W, Vermylen J, Verstraete M (1970) Outcome of recent thromboembolic occlusions of limb arteries treated with streptokinase. Br Med J 4: 639–644

Armon MP, Yusuf SW, Whitaker SC, Gregson RHS, Wenham PW, Hopkinson BR (1997) Results of 100 cases of pulse-spray thrombolysis for acute and subacute leg ischemia. Br J Surg 84:47–50

Atar S, Luo H, Nagai T, Siegel RJ (1999) Ultrasonic thrombolysis: catheter-delivered and transcutaneous applications. Eur J Ultrasound 9:39–54

Beard JD, Nyamekye I, Earnshaw JJ, Scott DJA, Thompson JF (1993) Intraoperative streptokinase: A useful adjunct to balloon-catheter embolectomy. Br J Surg 80:21–24

Belkin M, Donaldson MC, Whittemore AD, Polak JF, Grassi CJ, Harrington DP, Mannick JA (1990) Observations on the use of thrombolytic agents for thrombotic occlusion of infrainguinal vein grafts. J Vasc Surg 11:289–296

Berridge DC, Gregson RHS, Hopkinson BR, Makin GS (1991) Randomized trial of intra-arterial recombinant tissue plasminogen activator, intravenous recombinant tissue plasminogen activator and intra-arterial streptokinase in peripheral arterial thrombolysis. Br J Surg 78:988–995

Berridge DC, Gregson RHS, Makin GS, Hopkinson BR (1990) Tissue plasminogen activator in peripheral arterial thrombolysis. Br J Surg 77:179–182

Berridge DC, Makin GS, Hopkinson BR (1989) Local low dose intra-arterial thrombolytic therapy: the risk of stroke or major hemorrhage. Br J Surg 76:1230–1233

Blaisdell FW, Steele M, Allen RE (1978) Management of acute lower extremity arterial ischemia due to embolism and thrombosis. Surgery 84:822–834

Bookstein JJ, Bookstein FL (2000) Augmented experimental pulse-spray thrombolysis with tissue plasminogen activator, enabling dose reduction by one or more orders of magnitude. J Vasc Interv Radiol 11:299–303

Bookstein JJ, Valji K, Roberts AC (1994a) Pulsed versus conventional thrombolytic infusion techniques. Radiology 193:318–320

Bookstein JJ, Valji K, Roberts AC (1994b) Intrathrombic administration of lytic adjuncts. Radiology 193:324

Braithwaite BD, Buckenham TM, Galland RB, Heather BP, Earnshaw JJ, on behalf of the Thrombolysis Study Group (1997) Prospective randomised trial of high-dose bolus versus low-dose tissue plasminogen activator infusion in the management of acute limb ischaemia. Br J Surg 84:646–650

Braithwaite BD, Davies B, Birch PA, Heather BP, Earnshaw JJ (1998) Management of acute leg ischaemia in the elderly. Br J Surg 85:217–220

Braithwaite BD, Jones L, Yusuf SW, Dawson K, Berridge DC, Davies E, Bowyer R, Treska V, Earnshaw JJ, on behalf of the Thrombolysis Study Group (1995) Aspirin improves the outcome of intra-arterial thrombolysis with tissue plasminogen activator. Br J Surg 82:1357–1358

Buckenham TM, George CD, Chester JF, Taylor RS, Dormandy JA (1992) Accelerated thrombolysis using pulsed intra-thrombus recombinant human tissue type plasminogen activator (rt-PA). Eur J Vasc Surg 6:237–240

Cohen LH, Kaplan M, Bernhard VM (1986) Intraoperative streptokinase. An adjunct to mechanical arterial thrombectomy in the management of acute ischemia. Arch Surg 121:708–715

Comerota AJ (1988) Complications of thrombolytic therapy. In: Comerota AJ (ed) Thrombolytic therapy. Grune and Stratton, Orlando, pp 255–282

Comerota AJ, Koneti Rao A, Throm RC, Skibinski CI, Beck GJ, Ghosh S, Sun L, Curl GR, Ricotta JJ, Graor RA, Flinn WC, Roedersheimer RL, Alexander JB (1993) A prospective, randomized, blinded, and placebo-controlled trial of intra-operative intra-arterial urokinase infusion during lower extremity revascularization: Regional and systemic effects. Ann Surg 218:534–543

Comerota AJ, Weaver FA, Hosking JD, Froehlich J, Folander H, Sussman B, Rosenfield K (1996) Results of a prospective, randomized trial of surgery versus thrombolysis for occluded lower extremity bypass grafts. Am J Surg 172:105–112

Comerota AJ, White JV, Grosh JD (1989) Intraoperative intra-arterial thrombolytic therapy for salvage of limbs in patients with distal arterial thrombosis. Surg Gynecol Obstet 169:283–289

Cragg AH, Smith TP, Corson JD, Nakagawa N, Castaneda F, Kresowik TF, Sharp WJ, Shamma A, Berbaum KS (1991) Two urokinase dose regimens in native arterial and graft occlusions: initial results of a prospective, randomized clinical trial. Radiology 178:681–686

Davidian MM, Powell A, Benenati JF, Katzen BT, Becker GJ, Zemel G (2000) Initial results of reteplase in the treatment of acute lower extremity arterial occlusions. J Vasc Interv Radiol 11:289–294

Dawson K, Armon A, Braithwaite BD, Galland R, Kendrick R, Downes M, Buckenham T, Al-Kutoubi A, Berridge D, Earnshaw JJ, Hamilton G (1996) Stroke during intra-arterial thrombolysis: a survey of experience in the UK. Br J Surg 83:568 (Abstract)

Dotter CT, Rösch J, Seaman AJ (1974) Selective clot lysis with low-dose streptokinase. Radiology 111:31–37

Earnshaw JJ (1994) Intravenous thrombolysis. In: Earnshaw JJ, Gregson RHS (eds) Practical peripheral arterial thrombolysis. Butterworth-Heinemann, London, pp 28–37

Earnshaw JJ, Cosgrove C, Wilkins DC, Bliss BP (1990) Acute limb ischaemia: the place of intravenous streptokinase. Br J Surg 77:1136–1139

Earnshaw JJ, Westby JC, Makin GS, Hopkinson BR (1986) The systemic fibrinolytic effect of BRL 26921 during the treatment of acute peripheral arterial occlusions. Thromb Haemost 55:259–262

Elliott JM, Cross DB, Cederholm-Williams SA, White HD (1993) Neutralizing antibodies to streptokinase four years after intravenous thrombolytic therapy. Am J Cardiol 71:640–645

Fogarty TJ, Cranley JJ, Krause RJ, Strasser ES, Hafner CD (1963) A method for extracting arterial emboli and thrombi. Surg Gynecol Obstet 2:241–244

Galland RB, Earnshaw JJ, Baird RN, Lonsdale RJ, Hopkinson BR, Giddings AE, Dawson KJ, Hamilton G (1993) Acute limb deterioration during intra-arterial thrombolysis. Br J Surg 80:1118–1120

Gardiner GA Jr, Koltun W, Kandarpa K, Whittemore A, Meyerovitz MF, Bettmann MA, Levin DC, Harrington DP (1986) Thrombolysis of occluded femoropopliteal grafts. Am J Roentgenol 147:621–626

Giddings AE, Quraishy MS, Walker WJ (1993) Long-term results of a single protocol for thrombolysis in acute lower-limb ischaemia. Br J Surg 80:1262–1265

Görich J, Rilinger N, Sokiranski R, Kramer S, Mickley V, Schütz A, Brambs H-J, Pamler R (1998) Mechanical thrombolysis of acute occlusion of both the superficial and the deep femoral arteries using a thrombectomy device. Am J Roentgenol 170:1177–1180

Graor RA, Olin J, Bartholomew JR, Ruschhaupt WF, Young JR. (1990) Efficacy and safety of intrarterial local infusion of streptokinase, urokinase, or tissue plasminogen activator for peripheral arterial occlusion: a retrospective review. J Vasc Med Rev 2:310–315

Graor RA, Risius B, Young JR, Lucas FV, Beven EG, Hertzer NR, Krajewski LP, O'Hara PJ, Olin J, Ruschhaupt WF (1988) Thrombolysis of peripheral arterial bypass grafts: surgical arterial embolectomy compared with thrombolysis. A preliminary report. J Vasc Surg 7:347–355

Hamm CW, Steffen W, Terres W, de Scheerder I, Reimers J, Cumberland D, Siegel RJ, Meinertz T (1997) Intravascular therapeutic ultrasound thrombolysis in acute myocardial infarctions. Am J Cardiol 80:200–204

Hess H, Ingrisch H, Mietaschk A, Rath H (1982) Local low-dose thrombolytic therapy of peripheral arterial occlusions. N Engl J Med 307:1627–1630

Hess H, Mietaschk A, Bruckl R (1987) Peripheral arterial occlusions: a 6-year experience with local low-dose thrombolytic therapy. Radiology 163:753–758

Heymans S, Vanderschueren S, Verhaeghe R, Stockx L, Lacroix H, Nevelsteen A, Laroche Y, Collen D (2000) Outcome and one year follow-up of intra-arterial staphylokinase in 191 patients with peripheral arterial occlusion. Thromb Haemost 83:666–671

ISIS-3 (Third International Study of Infarct Survival) Collaborative Group (1992) ISIS-3: a randomised comparison of streptokinase vs tissue plasminogen activator vs anistreplase and of aspirin plus heparin vs aspirin alone among 41 299 cases of suspected acute myocardial infarction. Lancet 339:753–770

Juhan C, Haupert S, Miltgen G, Girard N, Dulac P (1991) A new intra arterial rt-PA dosage regimen in peripheral arterial occlusion: bolus followed by continuous infusion. Thromb Haemost 65:635 (letter)

Jørgensen B, Tønnesen KH, Bülow J, Dalsgård Nielsen J, Jørgensen M, Holstein P, Andersen E (1989) Femoral artery recanalisation with percutaneous angioplasty and segmentally enclosed plasminogen activator. Lancet i:1106–1108

Kakkar VV, Flanc C, O'Shea MJ, Flute PT, Howe CT, Clarke MB (1969) Treatment of deep-vein thrombosis with streptokinase. Br J Surg 56:178–183

Kandarpa K, Chopra PS, Aruny JE, Polak JF, Donaldson MC, Whittemore AD, Mannick JA, Goldhaber SZ, Meyerovitz MF (1993) Intraarterial thrombolysis of lower extremity occlusions: Prospective, randomized comparison of forced periodic infusion and conventional slow continuous infusion. Radiology 188:861–867

Katzen BT, Edwards KC, Albert AS, van Breda A (1984) Low-dose direct fibrinolysis in peripheral vascular disease. J Vasc Surg 1:718–722

Kröger K, Hwang I, Rudofsky G (1998) Recanalization of chronic peripheral arterial occlusions by alternating intra-arterial rt-PA and PGE1. VASA 27:20–23

Lammer J, Pilger E, Neumayer K, Schreyer H (1986) Intraarterial fibrinolysis: long-term results. Radiology 161:159–163

Lonsdale RJ, Berridge DC, Earnshaw JJ, Harrison JD, Gregson RHS, Wenham PW, Hopkinson BR, Makin GS (1992) Recombinant tissue-type plasminogen activator is superior to streptokinase for local intra-arterial thrombolysis. Br J Surg 79:272–275

Lonsdale RJ, Heptinstall S, Westby JC, Berridge DC, Wenham PW, Hopkinson BR, Makin GS (1993a) A study of the use of the thromboxane A_2 antagonist, sulotroban, in combination with streptokinase for local thrombolysis in patients with recent peripheral arterial occlusions: Clinical effects, platelet function and fibrinolytic parameters. Thromb Haemost 69:103–111

Lonsdale RJ, Whitaker SC, Berridge DC, Earnshaw JJ, Gregson RH, Wenham PW, Hopkinson BR, Makin GS (1993b) Peripheral arterial thrombolysis: intermediate-term results. Br J Surg 80:592–595

Luo H, Birnbaum Y, Fishbein MC, Peterson TM, Nagai T, Nishioka T, Siegel RJ (1998) Enhancement of thrombolysis in vivo without skin and soft tissue damage by transcutaneous ultrasound. Thromb Res 89:171–177

Mavor GE, Walker MG, Dhall DP (1972) Routine operative arteriography in arterial embolectomy. Br J Surg 59:482–484

McNamara TO, Fischer JR (1985) Thrombolysis of peripheral arterial and graft occlusions: improved results using high-dose urokinase. Am J Roentgenol 144:769–775

Meyerovitz MF, Didier D, Vogel JJ, Soulier-Parmeggiani L, Bounameaux H (1995) Thrombolytic therapy compared with mechanical recanalization in non-acute peripheral arterial occlusions: a randomized trial. J Vasc Interv Radiol 6:775–781

Meyerovitz MF, Goldhaber SZ, Reagan K, Polak JF, Kandarpa K, Grassi CJ, Donovan BC, Bettmann MA, Harrington DP (1990) Recombinant tissue-type plasminogen activator versus urokinase in peripheral arterial and graft occlusions: a randomized trial. Radiology 175:75–78

Nackman GB, Walsh DB, Fillinger MF, Zwolak RM, Bech FR, Bettmann MA, Cronenwett JL (1997) Thrombolysis of occluded infrainguinal vein grafts: predictors of outcome. J Vasc Surg 25:1023–131

Nilsson L, Albrechtsson U, Jonung T, Ribbe E, Thorvinger B, Thorne J, Åstedt B, Norgren L (1992) Surgical treatment versus thrombolysis in acute arterial occlusion: a randomised controlled study. Eur J Vasc Surg 6:189–193

Nyamekye I, Beard J, Gaines P (1993) The management of stroke during intraarterial thrombolysis. Br J Radiol 66:781–782

Odavic R (1995) Peripheral thrombolysis: Boehringer database. Presented at the Thrombolysis Study Group Meeting, Gloucester, UK, September, 1995

Ouriel K, Kandarpa K, Schuerr DM, Hultquist M, Hodkinson G, Wallin B (1999) Prourokinase versus urokinase for recanalization of peripheral occlusions, safety and efficacy: the PURPOSE trial. J Vasc Interv Radiol 10:1083–1091

Ouriel K, Shortell CK, Azodo MVU, Guiterrez OH, Marder VJ (1994a) Acute peripheral arterial occlusion: Predictors of success in catheter-directed thrombolytic therapy. Radiology 193:561–566

Ouriel K, Shortell CK, DeWeese JA, Green RM, Francis CW, Azodo MVU, Gutierrez OH, Manzione JV, Cox C, Marder VJ (1994b) A comparison of thrombolytic therapy with operative revascularization in the initial treatment of acute peripheral arterial ischemia. J Vasc Surg 19:1021–1030

Ouriel K, Veith FJ (1998) Acute lower limb ischemia: determinants of outcome. Surgery 124:336–41

Ouriel K, Veith FJ, Sasahara AA, for the Thrombolysis or Peripheral Arterial Surgery (TOPAS) Investigators (1998) A comparison of recombinant urokinase with vascular surgery as initial treatment for acute arterial occlusion of the legs. N Engl J Med 338:1105–1111

Ouriel K, Veith FJ, Sasahara AA, for the TOPAS Investigators (1996) Thrombolysis or peripheral arterial surgery: Phase I results. J Vasc Surg 23:64–75

Paterson IS, Darby S, Smith FCT, Tsang GMK, Hamer JD, Shearman CP (1993) Thrombolysis of occluded infrainguinal vein grafts before operative intervention. Br J Surg 80:529 (Abstract)

Peltonen S, Lassila R, Lepäntalo M (1996) Increased circulating plasminogen activator inhibitor-1 in patients with patent femorodistal venous bypass. Thromb Res 82:369–377

Philipp CS, Cisar LA, Kim HC, Wilson AC, Saidi P, Kostis JB (1997) Association of hemostatic factors with peripheral vascular disease. Am Heart J 134:978–984

Plecha FR, discussion in: Blaisdell FW, Steele M, Allen RE (1978) Management of lower extremity arterial ischemia due to embolism and thrombosis. Surgery 84:822–834

Ridker PM, Cushman M, Stampfer MJ, Tracy RP, Hennekens CH (1998) Plasma concentration of C-reactive protein and risk of developing peripheral vascular disease. Circulation 97:425–428

Rosenschein U, Gaul G, Erbel R, Amann F, Velasguez D, Stoerger H, Simon R, Gomez G, Troster J, Bartorelli A, Pieper M, Kyriakides Z, Laniado S, Miller HI, Cribier A, Fajadet J (1999) Percutaneous transluminal therapy of occluded saphenous vein grafts: can the challenge be met with ultrasound thrombolysis? Circulation 99:26–29

Roy S, Laerum F, Brosstad F (1998) Quantitative evaluation of selective thrombolysis techniques: influence of catheter characteristics and delivery parameters. Cathet Cardiovasc Diagn 43:111–119

Sanchez LA, Gupta SK, Veith FJ, Goldsmith J, Lyon RT, Wengerter KR, Panetta TF, Marin ML, Cynamon J, Berdejo G, et al (1991) A ten-year experience with one hundred fifty failing or threatened vein and polytetrafluoroethylene arterial bypass grafts. J Vasc Surg 14:729–36

Sandbaek G, Staxrud LE, Rosen L, Bay D, Stiris M, Gjolberg T (1996) Factors predicting the outcome of intraarterial thrombolysis in peripheral arterial and graft occlusions. Acta Radiol 37:299–304

Schweizer J, Altman E, Stosslein F, Florek F, Kaulen R (1996) Comparison of tissue plasminogen activator and urokinase in the local infiltration of thrombolysis of peripheral arterial occlusions. Eur J Radiol 22:129–132

Semba CP, Murphy TP, Bakal CW, Calis KA, Matalon TA, and the Advisory Panel (2000) Thrombolytic therapy with use of alteplase (rt-PA) in peripheral arterial occlusive disease: review of the clinical literature. J Vasc Interv Radiol 11:149–161

Spence LD, Hartnell GG, Reinking G, McEniff N, Gibbons G, Pomposelli F, Clouse ME (1997) Thrombolysis of infrapopliteal bypass grafts: efficacy and underlying angiographic pathology. Am J Roentgenol 169:717–721

Sullivan KL, Gardiner GA, Shapiro MJ, Bonn J, Levin DC (1989) Acceleration of thrombolysis with a high-dose transthrombus bolus technique. Radiology 173:805–808

Sullivan KL, Gardiner GA Jr, Kandarpa K, Bonn J, Shapiro MJ, Carabasi RA, Smullens S, Levin DC (1991) Efficacy of thrombolysis in infrainguinal bypass grafts. Circulation 83 [Suppl I]:I-99–I-105

Tawes RL Jr, Harris EJ, Brown WH, Shoor PM, Zimmerman JJ, Sydorak GR, Beare JP, Scribner RG, Fogarty TJ (1985) Arterial thromboembolism. A 20-year perspective. Arch Surg 120:595–599

Tepe G, Schott U, Erley CM, Albes J, Claussen CD, Duda SH (1999) Platelet glycoprotein IIb/IIIa receptor antagonist used in conjunction with thrombolysis for peripheral arterial thrombosis. AJR Am J Roentgenol 172:1343–1346

The STILE investigators (1994) Results of a prospective randomized trial evaluating surgery versus thrombolysis for ischemia of the lower extremity – the STILE trial. Ann Surg 220:251–268

Tønnesen KH, Holstein P, Andersen E (1991) Femoro-popliteal artery occlusions treated by percutaneous transluminal angioplasty and enclosed thrombolysis: results in 55 patients. Eur J Vasc Surg 5:429–434

Valji K, Bookstein JJ, Roberts AC, Sanchez RB (1993) Occluded peripheral arteries and bypass grafts: lytic stagnation as an end point for pulse-spray pharmacomechanical thrombolysis. Radiology 188:389–394

Valji K, Roberts AC, Davis GB, Bookstein JJ (1991) Pulsed-spray thrombolysis of arterial and bypass graft occlusions. Am J Roentgenol 156:617–621

van Breda A, Katzen BT, Deutsch AS (1987) Urokinase versus streptokinase in local thrombolysis. Radiology 165:109–111

Vanderschueren S, Stockx L, Wilms G, Lacroix H, Verhaeghe R, Vermylen J, Collen D (1995) Thrombolytic therapy of peripheral arterial occlusion with recombinant staphylokinase. Circulation 92:2050–2057

Verstraete M, Hess H, Mahler F, Mietaschk A, Roth FJ, Schneider E, Baert AL, Verhaeghe R (1988) Femoro-popliteal artery thrombolysis with intra-arterial infusion of recombinant tissue-type plasminogen activator – report of a pilot trial. Eur J Vasc Surg 2:155–159

Verstraete M, Vermylen J, Donati MB (1971) The effect of streptokinase infusion on chronic arterial occlusions and stenoses. Ann Intern Med 74:377–382

Wagner HJ, Starck EE (1992) Acute embolic occlusions of the infrainguinal arteries: percutaneous aspiration embolectomy in 102 patients. Radiology 182:403–407

Walker WJ, Giddings AE (1985) Low-dose intra-arterial streptokinase: benefit versus risk. Clin Radiol 36:345–353

Walker WJ, Giddings AEB (1988) A protocol for the safe treatment of acute lower limb ischaemia with intra-arterial streptokinase and surgery. Br J Surg 75:1189–1192

Ward AS, Andaz SK, Bygrave S (1994) Thrombolysis with tissue-plasminogen activator: Results with a high-dose transthrombus technique. J Vasc Surg 19:503–508

Weaver FA, Comerota AJ, Youngblood M, Froehlich J, Hosking JD, Papanicolaou G, and the STILE Investigators (1996) Surgical revascularization versus thrombolysis for nonembolic lower extremity native artery occlusions: results of a prospective randomized trial. J Vasc Surg 24:513–521

Wholey MH, Maynar MA, Wholey M, Pulido-Duque JM, Reyes R, Jarmolowski CR, Castaneda WR (1998) Comparison of thrombolytic therapy of lower-extremity acute, subacute, and chronic arterial occlusions. Cathet Cardiovasc Diagn 44:159–169

Working Party on Thrombolysis in the Management of Limb Ischemia (1998) Thrombolysis in the management of lower limb peripheral arterial occlusion – a consensus document. Am J Cardiol 81:207–218

Yusuf SW, Whitaker SC, Gregson RHS, Wenham PW, Hopkinson BR, Makin GS (1994) Experience with pulse-spray technique in peripheral thrombolysis. Eur J Vasc Surg 8:270–275

CHAPTER 15

Thrombotic Cerebrovascular Disease

G.J. DEL ZOPPO

A. Introduction

The first report of the clinical use of a plasminogen activator in cerebrovascular thrombosis appeared in 1958, which has been more recently followed by a small number of level III phase I/II clinical recanalization efficacy trials of acute thrombotic/ischemic stroke (SACKETT 1983; DEL ZOPPO et al. 1988, 1992; MORI et al. 1988; THERON et al. 1989; MATSUMOTO and SATOH 1991; ZEUMER et al. 1993; VON KUMMER and HACKE 1992) and level I phase III disability outcome trials of acute ischemic stroke (SACKETT 1983; MORI et al. 1992; YAMAGUCHI 1993; DONNAN et al. 1995, 1996; MULTICENTRE ACUTE STROKE TRIAL-ITALY (MAST-I) GROUP 1995; HACKE et al. 1995, 1998b; THE NATIONAL INSTITUTE OF NEUROLOGICAL DISORDERS AND STROKE TPA STROKE STUDY GROUP 1995; MULTICENTER ACUTE STROKE TRIAL-EUROPE STUDY GROUP 1996; DEL ZOPPO et al. 1998). Those studies, together with intervening anecdotal experience and experimental model work, have rested on the premise that restoration of flow in an occluded cerebral artery may facilitate the return of neurological function (WARDLAW and WARLOW 1992). Safety concerns, difficult to model adequately, are of central importance. With the exception of several small level I studies (MORI et al. 1992; YAMAGUCHI et al. 1993) and level III dose-finding (DEL ZOPPO et al. 1992) trials, an excess of symptomatic intracerebral hemorrhages has accompanied the use of plasminogen activators in focal cerebral ischemia (HACKE et al. 1995, 1998b; THE NATIONAL INSTITUTE OF NEUROLOGICAL DISORDERS AND STROKE TPA STROKE STUDY GROUP 1995; DEL ZOPPO et al. 1998). However, a compilation of recent experience has suggested that functional outcome may be improved with adherence to strict guidelines. Meta-analyses, based upon less recent experiences, have suggested the value of thrombolytics to improve disability outcome and mortality following ischemic stroke, and a significant association with symptomatic hemorrhage (WARDLAW and WARLOW 1992; WARDLAW et al. 1997). The conditions for the clinical use of plasminogen activators in acute cerebral ischemia continue in evolution.

B. Focal Cerebral Ischemia

I. Considerations for Tissue Salvage

The use of plasminogen activators in acute ischemic stroke is based on the observation that 80%–90% of focal cerebral ischemic events within 8–24 h of symptom onset are due to atherothrombotic or thromboembolic occlusions (SOLIS et al. 1977). Angiography-based studies of symptomatic carotid territory arterial thrombosis have documented occlusion in 76% of patients at 6 h (FIESCHI et al. 1989), 81% at 5.7 ± 1.4 h (DEL ZOPPO et al. 1992), 59% at 24 h (SOLIS et al. 1977), 57% at 3 days (FIESCHI and BOZZAO 1969), and 44% at 7 days (IRINO et al. 1977) after symptom onset. YAMAGUCHI et al. (1984) indicated that spontaneous recanalization is a relatively frequent occurrence following cerebral thromboembolism, and MORI et al. (1992) demonstrated that 17% of patients entered within 3.6 ± 1.1 h of symptom onset had undergone recanalization 2 h later. Therefore, a 20% frequency of spontaneous recanalization would be expected within 6 h of symptom onset in patients with thrombotic stroke. Plasmin generation would be expected to augment this frequency.

Elements important to outcomes in acute ischemic stroke include the integrity of collateral anastomoses, time from symptom onset to evaluation, occlusion location, cellular vulnerability, and local flow characteristics. Although limited, angiography-based studies of reperfusion using plasminogen activators have suggested that clinical outcome may be enhanced by the presence of patent collateral anastomoses (VON KUMMER and HACKE 1992). These include the circle of Willis, the retrograde ophthalmic circulation, and leptomeningeal anastomoses. Lenticulostriatal arterial branches of the middle cerebral artery (MCA), considered "end-arterioles," receive no apparent significant collateral protection from any source. ASTRUP et al. (1981) postulated a zone of neuronal tissue peripheral to an irreversibly ischemic core (infarct) that is metabolically injured and metastable, and functionally recoverable if sufficient blood flow could be restored within a short time. The existence of a "penumbra" has been implied by human PET studies (POWERS 1993) and experimental preparations (GARCIA et al. 1993). This concept is consistent with a finite interval after cessation of blood flow beyond which irreversible ischemic neuronal damage occurs (ASTRUP et al. 1977, 1981; SKYHOJ-OLSEN et al. 1983). Experimental preparations have suggested that certain regions of limited vascular flow and certain neurone populations are particularly sensitive to ischemia. For instance, blood cellular transit through the corpus striatum is substantially less than the cortex (FENSTERMACHER et al. 1991), and is differentially sensitive to ischemia. Such regional or cellular "selective vulnerabilities" may place unknown, but strict limits on tissue recoverability. In addition, local flow conditions may contribute to variation in the distribution or delivery of nutrients or agents. These are likely to reflect the overall vascular anatomy, as well as the state of the microvasculature. It remains unproven, but seems likely, that retrograde delivery of a plasminogen activator may take advantage of collateral vessels when the agents are delivered systemically. In an individual patient, these theoretical considerations may combine to define an unknown, but individual interval of tolerated ischemia manifested clinically as a limited interval for treatment (BARON et al. 1995).

In the cerebral arterial tree occlusion location may be important to the facility with which recanalization occurs. Several prospective thrombolytic intervention trials suggest that proximal internal carotid artery (ICA) occlusions are less likely to undergo recanalization than MCA branch lesions following either intravenous or regional infusion of a plasminogen activator (MORI 1991; DEL ZOPPO et al. 1992; MORI et al. 1992; LEE et al. 1994; SASAKI et al. 1995; ENDO et al. 1998). Thrombus volume and location underlie the relative resistance to dissolution by therapeutic thrombolysis, with proximal atheroma-based ICA thrombi being less responsive than distal embolic occlusions (DEL ZOPPO et al. 1986, 1988; MORI et al. 1988; MORI 1991). Exposure of the proximal (downstream) surface of the thrombotic occlusion to an agent depends upon the proximity of the occlusion to the nearest bifurcation in the anterograde direction. Lower frequencies of arterial recanalization have accompanied systemic or regional delivery of plasminogen activators due to the lack of flow proximal to the fixed occlusion, while consistently higher recanalization frequencies have been associated with directed intra-arterial infusion. On this basis, angiography-based studies have excluded patients with proximal ICA occlusions. This distinction has not been made for symptom-based studies.

II. Hemorrhagic Transformation

All plasminogen activators carry the risk of intracranial hemorrhage (ALDRICH et al. 1985). The central safety issue is that exposure of the ischemic bed to altered hemostasis and platelet reactivity engendered by plasminogen activators may increase the size or severity of naturally occurring hemorrhagic transformation.

Hemorrhagic transformation includes the development of hemorrhagic infarction (HI) or of parenchymal hematoma (PH) in relationship to focal ischemic injury (PESSIN et al. 1990, 1991). HI refers to petechial or confluent petechial hemorrhage confined to the ischemic zone, while PH consists of a homogeneous, discrete mass of blood (coagulum) initiated in the ischemic zone that may extend to the ventricle or appear isolated. In either case, PH may contribute to local swelling and shift of midline structures. Clinical deterioration caused by intracranial hemorrhage is most often associated with PH, although several factors may contribute to deterioration in the presence of HI, particularly when it is associated with large infarcts.

The radiographic distinction between severe HI and PH may be difficult. HI is more commonly associated with cardiogenic cerebral embolism than thrombosis in situ. YAMAGUCHI et al. (1984), in an angiographic study, reported HI in 37% of 120 cases of cardiogenic cerebral embolism, but in 2% of 105 cases of cerebral thrombosis in the carotid territory. HI was observed in 46% of 140 patients with carotid territory embolic stroke, while PH was observed in 16 (11%) patients (OKADA et al. 1989). In a separate study, HI occurred in 43% of 65 patients whereas PH occurred in 14% (HORNIG et al. 1986). Using CT scanning at admission and between day 5 and day 9 after stroke, TONI et al. (1996) arrived at similar results: 43% HI and 11% PH respectively. PH, which is most often symptomatic (PESSIN et al. 1990), seems to accompany acute carotid territory embolism in the setting of anticoagulant treatment (MEYER et al. 1963; KOLLER 1982; FURLAN et al. 1982; DRAKE and SHIN 1983; BABIKIAN et al. 1989).

The long-standing concept that hemorrhagic transformation may result from arterial reperfusion is not supported by angiographic studies (DEL ZOPPO et al. 1988, 1992; MORI et al. 1988, 1992). The finding of hemorrhagic transformation with persistent occlusion of the primary artery by Ogata and co-workers has suggested that hemorrhage may occur from other vascular sources (e.g., collateral channels) (OGATA et al. 1989; FISHER and ADAMS 1951, 1987). From those studies, the incidence of symptomatic hemorrhagic transformation has not exceeded 10%.

Several factors contribute to thrombolysis-related hemorrhage in acute ischemic stroke. One prospective study of recombinant tissue plasminogen activator (tPA, duteplase) in angiographically-defined cerebral arterial occlusion demonstrated that increased time from symptom onset to treatment was associated with a significant increase in all hemorrhage types (DEL ZOPPO et al. 1992). A post-hoc analysis of two symptom-based studies with tPA (alteplase) suggested that symptomatic hemorrhage was dependent upon diastolic hypertension, low body mass, age, increased tPA dose, and signs of ischemic injury at baseline (LEVY et al. 1994; HACKE et al. 1995; LARRUE et al. 1997). Those elements are also common contributors to the 0.95% overall frequency of intracerebral hemorrhage (ICH) observed in myocardial infarction patients treated with fibrinolytic agents (SOBEL 1994; SLOAN et al. 1995; LEVINE et al. 1995; GURWITZ et al. 1998) compared to untreated patients (0.02% ICH) (KASE et al. 1992; DEL ZOPPO and MORI 1992). UEDA et al. (1994) demonstrated that symptomatic intracerebral hemorrhage in patients with stroke treated with intra-arterial urokinase (UK) or tPA was significantly associated with low regional cerebral blood flow (rCBF). This is in keeping with the concept that prolonged rCBF reduction is associated with most severe tissue injury. The European Cooperative Acute Stroke Study (ECASS) indicated that symptomatic hemorrhage following intravenous tPA (alteplase) infusion was significantly associated with evidence of ischemic injury on the initial CT scan (HACKE et al. 1995).

III. Neurological Outcome

Approaches to neurological outcome measurements also evolve. Mortality, an outcome appropriate for cardiovascular studies, has much less utility in acute stroke trials because 1-year mortality generally does not exceed 15%, of which the majority are due to cardiovascular causes (SCANDINAVIAN STROKE STUDY GROUP 1985). To date, large-scale mortality outcome trials of stroke treatments, similar to the International Stroke Trial (IST) (INTERNATIONAL STROKE TRIAL COLLABORATIVE GROUP 1997) have not been developed for thrombolysis owing in part both to the need to assess neurologic status in a detailed way and to the contribution of hemorrhage to mortality in studies using thrombolytic agents.

It is accepted that stroke patients may display progressive improvement in the absence of treatment according to scoring instruments based upon the neurological examination [e.g., Scandinavian Stroke Scale (SCANDINAVIAN STROKE STUDY GROUP 1985)]. In the development of acute stroke interventions with plasminogen activators, the outcomes of recanalization frequency, regional blood flow, and measures of neurological function and disability have been employed. None of the neurological scoring instruments (including the NIH stroke scale (NIHSS) (BROTT et al. 1989, 1992; HALEY et al. 1992), the Hemispheric Stroke Scale (HSS) (MORI et al. 1992), or modifications of specific scales (YAMAGUCHI 1993)) used in acute ischemic stroke trials have been prospectively validated for long-term outcome (HANTSON et al. 1994), although inter-observer correlations for some scales have been published (LYDEN et al. 1994). Functional outcome as measured by such scoring instruments has been superseded by disability outcome measures, including the Rankin scale modified for mortality (RANKIN 1957), and the Barthel index (LOEWEN and ANDERSON 1988; MAHONEY and BARTHEL 1965). These disability indices have not yet been prospectively validated as outcome measures in acute stroke efficacy trials, but their use is broadening. Plasminogen activator trials which have employed disability outcome measures were also intended to validate their utility and appropriateness for acute stroke outcome assessment, in addition to demonstrating efficacy of the plasminogen activator (HACKE et al. 1995, 1998b; THE NATIONAL INSTITUTE OF NEUROLOGICAL DISORDERS AND STROKE TPA STROKE STUDY GROUP 1995; DEL ZOPPO et al. 1998).

In the absence of angiography, neurological status is a relatively poor correlate for cerebral arterial occlusion location (FIESCHI et al. 1989; SAITO et al. 1987). In general, MCA trunk (M1 segment) occlusion is associated with a very poor outcome typified by persistent disabling defects in 35 of 40 (87%) untreated patients with angiographic or post-mortem confirmation of the arterial occlusion (FISHER 1976). Substantial neurologic improvements in patients with carotid territory infarction and baseline Canadian Neurological Scale scores ≥6.5 (normal = 10.0) were noted (FIESCHI et al. 1989; CÔTÉ et al. 1986). SAITO et al. (1987) suggested that clinical outcome is directly related to occlusion site in the MCA territory. Specifically, 5 of 16 patients (31%) with distal M1 segment and M2 segment occlusions, and 5 of 17 patients (29%) with proximal M1 segment occlusions had fatal outcomes at 3 months, while the 3-month mortality among M3 segment (branch) occlusions was 14% (1 of 7 patients).

However, in a population of stroke patients not selected by vascular diagnostic techniques, those differences may not be apparent.

The clinical outcome of brain stem infarction secondary to basilar artery occlusion follows one of two profiles (KUBIK and ADAMS 1946; BIEMOND 1951; CASTAIGNE et al. 1974; KARP and HURTIG 1974; FISHER 1977; ARCHER and HORENSTEIN 1977; CAPLAN 1983, 1986; ZEUMER et al. 1982, 1983; PESSIN and CAPLAN 1986). One profile of vertebrobasilar territory ischemia is characterized by high mortality or severe disability (ARCHER and HORENSTEIN 1977). The majority of patients with basilar artery occlusion in one series succumbed within two days to five weeks (KUBIK and ADAMS 1946), while in another all nine patients with bilateral distal vertebral artery occlusion died (CAPLAN 1983). Among 22 consecutive patients with vertebral/basilar occlusions treated with antiplatelet agents and/or anticoagulants, 86% succumbed (HACKE et al. 1988). A separate, more benign course has been described in which patients present with only minor or transient brain stem symptoms (FIELDS et al. 1966; FISHER 1970; CAPLAN 1979; ASPLUND et al. 1980; BOGOUSSLAVSKY and REGLI 1985; PESSIN and CAPLAN 1986; BIEDERT et al. 1987). They would not be considered candidates for plasminogen activators.

In general, therefore, patients with milder fixed deficits are more likely to fare better than those with severe deficits. Mortality or ultimate neurologic status is generally a poor indicator of the anatomic location of the occlusion. Although neurologic deficits reflect occlusion location, the clinical deficits do not invariably correlate. Disability outcomes are now considered "standards" for outcome measures, although thresholds for efficacy assessment have not been validated.

C. Plasminogen Activators in Focal Cerebrovascular Ischemia

Studies of the efficacy of plasminogen activators in focal cerebral ischemia have focused on the hypotheses that (1) fibrinolytic substances facilitate cerebral arterial recanalization, which is necessary for neurological recovery, or (2) fibrinolytic substances may achieve neurological recovery. Two approaches have addressed these hypotheses: (1) angiography-based studies which define occlusion location and assess vascular outcome and related clinical benefits; and (2) symptom-based studies which assess clinical outcome only, without corroboration by vascular diagnosis or outcome. Among published experiments in acute cerebral ischemia, attention has focused on the use of streptokinase (SK), urinary-type plasminogen activator (uPA or urokinase = UK), tPA, and recombinant pro-urokinase (rpro-UK, sc-uPA) by local or regional intra-arterial infusion, which require angiography. Systemic infusion clinical outcome studies without angiographic confirmation have employed fibrinolysin, thrombolysin, SK, or tPA, while more recent angiography-based systemic infusion studies have uniformly used tPA.

I. Completed Stroke

There is no role for fibrinolytic agents in completed stroke or "stroke in evolution." Recanalization of patients with stable symptoms of completed stroke and angiographically documented intracranial ICA or MCA occlusions (SUSSMAN and FITCH 1958), or common carotid artery occlusion (CLARKE and CLIFFTON 1960) with fibrinolysin has been described. Direct intra-arterial delivery (CLARKE and CLIFFTON 1960) produced superior recanalization to intravenous delivery (SUSSMAN and FITCH 1958). HERNDON et al. (1960) described clinical improvement in 8 of 13 (61%) patients with presumed cerebral arterial occlusion of the carotid or basilar artery system treated with fibrinolysin. Single instances of successful recanalization of carotid territory occlusion with plasmin (ATKIN et al. 1964) and thrombolysin (MEYER et al. 1961) early following symptom onset were reported.

The possibility that intravenous infusion of thrombolysin might safely achieve recanalization was prospectively tested in a controlled study comparing 20 treated and 20 placebo subjects with symptoms of carotid territory occlusion (MEYER et al. 1963). Only 16 of the 40 patients had angiographic evidence of arterial occlusion. Recanalization was clearly demonstrated in a few patients, but clinical outcome was equivocal. A prospective study of intravenous SK ($n = 37$) vs placebo ($n = 36$) in patients presenting up to 72 h after carotid territory stroke was undertaken by MEYER et al. (1964). Angiographic control demonstrated occlusion of an intracranial artery in only 35 patients. Recanalization seemed more frequent in the SK group, as was clinical evidence of intracranial hemorrhage (11%).

The availability of UK a decade later spawned two studies which addressed recanalization and neurological outcome. ARAKI et al. (1973) reported a low frequency of recanalization in patients with carotid territory occlusion treated with UK up to 30 days after symptom onset in a placebo-controlled study. No neurological benefit was observed by FLETCHER et al. (1976) among 31 patients who received UK intravenously up to 36 h after symptom onset. Intracerebral hemorrhage occurred in seven (32%) patients, of whom four succumbed (HANAWAY et al. 1976). Those studies and that of MEYER et al. (1964) supported the view that the plasminogen activator use in (completed) stroke was unsafe and not clinically efficacious. A general contraindication of the use of fibrinolytic agents in patients suffering stroke ensued (NIH CONSENSUS CONFERENCE 1980). But, those studies suffered several weaknesses (DEL ZOPPO et al. 1986):

1. The majority of patients were treated >6 h from symptom onset.
2. Occlusion of a suspected symptom-producing artery was not documented in most studies.
3. The study populations included patients with "stroke in evolution," lacunar infarction, and completed stroke.
4. Intracerebral hemorrhage as the cause of stroke symptoms could not be definitely excluded because CT scanning techniques were not yet available.

5. The frequency of hemorrhagic transformation and the vascular "natural history" of each atherothrombotic embolic stroke was not known.
6. The dose-rate/duration of UK and SK infusion was not standardized or uniformly controlled.

Thus, the true relationship of vascular occlusion, plasminogen activator dose and infusion schedule, recanalization, and neurological outcome could not be evaluated.

   Subsequent non-angiographic systemic infusion studies of very low-dose UK and tPA (duteplase) in ischemic stroke were driven by safety concerns. Two placebo-controlled studies suggested clinical benefits following UK which did not reach statistical significance (ABE et al. 1981a,b; OTOMO et al. 1985), while subsequent comparisons gave no advantage to UK or tPA over placebo (OTOMO et al. 1988; ABE et al. 1989). The interval from symptom onset to treatment varied from 3 days to 10 days in those trials, and clinical outcomes were not linked to vascular reperfusion (ATARASHI et al. 1985; OTOMO et al. 1985, 1988; DEL ZOPPO et al. 1991). The rather low dose systemic infusion protocols used in that group of prospective studies were intended to minimize hemorrhagic risk (incidence: 0–1.1%) (ABE et al. 1981a,b; ATARASHI et al. 1985; OTOMO et al. 1985). A similar but somewhat higher frequency of hemorrhagic transformation, exclusively HI, was observed in the UK/tPA comparisons (OTOMO et al. 1988; ABE et al. 1989).

II. Acute Stroke

The concept that arterial recanalization must occur within the early minutes to hours of focal cerebral ischemia for neurological recovery to be significant was introduced in the early 1980s (DEL ZOPPO et al. 1986). The effects of intravenous infusion and intra-arterial local delivery of fibrinolytic agents on arterial recanalization, hemorrhagic transformation, and neurological outcome have been addressed by phase I (level III) dose-rate finding, and larger phase III (level I) clinical trials in carotid territory and vertebrobasilar territory ischemia.

1. Carotid Territory Ischemia

a) Angiography-Controlled Trials

α) Intra-Arterial Local Infusion

Intra-arterial infusion studies have established the feasibility of cerebral arterial recanalization in acute stroke patients (ZEUMER 1985; ZEUMER et al. 1989). Reports of single cases or small case series (CLARKE and CLIFFTON 1960; MEYER et al. 1961; ATKIN et al. 1964; NENCI et al. 1983; MIYAKAWA 1984; ZEUMER et al. 1989) supported larger prospective open trials of regional or local infusion of UK or SK that documented recanalization of acute carotid artery territory and of vertebrobasilar artery territory occlusions following acute intervention

(DEL ZOPPO et al. 1988; MORI et al. 1988; HACKE et al. 1988; THERON et al. 1989; ZEUMER et al. 1989, 1993; MATSUMOTO and SATOH 1991; MÖBIUS et al. 1991).

Arterial recanalization has been reported in 45–100% of patients with carotid territory focal ischemia treated acutely with thrombolytic agents. Table 1 lists prospective trials comprising ten or more patients treated locally with SK, UK, tPA, or pro-UK, or a combination of UK or tPA together with Lys-plasminogen. Overall complete or partial recanalization following local or regional intra-arterial infusion among these 323 patients was 63%. In the majority of studies, patients were entered up to 6h from symptom onset, although exceptionally one patient received the plasminogen activator 3 weeks after the initial symptoms. DEL ZOPPO et al. (1988) demonstrated the feasibility of recanalization in 18 of 20 patients treated within 6h of apparent symptom onset, the safety of angiographic techniques applied in the acute phase, and the benign outcome of HI which occurred within 24h of symptom onset and thrombolysis. MORI et al. (1991) demonstrated that recanalization was associated with significant reduction in infarction volume by CT scan. Recanalization was less frequent among ICA occlusions than distal MCA

Table 1. Recanalization outcome: carotid territory ischemia, intra-arterial delivery, angiography-based

Reference	Agent	Patients n	Δ(T-0) h	Recanalization n	Recanalization (%)[a]
DEL ZOPPO et al. (1988)	SK/UK	20	1–24	18	90
MORI et al. (1988)	UK	22	0.8–7	10	45
THERON et al. (1989)	SK/UK	12	2–504	12	100
MATSUMOTO and SATOH (1991)	UK	40	1–24	24	60
ZEUMER et al. (1993)	tPA/UK	31	<4	29	94
OVERGAARD et al. (1993)	tPA	17	<6	12	71
BARNWELL et al. (1994)	UK	10	<10	7	70
LEE et al. 1994	UK	20	<24?	9	45
SASAKI et al. (1995)	tPA/UK	35	<8	16	46
FREITAG et al. (1996)	tPA/UK	22	<6	14	64
	PA + lys-plg	14	<6	12	86
GÖNNER et al. (1998)	UK	33	<4	19	58
ENDO et al. (1998)	tPA/UK	21	<6	8	38
DEL ZOPPO et al. (1998)	pro-UK	26	<6	15	58
	Placebo	14	<6	2	14
JAHAN et al. (1999)	UK	26	<7.5	11	42
FURLAN et al. (1999)	pro-UK + H	121	<6	80	66
	H only	59	<6	11	18
LEWANDOWSKI et al. (1999)	iv. + ia. tPA	17	<3	9	53
	plac. + ia. tPA	18	<3	5	28

Δ(T-0), interval from onset to treatment; H, heparin; lys-plg, lys-plasminogen; PA, UK or tPA; plac, placebo; pro-UK, recombinant pro-urokinase; SK, streptokinase; tPA, recombinant tissue plasminogen activator; UK, urokinase plasminogen activator.
[a] Percent of total number of patients with recanalization.

Table 2. Hemorrhagic transformation: carotid territory ischemia, intra-arterial delivery

Reference	Patients n	Hemorrhage		With deterioration	
		HI	PH	(%)[a]	(%)[b]
DEL ZOPPO et al. (1988)	20	4	0	20	0
MORI et al. (1988)	22	1	3	18	14
THERON et al. (1989)	12	1	2	25	17
MATSUMOTO and SATOH (1991)	40	9	4	32	10
ZEUMER et al. (1993)	31	6	0	19	0
OVERGAARD et al. (1993)	17	2	0	12	0
BARNWELL et al. (1994)	10	3	0	30	0
LEE et al. (1994)	20	5	4	20	15
SASAKI et al. (1995)	35	7	1	23	3
GÖNNER et al. (1998)	33	5	2	21	6
DEL ZOPPO et al. (1998)					
Pro-UK	26	11[c]		42	15
Placebo	14	1[c]		7	14
JAHAN et al. (1999)	26	7	3	27	12
FURLAN et al. (1999)					
pro-UK+H	108	27	11	25	10
H only	54	6	1	11	2
LEWANDOWSKI et al. (1999)					
iv.+ia. tPA	17	4	2	24	12
plac.+ia. tPA	18	1	1	6	6

See Table 1.
HI, hemorrhagic infarction or asymptomatic hemorrhage (see text); PH, parenchymatous symptomatic hemorrhage (see text).
[a] Percent of total number of patients who demonstrated hemorrhagic transformation.
[b] Percent with hemorrhagic transformation (most often PH) and deterioration.
[c] HI + PH.

obstructions (MORI 1991; MORI et al. 1988; LEE et al. 1994; SASAKI et al. 1995; ENDO et al. 1998).

Hemorrhagic transformation occurred in 19%–32% of treated patients (Table 2). Overall, the incidence of PH (mostly symptomatic) in 266 patients was 15% and of HI 17%.

The Prolyse (rpro-UK) in Acute Cerebral Thromboembolism (PROACT I) trial was the first randomized, double-blind, multicenter trial comparing the safety, recanalization frequency, and clinical efficacy of direct intra-arterial infusion of recombinant pro-UK with placebo in patients with symptomatic MCA occlusion of less than 6h duration (DEL ZOPPO et al. 1998). That phase II trial randomized 26 patients to intra-arterial infusion of 6mg of pro-UK and 14 patients to placebo infusion. A high dose heparin (100 IU/kg bolus, followed by 1000 IU/h infusion for 4h) administration to the first 16 patients (11 pro-UK and 5 placebo) resulted in a very high recanalization rate of 82% in the pro-UK leg (vs 0% in the placebo group), but also produced a high incidence of intracranial hemorrhage (73% vs 20% in the placebo group). Reduction of the heparin dosage to a bolus of 2000 IU, followed by a 500 IU/h infusion in the 24 successive patients resulted in a lower 40% recanalization rate in the pro-UK patients (vs 22% in the placebo group) and also of the incidence of intracerebral hemorrhage (20% and 0% respectively). Overall 24 h results in the high and the low heparin group yielded 58% and 14% recanalization ($p = 0.017$), 42% and 7% intracranial hemorrhage ($p = 0.03$), and a 15% and 7% (n.s.) frequency of intracranial hemorrhage with clinical deterioration, respectively, in the pro-UK and placebo groups. Mortality within 90 days was 27% in the pro-UK and 43% in the placebo group (n.s.). Clinical outcome at 90 days, as measured by the Barthel index, the modified Rankin scale, and the NIHSS, was better in the pro-UK than in the placebo group.

Based on PROACT I, a larger multicenter, randomized, open trial was performed in 180 patients with MCA occlusion of <6 h duration. The total dose of recombinant pro-UK was increased from 6 to 9 mg and the same low-dose heparin schedule used as in PROACT I in an attempt to improve recanalization while limiting symptomatic brain hemorrhage. The initial mean NIHSS score was 17 in the pro-UK ($n = 121$) as well as in the group receiving only heparin and no interventional procedure ($n = 59$), reflecting a higher baseline stroke severity than in any randomized acute stroke trial. The median time to initiation of pro-UK treatment was 5.3 h (FURLAN et al. 1999). The 2-hour TIMI grade 2 + 3 recanalization rates in the PROACT II study were 66% in the pro-UK group and 18% in control patients. Intracranial hemorrhage at 24 h occurred in 35% of the pro-UK patients and 13% of controls; neurological deterioration in 10% and 2% respectively. For the primary efficacy analysis 40% of pro-UK patients and 25% of the non-intervention group had a modified Rankin score of ≤2 at 90 days ($p = 0.04$). No significant outcome difference was observed for a modified Rankin score of 0 and 1.

In the EMS bridging trial the feasibility of early administistion of intravenous tPA followed by intraarterial tPA (higher reported recanalization rates). It was hoped that the combination of these approaches would further optimize the treatment of acute stroke. In this multicenter, double-blind, randomized, placebo-controlled study 17 patients were assigned to the i.v./i.a. tPA group and 18 to the placebo/i.a. tPA group. On angiography thrombotic occlusion were only found in 22/35 patients. Recanalization was significantly better in the i.v./i.a. group with TIMI grade 3 flow in 6/11 patients versus 1/10 in the placebo/i.a. group ($p = 0.05$). Despite these excellent angiographic results clinical outcome as assessed by NIHSS score, Berthel index, modified Rankin Scale and Glasgow Oucome Scale was similar in the two groups (LEWANDOWSKI et al. 1999).

β) Intravenous (Systemic) Infusion

Four prospective angiography-based trials designed to assess the effects of intravenous tPA on recanalization or neurological outcome have been reported (Table 3) (VON KUMMER et al. 1991; DEL ZOPPO et al. 1992; MORI et al.

1992; VON KUMMER and HACKE 1992; YAMAGUCHI 1993, 1995). The intravenous infusions lasted 60–90min and were initiated early (within 6h or 8h) after symptom onset.

A prospective open dose-escalation study demonstrated no dose-rate dependence of recanalization following intravenous tPA (duteplase) (DEL ZOPPO et al. 1992). Recanalization at 60min post-tPA infusion was not significantly different at any of nine dose-rates from 0.12 MIU/kg to 0.75 MIU/kg, and was 34% overall. On retrospective analysis, recanalization of distal MCA occlusions was significantly more frequent than ICA occlusions (DEL ZOPPO et al. 1992), an impression which has been supported by other trials (YAMAGUCHI 1991; MORI et al. 1992; LEE et al. 1994; SASAKI et al. 1995; ENDO et al. 1998). The presence of the hyperdense middle cerebral artery sign (HMCAS), consistent with thrombotic arterial occlusion of the M1 segment, was not associated with recanalization following tPA (WOLPERT et al. 1993). The lack of dose-rate response in addition to the 18% frequency of ICA occlusions suggests that that study might have been underpowered to demonstrate a recanalization effect, or that the dose-rates applied were below the optimum for recanalization of visible carotid territory occlusions.

MORI et al. (1992) compared 20MIU and 30MIU tPA (duteplase) with placebo in a three-arm, double-blind level I trial. Increased recanalization and significantly better 30-day clinical outcome according to the (inverted) HSS was observed in the 30MIU (~0.5 MIU/kg per 60min) cohort compared to the placebo or 20MIU cohorts. No significant change in outcome at 24h to 48h was observed. YAMAGUCHI et al. (1993) extended those observations in a multicenter, placebo-controlled trial comparing 20MIU tPA (duteplase) with matched placebo in 95 patients treated within 6h of symptom onset. When reperfusion grades 3 and 4 were combined, 10 of 47 (21%) patients treated with tPA, and 2 of 48 (4%) patients treated with placebo had successful reperfusion ($p < 0.05$). The tPA group had a more favorable 30-day outcome ($p = 0.04$) according to a weighted percent reduction of the HSS (MORI et al.

Table 3. Recanalization outcome: carotid territory ischemia, intravenous delivery, angiography-based

Reference	Agent	Patients n	Δ(T-0) h	Recanalization	
				n	(%)[a]
DEL ZOPPO et al. (1992)	tPA	93 (104)	<8	32	34
MORI et al. (1992)	tPA	19	<6	9	47
	C	12		2	17
VON KUMMER and HACKE (1992)	tPA	22	<6	13	59
YAMAGUCHI (1993)	tPA	47	<6	10	21
	C	46		2	4

Δ(T-0), interval from onset to treatment; tPA, recombinant tissue plasminogen activator; C, control.
[a] Percent of total number of patients who demonstrated recanalization.

1992). A subsequent prospective blinded comparison of 20 MIU vs 30 MIU per patient demonstrated equal neurological recovery (YAMAGUCHI 1995).

Thus, a 21%–53% early partial or complete recanalization frequency following systemic thrombolysis was reported in these studies with similar entry criteria. In the two placebo-controlled trials, recanalization in the setting of the plasminogen activator surpassed that expected from spontaneous recanalization. Improvement in neurological score at 30 days was noted, but there was no assessment of long-term disability. Neurological outcome at 24 h did not predict subsequent outcome at 30 days (MORI et al. 1992). The possibility that patent-established collaterals at treatment may be associated with favorable neurological outcome was addressed by VON KUMMER and HACKE (1992) in their prospective angiography-based single-dose tPA (alteplase, 100 mg) study.

The incidence of hemorrhagic transformation among the intravenous trials was not different from that reported for intra-arterial infusion in the carotid territory (Table 4). In the dose-rate escalation study, HI occurred in 21% of treated patients and PH in 11% (DEL ZOPPO et al. 1992). Of note, 4 of 21 HI patients (19%) deteriorated clinically, while 5 of 11 PH patients (45%) improved or remained unchanged within the hospital observation period, indicating that some overlap of these entities is possible. Overall, however, PH was typically lobar, usually confined to the region of ischemia, and associated with clinical deterioration (DEL ZOPPO et al. 1992; YAMAGUCHI et al. 1993). No relationship between hemorrhagic transformation and 60-min recanalization, tPA dose-rate, pretreatment exposure to antiplatelet agents, or pretreatment blood pressure was apparent in retrospective analysis (DEL ZOPPO et al. 1992). In the trial of MORI et al. (1992) neurological deterioration occurred only in the tPA group. YAMAGUCHI (1993) noted massive hemorrhage in 8% and 11% of

Table 4. Hemorrhagic transformation: carotid territory ischemia; intravenous delivery, angiography-based

Reference	Agent	Patients n	Hemorrhagic		With deterioration	
			HI	PH	(%)[a]	(%)[b]
DEL ZOPPO et al. (1992)	tPA	93 (104)	21	11	31	11
MORI et al. (1992)	tPA	19	8	2	53	11
	C	12	4	1	42	0
VON KUMMER and HACKE (1992)	tPA	22	9	3	10	14
YAMAGUCHI (1993)	tPA	47 (51)	20	4	47	8
	C	46 (47)	17	5	47	11

See Table 1.
HI, hemorrhagic infarction (see text); PH, parenchymatous hemorrhage (see text).
[a] Percent of total number of patients who demonstrated hemorrhagic transformation.
[b] Percent with hemorrhagic transformation (most often PH) and deterioration.

patients treated with tPA and placebo, respectively. The frequency of hemorrhagic transformation with neurological deterioration ranged from 9% to 11% among the other three trials (DEL ZOPPO et al. 1992; VON KUMMER and HACKE 1992; MORI et al. 1992). The frequency of PH ranged from 7% to 11%, which was not substantially different from literature sources.

γ) Single Photon-Emission Computed Tomography (SPECT)

^{14}C-HMPAO-SPECT provides a convenient non-invasive measure of regional cerebral blood flow (MASDEU and BRASS 1995; ALEXANDROV et al. 1997). Several prospective clinical trials of thrombolysis in acute stroke have employed SPECT to assess relative regional CBF changes. OVERGAARD et al. (1993) (The Danish Pilot Study) suggested that rCBF improvement by SPECT correlated well with patency by angiography ($p = 0.015$) following intravenous delivery of tPA (alteplase). In contrast, BAIRD et al. (1994) found no correlation between evidence of reperfusion by SPECT and that provided by digital subtraction angiography (DSA). However, patients without SPECT evidence of reperfusion had significantly higher mortality, less improvement in neurological score, and more functional disability than those with normal perfusion. GROTTA and ALEXANDROV (1998) found a correlation between improvement by SPECT and the NIHSS following treatment with tPA. As noted above, UEDA et al. (1994) found a significant correlation of reduced rCBF by SPECT with hemorrhagic transformation following intra-arterial infusion UK in acute thrombotic stroke.

b) Symptom-Based (Clinical Outcome) Trials

Symptom-based, intravenous (systemic) delivery acute stroke intervention trials have selected patients according to CT scan and clinical criteria, and evaluated therapeutic efficacy by clinical symptoms (neurological score) or disability outcome. Disability status as a sensitive and accurate marker of outcome in this setting has been tested only in the context of ongoing thrombolysis trials.

Five prospective studies examined the impact of low dose UK in several configurations on general clinical outcome after stroke when applied up to five days after stroke onset (Table 5). No benefit was demonstrated in any trial.

The experience of three symptom-based, randomized, placebo-controlled trials of SK in acute ischemic stroke is relevant to concerns about intracerebral hemorrhage (Table 4). Each study used a single intravenous infusion of 1.5 MIU SK as the active arm. HOMMEL et al. (1995) described early termination of the Multicenter Acute Stroke Trial – Europe (MAST-E) when 10-day ($p < 0.001$) and 6-month ($p < 0.01$) mortality in the SK group significantly exceeded that of the placebo group. MAST-E compared intravenous SK with placebo given within 6h of the onset of MCA territory symptoms (THE MULTICENTER ACUTE STROKE TRIAL-EUROPE STUDY GROUP 1996). The significantly higher incidence of symptomatic intracranial hemorrhage observed in the SK

Table 5. Clinical outcome studies: intravenous delivery

Reference	Agent	Patients n	Δ(T-0) h	Outcome	Improvement (%)	Hemorrhage HI	Hemorrhage PH	(%)[a]
Abe et al. (1981)	UK	57	<720	N[b]	70	0	0	
	C	56			47	0	0	
Atarashi et al. (1985)	UK	191	<120	N[b]	45	2	0	1.0
	C	94C			44	1	0	1.1
Otomo et al. (1985)	UK	176	<120	N[b]	52	2	0	1.1
	C	188			41	1	0	0.5
Otomo et al. (1988)	tPA	171	<120	N[b]	59	2	0	1.1
	UK	184			55	3	0	1.6
Abe et al. (1990)	tPA	145	<72	N[b]	66	3	0	2.0
	UK	77			45	6	0	7.8
Haley et al. (1993)	tPA	10	<1.5	N[c]	60[g]	0	0	
	C	10			10	0	0	
	tPA	4	1.5–3.0	N[c]	50	0	0	
	C	3			67	0	1	
The Mast-E Trial (1996)	SK	156	<6.0	M[d]	34[g]	63	33	21.0[g]
	C	154			18	57	4	3.0
ASK (Donnan et al. 1995, 1996)	SK	165	<4.0	M[e]	36[g]	33	23	13.2[g]
	C	163			21	23	5	3.1
Mast-I (1995)	SK	313	<6.0	M[d]	25[g]	60	21	6.7[g]
	C	309			12	27	2	0.7
ECASS-I (Hacke et al. 1995)	tPA	313	<6.0	D[f]	36	72	62	19.8[g]
	C	307			29	93	30	6.5

Table 5. *Continued*

Reference	Agent	Patients n	Δ(T-0) h	Outcome	Improvement (%)	Hemorrhage HI	PH	(%)[a]
NINDS (I) (1995)	tPA	144	0–3.0	N[c]	47	–	13	5.6[g]
	C	147			39	–	3	0.0
NINDS (II)	tPA	168	0–3.0	D[f]/N[c]	48	–	21	7.1[g]
	C	165			39	–	8	2.1
ECASS-II (1998)	tPA	409	0–6.0	N[c]	40	142	48	11.7[g]
	C	391			37	141	12	3.1
ATLANTIS, Part B (CLARK et al. 1999)		272	3–5	D[f]/N[c]	34	31	19	7.0
	C	275			32	13	3	1.1
ATLANTIS, Part A (CLARK et al. 2000)	tPA	71	≤6.0	D[f]/N[c]	40	9	8	11
	C	71			21	3	0	0

Δ(T-0), interval from onset to treatment; C, control; HI, hemorrhagic infarction, or asymptomatic intracranial hemorrhage (see text); PH, parenchymatous hemorrhage, or symptomatic intracranial hemorrhage (see text); SK, streptokinase; tPA, recombinant tissue plasminogen activator; UK, urokinase plasminogen activator.

[a] Percent with hemorrhagic transformation (most often PH) and deterioration.

Outcomes of note to each study:

N[b], neurological outcome.

N[c], neurological outcome according to scoring instrument.

M[d], early mortality.

M[e], mortality (3- or 6-month).

D[f], best disability outcome (3-month).

[g] Statistically significant.

group [33 of 156 (21%) vs 4 of 154 (3%); $p < 0.001$] was probably due to the severity of stroke patients entered into this trial. This was suggested by the high short-term mortality in the placebo group (18%), consistent with severe tissue injury. The Australia Streptokinase (ASK) trial was a symptom-based, randomized, placebo-controlled trial of SK administered within 4h of acute ischemic stroke (DONNAN et al. 1995, 1996). Interim analysis of adverse outcome after 300 patients were entered prompted the safety monitoring committee to advise termination of the >3h (to 4h) arm because of significant increased mortality or mortality and Barthel index (modified) <120 among patients in that arm. No apparent safety concern was observed for patients treated in the 0–3h window. Unfavorable outcome in those patients were 14/41 (34%) in the SK arm and 15/29 (52%) in the placebo group (DONNAN et al. 1996). Those findings underscore the relevance of the time from symptom onset to treatment, and the importance of discrete and ongoing risk/benefit assessment. They are also reminiscent of the earlier findings of MEYER et al. (1964) regarding the safety of SK in stroke. In a substudy of the ASK trial INFELD et al. (1996) performed HMPAO-SPECT analysis in 15 SK- and in 9 placebo-treated patients. SK was associated with a greater amount of nonnutritional reperfusion than was placebo ($p = 0.04$). This luxury perfusion was associated with poor functional outcome ($p = 0.02$). In another substudy, using HMPAO-SPECT and/or transcranial Doppler analysis, a larger number of patients demonstrating the combined endpoint of reperfusion or recanalization was seen in the SK (13/14; 93%) than in the placebo group (7/14; 50%, $p = 0.01$) but this effect did not influence overall clinical outcome (YASAKA et al. 1998).

The Multicentre Acute Stroke Trial – Italy (MAST-I) group also terminated their trial of acute intravenous SK ± aspirin (ASA) within 6h of symptom onset after only 622 patients were entered (MULTICENTRE ACUTE STROKE TRIAL-ITALY (MAST-I) GROUP 1995; TOGNONI and RONCAGLIONI 1995; CICCONE et al. 1998). An excess 10-day case fatality was associated with SK ± ASA ($p < 0.00001$), most particularly when SK was given with ASA ($p < 0.00001$), but, no differences in six-month outcomes were seen. These experiences overall have limited the study of SK in acute ischemic stroke. It must be noted that a proper dose-finding study to arrive at a safe SK dose was not performed, and the dose used in all trials was that adapted from recommended doses for acute MI [GRUPPO ITALIANO PER LO STUDIO DELLA STREPTOCHINASI NELL'INFARTO MIOCARDICO (GISSI) 1986, 1990].

The European Cooperative Acute Stroke Study (ECASS), a prospective phase III randomized, symptom-based, placebo-controlled study, compared clinical outcomes following intravenous infusion tPA (1.1 mg/kg, maximum 100 mg) with placebo (HACKE et al. 1995). From 75 European centers, 620 patients were randomized to tPA or placebo within 6h of symptom onset. The primary outcome events were disability outcome by Barthel index and modified Rankin scale at day 90 post-treatment. Prospectively applied rules sought to exclude patients by neuroradiographic criteria with evidence of large regions of hemispheric injury. A post hoc re-evaluation of baseline CT scans in all patients defined a population with hemispheric injury not apparent on admission. Re-evaluation of outcomes with consideration for those patients and others who should have been excluded yielded a "target population." For the target analyses 109 patients were excluded, of which 66 (61%) were to have been excluded by CT scan criteria, mostly because of the presence of major early infarct signs not appreciated on the initial scan. SCHRIGER et al. (1998) have analyzed the accuracy of CT scan readings by a sample of 103 physicians involved in the screening of stroke patients. The 54 test scans demonstrated intracerebral hemorrhage, acute infarction, intracerebral calcifications, old cerebral infarction, and normal findings. The average correct readings by 38 emergency physicians were 67%, by 29 neurologists 83%, and by 36 general radiologists also 83%. Initial results with CT angiography, involving spiral CT scanning with an intravenous injection of non-ionic contrast medium, may help to further refine the indication for thrombolytic therapy in stroke patients by identifying, among others, those in whom spontaneous lysis of the occluding thrombus has taken place and in whom the administration of thrombolytic agents might do more harm than good (WILDERMUTH et al. 1998).

Intention-to-treat analysis demonstrated no apparent difference between the tPA and placebo-treated groups with regard to Barthel index and modified Rankin scale. Analysis of the "target population" suggested a significant difference in median Barthel index and modified Rankin scale score favoring the tPA-treated group. In the latter analysis an 11%–12% absolute improvement in best outcome (modified Rankin scales scores of 0 and 1) resulted. Cumulative mortality ("intention-to-treat") at 90 days was 23% in the tPA group and 16% in the placebo group, which was not different. The "intention-to-treat" model demonstrated a significantly higher proportion of patients with intracerebral hemorrhage causing neurological deterioration or death (PH2) with tPA (19 of 313) compared to placebo (7 of 307). An excess of PH2 (7%) was associated with tPA in "target" analyses as well. Early neurological recovery (i.e., Scandinavian Stroke Scale), functional outcome in the combined Barthel index/Rankin Scale, and duration of in-hospital stay were favorably influenced by tPA exposure in both. There was no benefit in neurological status at 24 h. Median neurological and disability outcomes at 30 and at 90 days were better in the tPA-treated "target" patients only. A singular interpretation of these data is that the lack of benefit in functional outcome and neurological recovery was "driven" by a subgroup of patients with clinically significant cerebral injury at increased risk for hemorrhage or further deterioration after intravenous thrombolysis. The overall analysis suggested that when a subgroup of patients at high risk ("early infarct signs") were excluded, 90-day disability outcome would be significantly affected. Defining the subgroup or subgroups represented by the excluded patients is of major importance to the application of this approach within 6 h from symptom onset. The ECASS experience underscored the need to define subtle alterations on initial CT scan consistent with "major early infarct signs" so as to exclude patients at high risk for poor outcome.

Two post-hoc analyses of the ECASS trial examined whether clinical outcome were better if only patients within 3 h of stroke onset had been enrolled (STEINER et al. 1998) and whether using three of the four dichotomized end points evaluated in the NINDS trial would modify the original analysis of the ECASS trial (HACKE et al. 1998a). In ECASS 87 patients had been randomized within 3 h of symptoms. Differences in favor of tPA treatment were found in all primary and secondary outcome measures, except for mortality at day 30 (STEINER et al. 1998). Using the statistical methodology of NINDS for the three end points of modified Rankin score, Barthel index, and the NIHSS, outcome was more favorable in the tPA-treated group compared to placebo. A modified Rankin score of 0 or 1 was found in 36% of patients in the tPA group vs 28% in the placebo group ($p = 0.044$), Barthel indexes of 95 or 100 were determined in 44% and 38% ($p = 0.1$), and NIHSS of 0–1 in 36% and 22% ($p = 0.001$) of patients respectively. Thus, when applying the statistical approach of the NINDS study to the ECASS intention to treat data set, the outcome of tPA-treated patients was significantly improved (HACKE et al. 1998a). The limitations of such analyses are that they were performed post-hoc, i.e., did not use predefined criteria.

Following a series of pilot studies to assess the relative safety of tPA (alteplase), a two-part, four-armed placebo-controlled clinical outcome study of tPA with entry at 0–90 min or 91–180 min from symptom onset was completed by the NINDS (THE NATIONAL INSTITUTE OF NEUROLOGICAL DISORDERS AND STROKE TPA STROKE STUDY GROUP 1995). Earlier reports from that group suggested a low incidence of symptomatic intracerebral hematomas in 3 of 74 (4%) patients treated with tPA (alteplase), although the frequency of hemorrhagic transformation in one open dose-escalation study was 30% (6 of 20) (HALEY et al. 1992, 1993).

In Part 1, among 291 patients randomized to tPA (0.9 mg/kg) or placebo within the two treatment time intervals, a slight improvement of the neurological status according to the NIHSS at 24 h was observed in the tPA group ($n = 144$) compared to the placebo ($n = 147$) group (n.s.; Table 5) (THE NATIONAL INSTITUTE OF NEUROLOGICAL DISORDERS AND STROKE TPA STROKE STUDY GROUP 1995). In Part 2, a significant 11%–13% absolute improvement in Barthel index, modified Rankin scale score, Glasgow outcome scale score, and NIHSS with minimal or no disability (deficit) was observed at three months in the tPA ($n = 168$) over placebo ($n = 165$) recipients. No difference in mortality was observed. However, in the combined experience of both parts, the frequency of symptomatic hemorrhage was significantly greater among those patients treated with tPA (6.4%, $2p < 0.001$) than those who received placebo (0.6%). Overall, there was no difference in outcomes between patients treated within 90 min and those treated within 180 min. It was concluded that substantial benefit in best outcome (no or minimal disability) could be achieved among patients treated within 3h of symptom onset with this plasminogen activator. At one-year follow-up patients treated with tPA were more likely to have minimal or no disability than the patients given placebo (Kwiatkowski et al. 1999)

Symptomatic hemorrhagic transformation contributed significantly to mortality within the tPA group at three months. Of interest is the observation that the frequency of symptomatic hemorrhage in the placebo group (0.6%) was substantially less than that reported in the literature from previous prospective fibrinolysis studies (del Zoppo et al. 1992; Mori et al. 1992; Yamaguchi et al. 1993). While the reasons for the lower frequency are not clear, it cannot be attributed to earlier arrival to the hospital setting or CT evaluation, or to less severe initial neurological status. In angiography-based studies contrast extravasation or anticoagulant use may have contributed to CT findings at 24h (del Zoppo et al. 1992; Mori et al. 1992; von Kummer and Hacke 1992; Yamaguchi 1993), but this cannot be the case for CT scan-based studies (Hornig et al. 1986; Hacke et al. 1995). The entry of proximal MCA lesions (which are compatible with large hemispheric injury) in those studies (del Zoppo et al. 1992; Mori et al. 1992) may have selected for a higher frequency of severe hemorrhage, whereas the lack of vascular diagnosis in other studies may have allowed selection of a larger proportion of less severe ischemic lesions or lesions of another source.

A cost-effectiveness analysis, based on the NINDS trial data set, arrived at the conclusion that thrombolytic therapy with tPA for acute ischemic stroke is likely to result in a net cost savings to the health care system (Fagan et al. 1998). Interestingly, tPA-treated patients, who were hypertensive after randomization and received antihypertensive therapy were less likely to have a favorable outcome at three months ($p < 0.01$) than those who were hypertensive and did not receive antihypertensive drugs (Brott et al. 1998). The favorable results with tPA in the NINDS trial led to rapid approval, by the FDA, of this drug for the treatment of acute ischemic stroke and the establishment of Guidelines for Thrombolytic Therapy for Acute Stroke by the American Heart Association Science Advisory and Coordinating Committee on June 20, 1996 (Adams et al. 1996). This was soon followed by the application of tPA thrombolysis for acute stroke in university and community hospitals (Trouillas et al. 1996; Grond et al. 1998b; Chiu et al. 1998; Tanne et al. 1998; Katzan et al. 2000; Albers et al. 2000).

ECASS-II was a non-angiographic, randomized, double-blind trial of 800 patients in Europe, Australia, and New Zealand (Hacke et al. 1998b). CT scans were used to exclude patients with signs of major infarction. tPA ($n = 409$) or placebo ($n = 391$) were randomly assigned with stratification for time from symptom onset (0–3h or 3–6h). The dose of tPA (0.9mg/kg; maximum 90mg, given as a bolus of 10% of the total dose, followed by a 60min infusion of the remaining dose) was chosen to match that of the NINDS trial. The primary end point was the modified Rankin scale (mRS) at 90 days. A favorable outcome (mRS 0 or 1) was seen in 40.3% of the patients in the tPA and 36.6% in the placebo group ($p = 0.28$). In a post-hoc analysis, in which outcome was classified in terms of independence (mRS 0–2), 54.3% of the patients in the tPA and 46.0% in the placebo group were independent ($p = 0.024$). There were no differences between the cohorts treated 0–3 h or 3–6 h after the onset of stroke symptoms, and in the 30-day and 90-day mortality rates between the two treatment groups. Severe intracranial hemorrhage was significantly more common in the tPA (11.7%) than in the placebo group (3.1%). Although the overall results were not statistically significant, the findings of the ECASS-II trial are consistent with the positive trend and benefits seen in previous trials and are supported by the significant absolute difference of 8.3% in favor of tPA when the mRS was dichotomized for dependency (HACKE et al. 1998b).

ATLANTIS was originally designed to assess the efficacy and safety of intravenous tPA in patients with acute stroke of <6 h duration. Enrollment was halted because of safety concerns in patients treated 5 to 6 h after onset of symptoms. Results of this initial ATLANTIS Part A trial were reported separately (CLARK et al. 2000) and Part II initiated, restricting the time window after onset of symptoms to 3–5 h. In Part A no significant benefit on any of the planned efficacy end points at 30 and 90 days was seen. The risk of symptomatic intracerebral hemorrhage was increased with tPA treatment, particularly in those patients treated 5-6 h after onset of symptoms (CLARK et al. 2000). In the target population of Part B, 272 patients were randomized to tPA and 275 to placebo. In about 20% of patients treatment was started 3-4 h and in 80% 4-5 h after onset of symptoms. Primary endpoint was an excellent neurological recovery at day 90 (NIHSS score ≤1), secondary endpoints included excellent recovery on functional outcome measures (Barthel index, modified Rankin scale and Glasgow Outcome Scale) at days 30 and 90. In none of these assessment was there any significant difference between the tPA and the placebo group. In the first 10 days asymptomatic and symptomatic intracranial hemorrhage was significantly increased in the tPA group (CLARK et al. 1999). The authors concluded that their results do not support the use of intravenous tPA for stroke treatment beyond 3 hours.

Taken together, the ECASS and NINDS studies indicate the enormous importance of patient selection to avoid risk attendant to plasminogen activators (HACKE et al. 1995, 1998b; THE NATIONAL INSTITUTE OF NEUROLOGICAL DISORDERS AND STROKE TPA STROKE STUDY GROUP 1995). The interval from symptom onset to treatment to achieve clinical improvement varies individually (BARON et al. 1995). All three studies suggest that CT scans and neurologic scores at study entry do not completely identify those at risk for hemorrhage, although proper attention to the presence and extent of ischemic injury on initial CT scan is likely to address hemorrhagic risk. It is currently not possible to separate benefit from hemorrhagic risk in a given patient based upon simple clinical criteria. However, the three studies and their forbears indicate that outcome improvement is feasible within a longer interval prior to treatment. The experience of MAST-E and studies with other imaging methods suggest that poor neurological status at outset and reduced rCBF may reflect cerebral tissue injury which will produce significant hemorrhage with deterioration upon exposure to plasminogen activators. Criteria for patient treatment (in the experimental setting) require a short interval from symptom onset to treatment and the absence of apparent tissue injury on initial CT scan.

2. Vertebrobasilar Territory Ischemia

a) Angiography-Controlled Trials

α) Intra-Arterial (Local) Infusion

The potentially severe and fatal outcome following vertebral and/or basilar artery occlusion has prompted an aggressive approach to this disorder (Table 6). NENCI et al. (1983) reported that local directed infusion of either SK or UK within 6–96 h of symptom onset in four patients with vertebrobasilar ischemia resulted in complete recanalization without hemorrhage. A retrospective comparison of general outcome and recanalization efficacy in 43 patients who received either SK or UK with 22 patients who received conventional antiplatelet/anticoagulant therapy for vertebrobasilar ischemia suggested benefit with recanalization (HACKE et al. 1988). Of the 43 patients with technically successful interventional procedures, 14 of 19 who achieved complete recanalization survived, whereas only 4 of the 23 patients who received anti-

Table 6. Recanalization outcome and hemorrhagic transformation: vertebrobasilar territory intra-arterial delivery, angiography-based ischemia

Reference	Agent	Patients n	Recanalization		Hemorrhage		
			n	(%)[a]	HI	PH	(%)[b]
HACKE et al. (1988)	SK/UK	43	19	44	2	2	9
ZEUMER et al. (1989)	UK	7	7	100	1	0	14
MÖBIUS et al. (1991)	SK/UK	18	14	78	0	0	0
MATSUMOTO and SATOH (1991)	UK	10	4	40	0	1	10
ZEUMER et al. (1993)	tPA/UK	28	28	100	2	0	0
BRANDT et al. (1996)	UK	44	23	52	3?	3	7?
BECKER et al. (1996)	UK	12	10	83	–	2	17
WIJDICKS et al. (1997)	UK	9	7	78	–	1	11
CROSS et al. (1997)	UK	20	13	65	–	3	15
LEVY et al. (1999)	UK	10	6	60			

HI, hemorrhagic infarction (see text); PH, parenchymatous hemorrhage (see text); SK, streptokinase; tPA, recombinant tissue plasminogen activator; UK, urokinase plasminogen activator.
[a] Percent of total number of patients who demonstrated recanalization.
[b] Percent of total number of patients who demonstrated hemorrhagic transformation.

coagulants survived. Recanalization was significantly associated with survival ($p = 0.000007$) and with improved outcome ($p = 0.00005$).

ZEUMER et al. (1989, 1993) reported recanalization in 100% of patients. MÖBIUS et al. (1991) realized recanalization in 78% of patients who received UK or SK by direct infusion. In contrast, MATSUMOTO and SATOH (1991) reported only 40% recanalization in patients who received UK by regional infusion. Of 18 patients described by MÖBIUS et al. (1991) who received either UK or SK, 10 of 14 patients who displayed recanalization within the 2h therapeutic period had at worst moderate residual deficits, whereas all four patients who did not display recanalization died. Similarly, three of four patients with basilar artery occlusion treated by MATSUMOTO and SATOH (1991) who displayed recanalization had fair or good clinical outcome. Those with progressive symptoms at outset seemed to fare worse than those with sudden symptom onset. The report of BRANDT et al. (1996) comprises a total of 51 patients of whom 44 were treated with intra-arterial UK. Recanalization was achieved in 52%. Mortality was 46% in the recanalization group and 92% in the group without recanalization ($p = 0.0004$). BECKER et al. (1996) reported recanalization in 10 of 12 (83%) patients with direct infusion of UK by a microcatheter embedded into the face of the arterial thrombus. An initial 1- to 2-h infusion was followed by a prolonged low-dose infusion, intended to prevent rethrombosis. Of the patients who exhibited recanalization, six succumbed of whom three had rethrombosis. Coma and quadriparesis on presentation seemed to increase the risk; all three patients with these conditions died.

β) Intravenous (Systemic) Infusion

There is meager information about recanalization efficacy following intravenous delivery of fibrinolytic agents in vertebrobasilar territory ischemia (HERDERSCHEÊ et al. 1991; VON KUMMER et al. 1991; YAMAGUCHI 1991; GROND et al. 1998a).

III. Other Conditions

1. Retinal Vascular Occlusion

The potential use of fibrinolytic agents in retinal artery or retinal vein occlusion has been examined by KWAAN (1988). Potential benefit with visual recovery was suggested when retinal vein occlusion was treated with (intravenous) SK within two weeks of symptom onset. HATTENBACH et al. (1999) studied 58 patients with central retinal vein occlusion. Results were poor in patients admitted >11 days after onset of symptoms. In 45 patients with a duration of symptoms of <11 days, 23 were treated with 50 mg tPA i.v. and heparin and 22 with hemodilution. An advancement of 2 or more lines on the logarithmic visual acuity chart was found in 44% of patients in the tPA group but only in 14% of patients in the hemodilution group. FREITAG et al. (1993) have noted that partial recovery of form vision was possible in some patients with acute retinal artery occlusion up to 33h after symptom onset. This confirmed a similar previous experience with microcatheter delivery of UK (Schmidt et al. 1992). Weber et al. (1998) administered UK (100000–900000 IU) through a microcatheter into the ophthalmic artery over 10–90min in 17 patients with acute central retinal artery occlusion. Three patients recovered completely and regained visual acuity. Two additional patients showed a marked and six a slight improvement of visual acuity. In six patients thrombolytic therapy had no effect. Compared with a historical control series of 15 patients, UK-treated patients fared better ($p = 0.01$). Richard et al. (1999) treated 53 patients with central artery occlusion (n-46) or branch arterial occlusion ($n = 7$) with 10 to 20mg of tPA delivered through a microcatheter inserted in the ophthalmic artery. At 3 months visual acuity had improved in 66% of patients. No prospective, placebo-controlled trials have been performed in acute retinal vascular occlusion.

2. Dural Sinus Thrombosis

Dural sinus thrombosis is associated variably with headache, altered mental status, focal neurologic deficits, seizures, and death. Management has varied from observation to aggressive treatment. Despite anecdotal reports (Tsai et al. 1992; Higashida et al. 1994; Smith et al. 1994; Horowitz et al. 1995), the delivery of plasminogen activators for severe symptomatic cerebral vein thrombosis has not been studied prospectively. Horowitz et al. (1995) treated 13 patients with UK by selective catheterization via the jugular vein, which was accessed via the femoral vein. Patency of the sinus was achieved in 12/13 patients (92%) all of whom had good or excellent clinical outcome. Tsai et al. (1992) infused lower doses of UK directly into the occluded sinus of five patients, all of whom recovered completely without any residual deficits. Similar results were achieved in 6 patients reported by Philips et al. (1999). Smith et al. (1994) treated seven patients with symptoms for one week to six months; UK infusions ranging from 88h to 244h were required, but all patients improved. FREY et al. (1999) treated 12 patients with tPA delivered into the thrombus by microcatheter. Flow could be restored in 9 patients.

D. Ongoing Stroke Trials with Thrombolytic Agents

In pilot sudies it was shown that transcranial Doppler ultrasound can accelerate tPA-mediated thrombolysis (Behrens et al. 1999; Alexandrov et al. 2000). Further studies on this promising approach are ongoing.

The AUST was designed to test the hypothesis that intra-arterial UK plus anticoagulants in acute basilar artery occlusion will reduce morbidity and mortality at six months compared with the administration of anticoagulants alone. Two hundred patients from eight centers in Australia were to be enrolled.

E. Conclusion

Experience with fibrinolysis as a treatment for acute cerebrovascular ischemia indicates that recanalization of carotid and vertebrobasilar territory occlusion is technically feasible within 3–6 h of symptom onset. A composite of prospective studies suggests that intra-arterial direct infusion of a plasminogen activator can produce a substantially greater recanalization frequency than systemic intravenous delivery. Complete occlusions of the cervical ICA by in situ thrombosis appear more resistant than occlusions of the stem and major branches of the MCA to thrombolysis. The optimal plasminogen activator, its dose-rate, and delivery system have yet to be defined in either territory. Three phase III trials (ECASS, ECASS-II, and NINDS) indicate the range of outcomes achievable and central risks to the use of plasminogen activators in the setting of acute focal cerebral ischemia. Any potential benefit with tPA in a stroke population was nullified by the treatment of patients with subtle signs of ischemic injury on CT scan (ECASS). Treatment of acute ischemic stroke patients within 3 h of symptom onset was associated with improvement in outcome (NINDS). Phase III trials of intravenous SK in carotid territory stroke have suggested increased mortality in some patients possibly related to stroke severity and excessive SK dosage. Hemorrhagic transformation invariably accompanies the use of fibrinolytic agents and is significantly increased by delayed intervention, diastolic hypertension, dose, and presence of substantial ischemia on initial CT scan. Future efforts will be directed to the reduction of these risk factors.

Acknowledgements. This work has been supported in part by grant R01 NS 26945 of the National Institutes of Health.

List of Abbreviations and Acronyms

ASA	acetylsalicylic acid (aspirin)
ASK	Australian SK trial
ATLANTIS	Alteplase Thrombolysis for Acute Noninterventional Therapy in Ischemic Stroke
AUST	Australian Urokinase Stroke Trial
DSA	digital subtraction angiography
ECASS	European Cooperative Acute Stroke Study
EMS	Emergency Management of Stroke (bridging trial)
HI	hemorrhagic infarction
HMCAS	hyperdense middle cerebral artery sign
HMPAO	hexamethylpropyleneamine oxime
HSS	hemispheric stroke scale
ICA	internal carotid artery

ICH	intracerebral hemorrhage
IST	International Stroke Trial
MAST	Multicenter Acute Stroke Trial
MCA	middle cerebral artery
MIU	Mega (1 million) international units
MU	Mega (1 million) units
NIHSS	National Institute of Health Stroke Scale
NINDS	National Institute of Neurological Disorders and Stroke
PET	positron emission tomography
PH	parenchymal hematoma
PROACT	Prolyse (rpro-UK) for Acute Cerebral Thromboembolism
rCBF	regional cerebral blood flow
sc-uPA	single chain uPA, also called pro-urokinase
SK	streptokinase
SPECT	single photon emission computed tomography
STARS	the Standard Treatment with Alteplase to Reverse Stroke
tPA	tissue-type plasminogen activator (abbr. also used for recombinant tPA)
UK	urokinase
uPA	urokinase-type plasminogen activator, also called urokinase

References

Abe T, Kazama M, Naito I, Ueda M, Tanaka T, Higuchi H, Saso S, Kariyone S, Yoshimura M, Maekawa T, Morimatsu M, Moriyasu N, Takagi S, Maezawa H, Kase M, Kuwana N, Odakura T, Takasu K, Sugawara Y, Takahashi A, Okubo M, Ijichi H, Yamamoto S, Kikuchi H, Kito S, Tokito S, Iwasaki Y (1981a) Clinical evaluation for efficacy of tissue culture urokinase (TCUK) on cerebral thrombosis by means of multicenter double-blind study. Blood Vessel 12:321–341

Abe T, Kazawa M, Naito I, Kanaya H, Machino R, Igata A, Tazaki Y, Kameyama M, Goto F, Maezawa H, Watanabe H, Urushiyama K, Togi H, Sakuragawa N, Atarashi J, Kito S, Araki G, Onodera H, Otomo E, Ogawa N (1981b) Clinical effect of urokinase (60000 units/day) on cerebral infarction – Comparative study by means of multiple center double blind test. Blood Vessel 12:342–358

Abe T, Machii K, Kajiwara N, Kawai C, Kanmatsuse K, Kanbara H, Nobuyoshi M, Yamaguchi T, Osada H, Kawada H, Matsuo O (1989) Clinical usefulness of intravenous application of SM-9527 (duteplase; t-PA) in acute myocardial infarction: A multicentered, double-blind comparative study with intravenous administration of urokinase. Clin Eval 17:485–515

Abe T, Terashi A, Tohgi H, Sasoh S, Naito I (1990) Clinical efficacy of intravenous administration of SM-9527 (t-PA) in cerebral thrombosis. Clin Eval 18:39–69

Adams HP, Brott TG, Furlan AJ, Gomez CR, Grotta J, Helgason CM, Kwiatkowski T, Lyden PD, Marler JR, Torner J, Feinberg W, Mayberg M, Thies W (1996) Guidelines for thrombolytic therapy for acute stroke: a supplement to the guidelines for the management of patients with acute ischemic stroke. A statement for healthcare professionals from a special writing group of the stroke council, American Heart Association. Stroke 27:1711–1718

Albers GW, Bates VE, Clark WM, Bell R, Verro P, Hamilton SA (2000) Intravenous tissue-type plasminogen activator for treatment of acute stroke: the Standard Treatment with Alteplase to Reverse Stroke (STARS) study. JAMA 283:1145–1150

Aldrich MS, Sherman SA, Greenberg HS (1985) Cerebrovascular complications of streptokinase infusion. JAMA 253:1777–1779

Alexandrov AV, Demchuk AM, Felberg RA, Christou I, Barber PA, Burgin WS, Malkoff M, Wojner AW, Grotta JC (2000) High rate of complete recanalization and dramatic clinical recovery during tPA infusion when continuously monitored with 2-MHz transcranial doppler monitoring. Stroke 31:610–614

Alexandrov AV, Masdeu JC, Devous MD, Black SE, Grotta JC (1997) Brain single-photon emission CT with HMPAO and safety of thrombolytic therapy in acute ischemic stroke. Proceedings of the meeting of the SPECT Safe Thrombolysis Study Collaborators and the members of the Brain Imaging Council of the Society of Nuclear Medicine. Stroke 28:1830–1834

Araki G, Minakami K, Mihara H (1973) Therapeutic effect of urokinase on cerebral infarction. Rinsho to Kenkyu 50:3317–3326

Archer CT, Horenstein S (1977) Basilar artery occlusion. Clinical and radiological correlation. Stroke 8:383–387

Asplund K, Wester P, Fodstad H (1980) Long time survival after vertebral/basilar occlusion. Stroke 11:304

Astrup J, Siesjö BK, Symon L (1981) Thresholds in cerebral ischemia – the ischemic penumbra. Stroke 12(6):723–725

Astrup J, Symon L, Branston NM, Lassen N (1977) Cortical evoked potential and extracellular K+ and H+ at critical levels of brain ischemia. Stroke 8:51–57

Atarashi J, Otomo E, Araki G, Itoh E, Togi H, Matsuda T (1985) Clinical utility of urokinase in the treatment of acute stage of cerebral thrombosis: Multi-center double-blind study in comparison with placebo. Clin Eval 13:659–709

Atkin N, Nitzberg S, Dorsey J (1964) Lysis of intracerebral thromboembolism with fibrinolysin. Report of a case. Angiology 15:346–439

Babikian VL, Kase CS, Pessin MS, Norrving B, Gorelick PB (1989) Intracerebral hemorrhage in stroke patients anticoagulated with heparin. Stroke 29:1500–1503

Baird AE, Donnan GA, Austin MC, Fitt GJ, Davis SM, McKay WJ (1994) Reperfusion after thrombolytic therapy in ischemic stroke measured by single-photon emission computed tomography. Stroke 25:79–85

Barnwell SL, Clark WM, Nguyen TT, O'Neill OR, Wynn ML, Coull BM (1994) Safety and efficacy of delayed intraarterial urokinase therapy with mechanical clot disruption for thromboembolic stroke. Am J Neuroradiol 15:1817–1822

Baron JC, von Kummer R, del Zoppo GJ (1995) Treatment of acute ischemic stroke. Challenging the concept of a rigid and universal time window (Editorial). Stroke 26:2219–2221

Barr JD, Horowitz MB, Mathis JM, Sclabassi RJ, Yonas H (1995) Intraoperative urokinase infusion for embolic stroke during carotid endarterectomy. Neurosurgery 36:606–611

Becker KJ, Monsein LH, Ulatowski J, Mirski M, Williams M, Hanley DF (1996) Intraarterial thrombolysis in vertebrobasilar occlusion. AJNR Am J Neuroradiol 17:255–262

Behrens S, Daffertshofer M, Spiegel D, Hennerici M (1999) Low-frequency, low-intensity ultrasound accelerates thrombolysis through the skull. Ultrasound Med Biol 25:269–273

Biedert S, Betz H, Reuther R (1987) Directional C-W Doppler sonography in the diagnosis of basilar artery disease. Stroke 18:101–107

Biemond A (1951) Thrombosis of the basilar artery and the vascularization of the brainstem. Brain 74:300–317

Bogousslavsky J, Regli F (1985) Vertebrobasilar transient ischemic attacks in internal carotid artery occlusion or tight stenosis. Arch Neurol 42:64–68

Brandt T, von Kummer R, Müller-Küppers M, Hacke W (1996) Thrombolytic therapy of acute basilar artery occlusion. Variables affecting recanalization and outcome. Stroke 27:875–881

Brott T, Adams HP, Olinger CP, et al (1989) Measurements of acute cerebral infarction: A clinical examination scale. Stroke 20:864–870

Brott TG, Haley EC Jr, Levy DE, Barsan W, Broderick J, Sheppard GL, Spilker J, Kongable GL, Massey S, Reed R, Marler JR (1992) Urgent therapy for stroke. Part I. Pilot study of tissue plasminogen activator administered within 90 min. Stroke 23:632–640

Brott T, Lu M, Kothari R, Fagan SC, Frankel M, Grotta JC, Broderick J, Kwiatkowski T, Lewandowski C, Haley EC Jr, Marler JR, Tilley BC, for the NINDS rt-PA Stroke Study Group (1998) Hypertension and its treatment in the NINDS rt-PA Stroke Trial. Stroke 29:1504–1509

Caplan LR (1979) Occlusion of the vertebral or basilar artery: Follow-up analysis of some patients with benign outcome. Stroke 10:277–282

Caplan LR (1983) Bilateral distal vertebral artery occlusion. Neurology (Minneap) 33:552–558

Caplan LR (1986) Vertebrobasilar occlusive disease. In: Barnett HS, Mohr J, Stein B, Yatsu F (eds) Stroke: pathophysiology, diagnosis and management. Churchill Livingstone, New York, pp 549–619

Castaigne F, Lhermitte F, Gautier JC, Escouroller R, Derouesne C, der Agopain P, Popa C (1974) Arterial occlusions in the vertebro-basilar system. A study of 44 patients with post-mortem data. Brain 96:133–154

Chiu D, Krieger D, Villar-Cordova C, Kasner SE, Morgenstern LB, Bratina PL, Yatsu FM, Grotta JC (1998) Intravenous tissue plasminogen activator for acute ischemic stroke: feasibility, safety, and efficacy in the first year of clinical practice. Stroke 29:18–22

Ciccone A, Motto C, Aritzu E, Piana A, Candelise L, on behalf of the MAST-I Collaborative Group (1998) Risk of aspirin use plus thrombolysis after acute ischemic stroke: a further MAST-I analysis. Lancet 352:880

Clark WM, Albers GW, Madden KP, Hamilton S, for the Thromblytic Therapy in Acute Ischemic Stroke Study Investigators (2000) The rtPA (alteplase) 0- to 6-hour acute stroke trial, part A (A0276g): results of a double-blind, placebo-controlled, multicenter study. Stroke 31:811–816

Clark WM, Wissman S, Albers GW, Jhamandas JH, Madden KP, Hamilton S (1999) Recombinant tissue-type plasminogen activator (Alteplase) for ischemic stroke 3 to 5 hours after symptom onset. JAMA 282:2019–2026

Clarke RL, Cliffton EE (1960) The treatment of cerebrovascular thrombosis and embolism with fibrinolytic agents. Am J Cardiol 30:546–551

Coté R, Hachinski VC, Shurtell BL, Norris JW, Wolfson C (1986) The Canadian Neurological Scale: A preliminary study in acute stroke. Stroke 17:731–737

Cross DT, Moran CJ, Akins PT, Angtuaco EE, Diringer MN (1997) Relationship between clot location and outcome after basilar artery thrombolysis. AJNR 18:1221–1228

del Zoppo GJ (1988) Thrombolysis: New concepts in the treatment of stroke. In: Hennerici M, Sitzer G, Weger H-D (eds) Carotid artery plaques. Karger, Basel, pp 247–272

del Zoppo GJ, Ferbert A, Otis S, Brückmann H, Hacke W, Zyroff J, Harker LA, Zeumer H (1988) Local intra-arterial fibrinolytic therapy in acute carotid territory stroke: A pilot study. Stroke 19:307–313

del Zoppo GJ, Higashida RT, Furlan AJ, Pessin MS, Rowley HA, Gent M, and the PROACT Investigators (1998) PROACT: a phase II randomized trial of recombinant pro-urokinase by direct arterial delivery in acute middle cerebral artery stroke. Stroke 29:4–11

del Zoppo GJ, Mori E (1992) Hematologic causes of intracerebral hemorrhage and their treatment. In: Batjer HH (ed) Spontaneous intracerebral hemorrhage. Neurosurg Clin North Am 3:637–658

del Zoppo GJ, Pessin MS, Mori E, Hacke W (1991) Thrombolytic intervention in acute thrombotic and embolic stroke. Semin Neurol 11:368–384

del Zoppo GJ, Poeck K, Pessin MS, Wolpert SM, Furlan AJ, Ferbert A, Alberts MJ, Zivin JA, Wechsler L, Busse O, Greenlee R Jr, Brass L, Mohr JP, Feldmann E, Hacke W, Kase CS, Biller J, Gress D, Otis SM (1992) Recombinant tissue plasminogen activator in acute thrombotic and embolic stroke. Ann Neurol 32:78–86 del Zoppo GJ, Zeumer H, Harker LA (1986) Thrombolytic therapy in acute stroke: Possibilities and hazards. Stroke 17:595–607

Donnan GA, Davis SM, Chambers BR, Gates PC, Hankey GJ, McNeil JJ, Rosen D, Stewart-Wynne EC, Tuck RR (1995) Trials of streptokinase in severe acute ischemic stroke. Lancet 345:578–579

Donnan GA, Davis SM, Chambers BR, Gates PC, Hankey GJ, Mcneil JJ, Rosen D, Stewart-Wynne EG, Tuck RR, for the Australian Streptokinase (ASK) Trial Study Group (1996) Streptokinase for acute ischemic stroke with relationship to time of administration. JAMA 276:961–966

Drake ME, Shin C (1983) Conversion of ischemic to hemorrhagic infarction by anti-coagulant administration. Report of two cases with evidence from serial computed tomographic brain scans. Arch Neurol 40:44–46

Endo S, Kuwayama N, Hirashima Y, Akai T, Nishijima M, Takaku A (1998) Results of urgent thrombolysis in patients with major stroke and atherothrombotic occlusion of the cervical internal carotid artery. AJNR Am J Neuroradiol 19:1169–1175

Fagan SC, Morgenstern LB, Petitta A, Ward RE, Tilley BC, Marler JR, Levine SR, Broderick JP, Kwiatkowski TG, Frankel M, Brott TG, Walker MD (1998) Cost-effectiveness of tissue plasminogen activator for acute ischemic stroke. NINDS rt-PA Stroke Study Group. Neurology 50:883–890

Fenstermacher J, Nakata H, Tajima A, Lin S-Z, Otsuka T, Acuff V, Wei L, Bereczki D (1991) Functional variations in parenchymal microvascular systems within the brain. Magn Reson Med 19:217–220

Fields WS, Ratinov G, Weibel J, Campos RJ (1966) Survival following basilar artery occlusion. Arch Neurol 15:463–471

Fieschi C, Argentino C, Lenzi GL, Sacchetti ML, Toni D, Bozzao L (1989) Clinical and instrumental evaluation of patients with ischemic stroke within the first six hours. J Neurol Sci 91:311–321

Fieschi C, Bozzao L (1969) Transient embolic occlusion of the middle cerebral and internal carotid arteries in cerebral apoplexy. J Neurol Neurosurg Psychiatry 32:236–240

Fisher CM (1970) Occlusion of the vertebral arteries causing transient basilar symptoms. Arch Neurol 22:13–19

Fisher CM (1976) The natural history of middle cerebral artery trunk occlusion. In: Austin GM (ed) Microneurosurgical anastomoses for cerebral ischemia. Charles C. Thomas, Springfield, pp 146–154

Fisher CM (1977) Bilateral occlusion of basilar artery branches. J Neurol Neurosurg Psychiatry 40:1182–1189

Fisher CM, Adams RD (1951) Observations on brain embolism with special reference to the mechanism of hemorrhagic infarction. J Neuropathol Exp Neurol 10:92–94

Fisher CM, Adams RD (1987) Observations on brain embolism with special reference to hemorrhage infarction. In: Furlan AJ (ed) The heart and stroke: exploring mutual cerebrovascular and cardiovascular issues. Springer, Berlin Heidelberg New York, pp 17–36

Fletcher AP, Alkjaersig N, Lewis M, Tulevski V, Davies A, Brooks JE, Mardin WB, Landau WM, Raichle ME (1976) A pilot study of urokinase therapy in cerebral infarction. Stroke 7:135–142

Freitag HJ, Becker VU, Thie A, Tilsner V, Philapitsch A, Schwarz HP, Webhof U, Müller A, Zeumer H (1996) Lys-plasminogen as an adjunct to local intra-arterial fibrinolysis for carotid territory stroke – laboratory and clinical findings. Neuroradiology 38:181–185

Freitag H-J, Zeumer H, Knospe V (1993) Acute central retinal artery occlusion and the role of thrombolysis. In: del Zoppo GJ, Mori E, Hacke W (eds) Thrombolytic therapy in acute ischemic stroke II. Springer, Berlin Heidelberg New York, pp 103–105

Frey JL, Muro GJ, McDougall CG, Dean BL, Jahnke HK (1999) Cerebral venous thrombosis: combined intrathrombus rtPA and intravenous heparin. Stroke 30: 489–494

Furlan AJ, Cavalier SJ, Hobbs RE, Weinstein MA, Modic MI (1982) Hemorrhage and anticoagulation after nonseptic embolic brain infarction. Neurology 32:280–282

Furlan A, Higashida R, Wechsler L, Gent M, Rowley H, Kase C, Pessin M, Ahuja A, Callahan F, Clark WM, Silver F, Rivera F, for the PROACT Investigators (1999) Intra-arterial prourokinase for acute ischemic stroke. The PROACT II study: a randomized controlled trial. JAMA 282:2003–2011

Garcia JH, Yoshida Y, Chen H, Li Y, Zhang ZG, Liam J, Chen S, Chopp M (1993) Progression from ischemic injury to infarct following middle cerebral artery occlusion in the rat. Am J Pathol 142:623–635

Gönner F, Remonda L, Mattle H, Sturzenegger M, Ozdoba C, Lövblad KO, Baumgartner R, Bassetti C, Schroth G (1998) Local intra-arterial thrombolysis in acute ischemic stroke. Stroke 29:1894–1900

Grond M, Rudolf J, Schmulling S, Stenzel C, Neveling M, Heiss WD (1998a) Early intravenous thrombolysis with recombinant tissue-type plasminogen activator in vertebrobasilar ischemic stroke. Arch Neurol 55:466–469

Grond M, Stenzel C, Schmülling S, Rudolf J, Neveling M, Lechleuthner A, Schneweis S, Heiss W-D (1998b) Early intravenous thrombolysis for acute ischemic stroke in a community-based approach. Stroke 29:1544–1549

Grotta JC, Alexandrov AV (1998) tPA-associated reperfusion after acute stroke demonstrated by SPECT. Stroke 29:429–432

Gruppo Italiano Per Lo Studio Della Streptochinasi Nell'Infarto Miocardico (GISSI) (1986) Effectiveness of intravenous thrombolytic treatment in acute myocardial infarction. Lancet 1:397–401

Gruppo Italiano Per Lo Studio Della Streptochinasi Nell'Infarto Miocardico (GISSI) (1990) GISSI-2: A factorial randomized trial of alteplase versus streptokinase and heparin versus no heparin among 12 490 patients with acute myocardial infarction. Lancet 336:65–71

Gurwitz JH, Gore JM, Goldberg RJ, Barron HV, Breen T, Rundle AC, Sloan MA, French W, Rogers WJ, for the Participants in the National Registry of Myocardial Infarction 2 (1998) Risk for intracranial hemorrhage after tissue plasminogen activator treatment for acute myocardial infarction. Ann Intern Med 129:597–604

Hacke W, Bluhmki E, Steiner T, Tatlisumak T, Mahagne M-H, Sacchetti M-L, Meier D, for the ECASS Study Group (1998a) Dichotomized efficacy end points and global end-point analysis applied to the ECASS intention-to-treat data set: post hoc analysis of ECASS I. Stroke 29:2073–2075

Hacke W, Kaste M, Fieschi C, Toni D, Lesaffre E, von Kummer R, Boysen G, Bluhmki E, Höxter G, Mahagne M-H, Hennerici M, for the ECASS Study Group (1995) Intravenous thrombolysis with recombinant tissue plasminogen activator for acute hemispheric stroke. The European Cooperative Acute Stroke Study (ECASS). JAMA 274:1017–1025

Hacke W, Kaste M, Fieschi C, von Kummer R, Davalos A, Meier D, Larrue V, Bluhmki E, Davis S, Donnan G, Schneider D, Diez-Tejedor E, Trouillas P, for the Second European-Australasian Acute Stroke Study Investigators (1998b) Randomised double-blind placebo-controlled trial of thrombolytic therapy with intravenous alteplase in acute ischemic stroke (ECASS II). Lancet 352:1245–1251

Hacke W, Zeumer H, Ferbert A, Brückmann H, del Zoppo GJ (1988) Intra-arterial thrombolytic therapy improves outcome in patients with acute vertebrobasilar occlusive disease. Stroke 19:1216–1222

Haley EC Jr, Brott TG, Sheppard GL, Barsan W, Broderick J, Marler JR, Kongable GL, Spilker J, Massey S, Hansen CA, Torner JC (1993) Pilot randomized trial of tissue plasminogen activator in acute ischemic stroke. Stroke 24:1000–1004

Haley EC Jr, Levy DE, Brott TG, Sheppard GL, Wong MCW, Kongable GL, Torner JC, Marler JR (1992) Urgent therapy for stroke. Part II. Pilot study of tissue plasminogen activator administered 91–180 minutes from onset. Stroke 23:641–645

Hanaway J, Torack R, Fletcher AP, Landau WM (1976) Intracranial bleeding associated with urokinase therapy for acute ischemic hemispheral stroke. Stroke 7:143–146

Hantson L, De Weerdt W, De Keyser J, Diener HC, Franke C, Palm R, Van Orshoven M, Schoonderwalt H, De Klippel N, Herroelen L, Feys H (1994) The European Stroke Scale. Stroke 25:2215–2219

Hattenbach L-O, Wellermann G, Steinkamp GWK, Scharrer I, Koch FHJ, Ohrloff C (1999) Visual outcome after treatment with low-dose recombinant tissue plasminogen activator or hemodilution in ischemic central retinal vein occlusion. Ophthalmologica 213:360–366

Herderscheê D, Limburg U, Hijdra A, Koster PA (1991) Recombinant tissue plasminogen activator in two patients with basilar artery occlusion. J Neurol Neurosurg Psychiatry 54:71–73

Herndon RM, Meyer JS, Johnson JF, Landers J (1960) Treatment of cerebrovascular thrombosis with fibrinolysin. Preliminary report. Am J Cardiol 30:540–545

Higashida RT, Halbach VV, Barnwell SL, Dowd CF, Hieshima GB (1994) Thrombolytic therapy in acute stroke. J Endovasc Surg 1:4–15

Hommel M, Boissel JP, Cornu C, Boutitie F, Lees KR, Bessorn G, Leys D, Amarenio P, Bogart M, for the MAST Group (1995) Termination of trial of streptokinase in severe acute ischemic stroke (Letter). Lancet 345:57

Hornig CR, Dorndorf W, Agnoli AL (1986) Hemorrhagic cerebral infarction: A prospective study. Stroke 17:179–185

Horowitz M, Purdy P, Unwin H, Carstens G, Greenlee R, Hise J, Kopitnik T, Batjer H, Rollins N, Samson D (1995) Treatment of dural sinus thrombosis using selective catheterization and urokinase. Ann Neurol 38:58–67

Infeld B, Davis SM, Donnan GA, Lichtenstein M, Baird AE, Binns D, Mitchell PJ, Hopper JL (1996) Streptokinase increases luxury perfusion after stroke. Stroke 27:1524–1529

International Stroke Trial Collaborative Group (1997) The International Stroke Trial (IST): a randomised trial of aspirin, subcutaneous heparin, both, or neither among 19435 patients with acute ischemic stroke. Lancet 349:1569–1581

Irino T, Taneda M, Minami T (1977) Angiographic manifestations in post-recanalized cerebral infarction. Neurology 27:471–475

Jahan R, Duckwiler GR, Kidwell CS, Sayre JW, Gobin YP, Villablanca JP, Saver J, Starkman S, Martin N, Vinuela F (1999) Intraarterial thrombolysis for treatment of acute stroke: experience in 26 patients with long-term follow-up. AJNR Am J Neuroradiol 20:1291–1299

Jansen O, von Kummer R, Forsting M, Hacke W, Sartor K (1995) Thrombolytic therapy in acute occlusion of the intracranial internal carotid artery bifurcation. Am J Neuroradiol 16:1977–1986

Karp J, Hurtig H (1974) "Locked-in" state with bilateral midbrain infarcts. Arch Neurol 30:176–178

Kase CS, Pessin MS, Zivin JA, del Zoppo GJ, Furlan AJ, Buckley JW, Snipes RG, Littlejohn JK (1992) Intracranial hemorrhage after coronary thrombolysis with tissue plasminogen activator. Am J Med 92:384–390

Katzan IL, Furlan AJ, Lloyd LE, Frank JI, Harper DL, Hinchey JA, Hammel JP, Qu A, Sila CA (2000) Use of tissue-type plasminogen activator for acute ischemic stroke: the Cleveland area experience. JAMA 283:1151–1158

Koller RL (1982) Recurrent embolic cerebral infarction and anticoagulation. Neurology 32:283–285

Kubik CS, Adams RD (1946) Occlusion of the basilar artery: A clinical and pathologic study. Brain 69:73–121

Kwaan H-C (1988) Thromboembolic disorders of the eye. In: Comerota AC (ed) Thrombolytic therapy. Grune and Stratton, Orlando, pp 153–163

Kwiatkowski TG, Libman RB, Frankel M, Tilley BC, Morgenstern LB, Lu M, Broderick JP, Lewandowski CA, Marler JR, Levine SR, Brott T, for the National Institute of Neurological Disorders and Stroke Recombinant Tissue Plasminogen Activator Stroke Study Group (1999) Effects of tissue plasminogen activator for acute ischemic stroke at one year. N Engl J Med 340:1781–1787

Larrue V, von Kummer R, del Zoppo G, Bluhmki E (1997) Hemorrhagic transforma-
    tion in acute ischemic stroke. Potential contributing factors in the European Coop-
    erative Acute Stroke Study. Stroke 28:957–960
Lee BI, Lee BC, Park SC, Shon YH, Kim DI, Jung TS, Suh JH (1994) Intra-carotid
    thrombolytic therapy in acute ischemic stroke of carotid arterial territory. Yonsei
    Med J 35:49–61
Levine MN, Goldhaber SZ, Gore JM, Hirsh J, Califf RM (1995) Hemorrhagic compli-
    cations of thrombolytic therapy in the treatment of myocardial infarction and
    venous thromboembolism. Chest 108 [Suppl]:291S–301S
Levy DE, Brott TG, Haley EC Jr, Marler JR, Sheppard GL, Barsan W, Broderick JP
    (1994) Factors related to intracranial hematoma formation in patients receiving
    tissue-type plasminogen activator for acute ischemic stroke. Stroke 25:291–297
Lewandowski CA, Frankel M, Tomsick TA, Broderick J, Frey J, Clark W, Starkman S,
    Grotta J, Spilker J, Khoury J, Brott T, and the EMS Bridging Trial Investigators
    (1999) Combined intravenous and intra-arterial r-TPA versus intra-arterial
    therapy of acute ischemic stroke. Emergency Management of Stroke (EMS)
    Bridging Trial. Stroke 30:2598–2605
Loewen SC, Anderson BA (1988) Reliability of the modified motor assessment scale
    and the Barthel index. Phys Ther 68:1077–1081
Lyden P, Brott T, Tilley B, Welch KM, Mascha EJ, Levine S, Haley EC, Grotta J, Marler
    J for the NINDS TPA Stroke Study Group (1994) Improved reliability of the NIH
    Stroke Scale using video training. Stroke 25:2220–2226
Mahoney FJ, Barthel DW (1965) Functional evaluation. The Barthel index. Md Med J
    14:61–65
Masdeu JC, Brass LM (1995) SPECT imaging of stroke. J Neuroimaging 5 [Suppl
    1]:S14–S22
Matsumoto K, Satoh K (1991) Topical intraarterial urokinase infusion for acute stroke.
    In: Hacke W, del Zoppo GJ, Hirschberg M (eds) Thrombolytic therapy in acute
    ischemic stroke. Springer, Berlin Heidelberg New York, pp 207–212
Meyer JS, Gilroy J, Barnhart MI, Johnson JF (1963) Therapeutic thrombolysis in cere-
    bral thromboembolism. Neurology 13:927–937
Meyer JS, Gilroy J, Barnhart MI, Johnson JF (1964) Anticoagulants plus streptokinase
    therapy in progressive stroke. JAMA 189:373
Meyer JS, Herndon RM, Gotoh F, Tazaki Y, Nelson JN, Johnson JF (1961) Therapeu-
    tic thrombolysis. In: Millikan CH, Siekert RG, Whisnant JP (eds) Cerebral vascu-
    lar diseases, third Princeton conference. Grune and Stratton, New York, pp 160–177
Miyakawa T (1984) The cerebral vessels and thrombosis. Rinsho Ketsueki 25:
    1018–1026
Möbius E, Berg-Dammer E, Kühne D, Nahser HC (1991) Local thrombolytic therapy
    in acute basilar artery occlusion: Experience with 18 patients. In: Hacke W, del
    Zoppo GJ, Hirschberg M (eds) Thrombolytic therapy in acute ischemic stroke.
    Springer, Berlin Heidelberg New York, pp 213–215
Mori E (1991) Fibrinolytic recanalisation therapy in acute cerebrovascular throm-
    boembolism. In: Hacke W, del Zoppo GJ, Hirschberg M (eds) Thrombolytic
    therapy in acute ischemic stroke. Springer, Berlin Heidelberg New York, pp
    137–146
Mori E, Tabuchi M, Yoshida T, Yamadori A (1988) Intracarotid urokinase with throm-
    boembolic occlusion of the middle cerebral artery. Stroke 19:802–812
Mori E, Yoneda Y, Tabuchi M, Yoshida T, Ohkawa S, Ohsumi Y, Kitano K, Tsutsumi A,
    Yamadori A (1992) Intravenous recombinant tissue plasminogen activator in acute
    carotid artery territory stroke. Neurology 42:976–982
Multicentre Acute Stroke Trial-Italy (MAST-I) Group (1995) Randomised controlled
    trial of streptokinase, aspirin, and combination of both in treatment of acute
    ischemic stroke. Lancet 346:1509–1514
Nenci GG, Gresele P, Taramelli M, Agnelli G, Signorini E (1983) Thrombolytic therapy
    for thromboembolism of vertebrobasilar artery. Angiology 34:561–571

NIH Consensus Conference (1980) Thrombolytic therapy in treatment. Br Med J 280:1585–1587

Ogata J, Yutani C, Imakita M, Ishibashi-Ueda H, Saku Y, Minematsu K, Sawada T, Yamaguchi T (1989) Haemorrhagic infarct of the brain without a reopening of the occluded arteries in cardioembolic stroke. Stroke 20:876–883

Ohtaki M, Shinya T, Yamamura A, Minamida Y, et al (1993) Local fibrinolytic therapy using superselective catheterization for patients with acute cerebral embolism within 10 hours. In: Tomita M, Mchedlishvili A, Rosenblum W, Heiss W-D, Fukuuchi Y (eds) Microcirculatory stasis in the brain. Elsevier Science Publishers, Amsterdam, pp 539–545

Okada Y, Yamaguchi T, Minematsu K, Miyashita T, Sawada T, Sadoshima S, Fujishima M, Omae T (1989) Hemorrhagic transformation in cerebral embolism. Stroke 20:598–603

Otomo E, Araki G, Itoh E, Tohgi H, Matuda T, Atarashi J (1985) Clinical efficacy of urokinase in the treatment of cerebral thrombosis. Clinical Evaluation 13:711–751

Otomo E, Tohgi H, Hirai S, Terashi A, Takakura K, Araki G, Ito E, Matsuda T, Sawada T, Fujishine M, Ogawa Y (1988) Clinical efficacy of AK-124 (tissue plasminogen activator) in the treatment of cerebral thrombosis: Study by means of multi-center double blind comparison with urokinase. Yakuri To Chiryo 16:3775–3821

Overgaard K, Sperling B, Boysen G, Pedersen H, Gam J, Ellemann K, Karle A, Arlien-Søborg P, Olsen TS, Videback C, Knudsen JB (1993) Thrombolytic therapy in acute ischemic stroke. A Danish pilot study. Stroke 24:1439–1446

Pessin MS, Caplan LR (1986) Heterogeneity of vertebrobasilar occlusive disease. In: Kunze K, Zangemeister WM, Arlt A (eds) Clinical problems of brainstem disorders. Thieme, Stuttgart New York, pp 30–42

Pessin MS, del Zoppo GJ, Estol CJ (1990) Thrombolytic agents in the treatment of stroke. Clin Neuropharmacol 13:271–289

Pessin MS, Teal PA, Caplan LR (1991) Hemorrhagic infarction: guilt by association. AJNR 12:1123–1126

Philips MF, Bagley LJ, Sinson GP, Raps EC, Galetta SL, Zager EL, Hurst RW (1999) Endovascular thrombolysis for symptomatic cerebral venous thrombosis. J Neurosurg 90:65–71

Powers WJ (1993) The ischemic penumbra: Usefulness of PET. In: del Zoppo GJ, Mori E, Hacke W (eds) Thrombolytic therapy in acute ischemic stroke II. Springer, Berlin Heidelberg New York, pp 17–21

Rankin J (1957) Cerebral vascular accident in patients over the age of 60. II. Prognosis. Scott Med J 2:200–215

Richard G, Lerche R-C, Knospe V, Zeumer H (1999) Treatment of retinal arterial occlusion with local fibrinolysis using recombinant tissue plasminogen activator. Ophthalmology 106:768–773

Sackett DL (1983) Rules of evidence and clinical recommendations on the use of antithrombotic agents. Chest 89 [Suppl]:2S–3S

Saito I, Segawa H, Shiokawa Y, Taniguchi M, Tsutsumi K (1987) Middle cerebral artery occlusion: Correlation of computed tomography and angiography with clinical outcome. Stroke 18:863–868

Sasaki O, Takeuchi S, Koike T, Koizumi T, Tanaka R (1995) Fibrinolytic therapy for acute embolic stroke: intravenous, intracarotid, and intra-arterial local approaches. Neurosurgery 36:246–252

Scandinavian Stroke Study Group (1985) Multicenter trial of haemodilution in ischemic stroke – background and study protocol. Stroke 16:885–890

Schmidt D, Schumacher M, Wakhloo AK (1992) Microcatheter urokinase infusion in central retinal artery occlusion. Am J Ophthalmol 113:429–434

Schriger DL, Kalafut M, Starkman S, Krueger M, Saver JL (1998) Cranial computed tomography interpretation in acute stroke: physician accuracy in determining eligibility for thrombolytic therapy. JAMA 279:1293–1297

Skyhoj-Olsen T, Larsen B, Herring M, Skawer EB, Lassen NA (1983) Blood flow and vascular reactivity in collateral perfused brain tissue: Evidence of an ischemic penumbra. Stroke 14:332–341

Sloan MA, Price TR, Petito CK, Randall AMY, Solomon RE, Terrin ML, Gore J, Collen D, Kleiman N, Feit F, Babb J, Herman M, Roberts WC, Sopko G, Bovill E, Forman S, Knatterud GL, for the TIMI Investigators (1995) Clinical features and pathogenesis of intracerebral hemorrhage after rt-PA and heparin therapy for acute myocardial infarction: the thrombolysis in myocardial infarction (TIMI) II pilot and randomized clinical trial combined experience. Neurology 45:649–658

Smith TP, Higashida RT, Barnwell SL, Halbach VV, Dowd CF, Fraser KW, Teitelbaum GP, Hieshima GB (1994) Treatment of dural sinus thrombosis by urokinase infusion. AJNR Am J Neuroradiol 15:801–807

Sobel BE (1994) Intracranial bleeding, fibrinolysis, and anticoagulation. Causal connections and clinical implications. Circulation 90:2147–2152

Solis OJ, Roberson GR, Taveras JM, Mohr J, Pessin MS (1977) Cerebral angiography in acute cerebral infarction. Revist Interam Radiol 2:19–25

Steiner T, Bluhmki E, Kaste M, Toni D, Trouillas P, von Kummer R, Hacke W, for the ECASS Study Group (1998) The ECASS 3-hour cohort. Secondary analysis of ECASS data by time stratification. Cerebrovasc Dis 8:198–203

Sussman BJ, Fitch TSP (1958) Thrombolysis with fibrinolysin in cerebral arterial occlusion. JAMA 167:1705–1709

Tanne D, Bates VE, Verro P, Kasner SE, Binder JR, Patel SC, Mansbach HH, Daley S, Schultz LR, Karanjia PN, Scott P, Dayno JM, Vereczkey-Porter K, Benesch C, Book D, Coplin WM, Dulli D, Levine SR, and the t-PA Stroke Survey Group (1999) Initial clinical experience with IV tissue plasminogen activator for acute ischemic stroke: a multicenter survey. Neurology 53:424–427

The Multicenter Acute Stroke Trial-Europe Study Group (1996) Thrombolytic therapy with streptokinase in acute ischemic stroke. N Engl J Med 335:145–150

The National Institute of Neurological Disorders and Stroke tPA Stroke Study Group (1995) Tissue plasminogen activator for acute ischemic stroke. N Engl J Med 333: 1581–1587

Theron J, Courtheoux P, Casaseo A, Alachkar F, Notari F, Garen F, Maiza D (1989) Local intra-arterial fibrinolysis in the carotid territory. AJNR 10:753–765

Tognoni G, Roncaglioni MC (1995) Dissent: an alternative interpretation of MAST-I. Lancet 346:1515

Toni D, Fiorelli M, Bastianello S, Sacchetti ML, Sette G, Argentino C, Montinaro E, Bozzao L (1996) Hemorrhagic transformation of brain infarct: predictability in the first 5 hours from stroke onset and influence on clinical outcome. Neurology 46: 341–345

Trouillas P, Nighoghossian N, Getenet J-C, Riche G, Neuschwander P, Froment J-C, Turjman F, Jin J-X, Malicier D, Fournier G, Gabry AL, Ledoux X, Derex L, Berthezène Y, Adeleine P, Xie J, Ffrench P, Dechavanne M (1996) Open trial of intravenous tissue plasminogen activator in acute carotid territory stroke. Correlations of outcome with clinical and radiological data. Stroke 27:882–890

Tsai FY, Higashida RT, Matovich V, Alfieri K (1992) Acute thrombosis of the intracranial dural sinus: direct thrombolytic treatment. AJNR Am J Neuroradiol 13: 1137–1141

Ueda T, Hatakeyama T, Kumon Y, Sakaki S, Uraoka T (1994) Evaluation of risk of hemorrhagic transformation in local intra-arterial thrombolysis in acute ischemic stroke by initial SPECT. Stroke 25:298–303

von Kummer R, Forsting M, Sartor K, Hacke W (1991) Intravenous recombinant tissue plasminogen activator in acute stroke. In: Hacke W, del Zoppo GJ, Hirschberg M (eds) Thrombolytic therapy in acute ischemic stroke. Springer, Berlin Heidelberg New York, pp 161–167

von Kummer R, Hacke W (1992) Safety and efficacy of intravenous tissue plasminogen activator and heparin in acute middle cerebral artery. Stroke 23:646–652

Wardlaw JM, Warlow CP (1992) Thrombolysis in acute ischemic stroke: does it work? Stroke 23:1826–1839

Wardlaw JM, Warlow CP, Counsell C (1997) Systematic review of evidence on thrombolytic therapy for acute ischemic stroke. Lancet 350:607–614

Weber J, Remonda L, Mattle HP, Koerner U, Baumgartner RW, Sturzenegger M, Ozdoba C, Koerner F, Schroth G (1998) Selective intra-arterial fibrinolysis of acute central retinal artery occlusion. Stroke 29:2076–2079

Wijdicks EF, Nichols DA, Thielen KR, Fulgham JR, Brown RD Jr, Meissner I, Meyer FB, Piepgras DG (1997) Intra-arterial thrombolysis in acute basilar artery thromboembolism: the initial Mayo Clinic experience. Mayo Clin Proc 72:1005–1013

Wildermuth S, Knauth M, Brandt T, Winter R, Sartor K, Hacke W (1998) Role of CT angiography in patient selection for thrombolytic therapy in acute hemispheric stroke. Stroke 29:935–938

Wolpert SM, Bruckmann H, Greenlee R, Wechsler L, Pessin MS, del Zoppo GJ, and the rt-PA Acute Stroke Study Group (1993) The neuroradiologic evaluation of patients with acute stroke treated with recombinant tissue plasminogen activator. AJNR 14:3–13

Yamaguchi T (1991) Intravenous rt-PA in acute embolic stroke. In: Hacke W, del Zoppo GJ, Hirschberg M (eds) Thrombolytic therapy in acute ischemic stroke. Springer, Berlin Heidelberg New York, pp 168–174

Yamaguchi T (1993) Intravenous tissue plasminogen activator in acute thromboembolic stroke: A placebo-controlled, double-blind trial. In: del Zoppo GJ, Mori E, Hacke W (eds) Thrombolytic therapy in acute ischemic stroke. Springer, Berlin Heidelberg New York, pp 59–65

Yamaguchi T (1995) The Japanese tPA trial. In: Yamaguchi T, Mori E, Minematsu K, del Zoppo GJ (eds) Thrombolysis in acute ischemic stroke III. Springer, Tokyo

Yamaguchi T, Hayakawa T, Kiuchi H (1993) Intravenous tissue plasminiogen activator ameliorates the outcome of hyperacute embolic stroke. Cerebrovasc Dis 3:269–272

Yamaguchi T, Minematsu K, Choki J, Ikeda M (1984) Clinical and neuroradiological analysis of thrombotic and embolic cerebral infarction. Jpn Circ J 48:50–58

Yasaka M, O'Keefe GJ, Chambers BR, Davis SM, Infeld B, O'Malley H, Baird AE, Hirano T, Donnan GA (1998) Streptokinase in acute stroke: effect on reperfusion and recanalization. Australian Streptokinase Trial Study Group. Neurology 50:626–632

Zeumer H (1985) Survey of progress: Vascular recanalizing techniques in interventional neuroradiology. J Neurol 231:287–294

Zeumer H, Freitag H-J, Zarella F, Thie A, Arnig C (1993) Local intra-arterial fibrinolytic therapy in patients with stroke: Urokinase vs recombinant tissue plasminogen activator (rt-PA). Neuroradiology 35:159–162

Zeumer H, Freitag HJ, Grzyka U, Neunzig HP (1989) Local intra-arterial fibrinolysis in acute vertebrobasilar occlusion. Technical developments and recent results. Neuroradiology 31:336–340

Zeumer H, Hacke W, Kolmann HL, Poeck K (1982) Lokale Fibrinolyse bei Basilaristhrombose. Dtsch Med Wochenschr 107:728–731

Zeumer H, Hacke W, Ringelstein EB (1983) Local intraarterial thrombolysis in vertebrobasilar thromboembolic disease. AJNR 4:401–404

Thrombolytic Agents in Development,
Biochemistry, Pharmacology –
Efficacy in Animal Experiments
and First Clinical Trials

CHAPTER 16
Staphylokinase

H.R. LIJNEN and D. COLLEN

A. Introduction

Thrombolytic therapy, consisting of the intravenous administration of plasminogen activators, has become an important therapeutic approach in patients with thromboembolic disease. Thrombolytic agents that are approved for clinical use or are under clinical investigation may be classified as "fibrin-specific" or "non-fibrin-specific" (COLLEN and LIJNEN 1991). Non-fibrin-specific plasminogen activators such as streptokinase (SK), anisoylated plasminogen streptokinase activator complex (APSAC), and two-chain urokinase-type plasminogen activator (tc-uPA, urokinase) activate both circulating and fibrin-bound plasminogen, inducing extensive systemic activation of the fibrinolytic system. Extensive systemic activation of plasminogen (plasma concentration 1.5–2 μmol/l) to plasmin, will result in depletion of the physiological plasmin inhibitor, α_2-antiplasmin (plasma concentration 1 μmol/l). Excess plasmin may degrade several plasma proteins including fibrinogen, factor V, and factor VIII, and induce the so-called "lytic state" (see Chap. 7). In contrast, fibrin-specific plasminogen activators such as the physiological molecules tissue-type plasminogen activator (tPA) (see Chap. 8), single-chain urokinase-type plasminogen activator (sc-uPA) (see Chap. 9), the bacterial plasminogen activator staphylokinase (SAK), as well as the Vampire bat plasminogen activator (see Chap. 17), preferentially activate plasminogen at the fibrin surface. Once formed, plasmin associated with the fibrin clot is protected from rapid inhibition by α_2-antiplasmin and may thus efficiently degrade the fibrin of a thrombus. These molecular interactions are schematically represented in Fig. 1.

The fibrin-specific mechanism of action of the physiological plasminogen activators has triggered great interest in their use for thrombolytic therapy, based on the premise that fibrin-specific activators would have a higher efficacy for coronary patency translating into a higher reduction of mortality as compared with non-fibrin-specific plasminogen activators. This "open artery hypothesis" was conclusively confirmed by the Global Utilization of SK and tPA for Occluded Coronary Arteries (GUSTO) trial and its angiographic substudy in patients with acute myocardial infarction, showing that early and per-

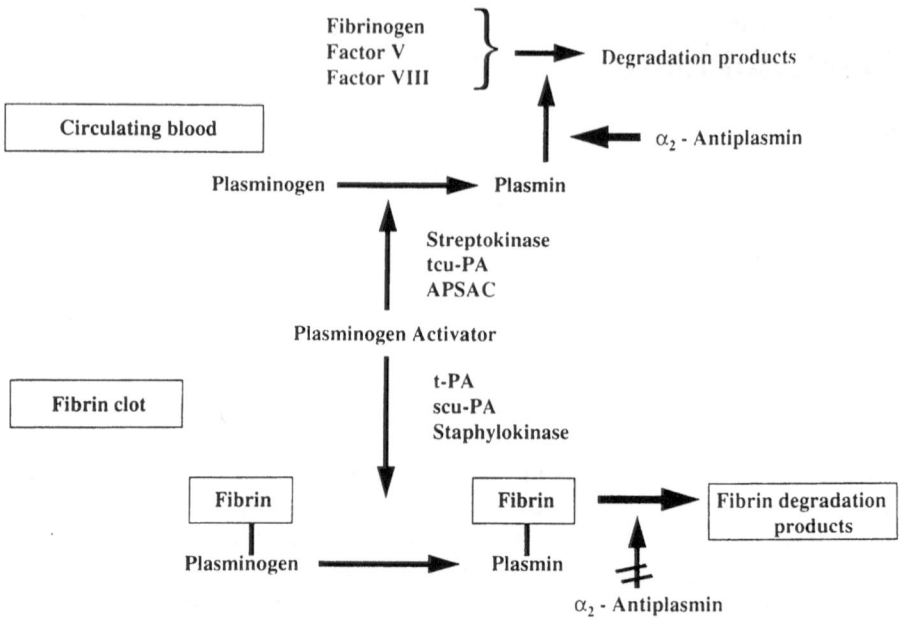

Fig. 1. Molecular interactions involved in the fibrin-specificity of plasminogen activation. Non-fibrin-specific plasminogen activators convert plasminogen to plasmin mainly in the circulation, resulting in depletion of α_2-antiplasmin and degradation of several plasma proteins. Fibrin-specific plasminogen activators mainly activate fibrin-bound plasminogen into fibrin-bound plasmin, which is protected from rapid inhibition by α_2-antiplasmin, and efficiently degrades fibrin

sistent coronary artery recanalization is the primary determinant of clinical benefit (Braunwald 1993).

At present, the thrombolytic potential of SAK, a highly fibrin-specific bacterial plasminogen activator, is under investigation (Collen and Lijnen 1994; Collen 1998b, 1999). In this chapter we will discuss the mechanisms involved in the fibrin-specificity of SAK, and its thrombolytic properties as evaluated in animal models of thrombosis and in (pilot) studies in patients with thromboembolic disease.

B. Gene and Protein Structure of Staphylokinase

I. Production of Staphylokinase

Staphylokinase, produced by certain strains of *Staphylococcus aureus* was shown to have profibrinolytic properties more than four decades ago (Lack 1948; Lewis and Ferguson 1951). Natural SAK has been purified from *Staphylococcus aureus* strains which were transformed with bacteriophages containing the SAK gene, or which had undergone lysogenic conversion to SAK production (WINKLER et al. 1965). More recently, natural SAK was purified from Tryptone Soy broth, conditioned by a selected *Staphylococcus aureus* strain, by batch adsorption to SP-Sephadex and chromatography on insolubilized active-site blocked plasmin (COLLEN et al. 1992a). However, the purification of SAK from lysogenic *Staphylococcus aureus* strains for detailed physical and biochemical studies as well as for the evaluation of its in vivo thrombolytic potential appeared to be elusive, mainly because of low expression levels and concomitant secretion of potent exotoxins.

The SAK gene has been cloned from the serotype B bacteriophage *sakøC* (SAKO et al. 1983), from the serotype F bacteriophage *sak*42D (BEHNKE and GERLACH 1987), from the ATCC 29213 strain (KIM et al. 1997), as well as from the genomic DNA (*sak*STAR) of a lysogenic *Staphylococcus aureus* strain (COLLEN et al. 1992a). It has been expressed under the control of the λ_{PR} promoter and its own translation signals in *Escherichia coli* (SAKO 1985) and also under the control of its natural promoter and translation signals in *Bacillus subtilis* (BEHNKE and GERLACH 1987, GERLACH et al. 1988) or *E. coli* (COLLEN et al. 1992a), resulting in accumulation of the gene product in the periplasmic space or in the culture medium, respectively. This mode of expression was, however, found to be suboptimal for the high yield production of mature SAK (COLLEN et al. 1993a).

Subsequently, recombinant plasmids were constructed with the signal sequence of the *sak*42D and the *sak*STAR genes replaced by an ATG start codon, which express SAK under the control of a tac promoter and two Shine-Dalgarno sequences in tandem (SCHLOTT et al. 1994a). Induction of transfected *E. coli* TG1 cells in a bacterial fermenter produced intracellular SAK (*sak*42D or *sak*STAR) representing 10%–15% of total cell protein. Highly purified recombinant SAK was purified from cytosol fractions by chromatography on SP-Sepharose and on phenyl-Sepharose, with yields of 50%–70% (approximately 200 mg/l fermentation broth) (SCHLOTT et al. 1994a). High level secretion of recombinant SAK from *Bacillus subtilis* using a P43 or P-sacB promoter (YE et al. 1999) or from *E. coli* using a tac promoter and an omp A signal sequence (LEE et al. 1998) has been reported. Large-scale preparation of the Δ10 form has been achieved by treating purified SAK with human plasminogen and purifying the truncated form by metal affinity and hydrophobic interaction chromatography (CHATTOPADHYAY et al. 1998).

II. Gene Structure of Staphylokinase

The SAK gene encodes a protein of 163 amino acids, with amino acid 28 corresponding to the NH_2-terminal residue of the full length mature protein. This coding sequence is preceded upstream by canonical Shine-Dalgarno, and –10 and –35 prokaryotic promoter sequences. Six nucleotide differences were found in the coding regions of the *sakøC*, *sak*42D, *sak*STAR, and ATCC 29213 genes, one of which constitutes a silent mutation. These affect the codons for amino acids 38, 60, 61, 63, and 70 (amino acids 11, 33, 34, 36, and 43 of the

TCA	AGT	TCA	TTC	GAC	AAA	GGA	AAA	TAT	AAA	AAA	GGC	GAT	GAC[42]
Ser	Ser	Ser	Phe	Asp	Lys	Gly	Lys	Tyr	Lys	Lys	Gly	Asp	Asp[14]

GCG	AGT	TAT	TTT	GAA	CCA	ACA	GGC	CCG	TAT	TTG	ATG	GTA	AAT[34]
Ala	Ser	Tyr	Phe	Glu	Pro	Thr	Gly	Pro	Tyr	Leu	Met	Val	Asn[28]

GTG	ACT	GGA	GTT	GAT	AGT	AAA	GGA	AAT	GAA	TTG	CTA	TCC	CCT[126]
Val	Thr	Gly	Val	Asp	Ser	Lys	Gly	Asn	Glu	Leu	Leu	Ser	Pro[42]

CAT	TAT	GTC	GAG	TTT	CCT	ATT	AAA	CCT	GGG	ACT	ACA	CTT	ACA[168]
His	Tyr	Val	Glu	Phe	Pro	Ile	Lys	Pro	Gly	Thr	Thr	Leu	Thr[56]

AAA	GAA	AAA	ATT	GAA	TAC	TAT	GTC	GAA	TGG	GCA	TTA	GAT	GCG[210]
Lys	Glu	Lys	Ile	Glu	Tyr	Tyr	Val	Glu	Trp	Ala	Leu	Asp	Ala[70]

ACA	GCA	TAT	AAA	GAG	TTT	AGA	GTA	GTT	GAA	TTA	GAT	CCA	AGC[252]
Thr	Ala	Tyr	Lys	Glu	Phe	Arg	Val	Val	Glu	Leu	Asp	Pro	Ser[34]

GCA	AAG	ATC	GAA	GTC	ACT	TAT	TAT	GAT	AAG	AAT	AAG	AAA	AAA[294]
Ala	Lys	Ile	Glu	Val	Thr	Tyr	Tyr	Asp	Lys	Asn	Lys	Lys	Lys[98]

GAA	GAA	ACG	AAG	TCT	TTC	CCT	ATA	ACA	GAA	AAA	GGT	TTT	GTT[336]
Glu	Glu	Thr	Lys	Ser	Phe	Pro	Ile	Thr	Glu	Lys	Gly	Phe	Val[112]

GTC	CCA	GAT	TTA	TCA	GAG	CAT	ATT	AAA	AAC	CCT	GGA	TTC	AAC[378]
Val	Pro	Asp	Leu	Ser	Glu	His	Ile	Lys	Asn	Pro	Gly	Phe	Asn[126]

TTA	ATT	ACA	AAG	GTT	GTT	ATA	GAA	AAG	AAA[408]
Leu	Ile	Thr	Lys	Val	Val	Ile	Glu	Lys	Lys[136]

Fig. 2. DNA sequence and deduced amino acid sequence of mature full-length staphylokinase (SakSTAR variant) (Collen et al. 1992b)

mature protein). Amino acid 38 is Lys in all four SAK moieties; amino acid 60 is Glu in variant ATCC 29213 but Asp in the other three variants; amino acid 61 is Ser in SakSTAR but Gly in the three other strains; amino acid 63 is Gly in SakSTAR and in Sak φC, but Arg in Sak42D and Glu in ATCC 29213. Finally, amino acid 70 is Arg in Sak42D but His in the three other variants (Sako and Tsuchida 1983; Behnke and Gerlach 1987; Collen et al. 1992b; Kim et al. 1997). The DNA sequence encoding the SakSTAR variant and the deduced amino acid sequence are shown in Fig. 2.

III. Protein Structure of Staphylokinase

Mature SAK consists of 136 amino acids in a single polypeptide chain without disulfide bridges (Sako and Tsuchida 1983; Behnke and Gerlach 1987; Collen et al. 1992b). The structure of SAK has been analyzed by X-ray scattering, dynamic light scattering, ultracentrifugation, ultraviolet circular dichroism spectroscopy, and by crystallography (Damaschun et al. 1993; Ohlenschläger et al. 1997, 1998). Crystallographic analysis revealed that

SAK folds into a compact ellipsoid structure with axial ratios of $1:0.55:0.49$. The longest axis of this ellipsoid has a length of 56 Å. The core of the protein is composed exclusively of hydrophobic amino acids. SAK is folded into a mixed five-stranded, slightly twisted β-sheet which wraps around a central α-helix and has two additional short two-stranded β-sheets opposing the central sheet (RABIJNS et al. 1997a,b). Upon interaction with microplasmin the convex surface of SAK formed by the side chains of strands $\beta3$, $\beta5$, $\beta1$, and $\beta2$ nestles against the 174(718) multiple-turn structure of the B-chain of microplasmin (chymotrypsinogen numbering; plasminogen numbering in parenthesis). Residues $Lys^{224(769)Plg}$-Glu^{19SAK}-$Arg^{175(719)Plg}$-Glu^{46SAK} participate in an extended network of salt bridges and charged hydrogen bonds, which is almost fully shielded from the bulk solvent through hydrophobic SAK residues. The strong buried salt bridge interactions of $Arg^{175(719)Plg}$ explain its importance in SAK binding as inferred from mutagenesis studies (JESPERS et al. 1998, 1999a,b). SAK amino acids Met^{26}, Tyr^{24}, and Tyr^{44} form an elongated hydrophobic patch that shields the network against water and contributes to the surface complementarity that SAK shows with microplasmin. This provides an explanation for the deleterious effect of replacing Met^{26} with residues of smaller side chains (LIJNEN et al. 1994).

Several molecular forms of SAK have been purified with slightly different M_r (16500–18000 on SDS-PAGE) and iso-electric points (SAKO 1985; GERLACH et al. 1988; COLLEN et al. 1992a; CHATTOPADHYAY et al. 1998). Lower M_r derivatives of mature SAK were obtained lacking the 6 ($\Delta6$) or the 10 ($\Delta10$) NH_2-terminal amino acids. Upon interaction with plasmin(ogen) in a buffer milieu, mature SAK (NH_2-terminal Ser-Ser-Ser-) is rapidly and quantitatively converted to the $\Delta10$ variant (NH_2-terminal Lys-Gly-Asp-). Purified SAK has a fairly high temperature stability (GASE et al. 1994). It was stable in liquid formulation after storage for more than 1.5 years at 4 °C (SINNAEVE et al. 1998).

C. Plasminogen Activation by Staphylokinase

I. Interaction with Plasmin(ogen)

SAK forms a 1:1 stoichiometric complex with plasminogen (KOWALSKA-LOTH and ZAKRZEWSKI 1975; LIJNEN et al. 1993a). Binding of SAK to plasminogen, as monitored by biospecific interaction analysis, occurs with an affinity constant of $1-2 \times 10^8 M^{-1}$, corresponding to a dissociation constant for the complex of 5–10 nmol/l. The binding data furthermore indicate that kringles 1–4 in the plasmin(ogen) A-chain do not contribute significantly to binding and that the active site in the plasmin molecule is not required for the high-affinity interaction (LIJNEN et al. 1994). Site-directed mutagenesis of nine solvent-exposed residues in microplasminogen led to the identification of one amino acid, Arg^{719} of which substitution by Ala strongly reduced the binding affinity and plasminogen activation potential of the SAK·microplasminogen complex (JESPERS et al. 1998).

A recombinant plasminogen mutant with the active site Ser[741] mutagenized to Ala (rPlg-Ala[741]) was not converted to a two-chain plasmin-like molecule by addition of SAK. In contrast, a preformed plasmin·SAK complex quantitatively converted rPlg-Ala[741] to its inactive two-chain derivative (Lijnen et al. 1991a). Two other recombinant plasminogen mutants, Arg[561]Ala, and Asp[646]Glu also did not result in the generation of plasmin activity after addition of SAK. The former mutant is a non-cleavable plasminogen (see Chap. 2), whereas in the latter mutant replacement of Asp[646], which is as Ser[741] part of the catalytic triad of amino acids of serine proteases, results in a non-activatable zymogen (Grella and Castellino 1997).

In equimolar mixtures of plasminogen and SAK, the active site titrant p-nitrophenyl-p'-guanidinobenzoate (NPGB) completely prevented active site exposure, whereas it reacted stoichiometrically with mixtures preincubated in its absence. Active site exposure was accompanied by quantitative conversion of plasminogen to plasmin (Collen et al. 1993b).

Taken together, these data indicate that SAK is not an enzyme, and that generation of an active site in its equimolar complex with plasminogen requires conversion of plasminogen to plasmin. Thus the plasmin·SAK complex is the active enzyme. This is in contrast with SK which produces a complex with plasminogen which exposes the active site in the plasminogen moiety without proteolytic cleavage (Reddy and Markus 1972).

Structure-function studies with recombinant SAK mutants, constructed according to a "clustered charge-to-alanine" scan revealed that replacement of clusters of charged amino acids in the regions comprising amino acids 11–14, 46–50, and 65–69 resulted in a 10–20-fold reduced binding to plasminogen (Silence et al. 1995; Lijnen et al. 1996). As a result, active site exposure in equimolar mixtures of these mutants with plasminogen, as monitored by titration with NPGB, was drastically impaired. The side chain of the amino acid at position 26 of SAK also appeared to be important for the initial binding to plasmin(ogen). Substitution of the unique Met residue in position 26 with Leu or Cys had little effect on the plasminogen activating potential of SAK. However, substitution with either Arg or Val resulted in a 10–20-fold reduced affinity for binding to plasminogen (Lijnen et al. 1994), and was associated with a total loss of activity (Schlott et al. 1994b).

Jespers et al. (1999a,b) have deduced a coherent docking model of the crystal structure of SAK on a homology-based model of microplasmin. SAK binding on microplasmin is primarily mediated by two surface-exposed loops, loops 174 and 215, at the rim of the active-site cleft, while the binding epitope of SAK on microplasmin involves several residues located in the flexible NH$_2$-terminal arm and in the five-stranded mixed β-sheet. Several SAK residues located within the unique α-helix and the $\beta2$ strand do not contribute to the binding epitope but are essential to induce plasminogen activating potential in the SAK·microplasmin complex. These residues form a topologically distinct activation epitope, which, upon binding of SAK to the catalytic domain of microplasmin, protrudes into a broad groove near the catalytic triad of microplasmin, thereby generating a competent binding pocket for microplasminogen.

This model was largely confirmed by crystallographic studies of the SAK·microplasmin complex suggesting that the NH_2-terminal tail of SAK may act as a flexible arm probing the environment for potential binding sites. This flexible arm could just reach into the active site of microplasmin, with Lys^{10} and Lys^{11} binding to the S_1 and S'_1 subsites of plasmin (PARRY et al. 1998). Indeed, deletion of Lys^{11} in $\Delta 10\,SAK$ resulted in a completely inactive molecule (GASE et al. 1996; SCHLOTT et al. 1998). Modeling studies suggested that Lys^{11} of $\Delta 10\,SAK$ could interact with the lysine binding site of kringle 5 of miniplasminogen (PARRY et al. 1998). Deletion of the COOH-terminal Lys^{136} also resulted in a molecule with strongly reduced plasminogen activation capacity (GASE et al. 1996).

II. Kinetics of Plasminogen Activation

Kinetic analysis revealed that the initial rate of plasmin generation in mixtures of plasminogen with catalytic amounts of SAK was not linear. The observed lag phase most likely corresponds to the time required for generation of active plasmin·SAK complex. Indeed, the initial activation rate was linear when catalytic amounts of preformed plasmin·SAK complex were used (LIJNEN et al. 1993a).

Activation of plasminogen by the preformed complex obeyed Michaelis-Menten kinetics, with K_m of $6.7\,\mu mol/l$ and k_{cat} of $1.8\,s^{-1}$, corresponding to a catalytic efficiency (k_{cat}/K_m) of $0.27\,\mu M^{-1}\,s^{-1}$. Similar catalytic efficiencies were observed for the activation of native plasminogen or of a plasminogen derivative lacking kringles 1–4 containing the lysine-binding sites (LIJNEN et al. 1993a). An independent kinetic analysis of the activation of plasminogen by the plasmin·SAK complex reported a comparable K_m value $(5\,\mu mol/l)$, but a much lower k_{cat} value $(0.03\,s^{-1})$ which would result in a 45-fold lower catalytic efficiency (SHIBATA et al. 1994).

In the "clustered charge-to-alanine" scan, three mutants were identified, with substitutions in the regions of amino acids 11–13 (SakSTAR13), 46–50 (SakSTAR48), and 65–69 (SakSTAR67) which virtually did not activate plasminogen. SakSTAR13 had a normal affinity for binding to plasminogen, but the plasmin·SakSTAR13 complex had a 14-fold reduced catalytic efficiency. SakSTAR48 and SakSTAR67 had a 10- to 20-fold reduced affinity for plasminogen and their complexes with plasmin had over a 20-fold reduced catalytic efficiency (SILENCE et al. 1995).

III. Mechanism of Plasminogen Activation

The kinetic data discussed above suggest the mechanism for plasminogen activation in a buffer milieu seen in Scheme 1. Plasminogen (Plg) and SAK produce an inactive 1:1 stoichiometric complex (Plg·SAK), which does not acti-

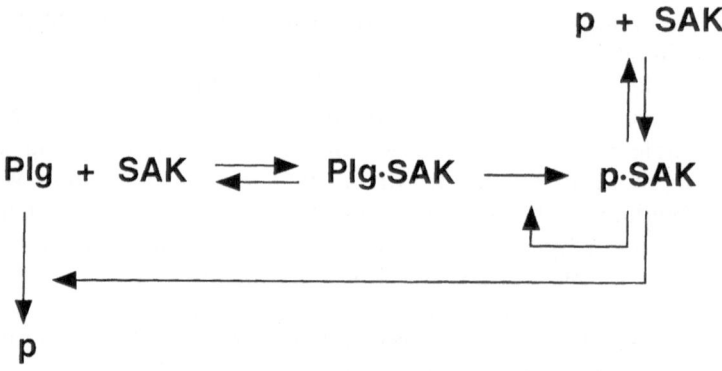

vate plasminogen. The activation reaction appears to be initiated by trace amounts of plasmin (p) which generates active plasmin·SAK complex (p·SAK). This is supported by the finding that plasminogen activation by catalytic amounts of SAK is enhanced by addition of traces of plasmin. Furthermore, it was shown that contamination of plasminogen with 30 ppm of plasmin is sufficient to explain the observed activation kinetics in equimolar mixtures of plasminogen and SAK (COLLEN et al. 1993b). In mixtures with excess plasminogen over SAK, generated p·SAK converts excess plasminogen to plasmin. In addition, kinetic analysis has revealed that generated p·SAK converts Plg·SAK to p·SAK at a rate which is several times higher than that of conversion of Plg to p, thus representing an efficient positive feed-back mechanism (SILENCE et al. 1995).

SAKHAROV et al. (1996) have shown that SAK has a much higher affinity for plasmin than for native plasminogen, indicating that the main pathway for plasmin generation in the above scheme is via activation of Plg by p·SAK formed by binding of SAK to p.

Interestingly, mono- and divalent ions modulate the activation of plasminogen by SAK. Cl⁻ showed the most striking inhibitory effect (64% inhibition at 10 mmol/l, but attenuated to 38% in the presence of a fibrin surface). Plasminogen activation was enhanced in the presence of Ca^{++} and fibrin (YARZABAL et al. 1999).

D. Fibrin-Specificity of Staphylokinase

I. Inhibition of Plasmin·SAK Complex by α_2-Antiplasmin

In purified systems, α_2-antiplasmin rapidly inhibits the plasmin·SAK complex (second-order inhibition rate constant of approximately $2 \times 10^6 M^{-1} s^{-1}$ (SAKAI et al. 1989; LIJNEN et al. 1991a). Addition of 6-aminohexanoic acid or of fibrin-like substances induces a more than 100-fold reduction of the inhibition rate of the plasmin·SAK complex by α_2-antiplasmin. Rapid inhibition by α_2-antiplasmin indeed requires the availability of the lysine-binding sites in the plasmin moiety of the complex (LIJNEN et al. 1992a). Fibrin, but not fibrinogen, reduces the inhibition rate of plasmin and the plasmin·SAK complex by α_2-antiplasmin by competing for interaction with the lysine-binding site(s).

Further analysis of the molecular interactions between plasmin·SAK and α_2-antiplasmin revealed that the activation rate of plasminogen by a mixture of plasmin·SAK and α_2-antiplasmin, which has no residual enzymatic activity, was indistinguishable from that by SAK alone. Furthermore, upon gel filtration of the mixture of plasmin·SAK and α_2-antiplasmin, SAK eluted with its own apparent M_r, whereas without addition of α_2-antiplasmin it eluted as a complex with plasmin (SILENCE et al. 1993a). These data indicate that neutralization of the plasmin·SAK complex by α_2-antiplasmin results in formation of an inactive plasmin-α_2-antiplasmin complex, and dissociation of functionally active SAK from the complex. SAK, which is converted to its $\Delta10$ derivative upon interaction with plasmin, is then recycled to other plasminogen molecules (SILENCE et al. 1993a).

II. Effect of Fibrin on Plasminogen Activation by Staphylokinase

SAK does not bind specifically to fibrin, and the initial rate of plasminogen activation by SAK is enhanced only two- to threefold by addition of fibrin (LIJNEN et al. 1991a). Cyanogen bromide-digested fibrinogen fragments FCB-2 and FCB-5 and plasmin-degraded crosslinked fibrin fragments DDE, DD, and E increased the k_{cat}/K_m ratio 10-fold, 5-fold, 30-fold, 38-fold and 8-fold, respectively (OKADA et al. 1996b). Experiments with purified fibrin clots revealed that SAK binds to plasminogen bound to partially degraded fibrin, but not to plasminogen bound to intact fibrin (SAKHAROV et al. 1996).

In equimolar mixtures of purified plasminogen and α_2-antiplasmin, no plasminogen activation is induced by catalytic amounts of SAK. However, addition of fibrin triggers activation of native plasminogen, but not of low M_r-plasminogen lacking the lysine-binding sites. These findings thus suggest that fibrin, by delaying inhibition of plasmin or plasmin·SAK by α_2-antiplasmin, facilitates generation of plasmin·SAK complex via a mechanism involving the lysine-binding sites of the plasmin moiety.

III. Molecular Mechanism of Fibrin-specificity in Plasma

The fibrin-specificity of SAK in human plasma has been explained by rapid inhibition of generated plasmin·SAK complex by α_2-antiplasmin (SAKAI et al. 1989; MATSUO et al. 1990; LIJNEN et al. 1991a), and by a more than 100-fold reduced inhibition rate at the fibrin surface (LIJNEN et al. 1992a), which may allow preferential plasminogen activation at the fibrin clot. However, SAK also dissociates in active form from the plasmin·SAK complex following neutralisation by α_2-antiplasmin, and is recycled to other plasminogen molecules

(SILENCE et al. 1993a). Thus, extensive systemic plasminogen activation with SAK would be expected in plasma, which is clearly in contradiction with its well established fibrin-specificity.

To elucidate this apparent paradox, the rate and extent of generation of plasmin·SAK complex in human plasma was monitored, both in the absence and the presence of fibrin (SILENCE et al. 1993b). In the absence of fibrin, no significant amounts of plasmin·SAK are generated because traces of plasmin are inhibited by α_2-antiplasmin; without plasmin·SAK complex, no significant plasminogen activation occurs. Traces of plasmin·SAK that may be formed will be rapidly inhibited by α_2-antiplasmin, whereby SAK dissociates from the complex (Fig. 3A). In the presence of fibrin, generation of the plasmin·SAK complex is facilitated because traces of fibrin-bound plasmin are protected from α_2-antiplasmin and, furthermore, inhibition of plasmin·SAK by α_2-

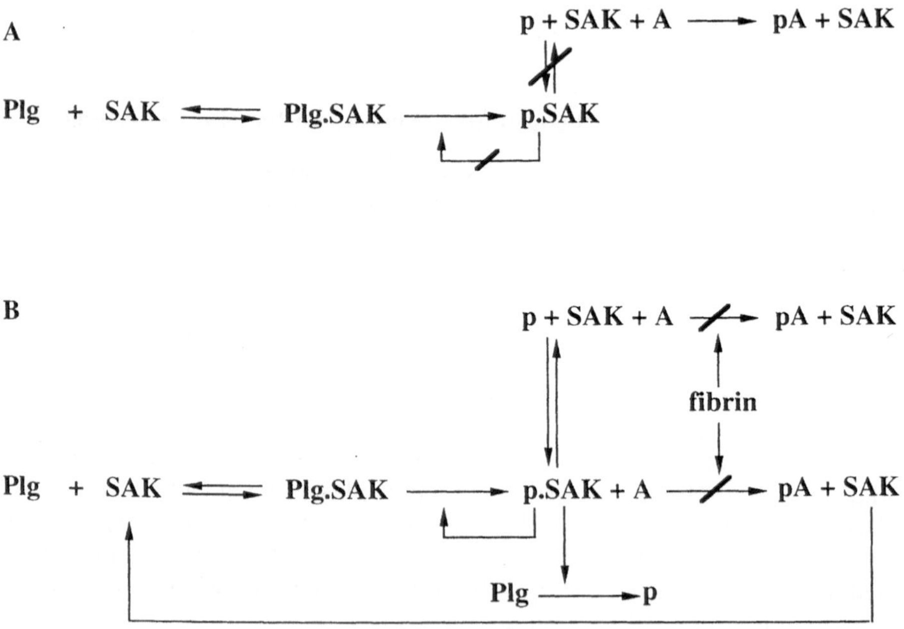

Fig. 3A,B. Molecular mechanism of fibrin-specificity of staphylokinase. **A** Plasminogen (*Plg*) and staphylokinase (*SAK*) produce an inactive stoichiometric complex (*Plg·SAK*). In the absence of α_2-antiplasmin (*A*) the activation reaction is initiated by traces of plasmin (*p*), which generate active plasmin·SAK complex (*p·SAK*). In a plasma milieu in the absence of fibrin, *A* inhibits formation of *p·SAK* by formation of inactive plasmin-α_2-antiplasmin complex (*pA*). **B** In a plasma milieu in the presence of fibrin, inhibition of traces of *p* by *A* is markedly delayed, thus allowing generation of active *p·SAK* complex. Inhibition of generated *p·SAK* at the fibrin surface by *A* is strongly delayed and thus conversion of *Plg·SAK* to *p·SAK* and activation of excess *Plg* to *p* can occur. Slow inhibition of *p·SAK* by *A* in the presence of fibrin results in generation of *pA*, whereby *SAK* dissociates from *p·SAK* and can be recycled to other *Plg* molecules generating *Plg·SAK*, or to fibrin-bound *p*

antiplasmin at the clot surface is delayed more than 100-fold. Thus, generated p·SAK may efficiently convert Plg·SAK to p·SAK and excess Plg to p. Recycling of SAK to fibrin-bound plasminogen, after slow neutralization of the plasmin·SAK complex, will result in more efficient generation of plasmin(ogen)·SAK complex (Fig. 3B). This mechanism is mediated via the lysine-binding sites of plasminogen and results in a significantly enhanced plasminogen activation at the fibrin surface.

E. Fibrinolytic Properties of Staphylokinase in Plasma In Vitro

I. Fibrinolytic Potency and Fibrin-specificity

The in vitro fibrinolytic properties of SAK were evaluated to some extent in the 1950s and 1960s (Sweet et al. 1965; Lewis et al. 1964). Using a clot lysis assay, Sweet et al. (1965) reported that SAK was a rapid and potent activator, but that a wide range of SAK-neutralizing activity was present in plasma.

Recombinant SAK (SakSTAR) induced dose-dependent lysis of ^{125}I-fibrin labeled human plasma clots submerged in citrated human plasma; 50% lysis in 2h of a 0.12-ml clot in 0.5ml plasma was obtained with 17nmol/l SAK, and was associated with only 5% plasma fibrinogen degradation. Corresponding values for SK were 68nmol/l and more than 90% fibrinogen degradation. In the absence of a fibrin clot, 50% fibrinogen degradation in human plasma in 2h required 790nmol/l SAK, but only 4.4nmol/l SK (Lijnen et al. 1991a). The fibrinolytic potency and fibrin-specificity of the preformed plasmin·SAK complex was very similar to that of SAK when added free (Lijnen et al. 1993b). This may be explained by inhibition of the preformed complex by α_2-antiplasmin, followed by recycling of SAK to other plasmin(ogen) molecules. The different M_r forms of SAK (mature SAK, $\Delta 6$, and $\Delta 10$ derivatives) have the same fibrinolytic and fibrinogenolytic potential in human plasma in vitro (Lijnen et al. 1992b). These results confirm and extend previous findings that in a human plasma milieu in vitro, SAK is more fibrinogen-sparing than SK (Sakai et al. 1989; Matsuo et al. 1990).

Ongoing clot lysis in human plasma induced by SAK could be more efficiently arrested by addition of antifibrinolytic amino acid, such as tranexamic acid than clot lysis with SK. The higher antifibrinolytic potency of tranexamic acid (which prevents binding of plasminogen to fibrin) towards SAK is most likely related to the requirement of fibrin-bound plasminogen for efficient lysis with SAK. These data suggest that tranexamic acid may be useful as an antidote to thrombolysis with SAK (Lijnen et al. 1995a).

The fibrinolytic activity of SAK towards platelet-rich (PRP; 400×10^9 platelets/l) and platelet-poor (PPP; $<5 \times 10^9$ platelets/l) plasma clots was compared in a human plasma milieu in vitro, which consisted of a 0.06ml ^{125}I-fibrin labeled plasma clot submerged in 0.5ml plasma. A 50% clot lysis in 2h, C_{50}, was obtained with 40 or 23 nmol/l SAK for PRP or PPP, respectively. With SK, no significant lysis of PRP clots was obtained with 440 nmol/l, whereas the C_{50} for PPP clots was 47 nmol/l (LIJNEN et al. 1992b). Similar results were reported using retracted or mechanically compressed plasma clots in vitro (HAUPTMANN and GLUSA 1995). This differential sensitivity to SAK and SK might result from alteration of the α_2-antiplasmin to plasminogen ratio in the clot during retraction. Indeed, in a plasma milieu in vitro, retracted blood clots are more sensitive to lysis with the fibrin-specific plasminogen activators tPA and sc-uPA than with the non-fibrin-specific agents SK and urokinase (SABOVIC et al. 1989). This phenomenon was explained by an enhanced systemic plasminogen activation with the non-fibrin-specific agents, which precluded recruitment of plasminogen from the surrounding plasma and thereby resulted in reduced clot lysis. Extrusion of non-fibrin-bound plasminogen from the clot, as a result of platelet-mediated retraction, also results in a reduced concentration of plasminogen, whereby the ratio of α_2-antiplasmin to plasminogen associated with the clot increases. Furthermore, during retraction additional α_2-antiplasmin is crosslinked to the C-terminal region of the α-chain of fibrin by Factor XIIIa (reviewed by AOKI and HARPEL 1984). Subsequent studies also revealed that PRP clots without normal clot retraction (Glanzmann thrombasthenia) were more resistant to lysis with SK than with SAK (LIJNEN et al. 1995b).

SAK did not have any effect on human platelets in vitro, as indicated by the absence of a significant effect of platelets on the rate of plasminogen activation by SAK, the lack of binding of SAK to platelets, and the absence of effects on platelet aggregation or disaggregation (ABDELOUAHED et al. 1997). In addition, therapeutic concentrations of SAK in plasma of patients with acute myocardial infarction treated with SAK did not affect platelet function, as revealed by unaltered platelet count, ADP- or collagen-induced platelet aggregation and ATP secretion (LIJNEN et al. 1995b). In another study, a very mild effect of SAK (at concentrations above 100 μg/ml) on collagen-induced platelet aggregation was reported (SUEHIRO et al. 1993).

In 24 patients with acute myocardial infarction, who were randomly assigned to receive either a double bolus of 15 mg of SAK 30 min apart or the classical scheme of accelerated tPA (maximum of 100 mg over 90 min, 10% administered as a bolus), baseline and 25 min and 90 min blood samples were drawn for the determination of the following markers of coagulation activation: fibrinopeptide A (FPA), prothrombin fragment 1 + 2 (F1 + 2), and of thrombin-antithrombin III complexes (TAT). In patients given SAK, FPA, F1 + 2, and TAT did not markedly increase during treatment ($p = 0.06$, $p = 0.4$ and $p = 0.03$ respectively). In contrast, during the administration of tPA the levels of FPA, F1 + 2, and TAT increased significantly over baseline ($p = 0.003$, $p < 0.0001$, and $p = 0.001$ respectively). These results demonstrate that SAK, in addition to being a fibrin-specific thrombolytic agent, also produces much less activation of the coagulation system than tPA (OKADA et al. 1996a).

II. Species-specificity

The comparative fibrinolytic and fibrinogenolytic properties of SAK were studied in human, baboon, rabbit, hamster, rat, and dog plasma in vitro. The plasma fibrinolytic systems of baboons, rabbits, and hamsters reacted comparably to the human system to SAK, the rat system appeared to be very resistant, whereas in the dog system SAK was very potent, but not fibrin-specific (LIJNEN et al. 1992c). The molecular basis of the marked interspecies variability in the response of plasma fibrinolytic systems to activation by SAK was studied using purified plasminogens and α_2-antiplasmins from different mammalian species (COLLEN et al. 1993c). The results indicate that this variability is determined mainly by the extent of complex formation of SAK with plasminogen, by the catalytic efficiencies of the complexes for the activation of autologous plasminogen, and by the rate of inhibition of these complexes by α_2-antiplasmin. The comparably high reactivity of SAK with both human and canine plasminogen may explain its high potency for clot lysis in these species, whereas the tenfold lower reactivity of the canine plasmin·SAK complex with α_2-antiplasmin may explain its markedly lower fibrin-specificity in the dog. Thus, the choice of the canine species for the initial in vivo evaluation of SAK (LEWIS et al. 1964; LEWIS and SHIRAKAWA 1964; KANAE 1986) may have produced misleading conclusions due to the unusually high sensitivity of the canine plasma fibrinolytic system to activation with SAK.

F. Thrombolytic Properties of Staphylokinase in Animal Models

I. Hamsters with Pulmonary Embolism

The thrombolytic properties of SAK (SakϕC) were compared with those of SK in hamsters with a pulmonary embolus produced from human or from hamster plasma. Continuous intravenous infusion of both agents over 1 h induced dose-dependent progressive clot lysis in the absence of significant systemic fibrinolytic activation. Both SAK and SK induced 50% clot lysis with a dose of less than 0.25 mg/kg. On a molar basis, however, SAK was less potent than SK (LIJNEN et al. 1991b).

The comparative thrombolytic properties of natural and recombinant SAK (SakSTAR) and of SK were also studied in hamsters with a pulmonary embolus consisting of a platelet-poor, a platelet-rich (300×10^9 platelets/l), or a platelet-enriched (1500×10^9 platelets/l) human plasma clot. The relative thrombolytic potencies, on a weight basis, of the SAK preparations vs SK were comparable in the platelet-poor and in the platelet-rich clot model, but five times higher in the platelet-enriched clot model (COLLEN et al. 1992c).

II. Rabbits with Jugular Vein Thrombosis

Continuous intravenous infusion over 4h of SAK (SakφC) or SK in rabbits with an autologous jugular vein clot produced dose-dependent clot lysis without significant systemic activation of the fibrinolytic system (Lijnen et al. 1991b). On a weight basis, SAK was about three times more potent than SK, but on a molar basis the thrombolytic potency of both agents was comparable.

III. Rabbits with Arterial Thrombosis

Helft and colleagues (1998) compared the effect of tPA and SAK in a rabbit model of femoral artery thrombosis. The main finding was that the infusion of SAK following a single bolus administration gave significantly higher blood flow values than the infusion of the same dose of tPA following a single bolus administration ($p < 0.05$). SAK was fibrin-specific and plasminogen-saving at doses below 0.5mg/kg. However, at higher doses (1 and 1.5mg/kg), which are above therapeutic doses, SAK significantly reduced fibrinogen levels in a dose- and time-dependent manner.

IV. Rabbit Embolic Stroke Model

In the NINDS trial, the administration of 90mg of tPA to patients with acute ischemic stroke of less than 3h duration resulted in improved clinical outcome at three months compared to controls (The National Institute of Neurological Disorders and Stroke rt-PA Stroke Study Group 1995). Vanderschueren et al. (1997) therefore compared the efficacy of SAK and of tPA in an experimental model of rabbit embolic stroke. Rabbits were administered a standardized ^{125}I-fibrin labeled autologous plasma clot into the internal carotid artery. After 15min, groups of 5–12 rabbits were given one of six thrombolytic regimens or a saline infusion: 1 or 2mg SAK/kg infused over 30 min, or 2mg SAK/kg given as a bolus, or 3 or 6mg/kg of tPA infused over 30 min, or 6mg/kg given as a bolus. Mean clot lysis after 60min was 4% in the saline group, 27–44% after SAK, and 15–34% after tPA. At the highest doses used, fibrinogen depletion was marginal with SAK but total with tPA. Marked prolongation of ear puncture and cuticle bleeding was only observed after bolus administration of tPA (Vanderschueren et al. 1997a).

V. Baboons with Venous and Arterial Thrombosis

Intravenous infusion of SK or SAK (SakSTAR) over 1h in doses up to 1mg/kg in baboons induced dose-dependent lysis of a ^{125}I-fibrin-labeled autologous jugular vein blood clot without systemic fibrinogen depletion (50% lysis requiring 0.14mg/kg SK and 0.058mg/kg SakSTAR, which are comparable amounts on a molar basis). In a baboon femoral arterial eversion graft model (a platelet-rich model), arterial recanalization was obtained more frequently and more persistently with SAK than with SK. Intravenous infusion of 1 mg/kg SK or 0.25 mg/kg SakSTAR in baboons given intravenous heparin (200 U bolus and 50 U per kg per hour) and aspirin (10 mg/kg) did not produce a significant prolongation of the median template bleeding time (COLLEN et al. 1993d).

G. Thrombolytic Properties of Staphylokinase in Patients

The preclinical animal experiments with SAK discussed above have yielded promising results in terms of a high thrombolytic potency towards venous and towards platelet-rich and retracted thrombi, a rapid onset of action, and a very high fibrin-specificity. These studies have thus formed the basis for the evaluation of the thrombolytic potential of SAK in patients.

I. Acute Myocardial Infarction (AMI)

Three dose-finding pilot studies used either a small bolus of 1 mg of SAK, followed by an infusion of 9 mg over 30 min (COLLEN and VAN DE WERF 1993; SCHLOTT et al. 1994a) or a double bolus of 20 mg SAK, 15 min apart, or an initial bolus of 20 mg, followed by a second bolus of 10 mg if the coronarography at 60 min revealed incomplete recanalization (VANDERSCHUEREN et al. 1996). TIMI grade 3 flow rates were similar to those obtained with accelerated tPA.

In the first randomized, multicenter trial SAK was administered IV in 48 patients and compared to accelerated weight-adjusted tPA (alteplase) given to 52 patients. All patients received 160 mg aspirin at entry (followed by 160 to 325 mg daily) and heparin (5000 U bolus in the first 25 patients, later on 10000 U bolus; followed by 1000 U/h and adjusted to APTT values of at least 1.5 times control). In the SAK group, the first 25 patients were given 10 mg of SAK infused over 30 min. TIMI grade 3 flow in this group was 50% at 90 min. In the subsequent 23 patients in the SAK group the dose was increased to 20 mg IV, given over 30 min. In this latter group TIMI grade 3 flow rates at 90 min of 77% compared favorably with those obtained in the accelerated tPA group (58%) (VANDERSCHUEREN et al. 1995a).

The second randomized, multicenter trial in patients with AMI treated within 6 h of onset of symptoms compared the administration of accelerated, weight-adjusted tPA (alteplase; 52 patients) with two bolus injections of 15 mg SAK, 30 min apart, in 50 patients. All patients received aspirin and heparin. TIMI grade 3 flow rates were achieved in 68% of patients treated with SAK versus 57% in patients treated with tPA. Double-bolus SAK was significantly more fibrin-specific than accelerated tPA with residual fibrinogen at 90 min of 105% ± 4.1% versus 68% ± 7.5% respectively ($p < 0.0001$). Thirteen patients in each group underwent PTCA of the culprit artery within the first 24 h because of suboptimal recanalization. TIMI grade 3 flow rates at 24h in the remaining patients was 100% in the SAK group, but only 79% in the accelerated tPA group. The administration of SAK did not induce allergic reactions, but resulted in the development of significant SAK-neutralizing antibodies (>5 µg/ml) and specific anti-SAK IgG developed in 73% of patients after 2 weeks (Vanderschueren et al. 1997b).

These finding were confirmed in a further, sequential, angiographic dose-finding study with the SAK42D variant. Eighty-two patients with AMI of <6 duration entered the CAPTORS trial. All patients received SAK42D as a 30-min infusion with 20% of the dose given over 1 min as a bolus. All patients received oral aspirin 160 to 325 mg and 325 mg daily thereafter. Heparin was given as an i.v. bolus followed by the infusion of 1000 U/h for patients ≥80 kg and 800 U/h for patients <80 kg. Coronary angiography was performed at 90 min. TIMI grade 3 (and TIMI grade 2 + 3) flow were 62% (76%) with the 15 mg dose, 65% (84%) with 30 mg and 63% (83%) with the 45 mg dose. Major bleeding occurred in 4 and moderate bleeding in 9 patients. The majority of these were related to vascular instrumentation and there was no clear relation between the extent of bleeding and the dose of SAK. Dose-related small to moderate decreases of fibrinogen, plasminogen and α_2-antiplasmin were observed (Armstrong et al. 2000). Thus, doses of approximately 30 mg of SAK produce TIMI grade 3 flow in patients with AMI that are equal or superior to accelerated tPA with a similar incidence of bleeding complications.

II. Peripheral Arterial Occlusion

Thirty patients with angiographically confirmed peripheral arterial occlusion were treated with intra-arterial SAK (SakSTAR), given as a 1-mg bolus followed by infusion of 0.5 mg per hour in 20 patients or as a 2-mg bolus followed by an infusion of 1 mg per hour in another 10 patients, all receiving heparin (Vanderschueren et al. 1995b). After 7.0 ± 0.7 mg SakSTAR infused over 8.7 ± 1.0 h, recanalization was complete in 25 patients (83%), partial in 2 and absent in 3 patients. Most of the patients underwent percutaneous transluminal angioplasty after SakSTAR-infusion. Major amputations were limited to two patients, both after failed thrombolysis, whereas three patients developed reocclusion within one month. Two major hemorrhagic complications occurred, including one fatal hemorrhagic stroke. However, intra-arterial SakSTAR did not induce a systemic fibrinolytic activation, and template bleeding times were not prolonged.

A recent, non-randomized study of 191 patients <80 years of age with peripheral arterial occlusion of <120 days duration reports on the effect of a 2 mg intra-arterial bolus injection of SAK or of SAK-mutants, followed by an infusion of 1 mg/h or 0.5 mg/h overnight. Ninety-nine patients presented with acute or subacute ischemia, 57 with severe claudication, 33 with chronic rest pain and 2 with gangrene. Occlusion occurred in 122 native arteries and in 69 grafts. Revascularization was complete in 83%, partial in 13% and absent in

4% after the administration of 12 ± 0.5 mg (mean $\pm$ SEM) SAK over $14 \pm$ 0.8 h. Complete restoration of patency was higher in proximal native arteries (95%) and grafts (89%) than in popliteal or more distal arteries (60%). Additional endovascular procedures were performed in 47% and subsequent elective bypass surgery in 23% of patients. Major bleeding occurred in 12%, one-month mortality was 3.1% and one-year mortality was 6.9%. Four patients had intracerebral hemorrhage (1.7%), one in a 85 year old woman (a protocol violation) and the other three in patients $\geq$74 years old. Amputations were performed within the first year in 8.6% of patients (HEYMANS et al. 2000).

SAK thus induced rapid and efficient restoration of vessel patency in the majority of patients with peripheral arterial occlusion and a high one-year amputation-free survival (84%), with an acceptable incidence of major bleedings, but with occasional fatal intracranial hemorrhages. These results compare favorably with data reported for other thrombolytic agents (see Chap. 14).

H. Pharmacokinetic Properties of Staphylokinase

I. In Animal Models

In hamsters and rabbits, ^{125}I-labeled SAK (SakϕC), was cleared from the circulation in a biphasic manner with an initial t1/2 of 1.8 min and a plasma clearance of 1.4 ml/min in hamsters, and corresponding values of 1.7 min and 14 ml/min in rabbits. The organ distribution of SAK documented a rapid uptake in muscle and kidney, but not in liver (LIJNEN et al. 1991b). The pharmacokinetics of SAK (SakSTAR) in hamsters were confirmed by antigen assays in plasma, and by measurement of SAK activity using a bio-immunoassay. The similar disappearance rate from plasma of SAK antigen and activity indicated that it is cleared from the circulation in an active form (LIJNEN et al. 1993c).

Following continuous intravenous infusion over 1 h in baboons, the plasma clearance of SakSTAR activity, determined from the infusion rate and the steady-state plasma level of SakSTAR activity, ranged between 45 and 62 ml/min for doses between 0.063 and 0.25 mg/kg (LIJNEN et al. 1993c).

II. In Patients

In five patients with AMI treated with an intravenous infusion of 10 mg SAK (SakSTAR) over 30 min, the concentration of SAK-related antigen in blood at the end of the infusion increased to between 0.9 and 1.7 μg/ml. The post-infusion disappearance of SAK-related antigen from plasma occurred in a biphasic manner with a t1/2α of 6.3 min and a t1/2β of 37 min, corresponding to a plasma clearance of 270 ml/min (COLLEN and VAN DE WERF 1993).

In the STAR trial of patients with acute myocardial infarction, SakSTAR antigen levels in 25 patients receiving 10 mg i.v. over 30 min were $0.56 \pm$

$0.06\,\mu g/ml$ at 25 min and $0.16 \pm 0.04\,\mu g/ml$ at 90 min, with corresponding levels of $1.9 \pm 0.22\,\mu g/ml$ and $0.42 \pm 0.06\,\mu g/ml$ in 23 patients receiving 20 mg SakSTAR over 30 min (Vanderschueren et al. 1995a).

Cysteine-linked polyethylene glycol derivatives of SAK have been constructed and exhibited half-clearance rates that were up to 10-fold slower than the parent molecule. Such PEGylated derivatives might thus yield suitable candidates for bolus treatment in patients with AMI (Vanwetswinkel et al. 2000).

I. Immunogenicity of Staphylokinase

I. In Animal Models

The immunogenicity of SAK (SakSTAR) as compared to SK was studied in rabbits, dogs, and baboons (Collen et al. 1992c, 1993d; Vanderschueren et al. 1994). Neutralizing activities were serially measured in plasma by an in vitro human plasma-based clot lysis time assay and were found to be IgG-related.

SAK appeared to be less immunogenic than SK in dogs, as evidenced by less rapid induction of antibody formation and resistance to clot lysis upon repeated administration (Collen et al. 1992c).

In four baboons with a ^{125}I-fibrin labeled clot in an extracorporeal arteriovenous loop, i.v. administration of $63\,\mu g/kg$ SAK over 1 h, repeated at weekly intervals, induced a progressive increase of SAK-neutralizing activity (from $0.05 \pm 0.1\,\mu g/ml$ at baseline to $4.8 \pm 1.5\,\mu g/ml$ at week 6), which was paralleled by a reduction of in vivo clot lysis (from $60 \pm 7\%$ to $8 \pm 3\%$). After temporary discontinuation of SAK administration, neutralizing activity reverted to baseline within seven weeks, whereafter the sensitivity of in vivo clot lysis to SAK was restored (Vanderschueren et al. 1994).

II. In Patients

Enzyme-linked immunosorbent assays, calibrated with affinospecific human antibodies, revealed 2.1–$65\,\mu g/ml$ (median $11\,\mu g/ml$) anti-SAK antibodies and 0.9–$370\,\mu g/ml$ (median $18\,\mu g/ml$) anti-SK antibodies in plasma from 100 blood donors. Corresponding values were 0.6–$100\,\mu g/ml$ (median $7.1\,\mu g/ml$) anti-SK antibodies and 0.4–$120\,\mu g/ml$ (median $7.3\,\mu g/ml$) anti-SAK antibodies in 104 patients with angina pectoris (Declerck et al. 1994).

In the first five patients with acute myocardial infarction given an intravenous infusion of 10 mg SAK over 30 min, neutralizing antibody titers against SAK (SakSTAR) were low at baseline and up to six days after infusion, but high titers (SAK-neutralizing titers of 12–$42\,\mu g/ml$ plasma) of antibodies, which did not cross-react with SK, were consistently demonstrable in plasma at 14–35 days (Collen and Van de Werf 1993). These observations were fully confirmed in the second pilot trial in five patients (Schlott et al. 1994a).

In the STAR trial, in 48 patients with acute myocardial infarction, SAK-neutralizing activity levels were low at baseline ($0.07 \pm 0.01\,\mu g/ml$) and during the first week after SAK administration ($1.5 \pm 0.39\,\mu g/ml$), but increased substantially from the second week on ($32 \pm 7.1\,\mu g/ml$) (VANDERSCHUEREN et al. 1995a). Also in patients with peripheral arterial occlusion, intra-arterial administration of SAK elicited high levels of SAK-neutralizing antibodies, which remained high for several months (VANDERSCHUEREN et al. 1994, 1995b).

III. Attempts to Reduce Immunogenicity

Considerable efforts have been undertaken to identify immunodominant epitopes in the SAK molecule (COLLEN et al. 1996a) to produce mutants by substitution of clusters of charged amino acids with alanine, and to analyze the SAK mutants for thrombolytic potency and immunogenicity (COLLEN et al. 1996b; reviewed in COLLEN 1998a,b). It soon became apparent that there exists a rather large species variability in the antibody response to SAK which implies that further studies in this field will require human or humanized systems (COLLEN et al. 1997a). Replacement of Lys^{74} with Ala or of Lys^{74}, Glu^{75}, and Arg^{77} with Ala produced variants with intact thrombolytic potencies which induced significantly less antibody formation in patients (COLLEN et al. 1997b). An elegant approach of mapping the most relevant epitopes of SAK that are antigenic for humans was undertaken by JENNÉ et al. (1998) who produced a phage-displayed library of SAK variants. These were selected for mutants that escaped binding to an affinity matrix derivatized with patient-specific polyclonal anti-SAK antibodies. Fifty-six escape variants were identified. Such studies may be valuable to guide efforts to reduce the immunogenicity of SAK using protein engineering techniques.

J. Conclusions

SAK is a profibrinolytic agent that forms a 1:1 stoichiometric complex with plasmin(ogen) which, following conversion to plasmin, activates other plasminogen molecules to plasmin. The plasmin·SAK complex, unlike the plasmin·SK complex, is rapidly inhibited by α_2-antiplasmin. In a plasma milieu, SAK is able to dissolve fibrin clots without associated fibrinogen degradation. This fibrin-specificity of SAK is the result of a higher affinity for fibrin-bound plasminogen, a reduced inhibition by α_2-antiplasmin of plasmin or plasmin· SAK complex bound to fibrin, and recycling of SAK from the plasmin·SAK complex following inhibition by α_2-antiplasmin. In several experimental animal models, SAK appeared to be equipotent to streptokinase for the dissolution of whole blood or plasma clots, and significantly more potent for the dissolution of platelet-rich or retracted thrombi.

The feasibility of fibrin-specific thrombolysis with recombinant SAK was demonstrated in patients with AMI and with peripheral arterial occlusion.

However, neutralizing antibodies against SAK were demonstrable from the third week on in most patients. Definition of the therapeutic benefit of recombinant SAK or of less immunogenic variants will require more larger scale randomized efficacy studies against other thrombolytic agents.

List of Abbreviations and Acronyms

APSAC	Anisoylated Plasminogen Streptokinase Activator Complex
AMI	Acute myocardial infarction
CAPTORS	Collaborative Angiographic Patency Trial Of Recombinant Staphylokinase
Δ6, Δ10 SAK	Staphylokinase cleavage product, lacking 6, respectively 10 NH_2-terminal amino acids
F1 + 2	prothrombin fragment F1 + 2
FPA	fibrinopeptide A
GUSTO	Global Utilization of Streptokinase and Tissue plasminogen activator for Occluded coronary arteries
NINDS	National Institute of Neurological Disorders and Stroke
NPGB	p-nitrophenyl-p'-guanidinobenzoate
uPA	urinary-type (or urokinase-type) plasminogen activator, also called urokinase
SAK	staphylokinase
sc-uPA	single chain uPA, also called pro-urokinase
SK	streptokinase
STAR	recombinant staphylokinase
TAT	thrombin-antithrombin complex
tc-uPA	two chain uPA, also called urokinase
tPA	tissue-type plasminogen activator

References

Abdelouahed M, Hatmi M, Helft G, Emadi S, Elalamy I, Samama MM (1997) Comparative effects of recombinant staphylokinase and streptokinase on platelet aggregation. Thromb Haemost 77:815–817

Aoki N, Harpel PC (1984) Inhibitors of the fibrinolytic enzyme system. Semin Thromb Hemost 10:24–41

Armstrong PW, Burton JR, Palisaitis D, Thompson CR, Van de Werf F, Rose B, Collen D, Teo KK (2000) Collaborative angiographic patency trial of recombinant staphylokinase (CAPTORS). Am Heart J 139:820–823

Behnke D, Gerlach D (1987) Cloning and expression in Escherichia coli, Bacillus subtilis, and Streptococcus sanguis of a gene for staphylokinase – a bacterial plasminogen activator. Molec Gen Genet 210:528–534

Braunwald E (1993) The open-artery theory is alive and well – again. N Engl J Med 329:1650–1652

Chattopadhyay D, Stewart JE, DeLucas LJ (1998) Large-scale preparation of the delta10 form of staphylokinase by in vitro processing of recombinant staphylokinase with purified human plasminogen. Appl Biochem Biotechnol 69:147–156

Collen D (1998a) Engineered staphylokinase variants with reduced immunogenicity. Fibrinolysis Proteolysis 12 [Suppl 2]:59–65

Collen D (1998b) Staphylokinase: a potent, uniquely fibrin-selective thrombolytic agent. Nat Med 4:279–284

Collen D (1999) The plasminogen (fibrinolytic) system. Thromb Haemost 82:259–270

Collen D, Bernaerts R, Declerck P, De Cock F, Demarsin E, Jenné S, Laroche Y, Lijnen HR, Silence K, Verstreken M (1996a) Recombinant staphylokinase variants with altered immunoreactivity. I: Construction and characterization. Circulation 94:197–206

Collen D, De Cock F, Demarsin E, Jenné S, Lasters I, Laroche Y, Warmerdam P, Jespers L (1997a) Recombinant staphylokinase variants with altered immunoreactivity. III: Species variability of antibody binding patterns. Circulation 95:455–462

Collen D, De Cock F, Stassen JM (1993d) Comparative immunogenicity and thrombolytic properties toward arterial and venous thrombi of staphylokinase and streptokinase in baboons. Circulation 87:996–1006

Collen D, De Cock F, Vanlinthout I, Declerck PJ, Lijnen HR, Stassen JM (1992c) Comparative thrombolytic and immunogenic properties of staphylokinase and streptokinase. Fibrinolysis 6:232–242

Collen D, De Mol M, Demarsin E, De Cock F, Stassen JM (1993a) Isolation and conditioning of recombinant staphylokinase for use in man. Fibrinolysis 7:242–247

Collen D, Lijnen HR (1991) Basic and clinical aspects of fibrinolysis and thrombolysis. Blood 78:3114–3124

Collen D, Lijnen HR (1994) Staphylokinase, a fibrin-specific plasminogen activator with therapeutic potential? Blood 84:680–686

Collen D, Moreau H, Stockx L, Vanderschueren S (1996b) Recombinant staphylokinase variants with altered immunoreactivity. II: Thrombolytic properties and antibody induction. Circulation 94:207–216

Collen D, Schlott B, Engelborghs Y, Van Hoef B, Hartmann M, Lijnen HR, Behnke D (1993b) On the mechanism of the activation of human plasminogen by recombinant staphylokinase. J Biol Chem 268:8284–8289

Collen D, Silence K, Demarsin E, De Mol M, Lijnen HR (1992a) Isolation and characterization of natural and recombinant staphylokinase. Fibrinolysis 6:203–213

Collen D, Stockx L, Lacroix H, Suy R, Vanderschueren S (1997b) Recombinant staphylokinase variants with altered immunoreactivity. IV: Identification of variants with reduced antibody induction but intact potency. Circulation 95:463–472

Collen D, Van de Werf F (1993) Coronary thrombolysis with recombinant staphylokinase in patients with evolving myocardial infarction. Circulation 87:1850–1853

Collen D, Van Hoef B, Schlott B, Hartmann M, Gührs KH, Lijnen HR (1993c) Mechanisms of activation of mammalian plasma fibrinolytic systems with streptokinase and with recombinant staphylokinase. Eur J Biochem 216:307–314

Collen D, Zhao ZA, Holvoet P, Marynen P (1992b) Primary structure and gene structure of staphylokinase. Fibrinolysis 6:226–231

Damaschun G, Damaschun H, Gast K, Misselwitz R, Zirwer D, Gührs K-H, Hartmann M, Schlott B, Triebel H, Behnke D (1993) Physical and conformational properties of staphylokinase in solution. Biochim Biophys Acta 1161:244–248

Declerck PJ, Vanderschueren S, Billiet J, Moreau H, Collen D (1994) Prevalence and induction of circulating antibodies against recombinant staphylokinase. Thromb Haemostas 71:129–133

Gase A, Birch-Hirschfeld E, Gührs K-H, Hartmann M, Vettermann S, Damaschun G, Damaschun H, Gast K, Misselwitz R, Zirwer D, Collen D, Schlott B (1994) The thermostability of natural variants of bacterial plasminogen-activator staphylokinase. Eur J Biochem 223:303–308

Gase A, Hartmann M, Gührs K-H, Röcker A, Collen D, Behnke D, Schlott B (1996) Functional significance of NH$_2$- and COOH-terminal regions of staphylokinase in plasminogen activation. Thromb Haemost 76:755–760

Gerlach D, Kraft R, Behnke D (1988) Purification and characterization of the bacterial plasminogen activator staphylokinase, secreted by a recombinant bacillus subtilis. Zbl Bakt Mikr Hyg 269:314–322

Grella DK, Castellino FJ (1997) Activation of human plasminogen by staphylokinase. Direct evidence that preformed plasmin is necessary for activation to occur. Blood 89:1585–1589

Hauptmann J, Glusa E (1995) Differential effects of staphylokinase, streptokinase and tissue-type plasminogen activator on the lysis of retracted human plasma clots and fibrinolytic plasma parameters in vitro. Blood Coagul Fibrinolysis 6:579–583

Helft G, Bara L, Bloch MF, Samama MM (1998) Comparative time course of thrombolysis induced by intravenous boluses and infusion of staphylokinase and tissue plasminogen activator in a rabbit arterial thrombosis model. Blood Coagul Fibrinolysis 9:411–417

Heymans S, Vanderschueren S, Verhaeghe R, Stockx L, Lacroix H, Nevelsteen A, Laroche Y, Collen D (2000) Outcome and one year follow-up of intra-arterial staphylokinase in 191 patients with peripheral arterial occlusion. Thromb Haemost 83:666–671

Jenné S, Brepoels K, Collen D, Jespers L (1998) High resolution mapping of the B cell epitopes of staphylokinase in humans using negative selection of a phage-displayed antigen library. J Immunol 161:3161–3168

Jespers L, Lijnen HR, Vanwetswinkel S, Van Hoef B, Brepoels K, Collen D, De Maeyer M (1999b) Guiding a docking mode by phage display: selection of correlated mutations at the staphylokinase-plasmin interface. J Mol Biol 290:471–479

Jespers L, Van Herzeele N, Lijnen HR, Van Hoef B, De Maeyer M, Collen D, Lasters I (1998) Arginine 719 in human plasminogen mediates formation of the staphylokinase:plasmin activator complex. Biochemistry 37:6380–6386

Jespers L, Vanwetswinkel S, Lijnen HR, Van Herzeele N, Van Hoef B, Demarsin E, Collen D, De Maeyer M (1999a) Structural and functional basis of plasminogen activation by staphylokinase. Thromb Haemost 81:479–485

Kanae K (1986) Fibrinolysis by staphylokinase in vivo. Biol Abstr 81:AB748 (Abstract 65436)

Kim S-H, Chun H-S, Han MH, Park N-Y, Suk K (1997) A novel variant of staphylokinase gene from Staphylococcus aureus ATCC 29213. Thromb Res 87:387–395

Kowalska-Loth B, Zakrzewski K (1975) The activation by staphylokinase of human plasminogen. Acta Biochim Pol 22:327–339

Lack CH (1948) Staphylokinase: an activator of plasma protease. Nature 161:559–560

Lee SJ, Kim IC, Kim DM, Bae KH, Byun SM (1998) High level secretion of recombinant staphylokinase into periplasm of Escherichia coli. Biotechnol Lett 20:113–116

Lewis JH, Ferguson JH (1951) A proteolytic enzyme system of the blood. III. Activation of dog serum profibrinolysin by staphylokinase. Am J Physiol 166:594–603

Lewis JH, Kerber CW, Wilson JH (1964) Effects of fibrinolytic agents and heparin on intravascular clot lysis. Am J Physiol 207:1044–1048

Lewis JH, Shirakawa M (1964) Effects of fibrinolytic agents and heparin on blood coagulation in dogs. Am J Physiol 207:1049–1052

Lijnen HR, Beelen V, Declerck PJ, Collen D (1993c) Bio-immuno assay for staphylokinase in blood. Thromb Haemostas 70:491–494

Lijnen HR, De Cock F, Matsuo O, Collen D (1992c) Comparative fibrinolytic and fibrinogenolytic properties of staphylokinase and streptokinase in plasma of different species in vitro. Fibrinolysis 6:33–37

Lijnen HR, De Cock F, Van Hoef B, Schlott B, Collen D (1994) Characterization of the interaction between plasminogen and staphylokinase. Eur J Biochem 224:143–149

Lijnen HR, Silence K, Hartmann M, Gührs K-H, Gase A, Schlott B, Collen D (1996) Fibrinolytic properties of staphylokinase mutants obtained by clustered charge-to-alanine mutagenesis. Fibrinolysis 10:177–182

Lijnen HR, Stassen JM, Collen D (1995a) Differential inhibition with antifibrinolytic agents of staphylokinase and streptokinase induced clot lysis. Thromb Haemostas 73:845–849

Lijnen HR, Stassen JM, Vanlinthout I, Fukao H, Okada K, Matsuo O, Collen D (1991b). Comparative fibrinolytic properties of staphylokinase and streptokinase in animal models of venous thrombosis. Thromb Haemostas 66:468–473

Lijnen HR, Van Hoef B, Collen D (1993a) Interaction of staphylokinase with different molecular forms of plasminogen. Eur J Biochem 211:91–97

Lijnen HR, Van Hoef B, Collen D. (1995b) Interaction of staphylokinase with human platelets. Thromb Haemostas 73:472–477

Lijnen HR, Van Hoef B, De Cock F, Okada K, Ueshima S, Matsuo O, Collen D (1991a) On the mechanism of fibrin-specific plasminogen activation by staphylokinase. J Biol Chem 266:11826–11832

Lijnen HR, Van Hoef B, Matsuo O, Collen D (1992a) On the molecular interactions between plasminogen-staphylokinase, α_2-antiplasmin and fibrin. Biochim Biophys Acta 1118:144–148

Lijnen HR, Van Hoef B, Smith RAG, Collen D (1993b) Functional properties of p-anisoylated plasmin-staphylokinase complex. Thromb Haemostas 70:326–331

Lijnen HR, Van Hoef B, Vandenbossche L, Collen D (1992b) Biochemical properties of natural and recombinant staphylokinase. Fibrinolysis 6:214–225

Matsuo O, Okada K, Fukao H, Tomioka Y, Ueshima S, Watanuki M, Sakai M (1990) Thrombolytic properties of staphylokinase. Blood 76:925–929

Ohlenschläger O, Ramachandran R, Flemming J, Gührs K-H, Schlott B, Brown LR (1997) NMR secondary structure of the plasminogen activator protein staphylokinase. J Biomol NMR 9:273–286

Ohlenschläger O, Ramachandran R, Gührs K-H, Schlott B, Brown LR (1998) Nuclear magnetic resonance solution structure of the plasminogen-activator protein staphylokinase. Biochemistry 37:10635–10642

Okada K, Lijnen HR, Moreau H, Vanderschueren S, Collen D (1996a) Procoagulant properties of intravenous staphylokinase versus tissue-type plasminogen activator. Thromb Haemost 76:857–859

Okada K, Ueshima S, Takaishi T, Yuasa H, Fukao H, Matsuo O (1996b) Effects of fibrin and α_2-antiplasmin on plasminogen activation by staphylokinase. Am J Hematol 53:151–157

Parry MAA, Fernandez-Catalan C, Bergner A, Huber R, Hopfner K-P, Schlott B, Gührs K-H, Bode W (1998) The ternary microplasmin-staphylokinase-microplasmin complex is a proteinase-cofactor-substrate complex in action. Nat Struct Biol 5:917–923

Rabijns A, Baeyens K, De Bondt HL, De Ranter C (1997a) Crystallization and preliminary x-ray analysis of recombinant staphylokinase. Proteins 27:160–161

Rabijns A, De Bondt HL, De Ranter C (1997b) Three-dimensional structure of staphylokinase, a plasminogen activator with therapeutic potential. Nat Struct Biol 4:357–360

Reddy KNN, Markus G (1972) Mechanism of activation of human plasminogen by streptokinase. Presence of active center in streptokinase-plasminogen complex. J Biol Chem 247:1683–1691

Sabovic M, Lijnen HR, Keber D, Collen D (1989) Effect of retraction on the lysis of human clots with fibrin specific and non-fibrin specific plasminogen activators. Thromb Haemostas 62:1083–1087

Sakai M, Watanuki M, Matsuo O (1989) Mechanism of fibrin-specific fibrinolysis by staphylokinase: participation of α_2-plasmin inhibitor. Biochem Biophys Res Comm 162:830–837

Sakharov DV, Lijnen HR, Rijken DC (1996) Interactions between staphylokinase, plasmin(ogen), and fibrin. Staphylokinase discriminates between free plasminogen and plasminogen bound to partially degraded fibrin. J Biol Chem 271:27912–27918

Sako T (1985) Overproduction of staphylokinase in Escherichia coli and its characterization. Eur J Biochem 149:557–563

Sako T, Sawaki S, Sakurai T, Ito S, Yoshizawa Y, Kondo I (1983) Cloning and expression of the staphylokinase gene of Staphylococcus aureus in E. coli. Molec Gen Genet 190:271–277

Sako T, Tsuchida N (1983) Nucleotide sequence of the staphylokinase gene from Staphylococcus aureus. Nucl Acids Res 11:7679–7693

Schlott B, Gührs KH, Hartmann M, Röcker A, Collen D (1998) NH2-terminal structural motifs in staphylokinase required for plasminogen activation. J Biol Chem 273:22346–22350

Schlott B, Hartmann M, Gührs K-H, Birch-Hirschfeld E, Gase A, Vetterman S, Collen D, Lijnen HR (1994b) Functional properties of recombinant staphylokinase variants obtained by site-specific mutagenesis of methionine 26. Biochim Biophys Acta 1204:235–242

Schlott B, Hartmann M, Gührs K-H, Birch-Hirschfeld E, Pohl HD, Vanderschueren S, Van de Werf F, Michoel A, Collen D, Behnke D (1994a) High yield production and purification of recombinant staphylokinase for thrombolytic therapy. Bio/technology 12:185–189

Shibata H, Nagaoka M, Sakai M, Sawada H, Watanabe T, Yokokura T (1994) Kinetic studies on the plasminogen activation by the staphylokinase-plasmin complex. J Biochem 115:738–742

Silence K, Collen D, Lijnen HR (1993a) Interaction between staphylokinase, plasmin(ogen) and α_2-antiplasmin. Recycling of staphylokinase after neutralization of the plasmin-staphylokinase complex by α_2-antiplasmin. J Biol Chem 268: 9811–9816

Silence K, Collen D, Lijnen HR (1993b) Regulation by α_2-antiplasmin and fibrin of the activation of plasminogen with recombinant staphylokinase in plasma. Blood 82:1175–1183

Silence K, Hartmann M, Gührs KH, Gase A, Schlott B, Collen D, Lijnen HR (1995) Structure-function relationships in staphylokinase as revealed by "clustered charge-to-alanine" mutagenesis. J Biol Chem 270:27192–27198

Sinnaeve P, Roelants I, Van de Werf F, Collen D (1998) Feasibility study of a liquid formulation of recombinant staphylokinase for coronary artery thrombolysis. Fibrinolysis Proteolysis 12:173–177

Suehiro A, Oura Y, Ueda M, Kakishita E (1993) Inhibitory effect of staphylokinase on platelet aggregation. Thromb Haemost 70:834–837

Sweet B, McNicol GP, Douglas AS (1965) In vitro studies of staphylokinase. Clin Sci 29:375–382

The National Institute of Neurological Disorders and Stroke rt-PA Stroke Study Group (1995) Tissue plasminogen activator for acute ischemic stroke. N Engl J Med 333:1581–1587

Vanderschueren S, Barrios L, Kerdsinchai P, Van den Heuvel P, Hermans L, Vrolix M, De Man F, Benit E, Muyldermans L, Collen D, Van de Werf F (1995a) A randomized trial of recombinant staphylokinase versus alteplase for coronary artery patency in acute myocardial infarction. Circulation 92:2044–2049

Vanderschueren S, Collen D, Van de Werf F (1996) A pilot study on bolus administration of recombinant staphylokinase for coronary artery thrombolysis. Thromb Haemost 76:541–544

Vanderschueren S, Dens J, Kerdsinchai P, Desmet W, Vrolix M, De Man F, Van den Heuvel P, Hermans L, Collen D, Van de Werf F (1997b) Randomized coronary patency trial of double-bolus recombinant staphylokinase versus front-loaded alteplase in acute myocardial infarction. Am Heart J 134:213–219

Vanderschueren SMF, Stassen JM, Collen D (1994) On the immunogenicity of recombinant staphylokinase in patients and in animal models. Thromb Haemostas 72:297–301

Vanderschueren S., Stockx L, Wilms G, Lacroix H, Verhaeghe R, Vermylen J, Collen D (1995b) Thrombolytic therapy of peripheral arterial occlusion with recombinant staphylokinase. Circulation 92:2050–205

Vanderschueren S, Van Vlaenderen I, Collen D (1997a) Intravenous thrombolysis with recombinant staphylokinase versus tissue-type plasminogen activator in a rabbit embolic stroke model. Stroke 28:1783–1788

Vanwetswinkel S, Plaisance S, Zhi-Yong Z, Vanlinthout I, Brepoels K, Lasters I, Collen D, Jespers L (2000) Pharmacokinetic and thrombolytic properties of cysteine-linked polyethylene glycol derivatives of staphylokinase. Blood 95:936–942

Winkler KC, DeWaart J, Grootsen C, Zegers BJM, Tellier NF, Vertegt CD (1965) Lysogenic conversion of staphylococci to loss of beta-toxin. J Gen Microbiol 39:321–333

Yarzabal A, Serrano RL, Puig J (1999) Modulation of staphylokinase-dependent plasminogen activation by mono- and divalent ions. Brazilian J Med Biol Res 32:39–43

Ye RQ, Kim JH, Kim BG, Szarka S, Sihota E, Wong SL (1999) High-level secretory production of intact, biologically active staphylokinase from Bacillus subtilis. Biotechnol Bioeng 62:87–96

CHAPTER 17

Desmodus rotundus (Common Vampire Bat) Salivary Plasminogen Activator

W.-D. Schleuning and P. Donner

A. Introduction

Already in the 1950s first attempts were made to treat acute myocardial infarction (AMI) by thrombolytic therapy (Fletcher et al. 1958, 1959). The acceptance of this therapeutic concept into general clinical practice, however, took more than a quarter of a century (Laffel and Braunwald 1984; Altschule 1985). Reasons for this unusually slow development are the various risks and side effects associated with the use of thrombolytic agents. Streptokinase (SK) – the primordial fibrinolytic agent – exhibits antigenicity, a most obvious shortcoming which is compounded by the presence of antibodies to SK due to previous streptococcal infections in most patients. Furthermore SK does not discriminate between clot-bound and circulating plasminogen, thereby promoting systemic plasminogen activation with the consequences of plasminogen, α_2-antiplasmin, and fibrinogen consumption and the degradation of clotting factors (Collen 1980). Urokinase (UK), the second "first generation" thrombolytic is not antigenic but likewise not specific for clot-bound plasminogen.

With the advent of the era of recombinant DNA-technology it became possible to produce the naturally occurring mammalian tissue-type plasminogen activator (tPA) (the second generation thrombolytic) in quantities sufficient for large-scale clinical trials and eventually for registration as one of the first drugs produced by recombinant DNA technology. tPA had previously been shown to exhibit the property of "fibrin specificity", i.e., its activity was strongly stimulated by fibrin (Camiolo et al. 1971; Hoylaerts et al. 1982), giving rise to the expectation that the negative effects of systemic plasminogen activation could be avoided. tPA fulfilled some but not all hopes of clinicians. Whereas it clearly exhibited fibrin specificity in vitro, there were good reasons to question whether this effect translated into a better tolerance of the drug in vivo. The results of the GUSTO (Global Utilization of Streptokinase and tPA for Occluded Coronary Arteries) trial (The GUSTO Investigators 1993) eventually demonstrated clear advantages of an accelerated tPA regimen over SK. A significantly lower total mortality was associated with the accelerated tPA regimen. To elucidate a significant difference between the two agents, however, the enrollment of large numbers of patients was required, and, even using the accelerated, front-loading tPA regimen, TIMI (Thrombolysis in Myocardial Infarction) grade 3 patency rates at 90 min were observed in only 53% of the patients. GUSTO also showed that early recanalization was essential to the survival benefit seen with tPA, and, that in general, the earlier the thrombolytic treatment was initiated, the better the results (THE GUSTO ANGIOGRAPHIC INVESTIGATORS 1993).

GUSTO nevertheless suggested that an aggressive bolus administration of an even more fibrin-dependent thrombolytic agent might further improve early and complete recanalization with a possible further increase in total survival after AMI. However, tPA, because of its short half-life and its limited fibrin specificity, cannot be considered the ideal candidate for this purpose. The administration of high doses of tPA to overcome its short half life is associated with the induction of severe plasminemia, coagulopathy, intracerebral hemorrhage, and stroke. Depletion of circulating (extrinsic) plasminogen with high plasma levels of tPA, a consequence of limited fibrin specificity, would also lower the efficacy of tPA on the clot surface through an effect known as "plasminogen steal." For these reasons there is an intensive search for thrombolytic agents that lack the evident disadvantages of first and second generation products.

In this context the vast cornucopia of opportunities offered by recombinant DNA technology has been exploited to its fullest possible extent. Numerous recombinant proteins such as staphylokinase (SAK), tPA deletion mutants, uPA/tPA hybrids, molecular chimeras between uPA and antifibrin monoclonal antibodies, hybrids between plasminogen and uPA and tPA respectively were produced and characterized in vitro and in some cases in vivo (VERSTRAETE and BACHMANN 1997; see also Chaps. 16, 18 and 19). So far none of these products have been proven clearly superior to tPA even though clinical trials with the finger-EGF deletion mutant reteplase [SEIFRIED et al. 1992; WILCOX, FOR THE INTERNATIONAL JOINT EFFICACY COMPARISON OF THROMBOLYTICS 1995; TOPOL, FOR THE GLOBAL USE OF STRATEGIES TO OPEN OCCLUDED CORONARY ARTERIES (GUSTO III) INVESTIGATORS 1997; BODE et al. 1996], with SAK (COLLEN 1998; VANDERSCHUEREN et al. 1997; ARMSTRONG et al. 2000), and with the so-called TNK-mutant of tPA (KEYT et al. 1994; CANNON et al. 1997; GIBSON et al. 1999) demonstrated at least equivalency.

We and others (GARDELL and FRIEDMAN 1993) have chosen a different path to generate a third generation fibrinolytic agent, initially starting from considerations derived from field- and evolutionary biology. Natural compounds used in medicine were selected by evolution for purposes obviously different from their therapeutic application. Many plant secondary metabolites that served as lead structures for modern drugs protect their hosts against foraging animals, or fungal, bacterial, or viral pathogens. An equally promising area for the search of pharmacologically active compounds are animal venoms. As the toxins contained in the venoms are usually proteins or peptides, only molecular cloning and heterologous expression can open this resource for exploitation. A bonanza of factors that interfere with the clotting system is provided by the digestive fluids of hematophageous animals such as leeches, ticks, hookworms, mosquitoes, and the only exclusively hematophageous mammals – the vampire bats. Building on observations from the 1930s (BIER 1932), HAWKEY (1966) discovered in the 1960s a highly potent plasminogen activator in the saliva of the common vampire *Desmodus rotundus*, which was later partially purified and characterized by CARTWRIGHT (1974). This enzyme was optimized by evolution to support the feeding habit of the animal, a functional target radically different from the physiological functions of other plasminogen activators, with their predominant role in wound healing. Already Cartwright had noticed the superior ability of the partially purified protein (which he named Desmokinase) to bring about lysis of preformed clots, and suggested its use in medicine. However, at the time it was difficult to imagine a method that would allow the generation of sufficient material for clinical testing. With this pioneering work in mind we decided to obtain vampire bat salivary plasminogen activators by cDNA cloning and heterologous gene expression.

B. Natural History of Vampire Bats

An excellent monograph on this subject has been published by GREENHALL and SCHMIDT (1988). Although vampire bats have been known by the inhabitants of the Americas for thousands of years, only during the last 150 years have they become the object of scientific studies. These were mainly directed to the control of the common vampire *Desmodus rotundus* which, as a carrier of rabies virus, had become the target of a veterinary health program in Panama and Trinidad. Much less is known about the white-winged vampire bat *Diaemus youngi* and the hairy-legged vampire bat *Diphylla ecaudata* because they have played no major role in transmission of disease. Vampire bats are not only unique among bats because of their diet but also because they exhibit a variety of interesting morphological, physiological, and behavioral adaptations related to their feeding habits. Using their razor blade sharp incisors they inflict a tiny superficial and apparently painless wound and subsequently lick the blood oozing from it. Returning to their caves they exhibit an unusual altruistic behavioral trait: they regurgitate blood in order to feed their kin that return from less successful foraging trips.

C. Biochemistry

I. Purification

The vampire bats used for our studies were collected from their nocturnal resting places in natural caves in the Guererro state of Mexico and maintained in captivity for several months. Saliva was collected by stimulating salivation by placing a small droplet of 1% pilocarpine nitrate in the buccal mucosa using a Pasteur pipette extended by a piece of silicone rubber tubing. Saliva was collected using the same device and placed in polystyrol tubes immersed in melting ice. The saliva was subsequently frozen, lyophilized, and shipped in dry ice for further processing. Saliva was fractionated by matrix-bound *Erythrina latissima* protease inhibitor (ETI) (Heussen et al. 1984). Three isoenzymes with molecular weights of 52 kd (DSPAα), 46 kd (DSPAβ), and 42 kd (DSPAγ) were identified (Gardell et al. 1989; Schleuning et al. 1992). All of these differed in their N-terminal amino acid sequences. The three forms which we named DSPA (*D. rotundus* Salivary Plasminogen Activator) could be further separated by hydrophobic interaction chromatography (Schleuning et al. 1992; Petri et al. 1992). We did not follow the nomenclature "BatPA" proposed by Gardell et al. (1989) for the following reasons:

1. There are several hundred species of bats but only three species of vampire bats. We believe that the name of the enzyme should contain a scientifically sound reference to its origin in addition to an unequivocal description of its function.
2. DSPA represents an evolutionary adaptation to a specific feeding habit and is as such clearly distinct from other plasminogen activators such as uPA and tPA, which have so far been found in all mammals investigated and are likely to occur also in bats. The designation BatPA does not take account of this distinctive characteristic and invites for a confusion with bat-tPA or uPA.

II. Cloning and Expression

RNA was isolated from *D. rotundus* salivary glands using the guanidinium isothiocyanate method and used as a template for cDNA synthesis and cloned into λgt10 and λZAP vector according to standard procedures (Maniatis et al. 1989). Candidate clones were isolated after screening the cDNA library using human tPA cDNA as a probe. Hybridizing clones were partially sequenced and found to correspond to four distinct forms (α_1, α_2, β, and γ). The cDNA sequences of the two largest forms (α_1 and α_2) were closely related (80 differences for a total of 2245 nucleotides). DSPAα_2 cDNA exhibited six nucleotide differences with a sequence published previously (Gardell et al. 1989). DSPAβcDNA was shortened by an internal 138 nucleotide deletion but displayed otherwise only one nucleotide difference when compared to DSPAα_2 cDNA. DSPAγ cDNA exhibited a 249 nucleotide long deletion and differed from DSPAα1 and β-cDNA in 54 and 23 positions respectively. When the sequences of DSPAs were aligned with tPA it became clear that they exhibited a modular structure characteristic of tPA and uPA. DSPAα_1 and DSPAα_2 consist of an array of known structural motifs: finger (F), epidermal growth factor (E), kringle (K), and protease (P). The formulas for DSPAβ and DSPAγ are EKP and KP respectively (Fig. 1). Remarkably, the K motif resem-

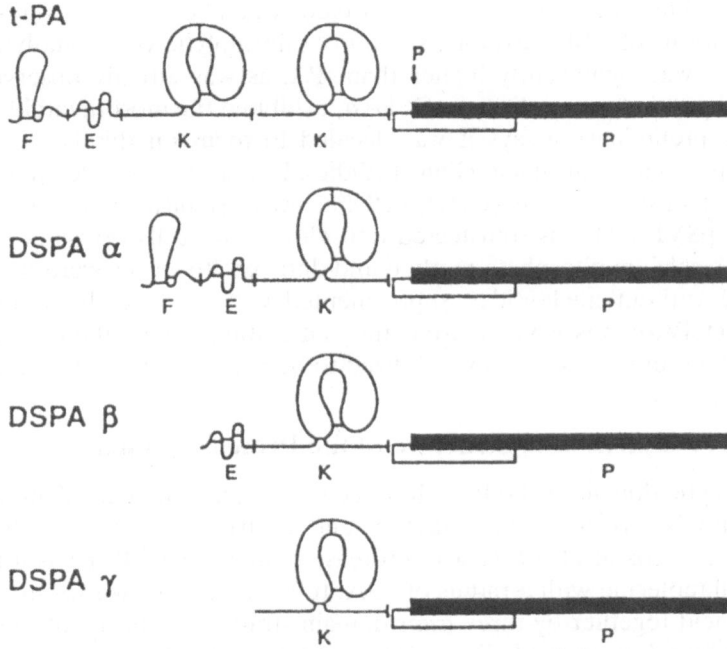

Fig. 1. Putative domain structures of tissue-type plasminogen activator (t-PA), DSPAα, β and γ. The *arrow* indicates a plasmin sensitive cleavage site. *F*, finger domain; *E*, epidermal growth factor domain; *K*, kringle domain; *P*, protease domain. (From PETRI et al. 1992, with permission)

bled more closely kringle 1 than kringle 2 of tPA. DSPAβ is closely related to DSPAα₂ whereas DSPAγ has diverged somewhat more, but still resembles α_2 more than α_1. All DSPA forms exhibit a potential N-glycosylation site in the protease domain. A further N-glycosylation site was also found at residue Asn[117] in the kringle of DSPAα₁ at the same position as in tPA (GOHLKE et al. 1997), whereas in α_2 and β it was located at a different position in the kringle. At residue Thr[61] a postranslational fucosylation site was identified (GOHLKE et al. 1996), similar to that of tPA (HARRIS et al. 1991) and uPA (BUKO et al. 1991). N-Glycosylation is not a prerequisite for the fibrinolytic activity of DSPA isoforms. No DSPAα₂ was identified in saliva. The structural features as well as the results of Southern blot experiments clearly indicated that the four forms of DSPA are products of different genes and not generated by differential mRNA splicing as suggested by other authors (GARDELL et al. 1989).

III. DSPA Gene Expression in Heterologous Host Cells

For a preliminary characterization, expression plasmids coding for DSPAα₁, α_2, β, and γ were transiently transfected into COS-1 cells. Supernatants of the transfected cells were analyzed by radial lysis in plasminogen containing casein- and fibrin-agar plates (details are described below). It was shown that all four forms of DSPA exhibited a ratio of fibrinolytic vs. caseinolytic activity which was significantly higher than tPA, as was already observed with material purified from saliva. As DSPAα_1 exhibited the most favorable profile in these preliminary assays it was decided to focus on this isoenzyme for further pharmacological and clinical studies. To this end a stable recombinant Chinese hamster ovarian (CHO)-cell line was established. The expression plasmid pSVPA 11 was transfected into dhfr⁻ CHO cells (Petri et al. 1995) using the calcium phosphate method and dhfr⁺ positive cells were selected in a MEM without nucleosides, supplemented with 2.5% dialyzed fetal calf serum. DSPAα_1 was isolated from the cell culture supernatants by affinity chromatography using ETI immobilized to Sepharose (Petri et al. 1995).

IV. Three-dimensional Structure of the Protease Domain

The catalytic domain of DSPAα_1 has been crystallized in a covalent complex with Glu-Gly-Arg-chloromethylketone and its structure solved at 2.9 Å resolution (Renatus et al. 1997). The protease domain of DSPAα1 is a roughly spherical molecule with a radius of 25 Å. It is folded into two six-stranded β-barrels held together by three *trans*-domain straps, one short "intermediate" and one long C-terminal helix, and several surface loops. It is very similar to other serine proteases, in particular to human two-chain tPA (Renatus et al. 1997). In the course of activation of a serine protease, the peptide bond between Arg[15] (or Lys[15]) and Ile[16] (or Val[16]) (chymotrypsin numbering) is cleaved giving rise to a new N-terminal Ile/Val[16]-X-X-Gly[19]. This segment inserts into the body of the proteinase, allowing formation of a salt bridge between the free N-terminus and the carboxylate group of Asp[194], causing a conformational change and creating the functional substrate binding site. DSPA is unusual in that is does not contain a cleavable activation peptide. The activation pocket which normally receives the N-terminal Ile[16] is occupied by the side chain of Lys[156], whose distal ammonium group creates an internal salt bridge with Asp[194] upon binding to fibrin and stabilizes DSPA in an enzymatically active form. This salt bridge does not exist in the enzymatically inactive form, leading to a disordered activation domain. In the absence of fibrin, the equilibrium between these two forms is thought to lean toward the enzymatically inactive form (Renatus et al. 1997).

V. Enzymology

"Fibrin specificity" is generally understood as a property of plasminogen activators that exhibit a higher catalytic efficiency in the presence of fibrin than in its absence. Indeed, whereas the catalytic efficiency of SK or UK is unaffected by fibrin, tPA catalyzed generation of plasmin is considerably stimulated in its presence and to a lesser extent also by fibrinogen, fibrin degradation products, β-amyloid, and other less well defined cofactors (Hoylaerts et al.

1982; SUENSON et al. 1990; WEITZ et al. 1991; KINGSTON et al. 1995). The underlying mechanism of this "cofactor dependence" has been intuitively attributed to "fibrin binding" by most authors but, as we will see later, there are more complex protein-protein interactions that contribute to this phenomenon.

The fibrin dependence of DSPA and tPA activity was compared using a modification of Astrup's radial fibrinolysis assay (ASTRUP and MÜLLERTZ 1954). Agar plates were prepared containing plasminogen and casein instead of fibrin. Wells of equal size were punched and filled with equimolar amounts of DSPA and tPA. As is clearly seen from Fig. 2, the caseinolytic activity of DSPAα_1 was 500 times lower than of that of tPA. Using agar plates containing plasminogen and fibrin, however, DSPA and tPA in equimolar concentration produced similar Lysis zones (data not shown).

In order to investigate the molecular basis of this striking fibrin dependence, the fibrin binding properties of all four forms of DSPA and a series of muteins' were studied and subjected to a detailed enzymological analysis (BRINGMANN et al. 1995a) using an assay in which plasminogen activation is measured by the determination of the generated plasmin in real time. Briefly the results are as follows:

1. Fibrin binding of DSPAs is exclusively dependent on the presence of a finger region, consequently DSPAα_1 and α_2 but not DSPAβ and γ bind to fibrin.
2. None of the DSPAs contain a lysine binding site.
3. All DSPAs are single chain molecules displaying substantial amidolytic activity but are almost inactive in a plasminogen activation assay in the absence of fibrin.
4. Upon addition of fibrin the catalytic efficiency (k_{cat}/k_m) of plasminogen activation by DSPAα_1 increases 105 000-fold whereas the corresponding value for tPA is only 550.

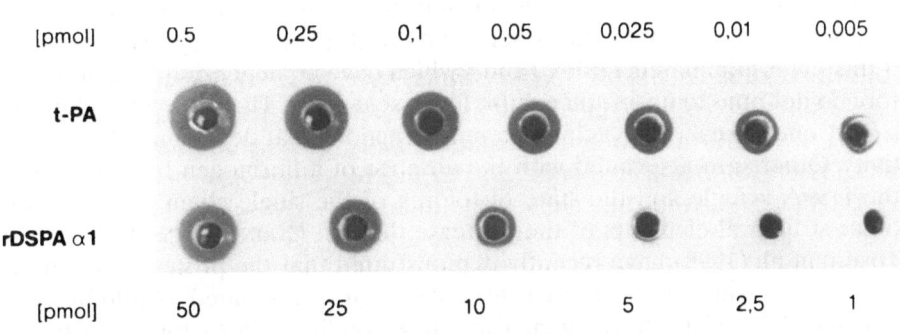

Fig. 2. Fibrin dependency of DSPAα_1 in comparison to human tissue-type plasminogen activator (tPA). 500-fold higher molar amounts (as indicated) of DSPAα_1 are required to achieve the same degree of fibrin independent lysis as tPA. (From SCHLEUNING et al. 1992, with permission)

5. The k_{cat}/k_m of DSPAα_1 and tPA in the presence of fibrin is almost equal but decreases dramatically with the loss of the finger and the kringle domain.
6. The ratio of the bimolecular rate constants of plasminogen activation in the presence of fibrin vs. fibrinogen (fibrin selectivity) of DSPAα_1, α_2, β, and γ was found to be 13000, 6500, 250, and 90 respectively. The corresponding value for tPA is only 72.

BERGUM and GARDELL (1992) have performed similar investigations with equivalent results and demonstrated in addition that tPA in contrast to DSPAα_2 (BatPA) is stimulated by the fragment X polymer, a fibrin degradation product which is generated during fibrinolysis.

These results establish a new paradigm for the molecular basis of "fibrin specificity." This term is unfortunately ambiguous because it is mostly understood without reference to other cofactors. Ideally the most "fibrin-specific" plasminogen activator would activate clot-bound plasminogen without affecting circulating plasminogen levels. It has to be taken into account however that circulating fibrinogen is a relatively potent cofactor of tPA-mediated plasminogen activation. A more meaningful quantitative parameter than "fibrin specificity" is therefore "fibrin selectivity," the quotient of the stimulatory effect of fibrin vs. fibrinogen, which ascribes to a plasminogen activator more accurately the preference for clot bound vs. circulating plasminogen. As mentioned above this factor is 13000 for DSPA whereas it is only 72 for tPA. What is the molecular basis of this striking difference? A plausible explanation was recently put forward by STEWART et al. (1998). Using light scattering spectroscopy these authors characterized two fibrin binding sites on tPA, one high affinity site associated with the finger and a low affinity site associated with kringle 2. The kringle 2 binding site is also able to react with fibrinogen and fibrin fragments. As this site is missing in DSPA, it cannot react with fibrinogen or fibrin fragments, hence the striking superiority of DSPA in fibrin selectivity. Another distinguishing structural feature of DSPA is the absence of a plasmin sensitive cleavage site in the peptide connecting the kringle and the protease domain: if such a site is introduced, fibrin selectivity decreases by a factor of ten. Likewise the fibrin selectivity of tPA increases by a similar factor if this site is eliminated. DSPAβ and γ which have no finger domain and therefore do not bind to fibrin still exhibit fibrin selectivity. Therefore fibrin binding is only one of several protein-protein interactions that determine fibrin selectivity. Others are associated with the absence of a fibrinogen binding site on the DSPA-kringle and the state of folding of the single chain molecule and other structural elements of the protease domain (BRINGMANN et al. 1995b). TOSCHI et al. (1998) have recently demonstrated that the protease domain of DSPAα_1 by itself exhibits fibrin selectivity, i.e., it is stimulated 32-fold by fibrin but only 1.5-fold by fibrinogen. The corresponding figures for the protease portion of tPA are 6 and 3 respectively.

DSPAα_2 (BatPA) is also slightly less susceptible than tPA to inactivation by plasminogen activator inhibitor 1 (PAI-1). The k_{ass} values for the interaction between PAI-1 and DSPAα_2 and two chain-tPA (in the presence of fibrinogen) are 4.4×10^6 and $13.1 \times 10^6 \text{M}^{-1} \text{sec}^{-1}$, respectively (GARDELL et al. 1990). In a similar assay system, but in the presence of fibrin, GRUBER (1995) compared DSPAα_1 ($k_{ass} = 0.7 \times 10^6 \text{M}^{-1} \text{sec}^{-1}$) and two-chain-tPA ($k_{ass} = 2.5 \times 10^6 \text{M}^{-1} \text{sec}^{-1}$).

D. Pharmacology

I. In Vitro Clot Lysis

Whole blot clot lysis (BEER et al. 1994) is employed to determine the activity of a fibrinolytic agent in a "semi-natural" environment, i.e., in the presence of all the cofactors, inhibitors known or unknown that are also likely to be present in vivo. Whole human blood clots were generated *in vitro*, aged for 1 h and placed in autologous plasma in the presence of varying concentrations of either tPA or DSPAα_1. The thrombolytic potential of both PAs, expressed as percent change in clot wet weight and the plasma fibrinogen content was determined after 6 h of incubation. tPA and DSPAα_1 exhibited a similar thrombolytic profile (Fig. 3). In contrast to tPA, however, only minimal fibrinogen degradation was observed in the DSPA-containing sample (SCHLEUNING et al. 1992).

We have also compared the fibrin selectivity of tPA and its mutein TNK-tPA with that of DSPAα_1. Kinetic analysis of plasminogen activation revealed that TNK-tPA is about eight times more fibrin selective than wild type tPA. However, the fibrin selectivity of DSPAα_1 is still 12 times higher than that of TNK-tPA.

At low concentrations (1–5 nmol/l) DSPAα_1 clearly exhibits a higher efficacy than TNK-tPA and tPA. At 5–25 nmol/l tPA's efficacy is very similar to that of DSPAα_1. The induction of clot lysis by 50 nmol/l tPA is however less efficacious and at 100 nmol/l tPA only incomplete clot lysis is achieved due to "plasminogen steal." The "plasminogen steal" effect is not observed at any concentration during DSPAα_1-mediated clot lysis, but flagrant at concentrations above 100 nmol/l TNK-tPA. SAKHAROV et al. (1999) investigated the lysis of compacted crosslinked human plasma clot in the presence of 9 different plasminogen activators. They found that the fibrin-selective PAs staphylokinase, TNK-tPA and DSPA induced rapid lysis in concentration ranges of 80-, 260- and 3500-fold respectively, much wider than that for tPA (35-fold). The phenomenon of smaller concentration ranges might be due to excessive activation of circulating plasminogen by less fibrin selective PAs followed by inactivation of plasmin by α_2-antiplasmin. However in terms of speed of lysis these three PAs exceeded tPA only slightly.

DSPAα_1- and SAK-mediated clot lysis is not accompanied by fibrinogenolysis even at high concentrations. In contrast, complete clot lysis induced by tPA, in particular at higher concentrations (25–100 nmol/l), is compromised by severe fibrinogen degradation. Even at 10 nmol/l tPA the level

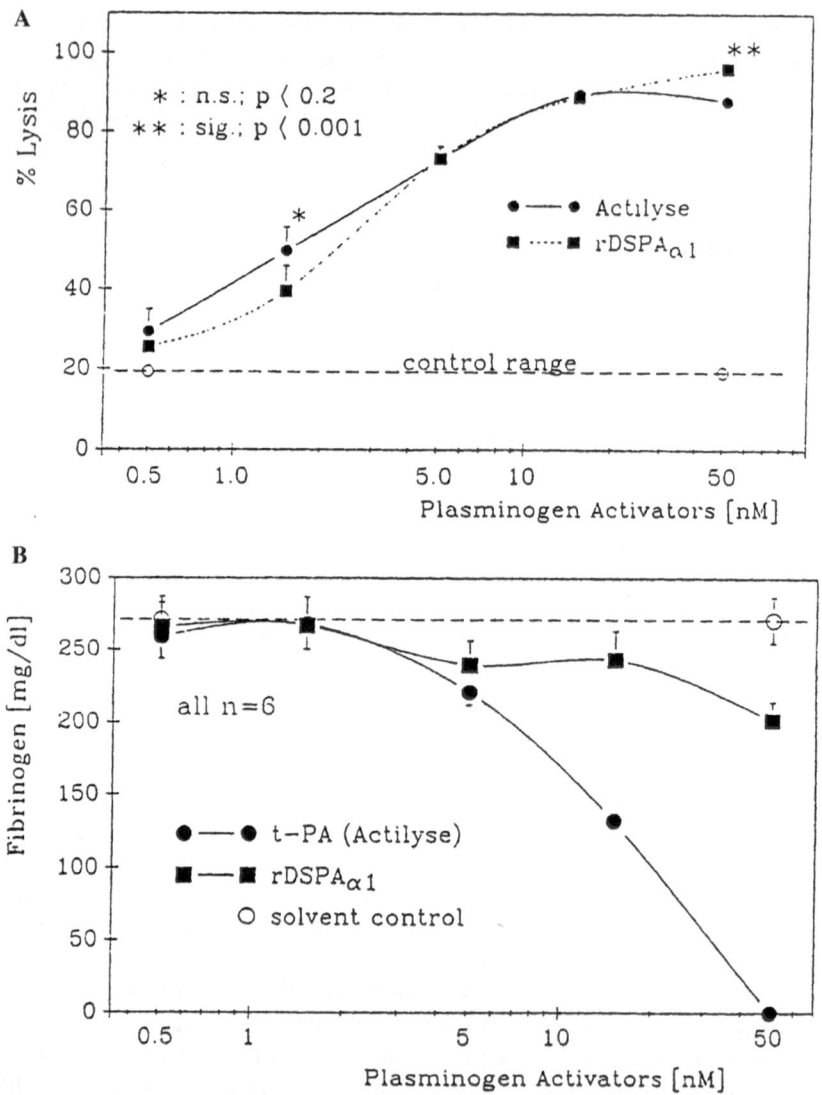

Fig. 3A,B. Human whole blood clot lysis in vitro. **A** Concentration response curves for tissue-type plasminogen activator (Actilyse) and DSPAα_1. **B** Fibrinogen concentrations after 6 h of incubation. (From Schleuning et al. 1992, with permission)

of functional fibrinogen is diminished by 70%. Whereas TNK-tPA is clearly more fibrin-selective than tPA, at elevated concentrations complete clot lysis is only achieved at the expense of notable fibrinogenolysis, indicating that the fibrin selectivity of TNK-tPA is still unsatisfactory. The consumption of α_2-antiplasmin corroborates our findings concerning fibrinogenolysis. α_2-Antiplasmin is virtually undetectable in plasma samples containing 10 nmol/l and higher concentrations of tPA and TNK-(tPA), while DSPAα_1-samples still contain measurable amounts of α_2-antiplasmin even at 100 nmol/l. Therefore, DSPAα_1 is a plasminogen activator with strict fibrin cofactor requirement and more potent than tPA and TNK-tPA at low doses (BRINGMANN et al. 1995a).

HARE and GARDELL (1992) compared DSPAα_2 (BatPA), tPA, and SK in a plasma clot lysis model using radiolabeled fibrinogen. Excessive fibrinogen degradation was only observed in the tPA- and SK-containing samples. These authors also demonstrated that the addition of a clot lysate did not stimulate DSPAα_2 activity, indicating that soluble fibrin degradation products do not influence DSPA activity.

II. Carotid and Femoral Artery Thrombosis in Rabbits

Thrombosis was induced by a copper coil inserted in the common carotid artery of groups of six anesthetized rabbits. Heparin and aspirin were given in addition to the thrombolytics. These experiments suggested that DSPA was at least as efficacious as tPA and probably 2–3 times more potent (MUSCHICK et al. 1993). With regard to hemostatic parameters, DSPAα_1 differed from tPA, as it was demonstrated to cause no significant fibrinogenolysis nor plasminogen depletion. α_2-Antiplasmin plasma levels decreased to a lesser extent than with tPA. The plasma half-life of DSPAα_1 exceeded that of tPA at all doses (Fig. 4).

In a rabbit model of femoral arterial thrombosis DSPAα_2 (Bat PA) was evaluated and compared with tPA (GARDELL et al. 1991). A thrombus was formed by injecting autologous whole blood, Ca^{++}, and thrombin into an isolated segment of the femoral artery. Following a 60-min aging period, DSPAα_2 or tPA were given by bolus intravenous injections and blood flow restoration was measured with an electromagnetic flow probe. At 14 nmol/kg and 42 nmol/kg, tPA reperfused 15% and 78% of the rabbits, respectively whereas at 4.7, 8.1, 14, and 42 nmol/kg, DSPAα_2 reperfused 0, 50, 75, and 80%. The thrombolytic efficacy of DSPAα_2 at 14 nmol/kg was comparable to a threefold higher dose of tPA (42 nmol/kg). Although the incidence of reperfusion by DSPAα_2 was not significantly different at 42 nmol/kg and 14 nmol/kg, other indices of efficacy such as median time to reperfusion and residual thrombus mass were significantly improved using the threefold higher dose.

Analysis of serial blood samples revealed that the administration of tPA (42 nmol/kg) but not of DSPA at the same dose elicited a marked decrease in the levels of plasma fibrinogen. This finding confirms the remarkable fibrin selectivity already noted using purified reagents as well as whole blot clot lysis.

III. Myocardial Infarction in Dogs

Copper coil-induced AMI in dogs is widely considered as one of the most meaningful models of human AMI. Therefore the thrombolytic properties of tPA and DSPAα_1 were compared in this system (WITT et al. 1994b). A copper

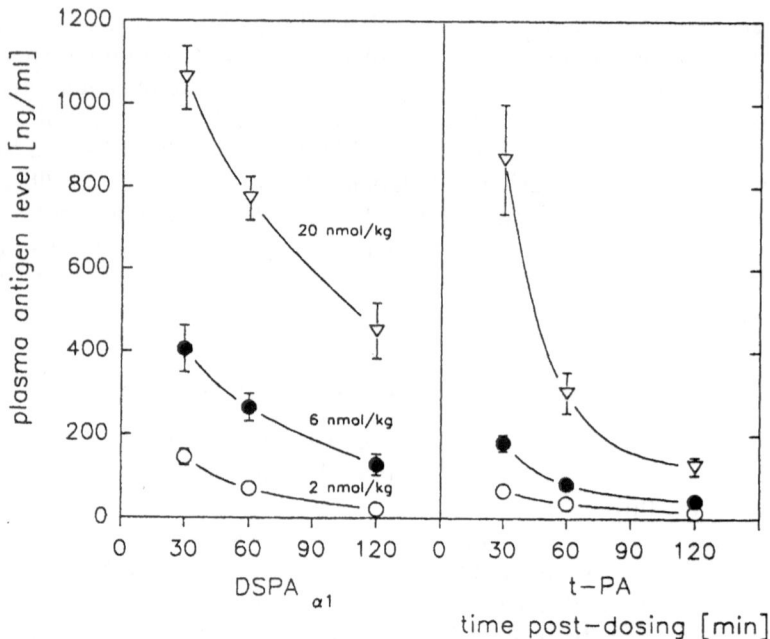

Fig. 4. Plasma antigen levels (mean ± SEM, $n = 6$–7) of DSPAα_1 and tissue-type plasminogen activator (tPA) following intravenous bolus injection at 30, 60, and 120 min. (From Muschick et al. 1993, with permission)

coil was inserted into a coronary artery of an anesthetized animal leading to the local formation of a thrombus. All dogs received heparin. Whereas control animals did not reperfuse within 180 min, intravenous bolus administration of DSPAα_1 at 25, 50, and 100 µg/kg resulted in 100% incidence of recanalization within 37, 23, and 18 min, respectively. tPA at 63 and 125 µg/kg reopened the coronary arteries in 33% and 50% of cases within 40 min. Eighty-three percent of the arteries were still patent 3 h after 50 and 100 µg/kg DSPAα_1, whereas only 20% (1 of 5) of all coronaries originally recanalized with both doses of tPA were still open at 3 h (Fig. 5). The clearance of DSPAα_1 was lower compared with tPA due to a prolonged terminal half-life.

IV. Experimental Pulmonary Embolisms in Rats

The thrombolytic properties of DSPAα_1 and tPA were compared in a rat model of pulmonary embolism (Witt et al. 1992). Whole blood clots, produced *in vitro* and labeled with ^{125}I-fibrinogen, were embolized into the lungs of anesthetized rats. Thrombolysis was calculated from the difference between initial clot radioactivity and that remaining in the lungs after 60 min. Blood was sampled for γ-counting, measurement of hemostatic factors, and plasminogen activator antigen levels. Thrombolysis using DSPAα_1 was significantly faster

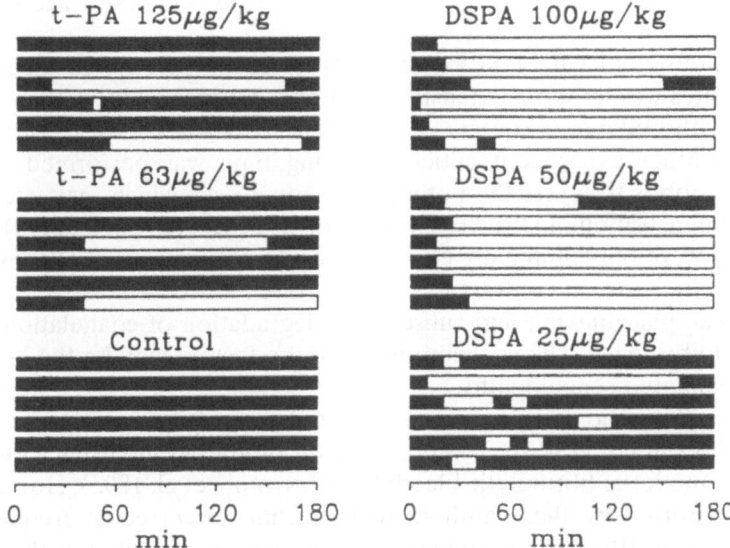

t–PA 125μg/kg   DSPA 100μg/kg

t–PA 63μg/kg   DSPA 50μg/kg

Control   DSPA 25μg/kg

0   60   120   180   0   60   120   180
        min   min

Fig. 5. Patency profiles of coronary arteries in individual dogs after intravenous bolus administration of different amounts (as indicated) of DSPAα_1, tissue-type plasminogen activator (tPA), or solvent (control). Each *bar* represents the angiographically defined patency status of a single dog coronary artery throughout the observation period of 180 min after dosing. *Solid bars* reflect periods of coronary occlusion; *open bars* depict periods of arterial reflow. (From Witt et al. 1994b, with permission)

than with tPA. Moreover tPA significantly decreased fibrinogen, plasminogen, and α_2-antiplasmin. Compared with tPA, DSPAα_1 was clearly the more potent and more clot selective ("fibrin selective") thrombolytic agent.

V. Bleeding Models

The propensity of bleeding induction by DSPAα_1 as compared to tPA and SK was studied in a venipuncture model in rat mesenteric arcade veins. Venipuncture bleeding time (5 min after thrombolysis induction) was prolonged >3 min in all of six animals given 30 nmol/kg tPA but only in one of six treated with 300 nmol/kg DSPA. One hour after administration of the thrombolytic agents, venipuncture bleeding time with tPA was still prolonged to 7.7 ± 3.5 min but only to 2.5 ± 1.8 min after DSPAα_1. Rebleeding, occurred in three of six animals treated with 30 nmol/kg tPA, in none up to 100 nmol/kg DSPA, but in five of six rats given 300 nmol/kg DSPA (Gulba et al. 1995).

In aspirin-pretreated rabbits, the administration of 14 or 42 nmol/kg DSPAα_2 increased the template bleeding time to a similar extent as did the administration of 42 nmol/kg of tPA. However, the cuticle bleeding times in the same animals were only prolonged after the injection of tPA. Even when the rabbits were pretreated with aspirin, there was only mild and transient bleeding observed with DSPAα_2. Furthermore, the severity of bleeds from surgical sites was far more extensive in rabbits that received tPA than in those treated with DSPAα_2 (Gardell et al. 1991).

In another experiment cuticle bleeding time was performed in anesthetized rabbits to assess the potential bleeding risks which may occur after bolus administration of DSPAα_2 (BatPA) or tPA (Mellott et al. 1995). Bleeding times were only minimally elevated with DSPAα_2 whereas tPA exhibited a 6.2-fold increase. In contrast to DSPAα_2, tPA induced a massive activation of systemic plasminogen and subsequent degradation of coagulation-factors VIII and fibrinogen. The consumption of these factors may be the key event in tPA-mediated coagulopathy.

Using the same dosages, a severe degradation of factor VIII and fibrinogen by tPA but not DSPAα_2 was observed by another group using a rabbit ear puncture model of fibrinolytic bleeding (Montoney et al. 1995). However the finding reported by these authors that bleeding occurred as frequently in DSPAα_2- as in tPA-treated animals is surprising and puzzling in the light of the observed factor VIII and fibrinogen degradation and in contrast to the results obtained by Mellott et al. (1995).

It has been suggested, that the bleeding time is the major risk indicator for severe hemorrhage in patients treated with thrombolytic agents (Gimple et al. 1989; Hirsch and Goldhaber 1990). If this assessment is correct, the results of the bleeding models suggest that thrombolysis with DSPAα_1 is superior in terms of therapeutic safety to other agents currently approved for clinical use. However, there are examples in which bleeding occurred despite only moderately prolonged bleeding time, as there are numerous examples of patients with massive prolongation of bleeding times that do not bleed. Thus a definitive statement on improved safety associated with the therapeutic use of DSPAα_1 will have to await the results of clinical trials.

E. Pharmacokinetics

A sandwich ELISA-system with affinity-purified and peroxidase-labeled DSPAα_1 antibodies raised in rabbits was developed and exhibited a limit of quantification of 3 ng/ml in undiluted spiked plasma. Accuracy was 98%–108% and precision accounted for 3%–9.5%. No cross-reactivity with human tPA or endogenous matrix constituents interfering with assay results was observed (Hildebrand et al. 1995). After i.v. administration of DSPAα_1 at 1 and 3 mg/kg in cynomolgus monkeys antigen levels in plasma were dose-dependent in both genders and exhibited a triphasic disposition profile with half-lives of 0.04–0.26 h, 0.6–3 h, and 4–8.5 h. The mean residence time of DSPAα_1 ranged from 3 to 9 h and total clearance was approximately 2 ml/min/kg independent of sex and dose. Several toxicokinetic studies with single or repeated administration of unlabeled DSPA were monitored by ELISA and FCLA (fibrin clot lysis assay) to measure antigen and activity levels. The dose range used was 1–30 mg/kg.

In rats and monkeys DSPAα_1 was characterized by long terminal half-lives of 1–2 h and 5–8 h with bi- or triphasically declining plasma levels. The terminal phase represented a partial AUC of 42–57% in monkeys. Total clearance accounted for 6–11 ml/min per kg and approx. 2 ml/min per kg in rats and monkeys, respectively. The volume of distribution in the central compartment was 0.5–0.1 l/kg in both species. In both species plasma antigen and activity levels exhibited a linear correlation with a slope close to 1 over the dose range of 1–30 mg/kg. In terms of distribution in rats radiolabel indicative for ^{125}I-DSPAα_1 was found in highly perfused organs and tissues. By means of allometric extrapolation a total clearance of approx. 1 ml/min per kg was predicted for humans (HILDEBRAND et al. 1996).

DSPAα_1 therefore displayed an advantageous pharmacokinetic and pharmacodynamic profile, especially due to its low total clearance, its long terminal half-life, and the full fibrinolytic activity of antigen present in the plasma, as compared to other established fibrinolytics (e.g., tPA). Animal data encourage a therapeutic dosage scheme with an i.v. bolus in humans.

These data confirm previous observations made in dogs in the context of a coronary thrombolysis model (WITT et al. 1994b). The half-life of DSPAα_1 in dogs was greatly prolonged compared with tPA, with the mean residence time being about 40–50 times longer. The clearance of DSPAα_1 was about five to nine times slower than that of tPA, TNK-tPA being cleared about half as fast as wild type tPA. This may translate into a minor advantage of TNK-tPA over tPA in relative potency and patency (including reocclusion). Reteplase, another mutant consisting of the kringle 2 and protease domains of human tPA, has a half-life of 13 min and a clearance of ~5 ml/min/kg in dogs. It is being pursued in clinical trials as a less costly alternative to tPA. Results in a canine model of coronary thrombosis as well as in patients with AMI indicate that the half-life achieved with reteplase necessitates a second bolus application to avoid early reocclusion and to achieve a satisfying degree of reperfusion (MARTIN et al. 1991; WILCOX, FOR THE INTERNATIONAL JOINT EFFICACY COMPARISON OF THROMBOLYTICS 1995; SMALLING et al. 1995; BODE et al. 1996; TOPOL, FOR THE GLOBAL USE OF STRATEGIES TO OPEN OCCLUDED CORONARY ARTERIES [GUSTO III] INVESTIGATORS 1997).

The thrombolysis experiments performed with DSPAα_2 in a canine model of arterial thrombosis by MELLOT et al. (1992) are relevant to pharmacokinetics and are therefore described here in further detail. The reperfusion incidences after the administration of tPA and DSPAα_2 at 14 nmol/kg were 50% and 88%, respectively . The mean times to reperfusion were not significantly different in the DSPAα_2 and tPA treatment groups. All animals reoccluded during the 4-h trial; however, the mean time to reocclusion in the dogs treated with DSPAα_2, 144 min, was significantly delayed relative to those treated with tPA, 37 min.

The approximately six times longer mean residence time of DSPAα_2 relative to tPA was a likely contributor to the delayed reocclusion and superior thrombolytic efficacy in the DSPA treatment group. The clearance profile for tPA was monoexponential, with a t1/2 of 2.4 min. The clearance profile for DSPAα_2 was biexponential, with a t1/2α of ~1 min and a t1/2β of ~20 min. Interestingly, the steady-state volume of distribution displayed by DSPAα_2 was 16 times greater than that of tPA. It was proposed that some DSPA was initially sequestered in an extraplasma compartment following administration and subsequently released back slowly into circulation; however no experimental data were provided to support this suggestion.

The fibrinogen levels were essentially unaffected following the administration of DSPAα_2 or tPA. It is also noteworthy that the mean residence time of DSPAα_2, albeit significantly longer than that of tPA, is still 6–8 times shorter than that of DSPAα_1. Indeed, this may be the reason why experiments conducted with DSPA$\alpha_{2\alpha}$ appear to show a stronger reocclusion tendency than those performed with DSPAα_1.

F. Toxicology

A detailed toxicological study in rats and monkeys revealed no negative result that would preclude the use of the substance in humans. The induction of an immunological memory but no antibody formation was observed after single bolus injections of 1 mg/kg body weight. Only after repeated injections of DSPAα_1 into rats was antibody formation observed. However no effect on the fibrinolytic potential in these animals nor any crossreactivity with human tPA was found (Witt et al. 1994a). These data indicate that antibody formation is no major concern restricting therapeutic applications for at least two successive treatments.

G. Phase I Clinical Studies

A phase 1 clinical trial with healthy volunteers was performed in order to determine the pharmacokinetics and the safety and tolerability of DSPAα_1. Subtherapeutic doses of 0.01, 0.03, and 0.05 mg DSPAα_1 were applied by intravenous bolus administration. The study was designed as an open uncontrolled interindividual group comparison. The pharmacokinetic parameters were determined by measuring DSPAα_1 antigen by ELISA up to 36 h after injection. Cardiovascular monitoring was extended over 1 h and laboratory parameters such as DSPAα_1 activity (fibrin clot lysis assay) hemodynamics, and adverse events were monitored up to 24 h post injection. The development of DSPAα_1 antibodies was checked for up to three month after the injection.

No clinically relevant changes or trends in hematology and clinical chemistry, no negative clinical effects on blood coagulation were observed, and no formation of antibodies could be detected. Therefore there were no safety-related concerns in the dose range tested and the pharmacokinetic prediction from animal studies was confirmed in humans. Based on the available pharmacokinetic data, and in comparison to pharmacokinetic and clinical data of tPA, the therapeutically effective dose of DSPAα_1 was predicted to be 0.5 mg/kg body weight (M. Mahler and T.W. Seifer, personal communication).

H. Phase Ib Clinical Studies

A phase Ib clinical study on AMI patients has been initiated but no data on the outcome have been published yet.

I. Conclusions and Perspectives

The underlying cause of AMI is the acute thrombotic occlusion of a coronary vessel. The re-establishment of coronary blood flow by thrombolytic therapy or surgical intervention in the early phase of this event is imperative to limit the ischemic damage to the myocardium. Numerous multi-center, double-blind studies have shown that early mortality of AMI can be significantly reduced by either method. The positive effect of coronary recanalization also translates into a reduction of long term mortality. Thrombolytic therapy relies on various drugs and various regimens all of which have the target to minimize the speed of reopening of the coronary vessel and the associated risks – mainly hemorrhagic complications – which in its most dreaded form, the life threatening intracranial bleeding, currently can still attain 1% (DE JAEGERE et al. 1992; LEVINE et al. 1995; GURWITZ et al. 1998). However, even the most aggressive regimens applied in clinical practice today achieve TIMI grade 3-patency rates that are relevant for the prognosis of the patient (TOPOL 1993) only in about 60% of the occluded vessels, and 20%–30% of AMI patients do not respond to thrombolytic therapy at all. Moreover, acute reocclusion occurs in about 10% of the patients who initially reperfuse and the incidence of mortality is still intolerably high (approximately 10% in the GISSI-2 [GRUPPO ITALIANO PER LO STUDIO DELLA STREPTOCHINASI NELL'INFARTO MIOCARDICO 1986] and ISIS-3 trials [ISIS 3 (THIRD INTERNATIONAL STUDY OF INFARCT SURVIVAL) COLLABORATIVE STUDY GROUP (1992)] and 7% in the GUSTO trial [THE GUSTO INVESTIGATORS 1993]). This limited efficacy and the number of hemorrhagic side effects has lead to a renaissance of the concept of mechanical recanalization, which however is not applicable to large patient populations for logistic and economic reasons. Hence if a further decrease in mortality from AMI is to be achieved there is the need for a new class of more efficacious and safer thrombolytics. Introduction of a fibrinolytic agent that is associated with a minimal bleeding risk would probably improve overall survival among AMI patients by increasing the number of patients eligible for therapy. A high number of patients today are excluded from the benefits of the treatment because of the risk of stroke or other complications. A plasminogen activator which is strictly fibrin selective and not activated by fibrinogen, fibrin degradation products, or denatured protein spares plasminogen and clotting factors. In contrast to tPA, DSPAα_1 is not activated by β-amyloid peptides adding another important safety feature, because deposits consisting of such material are frequently found in the vasculature of older patients (Kingston et al. 1995). DSPA, therefore, could present an advantageous feature when treating elderly stroke patients with thrombolytics. In all animal models tested so far DSPAα_1 and DSPAα_2 have proven to be more efficacious and to be associated with fewer undesirable effects than any other thrombolytic currently in use. Moreover the favorable pharmacokinetic profile of DSPAα_1 permits using a single i.v. bolus regimen, which is much easier to apply than the combined front loading and infusion regimens which are currently most widely used. This feature is particularly attractive because it opens the possibility to commence thrombolytic therapy in the emergency vehicle or even in the home of the patient and thus broadens the current scope of thrombolytic therapy.

List of Abbreviations and Acronyms

AMI   acute myocardial infarction
AUC   area under the curve
CHO   Chinese hamster ovary (cells)
DSPA   *Desmodus rotundus* Salivary Plasminogen Activator
ETI   *Erythrina latissima* trypsin inhibitor
GUSTO   Global Utilization of Streptokinase and t-PA for Occluded coronary arteries
MEM   modified Eagle's medium
SAK   staphylokinase
SK   streptokinase
tc-tPA   two chain tPA
TIMI   thrombolysis in myocardial infarction
tPA   tissue-type plasminogen activator
UK   urokinase (two-chain uPA)
uPA   urinary-type plasminogen activator

References

Altschule MD (1985) The coronary occlusion story. Prolonged neglect of early clinicopathologic findings and of the experimental animal physiology they stimulated. Chest 87:81–84
Armstrong PW, Burton JR, Palisaitis D, Thompson CR, Van de Werf F, Rose B, Collen D, Teo KK (2000) Collaborative angiographic patency trial of recombinant staphylokinase (CAPTORS). Am Heart J 139:820–823
Astrup T, Müllertz S (1954) The fibrin plate method for estimating fibrinolytic activity. Arch Biochem Biophys 40:346–351
Beer JH, Kläy H-P, Herren T, Haeberli A, Straub PW (1994) Whole blood clot lysis: Enhanced by exposure to autologous but not to homologous plasma. Thromb Haemost 71:622–626

Bergum PW, Gardell SJ (1992) Vampire bat salivary plasminogen activator exhibits a strict and fastidious requirement for polymeric fibrin as its cofactor, unlike human tissue-type plasminogen activator. J Biol Chem 267:17726–17731

Bier O (1932) Action anticoagulante et fibrinolytique de l'extrait des glandes salivaires d'une chauve-souris hematophage *Desmodus rufus*. C R Soc Biol (Paris) 110:129

Bode C, Smalling RW, Berg G, Burnett C, Lorch G, Kalbfleisch JM, Chernoff R, Christie LG, Feldman RL, Seals AA, Weaver WD, for the RAPID II Investigators (1996) Randomized comparison of coronary thrombolysis achieved with double-bolus reteplase (recombinant plasminogen activator) and front-loaded, accelerated alteplase (recombinant tissue plasminogen activator) in patients with acute myocardial infarction. Circulation 94:891–898

Bringmann P, Gruber D, Liese A, Toschi L, Krätzschmar J, Schleuning W-D, Donner P (1995b) Structural features mediating fibrin selectivity of vampire bat plasminogen activators. J Biol Chem 270:25596–25603

Bringmann P, Liese A, Toschi L, Petri T, Schleuning W-D, Donner P (1995a) Desmodus salivary plasminogen activator $\alpha 1$ is more potent and fibrin-selective in vitro than tPA, and TNK-(tPA). Thromb Haemostas 73:1129 (Abstract 876)

Buko AM, Kentzer EJ, Petros A, Menon G, Zuiderweg ERP, Sarin VK (1991) Characterization of a posttranslational fucosylation in the growth factor domain of urinary plasminogen activator. Proc Natl Acad Sci USA 88:3992–3996

Camiolo SM, Thorsen S, Astrup T (1971) Fibrinogenolysis and fibrinolysis with tissue plasminogen activator, urokinase, streptokinase-activated human globulin, and plasmin. Proc Soc Exp Biol Med 138:277–280

Cannon CP, McCabe CH, Gibson CM, Ghali M, Sequeira RF, McKendall GR, Breed J, Modi NB, Fox NL, Tracy RP, Love TW, Braunwald E, and the TIMI 10A Investigators (1997) TNK-tissue plasminogen activator in acute myocardial infarction. Results of the thrombolysis in myocardial infarction (TIMI) 10A dose-ranging trial. Circulation 95:351–356

Cartwright T (1974) The plasminogen activator of vampire bat saliva. Blood 43:317–326

Collen D (1980) On the regulation and control of fibrinolysis. Thromb Haemostas 43:77–89

Collen D (1998) Staphylokinase: a potent, uniquely fibrin-selective thrombolytic agent. Nat Med 4:279–284

De Jaegere PP, Arnold AA, Balk AH, Simoons ML (1992) Intracranial hemorrhage in association with thrombolytic therapy: Incidence and clinical predictive factors. JACC 19:289–294

Fletcher AP, Alkjaersig N, Smyrniotis FE, Sherry S (1958) Treatment of patients suffering from early myocardial infarction with massive and prolonged streptokinase therapy. Trans Assoc Am Physicians 71:287–296

Fletcher AP, Sherry S, Alkjaersig N, Smyrniotis FE, Jick S (1959) The maintenance of a sustained thrombolytic state in man. II. Clinical observations on patients with myocardial infarction and other thromboembolic disorders. J Clin Invest 38: 1111–1119

Gardell SL, Duong LT, Diehl RE, York JD, Hare TR, Register RB, Jacobs JW, Dixon RAF, Friedman PA (1989) Isolation, characterization and cDNA cloning of a vampire bat salivary plasminogen activator. J Biol Chem 264:17947–17952

Gardell SJ, Friedman PA (1993) Vampire bat salivary plasminogen activator. In: Lorand L, Mann KG (eds) Proteolytic enzymes in coagulation, fibrinolysis and complement activation. Methods in Enzymology, vol 223. Academic Press, pp 233–249

Gardell SJ, Hare TR, Bergum PW, Cuca GC, O'Neill-Palladino L, Zavodny SM (1990) Vampire bat plasminogen activator is quiescent in human plasma in the absence of fibrin unlike human tissue plasminogen activator. Blood 76:2560–2564

Gardell SJ, Ramjit DR, Stabilito II, Fujita T, Lynch JJ, Cuca GC, Jain D, Wang S, Tung J-S, Mark GE, Shebuski RJ (1991) Effective thrombolysis without marked plasminemia after bolus intravenous administration of vampire bat salivary plasminogen activator in rabbits. Circulation 84:244–253

Gibson CM, Cannon CP, Murphy SA, Adgey AAJ, Schweiger MJ, Sequeira RF, Grollier G, Fox NL, Berioli S, Weaver WD, Van de Werf F, Braunwald E, for the TIMI 10B Investigators (1999) Weight-adjusted dosing of TNK-tissue plasminogen activator and its relation to angiographic outcomes in the thrombolysis in myocardial infarction 10B trial. Am J Cardiol 84:976–980

Gimple LW, Gold HK, Leinbach RC, Coller BS, Werner W, Yasuda T, Johns JA, Ziskind AA, Finkelstein D, Collen D (1989) Correlation between template bleeding times and spontaneous bleeding during treatment of acute myocardial infarction with recombinant tissue-type plasminogen activator. Circulation 80:581–588

Gohlke M, Baude G, Nuck R, Grunow D, Kannicht C, Bringmann P, Donner P, Reutter W (1996) O-linked L-fucose is present in *desmodus rotundus* salivary plasminogen activator. J Biol Chem 271:7381–7386

Gohlke M, Nuck R, Kannicht C, Grunow D, Baude G, Donner P, Reutter W (1997) Analysis of site-specific N-glycosylation of recombinant desmodus rotundus salivary plasminogen activator rDSPAα-1 expressed in Chinese hamster ovary cells. Glycobiol 7:67–77

Greenhall AM, Schmidt U (1988) Natural history of vampire bats. CRC Press, Boca Raton, Florida

Gruber D (1995) Studien zum Wirkmechanismus neuer Plasminogen-Aktivatoren aus dem Speichel der Vampir-Fledermaus Desmodus rotundus, Dissertation, Freie Universität Berlin

Gruppo Italiano per lo Studio della Streptochinasi nell'Infarto Miocardico (GISSI) (1986) Effectiveness of intravenous thrombolytic treatment in acute myocardial infarction. Lancet 1:397–401

Gulba DC, Praus M, Witt W (1995) DSPA alpha – Properties of the plasminogen activators of the vampire bat *Desmodus rotundus*. Fibrinolysis 9 [Suppl 1]:91–96

Gurwitz JH, Gore JM, Goldberg RJ, Barron HV, Breen T, Rundle AC, Sloan MA, French W, Rogers WJ, for the Participants in the National Registry of Myocardial Infarction 2 (1998) Risk for intracranial hemorrhage after tissue plasminogen activator treatment for acute myocardial infarction. Ann Intern Med 129:597–604

Hare TR, Gardell SJ (1992) Vampire bat plaminogen activator promotes robust lysis of plasma clots in a plasma milieu without causing fluid phase plasminogen activation. Thromb Haemost 68:165–169

Harris RJ, Leonard CK, Guzzetta AW, Spellman MW (1991) Tissue plasminogen activator has an O-linked fucose attached to threonine-61 in the epidermal growth factor domain. Biochemistry 30:2311–2314

Hawkey C (1966) Plasminogen activator in the saliva of the vampire bat *Desmodus rotundus*. Nature 211:434–435

Heussen G, Joubert F, Dowdle EB (1984) Purification of human tissue plasminogen activator with *Erythrina* trypsin inhibitor. J Biol Chem 259:11635–11638

Hildebrand M, Bhargava AS, Bringmann P, Schütt A, Verhallen P (1996) Pharmacokinetics of the novel plasminogen activator *desmodus rotundus* plasminogen activator in animals and extrapolation to man. Fibrinolysis 10:269–276

Hildebrand M, Bunte T, Bringmann P, Schütt A (1995) Development of an ELISA for the measurement of DSPAα1 (*Desmodus rotundus* salivary plasminogen activator) in plasma and its application to investigate pharmacokinetics in monkeys. Fibrinolysis 9:107–111

Hirsch DR, Goldhaber SZ (1990) The bleeding time: Its potential utility among patients receiving thrombolytic therapy. Am Heart J 119:158–167

Hoylaerts M, Rijken DC, Lijnen HR, Collen D (1982) Kinetics of the activation of plasminogen by human tissue plasminogen activator. J Biol Chem 257:2912–2919

ISIS 3 (Third International Study of Infarct Survival) Collaborative Study Group (1992) ISIS 3: A randomised comparison of streptokinase vs tissue plasminogen activator vs anistreplase and of aspirin plus heparin vs aspirin alone among 41 299 cases of suspected acute myocardial infarction. Lancet 339:753–770

Keyt BA, Paoni NF, Refino CJ, Berleau L, Nguyen H, Chow A, Lai J, Peña L, Pater C, Ogez J, Etcheverry T, Botstein D, Bennett WF (1994) A faster-acting and more potent form of tissue plasminogen activator. Proc Natl Acad Sci USA 91: 3670–3674

Kingston IB, Castro MJM, Anderson S (1995) In vitro stimulation of tissue-type plasminogen activator by Alzheimer amyloid β-peptide analogues. Nat Med 1:138–142

Krätzschmar J, Haendler B, Langer G, Boidol W, Bringmann P, Alagon A, Donner P, Schleuning W-D (1991) The plasminogen activator family from the salivary gland of the vampire bat *Desmodus rotundus*: cloning and expression. Gene 105:229–237

Laffel GL, Braunwald E (1984) A new strategy for the treatment of acute myocardial infarction. N Engl J Med 311:710–717

Levine MN, Goldhaber SZ, Gore JM, Hirsh J, Califf RM (1995) Hemorrhagic complications of thrombolytic therapy in the treatment of myocardial infarction and venous thromboembolism. Chest 108 [Suppl]:291S–301S

Maniatis T, Fritsch EF, Sambrook J (1989) Molecular cloning, a laboratory manual. Cold Spring Harbor Laboratory Press, Cold Spring Harbor, New York

Martin U, Sponer G, König R, Smolarz A, Meyer-Sabellek W, Strein K (1991) Double bolus administration of the novel recombinant plasminogen activator BM 06.022 improves coronary blood flow after reperfusion in a canine model of coronary thrombosis. Blood Coagul Fibrinolysis 3:139–147

Mellott MJ, Ramjit DR, Stabilito II, Hare TR, Senderak ET, Lynch JJ Jr, Gardell SJ (1995) Vampire bat salivary plasminogen activator evokes minimal bleeding relative to tissue-type plasminogen activator as assessed by a rabbit cuticle bleeding time model. Thromb Haemost 73:478–483

Mellott MJ, Stabilito II, Holohan MA, Cuca GC, Wang S, Li P, Barrett JS, Lynch JJ, Gardell SJ (1992) Vampire bat salivary plasminogen activator promotes rapid and sustained reperfusion without concomitant systemic plasminogen activation in a canine model of arterial thrombosis. Arterioscler Thromb 12:112–221

Montoney M, Gardell SJ, Marder VJ (1995) Comparison of the bleeding potential of vampire bat salivary plasminogen activator versus tissue plasminogen activator in an experimental rabbit model. Circulation 91:1540–1544

Muschick P, Zeggert D, Donner P, Witt W (1993) Thrombolytic properties of *Desmodus* (vampire bat) salivary plasminogen activator DSPAα1, alteplase and streptokinase following intravenous bolus injection in a rabbit model of carotid artery. Fibrinolysis 7:284–290

Neuhaus KL, von Essen R, Vogt A, König R, Riess M, Appel KF, Meyer-Sabellek W (1991) GRECO-study Group. Dose-ranging study of a novel recombinant plasminogen activator in patients with acute myocardial infarction: results of the GRECO study. Circulation [Suppl II]: II-573 (Abstract)

Petri T, Baldus B, Boidol W, Bringmann P, Cashion L, Donner P, Haendler B, Krätzschmar J, Langer G, Siewert G, Witt W, Schleuning W-D (1992) Novel plasminogen activators from the vampire bat *Desmodus rotundus*. In: Spier RE, Griffiths JB, MacDonald C (eds) Animal Technology: Developments, Processes and Products. Butterworth-Heinemann Ltd, Oxford, pp 599–604

Petri T, Langer G, Bringmann P, Cashion L, Shallow S, Schleuning W-D, Donner P (1995) Production of vampire bat plasminogen activator DSPAα1 in CHO and insect cells. J Biotechnol 39:75–83

Renatus M, Stubbs MT, Huber R, Bringmann P, Donner P, Schleuning WD, Bode W (1997) Catalytic domain structure of vampire bat plasminogen activator: a molecular paradigm for proteolysis without activation cleavage. Biochemistry 36: 13483–13493

Sakharov DV, Barrett-Bergshoeff M, Hekkenberg RT, Rijken DC (1999) Fibrin-specificity of a plasminogen activator affects the efficiency of fibrinolysis and responsiveness to ultrasound: comparison of nine plasminogen activators in vitro. Thromb Haemost 81:605–612

Schleuning W-D, Alagon A, Boidol W, Bringmann P, Petri T, Krätzschmar J, Haendler B, Langer G, Baldus B, Witt W, Donner P (1992) Plasminogen activators from the saliva of *Desmodus rotundus* (common vampire bat): Unique fibrin specificity. Ann NY Acad Sci 667:395–403

Seifried E, Müller MM, Martin U, König R, Hombach U (1992) Bolus application of a novel plasminogen activator in acute myocardial infarction patients: pharmacokinetics and effects on the hemostatic system. Ann NY Acad Sci 667:417–420

Smalling RW, Bode C, Kalbfleisch J, Sen S, Limbourg P, Forycki F, Habib G, Feldman R, Hohnloser S, Seals A (1995) More rapid, complete, and stable coronary thrombolysis with bolus administration of reteplase compared with alteplase infusion in acute myocardial infarction. Circulation 91:2725–2732

Stewart RJ, Fredenburgh JC, Weitz JI (1998) Characterization of the interactions of plasminogen and tissue and vampire bat plasminogen activators with fibrinogen, fibrin, and the complex of D-dimer noncovalently linked to fragment E. J Biol Chem 273:18292–18299

Suenson E, Bjerrum P, Holm A, Lind B, Meldal M, Selmer J, Peterson LC (1990) The role of fragment X polymers in the fibrin enhancement of tissue plasminogen activator-catalyzed plasmin formation. J Biol Chem 265:22228–22237

The GUSTO Angiographic Investigators (1993) The effects of tissue plasminogen activator, streptokinase, or both on coronary-artery patency, ventricular function, and survival after acute myocardial infarction. N Engl J Med 329:1615–1622

The GUSTO Investigators (1993) An international randomized trial comparing four thrombolytic strategies for acute myocardial infarction. N Engl J Med 329:673–682

Topol EJ (1993) Validation of the early open infarct vessel hypothesis. Am J Cardiol 72 [Suppl G]:40G–45G

Topol EJ, for the Global Use of Strategies to Open Occluded Coronary Arteries (GUSTO III) Investigators (1997) A comparison of reteplase with alteplase for acute myocardial infarction. N Engl J Med 337:1118–1123

Toschi L, Bringmann P, Petri T, Donner P, Schleuning WD (1998) Fibrin selectivity of the isolated protease domains of tissue-type and vampire bat salivary gland plasminogen activators. Eur J Biochem 252:108–112

Verstraete M, Bachmann F (1997) Thrombolytic agents in development. In: Sasahara AA, Loscalzo J, Weitz J, Hirsh J, Willerswon JT, Benedict CR, Verstraete M (eds) Advances in therapeutic agents for thrombosis and thrombolysis. Marcel Dekker, New York, pp 639–647

Vanderschueren S, Dens J, Kerdsinchai P, Desmet W, Vrolix M, De Man F, Van den Heuvel P, Hermans L, Collen D, Van de Werf F (1997) Randomized coronary patency trial of double-bolus recombinant staphylokinase versus front-loaded alteplase in acute myocardial infarction. Am Heart J 134:213–219

Weitz JI, Leslie B, Ginsberg J (1991) Soluble fibrin degradation products potentiate tissue plasminogen activator-induced fibrinogen proteolysis. J Clin Invest 87:1082–1090

Wilcox RG, for the International Joint Efficacy Comparison of Thrombolytics (1995) Randomised, double-blind comparison of reteplase double-bolus administration with streptokinase in acute myocardial infarction (INJECT): trial to investigate equivalence. Lancet 346:329–336

Witt W, Baldus B, Bringmann P, Cashion L, Donner P, Schleuning W-D (1992) Thrombolytic properties of *Desmodus rotundus* (vampire bat) salivary plasminogen activator in experimental pulmonary embolism in rats. Blood 79:1213–1217

Witt W, Kirchhoff D, Woy P, Zierz R, Bhargava AS (1994a) Antibody formation and effects on endogeneous fibrinolysis after repeated administration of DSPAα1 in rats. Fibrinolysis (Abstract 182)

Witt W, Maass B, Baldus B, Hildebrand M, Donner P, Schleuning W-D (1994b) Coronary thrombolysis with *Desmodus* salivary plasminogen activator in dogs. Circulation 90:421–426

CHAPTER 18

Thrombus-Targeting of Plasminogen Activators

C. Bode, K. Peter, M.S. Runge, and E. Haber

A. Introduction

Large-scale studies with mortality endpoints comparing thrombolytic therapy with placebo in patients with acute myocardial infarction (AMI) have documented the benefit of timely dissolution of coronary arterial thrombi by intravenous infusion of plasminogen activators [Gruppo Italiano per lo Studio della Streptochinasi nell'Infarto miocardico (GISSI) 1986; Wilcox et al. 1988; Aims Trial Study Group 1988; The ISIS-2 (Second International Study of Infarct Survival) Collaborative Group 1988]. The GUSTO-1 trial (The GUSTO Investigators 1993), comparing four thrombolytic strategies, showed a small but noteworthy improvement (1.0%) in survival among patients treated with recombinant tissue-type plasminogen activator (tPA) compared with streptokinase (SK). Equally important with this incremental improvement in survival was the finding that there is a direct correlation between early and complete reperfusion and survival (The GUSTO Angiographic Investigators 1993). These data are consistent with the hypothesis that fibrin-specific thrombolytic agents achieve lower mortality by lysing coronary thrombi more rapidly and more completely. The results of the GUSTO trial thus appear to give reasonable direction to researchers involved in the design of plasminogen activators with improved potency and specificity.

Significant limitations, even of the most advanced thrombolytic regimens (The GUSTO Angiographic Investigators 1993; Bode et al. 1996) at present fail to achieve some reperfusion in 15%–20% and complete reperfusion in 35%–45% of patients within 90 min after start of therapy. The failure rate at earlier time points is even higher. In addition, an early reocclusion rate of 5%–15% limits the benefits of therapy in initially successfully treated patients. Furthermore, 3%–10% of patients experience bleeding episodes that necessitate transfusions and in 0.3%–0.7% of patients intracerebral hemorrhage occurs. It has to be emphasized that the above data refer to patients who have been carefully selected for thrombolytic therapy and who constitute only a "good risk" fraction of all patients with AMI.

Several lines of investigation are currently being pursued in the hope of enhancing the efficacy, speed, and freedom from side effects of thrombolytic therapy, thus making it more effective and more widely applicable. This chapter describes novel approaches for developing agents that dissolve clots or prevent clot formation that are based on the use of antibodies to target the agents to specific components of the thrombus.

B. The Principle of Antibody Targeting

Antibody targeting for the treatment or prevention of thrombosis entails the engineering of a bifunctional molecule that contains both a highly specific antigen-binding site which concentrates the molecule at the desired target, and an effector site which either initiates thrombolysis or prevents additional thrombus formation. The repertoire of potential antibody specificities is extremely large and allows for the selection of monoclonal antibodies that can differentiate among very similar antigens. Thus, it is possible to target to fibrin but not fibrinogen or to recognize the activated form of the platelet receptor glycoprotein (GP) IIb/IIIa but not its inactive isomer. Obviously, the effectiveness of a clot-specific antibody is dependent on a lack of cross-reactivity between the desired epitope and all non-thrombus epitopes on endothelial or other vascular cells.

I. Antifibrin Antibodies

Our most extensive experience has been gained with fibrin-specific antibody 59D8. The antibody was raised by immunizing Balb/C mice with a synthetic peptide containing the sequence of the thrombin-generated amino terminus of the fibrin β-chain, coupled to a carrier protein (Hui et al. 1983). Monoclonal antibody 59D8 binds to fibrin with high affinity (0.77 nmol/l) but does not bind to fibrinogen. In contrast, the binding constant of tPA for fibrin is 0.16 μmol/l. Other groups have used different antibodies specific for fibrin.

II. Antiplatelet Antibodies

Platelets undergo structural changes as they aggregate and are incorporated into a thrombus. The GP IIb/IIIa complex on the platelet cell membrane functions as a receptor for fibrinogen. As such the complex has a role in cross-linking platelets not only to one another but also to fibrin, thereby stabilizing the thrombus. Antibody 7E3 (abciximab) binds to and inhibits GP IIb/IIIa, thus inhibiting platelet aggregation. The antibody has been shown to be a useful adjunct to thrombolysis, PTCA and stenting (Coller et al. 1995; Coller 1998; The EPIC Investigators 1994; The EPILOG Investigators 1997; The EPISTENT Investigators 1998; reviewed in Adgey 1998; Bates and Weitz 1999; Califf 2000). For our purposes, 7E3 provides a model vehicle for targeting plasminogen activators or antithrombin agents to an epitope expressed on platelet-rich, arterial thrombi. The disadvantage of 7E3 as a targeting antibody is that it interacts with GP IIb/IIIa on both activated and resting platelets and thus targets the thrombus only with relative selectivity. Others have used the monoclonal antibody Fab-9 directed against GP IIb/IIIa that was subsequently optimized by mutagenesis and affinity maturation, using phage display techniques (SMITH et al. 1995).

III. Chemical Conjugates or Recombinant Fusion Proteins

Chemical conjugation between antibody and effector molecule is the fastest and – although challenging – easiest method to create a model bifunctional molecule. Recombinant fusion proteins have the advantage of a uniform and large scale preparation of protein material and they avoid the low-yield of chemical cross-linking. Also, recombinant fusion molecules are not limited like chemical conjugates to recombination of whole proteins or domains that can be generated by enzymatic digestion. Rather, the needed regions of proteins can be fused and, if necessary, cleavage sites for different enzymes can be introduced (see below) to endow the molecules with additional specificity. Thus, while chemical conjugates are suitable to explore the general concept, only recombinant fusion molecules can be tailored with suitable sophistication in order to endow them as powerful new therapeutic agents.

C. Synthesis, Purification, and Characterization of Chemical Antibody-Plasminogen Activator Conjugates

I. Synthesis

The method for chemically cross-linking plasminogen activators to antibodies has initially been established in our laboratories (BODE et al. 1985). The general approach is to generate stable disulfide or thio-ester bonds between the two proteins of interest. One of the most useful strategies is to link two proteins using the cross-linking reagent, N-succinimidyl 3-(2-pyridyldithio)propionate (SPDP). This is a heterobifunctional reagent: the attachment and cross-linking steps of the reaction are separate and it is possible to use either the same or different functional groups on each protein. In order to prevent the formation of aggregates and to increase the yield of the 1:1 conjugate it is essential to establish conditions that allow modification of a single amino group on each protein. Alternatively, free sulfhydryl groups can be generated on one of the proteins under reducing conditions. This approach has been very effectively used in coupling Fab′-fragments of the targeting antibody using the reduced inter-heavy chain disulfide bonds of the hinge-region as reactive sulfhydryl groups and reacting these with SPDP-modified plasminogen activator. A third commonly used approach is to modify the proteins to be coupled with two different cross-linking reagents. It is thus possible to control conditions such that the favored reaction is between the two desired species.

II. Purification

Methods to purify plasminogen activator-antibody conjugates have been reviewed by BODE et al. (1992). Although the methodology described in this section was first developed for purification of chemically coupled conjugates, similar methods are used for purification of recombinant hybrid molecules. In principle, purification strategies are employed that select for size (a 1:1 molar ratio of the antibody and plasminogen activator is the desired product) and also for biological activity of both the antibody and the enzyme. Thus, gel filtration is usually implemented to separate the desired 1:1 conjugate from uncoupled reactants as well as higher molecular weight polymers. Yet even a very pure peak that contains a 1:1 molar ratio of antibody and plasminogen activator, as assessed by SDS-polyacrylamide gel electrophoresis, does not guarantee that all conjugates contain biological activity for both the antibody and plasminogen activator. Sequential affinity chromatography with appropriate immobilized ligands can be used to select for biologically active conjugates or fusion proteins. The first affinity step selecting for active UK or tPA involves benzamidine immobilized on Sepharose, that specifically binds to (and inhibits) the intact active center of serine proteases. The benzamidine-Sepharose column does not retain uncoupled antibody or Fab'-fragment or inactivated enzyme, which are washed out. Elution by pH-change yields both conjugated material with an active plasminogen activator domain and free plasminogen activator, that was not coupled initially.

The second affinity step selects for molecules with an intact antibody portion. For example, antifibrin antibody 59D8 can be bound to a Sepharose matrix to which the peptide is coupled against which the antibody was initially raised (β-peptide). When the eluate from the benzamidine-Sepharose column purification step is further purified on a β-peptide-column, any conjugates with antibody unable to bind to its antigen will not bind to the column. Also, uncoupled plasminogen activator will not bind to the β-peptide-Sepharose column. The eluted final product from this column will thus be a 1:1 conjugate containing biologically active plasminogen activator and antibody domains. For antibodies or plasminogen activators [e.g., single-chain urinary-type plasminogen activator (scuPA)] for which suitable ligand-Sepharose columns are not available, immunoaffinity chromatography, using antibody-Sepharose columns, have been used successfully. Ideally, the immobilized antibody only recognizes and binds to the active form of plasminogen activator or antibody in the conjugate or fusion protein. If such an antibody is unavailable, less efficient alternatives include the use of immobilized antibodies which cannot distinguish active from inactive antibody or plasminogen activator or the use of immobilized protein A or G.

III. Characterization

Characterization of chemical conjugates as well as of recombinant fusion proteins includes assessment of molecular size by SDS gel electrophoresis under reducing and non-reducing conditions, antigenicity studies by Western-blot and analysis of the functional status of the antibody combining site (compared to native antibody) and of the plasminogen activator enzymatic properties (compared to native plasminogen activator). The focus of this review will be on functional characterization in terms of enhancement of fibrinolytic and thrombolytic potency of conjugate vs native plasminogen activator in vitro and in vivo.

Further insight can be gained by comparing fully functional conjugates to conjugates in which one partner is inactive. Control conjugates with plasminogen activator (e.g., UK) coupled to irrelevant antibodies (e.g., specific for digoxin) have been used to discriminate between enhancement of activity due to increase in molecular weight (and thus reduced clearance of conjugate vs native plasminogen activator) and enhancement due to targeting of the enzyme by the specific antibody combining site.

1. Antifibrin-UK Conjugates

Chemical conjugates between antifibrin antibody 59D8 and UK were 100–250 times as potent as uncoupled UK and 10 times as potent as tPA in an in vitro fibrinolytic screening assay with purified components (fibrin-Sepharose assay) (BODE et al. 1985, 1987b). The assay assesses the lysis of trace labeled fibrin immobilized on Sepharose after an activator solution has been applied to the column, followed by a washing step and the addition of plasminogen. Thus, fibrinolysis is heavily dependent on fibrin binding which makes the assay suitable for screening but clearly additional, more physiologic assays were required to assess the potential therapeutic effects of such conjugates.

In an assay that monitored the lysis of fresh human thrombus in a human plasma bath, the UK-antifibrin conjugate proved to be 2.2–4.4 times as efficient as UK alone. In both assays, conjugates of whole (bivalent) antifibrin IgG and those of (monovalent) antifibrin Fab fragments performed equally well which was important to find out in view of the construction of recombinant molecules (BODE et al. 1987a).

2. Antifibrin-tPA Conjugates

Tissue-type plasminogen activator is more difficult to handle than UK in conjugation experiments because it is less soluble at higher concentrations. The tPA-antifibrin conjugate was 10 times as efficient as uncoupled t-PA in the in vitro screening assay with purified components (fibrin-Sepharose assay) and it was 3.2 times more potent than tPA in the human plasma clot assay (RUNGE et al. 1988). In vivo, in a rabbit jugular vein model of thrombolysis, the conjugate proved to be 3.0–9.6 times as potent as tPA alone; the greatest benefit was observed at the lowest tPA concentrations (RUNGE et al. 1987). Both, in vitro and in vivo, the tPA antifibrin conjugate increased fibrinolytic potency without sacrificing specificity: fibrinogen, plasminogen and α_2-antiplasmin consumption were lower relative to that obtained with tPA alone, indicating that the increase in potency was the result of an increase in selectivity.

3. Antifibrin-scuPA Conjugates

Although scuPA has no affinity for fibrin, it is a relatively fibrin-selective agent through other mechanisms (see Chap. 9). The prospect of being able to improve the plasminogen-activating properties of scuPA by adding fibrin affinity, i.e., through a second mechanism of fibrin selectivity, appeared particularly interesting. In an assay with purified components (fibrin-Sepharose assay) the fibrinolytic potency of the conjugate proved to be 230 times greater than that of scuPA, 420 times greater than that of UK, and 33 times greater than that of tPA (BODE et al. 1990). Subsequent studies in vitro in human plasma showed enhanced lysis in terms of speed, completeness and specificity. In vivo, in the rabbit jugular vein model, antifibrin-scuPA was 29 times more efficacious than its natural counterpart.

4. Antiplatelet-UK Conjugates

Arterial thrombi contain a high concentration of activated platelets, and platelet-rich thrombi appear to be particularly resistant to thrombolysis. Thus, platelets may be a major reason for the limited success rate of thrombolytic therapy in patients with AMI and they also appear to play a major role in early reocclusion after initially successful thrombolysis. Antibody 7E3, which binds specifically to the platelet GP IIb/IIIa receptor which mediates platelet aggregation, has been shown to be a useful adjunct to thrombolysis, PTCA, and stenting (see above). It was of particular interest to determine whether the targeting of a plasminogen activator to the GP IIb/IIIa receptor through 7E3 would enhance the antibody's ability to block platelet aggregation. In addition to blocking access of the GP IIb/IIIa receptor to fibrinogen by virtue of 7E3's binding to the receptor, the high local concentration of plasmin achieved through targeting the plasminogen activator to platelets was thought to lyse fibrin and (locally) fibrinogen responsible for aggregating the platelets. A UK-7E3 Fab' conjugate bound to purified GP IIb/IIIa and intact activated platelets and also exhibited plasminogen activator activity. At the concentrations tested, UK showed very little activity against the clots, whereas the conjugate was 970 times more active. An equimolar mixture of UK and 7E3 was no more effective than UK alone (BODE et al. 1991a). The rate of lysis was dependent upon the concentration of platelets in the clot and no enhancement in lysis by the conjugate over UK was apparent in platelet-free clots. On the other hand, the conjugate was substantially more potent in inhibiting platelet aggregation when compared to antibody 7E3 alone. Thus, UK targeted to GP IIb/IIIa by conjugation to antibody 7E3 accounted for enhanced lysis and improved inhibition of platelet aggregation when compared to the parent molecules alone.

5. Antiplatelet-scuPA Conjugates

Similar results were obtained for a conjugate between 7E3 and scuPA. No advantage in using scuPA over UK was observed (BODE et al. 1991b). Using different antibodies, DEWERCHIN et al. (1991) confirmed and extended our observations by showing that an antiplatelet-scuPA conjugate was more effective than scuPA in lysing platelet-rich thrombi in vivo in a hamster model of pulmonary embolism.

D. Targeting with Bifunctional Antibodies

I. Principal Considerations

In naturally occurring bivalent IgG antibodies both antigen combining sites are identical in structure and specificity. An alternative to a chemical conjugate or a fusion protein that contains segments of two proteins of different function is a bifunctional antibody that is able to bind both thrombus (e.g., fibrin) and an effector protein (e.g., a plasminogen activator) without diminishing its physiologic function (e.g., enzyme activity). Such a bifunctional antibody would serve as an "adapter molecule" bringing the plasminogen activator into close proximity with fibrin, without the need to manipulate the enzyme chemically. Also, a bispecific antibody can be used by itself to enhance the potency of endogenous enzymes e.g., tPA. By using chemical methods it is possible either to cross-link intact antibodies of different specificities or to produce bispecific (Fab')2 molecules. A better method for producing bispecific antibodies on a larger scale is the hybrid-hybridoma approach, that is somatic cell fusion of two monoclonal antibody-producing hybridoma cell lines.

II. Bispecific IgG (Intact Antibodies)

It was anticipated that an antibody specific for both fibrin and tPA would bind to the fibrin matrix of a thrombus and also to circulating tPA, thereby increasing the concentration of tPA at the surface of thrombus. The feasibility of this approach was first tested by chemically cross-linking SPDP-modified antifibrin antibody 59D8 with an iminothiolane-modified antibody specific for tPA (TCL8, K_D of 5×10^{-9}). In an in vitro quantitative fibrinolysis assay, the relative fibrinolytic potency of tPA bound to the bispecific antibody was 13 times greater than that of tPA and 200 times greater than that of UK. When fibrin was treated with the bispecific antibody before being mixed (loaded) with tPA, the relative fibrinolytic potency of tPA was enhanced 14-fold (BODE et al. 1989). This capture of tPA also occurred when the concentration of tPA present in the assay was less than the concentration of tPA present in normal human plasma. In a human plasma clot assay, samples containing both the bispecific antibody and tPA exhibited significantly more lysis than did samples containing tPA alone. In spite of the increased clot lysis effected by the bispecific antibody, there was no significant increase in fibrinogen or α_2-antiplasmin consumption at equal tPA concentrations. The ability of the bispecific antibody to concentrate exogenous tPA in vivo was examined in the rabbit jugular vein model. Systemic infusion of a small amount of tPA produced no significant increment in thrombolysis over the level of spontaneous lysis (14%); however, simultaneous infusion of tPA and bispecific antibody resulted in 42% lysis. These results suggest that a bispecific antibody can enhance thrombolysis by capturing endogenous or exogenous tPA.

Other examples of the same approach include the synthesis of antiplatelet-anti-tPA bispecific antibodies (RUEF et al. 1999) and of antifibrin-anti-UK bispecific antibodies (CHARPIE et al. 1990).

III. Bispecific (Fab')2

As a potential pharmacologic agent, a bispecific (Fab')2 has several advantages over a bispecific IgG:

1. The (Fab')2 lacks the Fc region of the IgG, which is responsible for non-specific effector functions such as complement activation.
2. The molecular weight of the (Fab')2 is just 100 kD as opposed to 300 kD for IgG.
3. The (Fab')2 complex does not contain cross-linking residues.
4. The final product is uniform because the constituent Fab' molecules have only one alternative to form (Fab')2.

A bispecific (Fab')2 molecule was constructed by linking the monovalent Fab' from monoclonal antifibrin antibody 59D8 to the Fab' from monoclonal anti-tPA antibody TCL8 by means of inter-heavy-chain disulfide bonds. An immunochemical complex composed of the bispecific (Fab')2 molecule bound to tPA was then generated and purified. Its molecular weight was 170 kD, corresponding to the sum of (Fab')2 and tPA. The tPA-bispecific (Fab')2 complex was 8.6 times more efficient in fibrinolysis than tPA alone and 94 times more potent than UK (RUNGE et al. 1990). This enhancement in the fibrinolytic potency suggests that this (compared to bispecific IgG) pharmacologically preferable molecule is capable of binding both fibrin and tPA with similarly high affinity as bispecific IgG.

IV. Bispecific Antibodies by Hybrid-Hybridoma Technique

An alternative method that yields larger quantities of uniform and stable bispecific antibody is the hybrid-hybridoma technique. A thymidine kinase-deficient clone from the hybridoma producing the antifibrin antibody was subjected to somatic cell fusion with a hypoxanthine guanine phosphoribosyl transferase-deficient clone of the hybridoma producing the anti-tPA antibody. Surviving clones were selected in hypoxanthine aminopterin thymidine medium. The resulting bispecific antibodies were as effective as their chemically synthesized counterparts in concentrating tPA to fibrin (BRANSCOMB et al. 1990).

V. Double Targeting

In an effort to increase further the potency of plasminogen activators, UK was chemically linked to a bispecific antibody or bispecific (Fab')2 with specificity for fibrin (by means of a Fab'-fragment from antibody 59D8) and platelets (by means of a Fab'-fragment from antibody 7E3) (RUEF 1999). These ternary complexes were tested in several assays that revealed an increase of fibrinolytic potency of double-targeted UK over single antibody-UK conjugates, but only if the epitope targeted by the UK-conjugate was present in the clot. Assays with either immobilized platelets, GP IIb/IIIa or fibrin revealed that the conjugate activated plasminogen 10-, 58- and 13-fold more effectively than free UK ($p < 0.0001$ for each). In vitro clot lysis of platelet-rich and fibrin-rich plasma clots demonstrated an up to 5-fold higher potency of the conjugate compared to the parent molecule ($p < 0.0001$). In vitro platelet aggregation was effectively inhibited by the hybrid molecule, whereas UK had no effect. Binding of the conjugate to fibrin-Sepharose was 13-fold higher than that of uncoupled UK (RUEF et al. 1999). Thus the bispecific antifibrin-antiplatelet UK conjugate has the ability to lyse both fibrin-rich and platelet-rich thrombi with high efficacy and also inhibits platelet aggregation that occurs regularly on the surface of a fresh thrombus.

E. Expression and Characterization of Recombinant Antifibrin-Plasminogen Activator Fusion Proteins

I. Antifibrin-SK Constructs

In order to add fibrin selectivity to SK, GOLDSTEIN et al. (1996) have constructed a chimeric SK molecule that consists of the Fab'-fragment of the antifibrin antibody 59D8 and a full length SK sequence 1–414. An expression plasmid containing the cDNA encoding the heavy-chain variable region from 59D8 and the coding region of a genomic clone of SK was electroporated into a 59D8 light-chain producing hybridoma cell line. The chimeric SK was purified by affinity chromatography over the immobilized octapeptide ligand for 59D8. The 59D8-SK fusion protein increased clot lysis only twofold compared to SK but exhibited changed activator properties. It was relatively inactive in human plasma but lysed clots slowly and completely, whereas SK lysed clots rapidly but incompletely (GOLDSTEIN et al. 1996).

II. Antifibrin-tPA Constructs

Following successful construction of chemically conjugated antibody-plasminogen activator hybrids, recombinant DNA methodology was used to produce similar recombinant hybrid molecules. In theory, producing antibody-plasminogen activator hybrids by recombinant methods should allow much more flexibility in the design of hybrid molecules, as well as improve purity and yield. A number of different recombinant antibody-plasminogen activator hybrid molecules have been constructed. The first such molecule was an antifibrin (59D8) tPA fusion protein (Schnee et al. 1987; Love et al. 1989, 1993). It was found not to be an ideal construct, because the activity of tPA depends largely on fibrin-stimulation and this mechanism is impaired by the presence of an antibody at the N-terminus (Love et al. 1994).

Smith and collaborators (1995) used a different approach. They replaced amino acid sequence 63–71 in the epidermal growth factor (EGF) region of human tPA with a peptide containing the HCDR3 region of Fab-9, an antibody with nanomolar affinity for β_3-integrins. The modified activator, LG-tPA, had full enzymatic activity, and the presence of fibrin enhanced plasminogen activation by the modified tPA to the same degree as wild-type tPA. LG-tPA bound in a specific and saturable fashion to GP IIb/IIIa exhibited a K_D of approximately 0.9 nmol/l. It also bound to integrin $\alpha_v\beta_3$ receptor for vitronectin (Smith et al. 1995).

III. Antifibrin-scuPA Constructs

Pro-UK (scuPA) was found more suitable for integration into fusion proteins for the following reasons: first, scuPA does not require fibrin binding for activation, second, scuPA is a relatively fibrin-specific plasminogen activator even though it does not bind to fibrin (so we could anticipate enhanced specificity after endowing the molecule with high fibrin affinity through the antibody), and third, scuPA is resistant to inactivation by plasminogen activator inhibitor-1 (PAI-1). An expression plasmid containing DNA coding for the antibody 59D8 heavy chain variable and first constant domains and the catalytic domain of scuPA was transfected into a "heavy chain loss variant" of the hybridoma cell originally secreting antibody 59D8. The light chain of antibody 59D8, which was still produced in the variant hybridoma cell line, assembled with the chimeric molecule (heavy chain and plasminogen activator) within the variant hybridoma cells. A molecule was secreted that contained both an antigen binding site of predefined specificity (from antibody 59D8) and a catalytic site capable of cleaving plasminogen (from scuPA). In the original construct only the Fab part of the antibody had been included in the molecule in an effort to limit the mass of the chimera to its essential components, but further refinements towards a single-chain Fv molecule containing only the light- and heavy-chain variable region of the antibody as well as the introduction of human sequences (instead of murine) in order to limit antigenicity are under way. In a similar vein, the kringle and growth factor domains of scuPA were omitted and only the sequence corresponding to low molecular weight (LMW), 32 kD scuPA was used (Stump et al. 1986). The recombinant 59D8-scuPA fusion molecule that was used in the assays and experiments described below contained the heavy chain from residues 1–351 and a native light chain of antibody 59D8 and, in contiguous peptide sequence on the heavy chain, residues 144–411 of scuPA (Runge et al. 1991).

r-scuPA-59D8 was characterized in functional assays comparing it to the two parent molecules. The scuPA part did not differ from native scuPA in terms of percentage of uncleaved single-chain material (95% for both), specific amidolytic activity after conversion to two-chain UK by plasmin-Sepharose (85000 U/mg for scuPA and 83,900 U/mg for the 32-kD scuPA portion of the 103-kD fusion molecule, on a molar basis), or the ability to convert plasminogen to plasmin (K_M = 9.1 µmol/l vs 16.6 µmol/l). The fibrin binding ability also did not differ between the parent antibody molecule 59D8 and r-scuPA-59D8 as assessed by serial dilution assays.

In human plasma clot lysis assays r-scuPA-59D8 was six times more potent than scuPA and at equivalent thrombolytic doses more fibrin-specific as exhibited by less consumption of fibrinogen and α_2-antiplasmin. In vivo, the molecule was tested first in the rabbit jugular vein model (in situ formation of a human clot in the jugular vein of a rabbit and infusion of thrombolytic agent into the contralateral ear vein in simulation of a systemic infusion). Compared with native scuPA, the r-scuPA-59D8 fusion molecule showed a 20-fold increase in potency over a wide dose-response range. Only when clot lysis was nearly complete (in excess of 83%) was a decrease in fibrinogen concentration measured. A thrombolysis model in which thrombi are preformed in vivo in juvenile baboons was developed to compare the potencies of r-scuPA-59D8, scuPA, and tPA in lysing nonocclusive ^{111}In-labeled platelet-rich arterial type thrombi and ^{125}I-labeled fibrin-rich venous-type thrombi. Systemic infusion of 1.9 nmol/kg r-scuPA-59D8 produced thrombolysis that was comparable to that obtained with much higher doses of tPA (14.2 nmol/kg) and scuPA (28.5 nmol/kg). When steady-state plasma concentrations were normalized, r-scuPA-59D8 lysed thrombi six times more rapidly than scuPA and tPA ($p < 0.001$) and reduced the rate of new thrombus formation far more than comparable doses of the other activators. At equivalent thrombolytic doses r-scuPA-59D8 produced fewer antihemostatic effects than either tPA or scuPA. Template bleeding time measurements were shorter (3.5 ± 0.12 min for r-scuPA-59D8 vs 5.3 ± 0.36 and 5.2 ± 0.04 min for tPA and scuPA, respectively; $p < 0.05$), and α_2-antiplasmin consumption and D-dimer generation were significantly lower ($p < 0.05$). Because template bleeding times may have some association with the risk of hemorrhage in a clinical situation (GIMPLE et al. 1989), r-scuPA-59D8 may not only be more potent than other plasminogen activators but has the potential for greater safety as well (RUNGE et al. 1996).

Collen and co-workers (DEWERCHIN et al. 1989, 1992; HOLVOET et al. 1991, 1992, 1993ab; VANDAMME et al. 1992) have pursued similar studies with a different fibrin-specific antibody and reached similar conclusions.

YANG et al. (1994) have modified the scuPA catalytic domain of the r-scuPA-59D8 molecule described above. ScuPA has many cleavage sites (see Chap. 4); among these there is a thrombin-sensitive site between Arg[156] and Phe[157] that results in an inactivated molecule upon cleavage. Between Lys[158] and Ile[159] is the plasmin-sensitive site that results, after cleavage, in the enzymatically active two-chain UK. The deletion of Phe[157] and Lys[158] creates a thrombin-sensitive cleavage site that results, after cleavage, in an active two-chain plasminogen activator. Thus, upon activation by thrombin, this molecule converts plasminogen to plasmin and effects efficient clot lysis. Activation of the molecule can be inhibited by hirudin and the heparin/antithrombin complex, both thrombin inhibitors. These observations suggest that the thrombin-activatable form of r-scuPA-59D8 has the potential to lyse selectively fresh clots (which are thrombin rich) more effectively than older clots (which are poorer in thrombin). In a clinical situation like AMI a fresh coronary thrombus is most likely the causative agent that needs to by lysed whereas older thrombi that function as hemostatic plugs are best left intact in order to prevent hemorrhage. Thus, this variant of r-scuPA-59D8 may add extra safety features to the concept of antibody targeting in a clinical setting.

F. Other Approaches of Targeting Plasminogen Activators to Thrombi

Several approaches have been used to target plasminogen, tPA, scuPA, or staphylokinase to fibrin, platelets, P-selectin, or annexin V without the use of antibodies or Fab fragments. Yamada et al. (1996) have substituted the amino sequence 148–151 of a loop in kringle 1 of the human tPA with the integrin-specific sequence Arg-Gly-Asp (RGD) by site-directed mutagenesis. The mutant was expressed in COS-1 cells and purified by lysine-Sepharose affinity chromatography. It maintained full enzymatic activity compared to wild-type tPA and bound in a specific and saturable fashion to GP IIb/IIIa with a K_D of approximately 2 nmol/l.

Dawson et al. (1994) have altered the plasminogen cleavage site Arg^{561}–Val^{562} by substituting the P_3, P_2, and P_1' residues with sequences from thrombin cleavable proteins in an attempt to target the plasminogen to clot-bound thrombin. Plasminogen variants (100 μg/ml) with thrombin cleavage sites from fibrinogen, the thrombin receptor, factor XIII, and factor XI were 50% cleaved by thrombin (85 NIH units/ml) in 28 h, 2.5 h, 5.7 min, and 3 min respectively. When 40 μg/ml of factor XI-plasminogen mutant was added to citrated human plasma clotted by an activated partial thromboplastin reagent (aPTT) and Ca^{++} the clots formed were rapidly lysed.

P-selectin is expressed on the surface of activated platelets (Peter et al. 1999). It has been shown to be involved in the pathogenesis of reperfusion injury of the myocardium and post-ischemic no-reflow. Fujise et al. (1997) have therefore constructed fusion proteins consisting of amino acids 1–121 (P-selectin lectin domain) or 1–280 (P-selectin lectin domain, EGF-domain, and first two CR domains) fused to a PAI-1 resistant form of tPA. Both constructs were equally effective as wild-type tPA in an in vitro clot lysis assay. In a rat mesenteric artery cyclic flow variation model (CFV) wild-type tPA and the two P-selectin constructs reduced CFV to a similar extent. So far these constructs have not been tested in a model of coronary artery thrombosis.

WAN and collaborators (2000) also have engineered a recombinant chimeric UK construct which consists of a humanized monoclonal antibody (SZ-51Hu) directed against P-selectin and the UK sequence 1-411. In human plasma clots containing ^{125}I-fibrin and varying concentrations of human platelets clots lysis was 4 to 8 times enhanced by the conjugate compared to the parent UK. In a hamster pulmonary embolism model with clots prepared from fresh human platelet-rich plasma containing ^{125}I-fibrinogen the thrombolytic activity of the fusion protein was 4 times higher than that of 2000 IU/kg of UK. Fibrinogen breakdown using the construct was minimal.

The peptide Gly-Pro-Arg (GPR) which corresponds to the amino-terminal portion of the fibrin α-chain after release of the fibrinopeptide A prevents the polymerization of fibrin monomers. This peptide also binds to fibrinogen and to fibrin fragment D. Addition of proline to the tripeptide significantly increases binding and the inhibitory activity. HUA et al. (1996) have examined whether a construct of Gly-Pro-Arg-Pro fused to LMW-UK (Leu^{144}–Leu^{411}) improved its efficacy as a thrombolytic agent. The construct exhibited a sixfold greater affinity for fibrin and had a two- to threefold greater fibrinolytic potency in in vitro clot lysis assays. Fibrinogen degradation was much lower during clot lysis compared to that produced by wild-type LMW-UK.

Annexin V is a human protein that binds with high affinity (K_D of 7 nmol/l) to the abundant phosphatidyl serine molecules exposed on activated platelets. Binding to quiescent platelets in vitro is minimal. Maximally stimulated platelets contain approximately 200000 annexin V binding sites, substantially exceeding the number of approximately 25000 GP IIb/IIIa binding sites. TAIT et al. (1995) have fused full length scuPA (amino acids 1–411) or LMW-scuPA (amino acids 144–411) to full length annexin V. Both constructs, after activation by plasmin, had similar amidolytic activities and activated plasminogen marginally better than wild-type scuPA (K_M values of 5–6 μmol/l for the two former and 13 μmol/l for the wild type form). In vitro both constructs lysed platelet-rich clots as well but not better than LMW-scuPA. No in vivo experiments have been published so far.

VAN ZYL and collaborators (1997) produced a multivalent staphylokinase construct. Staphylokinase, a highly fibrin specific thrombolytic agent (see Chap. 16) was fused via a factor Xa cleavable linker to the cDNA of an antithrombotic peptide of 29 amino acids comprising: (1) an Arg-Gly-Asp (RGD) sequence to prevent binding of fibrinogen to platelets, (2) a portion of fibrinopeptide A, an inhibitor of thrombin, and (3) the tail of hirudin, a potent direct thrombin inhibitor. The construct was expressed in *E. coli* and purified using metal affinity chromatography. The fibrinolytic potential of PLATSAK (*Pl*atelet-*Anti*thrombin-*Staphylok*inase) was slightly lower than that of the parent molecule. PLATSAK markedly lengthened the thrombin time and the aPTT, thereby indicating inhibition of thrombin activity. It had a negligible effect on platelet aggregation, possibly due to inaccessibility of the RGD peptide in the tertiary structure of PLATSAK.

G. Targeting the Trombin Inhibitor Hirudin to Fibrin

I. General Considerations

The intent of this design is the inhibition of further fibrin deposition at the site of thrombosis while avoiding systemic anticoagulation. Antibody 59D8 is a particularly attractive targeting agent for a thrombin inhibitor because the epitope on fibrin, that the antibody binds to, only becomes exposed after thrombin has cleaved fibrinopeptide B off the fibrinogen β-chain. Thus, the inhibitor is concentrated at the very site of thrombin action.

II. Chemical Conjugates

Chemical conjugates of the antibody fragment 59D8-Fab' or of intact 59D8 and hirudin were constructed, purified, and tested in an in vitro assay measuring the deposition of fibrin on the surface of a standard clot. The 59D8-Fab'-hirudin conjugate was ten times more potent than hirudin alone or a mixture of hirudin and 59D8-Fab' in inhibiting fibrin deposition, presumably because of antithrombin concentration on the surface of the clot (BODE et al. 1994). The potency of fibrin-targeted hirudin was also compared in vivo with that of uncoupled recombinant hirudin in a baboon model of thrombus formation. ^{111}In-labeled platelet deposition was measured in a synthetic graft segment of an extracorporeal arteriovenous shunt in control animals and in animals receiving either fibrin-targeted hirudin or hirudin. Fibrin-targeted hirudin was again ten times more potent than hirudin in inhibiting platelet deposition and thrombus formation ($p < 0.05$) (BODE et al. 1997). These data indicate that targeting a thrombin inhibitor and presumably also other anticoagulants to an epitope present in thrombi such as fibrin or the platelet IIb/IIIa receptor results in significantly increased antithrombotic potency.

III. Recombinant Fusion Protein

A recombinant version of the 59D8-hirudin conjugate has been produced by methods similar to those used for the development of r-scuPA-59D8. However, a problem specific to hirudin had to be overcome. Because hirudin needs a free N- as well as a free C-terminus to be fully active, a fusion protein would be inactive. The problem was ultimately overcome by inserting a recognition sequence for factor Xa, which cleaves the C-terminus of the recognition sequence, thus liberating active hirudin (PETER et al. 2000). Because factor Xa is a constituent of thrombi, this approach may also add specificity to this approach in limiting the action of hirudin to the surface of a thrombus and thus decreasing the risk of bleeding complications.

A similar construct (M23) was engineered by LIJNEN et al. (1995). These authors fused the C-terminal amino acids 53–65 of hirudin via a 14-amino acid linker to the C-terminus of a 40 kD fragment (Ser^{47}–Leu^{411}) of scuPA and expressed the construct in *E. coli*. The thrombin inhibitory activity of the construct was about 20 times lower than that of hirudin, but still substantial. The chimera had a slightly lower plasminogen activating potential than parent scuPA (catalytic efficiency 0.55 vs 0.73 $\mu M^{-1} s^{-1}$ respectively) but a higher fibrin specificity. Fibrinogenolysis was lower when the construct was added to normal human plasma compared to scuPA. In the thrombotically occluded femoral artery of dogs M23 and scuPA (saruplase) exhibited similar lysis (reperfusion) and reocclusion rates, and time to reperfusion (SCHNEIDER et al. 1997). The efficacy of these two compounds to lyse venous ^{125}I-fibrin-labeled thrombi in dogs also was comparable, but M23 produced significantly less proteolysis and did not result in an increase of template bleeding times, whereas saruplase produced a very large increase in bleeding times and of the aPTT. Further modification of M23 by the introduction of a second linker (FLLRNP) from the human thrombin receptor conferred about a tenfold increase in anticoagulant activity to the new construct M37 compared with M23 (WNENDT et al. 1997).

H. Conclusion and Outlook

Antibody-targeting is an attractive approach to enhance the potency and specificity of effector molecules. The approach has been used in cancer research (HUANG et al. 1997) and was first applied to the field of thrombolysis in the 1980s (BODE et al. 1985). Since then, chemical conjugates have been substituted by recombinant fusion proteins and the tools of molecular biology have allowed to minimize the size of the fusion partners as well as to overcome largely the antigenicity of the antibody fragments by "humanization," a problem avoidable in the future by screening human antibody libraries. Promising results have been obtained in vitro and were substantiated by in vivo experiments in rabbits and non-human primates. Great efforts are currently undertaken to simplify the expression of fusion proteins by developing single-chain antibodies, both specific for fibrin, platelets, or to the pulmonary vasculature (MUZYKANTOV et al. 1996) and in developing other means (such as the incorporation of small peptide into a protein loop) for targeting plasminogen activators to fibrin (HUA et al. 1996) or platelet constituents (YAMADA et al. 1996) in a thrombus. The impressive results of antibody application in the treatment of acute coronary syndromes and as adjunctive therapy to intracoronary interventions with the GP IIb/IIIa inhibitor c7E3 show the promise of antibody treatment alone which may well be enhanced by using this antibody (as described above) or a single-chain Fv-fragment of an even more specific antiplatelet-antibody. Similar applications with different antibodies, such as against P-selectin, may also prove useful (FUJISE et al. 1997). Only the surface of the potential of the general concept has been scratched at present. Confirmation of the concept of antibody targeting in cardiovascular disease now awaits clinical trials.

List of Abbreviations and Acronyms

AMI	acute myocardial infarction
aPTT	activated partial thromboplastin time
CFV	cyclic flow variation
EGF	epidermal growth factor
GP	glycoprotein
GUSTO	Global Utilization of Streptokinase and Tissue plasminogen activator for Occluded coronary arteries
LMW	low molecular weight
PAI-1	plasminogen activator inhibitor type 1
PLATSAK	*pl*atelet-*a*nti*t*hrombin-*st*aphylo*k*inase
PTCA	Percutaneous Transluminal Coronary Angioplasty
SPDP	*N*-succinimidyl 3-(2-pyridyldithio)propionate
scuPA	single-chain urinary-type plasminogen activator, also called pro-UK
SK	streptokinase
tPA	tissue-type plasminogen activator
UK	urokinase

References

Adgey AA (1998) An overview of the results of clinical trials with glycoprotein IIb/IIIa inhibitors. Eur Heart J 19 [Suppl D]:D10–D21

AIMS Trial Study Group (1988) Effect of intravenous APSAC on mortality after acute myocardial infarction. Preliminary report of a placebo-controlled clinical trial. Lancet 1:545–549

Bates SM, Weitz JI (1999) Prevention of activation of blood coagulation during acute coronary ischemic syndromes: beyond aspirin and heparin. Cardiovasc Res 41:418–432

Bode C, Hanson SR, Schmedtje JF Jr, Haber E, Mehwald P, Kelly AB, Harker LA, Runge MS (1997) Antithrombotic potency of hirudin is increased in nonhuman primates by fibrin targeting. Circulation 95:800–804

Bode C, Hudelmayer M, Mehwald P, Bauer S, Freitag M, von Hodenberg E, Newell JB, Kübler W, Haber E, Runge MS (1994) Fibrin-targeted recombinant hirudin inhibits fibrin deposition on experimental clots more efficiently than recombinant hirudin. Circulation 90:1956–1963

Bode C, Matsueda GR, Hui KY, Haber E (1985) Antibody-directed urokinase: a specific fibrinolytic agent. Science 229:765–767

Bode C, Meinhardt G, Runge MS, Freitag M, Nordt T, Arens M, Newell JB, Kübler W, Haber E (1991a) Platelet-targeted fibrinolysis enhances clot lysis and inhibits platelet aggregation. Circulation 84:805–813

Bode C, Nordt T, Detlefs-Bartels U, Runge MS, Arens M, Kübler W, Haber E (1991b) Targeting of single-chain urokinase plasminogen activator by conjugation to an antiplatelet antibody results in enhanced clot lysis. Trans Assoc Am Physicians 104:29–31

Bode C, Runge MS, Branscomb EE, Newell JB, Matsueda GR, Haber E (1989) Antibody-directed fibrinolysis. An antibody specific for both fibrin and tissue plasminogen activator. J Biol Chem 264:944–948

Bode C, Runge MS, Haber E (1992) Purifying plasminogen activator antibody conjugates. Bioconjug Chem 3:269–272

Bode C, Runge MS, Newell JB, Matsueda GR, Haber E (1987a) Thrombolysis by a fibrin-specific antibody Fab'-urokinase conjugate. J Mol Cell Cardiol 19:335–341

Bode C, Runge MS, Newell JB, Matsueda GR, Haber E (1987b) Characterization of an antibody-urokinase conjugate. A plasminogen activator targeted to fibrin. J Biol Chem 262:10819–10823

Bode C, Runge MS, Schönermark S, Eberle T, Newell JB, Kübler W, Haber E (1990) Conjugation to antifibrin Fab' enhances fibrinolytic potency of single-chain urokinase plasminogen activator. Circulation 81:1974–1980

Bode C, Smalling RW, Berg G, Burnett C, Lorch G, Kalbfleisch JM, Chernoff R, Christie LG, Feldman RL, Seals AA, Weaver WD, for the RAPID II Investigators (1996) Randomized comparison of coronary thrombolysis achieved with double-bolus reteplase (recombinant plasminogen activator) and front-loaded, accelerated alteplase (recombinant tissue plasminogen activator) in patients with acute myocardial infarction. Circulation 94:891–898

Bode C, Wein M, Nordt T, Runge MS, Ruef J, Weber R, v. Hodenberg E, Kübler W, Haber E (1992) A bispecific antifibrin-antiplatelet UK conjugate enhances plasminogen activation at both targets. J Am Coll Cardiol 19 [Suppl]:393-A (Abstract)

Branscomb EE, Runge MS, Savard CE, Adams KM, Matsueda GR, Haber E (1990) Bispecific monoclonal antibodies produced by somatic cell fusion increase the potency of tissue plasminogen activator. Thromb Haemost 64:260–266

Califf RM (2000) Combination therapy for acute myocardial infarction: fibrinolytic therapy and glycoprotein IIb/IIIa inhibition. Am Heart J 139 [Suppl]:S33–S37

Charpie JR, Runge MS, Matsueda GR, Haber E (1990) A bispecific antibody enhances the fibrinolytic potency of single-chain urokinase. Biochemistry 29:6374–6378

Coller BS (1998) Monitoring platelet GP IIb/IIIa antagonist therapy. Circulation 97:5–9

Coller BS, Anderson K, Weisman HF (1995) New antiplatelet agents: platelet IIb/IIIa antagonists. Thromb Haemost 74:302–308

Dawson KM, Cook A, Devine JM, Edwards RM, Hunter MG, Raper RH, Roberts G (1994) Plasminogen mutants activated by thrombin. Potential thrombus-selective thrombolytic agents. J Biol Chem 269:15989–15992

Dewerchin M, Lijnen HR, Stassen JM, De Cock F, Quertermous T, Ginsberg MH, Plow EF, Collen D (1991) Effect of chemical conjugation of recombinant single-chain urokinase-type plasminogen activator with monoclonal antiplatelet antibodies on platelet aggregation and on plasma clot lysis in vitro and in vivo. Blood 78: 1005–1018

Dewerchin M, Lijnen HR, Van Hoef B, De Cock F, Collen D (1989) Biochemical properties of conjugates of urokinase-type plasminogen activator with a monoclonal antibody specific for cross-linked fibrin. Eur J Biochem 185:141–149

Dewerchin M, Vandamme AM, Holvoet P, De Cock F, Lemmens G, Lijnen HR, Stassen JM, Collen D (1992) Thrombolytic and pharmacokinetic properties of a recombinant chimeric plasminogen activator consisting of a fibrin fragment D- dimer specific humanized monoclonal antibody and a truncated single-chain urokinase. Thromb Haemost 68:170–179

Fujise K, Revelle BM, Stacy L, Madison EL, Yeh ETH, Willerson JT, Beck PJ (1997) A tissue plasminogen activator/P-selectin fusion protein is an effective thrombolytic agent. Circulation 95:715–722

Gimple LW, Gold HK, Leinbach RC, Coller BS, Werner W, Yasuda T, Johns JA, Ziskind AA, Finkelstein D, Collen D (1989) Correlation between template bleeding times and spontaneous bleeding during treatment of acute myocardial infarction with recombinant tissue-type plasminogen activator. Circulation 80:581–588

Goldstein J, Matsueda GR, Shaw S-Y (1996) A chimeric streptokinase with unexpected fibrinolytic selectivity. Thromb Haemost 76:429–438

Gruppo Italiano per lo Studio della Streptochinasi nell'Infarto Miocardico (GISSI) (1986) Effectiveness of intravenous thrombolytic treatment in acute myocardial infarction. Lancet 1:397–402

Holvoet P, Dewerchin M, Stassen JM, Lijnen HR, Tollenaere T, Gaffney PJ, Collen D
(1993a) Thrombolytic profiles of clot-targeted plasminogen activators: Parameters
determining potency and initial and maximal rates. Circulation 87:1007–1016
Holvoet P, Laroche Y, Lijnen HR, Van Cauwenberge R, Demarsin E, Brouwers E,
Matthyssens G, Collen D (1991) Characterization of a chimeric plasminogen activator consisting of a single-chain Fv fragment derived from a fibrin fragment D-
dimer-specific antibody and a truncated single-chain urokinase. J Biol Chem
266:19717–19724
Holvoet P, Laroche Y, Lijnen HR, Van Hoef B, Brouwers E, De Cock F, Lauwereys
M, Gansemans Y, Collen D (1992) Biochemical characterization of single-chain
chimeric plasminogen activators consisting of a single-chain Fv fragment of a
fibrin-specific antibody and single-chain urokinase. Eur J Biochem 210:945–
952
Holvoet P, Laroche Y, Stassen JM, Lijnen HR, Van Hoef B, De Cock F, Van Houtven
A, Gansemans Y, Matthyssens G, Collen D (1993b) Pharmacokinetic and thrombolytic properties of chimeric plasminogen activators consisting of a single-chain
Fv fragment of a fibrin-specific antibody fused to single-chain urokinase. Blood
81:696–703
Hua Z-C, Chen X-C, Dong C, Zhu D-X (1996) Characterization of a recombinant
chimeric plasminogen activator composed of Gly-Pro-Arg-Pro tetrapeptide and
truncated urokinase-type plasminogen activator expressed in *Escherichia coli.*
Biochem Biophys Res Commun 222:576–583
Huang X, Molema G, King S, Watkins L, Edgington TS, Thorpe PE (1997) Tumor infarction in mice by antibody-directed targeting of tissue factor to tumor vasculature.
Science 275:547–550
Hui KY, Haber E, Matsueda GR (1983) Monoclonal antibodies to a synthetic
fibrin-like peptide bind to human fibrin but not fibrinogen. Science 222:1129–
1132
Lijnen HR, Wnendt S, Schneider J, Janocha E, Van Hoef B, Collen D, Steffens GJ
(1995) Functional properties of a recombinant chimeric protein with combined
thrombin inhibitory and plasminogen-activating potential. Eur J Biochem 234:
350–357
Love TW, Quertermous T, Runge MS, Michelson KD, Matsueda GR, Haber E (1994)
Attachment of an antifibrin antibody to the amino terminus of tissue-type plasminogen activator impairs stimulation by fibrin. Fibrinolysis 8:326–332
Love TW, Quertermous T, Zavodny PJ, Runge MS, Chou C-C, Mullins D, Huang PL,
Schnee JM, Kestin AS, Savard CE, Michelson KD, Matsueda GR, Haber E (1993)
High-level expression of antibody-plasminogen activator fusion proteins in
hybridoma cells. Thromb Res 69:221–229
Love TW, Runge MS, Haber E, Quertermous T (1989) Recombinant antibodies pos-
sessing novel effector functions. Methods Enzymol 178:515–527
Muzykantov VR, Barnathan ES, Atochina EN, Kuo A, Danilov SM, Fisher AB (1996)
Targeting of antibody-conjugated plasminogen activators to the pulmonary vas-
culature. J Pharmacol Exp Ther 279:1026–1034
Peter K, Graeber J, Kipriyanov S, Zewe-Welschof M, Runge MS, Kübler W, Little M,
Bode C (2000) Construction and functional evaluation of a single-chain antibody
fusion protein with fibrin targeting and thrombin inhibition after activation by
factor Xa. Circulation 101:1158–1164
Peter K, Straub A, Kohler B, Volkmann M, Schwarz M, Kübler W, Bode C (1999)
Platelet activation as a potential mechanism of GP IIb/IIIa inhibitor-induced
thrombocytopenia. Am J Cardiol 84:519–524
Ruef J, Nordt TK, Peter K, Runge MS, Kübler W, Bode C (1999) A bispecific antifibrin-antiplatelet urokinase conjugate (BAAUC) induces enhanced clot lysis and
inhibits platelet aggregation. Thromb Haemost 82:109–114
Runge MS, Bode C, Matsueda GR, Haber E (1987) Antibody-enhanced thrombolysis:
Targeting of tissue plasminogen activator in vivo. Proc Natl Acad Sci USA
84:7659–7662

Runge MS, Bode C, Matsueda GR, Haber E (1988) Conjugation to an antifibrin monoclonal antibody enhances the fibrinolytic potency of tissue plasminogen activator in vitro. Biochemistry 27:1153–1157

Runge MS, Bode C, Savard CE, Matsueda GR, Haber E (1990) Antibody-directed fibrinolysis: a bispecific (Fab')2 that binds to fibrin and tissue plasminogen activator. Bioconjug Chem 1:274–277

Runge MS, Harker LA, Bode C, Ruef J, Kelly AB, Marzec UM, Allen E, Caban R, Shaw S-Y, Haber E, Hanson SR (1996) Enhanced thrombolytic and antithrombotic potency of a fibrin-targeted plasminogen activator in baboons. Circulation 94:1412–1422

Runge MS, Quertermous T, Zavodny PJ, Love TW, Bode C, Freitag M, Shaw S-Y, Huang PL, Chou C-C, Mullins D, Schnee JM, Savard CE, Rothenberg ME, Newell JB, Matsueda GR, Haber E (1991) A recombinant chimeric plasminogen activator with high affinity for fibrin has increased thrombolytic potency in vitro and in vivo. Proc Natl Acad Sci USA 88:10337–10341

Schnee JM, Runge MS, Matsueda GR, Hudson NW, Seidman JG, Haber E, Quertermous T (1987) Construction and expression of a recombinant antibody-targeted plasminogen activator. Proc Natl Acad Sci USA 84:6904–6908

Schneider J, Hauser R, Hennies HH, Korioth J, Steffens G, Wnendt S (1997) A novel chimaeric derivative of saruplase, rscu-PA-40 kDA/Hir, binds to thrombin and exerts thrombus-specific fibrinolysis in arterial and venous thrombosis in dogs. Thromb Haemost 77:535–539

Smith JW, Tachias K, Madison EL (1995) Protein loop grafting to construct a variant of tissue-type plasminogen activator that binds platelet integrin $\alpha_{IIb}\beta_3$. J Biol Chem 270:30486–30490

Stump DC, Lijnen HR, Collen D (1986) Purification and characterization of a novel low molecular weight form of single-chain urokinase-type plasminogen activator. J Biol Chem 261:17120–17126

Tait JF, Engelhardt S, Smith C, Fujikawa K (1995) Prourokinase-annexin V chimeras. Construction, expression, and characterization of recombinant proteins. J Biol Chem 270:21594–21599

The EPIC Investigators (1994) Use of a monoclonal antibody directed against the platelet glycoprotein IIb/IIIa receptor in high-risk coronary angioplasty. N Engl J Med 330:956–961

The EPILOG Investigators (1997) Platelet glycoprotein IIb/IIIa receptor blockade and low-dose heparin during percutaneous coronary revascularization. N Engl J Med 336:1689–1696

The EPISTENT Investigators (1998) Randomised placebo-controlled and balloon-angioplasty-controlled trial to assess safety of coronary stenting with use of platelet glycoprotein-IIb/IIIa blockade. Lancet 352:87–92

The GUSTO Angiographic Investigators (1993) The effects of tissue plasminogen activator, streptokinase, or both on coronary-artery patency, ventricular function, and survival after acute myocardial infarction. N Engl J Med 329:1615–1622

The GUSTO Investigators (1993) An international randomized trial comparing four thrombolytic strategies for acute myocardial infarction. N Engl J Med 329:673–682

The ISIS-2 (Second International Study of Infarct Survival) Collaborative Group (1988) Randomised trial of intravenous streptokinase, oral aspirin, both, or neither among 17 187 cases of suspected acute myocardial infarction: ISIS-2. Lancet 2: 349–360

Vandamme AM, Dewerchin M, Lijnen HR, Bernar H, Bulens F, Nelles L, Collen D (1992) Characterization of a recombinant chimeric plasminogen activator composed of a fibrin fragment-D-dimer-specific humanized monoclonal antibody and a truncated single-chain urokinase. Eur J Biochem 205:139–146

van Zyl WB, Pretorius GHJ, Hartmann M, Kotze HF (1997) Production of a recombinant antithrombotic and fibrinolytic protein, PLATSAK, in Escherichia coli. Thromb Res 88:419–426

Wan H, Liu Z, Xia X, Gu J, Wang B, Liu X, Zhu M, Li P, Ruan C (2000) A recombinant antibody-targeted plasminogen activator with high affinity for activated platelets increases thrombolytic potency in vitro and in vivo. Thromb Res 97:133–141

Wilcox RG, von der Lippe G, Olsson CG, Jensen G, Skene AM, Hampton JR, for the ASSET Study Group (1988) Trial of tissue plasminogen activator for mortality reduction in acute myocardial infarction. Anglo-Scandinavian Study of Early Thrombolysis (ASSET). Lancet 2:525–530

Wnendt S, Janocha E, Steffens GJ, Strassburger W (1997) A strong thrombin-inhibitory prourokinase derivative with sequence elements from hirudin and the human thrombin receptor. Protein Eng 10:169–173

Yamada T, Shimada Y, Kikuchi M (1996) Integrin-specific tissue-type plasminogen activator engineered by introduction of the Arg-Gly-Asp sequence. Biochem Biophys Res Commun 228:306–311

Yang W-P, Goldstein J, Procyk R, Matsueda GR, Shaw S-Y (1994) Design and evaluation of a thrombin-activable plasminogen activator. Biochemistry 33:2306–2312

The Hunt for the Ideal Thrombolytic Agent: Mutants of tPA and uPA, Chimera of Both Molecules, Fibrolase

M. VERSTRAETE

A. Mutants and Variants of Single-Chain Urokinase-Type Plasminogen Activator

Some approaches to improve the thrombolytic profile of recombinant single-chain urokinase-type plasminogen activator (rsc-uPA, prourokinase) are briefly summarised below.

I. Plasmin-Resistant Mutants of sc-uPA

Because clot lysis with sc-uPA in a plasma milieu is associated with a higher degree of fibrin specificity compared to two-chain urokinase-type plasminogen activator (tc-uPA, urokinase), several investigators have constructed mutants of recombinant sc-uPA in which the plasmin cleavage site was destroyed by site-specific mutagenesis. Such plasmin-resistant recombinant mutants of sc-uPA (e.g., rsc-uPA Lys[158]Ala, rsc-uPA Lys[158]Gly) have, however, a thrombolytic potency in rabbits with jugular vein thrombosis that is about five times lower than that of sc-uPA (COLLEN et al. 1989a). The thrombolytic potency of such mutants thus appears to be too low to allow their efficient use in humans.

II. Low-Molecular-Weight sc-uPA

A low molecular weight derivative of human sc-uPA, sc-uPA-32k, lacking the amino-terminal 143 residues, was purified from cell cultures (STUMP et al. 1986) and subsequently produced by recombinant DNA technology (rsc-uPA-32k) (LIJNEN et al. 1988). In a rabbit jugular vein thrombosis model, comparable clot lysis to 33kDa tc-uPA was obtained with rsc-uPA-32k, but was associated with less pronounced systemic fibrinogen breakdown (LIJNEN et al. 1988). Unfortunately, clustered charge-to-alanine mutants of rsc-uPA-32k designed to eliminate charged regions with the highest solvent accessibility did not exhibit significantly improved functional, fibrinolytic or pharmacokinetic properties (UESHIMA et al. 1994).

III. Mutant of sc-uPA Resistant to PAI-1

Deletion of the amino acid sequence Arg[179]-Ser[184] in uPA (which is homologous to the plasminogen activator inhibitor (PAI-1) binding site of tPA) resulted in a urokinase mutant which was resistant to inhibition by PAI-1 (Adams et al. 1991). This mutant has, however, not been evaluated in in vivo experiments.

B. Mutants of Tissue-Type Plasminogen Activator (tPA)

It is well documented for a large number of tPA variants that the type and extent of glycosylation as well as the presence or absence of the various non-catalytic domains affect the pharmacokinetics of the protein. Thus several mutants of recombinant tissue-type plasminogen activator (rtPA, alteplase) have been constructed with interesting properties, including slower clearance from the circulation, more selective binding to fibrin, stronger stimulation by fibrin, and resistance to plasma protease inhibitors (Larsen et al. 1989, 1991; Madison et al. 1989, 1993; Madison and Sambrook 1993; Keyt et al. 1994; Ke et al. 1997; Strandberg and Madison 1995; Tachias and Madison 1997) (see also Chap. 2).

I. Mutants with Modified Epidermal Growth Factor Domain

A modified tPA (E6010, monteplase, Cleactor; Eisai Co., Japan) constructed by substituting only one amino acid in the epidermal growth factor domain ($Cys^{84} \rightarrow Ser$) has a prolonged half-life of more than 20 min, compared with 6 min for wild-type tPA (Suzuki et al. 1991). Monteplase has been used successfully for bolus administration in femoral artery (Suzuki et al. 1991, 1994) and coronary artery (Saito et al. 1994, 1995; Suzuki et al. 1995) thrombosis models in dogs. In a prospective randomized, double-blind multicenter trial monteplase was compared to tPA (tisokinase) in 199 patients with acute myocardial infarction (AMI), treated within 6 h of onset of symptoms (Kawai et al. 1997). Monteplase was given IV in a dose of 0.22 mg/kg over 2 min and tPA was administered IV in an unusually small dose of 28.8 mg (10% over 1–2 min, the rest over 60 min). Primary endpoint in this study was the recanalization rate of the infarct-related artery at 60 min. TIMI grade 3 flow rates at 30, 45, and 60 min were 25%, 43%, and 53% for monteplase and 9%, 20%, and 36% for tPA respectively. From this study it is difficult to ascertain whether monteplase is superior to tPA since tisokinase was given in doses which were only 25%–29% of those commonly used in the United States and in Europe.

Also, other point mutations in the epidermal growth factor region can considerably prolong the circulating half-life of tPA (Browne et al. 1993; Bassel-Duby et al. 1992).

Deletion of the entire growth factor domain (amino acids 51–87) resulted in a tPA mutant which was cleared four to ten times more slowly than wild-type tPA in rat, rabbit, and guinea pig models and maintained the same specific activity as tPA (Browne et al. 1988, 1989; Lucore et al. 1989; Johannesen et al. 1990).

II. Mutants with Modified or Deleted Finger Domain

Deletion of the finger domain (amino acids 4–50) increased the circulating half-life of tPA 20-fold in a rat model (Larsen et al. 1989). Mutants of tPA containing 3–6 contiguous substitutions in the finger domain (amino acids 7–9, 10–14, 15–19, 28–33, and 37–42) were constructed by Yahara et al. (1992), and exhibited a 6- to 12-times longer half-life in rabbits.

III. Deletion of the Finger and Epidermal Growth Factor Domains

A variant of tPA (ΔFE1X) lacks both the finger and growth factor domains and has an amino acid substitution of Gln at the N-linked glycosylation site Asn[117] which prevents clearance by the mannose receptor (Otter et al. 1992). This molecule is known as novel plasminogen activator (nPA, also lanoteplase), coded SUN9216 or BMS-200980. Produced by recombinant techniques in hamster ovary cells, this mutant of tPA displays a 22-fold enhanced circulating half-life and 9.6-fold increased thrombolytic potency compared to wild-type tPA in the rabbit (Larsen et al. 1991). It is suitable for administration as a single rapid bolus injection, rather than IV infusion over several hours. A single bolus injection of 1 mg/kg 30 min after occlusion of the middle cerebral artery in rats was effective in recanalizing the vessel and reducing the area of cerebral infarction. Blood flow in the middle cerebral artery was restored in 32% of rats treated with SUN9216, in 59% of rats given SUN9216 and a thromboxane receptor inhibitor (vapiprost) but in none of the saline-treated rats. The area of cerebral infarction in rats perfused with SUN9216 alone or combined with vapiprost was significantly reduced compared with that in the control group. The time of reopening the middle cerebral artery was 38 min and 21 min in rats treated with SUN9216 alone or in combination with vapiprost, respectively. No bleeding in the cerebrum was noted in rats treated with the combination (Unemura et al. 1993, 1994). Adding an endothelin receptor antagonist (FR 139317) offered no extra advantage over SUN9216 on its own, either in reopening the middle cerebral artery or in reducing the size of cerebral infarction (Unemura et al. 1995).

In the double-blind, angiographic InTIME trial, 602 patients with AMI, presenting within 6 h of symptom onset were randomized to receive one of four doses of lanoteplase and an alteplase placebo, or accelerated alteplase and a lanoteplase placebo (Den Heijer et al. 1998). Lanoteplase, 15, 30, 60, or 120 kU/kg (not to exceed 12 MU), was given as an IV bolus over 2–4 min. Alteplase or its matching placebo was administered according to the accelerated regimen: a 15 mg bolus, followed by 0.75 mg/kg (not to exceed 50 mg) over 30 min, then 0.5 mg/kg (not to exceed 35 mg) over 60 min. Aspirin, 150–325 mg/day and heparin, IV bolus of 5000 U were initiated prior to thrombolytic treatment. Heparin was continued at a dose of 1000 U/h for at least 48 h and adjusted to an activated partial thromboplastin time (APTT) of 60–85 s. At 60 min the following angiographic TIMI grade 3 flow rates had been achieved with 15, 30, 60, and 120 kU/kg of lanoteplase and alteplase respectively: 24%, 29%, 44%, 47%, and 37%. At 90 min TIMI 3 flow rates were 26%, 32%, 47%, 57%, and 46%, respectively. Major bleeding, defined as hemorrhagic stroke or bleeding associated with hemodynamic compromise requiring transfusion occurred in 1.5% of the lanoteplase-treated patients and in 5.6% of those receiving alteplase. Thirty day mortality was 3.1% in the lanoteplase and 6.5% in the alteplase group (DEN HEIJER et al. 1998). None of these differences were statistically significant, due to the relatively small number of patients in each group. Noteworthy is the low TIMI grade 3 patency rate of 46% with alteplase, which was lower in this medium-sized trial than in most larger trials using the accelerated alteplase dosing scheme.

The highest dose of lanoteplase produced a proteolytic state comparable to that produced by alteplase. However, the 60 kU/kg dose which was similarly efficacious as alteplase with respect to patency rates resulted in a lesser fibrinogen breakdown and plasminogen and α_2-antiplasmin consumption (KOSTIS et al. 1999). A pharmacokinetic substudy of InTIME in 31 AMI patients revealed a two-compartment elimination profile with a plasma clearance of 51 $\pm$ 16 ml/min (mean $\pm$ SD), a t/2α of 37 $\pm$ 11 min and a t/2β of 586 $\pm$ 278 ml/min (LIAO et al. 1997). Patients experiencing an AMI often have elevated PAI-1 plasma levels at admission (ALMÉR and ÖHLIN 1987; SANE et al. 1991; THÖGERSEN et al. 1998). During thrombolytic therapy with tPA, PAI-1 levels fall to values around zero, but exhibit a rebound to high levels one to several hours later (RAPOLD et al. 1991). It also has been demonstrated in vitro that tPA stimulates PAI-1 expression in endothelial cell cultures (SHI et al. 1996). Interestingly, this rebound phenomenon was significantly attenuated during thrombolytic therapy with lanoteplase (OGATA et al. 1998).

IV. Deletion of the Epidermal Growth Factor and Finger Domains and of Glycosylation Sites

A molecule lacking the finger- and the epidermal growth factor regions and with all three glycosylation sites removed (Asn^{117}Gln, Asn^{1184}Gln, Asn^{448}Gln) (LARSEN et al. 1989) was found to have an initial half-life of 14 min in rabbits and dogs and was more effective than wild-type tPA in a canine model of coronary thrombosis (CAMBIER et al. 1988).

V. Mutants Consisting of Kringle 2 and the Protease Domain

Mutants consisting only of the kringle 2 and the protease domain of tPA were reported to display a 2- to 20-fold prolonged circulating half-life and 3- to 5-fold higher thrombolytic potency in a variety of thrombosis models in different animals (JACKSON et al. 1990, 1992; FELEZ et al. 1990; NICOLINI et al. 1992; BURCK et al. 1990; BODE et al. 1995; NEUHAUS et al. 1994).

Reteplase (rPA, BM 06.022, Rapilysin, Retavase, Ecokinase; Boehringer Mannheim) is the best studied single-chain, non-glycosylated deletion variant of tPA (reviewed in MARTIN et al. 1999). Its structure is depicted in Fig. 1.

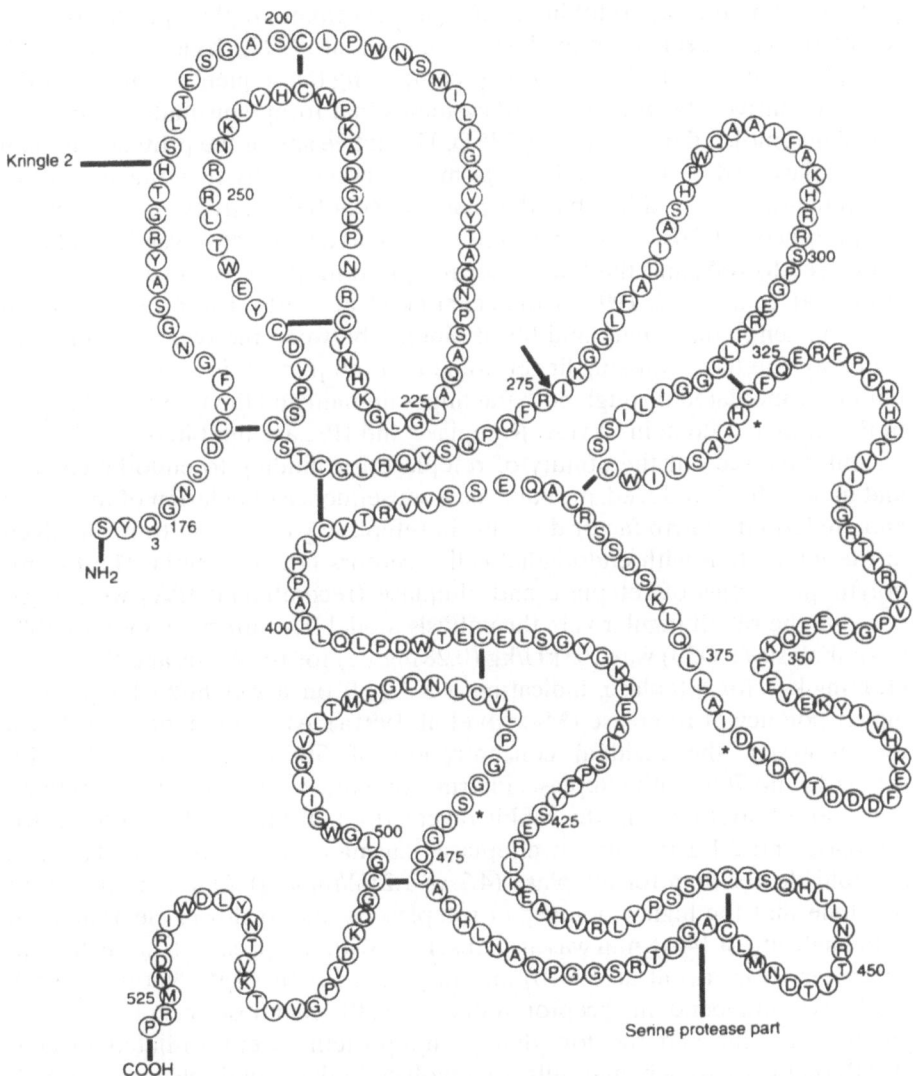

Fig. 1. Diagram of the primary structure of reteplase (amino acids Ser^1-Gln^3 and Gly^{176}-Pro^{527} of tPA) with domain organization and sites of post-translational modification. The amino acids are represented by their *single letter codes*, *black bars* indicate disulfide intra- and interchain bridges and the *asterisks* indicate the active site residues in the protease part. The *arrow* indicates the plasmin cleavage site

Expression in a prokaryotic host, *E. coli*, provides for the absence of glycosylation without the need for point mutation at the sites of post-translational modification. Moreover, expression in bacteria is generally considered preferable to that in mammalian cells since many of the complicating virological and related safety issues associated with production in mammalian cell culture are avoided. Production of reteplase in *E. coli* leads to the accumulation of inactive protein aggregates (inclusion bodies) inside the cells. The isolation of the inclusion bodies, the refolding, and the chromatographic purification of reteplase have been described (KOHNERT et al. 1992; STERN et al. 1989). Reteplase consists of 355 amino acids comprising the sequences 1 to 3 and 176 to 527 of native tPA. It contains 18 cysteines that form 9 disulfide bridges, the calculated molecular weight is 39 575 D. The active site of the protease domain of reteplase and of tPA, and their plasminogenolytic activity in the absence of a stimulator, do not differ but the plasminogenolytic activity of reteplase in the presence of CNBr fragments of fibrinogen as a stimulator was 4-fold lower compared to tPA, and the binding of reteplase to fibrin was five times lower (KOHNERT et al. 1992, 1993; STÜRZEBECHER et al. 1991). These differences in plasminogenolytic activity and fibrin binding between the two molecules are probably due to the missing finger domain in reteplase. It is known that fibrin binding is mediated through both the finger domain and the lysine binding site in the kringle 2 domain of tPA. Reteplase and tPA are inhibited by PAI-1 to a similar degree but the affinity of reteplase for binding to endothelial cells and monocytes is reduced, probably as a consequence of deletion of the finger and epidermal growth factor domains in reteplase which seem to be involved in the interaction with endothelial cell receptors (HAJJAR 1991). The thrombolytic properties of reteplase and alteplase (recombinant tPA) were compared in the rabbit jugular vein thrombosis model. The effective dose for 50% thrombolysis (ED50) was 163 kU/kg (0.28 mg/kg) for reteplase, and 871 kU/kg (1.09 mg/kg) for alteplase, indicating a 5.3 (3.9 on a weight/kg basis) -fold higher potency of reteplase (MARTIN et al. 1991a). At equipotent doses (50% thrombolysis), the residual concentration of fibrinogen was 74% with reteplase and 76% with alteplase. Pharmacokinetic analysis of plasma activity at a dose of 400 kU/kg in the rabbit revealed a half life of 18.9 ± 1.5 min for reteplase and 2.1 ± 0.1 min for alteplase. Plasma clearance for reteplase was fourfold slower than for alteplase (4.7 vs 1.2 ml/min/kg). One may therefore conclude that the higher potency of reteplase is due to slower clearance. An initial half-life of 14–18 min was also observed with reteplase in healthy human volunteers (MARTIN et al. 1991b) and in patients with AMI (NEUHAUS et al. 1994). Two independent receptor systems, i.e., the mannose receptor on liver endothelial cells and the low density lipoprotein receptor-related protein (LRP) on liver parenchymal cells, are involved in the hepatic clearance of tPA (MARTIN et al. 1993; CAMANI and KRUITHOF 1994). Organ failure studies were conducted to elucidate the metabolic capacity of the liver and kidneys. Bilateral nephrectomy in rats resulted in a longer half-life, an increase in liver uptake and a reduced clearance rate compared with tPA (MARTIN et al. 1993). Nevertheless, plasma activity concentration decreased over time indicating that there are compensatory mechanisms. In addition to the liver, blood plasma was identified as contributing to inactivation of reteplase by inhibitors, i.e., C1-inactivator, α_2-antiplasmin and α_1-antitrypsin (Rijken et al. 1994). In the healthy rat, the relative contribution of this inhibition pathway to total clearance of reteplase was calculated to be 32%. Results of the functional studies suggest that the catabolic role of the kidneys might be higher than proposed by the uptake studies (20%), potentially via filtration as shown for other mutants of tPA (Larsen et al. 1989). Further studies with varying degrees of renal dysfunction indicated that slight impairment of renal function does not significantly influence the pharmacokinetic properties of reteplase. Chemically induced hepatic failure only marginally (by 14%) reduced the clearance rate indicating that liver failure can be better compensated for than kidney failure. In studies with renal and hepatic failure in rats, reteplase was still inactivated in plasma (Martin et al. 1993).

Dose-escalating studies were conducted in healthy male volunteers (Martin et al. 1991b). At the doses studied, i.e., up to 5.5 MU of reteplase, there was no reduction in plasma fibrinogen or prolongation of the clotting times (thrombin time, prothrombin time, APTT). Plasminogen fell to a nadir of 86% and α_2-antiplasmin to 70% of baseline, α_2-macroglobulin and antithrombin III remained unchanged. The decreases most probably reflect unspecific systemic formation of plasmin by reteplase. Indirect evidence of the fibrinolytic action of reteplase was obtained by a dose-dependent increase of fibrin D-dimers up to 1147 ± 380 ng/ml and of fibrin degradation products. The reduction in PAI-I activity indicated normal interaction of reteplase with PAI-1. The bleeding time, platelet count and platelet aggregation were in the normal range after administration of reteplase. Evaluation of hemodynamic and clinicochemical parameters of safety did not reveal relevant changes after the 2 min IV bolus injection of reteplase. Formation of antibodies to reteplase was not found in samples taken up to 1 year post administration. The area under the curve (AUC) values for reteplase activity showed a dose-dependent linear increase beginning at doses of 2.2 MU reteplase. At the highest dose studied, i.e., 5.5 MU, the total plasma clearance for reteplase activity was 306 ± 40 ml/min and the half life was 14.4 ± 1.1 min.

Dose-ranging studies of bolus reteplase were performed in an open-label sequential multicenter trial coded GRECO (Neuhaus et al. 1994). With a dose of 10 MU of reteplase, a patent infarct-related coronary artery (TIMI grade 3 flow) was obtained at 30 min in 46%, at 60 min in 48%, at 90 min in 52%, and at 24 h to 48 h in 88% of patients with AMI. With 15 MU a higher angiographic patency rate at the same time intervals was obtained (38%, 58%, 69%, and 85%). Because there was a 20% (10 MU) and 12.5% (15 MU) reocclusion rate between the 30 min and 90 min angiogram, the administration of a second smaller bolus of reteplase (5 MU) 30 min after the initial bolus (10 MU) was investigated in an open uncontrolled study coded GRECO DB (Tebbe et al. 1993). TIMI grade 3 flow rates reached 50% at 60 min, 58% at 90 min, and 84% at 24–48 h. Only 1 of the 50 patients studied had reocclusion in the first 24–48 h. In a randomized, open label controlled study (coded RAPID-1) in 606 patients with AMI, different bolus doses of reteplase (single dose of 15 MU, 10 MU and 5 MU 30 min later, 10 MU and 10 MU 30 min later) were compared with the conventional dose regimen of alteplase (100 mg over 3 h) (Smalling et al. 1995). TIMI-3 patency rates at 90 min were obtained with the given reteplase regimen in 41%, 46%, 63%, respectively, and in 49% of patients treated with alteplase. The difference between the 10 MU + 10 MU reteplase and alteplase arms is significant ($p < 0.05$).

The RAPID-2 study was a randomized, open-label, angiographic study in 324 patients with AMI (Bode et al. 1996). It was designed to compare the effect of 10 + 10 U reteplase with that of accelerated, front-loaded, dose alteplase (100 mg over 90 min) on the TIMI flow rate of the infarct-related coronary artery 90 min after the initiation of thrombolytic therapy. There was no age limit and patients were recruited up to 12 h after onset of symptoms, all received aspirin. The heparin regimen consisted of a 5000 IU IV bolus that was administered before thrombolytic therapy followed by an infusion of 1000 IU per hour for at least 24 h. In this study, reteplase achieved earlier and more complete reperfusion than accelerated-dose alteplase. TIMI grade 2 or 3 and TIMI grade 3 flow rates of the infarct-related artery at 90 min were significantly higher for reteplase relative to the alteplase control (83% vs 73% and 60% vs 45% respectively). At 60 min, both the TIMI grade 2 or 3 and the TIMI grade 3 flow rates were significantly higher for reteplase as compared to alteplase. Reteplase treated patients required significantly fewer additional coronary interventions within the first 6 h of treatment (14% vs 26.5%). As expected in a trial of this size, there was no significant differences between the reteplase and alteplase groups with respect to 35-day mortality (4.1% vs 8.4%) and hemorrhagic stroke (1.2% vs 1.9%).

Two mortality trials with reteplase were conducted in patients with AMI. The INJECT study was designed to determine whether reteplase was at least as effective in mortality reduction (within 1% of fatality rate) as a standard streptokinase regimen (International Joint Efficacy Comparison of Thrombolytics 1995). In this double-blind study, 3004 patients were randomized to a double bolus of 10 + 10 U of reteplase, 30 min apart, and 3006 patients were randomized to 1.5 MU of streptokinase over 60 min. Treatment could be started up to 12 h from onset of symptoms. All patients received IV heparin for at least 24 h and aspirin. The 35-day mortality rate was 9.0% in the reteplase group and 9.5% in the streptokinase group, a non-significant difference (95% CI: −1.98% to 0.96%). At six months, mortality rates were 11% for reteplase and 12% for streptokinase, a difference of −1% (95% CI: −2.65% to 0.59%). Bleeding events were similar in the two groups (0.7% for reteplase and 1% for streptokinase). The in-hospital stroke rates were 1.2% for reteplase and 1% for streptokinase. The incidence of recurrent myocardial infarction was similar in the two groups. In the GUSTO III trial 15,059 patients with AMI from 807 hospitals in 20 countries were randomly assigned in a 2:1 ratio to receive reteplase, two bolus doses of 10 MU each given 30 min apart, or the accelerated tPA (alteplase) regimen [The Global Use of Strategies

TO OPEN OCCLUDED CORONARY ARTERIES (GUSTO III) INVESTIGATORS 1997].
The mortality rate at 30 days was 7.5% for reteplase and 7.2% for alteplase
(n.s.). Stroke occurred in 1.6% of patients treated with reteplase and in 1.8%
of those treated with alteplase (n.s.). The combined endpoints of death or non-
fatal, disabling stroke were 7.9% in both groups. Thus, in this large study,
reteplase, although easier to administer than alteplase did not result in a survival benefit in patients with AMI.

A preliminary pilot study (SPEED TRIAL = GUSTO-4), using two
boluses of low-dose reteplase (5 + 5 U) combined with full-dose abciximab and
slightly reduced, bodyweight-adjusted heparin doses in AMI resulted in
enhanced early reperfusion (OHMAN et al. 1998; CALIFF 1999; BODE et al. 1999).

In phase B of the SPEED trial 115 patients received two bolus doses of
5 U of reteplase, 30 min apart, plus the standard dose of abciximab, and 109
patients were randomized to high dose reteplase (2 × 10 MU 30 min apart)
(TOPOL et al. 2000). The primary endpoint, TIMI grade 3 flow, was 54% in
the combination therapy group and 47% in patients receiving reteplase only.
TIMI grade 3 flow was dependent on the dose of the initial heparin bolus
administered in the combination group (61% for 60 U/kg and 51% for 40 U/kg
body weight). Major bleeding rates were higher in the abciximab-reteplase
5 + 5 MU group (9.8%) than in the reteplase 10 + 10 MU group (3.7%).

Reteplase caused greater ADP- and thrombin-induced platelet aggregation and greater GP IIb/IIIa expression than alteplase in the GUSTO-III trial
(GURBEL et al. 1998).

Molecular markers of coagulation and fibrinolysis were serially examined
for up to 5 days in a prospective, randomized study (HOFFMEISTER et al.
2000). Fifty patients with AMI were either treated with double bolus (10 +
10 MU) reteplase or with front-loaded alteplase (up to 100 mg) within 6 hours
of symptom onset. Thirty apparently healthy persons served as controls. Para-
doxical thrombin activation at 3 hours after initiation of therapy was comparable between reteplase and alteplase. Reteplase and alteplase caused
significantly elevated kallikrein activity at 3 hours after administration (65 ±
U/l and 72 ± 8 U/l, respectively) compared to controls (30 ± 1 U/l; $p < 0.01$).
Fibrin specificity was less for reteplase with a decrease in fibrinogen at 3 hours
to 1.22 ± 0.27 g/l versus 2.24 ± 0.28 g/l for alteplase ($p < 0.05$). D-Dimer levels
at 3 hours were higher after reteplase (5460 ± 610 ng/ml) versus alteplase (3440
± 680 ng/ml) ($p < 0.05$). These results demonstrate a lesser fibrin specificity of
reteplase compared with front-loaded alteplase. Both treatment schemes produced a moderate procoagulant effect.

VI. Mutant Consisting of Kringle 2 with Point Mutation in Linkage to Protease Light Chain

YM866 (pamiteplase, Solinase, Yamanouchi Pharmaceuticals, Japan) is a novel
modified tPA with deletion of the kringle-1 domain and a point mutation at

the cleavage site of single-chain tPA (del 92–173, Arg^{275} Glu) (Kawauchi et al. 1991). The replacement of Arg by Glu at position 275 renders this tPA mutant resistant to cleavage by plasmin. Pamiteplase possesses the same affinity for fibrin and specific activity of tPA in vitro (Katoh et al. 1991). Furthermore, the inhibition of pamiteplase and tPA activities by plasminogen activator inhibitor-1 (PAI-1) is comparable. The plasma clearance of pamiteplase estimated from pharmacokinetic studies in rats was seven times slower than that of tPA (mean residence time: pamiteplase, 62 min; tPA, 9 min) (Katoh et al. 1991). Pamiteplase administered by IV bolus injection produced a greater thrombolytic effect than tPA in a canine model of coronary artery thrombosis (Kawasaki et al. 1993).

Pamiteplase was compared with that of recombinant tPA (tPA) in rabbits with experimental jugular vein thrombosis (Kawasaki et al. 1994). Thrombus was induced in an isolated segment of the jugular vein from a mixture of whole autologous blood and thrombin. Then, 30 min after the induction of thrombus, pamiteplase was administered by IV bolus injection, while tPA was given by the same method or by a 60 min IV infusion. Thrombi were removed 70 min after IV bolus injection or 10 min after the termination of IV infusion, and thrombus size was measured by total protein content in the isolated thrombus. Both pamiteplase and tPA exhibited dose-dependent thrombolysis; the thrombolytic activity of pamiteplase, however, was four times greater than that of tPA. The improved thrombolytic activity of pamiteplase correlated with its relatively higher antigen levels in plasma, which result from its prolonged biological half-life. Depletion of plasma fibrinogen to less than 20% of baseline levels was observed in all groups. Template bleeding time was not significantly altered in any group.

A pharmacokinetic study over a wide range of pamiteplase dosages in human volunteers revealed a t/2 of biologically active pamiteplase of 30–47 min after the administration of a single dose of 0.5–4 mg of the mutant (Hashimoto et al. 1996). In a dose-finding early phase II study 157 patients with AMI were treated with 0.05, 0.1, 0.2, or 0.3 mg/kg of pamiteplase. TIMI grade 3 flow rates were achieved after 60 min in 42%, 57%, 63%, and 54% of patients respectively. Adverse effects occurred in 7% of patients receiving 0.2 mg/kg, but in 17% of those given the highest dose of 0.3 mg/kg (Sumiyoshi et al. 1996). In a double-blind multicenter trial pamiteplase, given as an IV bolus, was compared with tPA (tisokinase) in patients with AMI. Pamiteplase was administered in a dosage of 0.1 mg/kg and tisokinase in the low dosage commonly used in Japan of 14.4 MU (28.8 mg). TIMI grade 3 flow rates after 30, 45, and 60 min were 25%, 36%, and 50% with pamiteplase and 16%, 26%, and 48% respectively with tisokinase (Sumiyoshi et al. 1996). As stated above for E6010, it is difficult to determine whether the use of approximately three- to four-fold lower tPA doses than those administered in Europe and the United States permits a judgement on the comparative efficacy of pamiteplase and of tPA.

VII. Mutant with Modifications in Kringle 1

The mutant Asn^{117}Glu of human tPA, termed long-acting tPA (LA-tPA), eliminates an N-linked high mannose oligosaccharide carbohydrate side-chain in the kringle 1 region of tPA that is recognised by receptors on liver Kupffer cells. In a rabbit model of pulmonary embolism LA-tPA had a fourfold longer circulating half-life than wild-type tPA and brought about more rapid thrombolysis without increased systemic fibrinogenolysis (LAU et al. 1987). In the dog, the hepatic clearance of LA-tPA was approximately six times slower than that of wild-type tPA (α-phase 8.9 ± 1.7 min v 2.8 ± 0.6 min). When equal bolus doses of LA-tPA and native tPA were given to dogs subjected to copper-coil induced coronary thrombosis, LA-tPA resulted in slightly faster thrombolysis of fresh thrombi (<30 min old) than native tPA, but was associated with a high reocclusion rate (EIDT et al. 1991).

VII. Recombinant TNK-tPA

A tPA mutant in which Thr^{103} is substituted by Asn (coded T-tPA) was found to have a prolonged half-life (KEYT et al. 1994). By mutagenizing the sequence Lys^{296}-His-Arg-Arg (which binds tightly to PAI-1) to a tetra-alanine substitution at position 296–299 (code K-tPA), resistance to PAI-1 was achieved (BENNETT et al. 1991; PAONI et al. 1992, 1993). Recent work, using wild-type and PAI-1 deficient mice, demonstrated that PAI-1 is a major determinant of the resistance of platelet-rich arterial thrombi to pharmacological concentrations of wild-type tPA (ZHU et al. 1999). The double mutant T103N, KHHR(296–299)AAAA (code TK-tPA) had an increased potency on platelet-rich arterial thrombi (rich in PAI-1) in a rabbit arteriovenous shunt model (REFINO et al. 1993). Additional substitution in this mutant of Asn^{117} by Gln (N-tPA) resulted in a tPA variant with 8-fold slower clearance, and 200-fold enhanced resistance to PAI-1 (GUZETTA et al. 1993; KEYT et al. 1994). These three combinations in a single molecule are referred to as TNK-tPA (Fig. 2).

The variant is more potent than wild-type tPA. TNK-tPA (tenecteplase, Inkase Genentech Inc. Metalyse, Boehninger Ingelheim) had a substantially slower in vivo clearance in rabbits, near-normal fibrin binding and plasma clot lysis activity, resistance to PAI-1 and enhanced fibrin-specificity (KEYT et al. 1994). To investigate the higher fibrin specificity of TNK-tPA the kinetics of Glu-plasminogen activation for t-PA (alteplase) and TNK-tPA in the presence of fibrin, the fibrin degradation product DDE and fibrinogen were compared. Although these two activators have similar catalytic efficiency in the presence of fibrin, the catalytic efficiency of TNK-tPA is 15-fold lower than that for tPA in the presence of DDE or fibrinogen (STEWART et al. 2000). In vivo models of fibrinolysis in rabbits indicate that TNK-tPA (by bolus) achieves 50% lysis of a whole blood clot in one-third the time required by an equivalent dose of tPA (as an infusion). In the same model, the TNK variant is 8- and 13-fold more

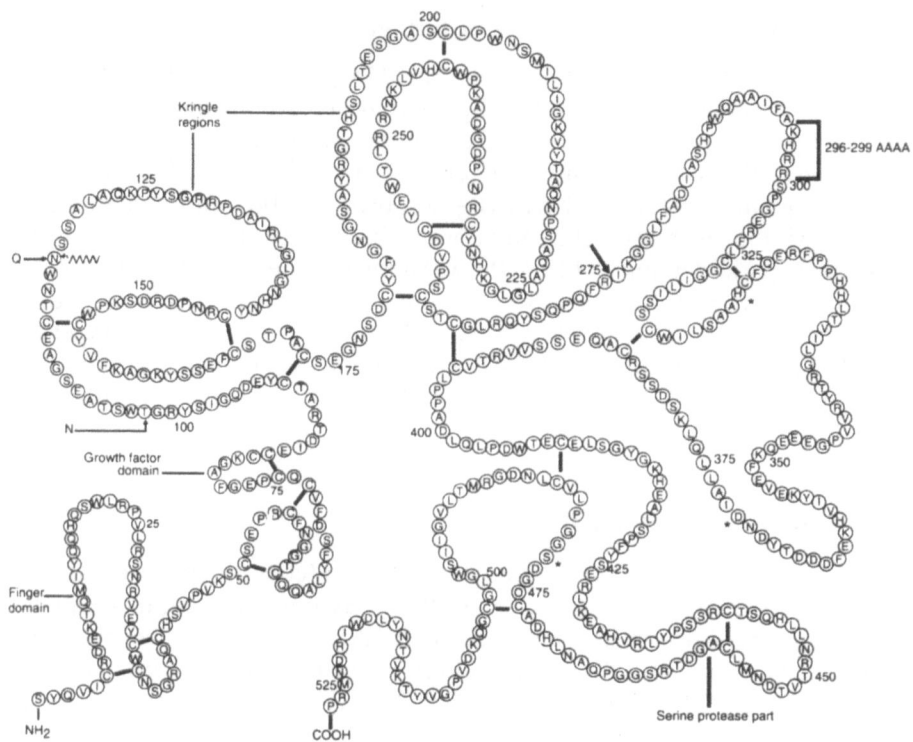

Fig. 2. Schematic representation of the primary structure of rtPA-TNK (substitution in rtPA of Thr[103] by Asn, of Asn[117] by Gln and of Lys[296]-His-Arg-Arg by Ala-Ala-Ala-Ala). The amino acids are represented by their *single letter symbols*, *black bars* indicate disulfide bonds and the *asterisks* indicate the active site residues in the protease part. The *arrow* indicates the plasmin cleavage site

potent than tPA toward whole blood clots and platelet-enriched clots, respectively. TNK-tPA conserves fibrinogen and, because of its slower clearance and normal clot lysis activity, is effective as a bolus at a relatively low dose (KEYT et al. 1994). Similar results have been obtained in a platelet-rich arterial eversion graft thrombosis model in dogs (COLLEN et al. 1994).

The clearance of TNK-tPA was evaluated in rabbits following IV injections. The AUC for TNK-tPA was 8.4-fold greater than for tPA which resulted in clearance values of 1.9 ml/min/kg and 16.1 ml/min/kg for TNK-tPA and tPA, respectively. The data were fitted to bi-exponential and tri-exponential equations for TNK-tPA and tPA, respectively. Clearance of TNK-tPA exhibited an α and β phase with calculated half-life values of approximately 9 and 31 min. Wild-type tPA exhibited α, β, and γ phases of clearance with calculated half-life values of 1.3, 6, and 33 min, respectively. The rates of elimination (terminal phase of clearance) for the two molecules were equivalent as indicated by similar slopes of the clearance curves after 2 h. However, the early phase of clearance was very different. At 15 min after injection, 50% of TNK-tPA remained in the circulation compared with only 1% of tPA remaining at the same time (BENEDICT et al. 1995). The plasma pharmacokinetics of TNK-tPA were also characterized in male beagle dogs following doses of 0.3, 2, 10, and 30 mg/kg. In this study, the pharmacodynamic effect of TNK-tPA on fibrinogen, α_2-antiplasmin, and fibrinogen degradation products was also investigated. Following a dose of 0.3 mg/kg, the initial and the terminal half-life were approximately 3 and 40 min, respectively, and the clearance was 17 ± 3 ml/min/kg. The estimated clearance decreased with increasing dose. At a dose of 30 mg/kg, the plasma clearance was 3.5 ± 0.9 ml/min/kg. This may be indicative of a saturation of the clearance in dogs. TNK-tPA produced a dose-dependent effect on the hematologic parameters that was consistent with the pharmacologic action of a thrombolytic agent. There was a transient decrease in the fibrinogen and α_2-antiplasmin concentrations and an increase in the fibrinogen degradation products. The hematologic parameters had returned to baseline levels by 24 h (McCLUSKEY et al. 1996).

Co-administration of heparin and aspirin does not seem to affect the pharmacokinetics of TNK-tPA. The elimination of TNK-tPA appears to be mediated hepatically and the clearance may be saturable at high doses (i.e., >10 mg/kg) (McCLUSKEY et al. 1996).

TNK-tPA was investigated in an open-label, dose-ranging, pilot trial of single-bolus, IV TNK-tPA in patients with AMI (TIMI 10 A) (CANNON et al. 1997). Patients were treated with one of eight ascending doses of TNK-tPA (5, 7.5, 10, 15, 20, 30, 40, or 50 mg), administered as an IV bolus. All patients received oral aspirin and IV heparin therapy titrated to an activated partial thromboplastin time (APTT) of 55–85 s. The primary goal of TIMI 10 A was to establish the pharmacokinetic profile of TNK-tPA in AMI patients following bolus administration. Overall the clearance of TNK-tPA was 151 ± 56 ml/min which is significantly slower compared with the clearance of 572 ± 132 ml/min for native tPA. The initial plasma half-life was from 11–20 min as compared with a half-life of 3.5 min for native tPA. There was a slight decrease in clearance with increasing dose over the 5–50 mg dose range studied (CANNON et al. 1997; MODI et al. 1998).

A comparison of the bolus dose of TNK-tPA with the accelerated 90 min infusion of native tPA demonstrated a similar plasma concentration profile for the two regimens. Based on pharmacokinetic analysis, it appeared that a 30 mg bolus dose of TNK-tPA could provide a similar plasma exposure to 100 mg of "accelerated infusion" tPA.

There was no decrease seen in fibrinogen and only a slight decrease of plasminogen concentrations as the dose of TNK-tPA was increased from 5 to 50 mg. Similarly, the decreases seen in the α_2-antiplasmin levels to approximately 20% below baseline were much less than the 40%–45% decrease seen with native tPA. Although at even higher doses of TNK-tPA, a further drop in α_2-antiplasmin might be expected, it is clear from these data that a generalized proteolytic effect is not induced. In a recent pilot study it was shown that a single bolus of 30–40 mg of TNK-tPA resulted in a much smaller increase of thrombin-antithrombin complexes 2 h after thrombolysis and a much smaller rebound of PAI-1 in the period 4–24 h post-treatment than after the administration of 1.5 MU of streptokinase or 100 mg of tPA (ANDREOTTI et al. 1999).

TNK-tPA also produced encouraging results with respect to angiographic recanalization rates. Already with the low doses of 7.5–15 mg good 90 min patency rates were seen. The best TIMI-3 flow rates in the TIMI 10 A trial were seen with 30, 40, and 50 mg (57%–64%) which compares favorably with the 54% of tPA treated patients in GUSTO.

In the initial phase I study there were 4 deaths in 113 patients (3.5%). No patient experienced a stroke and there were no cases of intracerebral hemorrhage in the study. Six patients (5.3%) experienced a major hemorrhage and four (3.5%) of those occurred at sites of vascular access following catheterization. This is in keeping with the lower rate of surgical site hemorrhage noted in animal studies and is lower than in studies using recombinant tPA. There were no episodes of anaphylaxis and no evidence of anti-TNK-tPA antibody formation in any of the patients. Overall, TNK-tPA appeared to be well tolerated.

Two phase II trials have been conducted so far. The TIMI 10B trial was designed to compare angiographic recanalisation of a bolus of 30, 40, and 50 mg of TNK-tPA and of the accelerated t-PA regimen. The ASSENT trial was designed as a safety study to evaluate the effect of 30, 40, and 50 mg of TNK-tPA only on mortality and the incidence of intracerebral hemorrhage in a large cohort of 3325 patients with AMI.

The initial study protocol for the TIMI 10B multicenter trial randomized patients with AMI, within 12 h of onset of symptoms, to a bolus of 30 or 50 mg of TNK-tPA, given over 5–10 s, or the accelerated tPA regimen in a 1:1:1 ratio (CANNON et al. 1998). All patients received 150–325 mg of aspirin and β-blockers were recommended. The protocol offered the following guidelines for the administration of heparin: an initial bolus of 5000 U, followed by a continuous IV infusion of 1000 U/h for patients weighing >67 kg or of 800 U/h for those ≤67 kg for 48–72 h. Heparin doses were to be adjusted according to APTTs with a target range of 55–80 s. The primary end point was the rate of TIMI grade 3 flow at 90 min. Secondary end points included TIMI grade 3 flow at 60 and 75 min, TIMI grade 2 or 3 flow and TIMI frame counts at all time points, pharmacokinetics, coagulation parameters, recurrent MI, and serious bleeding. Early on in the trial it was noted that the heparin doses in some patients in the 50 mg TNK group who experienced intracranial hemorrhage were higher than the suggested doses. There were three cases (3.8%) of intracranial hemorrhage in the first 78 patients; however this was balanced by an unusually low mortality rate in this group of also 3.8%. On the recommendation of the Safety Monitoring Board, the 50 mg TNK-tPA arm was replaced by 40 mg, and at the same time the initial bolus of heparin for patients ≤67 kg was reduced to 4000 U and recommended doses (see above) were declared mandatory. TIMI grade 3 flow rates at 90 min for 30, 40, 50 mg TNK-tPA and for accelerated tPA were 54%, 63%, 66%, and 63% respectively. The performance of corrected TIMI frame counts (CTFC) has been shown to be a useful method to add additional prognostic information by segregating patients with TIMI grade 3 flow into lower- and higher-risk groups (GIBSON et al. 1999). In the four TIMI 10B groups CTFCs of <40 (faster flow) were: 53%, 63%, 66%, and 61% (40 mg TNK-tPA vs tPA: $p = 0.07$). When TIMI grade 3 flow rates were analyzed in function of the TNK-tPA dose/kg it was noted that TIMI grade 3 flow was 62%–63% for doses ≥0.5 mg/kg, but 51%–54% at doses lower than this ($p = 0.028$ across quintiles). However, Table 1 illustrates that bleeding complications were also correlated with dose/kg and intensity of heparin administration. The incidence of intracerebral hemorrhage and of other serious bleeds was particularly high in patients receiving >0.55 mg/kg of TNK-tPA and the non-reduced heparin doses. Reduction of the heparin dose still resulted in an unacceptable high incidence of cerebral hemorrhage in the >0.55 mg/kg TNK-tPA group, but was nil in patients receiving less than 0.55 mg/kg. Mortality at 30 days was 4.9% overall and reinfarction was 5.4% overall, both without significant differences between TNK-tPA doses and tPA (CANNON et al. 1998).

   The results of the ASSENT-1 phase II safety trial have recently been published (VAN DE WERF et al. 1999). Some data have been discussed in review articles or abstracts (GIUGLIANO et al. 1997; WHITE and VAN DE WERF 1998; FOX et al. 1999). In the original protocol two treatment arms were planned: 30 mg and 50 mg of TNK-tPA only. The primary end points for the ASSENT-1 study were safety, including intracerebral hemorrhage and serious/life threatening bleeding complications, death, total stroke, recurrent MI, cardiogenic shock, anaphylaxis, pulmonary edema, revascularization procedures. Early on in this study it was noted that the incidence of intracerebral hemorrhage was unacceptably high in the 50 mg arm, and as in the TIMI 10B study, this arm was replaced by a 40 mg arm and heparin doses were adjusted as in the amended TIMI 10B protocol. A total of 3325 patients were enrolled. With the 40 mg

Table 1. Intracerebral hemorrhage and serious bleeding complications during thrombolytic treatment with TNK-tPA in function of dose/weight administered

	TNK-tPA dose/weight (mg/kg)				Accelerated tPA
	0.35–0.42	>0.42–0.48	>0.48–0.55	>0.55	
All patients					
Intracerebral h (%)	0.9	0.9	0	5.6	1.9
Serious bleeding (%)	1.9	0.9	4.5	12.0	8.5
Reduced heparin					
Intracerebral h (%)	0	0	0	4.6	1.2
Serious bleeding (%)	0	0	6.1	6.1	2.3

Date compiled from CANNON et al. 1998.

dose an intracranial hemorrhage rate of 0.62% was observed. Mortality at 30 days was 5.6% (Van de Werf et al. 1999). This incidence was considered acceptable because 15% of the patients in the ASSENT-1 trial were over the age of 75 years. Serious bleeding complications requiring transfusions were 1.3%–1.4% in both the TIMI 10B and ASSENT-1 trial in the 40 mg TNK-tPA arm, compared to 7% in the accelerated t-PA arm.

Based on the patency and safety data from the Phase II trials, the ASSENT-2 double-blind randomized phase III trial was designed to compare tenecteplase to the accelerated infusion of alteplase (Assessment of the Safety and Efficacy of a New Thrombolytic (ASSENT-2) Investigators 1999). A simple five increment, weight-adjusted dosing regimen was devised, based on a target dose/weight of 0.53 mg/kg (Wang-Clow et al. 1998). A total of 16949 patients with ST-segment elevation and symptom onset of ≤6 h entered the trial. The 8461 patients randomized to tenecteplase were to receive a single bolus over 5–10 s. Patients who weighed <60 kg were given 30 mg of tenecteplase; in the weight ranges 60–69.9 kg, 70–79.9 kg, 80–89.9 kg, and >90 kg the following doses were administered: 35 mg, 40 mg, 45 mg, and 50 mg. All patients randomized to tenecteplase also received a placebo alteplase infusion. The 8488 patients assigned to the alteplase group were given the classical accelerated dosage scheme (see Table 2) and a tenecteplase bolus placebo. All patients received 150–325 mg of aspirin orally and IV heparin (bolus of 4000 U and infusion of 800 U/h for those weighing ≤67 kg, 5000 U bolus and infusion of 1000 U/h for those weighing >67 kg), adjusted to maintain an APTT of 50–75 s.

The primary endpoint was all-cause mortality at 30 days. Secondary endpoints included net clinical benefit, defined as absence of death or non-fatal stroke at 30 days, major nonfatal cardiac events in hospital, and stroke. Covariate-adjusted 30-day mortality was almost identical in the two groups – 6.18% for tenecteplase and 6.15% for alteplase. Subgroup analyses revealed no differences between the two treatment groups except in patients treated >4 h after onset of symptoms, where the more fibrin-specific tenecteplase resulted in a lower 30-day mortality of 7.0% (alteplase 9.2%; $p < 0.02$). Total stroke rates were slightly higher in the tenecteplase group (1.78% vs 1.66%; n.s.) but the incidence of intracerebral hemorrhage was identical in the two groups (0.93% and 0.94%). Bleeding complications were fewer with tenecteplase (total bleeding 26.4% vs 28.95%; $p = 0.0003$). Major bleeding occurred in 4.7% in the tenecteplase group and in 5.9% in the alteplase group ($p = 0.0002$) (Assessment of the Safety and Efficacy of a New Thrombolytic (ASSENT-2) Investigators 1999).

In a substudy of ASSENT-2 the speed of recanalization by means of enzymatic assay, based on the time-dependent interconversion of isoforms of creatinine kinase mediated by carboxypeptidase N was investigated. Early recanalization within 40 min of onset of treatment occurred in 56% of patients treated with tPA and in 76 of those treated with TNK-tPA (Binbrek et al. 2000).

Table 2. Characteristics of tPA and newer mutants in clinical use

	Alteplase (tPA or rtPA)	Reteplase (rPA)	Tenecteplase (TNK-tPA)	Lanoteplase (nPA, SUN 9216)	Monteplase (E 6010)	Pamiteplase (YM 866)
Genetic modification	None	Deletion of F and E domain	$Thr^{103} \rightarrow Asn$, $Asn^{117} \rightarrow Gln$, Lys^{296} His-Arg-Arg $\rightarrow$ Ala$_4$ substitution	Deletion of F and E domain, Asn^{117} Gln substitution	Cys^{84}->Ser substitution	Deletion of K1; $Arg^{275} \rightarrow$ Glu substitution
Fibrin specificity	++	+	+++	+	++	++
Plasma half-clearance rate (min)	~5	~16	~20	~37	>20	~35
PAI-1 resistance	0	0	+	+	0	0
Dose	15 mg bolus, then 0.75 mg/kg (max. 50 mg) during 30 min, then 0.5 mg/kg (max. 35 mg) during 60 min.	10 + 10 MU double bolus 30 min apart	0.53 mg/kg single bolus	120 kU/kg single bolus	0.22 mg/kg single bolus	0.1 to 0.2 mg/kg single bolus

F, finger; E, epidermal growth factor; K, kringle.

Tenecteplase thus represents an attractive alternative to alteplase with respect to ease of administration, which facilitates prehospital thrombolysis, and a lower risk of non-cerebral hemorrhage.

IX. Unglycosylated Protease Domains of tPA

The glycosylated forms of the protease domain of tPA lack fibrin binding and have a reduced plasmin-generating activity (Dodd et al. 1986; Rijken 1986). Kohnert et al. (1996) have shown that the unglycosylated protease domain of tPA produced by *Escherichia coli* has plasmin-generating activity in a buffer milieu and is an effective fibrinolytic agent in a plasma containing a dynamic (closed circuit under 10 mbar pressure) in vitro clot lysis system, probably by converting single-chain tPA and uPA to their more active two-chain forms. In a rabbit model of jugular vein thrombosis the unglycosylated tPA protease domain induced significant thrombolysis (Martin et al. 1996). Effective doses of the protease domain were associated with a systemic lytic state. However, despite the significant fibrinogen degradation (67%, with tPA 21%) occurring during the 90 min infusion of 2 mg/kg of protease, thrombolytic efficacy and blood losses from rebleeding from fresh wounds were similar to those produced by a 90 min infusion of 1.4 mg/kg of tPA.

C. Chimeric Plasminogen Activators

Recombinant chimeric plasminogen activators have been constructed using different regions of tPA and sc-uPA, although several alternative combinations have been evaluated to some extent (Collen et al. 1989b, 1991). The thrombolytic properties and fibrin-specificity of these chimeras were similar but not superior to those of the parent molecules. In general, the combination of the A-chain of tPA, which confers fibrin-affinity to the molecule, with the enzymatic properties of sc-uPA did not improve the thrombolytic potency of the chimeras. There may be one exception: a chimera consisting of the two kringle domains of tPA (amino acids Ser^1-Gln^3 and Asp^{87}-Phe^{274}) and the serine protease domain of sc-uPA (amino acids Ser^{138}-Leu^{411}; Fig. 3). In vivo evaluation in animal models of thrombosis of this molecule ($K_1K_2P_u$) indicated a markedly enhanced thrombolytic potency towards venous and arterial thrombi (Collen et al. 1991). The 6- to 20-fold prolonged half-life of $K_1K_2P_u$ and its 3- to 16-fold enhanced thrombolytic activity in hamster, rabbit, and baboon models of thrombolysis compared with tPA or sc-uPA suggested that the total amount of material required for thrombolytic therapy may be significantly reduced, and that its administration by bolus injection may be effective (Collen et al. 1991).

Other recombinant chimeric plasminogen activators which have been evaluated in animal models of thrombosis include FK_2tu-PA and K_2tu-PA (F and K_2 domains or K_2 domain only of tPA, linked to the protease domain of

Fig. 3. Schematic representation of the primary structure of $K_1K_2P_u$ (amino acids Ser^1-Gln^3 and Asp^{87}-Phe^{274} of tPA and Ser^{138}-Leu^{411} of sc-uPA). The amino acids are represented by their *single letter symbols*, *black bars* indicate disulfide bonds and the *asterisks* indicate the active site residues in the protease part. The *arrow* indicates the plasmin cleavage site

sc-uPA). In a rabbit jugular vein thrombosis model, the thrombolytic activity of K_2tu-PA was reported to be significantly higher than that of both tPA and sc-uPA, whereas systemic effects were not different (AGNELLI et al. 1992). In contrast, hybrids containing the growth factor domain of uPA and the K_2 and protease domain of tPA were found to have a prolonged half-life but were virtually inactive in a rabbit jugular vein thrombosis model (ASSELBERGS et al. 1993).

An acylated recombinant chimera consisting of the fibrin-binding domains of plasminogen covalently linked to the protease domain of tPA, was more potent and more fibrin-selective than tPA upon bolus administration in a guinea pig pulmonary embolism model (ROBINSON et al. 1992).

A hybrid consisting of the single kringle of urokinase inserted immediately before the double kringle of tPA has been constructed (LEE et al. 1988). Insertion of the uPA kringle domain prolonged the half-life of the hybrid compared with wild-type tPA (FU et al. 1988). This molecule induced coronary thrombolysis in dogs at markedly lower doses than those required with tPA and prolonged the time to reocclusion (WEINHEIMER et al. 1991).

None of these many chimeric plasminogen activators has ever been evaluated in clinical trials.

D. Fibrolase

A fibrinolytic enzyme present in *Agkistrodon contortrix contortrix* (southern copperhead snake) venom has been purified (RETZIOS and MARKLAND 1988). The enzyme, fibrolase, has a molecular weight of 22 891 D and is composed of a single polypeptide chain with 203 amino acids. It is a metalloproteinase (1 mole zinc per mole of enzyme) which exhibits direct fibrinolytic activity and does not activate plasminogen or protein C, nor does it require any blood-borne components for activity. The enzyme degrades the α- and β-chains of fibrin and fibrinogen (GUAN et al. 1991; TRIKHA et al. 1995). Expression of recombinant fibrolase has been obtained from yeast (LOAYZA et al. 1995). The combination of fibrolase and monoclonal antibodies against GPIIb/IIIa produced rapid and sustained thrombolysis in the canine carotid arterial thrombosis model (MARKLAND et al. 1994).

E. Other Fibrinolytic Agents

Several hundreds of other mutants, chimeras, and conjugates of thrombolytically active substances have been developed. Most of these have not even reached the stage of animal experimentation and none are in clinical development. Those other thrombolytic agents which have shown a potential for or are in clinical development are discussed in Chaps. 16 (staphylokinase), 17 (vampire bat plasminogen activator), and 18 (fibrin-targeting of plasminogen activators).

List of Abbreviations and Acronyms

AMI	acute myocardial infarction
APTT	activated partial thromboplastin time
ASSENT	Assessment of the Safety and Efficacy of a New Thrombolytic agent
AUC	area under the curve
CTFC	corrected TIMI frame count
GRECO	German RECOmbinant plasminogen activator study
GRECO DB	German RECOmbinant plasminogen activator Double Bolus study
GUSTO	Global Utilization of Streptokinase and Tissue plasminogen activator for Occluded coronary arteries
INJECT	International Joint Efficacy Comparison of Thrombolytics
InTIME	Intravenous nPA for Treatment of Infarcting Myocardium Early
IV	intravenous(ly)
MU	Mega (1 million) units
nPA	novel plasminogen activator (also designated as SUN9216, BMS-200980 or lanoteplase)

PAI-1   Plasminogen Activator Inhibitor type-1
RAPID   Reteplase vs Alteplase Patency Investigation During acute
               myocardial infarction
sc-uPA   single chain uPA, also called pro-urokinase
SPEED   Strategies for Patency Enhancement in the Emergency
               Department
tc-uPA   two chain uPA, also called urokinase
tPA   tissue-type plasminogen activator
uPA   urinary-type (or urokinase-type) plasminogen activator, also
               called urokinase

References

Adams DS, Griffin LA, Nachajko WR, Reddy VB, Wei C (1991) A synthetic DNA
    encoding a modified human urokinase resistant to inhibition by serum plasmino-
    gen activator inhibitor. J Biol Chem 266:8476–8482
Agnelli G, Pascucci C, Colucci M, Nenci GG, Mele A, Bürgi R, Heim J (1992) Throm-
    bolytic activity of two chimeric recombinant plasminogen activators (FK$_2$tu-PA
    and K$_2$tu-PA) in rabbits. Thromb Haemost 68:331–335
Almér L-O, Öhlin H (1987) Elevated levels of the rapid inhibitor of plasminogen acti-
    vator (t-PAI) in acute myocardial infarction. Thromb Res 47:335–339
Andreotti F, De Marco E, Ianni A, van de Greef W, Biasucci LM, Rebuzzi AG, Maseri
    A (1999) TNK-tPA is devoid of procoagulant effects in sharp contrast with strep-
    tokinase and rt-PA. J Am Coll Cardiol 33:325 (Abstract)
Asselbergs FAM, Bürgi R, Hamerman J, Heim J, Van Oostrum J, Agnelli G (1993) A
    hybrid plasminogen activator binds to the u-PA receptor and has a reduced throm-
    bolytic potency in vivo. Thromb Haemost 69:50–55
Assessment of the Safety and Efficacy of a New Thrombolytic (ASSENT-2) Investiga-
    tors (1999) Single-bolus tenecteplase compared with front-loaded alteplase in
    acute myocardial infarction: the ASSENT-2 double-blind randomised trial. Lancet
    354:716–722
Bassel-Duby R, Jiang NY, Bittick T, Madison E, McGookey D, Orth K, Shohet R,
    Sambrook J, Gething M J (1992) Tyrosine 67 in the epidermal growth factor-
    like domain of tissue- type plasminogen activator is important for clearance by a
    specific hepatic receptor. J Biol Chem 267:9668–9677
Benedict CR, Phil D, Refino CJ, Keyt BA, Pakala R, Paoni NF, Thomas R, Bennett WF
    (1995) New variant of human tissue plasminogen activator (TPA) with enhanced
    efficacy and lower incidence of bleeding compared with recombinant human TPA.
    Circulation 92:3032–3040
Bennett WF, Paoni NF, Keyt BA, Botstein D, Jones AJS, Presta L, Wurm FM, Zoller
    MJ (1991) High resolution analysis of functional determinants on human tissue-
    type plasminogen activator. J Biol Chem 266:5191–5201
Binbrek A, Rao N, Absher PM, Van de Werf F, Sobel BE (2000) The relative rapidity of
    recanalization induced by recombinant tissue-type plasminogen activator (r-tPA)
    and TNK-tPA, assessed with enzymatic methods. Coron Artery Dis 11:429–435
Bode C, Cantor WJ, Ohman EM, Betriu A, Forycki F, Morocutti G, Ardissino D,
    Aylward PE, Califf RM, Topol EJ (1999) Importance of antithrombin therapy in
    combination with abciximab and low-dose fibrinolytic therapy: preliminary results
    of the GUSTO-4 pilot study. Eur Heart J 20 [Suppl]:286 (Abstract)
Bode C, Kohler B, Smalling RW, Runge MS (1995) Reteplase (r-PA): a novel recom-
    binant plasminogen activator. Fibrinolysis 9 [Suppl 1]:97–99
Bode C, Smalling RW, Berg G, Burnett C, Lorch G, Kalbfleisch JM, Chernoff R, Christie
    LG, Feldman RL, Seals AA, Weaver WD, for the RAPID II Investigators (1996)

Randomized comparison of coronary thrombolysis achieved with double-bolus reteplase (recombinant plasminogen activator) and front-loaded, accelerated alteplase (recombinant tissue plasminogen activator) in patients with acute myocardial infarction. Circulation 94:891–898

Browne MJ, Carey JE, Chapman CG, Dodd I, Esmail AF, Lawrence GMP, McMurdo L, Williamson I, Wilson S, Robinson JH (1993) Protein engineering and comparative pharmacokinetic analysis of a family of novel recombinant hybrid and mutant plasminogen activators. Fibrinolysis 7:357–364

Browne MJ, Carey JE, Chapman CG, Tyrrell AWR, Entwisle C, Lawrence GMP, Reavy B, Dodd I, Esmail A, Robinson JH (1988) A tissue-type plasminogen activator mutant with prolonged clearance in vivo. Effect of removal of the growth factor domain. J Biol Chem 263:1599–1602

Burck PJ, Berg DH, Warrick MW, Berg DT, Walls JD, Jaskunas SR, Crisel RM, Weigel B, Vlahos CJ, McClure DB (1990) Characterization of a modified human tissue plasminogen activator comprising a kringle-2 and a protease domain. J Biol Chem 265:5170–5177

Califf RM (1999) Glycoprotein IIb/IIIa blockade and thrombolytics: early lessons from the SPEED and GUSTO IV trials. Am Heart J 138:S12–S15

Camani C, Kruithof EKO (1994) Clearance receptors for tissue-type plasminogen. Int J Hematol 60:97–109

Cambier P, Van de Werf F, Larsen GR, Collen D (1988) Pharmacokinetics and thrombolytic properties of a nonglycosylated mutant of human tissue-type plasminogen activator, lacking the finger and growth factor domains, in dogs with copper coil-induced coronary artery thrombosis. J Cardiovasc Pharmacol 11:468–472

Cannon CP, Gibson CM, McCabe CH, Adgey AAJ, Schweiger MJ, Sequeira RF, Grollier G, Giugliano RP, Frey M, Mueller HS, Steingart RM, Weaver WD, Van de Werf F, Braunwald E, for the Thrombolysis in Myocardial Infarction (TIMI) 10B Investigators (1998) TNK-tissue plasminogen activator compared with front-loaded alteplase in acute myocardial infarction. Results of the TIMI 10B trial. Circulation 98:2805–2814

Cannon CP, McCabe CH, Gibson CM, Ghali M, Sequeira RF, McKendall GR, Breed J, Modi NB, Fox NL, Tracy RP, Love TW, Braunwald E, and the TIMI 10 A Investigators (1997) TNK-tissue plasminogen activator in acute myocardial infarction. Results of the thrombolysis in myocardial infarction (TIMI) 10 A dose-ranging trial. Circulation 95:351–356

Collen D, Lu HR, Lijnen HR, Nelles L, Stassen JM (1991) Thrombolytic and pharmacokinetic properties of chimeric tissue-type and urokinase-type plasminogen activators. Circulation 84:1216–1234

Collen D, Mao J, Stassen JM, Broeze R, Lijnen HR, Abercrombie D, Puma P, Almeda S, Vovis G (1989a) Thrombolytic properties of Lys-158 mutants of recombinant single chain urokinase-type plasminogen activator in rabbits with jugular vein thrombosis. J Vasc Med Biol 1:46–49

Collen D, Stassen JM, Demarsin E, Kieckens L, Lijnen HR, Nelles L (1989b) Pharmacokinetics and thrombolytic properties of chimaeric plasminogen activators consisting of the NH_2-terminal region of human tissue-type plasminogen activator and the COOH-terminal region of human single chain urokinase-type plasminogen activator. J Vasc Med Biol 1:234–240

Collen D, Stassen J-M, Yasuda T, Refino C, Paoni N, Keyt B, Roskams T, Guerrero JL, Lijnen HR, Gold HK, Bennett WF (1994) Comparative thrombolytic properties of tissue-type plasminogen activator and of plasminogen activator inhibitor-1-resistant glycosylation variant, in a combined arterial and venous thrombosis model in the dog. Thromb Haemost 72:98–104

den Heijer P, Vermeer F, Ambrosioni E, Sadowski Z, López-Sendón JL, von Essen R, Beaufils P, Thadani U, Adgey J, Pierard L, Brinker J, Davies RF, Smalling RW, Wallentin L, Caspi A, Pangerl A, Trickett L, Hauck C, Henry D, Chew P, on behalf of the InTIME Investigators (1998) Evaluation of a weight-adjusted single-bolus plasminogen activator in patients with myocardial infarction. A double-blind, randomized angiographic trial of lanoteplase versus alteplase. Circulation 98:2117–2125

Dodd I, Fears R, Robinson JH (1986) Isolation and preliminary characterization of active B-chain of recombinant tissue-type plasminogen activator. Thromb Haemost 55:94–97.

Eidt JF, McNatt J, Wydro RM, Yao SK, Allson P, Garramone S, Peters L, Livingstone D, Buja LM, Willerson JT (1991) Coronary thrombolysis with a variant of human tissue-type plasminogen activator in a canine preparation: slightly enhanced thrombolysis and prolonged time to reocclusion. Coron Art Dis 2:931–940

Felez J, Miles SA, Plescia J, Plow EF (1990) Regulation of plasminogen receptor expression on human monocytes and monocytoid cell lines. J Cell Biol 111:1673–1683

Fox NL, Cannon CP, Berioli S, Wang-Clow F, Danays T, Sarelin H, Braunwald E, Van de Werf F (1999) Rates of serious bleeding events requiring transfusion in AMI patients treated with TNK-tPA. J Am Coll Cardiol 33 [Suppl]:353 (Abstract)

Fu KP, Lee S, Hetzel N, Fenichel R, Hum HW, Speth J, Kalyan N, Hung PP (1988) Clearance of a novel recombinant tissue plasminogen activator in rabbits. Thromb Res 50:679–685

Gibson CM, Murphy SA, Rizzo MJ, Ryan KA, Marble SJ, McCabe CH, Cannon CP, Van de Werf F, Braunwald E, for the Thrombolysis In Myocardial Infarction (TIMI) Study Group (1999) Relationship between TIMI frame count and clinical outcomes after thrombolytic administration. Circulation 99:1945–1950

Giugliano RP, Cannon CP, McCabe CH, Van de Werf F, Braunwald E (1997) Lower dose heparin with thrombolysis is associated with lower rates of intracranial hemorrhage: results from TIMI 10B and ASSENT. Circulation 96 [Suppl]:I-535 (Abstract)

Guan AL, Retzios AD, Henderson GN, Markland FS Jr. (1991) Purification and characterization of a fibrinolytic enzyme from venom of the Southern copperhead snake (Agkistrodon contortrix contortrix). Arch Biochem Biophys 289:197–207

Gurbel PA, Serebruany VL, Shustov AR, Bahr RD, Carpo C, Ohman EM, Topol EJ, for the GUSTO-III Investigators (1998) Effects of reteplase and alteplase on platelet aggregation and major receptor expression during the first 24 hours of acute myocardial infarction treatment. J Am Coll Cardiol 31:1466–1473

Guzzetta AW, Basa LJ, Hancock WS, Keyt BA, Bennett WF (1993) Identification of carbohydrate structures in glycoprotein peptide maps by the use of LC/MS with selected ion extraction with special reference to tissue plasminogen activator and a glycosylation variant produced by site directed mutagenesis. Anal Chem 65:2953–2962

Hajjar KA (1991) The endothelial cell tissue plasminogen activator receptor. J Biol Chem 266:21962–21970

Hashimoto K, Oikawa K, Miyamoto I, Hayamizu K, Abe Y (1996) Phase I study of novel modified tissue-type plasminogen activator, YM866. J Med Pharm Sci 36:623–646

Hoffmeister HM, Kastner C, Szabo S, Beyer ME, Helber U, Kazmaier S, Baumbach A, Wendel HP, Heller W (2000) Fibrin specificity and procoagulant effect related to the kallikrein-contact phase system and to plasmin generation with double-bolus reteplase and front-loaded alteplase thrombolysis in acute myocardial infarction. Am J Cardiol 86:263–268

International Joint Efficacy Comparison of Thrombolytics (1995) Randomised, double-blind comparison of reteplase double-bolus administration with streptokinase in acute myocardial infarction (INJECT): trial to investigate equivalence. Lancet 346:329–336; erratum in Lancet 1995; 329:980

Jackson CV, Crowe G, Craft TJ, Sundboom JL, Grinnell BW, Bobbitt JL, Burck PJ, Quay JF, Smith GF (1990) Thrombolytic activity of a novel plasminogen activator, LY210825, compared with recombinant tissue-type plasminogen activator in a canine model of coronary artery thrombosis. Circulation 82:930–940

Jackson CV, Frank JD, Craft TJ, Sundboom JL, Smith GF (1992) Comparison of the thrombolytic activity of the novel plasminogen activator, LY210825, to anisoylated plasminogen-streptokinase activator complex in a canine model of coronary artery thrombolysis. J Pharmacol Exp Ther 260:64–70

Johannessen M, Diness V, Pingel K, Petersen LC, Rao D, Lioubin P, O'Hara P, Mulvihill E (1990) Fibrin affinity and clearance of t-PA deletion and substitution analogues. Thromb Haemost 63:54–59

Katoh M, Suzuki Y, Miyamoto I, Watanabe T, Mori K, Arakawa H, Gushima H (1991) Biochemical and pharmacokinetic properties of YM866, a novel fibrinolytic agent. Thromb Haemost 65:1193 (Abstract)

Kawai C, Yui Y, Hosoda S, Nobuyoshi M, Suzuki S, Sato H, Takatsu F, Motomiya T, Kanmatsuse K, Kodama K, Yabe Y, Minamino T, Kimata S, Nakashima M, on behalf of the E6010 Study Group (1997) A prospective, randomized, double-blind multicenter trial of a single bolus injection of the novel modified t-PA E6010 in the treatment of acute myocardial infarction: comparison with native t-PA. J Am Coll Cardiol 29:1447–1453

Kawasaki T, Katoh M, Kaku S, Gushima H, Takenaka T, Yui Y, Kawai C (1993) Thrombolytic activity of a novel modified tissue-type plasminogen activator, YM866, in a canine model of coronary artery thrombosis. Jpn J Pharmacol 63:9–16

Kawasaki T, Kaku S, Sakai Y, Takenaka T (1994) Thrombolytic activity of YM866, a novel modified tissue-type plasminogen activator, in a rabbit model of jugular vein thrombosis. Drug Devel Res 33:33–38

Kawauchi Y, Morinaga T, Yokota M, Kinoshita A, Kawamura K, Suzuki Y, Takayama M, Furuichi K, Gushima H (1991) Gene construction and large scale production of a novel thrombolytic agent (YM866) in CHO cells. Thromb Haemost 65:1193 (Abstract)

Ke S-H, Tachias K, Lamba D, Bode W, Madison EL (1997) Identification of a hydrophobic exosite on tissue type plasminogen activator that modulates specificity for plasminogen. J Biol Chem 272:1811–1816

Keyt BA, Paoni NF, Refino CJ, Berleau L, Nguyen H, Chow A, Lai J, Peña L, Pater C, Ogez J, Etcheverry T, Botstein D, Bennett WF (1994) A faster-acting and more potent form of tissue plasminogen activator. Proc Natl Acad Sci USA 91: 3670–3674

Kohnert U, Hellerbrand K, Martin U, Stern A, Popp F, Fischer S (1996) The recombinant E. coli-derived protease-domain of tissue-type plasminogen activator is a potent and fibrin-specific fibrinolytic agent. Fibrinolysis 10:93–102

Kohnert U, Horsch B, Fischer S (1993) A variant of tissue plasminogen activator (t-PA) comprised of the kringle 2 and the protease domain shows a significant difference in the in vitro rate of plasmin formation as compared to the recombinant human t-PA from transformed Chinese hamster ovary cells. Fibrinolysis 7:365–372

Kohnert U, Rudolph R, Verheijen JH, Jacoline E, Weening-Verhoeff EJ, Stern A, Opitz U, Martin U, Lill H, Prinz H, Lechner M, Kresse G-B, Buckel P, Fischer S (1992) Biochemical properties of the kringle 2 and protease domains are maintained in the refolded t-PA deletion variant BM 06.022. Protein Eng 5:93–100

Kostis JB, Liao W-C, Beierle FA, Ens GE, Tracy RP, Hauck CJ, Abbud ZA, Smith RA (1999) Single bolus regimen of lanoteplase (nPA) in acute myocardial infarction: hemostatic evaluation vs. tPA in the InTIME Study. J Am Coll Cardiol 33 [Suppl]:325A (Abstract)

Larsen GR, Metzger M, Henson K, Blue Y, Horgan P (1989) Pharmacokinetic and distribution analysis of variant forms of tissue-type plasminogen activator with prolonged clearance in rat. Blood 73:1842–1850

Larsen GR, Timony GA, Horgan PG, Barone KM, Henson KS, Angus LB, Stoudemire JB (1991) Protein engineering of novel plasminogen activators with increased thrombolytic potency in rabbits relative to activase. J Biol Chem 266:8156–8161

Lau D, Kuzma G, Wei C, Livingston D J, Hsiung N (1987) A modified human tissue plasminogen activator with extended half-life in vivo. Bio/Technology 5:953–958

Lee SG, Kalyan N, Wilhelm J, Hum W, Rappaport R, Cheng S, Dheer S, Urbano C, Hartzell RW, Ronchetti Blume M, Levner M, Hung PP (1988) Construction and expression of hybrid plasminogen activators prepared from tissue-type plasminogen activator and urokinase-type plasminogen activator genes. J Biol Chem 263:2917–2924

Liao W-C, Beierle FA, Stouffer BC, Dockens RC, Abbud ZA, Tay LK, Knaus DM, Raymond RH, Chew PH, Kostis JB (1997) Single bolus regimen of lantoteplase (nPA) in acute myocardial infarction: pharmacokinetic evaluation from inTIME-I Study. Circulation 96 [Suppl]:I-260–I-261 (Abstract)

Lijnen HR, Nelles L, Holmes WE, Collen D (1988) Biochemical and thrombolytic properties of a low molecular weight form (comprising Leu144 through Leu411) of recombinant single-chain urokinase-type plasminogen activator. J Biol Chem 263:5594–5598

Loayza SL, Trikha M, Markland FS, Riquelme P, Kuo J (1994) Resolution of isoforms of natural and recombinant fibrolase, the fibrinolytic enzyme from Agkistrodon contortrix contortrix snake venom, and comparison of their EDTA sensitivities. J Chromatogr B Biomed Appl 662:227–243

Lucore CL, Fujii S, Sobel BE (1989) Dependence of fibrinolytic activity on the concentration of free rather than total tissue-type plasminogen activator in plasma after pharmacologic administration. Circulation 79:1204–1213

Madison EL, Goldsmith EJ, Gerard RD, Gething MJH, Sambrook JF (1989) Serpin-resistant mutants of human tissue-type plasminogen activator. Nature 339:721–724

Madison EL, Kobe A, Gething M-J, Sambrook JF, Goldsmith EJ (1993) Converting tissue plasminogen activator to a zymogen: a regulatory triad of Asp-His-Ser. Science 262:419–421

Madison EL, Sambrook JF (1993) Probing structure-function relationships of tissue-type plasminogen activator by oligonucleotide-mediated site-specific mutagenesis. Methods Enzymol 223:249–271

Markland FS, Friedrichs GS, Pewitt SR, Lucchesi BR (1994) Thrombolytic effects of recombinant fibrolase or APSAC in a canine model of carotid artery thrombosis. Circulation 90:2448–2456

Martin U, Fischer S, Kohnert U, Opitz U, Rudolph R, Sponer G, Stern A, Strein K (1991a) Thrombolysis with an Escherichia coli-produced recombinant plasminogen activator (BM 06.022) in the rabbit model of jugular vein thrombosis. Thromb Haemost 65:560–564

Martin U, Kaufmann B, Neugebauer G (1999) Current clinical use of reteplase for thrombolysis. A pharmacokinetic-pharmacodynamic perspective. Clin Pharmacokinet 36:265–276

Martin U, Köhler J, Sponer G, Strein K (1992) Pharmacokinetics of the novel recombinant plasminogen activator BM 06.022 in rats, dogs, and non-human primates. Fibrinolysis 6:39–43

Martin U, Kohnert U, Hellerbrand K, Stern A, Popp F, Doerge L, Stegmeier K, Müller-Beckmann B, Fischer S (1996) Effective thrombolysis by a recombinant Escherichia coli-produced protease domain of tissue-type plasminogen activator in the rabbit model of jugular vein thrombosis. Fibrinolysis 10:87–92

Martin U, Sponer G, Strein K (1993) Influence of hepatic and renal failure on pharmacokinetic properties of the novel recombinant plasminogen activator BM 06.022 in rats. Drug Metab Disp 21:236–241

Martin U, van Möllendorf E, Akpan W, Kientsch-Engel R, Kaufmann B, Neugebauer G (1991b) Dose-ranging study of the novel recombinant plasminogen activator BM 06.022 in healthy volunteers. Clin Pharmacol Ther 50:429–436

McCluskey ER, Keyt BA, Refino CJ, Modi NB, Zioncheck TF, Bussiere JL, Love TW (1997) Biochemistry, pharmacology, and initial clinical experience with TNK-tPA. In: Sasahara AA, Loscalzo J (eds) New therapeutic agents in thrombosis and thrombolysis. Marcel Dekker, New York, pp 475–493

Modi NB, Eppler S, Breed J, Cannon CP, Braunwald E, Love TW (1998) Pharmacokinetics of a slower clearing tissue plasminogen activator variant, TNK-tPA, in patients with acute myocardial infarction. Thromb Haemost 79:134–139

Neuhaus K-L, von Essen R, Vogt A, Tebbe U, Rustige J, Wagner H-J, Appel K-F, Stienen U, König R, Meyer-Sabellek W (1994) Dose finding with a novel recombinant plasminogen activator (BM 06.022) in patients with acute myocardial infarction: results of the german recombinant plasminogen activator study. J Am Coll Cardiol 24:55–60

Nicolini FA, Nichols WW, Mehta JL, Saldeen TGP, Schofield R, Ross M, Player DW, Pohl GB, Mattsson C (1992) Sustained reflow in dogs with coronary thrombosis with K2P, a novel mutant of tissue-plasminogen activator. J Am Coll Cardiol 20:228–235

Ogata N, Ogawa H, Ogata Y, Numata Y, Morigami Y, Suefuji H, Soejima H, Sakamoto T, Yasue H (1998) Comparison of thrombolytic therapies with mutant tPA (lanoteplase/SUN9216) and recombinant tPA (alteplase) for acute myocardial infarction. Jpn Circ J 62:801–806

Ohman EM, Lincoff AM, Bode C, Bachinsky WB, Ardissino D, Betriu A, Schildcrout JS, Oliverio R, Barnathan E, Sherer J, Sketch MS, Topol EJ (1998) Enhanced early reperfusion at 60 minutes with low-dose reteplase combined with full-dose abciximab in acute myocardial infarction: preliminary results from the GUSTO-4 pilot (SPEED) dose-ranging trial. Circulation 98 [Suppl I]:504 (Abstract)

Otter M, Zocková P, Kuiper J, Van Berkel TJC, Barrett-Bergshoeff MM, Rijken DC (1992) Isolation and characterization of the mannose receptor from human liver potentially involved in the plasma clearance of tissue-type plasminogen activator. Hepatology 16:54–59

Paoni NF, Refino CJ, Brady K, Peña LC, Nguyen HV, Kerr EM, van Reis R, Botstein D, Bennett WF (1992) Protein Eng 5:259–266

Paoni NF, Chow AM, Pena LC, Keyt BA, Zoller MJ, Bennett WF (1993) Making tissue-type plasminogen activator more fibrin specific. Protein Eng 6:529–534

Rapold HJ, Grimaudo V, Declerck PJ, Kruithof EKO, Bachmann F (1991) Plasma levels of plasminogen activator inhibitor type 1, β-thromboglobulin, and fibrinopeptide A before, during, and after treatment of acute myocardial infarction with alteplase. Blood 78:1490–1495

Refino CJ, Paoni NF, Keyt BA, Pater CS, Badillo JM, Wurm FM, Ogez J, Bennett WF (1993) A variant of t-PA (T103N, KHRR 296–299 AAAA) that, by bolus, has increased potency and decreased systemic activation of plasminogen. Thromb Haemost 70:313–319

Retzios AD, Markland FS Jr (1988) A direct-acting fibrinolytic enzyme from the venom of agkistrodon contortrix contortrix: effects on various components of the human blood coagulation and fibrinolysis systems. Thromb Res 52:541–552

Robinson JH, Browne MJ, Carey JE, Chamberlain PD, Chapman CG, Cronk DW, Dodd I, Entwisle C, Esmail AF, Kalindjian SB, Lawrence GMP, McMurdo L, Mitchell D, Smith RAG, Wilson S (1992) A recombinant, chimeric enzyme with a novel mechanism of action leading to greater potency and selectivity than tissue-type plasminogen activator. Circulation 86:548–552

Rijken DC, Groeneveld E (1986) Isolation and functional characterization of the heavy and light chains of human tissue-type plasminogen activator. J Biol Chem 261: 3098–3102

Rijken DC, Groeneveld E, Barrett-Bergshoeff MM (1994) In vitro stability of a tissue-type plasminogen activator mutant, BM 06.022, in human plasma. Thromb Haemost 72:906–911

Saito M, Suzuki S, Yui Y, Kawai C (1994) A novel modified tissue-type plasminogen activator (t-PA), E6010, gradually increases coronary blood flow after thrombolysis compared with native t-PA, urokinase and balloon catheter occlusion-reperfusion. Jpn J Pharmacol 66:17–23

Sane DC, Stump DC, Topol EJ, Sigmon KN, Kereiakes DJ, George BS, Mantell SJ, Macy E, Collen D, Califf RM (1991) Correlation between baseline plasminogen activator inhibitor levels and clinical outcome during therapy with tissue plasminogen activator for acute myocardial infarction. Thromb Haemost 65:275–279

Saito M, Suzuki S, Yui Y, Kawai C (1995) A novel modified tissue-type plasminogen activator (t-PA), E6010, reduces reperfusion arrhythmias induced after coronary thrombolysis–comparison of native t-PA and urokinase. Jpn Circ J 59:556–564

Shi GY, Hsu CC, Chang BI, Tsai CF, Han HS, Lai MD, Lin MT, Chang WC, Wing LYC, Jen CJ, Tang MJ, Wu HL (1996) Regulation of plasminogen activator inhibitor activity in endothelial cells by tissue-type plasminogen activators. Fibrinolysis 10:183–191

Smalling RW, Bode C, Kalbfleisch J, Sen S, Limbourg P, Forycki F, Habib G, Feldman R, Hohnloser S, Seals A (1995) More rapid, complete, and stable coronary thrombolysis with bolus administration of reteplase compared with alteplase infusion in acute myocardial infarction. Circulation 91:2725–2732

Stern A, Kohnert U, Rudolph R, et al (1989) Gewebs-Plasminogenaktivator-Derivat. Eur Patent Appl 38217

Stewart RJ, Fredenburgh JC, Leslie BA, Keyt BA, Rischke JA, Weitz JI (2000) Identification of the mechanism responsible for the increased fibrin specificity of TNK-tissue plasminogen activator relative to tissue plasminogen activator. J Biol Chem 275:10112–10120

Stump DC, Lijnen HR, Collen D (1986) Purification and characterization of a novel low molecular weight form of single-chain urokinase-type plasminogen activator. J Biol Chem 261:17120–17126

Strandberg L, Madison EL (1995) Variants of tissue-type plasminogen activator with substantially enhanced response and selectivity toward fibrin co-factors. J Biol Chem 270:23444–23449

Stürzebecher J, Neumann U, Kohnert U, Kresse GB, Fischer S (1991) Mapping of the catalytic site of CHO-t-PA and the t-PA variant BM 06.022 by synthetic inhibitors and substrates. Prot Sci 1:1007–1013

Sumiyoshi T, Kawai C, Hosoda S, Aoki N, Takano T, Kanmatsuse K, Motomiya T, Yui Y, Minamino T, Kodama K, Haze K, Satoh H, Nobuyoshi M (1996) Clinical evaluation of intravenous bolus injection of YM866 (modified tissue-plasminogen activator) in patients with acute myocardial infarction : multicenter early phase 2 study. Medicine and Pharmacy 36:1–30

Suzuki S, Saito M, Suzuki N, Kato H, Nagaoka N, Yoshitake S, Mizuo H, Yuzuriha T, Yui Y, Kawai C (1991) Thrombolytic properties of a novel modified human tissue-type plasminogen activator (E6010): a bolus injection of E6010 has equivalent potency of lysing young and aged canine coronary thrombi. J Cardiovasc Pharmacol 17:738–746

Suzuki S, Saito M, Yui Y, Kawai C (1995) A novel modified t-PA, E-6010, induces faster recovery of ventricular function after coronary thrombolysis than native t-PA in a canine thrombosis model. Jpn Circ J 59:205–212

Suzuki N, Suzuki S, Nagaoka N, Mizuo H, Yuzuriha T, Yoshitake S, Kanmatsuse K (1994) Thrombolysis of canine femoral artery thrombus by a novel modified tissue-type plasminogen activator (E6010). Jpn J Pharmacol 65:257–263

Tachias K, Madison EL (1997) Variants of tissue-type plasminogen activator that display extraordinary resistance to inhibition by the serpin plasminogen activator inhibitor type 1. J Biol Chem 272:14580–14585

Tebbe U, von Essen R, Smolarz A, Limbourg P, Rox J, Rustige J, Vogt A, Wagner J, Meyer-Sabellek W, Neuhaus K-L (1993) Open, non-controlled dose finding study with a novel recombinant plasminogen activator (BM 06.022) given as double bolus in patients with acute myocardial infarction. Am J Cardiol 72:518–524

The Global Use of Strategies to Open Occluded Coronary Arteries (GUSTO III) Investigators (1997) A comparison of reteplase with alteplase for acute myocardial infarction. N Engl J Med 337:1118–1123

Thögersen AM, Jansson J-H, Boman K, Nilsson TK, Weinehall L, Huhtasaari F, Hallmans G (1998) High plasminogen activator inhibitor and tissue plasminogen activator levels in plasma precede a first acute myocardial infarction in both men and women. Evidence for the fibrinolytic system as an independent primary risk factor. Circulation 98:2241–2247

Topol EJ, Ohman EM, Barnathan E, Califf RM, Lee KL, Lincoff AM, Scherer J, Scott R, Sketch MH, Weisman HF, Cantor W, Evans B, Harris J, Kirby J, Miller J, Minshall D, Oliverio RM, Stebbins AL, Webb L, Wittlief J, Moliterno DJ, Ivanc T, Damaraju L, Hadley K, Kukulewicz L, et al. (2000) Trial of abciximab with and without low-dose reteplase for acute myocardial infarction. Strategies for Patency Enhancement in the Emergency Department (SPEED) Group. Circulation 101:2788–2794

Trikha M, Schmitmeier S, Markland FS (1994) Purification and characterization of fibrolase isoforms from venom of individual southern copperhead (Agkistrodon contortrix contortrix) snakes. Toxicon 32:1521–1531

Ueshima S, Holvoet P, Lijnen HR, Nelles L, Seghers V, Collen D (1994) Expression and characterization of clustered charge-to-alanine mutants of low M_r single-chain urokinase-type plasminogen activator. Thromb Haemost 71:134–140

Unemura K, Ishiye M, Kasuge K, Nakashima M (1995) Effect of combination of a tissue-type plasminogen activator and an endothelin receptor antagonist FR 139317, in the rat cerebral infarction model. Eur J Pharmacol 275:17–21

Unemura K, Toshima Y, Nakashima M (1994) Thrombolytic efficacy of a modified tissue-type plasminogen activator, SUN9216, in the rat middle cerebral artery thrombosis model. Eur J Pharmacol 262:27–31

Unemura K, Wada K, Uematsu T, Nakashima M (1993) Evaluation of the combination of a tissue-type plasminogen activator, SUN9216 and a thromboxane A_2 receptor antagonist, vapiprost, in a rat middle cerebral artery thrombosis model. Stroke 24:1077–1082

Van de Werf F, Cannon CP, Luyten A, Houbracken K, McCabe CH, Berioli S, Bluhmki E, Sarelin H, Wang-Clow F, Fox NL, Braunwald E, for the ASSENT-1 Investigators (1999) Safety assessment of single-bolus administration of TNK tissue-plasminogen activator in acute myocardial infarction: the ASSENT-1 trial. Am Heart J 137:786–791

Wang-Clow F, Fox NL, Berioli S, Danays T, Bluhmki E, Modi N, Tanswell P, Stump D, Gibson CM, Van de Werf F (1998) A simple, incremental weight-adjusted dosing scheme for TNK-tPA, a bioengineered variant of the natural t-PA molecule. Circulation 98 [Suppl I]:I-280 (Abstract)

Weinheimer CJ, James HL, Kalyan NK, Wilhelm J, Lee SG, Hung PP, Sobel BE, Bergmann SR (1991) Induction of sustained patency after clot-selective coronary thrombolysis with Hybrid-B, a genetically engineered plasminogen activator with a prolonged biological half-life. Circulation 83:1429–1436

White HD, Van de Werf FJJ (1998) Thrombolysis for acute myocardial infarction. Circulation 97:1632–1646

Yahara H, Matsumoto K, Maruyama H, Nagaoka T, Ikenaka Y, Yajima K, Fukao H, Ueshima S, Matsuo O (1992) Recombinant variants of tissue-type plasminogen activator containing amino acid substitutions in the finger domain. Thromb Haemost 68:672–677

Zhu Y, Carmeliet P, Fay WP (1999) Plasminogen activator inhibitor-1 is a major determinant of arterial thrombolysis resistance. Circulation 99:3050–3055

Agents which Increase Synthesis and Release of Tissue-Type Plasminogen Activator

T. KOOISTRA, and J.J. EMEIS

A. Introduction

I. The Plasminogen/Plasmin System

The plasminogen/plasmin system represents a highly regulated enzyme cascade that generates localized proteolysis. The system plays a key role in an extraordinary number of normal biological and pathophysiological contexts such as fibrinolysis, cell migration, and tissue remodeling. Plasminogen, present in high concentrations in blood and other body fluids, constitutes an abundant source of latent proteolytic activity. Conversion of the inactive proenzyme plasminogen into the active serine proteinase plasmin is catalyzed by plasminogen activators. The two main physiological plasminogen activators are tissue-type and urokinase-type plasminogen activator (tPA and uPA, respectively). Consistent with the important role of the plasminogen/plasmin system in so many processes, the action of tPA and uPA is well controlled by specific interactions between the various components, the synthesis and/or release of tPA and uPA, and the presence of specific protease inhibitors.

In this chapter we will mainly concentrate on the regulation of the synthesis and release of tPA into the circulation, with particular reference to the possibilities to modulate tPA secretion pharmacologically. Where appropriate and possible, we will include relevant data on the regulation of uPA plasma levels. For reviews on the biochemical aspects of tPA and uPA or the role of the plasminogen/plasmin system in various (patho)physiological processes, the reader is referred to other reviews (this book; DANØ et al. 1985; SAKSELA and RIFKIN 1988; VASSALLI et al. 1991; COLLEN and LIJNEN 1994; BACHMANN 1994; DECLERCK et al. 1994; RIJKEN 1995).

II. Intravascular Fibrinolysis

The dissolution of the fibrin component of thrombi and hemostatic plugs is mediated by the intravascular plasminogen/plasmin system (COLLEN and LIJNEN 1991). Whereas inadequate dissolution of fibrin may result in the obstruction of a blood vessel, excessive premature fibrin degradation can lead to bleeding (LIJNEN and COLLEN 1989). Clearly, fibrin degradation (fibrinolysis) needs to be finely regulated in time, location, and extent. Studies with transgenic mice over- or under-expressing components of the fibrinolytic system have confirmed the importance of this system in fibrin clot surveillance and indicate that tPA and uPA both play a role in maintaining vascular patency. The role of the fibrinolytic system in thrombosis became particularly apparent when exposing mice with tPA or uPA deficiencies to inflammatory or traumatic challenges (CARMELIET and COLLEN 1995).

As regards the relative importance of tPA and uPA in intravascular fibrinolysis and thrombolysis, evidence from in vitro experiments suggests that tPA is the primary initiator of fibrin degradation (WUN and CAPUANO 1985, 1987), while uPA may be more involved in later stages of fibrin dissolution (GUREVICH 1988). tPA differs from other proteins involved in fibrinolysis and coagulation by being secreted into the circulation as an active enzyme, thus enabling it to trigger directly the fibrinolytic process. In the absence of fibrin, tPA has a low activity towards plasminogen. In the presence of fibrin, this activity is two orders of magnitude higher (COLLEN and LIJNEN 1991). Fibrin essentially increases the local plasminogen concentration by creating an additional interaction between tPA and its substrate. Fibrin thus fulfils a dual function, both as a stimulator of plasminogen activation and as a final substrate of generated plasmin. The high affinity of tPA for plasminogen in the presence of fibrin allows efficient activation on the fibrin clot, while plasminogen activation by tPA in plasma is rather inefficient. The fibrinolytic process thus seems to be triggered by and confined to fibrin.

An important point to note is that tPA is much more effective in inducing the lysis of fibrin (or of a thrombus) if present prior to fibrin formation than after fibrin formation has occurred (BROMMER 1984; ZAMARRON et al. 1984; Fox et al. 1985). This difference in potency can be several hundred-fold and explains at least in part the large therapeutic doses of tPA required in thrombolytic therapy to dissolve coronary artery thrombi associated with acute myocardial infarction. It also emphasizes the preventive potential of a proper endogenous fibrinolytic capacity to oppose the development of intravascular thrombosis.

III. Tissue-Type Plasminogen Activator (tPA) in the Circulation

tPA is cleared very rapidly from the circulation by the liver (KUIPER et al. 1988), its plasma half-life ranging from 1–2 min in rodents (DEVRIES et al. 1987) to about 5 min in humans (BOUNAMEAUX et al. 1985; CHANDLER et al. 1997). One of the consequences of this rapid clearance is that tPA must be secreted continuously into the blood in order to maintain a stable level of tPA. Another consequence is that a change in the rate of tPA secretion will be immediately translated into a change in the plasma concentration of tPA.

However, the plasma level of tPA activity is not merely the resultant of the rate of tPA secretion and the rate of tPA clearance. tPA activity is also controlled by specific plasminogen activator inhibitors (PAIs) of which PAI-1 is the physiologically relevant inhibitor in plasma (Sprengers and Kluft 1987; Loskutoff et al. 1988). The rate of tPA:PAI-1 complex formation will be a function of the plasma concentrations of the two proteins: the more tPA and/or PAI-1, the more complex will be formed. Because the clearance rate of the tPA:PAI-1 complex is slower than the clearance rate of free tPA (Brommer et al. 1988; Chandler et al. 1997), it follows that the higher the plasma PAI-1 level, the more tPA:PAI-1 complex will be in the circulation, resulting – because of the slower clearance of the complex – in a higher plasma level of tPA antigen (but not of tPA activity, which will decrease with increasing plasma PAI-1 levels) (Chandler et al. 1990). This may explain the reported association of increased plasma levels of tPA and of PAI-1 antigen with increased risk for the development of cardiovascular disease (Jansson et al. 1991, 1993; Ridker et al. 1993, 1994). This constellation might well represent an association of decreased tPA activity with increased risk (Meade et al. 1994). Attempts to increase tPA secretion in a pharmacological way to restore an impaired plasma fibrinolytic activity should therefore require a rather selective effect on tPA secretion without a concomitant effect on PAI-1 production.

IV. Involvement of the Endothelium in Plasma Fibrinolysis

It is generally accepted that tPA in the circulation originates mainly from the vascular endothelium (see reviews and references therein by Kooistra 1990; Kooistra et al. 1994; Emeis 1995a,b; Emeis et al. 1996). The expression of endothelial tPA is a function defined by vessel size and anatomical location as well as by the developmental stage of an organism. tPA expression disappeared from a number of larger arteries after animals reached maturity (Levin and Del Zoppo 1994; Levin EG et al. 1997, 2000). Chromaffin cells (Parmer et al. 1997) and mast cells (Bankl et al. 1999; Sillaber et al. 1999) might be other sources of some circulating tPA.

Vascular endothelial cells are ideally positioned to contribute to plasma levels of tPA. Cultured endothelial cells synthesize tPA in vitro and endothelial cells in vivo contain large amounts of tPA, as shown by histochemical (fibrin autography), immunohistochemical, and biochemical techniques. Compounds that stimulate the synthesis of tPA in cultured endothelial cells can also increase the plasma level of tPA in vivo in experimental animals. Interestingly, shear stress increases tPA synthesis and release in endothelial cell cultures (Kawai et al. 1996). Studies, using in situ hybridization methodology, also localized messenger RNA (mRNA) for tPA in vascular endothelial cells (Levin and Del Zoppo 1994). Since tPA mRNA, antigen and activity are all associated with the endothelium in vivo, and since tPA expression has not been demonstrated in any other major cell type, it is concluded that vascular endothelial cells are the major source of plasma tPA.

In contrast to tPA, the cellular source of plasma uPA is unknown (see review by Emeis et al. 1996). Immunocytochemical analysis for uPA in normal human tissue revealed that this protein was present in luminal endothelial cells and in medial smooth muscle cells (LUPU et al. 1995). However, signals for uPA mRNA were weak over smooth muscle cells and undetectable over endothelial cells. The absence of a signal over endothelial cells may be due to insufficient sensitivity of the in situ hybridization technique but it is also possible that intimal uPA is derived from the blood circulation and becomes bound to the endothelial uPA receptor (BARNATHAN et al. 1990).

Since endothelial cells are the main source of plasma tPA, any change in tPA secretion from the endothelium will result in a change in plasma tPA levels. And because a low plasma tPA activity is generally considered a risk factor for the development of cardiovascular disease (HAMSTEN and ERIKSSON 1994), it would be an attractive option to develop (pharmacological) tools to enhance tPA secretion (the other options are to decrease tPA clearance or PAI-1 synthesis, or to inhibit PAI-1 activity: see Sect. D).

The mechanism by which tPA is released from the endothelial cells is an important issue. Endothelial cells can release tPA in two ways. One route is known as constitutive secretion because the tPA synthesized is exported continuously. The other way involves the concentration and storage of tPA in specialized storage vesicles (KOOISTRA et al. 1994; EMEIS et al. 1997; PARMER et al. 1997; ROSNOBLET et al. 1999; SANTELL et al. 1999). The contents of these storage vesicles are delivered to the cell surface in response to specific extracellular stimuli and, therefore, export via this route is said to be regulated. Since both secretion pathways may represent important control points for regulation of plasma tPA levels, we will discuss the regulatory mechanisms underlying constitutive secretion and regulated secretion (also called acute release), their relative importance, and agents known to influence these pathways.

B. Regulation of Constitutive tPA Synthesis

I. General

Endothelial cells synthesize and store tPA in culture (KOOISTRA 1990; VAN HINSBERGH et al. 1991; KOOISTRA et al. 1994). A good approach to develop ways (drugs, procedures) that increase tPA secretion is therefore to study tPA synthesis and release in cultured human endothelial cells. Such studies have allowed the identification of a number of factors and signaling pathways to increase tPA secretion in human endothelial cells. In experimental animals, these developments have resulted in treatment schedules that increase tPA synthesis, tissue concentrations of tPA, acute release of tPA, as well as plasma levels of tPA. The objective of increasing plasma tPA levels in humans by means of increasing endothelial tPA synthesis obviously is a next step and may well become a real possibility within the next few years.

Many of the changes in tPA synthesis depend on modulation of gene transcription. The human (NY et al. 1984; FISHER et al. 1985; FRIEZNER DEGEN et al. 1986; BULENS et al. 1995; ARTS et al. 1997), rat (FENG et al. 1990), and mouse

(RICKLES et al. 1989) tPA genes have been isolated and characterized. The specific DNA sequences and transcription factors involved in the regulation of tPA gene expression are being elucidated and explain some of the species-specific differences in tPA gene regulation (HOLMBERG et al. 1995; LEONARDSSON and NY 1997; COSTA et al. 1998). Some of the pharmacological factors and intracellular signaling pathways that regulate tPA synthesis and release will be reviewed below.

II. Intracellular Signaling and Activation of Protein Kinase C

tPA production can be stimulated through multiple intracellular signaling pathways, and these pathways interact to potentiate or synergize the expression of tPA. Vasoactive compounds such as thrombin and histamine increase tPA synthesis in cultured human endothelial cells (LEVIN EG et al. 1984; VAN HINSBERGH et al. 1987; HANSS and COLLEN 1987; RYDHOLM et al. 1998), most likely via a pathway that involves activation of protein kinase C: the increase in tPA synthesis can be prevented by inhibition of protein kinase C activity (LEVIN and SANTELL 1988; SANTELL and LEVIN 1988), and direct activators of protein kinase C, such as diacylglycerol and the phorbol ester, 4-β-phorbol 12-myristate 13-acetate (PMA), stimulate tPA production in a manner similar to thrombin and histamine (LEVIN and SANTELL 1988; LEVIN EG et al. 1989; GRÜLICH-HENN and MÜLLER-BERGHAUS 1990; KOOISTRA et al. 1991a; ARTS et al. 1997). Experiments with transgenic mice carrying a 1.4-kb human tPA promoter fragment (including the first exon and part of the first intron) fused to the bacterial β-galactosidase gene show an increased expression of mouse tPA and bacterial β-galactosidase in various tissues after PMA treatment (THEURING et al. 1996). These experiments show for the first time that activation of protein kinase C can also activate the human tPA promoter in vivo and underline the value of cultured endothelial cells as a model system to study regulation of tPA synthesis.

Cyclic AMP (cAMP), which in itself does not affect tPA synthesis in human cells, has a potentiating effect on protein kinase C-dependent tPA stimulation (SANTELL and LEVIN 1988; KOOISTRA et al. 1991a). Low concentrations of alcohol (0.25–25 mmol/l) likewise enhance cAMP-dependent tPA gene transcription in human endothelial cells (MIYAMOTO et al. 1999). This may be, at least in part, responsible for the beneficial effect of moderate alcohol consumption in cardiovascular disease. Interestingly, cAMP alone is sufficient to increase tPA synthesis in rat endothelial cells (EMEIS et al. 1998). This species-specific difference in tPA regulation is due to a mutation in the tPA promoter: while the rat tPA promoter contains a perfect consensus cAMP response element, the same position in the human and mouse tPA promoter is converted to a PMA-response element by a one nucleotide substitution (FENG et al. 1990; HOLMBERG et al. 1995; COSTA et al. 1998). These differences in tPA regulation have now also been confirmed in vivo. Administration of the adenylate cyclase activators forskolin or cholera toxin to rats very effectively increased plasma tPA activity and antigen levels and tissue tPA mRNA and protein levels (VAN DEN EIJNDEN-SCHRAUWEN et al. 1995). In contrast, mice showed increased plasma and tissue concentrations of tPA after PMA administration (THEURING et al. 1996), but not after administration of cAMP-elevating compounds (T. Kooistra et al., unpublished data). These animal experiments also show that it is feasible, by increasing endothelial tPA synthesis, to increase plasma tPA levels pharmacologically.

III. Retinoids

Vitamin A and derivatives thereof (collectively called retinoids) effectively stimulate tPA synthesis in cultured human endothelial cells, without markedly influencing PAI-1 expression (KOOISTRA et al. 1991b, 1995; THOMPSON et al. 1991; BULENS et al. 1992; MEDH et al. 1992; KOOISTRA 1995). Studies in rats demonstrated that tPA production and retinoid status are also correlated in vivo. tPA activity levels in the plasma and tissue of rats undergoing retinoic acid treatment or hypervitaminosis A showed an increase of activity in both by 50% which lasted for at least 8 weeks. The plasma activities of uPA and PAI-1 were not changed in these rats. On the other hand, plasma tPA activity in vitamin A-deficient rats was found to be about 3-fold lower than control values (KOOISTRA et al. 1991b; VAN BENNEKUM et al. 1993). uPA activity was decreased to about 50% of the control value in the vitamin A-deficient rats, but PAI-1 activity was increased about twofold. This increase in plasma PAI-1 activity does not necessarily reflect an increased synthesis rate, but could be the result of the lower plasma levels of tPA and uPA. These in vivo findings underline the physiological relevance of retinoids in regulating tPA and, to a lesser extent, uPA, but also illustrate that under normal physiological conditions retinoid levels in plasma and tissues are sufficient to maintain almost maximal retinoid-dependent tPA expression in vivo. This would also explain why treatment of humans with isotretinoin (13-*cis* retinoic acid) maximally elevated plasma tPA antigen levels by only 50% (WALLNÖFER et al. 1993; DECLERCK et al. 1993; DOOTSON et al. 1995; BÄCK and NILSSON 1995).

The potential future usefulness of retinoids as a profibrinolytic drug will depend therefore on the availability and/or development of an appropriate synthetic derivative that is a more potent tPA stimulator than retinoic acid and has less side effects (BIESALSKI 1989). Two types of observations could be build upon to develop such retinoids with a superior benefit/risk ratio as compared with retinoic acid. Firstly, synthetic retinoids have been identified which are more potent than retinoic acid in stimulating tPA expression in cultured human endothelial cells and which can serve as lead compounds for further evaluation (KOOISTRA et al. 1991b; BULENS et al. 1992). For example, Ro 13–7410 (TTNPB), in which the terminal ring of retinoic acid is substituted with a methyl group and the conformational mobility of the side chain is restricted by the introduction of aromatic rings within the skeleton (see Fig. 1), stimulated tPA synthesis in cultured human endothelial cells about three

All-trans-retinoic acid

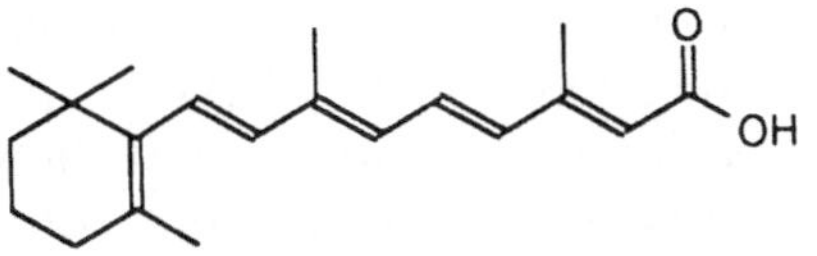

Ro 13-7410
(TTNPB)

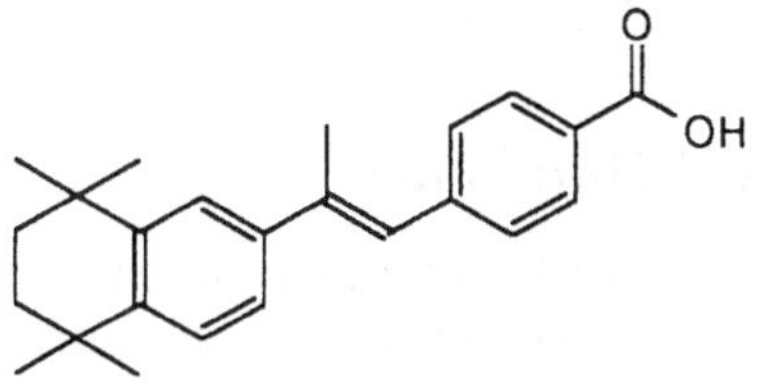

Fig. 1. Chemical structures of all-*trans*-retinoic acid and the synthetic retinoid TTNPB (Ro 13–7410)

times stronger than retinoic acid during a 48-h incubation period (KOOISTRA et al. 1991b; BULENS et al. 1992). Second, the discovery of nuclear retinoid receptors has provided a conceptual framework for the mechanism by which retinoids can regulate tPA gene expression. Two families of these receptors have been identified, the retinoic acid receptors (RARs) and the retinoid X/9-*cis* retinoic acid receptors (RXRs). Both RARs and RXRs are ligand-inducible transcription factors which modify the expression of specific genes by binding to specific DNA sequences, designated retinoic acid response elements (RAREs) (for reviews, see STUNNENBERG 1993; CHAMBON 1994). BULENS et al. (1995) showed the presence of a functional RARE (at position –7.3 kb) in the tPA promoter. Both RARs and RXRs consist of three subtypes, designated α, β, and γ, and due to alternative splicing and the use of different promoters, there are at least two different isoforms of each receptor subtype. Since RARs and RXRs function as either homo- or heterodimers in binding to the RAREs, a great variety of homo- and heterodimeric combinations is possible. Several lines of evidence suggest that each RAR/RXR subtype combination may specifically control the expression of a subset of retinoid target genes (NAGPAL et al. 1992), although a certain degree of redundancy may exist in the retinoid signalling pathway (CHAMBON 1993). Using subtype-specific retinoids and an antagonist with a high preference for RARα, KOOISTRA et al. (1995) identified RARα to be involved in the stimulation of tPA expression by retinoids in human endothelial cells. However, the relatively slow action of retinoic acid on tPA gene expression suggested a mechanism, in which the induction of tPA is a secondary response to activation of RARα by retinoic acid. Further research revealed that activation of RARα leads to the induction of RARβ2, which subsequently enhances tPA gene expression (LANSINK and KOOISTRA 1996). Since under normal physiological conditions retinoic acid concentrations in tissues are sufficient to maintain a steady RARβ expression

(KATO et al. 1992), future attempts to increase plasma levels of tPA in vivo should be directed at finding RARβ2-specific ligands with a higher transactivating capacity than retinoic acid. It should also be considered that retinoids are susceptible to rapid metabolism. All-*trans*-retinoic acid is rapidly metabolized in endothelial cells and slowly in hepatocytes, whereas 9-*cis*-retinoic acid degradation takes place rapidly in hepatocytes and slowly in endothelial cells (LANSINK et al. 1997).

IV. Steroid Hormones

Another class of hormones which, like retinoids, may act via nuclear receptors, are steroid hormones. A multihormone response enhancer has recently been identified in the tPA gene 7.1 kb upstream of the transcription initiation site (BULENS et al. 1997). Several reports have documented an apparent increase in tPA activity in women taking oral contraceptives (JESPERSEN and KLUFT 1985, 1986; GEVERS LEUVEN et al. 1987; SIEGBAHN and RUUSUVAARA 1988). This increase in tPA activity appeared not to be caused by an increase in tPA antigen, which, to the contrary, was significantly reduced. The explanation is assumed to be a reduction in PAI-1 antigen concentration. A similar finding has been observed with the anabolic steroid stanozolol (VERHEIJEN et al. 1984).

VAN KESTEREN et al. (1998) found that in male-to-female transsexual subjects, exogenous ethinylestradiol in combination with cyproterone acetate (an anti-androgen) significantly reduced plasma levels of tPA, uPA, and PAI-1 at 4 months of treatment. In contrast, there was no significant change in the plasma concentration of tPA, uPA, and PAI-1 upon treatment of a group of female-to-male transsexual subjects with testosterone for 4 months. The absence of a significant correlation between tPA antigen and testosterone was also found in two other studies. In one study, the effects of exogenous testosterone on the hemostatic system have been studied in a group of healthy men undergoing a clinical trial of hormonal male contraception (ANDERSON et al. 1995). In a second study, a possible correlation between fibrinolysis parameters and testosterone was investigated in normogonadic men and men with severe hypogonadotrophinic hypogonadism (testosterone <3 nmol/l) (CARON et al. 1989). Recent findings demonstrate that estrogen treatment of rodents results in increased plasma clearance rate of tPA via induction of the mannose receptor, which could explain the inverse relationship between estrogen status and plasma tPA concentrations observed in humans (LANSINK et al. 1999). In line with this, in vitro studies have shown that incubation of cultured human endothelial cells with ethinylestradiol (or progestagens) did not affect the secretion of tPA, and that the secretion of PAI-1 by human endothelial cells and hepatocytes was also unaffected by these sex steroids (KOOISTRA et al. 1990). A recent study of GILTAY et al. (2000) also supports increased hepatic clearance of tPA rather than increased tPA synthesis after the oral administration of ethinyl estradiol to male-to-female transsexual subjects. Whereas a decrease in basal plasma tPA levels after estrogen treatment was confirmed, no significant difference in increment in venous plasma tPA levels in response to occlusion of the upper arm was found, suggesting that estrogen did not alter tPA synthesis.

V. Sodium Butyrate and Other Inhibitors of Histone Deacetylase

Sodium butyrate and similar short chain fatty acids are strong and rather selective inducers of tPA expression in cultured human endothelial cells (KOOISTRA et al. 1987). Studies on the relationship between structure and tPA stimulatory activity revealed that a straight-chain C_4 monocarboxylate structure with a methyl group at one end and a carboxy moiety at the other seems to be required for the optimal induction of tPA in cultured endothelial cells. This makes it very likely that butyrate itself is active rather than a metabolic product, since other even-chain fatty acids are metabolized through the same pathway, but are much less effective.

Sodium butyrate is a pleiotropic agent with many different effects (KRUH 1982). One of the most evident changes brought about is the acetylation of histones via inhibition of the enzyme histone deacetylase (RIGGS et al. 1977). More recently, evidence has been provided that the tPA stimulatory effect of butyrate involves histone H_4 acetylation, and that this induction can be mimicked by specific, structurally unrelated, histone deacetylase inhibitors, such as trichostatin A (ARTS and KOOISTRA 1995; ARTS et al. 1995) and trapoxin (T Kooistra, unpublished data) (see Fig. 2 for structures). In accordance with the

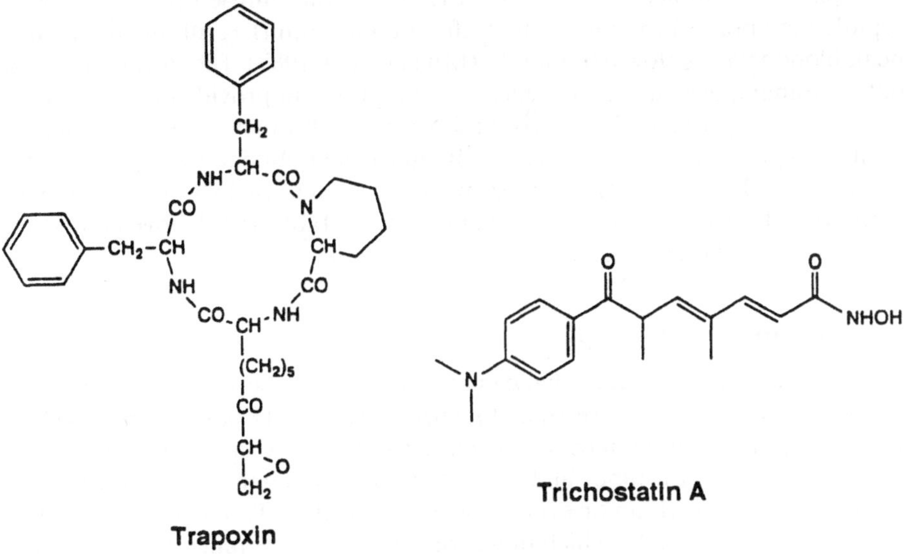

Trapoxin

Trichostatin A

Fig. 2. Chemical structures of the histone deacetylase inhibitors trichostatin A and trapoxin

Table 1. Plasma tPA and uPA in trichostatin A-treated rats

	tPA activity (ng/ml)	tPA antigen (ng/ml)	uPA activity (ng/ml)
Controls			
4h	0.81	3.15	1.00
8h	0.72	3.04	0.71
24h	0.70	2.70	0.73
Trichostatin A			
4h	0.60	3.21	0.89
8h	0.52	3.51	0.81
24h	1.55	3.50	0.84

Rats were treated with a single dose of trichostatin A (5 ml/kg i.v. = 1 mmol/kg) or vehicle (5 ml/kg i.v. of a 0.2% DMSO solution in saline). Blood samples were taken only once from each rat at the times indicated.

much lower Ki value of trichostatin A for histone deacetylase (Yoshida et al. 1990), much lower concentrations of trichostatin A (optimal concentration 1 μmol/l) than of butyrate (optimal concentration 3 mmol/l) were required for maximal induction of tPA expression. Trapoxin, an irreversible inhibitor of mammalian histone deacetylase (Kijima et al. 1993), was found to stimulate maximally tPA synthesis in cultured human endothelial cells already at a concentration of 10 nmol/l (T. Kooistra, unpublished data).

At present it is not known whether such agents are also effective in vivo. A possible effect will depend on prolonged exposure to a minimum concentration of the agent. From in vivo studies of butyrate, used in clinical trials to treat patients with acute leukaemia, it is known that infused butyrate is very rapidly metabolized (metabolic half-life of about 6 min), resulting in maximal peak blood levels below 0.05 mmol/l (Miller et al. 1987). The development of butyrate analogues that are as effective as butyrate but provide more sustained concentrations in vivo (Newmark and Young 1995) may create new opportunities to perform in vivo studies. Preliminary data obtained with trichostatin A in rats indicate that the tPA stimulatory effects of histone deacetylase inhibitors in vitro may also be relevant in vivo (Table 1; T. Kooistra and J.J. Emeis, unpublished data).

VI. Triazolobenzodiazepines

Another class of compounds that can enhance the synthesis of tPA in cultured human endothelial cells, are triazolobenzodiazepines (Kooistra et al. 1993). The most potent compounds, U-34599, U-41695, and U-51477 (see Fig. 3 for structures), showed a time- and concentration-dependent stimulatory effect on tPA synthesis, with no or even a lowering effect on PAI-1 production. The regulatory mechanism by which these triazolobenzodiazepines exert their tPA stimulatory action is not understood at present. No positive correlation was found between the ability of the various triazolobenzodiazepines to stimulate

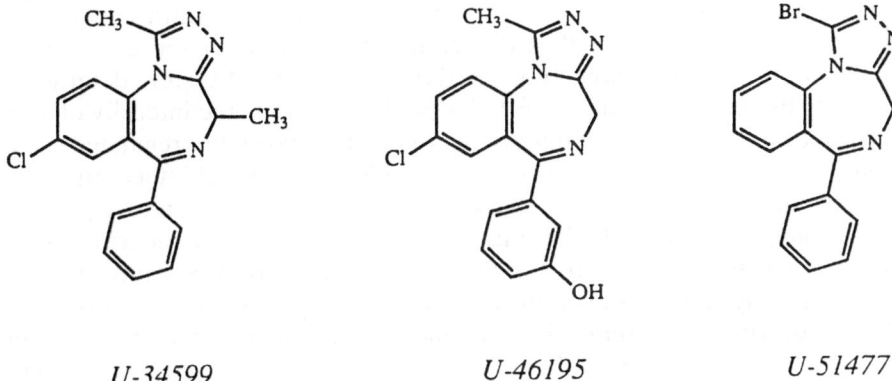

U-34599   *U-46195*   *U-51477*

Fig. 3. Chemical structures of triazolobenzodiazepines from the Upjohn Co

tPA synthesis and their affinity for the central-type benzodiazepine receptor. In fact, binding studies with specific ligands for central-type and peripheral-type benzodiazepine receptors provided no evidence for the presence of such receptors in cultured human endothelial cells (KOOISTRA et al. 1993).

A possible lead to the mode of action of triazolobenzodiazepines was suggested by the relationship between the tPA stimulatory activity of the various triazolobenzodiazepines and their platelet-activating factor (PAF) antagonist activity. However, several non-benzodiazepine PAF-antagonists did not stimulate tPA production, indicating that the three lead compounds shown in Fig. 3 provide a structural rather than a functional lead at present for finding other tPA stimulatory compounds. Triazolo "benzodiazepine fragments" and open ring triazolo and imidazolo benzodiazepines (benzophenones) related to nizofenone (claimed to be effective in subarachnoid haemorrhage (OHTA et al. 1986)) showed no stimulating effect on tPA synthesis in cultured human endothelial cells.

Further research should now be directed at testing the triazolobenzodiazepines in vivo to evaluate their profibrinolytic potential.

C. Regulated Secretion of tPA

I. General

Although it has been speculated that tPA may have intracellular functions, for instance in hormone processing (KRISTENSEN et al. 1985, 1986), its major site of action is extracellular. Of the two secretory pathways – constitutive secretion and regulated secretion – the former has been discussed above. In this pathway, newly synthesized tPA is directly, without storage in an intracellular compartment, transported out of the cell. The rate of constitutive tPA secretion is therefore determined largely by the rate of tPA synthesis. In the regulated pathway, newly synthesized tPA is first stored in an intracellular compartment, and secreted into the extracellular space only after appropriate stimulation of the cell. The rate of regulated tPA secretion is therefore determined not only by the size of the intracellular pool, but also by the intensity of stimulation of the cellular second messenger systems involved in regulated secretion. This results in further possibilities for pharmacological intervention.

In cultured human endothelial cells, the predominant secretory pathway is the constitutive one. Whether this is also true in vivo is still undecided, since the evidence for an in vivo role of the constitutive pathway is scanty and indirect. This may well, however, be due to the lack of proper methodology to detect constitutive secretion in vivo, and an important in vivo role for this pathway in tPA metabolism should by no means be disregarded. The existence of a regulated secretory pathway for tPA in man is, in contrast, well established (see Sect. C.IV.1).

In the following we will discuss the nature of the tPA storage compartment in endothelial cells, the circulatory regulation of tPA, and the cellular mechanisms involved in the regulated secretion of tPA from the endothelium. Finally, we will discuss compounds that affect the regulated secretion of tPA in man.

II. The Endothelial tPA Storage Compartment and the Circulatory Regulation of tPA

The association of functionally active tPA with human vascular endothelium in vivo has been demonstrated in a large number of functional and immuno-histochemical studies (reviewed in KRISTENSEN 1992; EMEIS et al. 1996). The precise intracellular localization of tPA could, however, not be established by these techniques. Using density gradient centrifugation, we have isolated from cultured human endothelial cells and from lung tissue a small, dense (density 1.11 g/ml) particle containing tPA. By confocal laser scanning immuno-microscopy, small tPA-containing particles were found throughout the cytoplasm of cultured human endothelial cells (EMEIS et al. 1997). It is thus likely that, similar to many other proteins that are secreted via the regulated pathway (e.g., hormones, pancreatic enzymes), tPA is stored in a secretory granule. Other storage granules have been described in endothelial cells as well, for example particles containing von Willebrand factor (the so-called Weibel-Palade bodies; for reviews see REINDERS et al. 1988; WAGNER 1990, 1993), protein S (BRETT et al. 1988), endothelin-1 (HARRISON et al. 1995), or tissue factor pathway inhibitor (LUPU C et al. 1995). Whether the tPA-containing particle is different from the Weibel-Palade body is still being discussed (EMEIS et al. 1997; ROSNOBLET et al. 1999). Secretion of the one protein is not necessarily linked with secretion of the other (VAN DEN EIJNDEN-SCHRAUWEN 1996). In neuroendocrine cells tPA is stored in dense core granules and colocalizes with ACTH. Pulse-chase experiments revealed that a portion of newly synthesized tPA is retained in neuroendocrine cells for at least 4 h and is released in response to 8-bromo-cAMP (SANTELL et al. 1999). In chromaffin cells, such as pheochromocytoma cell lines or bovine adrenal chromaffin cells, tPA is stored in dense storage vesicles, of identical density as those for norepinephrine. Treatment of chromaffin cells with secretagogues, such as nicotine, KCl or $BaCl_2$ resulted in co-release of tPA in parallel with catecholamines (PARMER et al. 1997).

The amounts of tPA stored in tissues are large in comparison to the amount circulating in the blood. It has been estimated (PADRÓ et al. 1990) that, in rats, the tissue stores of tPA are sufficient to maintain steady-state plasma levels of tPA for two days in the absence of protein synthesis. Whether comparable amounts of tPA are stored in human tissues is, for lack of reliable data on human tissue, not known, but it is likely that large stores of tPA are present in man as well. As shown in Table 2 for baboon tissues, primate tissues contain large amounts of stored tPA as well. The acute release of all available tPA

Table 2. Tissue-type plasminogen activator activity content of baboon tissues

Tissue	Extraction buffer	tPA activity (units/g)
Spleen	KSCN buffer	10[a]
Skin	KSCN	40
Stomach	KSCN	90
Lung	KSCN buffer	135[b]
Kidney cortex	KSCN buffer	190[a]
Adrenal	KSCN buffer	285
Small intestines	Camiolo's buffer	315
Fast skeletal muscle	Camiolo's buffer	340
Large intestines	Camiolo's buffer	375
Slow skeletal muscle	Camiolo's buffer	415
Cerebellum	Camiolo's buffer	425
Arteria femoralis	KSCN buffer	435
Cerebrum	Camiolo's buffer	440
Vena cava superior	Camiolo's buffer	460
Ureter	KSCN buffer	480
Oesophagus	Camiolo's buffer	480
Aorta	KSCN buffer	530
Heart muscle	Camiolo's buffer	565
Vena cava inferior	Camiolo's buffer	925
Urine bladder	Camiolo's buffer	970

Baboon tissues were extracted (PADRÓ et al. 1990) in Camiolo's buffer (CAMIOLO et al. 1982) and in potassium thiocyanate (KSCN: 0.5 mol/l KSCN in 0.1 mol/l Tris buffer pH = 7.4, 0.1% Tween-80; ERIKSSON et al. 1983).
Data are given, for the more efficient of the two extraction procedures used per tissue, as tPA activity per gram of wet tissue weight. One unit of activity equals the activity of one nanogram of recombinant human tPA.
Unpublished data of R.C. FRANZ and J.J. EMEIS (1994).
[a] These tissues contained in addition appreciable amounts of non-tPA-related (possibly uPA) activity. Kidney medulla could not reliably be estimated, due to excess uPA activity.
[b] Human lung: 700 units/gram.

from the vascular endothelium would result in increases in plasma levels ranging from nanograms per millilitre (in large vessels) to micrograms per millilitre (in capillaries), depending upon the endothelial surface/blood volume ratio of the vessel involved. For the whole body, assuming a blood volume of 5 l, the average increase would be between 1.4 and 2.8 µg/ml of blood (for calculations, see Emeis 1992; Schrauwen et al. 1994b).

One should realize, though, that local differences other than surface/volume ratios may also result in local differences in tPA metabolism. tPA is very heterogenously distributed in the vasculature, while even within a single vessel appreciable differences may exist between the various vessel segments (see, e.g., Keber 1983; Levin and Del Zoppo 1994; Levin et al. 1997; reviewed in Emeis et al. 1996). Regional differences in plasminogen activator secretion have also been noted (Chandler et al. 1992b; Jern et al. 1997a; Seeman-Lodding et al. 1997).

In a kinetic model of the circulatory regulation of tPA (Chandler 1990; Chandler et al. 1993, 1995, 1997) plasma tPA levels are determined by peripheral endothelial secretion of tPA, by complex formation between tPA and inhibitors, and by the clearance of tPA and tPA/inhibitor complexes (see also Sect. A.III above). Under steady-state conditions, the secretion of tPA is, according to the model, on average 0.07 ± 0.02 pmole l^{-1} s^{-1} (Chandler et al. 1995), or about 2 mg day^{-1} for a 70 kg individual. If the endothelial surface of the circulatory system is estimated at 750–1500 m^2 (Wolinski 1980), this would require a synthetic rate of 0.12–0.24 ng cm^{-1} day^{-1}, in good agreement with synthetic rates observed in vitro (Van Hinsbergh 1988). The steady-state level of tPA is, apart from secretion, determined by the interaction of tPA with its inhibitors (which aspect will not be discussed here) and by the clearance of tPA by the liver. A decrease in liver flow results in a decrease in clearance and an increase in plasma tPA level (Bounameaux et al. 1986; De Boer et al. 1992; Van Griensven et al. 1998). In contrast, eating a meal increases liver blood flow and decreases plasma tPA (Van Griensven 1995). Chandler et al. (1993, 1995) have shown that part of the increase in plasma tPA during the infusion of epinephrine is due to an epinephrine-induced decrease in liver blood flow. Similarly, part of the increase in tPA observed during exhaustive exercise must be ascribed to an (epinephrine-mediated) decrease in liver blood flow. The role of adrenergic agents in these processes will be discussed in more detail below (Sect. C.IV.2). The clearance of tPA is also dependent upon the plasma concentration of PAI-1. High levels of PAI-1 lead to rapid neutralisation of active tPA. On the other hand, tPA antigen will be cleared more slowly because the clearance rate of the tPA/PAI-1 complex is slower than that of active tPA (Chandler et al. 1997).

The important conclusion from these studies is that an acute change in plasma tPA level is not necessarily due to an acute change in tPA secretion. It might as well be due to a change in tPA clearance. Pharmacological compounds affecting liver blood flow could therefore be used to affect plasma tPA levels, but so far this principle has not found application in clinical practice.

III. Cellular Mechanisms Involved in Regulated tPA Secretion

Until recently, the mechanisms involved in the regulated secretion (acute release) of tPA were studied mainly in animal models (MARKWARDT 1983; TAPPY et al. 1984; EMEIS 1988, 1992; PROWSE and MACGREGOR 1988; KLÖCKING 1991; EMEIS and TRANQUILLE 1992; KOOISTRA et al. 1994). From these studies it was concluded that tPA can be released acutely (within seconds) from perfused animal vascular beds after stimulation with a variety of compounds or conditions such as acidosis. Compounds that consistently induced tPA secretion in the animal systems studied include thrombin, bradykinin, eledoisin, histamine, acetylcholine, epinephrine, PAF, endothelin, and calcium ionophore A-23187. A common denominator of these compounds is that they all induce calcium influx into endothelial cells by activating their cognate (G-protein coupled) seven-transmembrane-domain receptors on the endothelium, and that they all induce in addition prostaglandin synthesis and increase synthesis of nitric oxide in endothelial cells (EMEIS 1988, 1995a). Regulated tPA secretion requires extracellular calcium (TRANQUILLE and EMEIS 1991). In some studies (reviewed in KLÖCKING 1991; see also ZHU et al. 1989), compounds that increase cAMP also induced tPA release, while in other studies (TRANQUILLE and EMEIS 1993) such compounds were inactive or even inhibitory. Finally, acute release of von Willebrand factor (vWF) was found (when studied) to occur simultaneously with the release of tPA with all compounds (TRANQUILLE and EMEIS 1990), except ADP (SMALLEY et al. 1993).

Various attempts to demonstrate a similar acute secretion of tPA by stimulating human (or animal) endothelial cells in vitro consistently failed in the past. On hindsight, this was presumably due to the fact that the amount of tPA secreted by cultured endothelial cells was too low to be detected by the antigen assay methods then available, and to the fact that the (more sensitive) activity assays could not be applied because of the presence of excess PAI-1 in the conditioned media of cultured endothelial cells. BOOYSE et al. (1986) demonstrated, using fibrin autography, that cultured human endothelial cells would rapidly release tPA (and, in subcultured cells, uPA as well) upon stimulation with thrombin or the calcium ionophore A-23187.

The development of a very sensitive ELISA procedure for tPA (SCHRAUWEN et al. 1994a) finally made possible a detailed study, in cultured human umbilical vein endothelial cells (HUVEC), of the cellular mechanisms involved in regulated secretion (acute release) of tPA. From these studies the following conclusions regarding the regulated secretion of tPA from HUVEC in vitro were drawn (VAN DEN EIJNDEN-SCHRAUWEN et al. 1995; VAN DEN EIJNDEN-SCHRAUWEN 1996). After stimulation of HUVEC with thrombin, tPA was rapidly released; maximal release occurred in the first minute after stimulation, while secretion was complete by 3 min. The amount of tPA secreted was dose-dependently related to the thrombin concentration. The amount of tPA acutely secreted after maximal stimulation (generally induced by 1 NIH Unit of α-thrombin per ml) was linearly related to the synthetic rate of tPA of the cells (which was measured as the rate of constitutive secretion during the 30 min prior to stimulation). An increase in the intracellular calcium concentration $[Ca^{2+}]_i$ was necessary, but not sufficient, to induce regulated tPA secretion, while it did not matter whether the increase in $[Ca^{2+}]_i$ was derived from an intracellular or extracellular calcium pool (Van den Eijnden-Schrauwen et al. 1997). Apart from an increase in $[Ca^{2+}]_i$, activation of at least two (Pertussis toxin-sensitive) G-proteins was also involved in the induction of regulated tPA secretion. In permeabilized cells, regulated tPA secretion could in addition be induced by activating G-proteins with aluminium fluoride or GTPγS, without a measurable increase in $[Ca^{2+}]_i$. Activation of phospholipase A_2 or protein kinase C was not required for regulated secretion. In HUVEC, regulated tPA secretion could be evoked by thrombin, ADP, ATP, bradykinin, histamine, PAF, and the calcium ionophore ionomycin. These in vitro data thus fitted well with, and greatly extended, the observations made in perfused animal vascular systems. Together, the data suggest that activation of G-protein-coupled endothelial receptors, together with calcium influx, is necessary and sufficient to induce regulated tPA secretion (Van den Eijnden-Schrauwen 1996; Van den Eijnden-Schrauwen et al. 1997).

It therefore came as a surprise when it was found in further experiments that compounds that increased endothelial cAMP are also able to induce tPA secretion (Hegeman et al. 1998). Forskolin, dibutyryl-cAMP and prostacyclin induced, dose-dependently, acute release of tPA; vWF was released as well. In contrast to, for instance, thrombin-induced regulated secretion, the secretion induced by cAMP occurred more slowly, taking about 10 min to be completed, while secretion occurred without a measurable increase in $[Ca^{2+}]_i$. It is thus likely that at least two, possibly separately regulated, systems in HUVEC are capable of inducing regulated secretion of tPA (and vWF).

IV. Compounds Affecting Regulated tPA Secretion in Man

1. The Occurrence of Regulated tPA Secretion in Man

Provided that a sudden change in liver blood flow can be excluded (see Sect. C.II above), an abrupt increase in tPA level is most likely due to a sudden increase in regulated secretion (acute release) of tPA (Emeis 1988). Direct evidence for regulated tPA secretion in man in vivo has recently been presented (C. Jern et al. 1994; S. Jern et al. 1994). Using a technique in which differences in an arteriovenous tPA antigen concentration could be determined during intra-brachial drug infusion, S. Jern et al. (1994) demonstrated that muscarinic receptor stimulation gave rise (also after correction for changes in blood flow) to an increase in the tPA antigen concentration in the venous effluent, or, stated differently, to acute release of tPA. Similar techniques have previously been used. Cash (1978) studied, in ipsilateral return venous blood, the effect on fibrinolytic activity of an intrabrachial infusion of epinephrine (which increased activity) and DDAVP (which had no effect). Rosing et al. (1978), using a similar technique of intra-arterial drug infusion and sampling of venous return blood, found that an infusion of epinephrine gave an increase in fibrinolytic activity which was inhibited by propranolol (a β-adrenergic blocker) but not by the α-adrenergic blocker phentolamine. These authors also found a positive effect of nicotinic acid. At that time, tPA antigen or specific tPA activity assays were of course not yet in existence, so that the studies of C. JERN et al. (1994) and of S. JERN et al. (1994) are the first to formally demonstrate regulated tPA secretion in peripheral vessels in man.

Taken together, the reports of CHANDLER et al. (1993, 1995) on the effects of epinephrine infusion, in combination with the report of S. JERN et al. (1994) on arteriovenous differences in tPA antigen during acetylcholine infusion, and that of VAN DEN EIJNDEN-SCHRAUWEN et al. (1995) on regulated tPA secretion from cultured human endothelial cells, strongly support the statement that regulated tPA secretion is an important physiological mechanism in man. These reports also suggest that plasma tPA levels are upregulated by activating the process of regulated release from large tPA storage pools in endothelial cells. On the basis of the observations in cultured human endothelial cell systems and in animal models discussed above, a large number of compounds would in principle be suitable to stimulate regulated secretion. Only a few of these have, however, been used in man. The data available up till 1990 have been regularly reviewed (KWAAN et al. 1957; FEARNLEY 1970; DESNOYERS 1978; NILSSON et al. 1980; MARKWARDT 1983; PROWSE and CASH 1984; PROWSE and MACGREGOR 1988; EMEIS 1988, 1992, 1995a,b; KLÖCKING 1991; KOOISTRA et al. 1994). Therefore we will discuss these data in the following paragraph only in so far as they are relevant to recent observations.

A cautionary remark should be made regarding these older studies. The reader should be aware that the distinction between regulated and constitutive secretion was not always taken into consideration, so that increased constitutive secretion (synthesis) has, especially in cell culture studies, often been described as increased release, suggesting an increase in regulated secretion. Also, in the older literature, the effects of interventions on plasminogen activator(s) cannot always easily be distinguished from effects on the, at the time still unknown, plasminogen activator inhibitor(s).

Clinical studies to estimate the body's capacity to secrete tPA into the circulation have mostly used one of three procedures considered to reflect this capacity: venous occlusion, exercise, or the infusion or injection of DDAVP (Desmopressin: 1-deamino-8-D-arginine vasopressin, a V_2-receptor-binding vasopressin analogue). These three procedures, all supposedly inducing tPA secretion, have been critically discussed recently (EMEIS 1995b), and will not be dealt with here in detail. All three procedures certainly increase the plasma level of tPA antigen in healthy people.

As will be clear from the foregoing, however, the increase in tPA during exercise is largely determined by changes in liver blood flow, and by the release of epinephrine into the circulation, and will thus not necessarily reflect the size of the endothelial storage pool, or the capacity of the endothelium to release tPA. A more straightforward way would be to infuse known amounts of epinephrine (Chandler et al. 1992a).

The increase in tPA antigen in an occluded limb during venous occlusion is now thought to reflect the rate of constitutive tPA secretion into the occluded vessels, rather than secretory capacity (Keber 1988; Keber et al. 1990). Provided great care is taken to correct for hemoconcentration in, and protein loss from, the occluded vessels (see, e.g., Wieczorek et al. 1993), venous occlusion may therefore provide a useful estimate of the constitutive tPA secretory pathway. One should keep in mind, however, that large differences exist between vascular beds, for instance that in the leg and that in the arm (see, e.g., Keber 1983). The use of DDAVP will be discussed below in Sect. C.IV.4.

2. Adrenergic Agents

Intravenous isoproterenol, a β-adrenergic agent, dose-dependently increases tPA plasma levels by inducing the acute release of tPA (and uPA) when given either systemically (Chandler et al. 1995) or locally in the forearm (Stein et al. 1998). Phenylephrine, an α-adrenergic agent, also increases tPA plasma levels dose-dependently, but in this case by decreasing liver blood flow (Chandler et al. 1995). These data explain the many confusing observations (discussed in, e.g., Prowse and Cash 1984; Prowse and MacGregor 1988) on the role of adrenergic stimulation in the activation of the fibrinolytic system which suggested that tPA release was mediated by both α-adrenergic and β-adrenergic receptors. The dual role of adrenergic agents in inducing both increased secretion and reduced clearance finally explains these observations. During submaximal exercise, the effects of a reduction in blood flow prevail. During maximal exercise plasma epinephrine levels are increased, resulting in β-adrenergic stimulation of regulated tPA secretion, as well as in additional reduction in liver blood flow by an α-adrenergic mechanism (Chandler et al. 1993, 1995). All procedures that increase plasma epinephrine levels (stress, anxiety, exercise, etc.) are thus likely to increase tPA antigen levels (e.g., C. Jern et al. 1991, 1994). Norepinephrine has no effects on tPA levels.

The β-adrenergic receptor present on cultured human aortic and micro-vascular endothelial cells is a salbutamol-responsive, β_2-adrenergic receptor (Draijer 1996). This is in good agreement with the observations (see Prowse and MacGregor 1988) that salbutamol induces an increase in regulated secretion of tPA, and that, during local stimulation of the forearm with epinephrine, the release of tPA is inhibited by a β-blocker (Rosing et al. 1978; see also Stein et al. 1998). Since β-adrenergic stimulation of endothelial cells will result in increased levels of cAMP (Hopkins and Gorman 1981; Langerer and Van Hinsbergh 1991; Schafer et al. 1980), it is likely that the regulated secretion of tPA during β-adrenergic stimulation is driven by the cAMP-mediated mechanism described above. In perfused rat hind legs, epinephrine gave acute release of PA, which was inhibited by propranolol but not by phentolamine.

In this system, dibutyryl-cAMP similarly gave acute release of PA activity (ZHU et al. 1989).

3. Vasopressin

During many situations of stress (surgery, electroconvulsive therapy, apomorphine-induced nausea, to mention a few examples), not only epinephrine is increased, but arg-vasopressin (AVP) as well. The question therefore arises whether vasopressin might, like epinephrine, be involved in the increase in fibrinolytic activity during stress. As extensively reviewed by GRANT (1990, 1993), this is indeed the case, although the mechanism(s) involved are still not well defined. During major abdominal surgery, AVP concentrations will increase to well above 50pg/ml (GRANT et al. 1986), sufficient to induce increased plasma levels of tPA activity (GRANT et al. 1985) and antigen (HARIMAN et al. 1990). Modified electroconvulsive shock (HAMPTON et al. 1990), as well as postural hypotension in patients with progressive autonomic failure (GRANT 1993), both induce an increase in plasma AVP, but no epinephrine response. Still, increased tPA antigen levels could be demonstrated in these two situations, showing that during stress AVP can induce increased levels of tPA in the absence of increased epinephrine. In these latter experiments, no increase in factor VIIIC (or vWF) was noted, although this was the case in "unmodified" electroconvulsive shock (PINA-CABRAL and RODRIGUES 1974), which procedure increases the epinephrine level.

No direct effect of AVP on tPA release has so far been reported in isolated vessels or in cultured endothelial cells (GRANT 1990, 1993), suggesting that AVP might act indirectly on the regulated secretion of tPA, as has also been suggested for the AVP analogue DDAVP (see next paragraph). Another explanation might be that AVP increases tPA levels by reducing liver blood flow. This hypothesis would fit with the observation that, in the absence of an increase in epinephrine (for instance in modified electroconvulsive therapy and in progressive autonomic failure), AVP is unable to increase the plasma level of vWF. Plasma levels of vWF are, because of the long half-life of vWF, not sensitive to changes in liver blood flow. In the case where both epinephrine and AVP are increased (unmodified electroconvulsive therapy, surgery, etc.), this would result in an (epinephrine-mediated) acute release of tPA and vWF, an (epinephrine and AVP mediated) reduction in liver blood flow, and an additional increase in tPA. If only AVP is increased (modified electroconvulsive therapy, postural hypotension in progressive autonomic failure), only liver blood flow would be changed, and only tPA would increase.

4. DDAVP

The vasopressin analogue DDAVP was originally developed as an analogue which mimicked vasopressin's antidiuretic effect without having its vasopressin's vasoactive effects. DDAVP was subsequently shown to increase plasma levels of vWF, factor VIIIC, and tPA (for reviews, see MANNUCCI 1988;

Mariani et al. 1993; Lethagen 1994). In the field of hemostasis, DDAVP is mostly used to increase plasma levels of factor VIIIC/vWF, e.g., in hemophiliacs or in von Willebrand disease patients prior to minor surgery, or in blood donors prior to blood donation (Mannucci 1988). The application of DDAVP's potential to increase the plasma level of tPA has been largely restricted to evaluate the regulated secretion of tPA.

The mechanism by which DDAVP influences the plasma level of tPA and vWF is still a matter of controversy. Studies on the effect of DDAVP on the release of tPA and vWF in cultured human endothelial cells and in isolated perfused human umbilical cord blood vessels have proved uniformly negative. The original observation of Cash (1978) that infusion of DDAVP into a brachial artery did not result in measurable changes in tPA and vWF in the outflowing venous blood, while infusion of epinephrine did, led to the hypothesis that DDAVP induces the release, possibly from the pituitary, of an intermediary "plasminogen activator releasing hormone," which in turn would induce the release of tPA and vWF from the endothelium. Little evidence for the existence of a plasminogen activator releasing hormone has since been forthcoming, and the pituitary has been excluded as a source of this putative hormone (Juhan-Vague et al. 1984; see discussion in Emeis 1995b). Another possibility, namely that DDAVP affects liver blood flow, can also be excluded (Burggraaf et al. 1994). Hashemi et al. (1990, 1993) have suggested that the hypothetical intermediary compound is PAF, released from DDAVP-activated blood monocytes. This suggestion was based on the observation that supernatants from DDAVP-stimulated monocytes induced more vWF release (synthesis) in cultured endothelial cells than did supernatants from unstimulated monocytes, the effect of which was abrogated by a PAF-antagonist (Hashemi et al. 1993). PAF is, at least in experimental animals, a potent stimulus for the acute release of tPA and vWF (Emeis and Kluft 1985). However, in dogs, pretreatment with SR 27417, a PAF receptor antagonist, had no effect on the DDAVP-induced increases of FVIIIC, vWF, and tPA plasma levels (Bernat et al. 1997).

DDAVP does not work in patients with nephrogenic diabetes insipidus (DI), in whom the vasopressin type-2 (V_2)-receptor is inactivated by mutation (Davies 1992). The hemostatic effects of DDAVP on tPA and vWF are also absent in patients with congenital X-linked nephrogenic DI (Bichet et al. 1988; Brommer et al. 1990; Knoers et al. 1990; Kobrinski et al. 1985), although a few patients with non-X-linked nephrogenic DI show a normal clotting response to DDAVP (Brenner et al. 1988). It can thus be assumed that DDAVP acts, as regards tPA and vWF release, on a V_2-receptor. This conclusion is strengthened by the observation that, in rhesus monkeys, the DDAVP-induced increase in plasma levels of vWF and factor VIIIC (tPA was not studied in these monkeys) was inhibited by the V_2-receptor antagonist SK&F 105494 (Kinter et al. 1992). Three other V_2-agonists, also active on renal V_2-receptors, did not affect vWF and factor VIIIC, however (Kinter et al. 1992). The authors concluded that "dDAVP stimulation of clotting factor release is mediated by a low-affinity, V_2-like receptor mechanism." Animal studies in this field have in the past been greatly hampered by the still largely unexplored differences between species in their sensitivity to agonists, antagonists, and analogues of vasopressin (BURRELL et al. 1994; EMEIS 1995b). In man, vascular vasodilatory V_2-receptors have been identified in various vascular beds (see, e.g., HIRSCH et al. 1989; SCHWARTZ 1989; SUZUKI et al. 1989; TAGAWA et al. 1995; VALLOTTON 1991). Although the effect of DDAVP on forearm blood flow is not NO-dependent (VAN LIEBURG et al. 1995), it is very likely that forearm V_2-receptors are located on vascular endothelial cells, and might thus mediate tPA secretion.

The mechanism underlying DDAVP-induced secretion has been clarified further by KAUFMANN et al. (2000). Having previously demonstrated that vWF release from endothelial cells could be induced by cAMP-increasing agents (VISCHER and WOLLHEIM 1997), this group now shows that DDAVP can directly induce vWF release from cultured human lung microvascular endothelial cells by activating the V_2-receptor. Previous attempts to demonstrate this were apparently unsuccessful because the V_2-receptor is not normally present on cultured HUVECs. Transfection of HUVECs with the V_2-receptor cDNA conferred responsiveness to DDAVP to these cells. It is likely that tPA secretion can also proceed via this mechanism (see Section III).

WALL et al. (1997, 1998) demonstrated, in carefully conducted studies, that the infusion of DDAVP (70 ng/min) into the brachial artery of volunteers led to a large increase of forearm blood flow and to a release of tPA (but not of vWF). At baseline, the average net release rate of tPA across the whole forearm vascular bed was 6.7 ng/min. Infusion of DDAVP induced a massive regulated release of tPA with a peak after 15 min and a maximum net release rate of 178 ng/min. To determine whether the increase in flow rate was responsible for the observed tPA release additional experiments were performed with sodium nitroprusside infusions, which induced an even greater increase of forearm blood flow but no increase in tPA release. The lack of a tPA response in the contralateral control arm, as well as the unaltered systemic hemodynamic parameters during the local DDAVP infusion, confirm that the observed tPA response was independent of a central mechanism, but due to a local release mechanism.

5. Acetylcholine and Methacholine

Several reports describe increased fibrinolytic activity after the intra-arterial infusion of acetylcholine or methacholine (TESI 1975; S. JERN et al. 1994; C. JERN et al. 1997b; WALL et al. 1997; STEIN et al. 1998; DELL'OMO et al. 1999). The increase is due to a local release of tPA, and is mediated by muscarinic receptors, both in animals (EMEIS 1988) and in man (S. JERN et al. 1994). The subtype of the muscarinic receptor involved has not been determined.

6. Bradykinin and Substance P

Bradykinin and substance P are potent inducers of tPA secretion in animal systems (reviewed by EMEIS and TRANQUILLE 1992), while bradykinin also induces regulated secretion of tPA in HUVEC via the B_2-receptor (EMEIS et al., unpublished data). Intravenous bradykinin will increase plasma PA activity in man (NERI SERNERI et al. 1965; TESI and CARAMELLI 1972), and even more extensively in angiotensin-converting enzyme inhibitor-pretreated subjects (BROWN et al. 1997). As shown by BROWN et al. (1999) in the human forearm, bradykinin releases tPA through a local effect on the vaculature. Similarly, substance P will induce tPA release when infused into the forearm (FANCIULLACCI et al. 1993; NEWBY et al. 1997).

7. Coagulation Activation Products

In cultured human endothelial cells, thrombin is a potent stimulus for regulated tPA secretion (BOOYSE et al. 1986; VAN DEN EIJNDEN-SCHRAUWEN et al. 1995). The same is true for animal systems (EMEIS 1992). As can be understood, no in vivo data are available for man. In baboons, infusion of phospholipid vesicles plus factor X resulted in thrombin generation and massive, possibly thrombin-mediated, secretion of tPA (GILES et al. 1990). The potential role of coagulation activation products in regulated tPA secretion has been reviewed previously (EMEIS 1992).

8. cAMP

If, as proposed above, β-adrenergic agents cause increased regulated tPA secretion by increasing cAMP in endothelial cells, it would follow that other compounds that increase endothelial cAMP levels would also induce increased secretion of tPA, and increased plasma levels of tPA (provided liver blood flow is not strongly affected). Candidate compounds are prostacyclin (CASNOCHA et al. 1989; HOPKINS and GORMAN 1981; LANGERER and VAN HINSBERGH 1991; SCHAFER et al. 1980; HEGEMAN et al. 1998), calcitonin gene-related peptide (CROSSMAN et al. 1987), ADP and ATP (GRIESMACHER et al. 1992), and DDAVP. However, MANNUCCI (1974) found no effect of infusing the cAMP analogue dibutyryl-cAMP ($0.35\,mg\,kg^{-1}\,min^{-1}$) on PA activity in humans. No data are available on the other candidate compounds in man.

9. Miscellaneous Compounds

Nicotinic acid, given intra-arterially, directly induces secretion of PA activity from the forearm (ROSING et al. 1978), but the compound has not found wide acceptance, mainly due to the rapid induction of prolonged tachyphylaxis. The mechanism of action of nicotinic acid is fully unknown. For further details on nicotinic acid, see the reviews cited in Sect. C.IV.1.

Acetylsalicylic acid (aspirin) has been reported to reduce tPA secretion during venous occlusion (R.I. LEVIN et al. 1984). This observation has subsequently been both confirmed and denied several times (see references in R.I. LEVIN et al. 1989). The published observations are hard to interpret, because venous occlusion is now widely considered to reflect base-line constitutive secretion, rather than capacity for regulated secretion. Moreover, in only a few of the studies have the data been corrected for changes in hematocrit, making interpretation of the observations hazardous. Acetylsalicylic acid does not affect exercise-induced tPA changes (KEBER et al. 1987), nor does it affect DDAVP-induced changes in tPA (BROMMER et al. 1984).

10. Abnormal Release of tPA

Recently, a few studies have been published on tPA release from the perfused human forearm in patient populations. In contrast to a normal tPA release in response to methacholine in borderline hypertensive male patients (S. JERN et al. 1997), the same group of investigators subsequently reported (HRAFNKELS-DÓTTIR et al. 1998) an impaired capacity for tPA release in response to DDAVP in (non-smoking) essentially hypertensive male patients, though the response to methacholine was normal. The forearm vasodilator responses were in all instances normal, again demonstrating that release of tPA and vasodilatory response are not related. DELL'OMO et al. (1999) reported, in elderly smoking male hypertensives, a deficient tPA release and a deficient vasodilatory response to acetylcholine, which led these authors to postulate that vasodilatation induces tPA release (DELL'OMO et al. 1999). In view of the absence of tPA release during endothelium-independent vasodilation as induced by sodium nitroprusside, it is more likely, however, that in these patients the acetylcholine response itself was deficient. NEWBY et al. (1999) compared the effects of intrabrachial infusion of subsystemic, locally active doses of substance P, an endothelium-dependent vasodilator, in smokers and non-smokers (NEWBY et al. 1999). The increase in forearm blood flow and the release of tPA were found to be reduced in the smokers. Since these authors had previously demonstrated that inhibition of NO synthesis reduced tPA release (NEWBY et al. 1998), together these data suggest that endothelial dysfunction may hamper the optimal release of tPA.

D. Increase of tPA Activity by Inhibition of PAI-1

During the last several years a number of compounds have been developed which decrease PAI-1 activity or synthesis, or interfere with the tPA/PAI-1 complex formation and thus increase plasma tPA-mediated fibrinolytic activity. Some of these compounds have been tested in animals, but none has yet been used in man.

The flufenamic acid derivative AR-H029953XX, a low-molecular PAI-1 inhibitor, increases the concentration of free tPA by inhibiting the formation of a tPA/PAI-1 complex (BJÖRQUIST et al. 1998). Several diketopiperazine-based low-molecular weight inhibitors of PAI-1 have been evaluated. XR334, XR1853, and XR5082 inhibited the inhibition of tPA and of two-chain uPA by PAI-1. When injected into rats, a 32%–60% increase in ex vivo whole blood clot lysis was observed (CHARLTON et al. 1996). The derivative XR5118 was shown to bind to an area between amino acids 110 and 145 of the PAI-1 mol-

544 T. Kooistra and J.J. Emeis

ecule, which is known to bind to tPA. Systemic infusion of XR5118 resulted in a significant reduction of PAI-1 activity in rabbits (from 24 to 11 IU/ml). In a rabbit jugular vein thrombosis model, XR5118 inhibited thrombus growth (FRIEDERICH et al. 1997).

A third novel compound, T-686, (3E,4E)-3-benzylidene-4-(3,4,5-trimethoxy-benzylidene)-pyrrolidine-2,5-dione, attenuated the increase of PAI-1 after lipopolysaccharide injection into mice and reduced mortality of mice (MURAKAMI et al. 1997). It also decreased venous thrombus growth in a rat thrombosis model (OHTANI et al. 1997). It attenuated the PAI-1 antigen accumulation induced by TGFβ in conditioned medium from HUVEC cultures. In rabbits fed an atherogenic diet and exposed to vascular injury, T-686 attenuated the development of vascular lesions (VINOGRADSKY et al. 1997).

List of Abbreviations and Symbols

AVP   arginine-vasopressin
$[Ca^{2+}]_i$ intracellular calcium concentration
cAMP   adenosine cyclic 3'-5' monophosphate
DDAVP   1-deamino-8-D-arginine vasopressin
DI   diabetes insipidus
HUVEC   human umbilical vein endothelial cells
IBMX   isobutylmethylxanthine
PAI-1   plasminogen activator inhibitor type 1
PAF   platelet-activating factor
PMA   4-β-phorbol 12-myristate 13-acetate
RAR   RXR, retinoic acid receptor, retinoid X/9-cis retinoic acid receptor
RARE   retinoic acid response element
tPA   tissue-type plasminogen activator
TTNPB $p[(E)-2-(5,6,7,8-tetrahydro-5,5,8,8-tetramethyl-2-naphthyl)-1-propenyl]$ benzoic acid
uPA   urokinase-type plasminogen activator
vWF   von Willebrand factor

References

Anderson RA, Ludlam CA, Wu FCW (1995) Haemostatic effects of supraphysiological levels of testosterone in normal men. Thromb Haemost 74:693–697
Arts J, Herr I, Lansink M, Angel P, Kooistra T (1997) Cell-type specific DNA-protein interactions at the tissue-type plasminogen activator promoter in human endothelial and Hela cells in vivo and in vitro. Nucleic Acids Res 25:311–317
Arts J, Kooistra T (1995) Studies on the mechanism of sodium butyrate-stimulated t-PA expression in cultured human endothelial cells. Effects of trichostatin A and 2-deoxy-D-glucose. Fibrinolysis 9:293–297
Arts J, Lansink M, Grimbergen J, Toet KH, Kooistra T (1995) Stimulation of tissue-type plasminogen activator gene expression by sodium butyrate and trichostatin A in human endothelial cells involves histone acetylation. Biochem J 310:171–176
Bäck O, Nilsson, TK (1995) Retinoids and fibrinolysis. Acta Derm Venereol 75:290–292

Bachmann F (1994) The plasminogen-plasmin enzyme system. In: Hemostasis and Thrombosis: Basic Principles and Clinical Practice, Colman RW, Hirsh J, Marder VJ, Salzman EW. J.B. Lippincott, Philadelphia, pp. 1592–1622

Bankl HC, Grossschmidt K, Pikula B, Bankl H, Lechner K, Valent P (1999) Mast cells are augmented in deep vein thrombosis and express a profibrinolytic phenotype. Hum Pathol 30:188–194

Barnathan ES, Kuo A, Rosenfeld L, Karikó K, Leski M, Robbiati F, Nolli ML, Henkin J, Cines DB (1990) Interaction of single-chain urokinase-type plasminogen activator with human endothelial cells. J Biol Chem 265:2865–2872

Bernat A, Hoffmann P, Dumas A, Serradeil-le Gal C, Raufaste D, Herbert JM (1997) V_2 receptor antagonism of DDAVP-induced release of hemostasis factors in conscious dogs. J Pharmacol Exp Ther 282:597–602

Bichet DG, Razi M, Lonergan M, Arthus M-F, Papukna V, Kortas C, Barjon J-N (1988) Hemodynamic and coagulation responses to 1-desamino[8-D-arginine]vasopressin in patients with congenital nephrogenic diabetes insipidus. New Eng J Med 318:881–887

Biesalski HK (1989) Comparative assessment of the toxicology of vitamin A and retinoids in man. Toxicology 57:117–161

Björquist P, Ehnebom J, Inghardt T, Hansson L, Lindberg M, Linschoten M, Strömqvist M, Deinum J (1998) Identification of the binding site for a low-molecular-weight inhibitor of plasminogen activator inhibitor type 1 by site-directed mutagenesis. Biochemistry 37:1227–1234

Booyse FM, Bruce R, Dolenak D, Grover M, Casey LC (1986) Rapid release and deactivation of plasminogen activators in human endothelial cell cultures in the presence of thrombin and ionophore A23187. Semin Thromb Haemost 12:228–232

Bounameaux H, Stassen JM, Seghers C, Collen D (1986) Influence of fibrin and liver blood flow an the turnover and the systemic fibrinogenolytic effects of recombinant human tissue-type plasminogen activator in rabbits. Blood 67:1493–1497

Bounameaux H, Verstraete M, Collen D (1985) Biological and therapeutic properties of new thrombolytic agents. In: Collen D, Lijnen HR, Verstraete M (eds) Thrombolysis. Churchill Livingstone, Edinburgh, pp 86–91

Brenner B, Seligsohn U, Hochberg Z (1988) Normal response of factor VIII and von Willebrand factor to 1-deamino-8-D-arginine vasopressin in nephrogenic diabetes insipidus. J Clin Endocrinol Metab 67:191–193

Brett JG, Steinberg SF, de Groot PG, Nawroth PP, Stern DM (1988) Norepinephrine down-regulates the activity of protein S on endothelial cells. J Cell Biol 106:2109–2118

Brommer EJP (1984) The level of extrinsic plasminogen activator (t-PA) during clotting as a determinant of the rate of fibrinolysis; inefficiency of activators added afterwards. Thromb Res 34:109–115

Brommer EJP, Brink H, Derkx FHM, Schalekamp MADH, Stibbe J (1990) Normal homeostasis of fibrinolysis in nephrogenic diabetes insipidus in spite of defective V_2-receptor-mediated responses of tissue plasminogen activator release. Eur J Clin Invest 20:72–78

Brommer EJP, Derkx FHM, Barrett-Bergshoeff MM, Schalekamp MADH (1984) The inability of propranolol and aspirin to inhibit the response of fibrinolytic activity and factor VIII-antigen to infusion of DDAVP. Thromb Haemost 51:42–44

Brommer EJP, Derkx FHM, Schalekamp MADH, Dooijewaard G, Van der Klaauw MM (1988) Renal and hepatic handling of endogenous tissue-type plasminogen activator (t-PA) and its inhibitor in man. Thromb Haemost 59:404–411

Brown NJ, Gainer JV, Stein CM, Vaughan DE (1999) Bradykinin stimulates tissue plasminogen activator release in human vasculature. Hypertension 33:1431–1435

Brown NJ, Nadeau J, Vaughan DE (1997). Selective stimulation of tissue-type plasminogen activator (tPA) in vivo by infusion of bradykinin. Thromb Haemostas 77:522–525

Bulens F, Ibañez-Tallon I, Van Acker P, De Vriese A, Nelles L, Belayew A, Collen D (1995) Retinoic acid induction of human tissue-type plasminogen activator gene expression via a direct repeat element (DR5) located at –7 kilobases. J Biol Chem 270:7167–7175

Bulens F, Merchiers P, Ibañez-Tallon I, de Vriese A, Nelles L, Claessens F, Belayew A, Collen D (1997) Identification of a multihormone responsive enhancer far upstream from the human tissue-type plasminogen activator gene. J Biol Chem 272:663–671

Bulens F, Nelles L, Van den Panhuyzen N, Collen D (1992) Stimulation by retinoids of tissue-type plasminogen activator secretion in cultured human endothelial cells: relations of structure to effect. J Cardiovasc Pharmacol 19:508–514

Burggraaf J, Schoemaker HC, Kroon JM, Huisman L, Kluft C, Cohen AF (1994) Influence of 1-desamino-8-D-vasopressin on endogenous fibrinolysis, haemodynamics and liver blood flow in healthy subjects. Clin Sci 86:497–503

Burrell LM, Phillips PA, Rolls KA, Buxton BF, Johnston CI, Liu JJ (1994) Vascular responses to vasopressin antagonists in man and rat. Clin Sci 87:389–395

Camiolo SM, Siuta MR, Madeja JM (1982) Improved medium for extraction of plasminogen activator from tissues. Prep Biochem 12:297–305

Carmeliet P, Collen D, (1995) Gene targeting and gene transfer studies of the biological role of the plasminogen/plasmin system. Thromb Haemost 74:429–436

Caron P, Bennet A, Camare R, Louvet JP, Boneu B, Sié P (1989) Plasminogen activator inhibitor in plasma is related to testosterone in men. Metabolism 38:1010–1015

Cash JD (1978) Control mechanism of activator release. In: Davidson JF, Rowan RM, Samama MM, Desnoyers PC (eds) Progress in chemical fibrinolysis and thrombolysis, vol 3. Raven Press, New York, pp 65–75

Casnocha SA, Eskin SG, Hall ER, McIntyre LV (1989) Permeability of human endothelial monolayers: effect of vasoactive agonists and cAMP. J Appl Physiol 67:1997–2005

Chambon P (1993) The molecular and genetic dissection of the retinoid signaling pathway. Gene 135:223–228

Chambon P (1994) The retinoid signaling pathway: molecular and genetic analyses. Semin Cell Biol 5:115–125

Chandler WL (1990) A kinetic model of the circulatory regulation of tissue plasminogen activator. Thromb Haemost 66:321–328

Chandler WL, Alessi MC, Aillaud MF, Henderson P, Vague P, Juhan-Vague I (1997) Clearance of tissue plasminogen activator (TPA) and TPA/plasminogen activator inhibitor type 1 (PAI-1) complex: relationship to elevated TPA antigen in patients with high PAI-1 activity levels. Circulation 96:761–768

Chandler WL, Levy WC, Stratton JR (1995) The circulatory regulation of t-PA and u-PA secretion, clearance, and inhibition during exercise and during the infusion of isoproterenol and phenylephrine. Circulation 92:2984–2994

Chandler WL, Levy WC, Veith RC, Stratton JR (1993) A kinetic model of the circulatory regulation of tissue plasminogen activator during exercise, epinephrine infusion, and endurance training. Blood 81:3293–3302

Chandler WL, Loo SC, Mornin D (1992b) Adrenergic stimulation of regional plasminogen activator release in rabbits. Thromb Haemost 68:545–551

Chandler WL, Trimble SL, Loo S-C, Mornin D (1990) Effect of PAI-1 levels on the molar concentrations of active tissue plasminogen activator (t-PA) and t-PA/PAI-1 complex in plasma. Blood 76:930–937

Chandler WC, Veith RC, Fellingham GW, Levy WC, Schwartz RS, Cerqueira MD, Kahn SE, Larson VG, Cain KC, Beard JC, Abrass IB, Stratton JR (1992a) Fibrinolytic response during exercise and epinephrine infusion in the same subjects. J Amer Coll Cardiol 19:1412–1420

Charlton PA, Faint RW, Bent F, Bryans J, Chicarelli-Robinson I, Mackie I, Machin S, Bevan P (1996) Evaluation of a low molecular weight modulator of human plasminogen activator inhibitor-1 activity. Thromb Haemost 75:808–815

Collen D, Lijnen HR (1991) Basic and clinical aspects of fibrinolysis and thrombolysis. Blood 78:3114–3124

Collen D, Lijnen HR (1994) Fibrinolysis and the control of hemostasis. In: Stamatoyannopoulos GS, Nienhuis AW, Majerus PW, Varmus H (eds) The molecular basis of blood diseases. Saunders, Philadelphia, pp 725–752

Costa M, Shen Y, Maurer F, Medcalf RL (1998) Transcriptional regulation of the tissue-type plasminogen-activator gene in human endothelial cells: identification of nuclear factors that recognise functional elements in the tissue-type plasminogen-activator gene promoter. Eur J Biochem 258:123–131

Crossman D, McEwan J, MacDermot J, MacIntyre I, Dollery CT (1987) Human calcitonin gene-related peptide activates adenylate cyclase and releases prostacyclin from human umbilical vein endothelial cells. Br J Pharmacol 92:695–701

Danø K, Andreasen PA, Grøndahl-Hansen J, Kristensen P, Nielsen LS, Skriver L (1985) Plasminogen activators, tissue degradation, and cancer. Adv Cancer Res 44:139–266

Davies K (1992) Diabetes defect defined. Nature 359:434

De Boer A, Kluft C, Kroon MJ, Kasper FJ, Schoemaker HC, Pruis J, Breimer DD, Soons PA, Emeis JJ, Cohen AF (1992) Liver blood flow as a major determinant of the clearance of recombinant tissue-type plasminogen activator. Thromb Haemost 67:83–87

Declerck PJ, Boden G, Degreef H, Collen D (1993) Influence of oral intake of retinoids on the human plasma fibrinolytic system. Fibrinolysis 7:347–351

Declerck PJ, Juhan-Vague I, Felez J, Wiman B (1994) Pathophysiology of fibrinolysis. J Intern Med 236:425–432

Dell'Omo G, Ferrini L, Morale M, De Negri F, Melillo E, Carmassi F, Pedrinelli R (1999) Acetylcholine-mediated vasodilatation and tissue-type plasminogen activator release in normal and hypertensive men. Angiology 50:273–282

Desnoyers PC (1978) Indirect fibrinolytic agents. In: Markwardt F (ed) Fibrinolytics and antifibrinolytics. Springer, Berlin Heidelberg New York, pp 273–314

Devries SR, Fox KAA, Robison A, Rodriguez U, Sobel BE (1987) Determinants of clearance of tissue-type plasminogen activator (t-PA) from the circulation. Fibrinolysis 1:17–22

Dootson GM, Keidan J, Harper PL (1995) The influence of isotretinoin upon fibrinolysis in patients with acne. Brit J Dermatol 133:66–70

Draijer R (1996) Regulation of human endothelial barrier function by cyclic nucleotides, nitric oxide and protein kinases. Thesis, University of Leiden

Emeis JJ (1988) Mechanisms involved in short–term changes in blood levels of tPA. In: Kluft C (ed) Tissue–type plasminogen activator (tPA): physiological and clinical aspects, vol 2. CRC Press, Boca Raton, pp 21–35

Emeis JJ (1992) Regulation of the acute release of tissue–type plasminogen activator from the endothelium by coagulation activation products. Ann NY Acad Sci 667:249–258

Emeis JJ (1995a) The control of t-PA and PAI-1 secretion from the vessel wall. Vasc Med Rev 6:153–166

Emeis JJ (1995b) Normal and abnormal endothelial release of tissue-type plasminogen activator. In: Glas-Greenwalt P (ed) Fibrinolysis in disease. CRC Press, Boca Raton, pp 55–64

Emeis JJ, Kluft C (1985) PAF-acether-induced release of tissue-type plasminogen activator from vessel walls. Blood 66:86–91

Emeis JJ, Tranquille N (1992) On the role of bradykinin in secretion from vascular endothelial cells. Agents Actions 38 [Suppl II]:285–291

Emeis JJ, van den Eijnden-Schrauwen Y, Kooistra T (1996) Tissue-type plasminogen activator and the vessel wall: synthesis, storage and secretion. In: van Hinsbergh VWM (ed) Vascular control of haemostasis. Gordon and Breach Science Publishers SA, Amsterdam, pp 187–206

Emeis JJ, van den Eijnden-Schrauwen Y, Van den Hoogen CM, de Priester W, Westmuckett A, Lupu F (1997) An endothelial storage granule for tissue-type plasminogen activator. J Cell Biol 139:245–256

Emeis JJ, van den Hoogen CM, Diglio CA (1998) Synthesis, storage and regulated secretion of tissue-type plasminogen activator by cultured rat heart endothelial cells. Fibrinol Proteol 12:9–16

Fanciullacci M, Fedi S, Alessandri M, Pietrini U (1993) Substance P-induced fibrinolysis in the forearm of healthy humans. Experientia 49:242–244

Fearnley GR (1970) Pharmacological enhancement of fibrinolysis. In: Schor JM (ed) Chemical control of fibrinolysis and thrombolysis. Wiley, New York, pp 205–243

Feng P, Ohlsson M, Ny T (1990) The structure of the TATA-less rat tissue-type plasminogen activator gene. Species-specific sequence divergences in the promoter predict differences in regulation of gene expression. J Biol Chem 265:2022–2027

Fisher R, Waller EK, Grossi G, Thompson D, Tizard R, Schleuning W-D (1985) Isolation and characterization of the human tissue-type plasminogen activator structural gene including its 5' flanking region. J Biol Chem 260:11223–11230

Fox KAA, Robison AK, Knabb RM, Rosamond TL, Sobel BE, Bergmann SR (1985) Prevention of coronary thrombosis with subthrombolytic doses of tissue-type plasminogen activator. Circulation 72:1346–1354

Friederich PW, Levi M, Biemond BJ, Charlton P, Templeton D, van Zonneveld AJ, Bevan P, Pannekoek H, ten Cate JW (1997) Novel low-molecular-weight inhibitor of PAI-1 (XR5118) promotes endogenous fibrinolysis and reduces postthrombolysis thrombus growth in rabbits. Circulation 96:916–921

Friezner Degen SJ, Rajput B, Reich E (1986) The human tissue plasminogen activator gene. J Biol Chem 261:6972–6985

Gevers Leuven JA, Kluft C, Bertina RM, Hessel LW (1987) Effects of two low-dose oral contraceptives on circulating components of the coagulation and fibrinolytic systems. J Lab Clin Med 109:631–636

Giles AR, Nesheim ME, Herring SW, Hoogendoorn H, Stump DC, Heldebrant CM (1990) The fibrinolytic potential of the normal primate following the generation of thrombin in vivo. Thromb Haemost 63:476–481

Giltay EJ, Gooren LJG, Emeis JJ, Kooistra T, Stehouwer CDA (2000) Oral, but not transdermal administration of estrogens lowers tissue-plasminogen activator levels in humans without affecting its endothelial synthesis. Arterioscler Thromb Vasc Biol 20:1396–1403

Grant PJ (1990) Hormonal regulation of the acute haemostatic response to stress. Blood Coagul Fibrinolysis 1:299–306

Grant PJ (1993) Regulation of haemostasis: the role of arginine vasopressin. In: Mariani G, Mannucci PM, Cattaneo M (eds) Desmopressin in Bleeding Disorders. Plenum Press, New York, pp 1–10

Grant PJ, Davies JA, Tate GM, Boothby M, Prentice CRM (1985) Effects of physiological concentrations of vasopressin on haemostatic function in man. Clin Sci 69:471–476

Grant PJ, Tate GM, Davies JA, Williams NS, Prentice CRM (1986) Intra-operative activation of coagulation – a stimulus to thrombosis mediated by vasopressin? Thromb Haemost 55:104–107

Griesmacher A, Weigel G, David M, Horvath G, Mueller MM (1992) Functional implications of cAMP and Ca^{2+} on prostaglandin I_2 and thromboxane A_2 synthesis by human endothelial cells. Arterioscler Thromb 12:512–518

Grülich-Henn J, Müller-Berghaus G (1990) Regulation of endothelial tissue plasminogen activator and plasminogen activator inhibitor type 1 synthesis by diacylglycerol, phorbol ester, and thrombin. Blut 61:38–44

Gurevich V (1988) Pro-urokinase: physiochemical properties and promotion of its fibrinolytic activity by urokinase and by tissue plasminogen activator with which it has a complementary mechanism of action. Semin Thromb Haemost 14:100–115

Hampton KK, Grant PJ, Boothby M, Dean HG, Davies JA, Prentice CRM (1990) The effect of modified electroconvulsive therapy on vasopressin release and haemostasis in man. Blood Coagul Fibrinolysis 1:293–297

Hamsten A, Eriksson P (1994) Fibrinolysis and atherosclerosis: an update. Fibrinolysis 8 [Suppl 1]:253–262

Hanss M, Collen D (1987) Secretion of tissue-type plasminogen activator and plasminogen activator inhibitor by cultured human endothelial cells: modulation by thrombin, endotoxin and histamine. J Lab Clin Med 109:97–104

Hariman H, Grant PJ, Hughes JR, Booth NA, Davies JA, Prentice CRM (1990) Effect of physiological concentrations of vasopressin on components of the fibrinolytic system. Thromb Haemost 61:298–300

Harrison VJ, Barnes K, Turner AJ, Wood E, Corder R, Vane JR (1995) Identification of endothelin 1 and big endothelin 1 in secretory vesicles isolated from bovine aortic endothelial cells. Proc Natl Acad Sci USA 92:6344–6348

Hashemi S, Palmer DS, Aye MT, Ganz PR (1993) Platelet-activating factor secreted by DDAVP-treated monocytes mediates von Willebrand factor release from endothelial cells. J Cell Physiol 154:496–505

Hashemi S, Tackaberry ES, Palmer DS, Ganz PR (1990) DDAVP-induced release of von Willebrand factor from endothelial cells in vitro: the effect of plasma and blood cells. Biochem Biophys Acta 1052:63–70

Hegeman RJ, van den Eijnden-Schrauwen Y, Emeis JJ (1998) Adenosine 3':5'-cyclic monophosphate induces regulated secretion of tissue-type plasminogen activator and von Willebrand factor from cultured human endothelial cells. Thromb Haemost 79:853–858

Hirsch AT, Dzau VJ, Majzoub JA, Creager MA (1989) Vasopressin-mediated forearm vasodilation in normal humans. Evidence for a vascular vasopressin V_2 receptor. J Clin Invest 84:418–426

Holmberg M, Leonardsson G, Ny T (1995) The species-specific differences in the cAMP regulation of the tissue-type plasminogen activator gene between rat, mouse and human is caused by a one-nucleotide substitution in the cAMP-responsive element of the promoters. Eur J Biochem 231:466–474

Hopkins NK, Gorman RR (1981) Regulation of endothelial cyclic nucleotide metabolism by prostacyclin. J Clin Invest 67:540–546

Hrafnkelsdóttir T, Wall U, Jern C, Jern S (1998) Impaired capacity for endogenous fibrinolysis in essential hypertension. Lancet 352:1597–1598

Jansson JH, Nilsson TK, Olofsson BO (1991) Tissue plasminogen activator and other risk factors as predictors of cardiovascular events in patients with severe angina pectoris. Eur Heart J 12:157–161

Jansson JH, Olofsson BO, Nilsson TK (1993) Predictive value of tissue plasminogen activator mass concentration on long-term mortality in patients with coronary artery disease. A 7-year follow-up. Circulation 88:2030–2034

Jern C, Manhem K, Eriksson E, Tengborn L, Risberg B, Jern S (1991) Hemostatic responses to mental stress during the menstrual cycle. Thromb Haemost 66:614–618

Jern C, Seeman-Lodding H, Biber B, Winsö O, Jern S (1997a) An experimental multiple-organ model for the study of regional net release/uptake rates of tissue-type plasminogen activator in the intact pig. Thromb Haemost 78:1150–1156

Jern C, Selin L, Jern S (1994) In vivo release of tissue-type plasminogen activator across the human forearm during mental stress. Thromb Haemost 72:285–291

Jern C, Selin L, Tengborn L, Jern S (1997b) Sympathoadrenal activation and muscarinic receptor stimulation induce acute release of tissue-type plasminogen activator but not von Willebrand factor across the human forearm. Thromb Haemost 78:887–891

Jern S, Selin L, Bergbrant A, Jern C (1994) Release of tissue-type plasminogen activator in response to muscarinic receptor stimulation in human forearm. Thromb Haemost 72:588–594

Jern S, Wall U, Bergbrant A, Selin-Sjögren L, Jern C (1997) Endothelium-dependent vasodilation and tissue-type plasminogen activator release in borderline hypertension. Arterioscler Thromb Vasc Biol 17:3376–3383

Jespersen J, Kluft C (1985) Increased euglobulin fibrinolytic potential in women on oral contraceptives low in oestrogen – levels of extrinsic and intrinsic plasminogen activators, prekallikrein, factor XII, and C1-inactivator. Thromb Haemost 54:454–459

Jespersen J, Kluft C (1986) Inhibition of tissue-type plasminogen activator in plasma of women using oral contraceptives and in normal women during a menstrual cycle. Thromb Haemost 55:388–389

Juhan-Vague I, Conte-Devola B, Aillaud MF, Mendez C, Oliver C, Collen D (1984) Effects of DDAVP and venous occlusion on the release of tissue- type plasminogen activator and von Willebrand factor in patients with panhypopituitarism. Thromb Res 33:653–659

Kato S, Mano H, Kumazawa T, Yoshizawa Y, Kojima R, Masushige S (1992) Effect of retinoid status on α, β and γ retinoic acid receptor mRNA levels in various rat tissues. Biochem J 286:755–760

Kaufmann JE, Oksche A, Wollheim CB, Günther G, Rosenthal W, Vischer UM (2000) Vasopressin-induced von Willebrand factor secretion from endothelial cells involves V2 receptors and cAMP. J Clin Invest 106:107–116

Kawai Y, Matsumoto Y, Watanabe K, Yamamoto H, Satoh K, Murata M, Handa M, Ikeda Y (1996) Hemodynamic forces modulate the effects of cytokines on fibrinolytic activity of endothelial cells. Blood 87:2314–2321

Keber D (1983) The increase of leg fibrinolytic potential after reduction of hydrostatic stimulus. Thromb Haemost 50:731–734

Keber D (1988) Mechanism of tissue plasminogen activator release during venous occlusion. Fibrinolysis 2 [Suppl 2]:96–103

Keber D, Blinc A, Fettich J (1990) Increase of tissue plasminogen activator in limbs during venous occlusion: a simple haemodynamic model. Thromb Haemost 64:433–437

Keber I, Jereb M, Keber D (1987) Aspirin decreases fibrinolytic potential during venous occlusion, but not during acute physical activity. Thromb Res 46:205–212

Kijima M, Yoshida M, Sugita K, Horinouchi S, Beppu T (1993) Trapoxin, an antitumor cyclic tetrapeptide, is an irreversible inhibitor of mammalian histone deacetylase. J Biol Chem 268:22429–22435

Kinter LB, McConnell I, Goodwin BT, Campbell S, Huffman WF, Arthus M-F, Lonergan M, Bichet DG (1992) Vasopressin antagonist inhibition of clotting factor release in the rhesus monkey (Macaca mulatta). J Pharmacol Exp Therap 261:462–469

Klöcking H-P (1991) Pharmakologische Beeinflussung der Freisetzung von t-PA aus dem Gefässendothel. Hämostaseologie 11:76–88

Knoers N, Brommer EJP, Willems H, van Oost BA, Monnens LAH (1990) Fibrinolytic responses to 1-desamino-8-D-arginine-vasopressin in patients with congenital nephrogenic diabetes insipidus. Nephron 54:322–326

Kobrinsky NL, Doyle JJ, Israels ED, Winter JSD, Cheang MS, Walker RD, Bishop AJ (1985) Absent factor VIII response to synthetic vasopressin analogue (DDAVP) in nephrogenic diabetes insipidus. Lancet i:1293–1294

Kooistra T (1990) The use of cultured human endothelial cells and hepatocytes as an in vitro model system to study modulation of endogenous fibrinolysis. Fibrinolysis 4 [Suppl 2]:33–39

Kooistra T (1995) The potentials of retinoids as stimulators of endogenous tissue-type plasminogen activator. In: Glas-Greenwalt P (ed) Fibrinolysis in disease. CRC Press, Boca Raton, pp 237–245

Kooistra T, Bosma PJ, Jespersen J, Kluft C (1990) Studies on the mechanism of action of oral contraceptives with regard to fibrinolytic variables. Am J Obstet Gynecol 163:404–413

Kooistra T, Bosma PJ, Toet K, et al. (1991a) Role of protein kinase C and cAMP in the regulation of tissue-type plasminogen activator, plasminogen activator inhibitor 1 and platelet-derived growth factor mRNA levels in human endothelial cells. Possible involvement of protooncogenes c-jun and c-fos. Arterioscler Thromb 11:1042–1052

Kooistra T, Lansink M, Arts J, Sitter T, Toet K (1995) Involvement of retinoic acid receptor a in the stimulation of tissue-type plasminogen-activator gene expression in human endothelial cells. Eur J Biochem 232:425–432

Kooistra T, Opdenberg JP, Toet K, Hendriks HFJ, Van den Hoogen RM, Emeis JJ (1991b) Stimulation of tissue-type plasminogen activator synthesis by retinoids in cultured human endothelial cells and rat tissues in vivo. Thromb Haemost 65:565–572

Kooistra T, Schrauwen Y, Arts J, Emeis JJ (1994) Regulation of endothelial cell tPA synthesis and release. Int J Hematol 59:233–255

Kooistra T, Toet K, Kluft C, et al (1993) Triazolobenzodiazepines: a new class of stimulators of tissue-type plasminogen activator synthesis in human endothelial cells. Biochem Pharmacol 46:61–67

Kooistra T, Van den Berg J, Töns A, Platenburg G, Rijken DC, Van den Berg E (1987) Butyrate stimulates tissue-type plasminogen-activator synthesis in cultured human endothelial cells. Biochem J 247:605–612

Kristensen P (1992) Localization of components from the plasminogen activation system in mammalian tissues. Acta Pathol Microbiol Immunol Scand 100 (Suppl 29):1–27

Kristensen P, Hougaard DM, Nielsen LS, Danø K (1986) Tissue-type plasminogen activator in rat adrenal medulla. Histochemistry 85:431–436

Kristensen P, Nielsen LS, Grøndahl-Hansen J, Andresen PB, Larsson L-I, Danø K (1985) Immunocytochemical demonstration of tissue-type plasminogen activator in endocrine cells of the rat pituitary gland. J Cell Biol 101:305–311

Kruh J (1982) Effect of sodium butyrate, a new pharmacological agent, on cells in culture. Mol Cell Biochem 42:65–82

Kuiper J, Otter M, Rijken DC, Van Berkel TJC (1988) Characterization of the interaction in vivo of tissue-type plasminogen activator with liver cells. J Biol Chem 263:18220–18224

Kwaan HC, Lo R, McFadzean AJS (1957) On the production of plasma fibrinolytic activity within veins. Clin Sci 16:241–253

Langerer EG, van Hinsbergh VWM (1991) Norepinephrine and iloprost improve barrier function of human endothelial cell monolayers: role of cAMP. Am J Physiol 260:C2052–C2059

Lansink M, Jong M, Bijsterbosch M, Bekkers M, Toet K, Havekes L, Emeis J, Kooistra T (1999) Increased clearance explains lower plasma levels of tissue-type plasminogen activator by estradiol: evidence for potently enhanced mannose receptor expression in mice. Blood 94:1330–1336

Lansink M, Kooistra T (1996) Stimulation of tissue-type plasminogen activator expression by retinoic acid in human endothelial cells requires retinoic acid receptor β-2 induction. Blood 88:531–541

Lansink M, Van Bennekum AM, Blaner WS, Kooistra T (1997) Differences in metabolism and isomerization of all-*trans*-retinoic acid and 9-*cis*-retinoic acid between human endothelial cells and hepatocytes. Eur J Biochem 247:596–604

Leonardsson G, Ny T (1997) Characterisation of the rat tissue-type plasminogen activator gene promoter. Identification of a TAAT-containing promoter element. Eur J Biochem 248:676–683

Lethagen S (1994) Desmopressin (DDAVP) and haemostasis. Ann Haematol 69: 173–180

Levin EG, Banka CL, Parry GC (2000) Progressive and transient expression of tissue plasminogen activator during fetal development. Arterioscler Thromb Vasc Biol 20:1668–1674

Levin EG, Del Zoppo GJ (1994) Localization of tissue plasminogen activator in the endothelium of a limited number of vessels. Am J Pathol 144:855–861

Levin EG, Marotti KR, Santell L (1989) Protein kinase C and the stimulation of tissue plasminogen activator release from human endothelial cells. Dependence on the elevation of messenger RNA. J Biol Chem 264:16030–16036

Levin EG, Marzec U, Anderson J, Harker LA (1984) Thrombin stimulates tissue plasminogen activator release from cultured human endothelial cells. J Clin Invest 74:1988–1995

Levin EG, Santell L (1988) Stimulation and desensitization of tissue plasminogen activator release from human endothelial cells. J Biol Chem 263:9360–9365

Levin EG, Santell L, Osborn KG (1997) The expression of endothelial tissue plas-
    minogen activator in vivo – a function defined by vessel size and anatomic loca-
    tion. J Cell Sci 110:139–148
Levin RI, Harpel PC, Harpel JG, Recht PA (1989) Inhibition of tissue plasminogen
    activator activity by aspirin in vivo and its relationship to levels of tissue plas-
    minogen activator activity antigen, plasminogen activator inhibitor, and their com-
    plexes. Blood 74:1635–1643
Levin RI, Harpel PC, Weil D, Chang TS, Rifkin DB (1984) Aspirin inhibits vascular
    plasminogen activator activity in vivo. Studies utilizing a new assay to quantify
    plasminogen activator activity. J Clin Invest 74:571–580
Lijnen HR, Collen D (1989) Congenital and acquired deficiencies of components of
    the fibrinolytic system and their relation to bleeding or thrombosis. Fibrinolysis
    3:67–77
Loskutoff DJ, Sawdey M, Mimuro J (1988) Type 1 plasminogen activator inhibitor. Prog
    Haemostasis Thromb 9:87–115
Lupu C, Lupu F, Dennehy U, Kakkar VV, Scully MF (1995) Thrombin induces the
    redistribution and acute release of tissue factor pathway inhibitor from specific
    granules within human endothelial cells in culture. Arterioscler Thromb Vasc Biol
    15:2055–2062
Lupu F, Heim DA, Bachmann F, Hurni M, Kakkar VV, Kruithof EKO (1995) Plas-
    minogen activator expression in human atherosclerotic lesions. Arterioscler
    Thromb Vasc Biol 15:1444–1455
Mannucci PM (1974) Enhancement of plasminogen activator by vasopressin and
    adrenaline: a role of cyclic AMP? Thromb Res 4:539–549
Mannucci PM (1988) Desmopressin: a nontransfusional form of treatment for con-
    genital and acquired bleeding disorders. Blood 72:1449–1455
Mariani G, Mannucci PM, Cattaneo M (eds) (1993) Desmopressin in Bleeding Disor-
    ders. Plenum Press, New York
Markwardt F (1983) Gefässwand und Fibrinolyse. Arzneim-Forsch/Drug Res
    33:1370–1374
Meade TW, Howarth DJ, Cooper J, MacCallum PK, Stirling Y (1994) Fibrinolytic activ-
    ity and arterial disease [Letter]. Lancet 343:1442
Medh RD, Santell L, Levin EG (1992) Stimulation of tissue plasminogen activator pro-
    duction by retinoic acid: synergistic effect on protein kinase C-mediated activa-
    tion. Blood 80:981–987
Miller AA, Kurschel E, Osieka R, Schmidt CG (1987) Clinical pharmacology of
    sodium butyrate in patients with acute leukemia. Eur J Cancer Clin Oncol
    23:1283–1287
Miyamoto A, Yang S-X, Laufs U, Ruan X-L, Liao JK (1999) Activation of guanine
    nucleotide-binding proteins and induction of endothelial tissue-type plasminogen
    activator gene transcription by alcohol. J Biol Chem 274:12055–12060
Murakami J, Ohtani A, Murata S (1997) Protective effect of T-686, an inhibitor of plas-
    minogen activator inhibitor-1 production, against the lethal effect of lipopolysac-
    charide in mice. Jpn J Pharmacol 75:291–294
Nagpal S, Saunders M, Kastner P, Durand B, Nakshrati H, Chambon P (1992) Pro-
    moter context- and response element-dependent specificity of the transcriptional
    activation and modulating functions of retinoic acid receptors. Cell 70:1007–
    1019
Newby DE, Wright RA, Dawson P, Ludlam CA, Boon NA, Fox KA, Webb DJ (1998)
    The L-arginine/nitric oxide pathway contributes to the acute release of tissue plas-
    minogen activator in vivo in man. Cardiovasc Res 38:485–492
Newby DE, Wright RA, Ludlam CA, Fox KAA, Boon NA, Webb DJ (1997) An in vivo
    model for the assessment of acute fibrinolytic capacity of the endothelium. Thromb
    Haemost 78:1242–1248
Newby DE, Wright RA, Labinjoh C, Ludlam CA, Fox KAA, Boon NA, Webb DJ
    (1999) Endothelial dysfunction, impaired endogenous fibrinolysis, and cigarette

smoking: a mechanism for arterial thrombosis and myocardial infarction. Circulation 99:1411–1415

Newmark HL, Young CW (1995) Butyrate and phenylacetate as differentiating agents: practical problems and opportunities. J Cell Biochem 22 [Suppl]:247–253

Neri Serneri GG, Rossi Ferrini PL, Paoletti P, Panti A, D'Ayala Valva G (1965) Effects of bradykinin on coagulation and fibrinolysis, study in vivo and in vitro. Thromb Diathes Haemorrh 14:508–518

Nilsson IM, Hedner U, Pandolfi M (1980) Physiology of fibrinolysis. In: Kline DL, Reddy KNN (eds) Fibrinolysis. CRC Press, Boca Raton, pp 165–183

Ny T, Elgh F, Lund B (1984) The structure of the human tissue-type plasminogen activator gene: correlation of intron and exon structures to functional and structural domains. Proc Natl Acad Sci USA 81:5355–5359

Ohta T, Kikuchi H, Hashi K, Kudo Y (1986) Nizofenone administration in the acute stage following subarachnoid hemorrhage. Results of a multi-center controlled double-blind clinical study. J Neurosurg 64:420–426

Ohtani A, Murakami J Hirano-Wakimoto A (1997) T-686, a novel inhibitor of plasminogen activator inhibitor-1, inhibits thrombosis without impairment of hemostasis in rats. Eur J Pharmacol. 330:151–156

Padró T, van den Hoogen CM, Emeis JJ (1990) Distribution of tissue–type plasminogen activator (activity and antigen) in rat tissues. Blood Coag Fibrinolysis 1: 601–608

Parmer RJ, Mahata M, Mahata S, Sebald MT, O'Connor DT, Miles LA (1997) Tissue plasminogen activator (t-PA) is targeted to the regulated secretory pathway. Catecholamine storage vesicles as a reservoir for the rapid release of t-PA. J Biol Chem 272:1976–1982

Pina-Cabral JM, Rodrigues C (1974) Blood catecholamine levels, factor VIII and fibrinolysis after therapeutic electroshock. Br J Haematol 28:371–380

Prowse CV, Cash JD (1984) Physiologic and pharmacologic enhancement of fibrinolysis. Semin Thromb Haemost 10:51–60

Prowse CV, MacGregor IR (1988) Regulation of the plasminogen activator level in blood. In: Kluft C (ed) Tissue–type plasminogen activator (t-PA): physiological and clinical aspects, vol 2. CRC Press, Boca Raton, pp 49–66

Reinders JH, de Groot PG, Sixma JJ, van Mourik JA (1988) Storage and secretion of von Willebrand factor by endothelial cells. Haemostasis 18:246–261

Rickles RJ, Darrow AL, Strickland S (1989) Differentiation-responsive elements in the 5′ region of the mouse tissue plasminogen activator gene confer two-stage regulation by retinoic acid and cyclic AMP in teratocarcinoma cells. Mol Cell Biol 9:1691–1704

Ridker PM, Hennekens CH, Stampfer MJ, Manson JE, Vaughan DE (1994) Prospective study of endogenous tissue plasminogen activator and risk of stroke. Lancet 343:940–943

Ridker PM, Vaughan DE, Stampfer MJ, Manson JE, Hennekens CH (1993) Endogenous tissue-type plasminogen activator and risk of myocardial infarction. Lancet 341:1165–1168

Riggs MG, Whittaker RG, Neumann JR, Ingram VM (1977) n-Butyrate causes histone modification in Hela and Friend erythroleukemia cells. Nature 262:462–464

Rijken DC (1995) Plasminogen activators and plasminogen activator inhibitors: biochemical aspects. In: Lijnen HR, Collen D (eds) Fibrinolysis. Baillieres Clin Haematol 8 [no 2]:291–312

Rosing DR, Redwood DR, Brakman P, Astrup T, Epstein SE (1978) The fibrinolytic response of man to vasoactive drugs measured in arterial blood. Thromb Res 13:419–428

Rosnoblet C, Vischer UM, Gerard RD, Irminger J-C, Halban P, Kruithof EKO (1999) Storage of tissue-type plasminogen activator in Weibel-Palade bodies of human endothelial cells. Arterioscler Thromb Vasc Biol 19:1796–1803

Rydholm HE, Falk P, Eriksson E, Risberg B (1998) Thrombin signal transduction of the fibrinolytic system in human adult venous endothelium in vitro. Scand J Clin Lab Invest 58:347–352

Saksela O, Rifkin DB (1988) Cell-associated plasminogen activation: Regulation and physiological functions. Annu Rev Cell Biol 4:93–126

Santell L, Levin EG (1988) Cyclic AMP potentiates phorbol ester stimulation of tissue plasminogen activator release and inhibits secretion of plasminogen activator inhibitor-1 from human endothelial cells. J Biol Chem 263:16802–16808

Santell L, Marotti KR, Levin EG (1999) Targeting of tissue plasminogen activator into the regulated secretory pathway of neuroendocrine cells. Brain Res 816:258–265

Schafer AI, Gimbrone MA, Handin RI (1980) Endothelial cell adenylate cyclase: activation by catecholamines and prostaglandin I_2. Biochem Biophys Res Comm 96:1640–1647

Schrauwen Y, Emeis JJ, Kooistra T (1994a) A sensitive ELISA for human tissue–type plasminogen activator applicable to the study of acute release from cultured human endothelial cells. Thromb Haemost 71:225–229

Schrauwen Y, de Vries REM, Kooistra T, Emeis JJ (1994b) Acute release of tissue-type plasminogen activator (tPA) from the endothelium: regulatory mechanisms and therapeutic target. Fibrinolysis 8 [Suppl 2]:8–12

Schwartz J (1989) Vasodilation associated with V_2-type vasopressin activity: findings and implications. Mol Cell Endocrinol 64:133–136

Seeman-Lodding H, Haggmark S, Jern C, Jern S, Johansson G, Winso O, Biber B (1997) Aortic cross-clamping influences regional net release and uptake rates of tissue-type plasminogen activator in pigs. Acta Anaesthesiol Scand 41:1114–1123

Siegbahn A, Ruusuvaara L (1988) Age dependence of blood fibrinolytic components and the effects of low-dose oral contraceptives on coagulation and fibrinolysis in teenagers. Thromb Haemost 60:361–364

Sillaber C, Baghestanian M, Bevec D, Willheim M, Agis H, Kapiotis S, Füreder W, Bankl HC, Kiener HP, Speiser W, Binder BR, Lechner K, Valent P (1999) The mast cell as site of tissue-type plasminogen activator expression and fibrinolysis. J Immunol 162:1032–1041

Smalley DM, Fitzgerald JE, O'Rourke J (1993) Adenosine diphosphate stimulates the endothelial release of tissue-type plasminogen activator but not von Willebrand factor from isolated-perfused rat hind limbs. Thromb Haemost 70:1043–1046

Sprengers ED, Kluft C (1987) Plasminogen activator inhibitors. Blood 69:381–387

Stein CM, Brown N, Vaughan DE, Lang CC, Wood AJ (1998) Regulation of local tissue-type plasminogen activator release by endothelium-dependent and endothelium-independent agonists in human vasculature. J Am Coll Cardiol 32:117–122

Stunnenberg HG (1993) Mechanisms of transactivation by retinoic acid receptors. Bioessays 15:309–315

Suzuki S, Takeshita A, Imaizumi T, Hirooka Y, Yoshida M, Ando S, Nakamura M (1989) Biphasic forearm vascular responses to intraarterial arginine vasopressin. J Clin Invest 84:427–434

Tagawa T, Imaizumi T, Shiramoto M, Endo T, Hironaga K, Takeshita A (1995) V_2 receptor mediated vasodilation in healthy humans. J Cardiovasc Pharmacol 25:387–392

Tappy L, Hauert J, Bachmann F (1984) Effects of hypoxia and acidosis on vascular plasminogen activator release in the pig ear perfusion system. Thromb Res 33:117–124

Tesi M (1975) Aspects of clinical methodology in fibrinolytic therapy. In: Davidson JF, Samama MM, Desnoyers PC (eds) Progress in chemical fibrinolysis and thrombolysis, vol 1. Raven Press, New York, pp 255–265

Tesi M, Caramelli L (1972) Influence of the autonomic nervous system on fibrinolytic activity caused by bradykinin. In: Back N, Sicuteri F (eds) Vasopeptides. Plenum Press, New York, pp 209–220

Theuring F, Kooistra T, Schleuning W-D (1996) Pharmacological modulation of gene expression in t-PA-lacZ transgenic mice. In: Klug S, Thiel R (eds) Methods in developmental toxicology and biology. Blackwell, Berlin, pp 103–109

Thompson EA, Nelles L, Collen D (1991) Effect of retinoic acid on the synthesis of tissue-type plasminogen activator and plasminogen activator inhibitor-1 in human endothelial cells. Eur J Biochem 201:627–632

Tranquille N, Emeis JJ (1990) The simultaneous acute release of tissue–type plasminogen activator and von Willebrand factor in the perfused rat hindleg region. Thromb Haemost 63:454–458

Tranquille N, Emeis JJ (1991) On the role of calcium in the acute release of tissue–type plasminogen activator and von Willebrand factor from the rat perfused hindleg region. Thromb Haemost 66:479–483

Tranquille N, Emeis JJ (1993) The role of cyclic nucleotides in the release of tissue–type plasminogen activator and von Willebrand factor. Thromb Haemost 69:259–261

Vallotton MB (1991) The multiple faces of the vasopressin receptors. Mol Cell Endocrinol 78:C73–C76

Van Bennekum AM, Emeis JJ, Kooistra T, Hendriks HFJ (1993) Modulation of tissue-type plasminogen activator by retinoids in rat plasma and tissues. Am J Physiol 264:R931–937

Van den Eijnden-Schrauwen Y (1996) Acute release of tissue-type plasminogen activator from human endothelial cells. Thesis, University of Leiden. Allin/Verhagen, Katwijk

Van den Eijnden-Schrauwen Y, Atsma DE, Lupu F, de Vries REM, Kooistra T, Emeis JJ (1997) Involvement of calcium and G proteins in the acute release of tissue-type plasminogen activator and von Willebrand factor from cultured human endothelial cells. Arterioscler Thromb Vasc Biol 17:2177–2187

Van den Eijnden-Schrauwen Y, Kooistra T, de Vries REM, Emeis JJ (1995) Studies on the acute release of tissue-type plasminogen activator from human endothelial cells in vitro and in rats in vivo: evidence for a dynamic storage pool. Blood 85:3510–3517

Van Griensven JM (1995) Causes of variability in the pharmacokinetics of thrombolytic drugs. Thesis, University of Leiden. ICG Printing, Dordrecht

Van Griensven JMT, Koster RW, Burggraaf J, Huisman LG, Kluft C, Kroon R, Schoemaker RC, Cohen AF (1998) Effects of liver blood flow on the pharmacokinetics of tissue-type plasminogen activator (alteplase) during thrombolysis in patients with acute myocardial infarction. Clin Pharmacol Ther 63:39–47

Van Hinsbergh VWM (1988) Synthesis and secretion of plasminogen activators and plasminogen activator inhibitor by endothelial cells. In: Kluft C (ed) Tissue–type plasminogen activator (t-PA): physiological and clinical aspects, vol 2. CRC Press, Boca Raton, pp 3–20

Van Hinsbergh VWM, Kooistra T, Emeis JJ, Koolwijk P (1991) Regulation of plasminogen activator production by endothelial cells: role in fibrinolysis and local proteolysis. Int J Rad Biol 60:261–272

Van Hinsbergh VWM, Sprengers ED, Kooistra T (1987) Effect of thrombin on the production of plasminogen activators and PA inhibitor-1 by human foreskin microvascular endothelial cells. Thromb Haemost 57:148–153

Van Kesteren PJM, Kooistra T, Lansink M, van Kamp GJ, Asscheman H, Gooren LJG, Emeis JJ, Vischer UM, Stehouwer CDA (1998) The effects of sex steroids on plasma levels of marker proteins of endothelial cell functioning. Thromb Haemost 79:1029–1033

Van Lieburg AF, Knoers NV, Monnens LA, Smits P (1995) Effects of arginine vasopressin and 1-desamino-8-D-arginine vasopressin on forearm vasculature of healthy subjects and patients with a V2 receptor defect. J Hypertens 13:1695–1700

Vassalli J-D, Sappino A-P, Belin D (1991) The plasminogen activator/plasmin system. J Clin Invest 88:1067–1072

Verheijen JH, Rijken DC, Chang GTG, Preston FE, Kluft C (1984) Modulation of rapid plasminogen activator inhibitor in plasma by stanozolol. Thromb Haemost 51:396–397

Vinogradsky B, Bell SP, Woodcock-Mitchell J, Ohtani A, Fujii S (1997) A new butadiene derivative, T-686, inhibits plasminogen activator inhibitor type-1 production in vitro by cultured human vascular endothelial cells and development of atherosclerotic lesions in vivo in rabbits. Thromb Res 85:305–314

Vischer UM, Wollheim CB (1997) Epinephrine induces von Willebrand factor release from cultured endothelial cells: involvement of cyclic AMP-dependent signalling in exocytosis. Thromb Haemostas 77:1182–1188

Wagner DD (1990) Cell biology of von Willebrand factor. Annu Rev Cell Biol 6:217–246

Wagner DD (1993) The Weibel-Palade body: the storage granule for von Willebrand factor and P-selectin. Thromb Haemost 70:105–110

Wall U, Jern C, Jern S (1997) High capacity for tissue-type plasminogen activator release from vascular endothelium in vivo. J Hypertens 15:1641–1647

Wall U, Jern S, Tengborn L, Jern C (1998) Evidence of a local mechanism for desmopressin-induced tissue-type plasminogen activator release in human forearm. Blood 91:529–537

Wallnöfer AE, Van Griensven JMT, Schoemaker HC, Cohen AF, Lambert W, Kluft C, Meijer P, Kooistra T (1993) Effect of isotretinoin on endogenous tissue-type plasminogen activator (tPA) and plasminogen activator inhibitor 1 (PAI-1) in humans. Thromb Haemost 70:1005–1008

Wieczorek I, Ludlam CA, MacGregor IR (1993) Venous occlusion does not release von Willebrand factor, factor VII or PAI-1 from endothelial cells – the importance of consensus on the use of correction factors for haemoconcentration. Thromb Haemost 69:91

Wolinski H (1980) A proposal linking clearance of circulating lipoproteins to tissue metabolic activity as a basis for understanding atherogenesis. Circulation 47:301–311

Wun T-C, Capuano A (1985) Spontaneous fibrinolysis in whole human plasma. Identification of tissue activator-related protein as the major plasminogen activator causing spontaneous activity in vitro. J Biol Chem 260:5061–5066

Wun T-C, Capuano A (1987) Initiation and regulation of fibrinolysis in human plasma at the plasminogen activator level. Blood 69:1354–1362

Yoshida M, Kijima M, Akita M, Beppu T (1990) Potent and specific inhibition of mammalian histone deacetylase both in vivo and in vitro by trichostatin A. J Biol Chem 265:17174–17179

Zamarron C, Lijnen HR, Collen D (1984) Influence of exogenous and endogenous tissue-type plasminogen activator on the lysability of clots in a plasma milieu in vitro. Thromb Res 35:335–345

Zhu GJ, Abbadini M, Donati MB, Mussoni L (1989) Tissue-type plasminogen activator release in response to epinephrine in perfused rat hindlegs. Am J Physiol 256:H404–H410

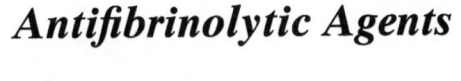

Antifibrinolytic Agents

CHAPTER 21

Structure, Pharmacology, and Clinical Use of Antifibrinolytic Agents

C.M. Samama, W. Dietrich, J. Horrow, and O. Taby, M.M. Samama

A. Introduction

Antifibrinolytic therapy constitutes an effective means to control or reduce bleeding and to limit or avoid blood transfusion in current medical practice. During the past 30 years it has been used as an established antidote in patients treated by thrombolytic agents and experiencing bleeding complications. Antifibrinolytic drugs have also been prescribed in patients developing clinical hyperfibrinolysis. However, use of these products has been limited to very specific situations involving small numbers of patients.

Up to the early 1980s, blood products have served as the primary treatment for patients with mild congenital and acquired disorders. There was no concern about the potential risk of transmitting infectious agents by homologous blood transfusions. More recently, the human immunodeficiency virus (HIV) epidemic has completely changed the practice of blood transfusion. Attention has focused on new techniques involving a more sparing use of blood, cell savers, autotransfusion, hemodilution, and use of hemostatic drugs.

The interest in antifibrinolytic agents dramatically increased when Royston et al. (1987) were able to demonstrate, that high doses of aprotinin reduced intraoperative blood loss and transfusions in cardiac surgery. Many studies have been performed in this setting and large trials have taken place in vascular, liver, and orthopedic surgery, studying different dosages and modes of administration. Furthermore, several additional studies have been performed with synthetic cost-effective agents, such as 6-amino-hexanoic acid (ε-amino caproic acid, EACA) and 4-amino-methyl-cyclohexane carboxylic acid (tranexamic acid, TA) in order to determine whether a similar efficacy on bleeding and transfusion requirements could be obtained as compared to aprotinin.

B. Pharmacology and Clinical Use of EACA and TA

Several pharmaceuticals inhibit the breakdown of cross-linked fibrin. Okamoto et al. (1959) reported efficient inhibition of plasmin activity associated with low toxicity in experimental animals with EACA. This compound strongly resembles the amino acid lysine (Fig. 1). Antifibrinolytic activity depends on a 7 Å distance between the amino- and carboxyl groups. Replacement of the 4-carbon chain with the spatially similar cyclohexane molecule in its "chair" (*trans*) form produces the cyclic compound TA. The more rigid molecular structure of the cyclohexane ring accounts for the seven to ten times greater potency of TA relative to EACA (OGSTON 1984).

EACA and TA bind to the lysine binding sites of plasminogen and plasmin, preventing these molecules from binding to C-terminal lysine residues of fibrin(ogen) (HOYLAERTS et al. 1981). In the process of binding to plasminogen, these compounds induce a conformational change of the plasminogen molecule and actually hasten its conversion to plasmin (THORSEN and MÜLLERTZ 1974; TAKADA et al. 1989). However, this conversion carries no physiologic consequence, since EACA and TA rapidly inhibit the action of plasmin.

Despite the known effects of antifibrinolytic drugs in preventing the breakdown of cross-linked fibrin, evidence that platelet dysfunction constitutes the principal hemostatic defect of cardiopulmonary bypass (CPB) (McKENNA et al. 1975; HARKER et al. 1980; HARKER 1986) prompted investigators to seek a platelet sparing effect of the antifibrinolytic compounds. Plasmin causes platelet mediator release and activation (ADELMAN et al. 1988). TA prevents platelet activation ex vivo and preserves platelet adenosine diphosphate content in patients undergoing CPB (SOSLAU et al. 1991).

How do antifibrinolytics improve hemostasis? Most likely, several pathways lead to this beneficial result: (1) platelet preservation, (2) decreased release of tissue-type plasminogen activator (tPA) by inhibiting formation of compounds which release tPA, and (3) inhibition of systemic fibrinolysis. A discussion of the clinical uses of the synthetic antifibrinolytics follows.

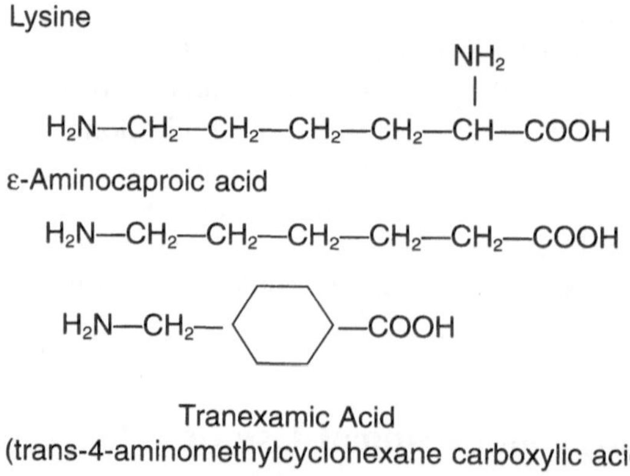

Lysine

$$NH_2$$
$$|$$
$$H_2N—CH_2—CH_2—CH_2—CH_2—CH—COOH$$

ε-Aminocaproic acid

$$H_2N—CH_2—CH_2—CH_2—CH_2—CH_2—COOH$$

Tranexamic Acid
(trans-4-aminomethylcyclohexane carboxylic acid)

Fig. 1. Chemical structures of the amino acid lysine, its deaminated product ε-aminocaproic acid (EACA), and tranexamic acid, which is formed by substituting a cyclohexane ring for four internal elements in the carbon chain

I. ε-Aminocaproic Acid (EACA)

1. Pharmacology

After administration of a loading dose of 150 mg/kg/30 min of EACA mean plasma concentrations in 27 patients undergoing extracorporeal circulation was 593 ± 154 mg/l. EACA concentration above 150 mg/l, which effectively inhibit plasma fibrinolytic activity (McNicol et al. 1962), were maintained by an infusion of 30 mg/kg per hour (Bennett-Guerrero et al. 1997). However, a recent pharmacokinetic study in 20 patients undergoing elective coronary surgery suggests that a loading dose of 50 mg/kg over 20 min before bypass, followed by 25 mg/kg per hour is sufficient to maintain plasma concentrations of EACA above 260 mg/l, which is twice the concentration needed to inhibit plasma fibrinolytic activity. Clearance of EACA during bypass fell from approximately 85 ml/min to 7 ml/min (Butterworth et al. 1999). EACA does not bind to other plasma proteins except for plasminogen. EACA does not cross the intact blood brain barrier (Ogston 1984) but inhibits fibrinolytic activity in the cerebrospinal fluid in the presence of subarachnoid hemorrhage. Between 70% and 90% of the drug appears in the urine within a few hours of administration, achieving a urinary drug concentration 50–100 times that of plasma (Verstraete 1985). Since EACA is so dependent upon renal excretion and glomerular filtration, the infusion dose should be adjusted in patients with an elevated serum creatinine or aprotinin should be chosen as drug of choice (Butterworth et al. 1999). The terminal elimination half-life is 2 h (Ogston 1984; Verstraete 1985).

2. Noncardiac Surgery

a) Urologic Surgery

Severe bleeding in the upper urinary tract presents a contraindication to EACA administration because of the inhibition of urokinase-mediated clot lysis, leading to obstruction of urinary flow (van Itterbeek et al. 1968; Wymenga and van der Boon 1998). For lower tract urinary bleeding, particularly after prostatectomy, many controlled trials attest to the drug's safety (Sharifi et al. 1986; Smith et al. 1984). However, a prophylactic role in bleeding from prostate surgery has not been demonstrated (Smith et al. 1984).

b) Oral Surgery

Saliva, a fluid with high fibrinolytic activity (Sindet-Pedersen et al. 1987) worsens the inherent coagulation defect in patients with hemophilia or von Willebrand disease (Stern and Catone 1975). Antifibrinolytic therapy, either parenterally or by mouthwash, provides routine prophylaxis against bleeding in this and other populations at risk (Blombäck et al. 1989). The combination of desmopressin and EACA may enhance the therapeutic effect (Williamson and Eggleston 1988).

c) Liver Transplantation

Orthotopic liver transplantation may be divided into three stages: prehepatic, anhepatic and recirculation stage. Fibrinolysis is altered by different mechanisms in each of these stages of surgery. Patients with liver disease commonly exhibit a decreased clearance of tPA resulting in increased circulating plasma levels of tPA. This phenomenon is accentuated during the anhepatic phase, when clearance falls still further. However, heavy blood losses with transfusion of packed red blood cells and stored whole blood (containing virtually no active tPA) represent an alternate extra-hepatic clearance mechanism and contributre to a partial correction of elevated tPA levels (CROOKSTON et al. 2000). KANG et al. (1987) successfully treated 20 of 97 patients undergoing liver transplantation with EACA, suppressing fibrinolytic activity, reducing blood losses with no thrombotic complications.

3. Use in Cardiac Surgery

Routine monitoring of anticoagulation during CPB and improvements in the biomaterials have decreased the coagulation activation. Nevertheless, the evidence for ongoing coagulation during CPB despite adequate heparin therapy (TANAKA et al. 1989) suggests that subsequent fibrinolysis occasionally explains excessive bleeding after operation.

Initial attempts to decrease bleeding after heart operations with EACA produced mixed results. Early reports lacked blinding and randomization, and often employed subtherapeutic or minimally effective doses of drug (Table 1). In well designed studies, prophylactic EACA decreased bleeding (DEL ROSSI et al. 1989; KARSKI et al. 1993; DAILY et al. 1994; TROIANOS et al. 1995, 1999). A recent meta-analysis of nine randomized studies of EACA vs controls revealed a mean reduction of blood losses with EACA of 35% ($p < 0.001$) (MUNOZ et al. 1999). Treated patients received less homologous blood, with no increase in adverse events.

4. Adverse Effects

Case reports of thrombosis from EACA accumulated following its introduction (BERGIN et al. 1966). A report of glomerular thrombosis in a bleeding patient with metastatic prostate carcinoma given EACA (CHARYTAN and PURTILO 1969) prompted an editorial branding EACA "a dangerous weapon" (RATNOFF 1969). While antifibrinolytic therapy favors the development of thrombosis in the presence of a consumptive coagulopathy, glomerular thrombosis could have resulted from the underlying coagulopathy without EACA therapy (GRALNICK 1971; COLUCCI et al. 1991). Several case reports document an association of cerebral thrombosis with EACA therapy for subarachnoid hemorrhage (SONNTAG and STEIN 1974; HOFFMAN and KOO 1979).

However, prospective studies of prophylactic EACA demonstrate its safety to the extent that this is possible with cohorts of moderate size (VINNICOMBE

Table 1. Studies of EACA to reduce bleeding after cardiac surgery

Year	Reference[a]	n^b	Blood loss reduction (%)	Structure	Dose of EACA	Timing of dose
1967	STERNS	240/100	34	Retrospective	≈5 g	After CPB
1970	GOMES	202/137	n.s.	Retrospective	Unknown	At sternotomy
1971	MIDELL	48/25	58	Prospective	125 mg/kg	Before CPB
1974	MCCLURE	12/18	42	Blinded; randomized	75 mg/kg	At sternotomy
1985	SAUSSINE	29/28	n.s.	Blinded; randomized	4 g	After protamine
1988	VANDER SALM	31/27	18	Blinded; randomized	5 g	After CPB
1989	DEL ROSSI	170/180	30	Blinded; randomized	5 g	Before incision
1993	KARSKI	125/91	25	Retrospective	10 or 15 g	Before operation
1994	DAILY	21/19	33	Blinded; randomized	10 g	Before incision
1994	AROM	100/100	34	Sequential	5 g	Before CPB
1995	TROIANOS	30/30	30	Blinded; randomized	125 mg/kg	Before CPB
1999	TROIANOS	34/33	23	Blinded; randomized	125 mg/kg	Before CPB

CPB, cardiopulmonary bypass; EACA, ε aminocaproic acid.
[a] First author.
[b] Number of patients in the treated group/number in the control group.

and Shuttleworth 1966; Smith et al. 1984; Sharifi et al. 1986; Del Rossi et al. 1989). A prudent clinician, however, will avoid its use in patients with fulminant consumptive coagulopathy unless otherwise anticoagulated.

II. Tranexamic Acid

1. Pharmacology

Administration of 10 mg/kg of TA produces plasma concentrations between 30 and 50 mg/l, decreasing to 5 mg/l 5 h later (Hoylaerts et al. 1981; Åstedt 1987). To achieve 80% inhibition of fibrinolytic activity requires concentrations of 10 mg/l (Ogston 1984; Åstedt 1987). With normal renal function, nearly 60% appears in the urine within 3 h, and over 90% within 24 h. A triexponential model of TA plasma disappearance demonstrates an overall biological elimination half-life of 2 h (Eriksson et al. 1974). TA, like EACA, binds negligibly to proteins other than plasminogen; unlike EACA, however, it passes readily into most tissues, including placenta, aqueous humor, joints, seminal fluid, and across the damaged blood-brain barrier. Oral administration results in more sustained tissue levels than intravenous administration. Rectal administration, employed to treat ulcerative colitis, results in minimal absorption and low plasma levels (Almer et al. 1992).

2. Nonsurgical Applications

Promyelocytic leukemia cells with the t (15;17) translocation express abnormally high levels of cell surface annexin II, a receptor for both plasminogen and tPA. The increased occupancy with annexin II on the surface of promyeloblasts leads to an abnormally high generation of plasmin which predisposes patients to bleeding complications (Menell et al. 1999). Plasmin cleaves (and partially inactivates) the thrombin-activatable fibrinolysis inhibitor (TAFI) resulting in a diminution of the fibrinolysis-inhibitory potential of the plasma. This mechanism contributes to the severity of the bleeding complications (Meijers et al. 2000). These have been successfully prevented by oral TA during initiation of chemotherapy of promyelocytic leukemia (Avvisati et al. 1989). Other reported successful applications include abruptio placenta (Svanberg et al. 1980), gastrointestinal bleeding (von Holstein et al. 1987), epistaxis (White 1988), hemothorax associated with malignant mesothelioma (De Boer et al. 1991), traumatic hyphema (Jerndal and Frisen 1976; Deans et al. 1992; Rahmani and Jahadi 1999; Rahmani et al. 1999), and even consumptive coagulopathy (Takada et al. 1990). In subarachnoid hemorrhage, a second episode of bleeding often complicates care, producing significant morbidity. Antifibrinolytic therapy in the form of EACA, or more commonly TA, successfully prevents rebleeding (Schisano 1978). However, concern that these drugs potentiate the development of cerebral vasospasm and the fact that TA did not result in a survival benefit in eight randomized studies (Roos et al. 1998) sharply curtailed its application.

Patients with advanced chronic renal failure typically exhibit platelet function abnormalities and a prolonged bleeding time. TA 20 mg/kg per day, given over a period of 6 days resulted in a shortening of the bleeding time in 26 of 37 patients (67%) and was associated with a significant improvement of platelet aggregation and secretion (MEZZANO et al. 1999). These findings indicate that enhanced fibrinolysis contributes to the defect in primary hemostasis in chronic advanced renal failure.

3. Noncardiac Surgery

Mouthwashes with TA successfully prevented excessive bleeding following oral surgery in anticoagulated patients (SINDET-PEDERSEN et al. 1989). TA, 40 mg/kg per hour, reduced homologous blood transfusion requirements in a randomized double-blind placebo-controlled group in orthotopic liver transplantation (BOYLAN et al. 1996). TA has even found application in reducing blood loss after total knee arthroplasty (HIIPPALA et al. 1995) in patients also receiving enoxaparin for the prophylaxis of frequently occurring postoperative deep venous thrombosis.

4. Cardiac Surgery

HORROW et al. (1990) demonstrated that prophylactic TA decreased bleeding by about 30% in a mixed population of patients undergoing cardiac surgery. Those initial observations in a small patient group have subsequently been confirmed (Table 2, Fig. 2) (SOSLAU et al. 1991; HORROW et al. 1991, 1995; KARSKI et al. 1995; MATSUZAKI et al. 1999). In a recently published trial three different doses of TA were compared (50 mg/kg, 100 mg/kg, and 150 mg/kg). Of the three doses tested the most cost-effective one to reduce bleeding was 100 mg/kg (KARSKI et al. 1998). The magnitude of savings with TA appears equivalent to that obtained in similar populations with EACA (DEL ROSSI et al. 1989) and with aprotinin (DIETRICH et al. 1992). Direct comparisons with low-dose aprotinin and with EACA revealed equivalent but somewhat lower hemostatic efficacy when compared to high-dose aprotinin (PUGH and WIELOGOSRSKI 1995; BOUGHENOU et al. 1995; PENTA DE PEPPO et al. 1995; MENICHETTI et al. 1996; LANDYMORE et al. 1997; MUNOZ et al. 1999).

5. Adverse Effects

As with EACA, prospective studies failed to implicate short term TA administration to various groups of patients as an accelerator of thrombosis (SOSLAU et al. 1991; HORROW et al. 1991, 1995; BOYLAN et al. 1992; KARSKI et al. 1993). Studies in normal patients revealed no thrombogenic effect of TA, mediated via platelet activation or elevated levels of factor VIII or von Willebrand factor (LETHAGEN and BJÖRLIN 1991).

Case reports, however, question the safety of long term administration. One patient with acute myelogenous leukemia suffered a retinal artery occlu-

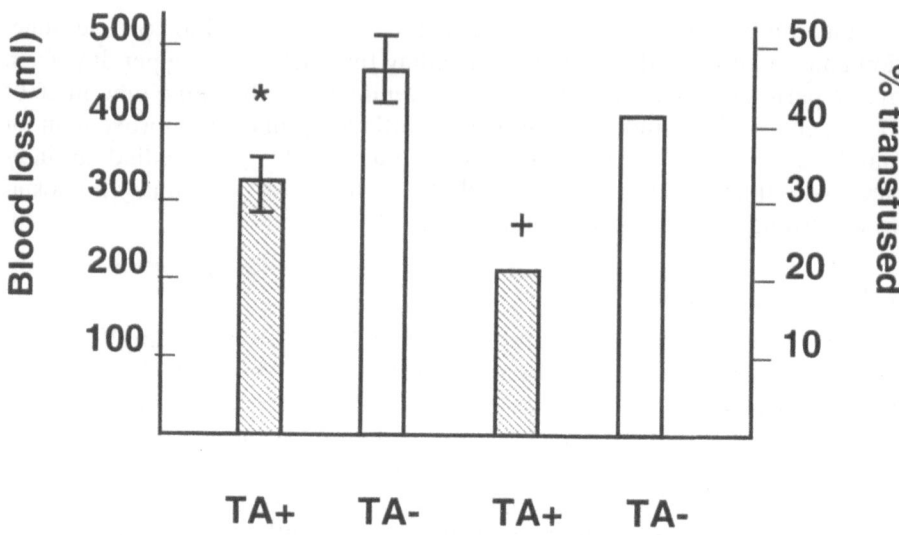

Fig. 2. Blood loss (*left vertical axis*) and percentage of patients receiving any allogeneic red cell transfusion within 5 days of operation (*right vertical axis*) achieved with administration of prophylaxic tranexamic acid. *$p < 0.0001$; ⁺$p = 0.011$. Means and SEM shown for blood loss. Adapted from Horrow et al. (1991) with permission

Table 2. Studies of tranexamic acid to reduce bleeding after cardiac surgery

Year	Reference[a]	N[b]	Blood loss reduction	Structure	Dose of TA	Timing of dose
1990	Horrow	18/20	34%	Blinded; randomized	10 mg/kg	Before incision
1991	Horrow	77/82	30%	Blinded; randomized	10 mg/kg	Before incision
1991	Soslau	8/9	36%	Blinded; randomized	7 mg/kg	Before and after CPB
1993	Karski	65/91	35%	Retrospective	10 g	Before CPB
1995	Karski	99/48	36%	Blinded; randomized	10 g or 20 g	Before CPB
1995	Horrow	97/51	31%	Blinded; randomized	5 to 40 mg/kg	Before incision

CPB, cardiopulmonary bypass; EACA, ε aminocaproic acid; TA, tranexamic acid.
[a] First author.
[b] Number of patients in the treated group/number in the control group.

sion after 5 days' treatment with TA (Parsons et al. 1988). An elderly woman given intravenous TA for 10 days after subarachnoid hemorrhage suffered a massive pulmonary embolism (Woo et al. 1989). Two young women treated for a year with oral TA for menorrhagia developed intracranial arterial thrombosis (Rydin and Lundberg 1976). A patient with idiopathic thrombocytopenic purpura developed deep vein thrombosis after 16 months of oral TA therapy (Endo et al. 1988). A woman with hereditary angioneurotic edema who received either EACA or TA over 5 years developed thrombosis of the left common carotid artery (Davies and Howell 1977).

C. Pharmacology of Aprotinin

Aprotinin is a naturally occurring serine protease inhibitor. It was independently discovered by Kraut et al. (1930) and Kunitz and Northrop (1936) as an inhibitor of kallikrein and trypsin. It is also known as basic pancreatic trypsin inhibitor because it was first isolated from bovine pancreas. It is present in other mammalian tissues such as lungs, parotid gland, spleen, liver, and seminal vesicles (Fritz and Wunderer 1983). For commercial purpose it is generally isolated from bovine lungs.

I. Structure

Aprotinin is a 58 amino-acid polypeptide with a molecular weight of 6.5 kD. Its structure consists of a single chain cross-linked by three disulfide bridges. The molecule is very basic with an isoelectric point of 10.5 (Kassel 1970). The three-dimensional structure of aprotinin has been obtained by X-ray crystallography (Huber et al. 1972). The arrangement of the molecule is very compact. This is the main reason why aprotinin is remarkably stable and resistant to heat, extreme pH, and proteolysis. Potency of aprotinin is expressed in kallikrein inhibitor units (KIU) or in protease inhibitor units (PIU); 1 KIU is equivalent to 8 PIU and 1 million KIU represent 140 mg of the pure inhibitor (Robert et al. 1996).

II. Pharmacology

Serine proteases inhibited by aprotinin and their inhibition constants (K_i) are reported in Table 3. Aprotinin has a broad inhibitory specificity but the most

Table 3. Serine proteases inhibited by aprotinin

Protease	K_i mol/l	Reference
Trypsin	6×10^{-14}	Vincent and Lazdunski (1972)
Urinary kallikrein	9×10^{-11}	Fritz and Wunderer (1983)
Plasmin	1×10^{-9}	Fritz and Wunderer (1983)
Chymotrypsin	9×10^{-9}	Fritz and Wunderer (1983)
Plasma kallikrein	3×10^{-8}	Fritz and Wunderer (1983)
Activated protein C	1.35×10^{-6}	Taby et al. (1990)
Urokinase	2.7×10^{-5}	Lottenberg (1988)
Tissue factor/factor VIIa	3×10^{-5}	Chabbat et al. (1993)

important physiological effects are its inhibition of plasmin and of kallikreins of different origin. The mechanism of inhibition is the same for all the proteases listed in Table 3. Aprotinin forms a 1:1 stoichiometric complex with the enzyme and blocks the active site of the latter. The amino acid lysine in position 15 is involved in the inhibitory activity of aprotinin (CHAUVET and ACHER 1967). It forms a covalent bond with the serine residue of the catalytic site as demonstrated by crystal structure analysis of the complex of aprotinin and bovine trypsin (RÜHLMANN et al. 1973).

While there is ample evidence of the inhibition of plasmin and kallikreins by aprotinin it has not been clearly determined whether inhibition of activated protein C (an inhibitor of blood coagulation) and of the tissue factor/factor VIIa complex (initiates the coagulation cascade) plays a role during therapy with aprotinin. K_i values are about two to three orders of magnitude lower than those for plasmin and kallikrein and the therapeutic aprotinin concentrations used may not be sufficient to inhibit the two latter enyzmes (TABY et al. 1990; CHABBOT et al. 1993).

III. Pharmacokinetics

The administration of aprotinin is intravenous because it is inactivated in the upper gastrointestinal tract (ROYSTON 1992). After injection, the drug distributes rapidly into the extracellular space. The distribution half-life is 0.7–2.5 h and the elimination half-life is 7–10h. Aprotinin exhibits linear pharmacokinetics over the dose range of $0.5–2 \times 10^6$ KIU. Its apparent volume of distribution is 261 (LEVY 1994). Plasma concentrations of aprotinin can be determined by enzyme-linked immunosorbent assay.

Studies in animals have shown that the main metabolic and excretory organ for this drug is the kidney. Approximately 90% of the dose appears in the kidney within a few hours after the injection and remains there for 12–14h (VERSTRATE 1985; WESTABY 1993). Aprotinin is filtered by the glomeruli and actively reabsorbed by the proximal tubules, where it remains until metabolized by renal lysosomes into small peptides or amino acids. Of a single dose, 25%–40% is excreted in the urine over 48h, predominantly as metabolites.

IV. Clinical Use

1. History

The drug was released to the market in Europe in the late 1950s and has, for nearly 40 years, been in clinical use for a variety of indications. Early trials with aprotinin in indications such as pancreatitis and septic shock do not satisfy current requirements of trial design and are therefore difficult to interpret.

The first report on the use of aprotinin to reduce bleeding and allogeneic blood requirement in open-heart surgery was published in the 1960s (TICE et al. 1964). Encouraging results on the use of aprotinin in pediatric open-heart surgery were published in 1985 (Popov-Cenic et al. 1985). At this time there was no generally accepted dosage regime for aprotinin and the drug was used in dosages between 100000 KIU and 1 Million (M) KIU per patient.

Royston et al. (1987) described a novel high-dosage regimen of aprotinin and reported remarkable reductions on bleeding and allogeneic blood requirement. His dosage ("Hammersmith dosage") consisted of a bolus of 2M KIU aprotinin prior to skin incision, a continuous infusion of 500000 KIU/h during the operation, and an additional bolus of 2M KIU given to the pump prime of the heart-lung machine. Since the publication of this study, the drug has been investigated in cardiac surgery in numerous studies. Unequivocally, all these studies showed a significant reduction of bleeding tendency during open-heart surgery by the use of high-dose aprotinin.

2. Mode of Action of Aprotinin (Clinical Point of View)

Platelet dysfunction is the main cause of nonsurgical bleeding after cardiopulmonary bypass (Harker et al. 1980). Platelets are activated mechanically by contact with the foreign surfaces of the heart-lung machine, the roller pumps, the suction lines, the oxygenator, and by ADP released from red blood cells (McKenna et al. 1975). Additionally, activation of the coagulation cascade, which occurs during CPB, also leads to platelet activation and platelet dysfunction. The contact of blood with the artificial, negatively charged surfaces of the extracorporeal circuit results in the activation of the contact phase of hemostasis. The Hageman factor (FXII) is converted into its active form and converts, in the presence of high molecular weight kininogen, prekallikrein to kallikrein, which again activates the Hageman factor and also converts single-chain urokinase to the active two-chain form (Ichinose et al. 1986; Hauert et al. 1989). Kallikrein leads via activation of Factor XI to the activation of the clotting cascade. The pivotal point of hemostatic activation during CPB is generation of thrombin (Winters et al. 1991). Thrombin not only converts fibrinogen into fibrin, but is also the most powerful platelet activator, activates the endothelium and fibrinolysis via the release of tPA from the endothelium, stimulates white blood cells, and has a mitogenic effect on vascular smooth muscle cells (Kanthou et al. 1992; Harker et al. 1995). Heparin is commonly used during CPB to inhibit thrombin activity. However, thrombin, which is clot-bound, is no longer accessible for the heparin/ATIII complex (Weitz et al. 1990). Thus, despite the presence of heparin, there always remains a residual thrombin activity (Dietrich 1996).

This process of hemostatic activation is controlled by amplification cascades of proteolytic enzymes and by physiologic inhibitors. The vast majority of those are mediated by serine proteases (Royston 1992). This whole body inflammatory response (Kirklin et al. 1983) is in part manifested by bleeding tendencies, which occur postoperatively despite proper surgical technique. The main hemostatic functions of platelets are the adhesion to damaged blood vessel walls, the aggregation to form a platelet plug, and the promotion of fibrin clots. Adhesion is primarily mediated by a specific receptor on the platelet surface – the glycoprotein receptor Ib. Platelet aggregation is initiated by the interaction of the glycoprotein receptor IIb-IIIa and fibrinogen. It is known that both surface glycoproteins are decreased during CPB. This reduction of surface glycoprotein may be responsible for the functional defects of platelets after CPB (EDMUNDS 1993; RINDER et al. 1991).

Aprotinin has a number of biochemical effects on this process of hemostatic activation. The drug exerts its inhibitory effect on the target serine protease by forming reversible stoichometric enzyme-inhibitor complexes. The most striking effect of aprotinin on the hemostatic system is the reduced fibrinolytic activation in patients treated with aprotinin. The concentration required to inhibit serine proteases varies from approximately 50 KIU/ml for plasmin to 200 KIU/ml for plasma kallikrein. Unequivocally, all studies report a reduction of D-dimers (BLAUHUT et al. 1991), fibrin degradation products (LU et al. 1991), fibrinolytic activation on fibrin plates (DIETRICH et al. 1990), and reduction of the plasmin-antiplasmin complexes (DIETRICH et al. 1995). On the other hand, there is no influence of aprotinin on the extrinsic pathway of fibrinolysis – no differences in tPA concentrations could be detected (DIETRICH et al. 1990). Unquestionably, aprotinin given in high dosages is a strong and effective antifibrinolytic drug.

The rationale for using the high aprotinin dosages was to achieve aprotinin plasma concentrations higher than 200 KIU/ml. These concentrations are sufficient to inhibit kallikrein activation (FRITZ and WUNDERER 1983) and, therefore, reduce contact phase activation of hemostasis. However, it could be shown that the plasma concentrations of aprotinin sufficient to inhibit kallikrein, are not maintained throughout the entire period of CPB (DIETRICH et al. 1990; FEINDT et al. 1993). MARX et al. (1991) could not demonstrate significant differences in Factor XII- or Factor XI-activity in patients treated with aprotinin compared to patients without aprotinin. On the other hand, in a model of simulated CPB it was possible to demonstrate reduced contact phase activation and reduced expression of tissue factor on monocytes by the use of aprotinin (WACHTFOGEL et al. 1993; KHAN et al. 1999).

Aprotinin prolongs the activated partial thromboplastin time (APTT) as well as the celite-activated clotting time (ACT) (DIETRICH et al. 1995) while the kaolin activated clotting time is not prolonged (DESPOTIS et al. 1995). It was speculated that this ACT prolongation is an artificial, in vitro effect of aprotinin and might be responsible for a reduction of heparin dosage during CPB which finally may lead to increased graft occlusion due to insufficient anticoagulation (COSGROVE et al. 1992). However, it was demonstrated that aprotinin binds to kaolin in the test tube and the celite ACT prolongation reflects a true anticoagulatory effect of aprotinin (DIETRICH and JOCHUM 1995). Additionally, there is evidence that clotting activation is reduced by aprotinin. Plasma levels of prothrombin fragment F1+2 (DIETRICH et al. 1995), thrombin/AT III complex (LU et al. 1991), fibrin monomers (DIETRICH et al. 1990), and fibrinopeptide A (MARX et al. 1991) are lower in aprotinin-treated patients. Though these results are not as uniform as aprotinin's effect on fibrinolysis (VERSTRAETE 1985), there is ample evidence that aprotinin not only acts as an antifibrinolytic agent but also as an anticoagulant (QUERESHI et al. 1992; DIETRICH 1996).

Platelet function is better preserved by high-dose aprotinin treatment (MOHR et al. 1992; VAN OEVEREN et al. 1990), probably due to reduced thrombin activity (DIETRICH et al. 1990; KAWASUJI et al. 1993; ORCHARD et al. 1993; SPANNAGL et al. 1994). The hypothesis about the mode of action of aprotinin in open heart surgery is that aprotinin inhibits contact phase activation and activation of the fibrinolytic system (DIETRICH 1996). Because thrombin and plasmin are both platelet stimulators, platelet function is preserved after CPB. The consequence of this better preserved platelet function is the reduced intra- and postoperative bleeding tendency (DIETRICH et al. 1995).

3. Efficacy of Aprotinin Treatment

Historically, patients with cardiac reoperations were the first group in whom the high-dose aprotinin dosage regimen was used (ROYSTON et al. 1987). These patients are at risk for perioperative bleeding and require more allogeneic blood transfusions than primary operations. In his original publication ROYSTON et al. (1987) described a small group of 22 patients having cardiac reoperations. Eleven patients were treated with high-dose aprotinin, while the other 11 patients served as control group. A reduction in postoperative bleeding tendency from 1509 ± 388 ml in the control group to 286 ± 48 ml in the aprotinin group was described. Four of 11 patients in the aprotinin group received a total of 5 units of blood compared to a total of 41 units in the control group. These encouraging results initiated several other controlled and placebo-controlled studies in cardiac surgery (VAN OEVEREN et al. 1987; BIDSTRUP et al. 1989; DIETRICH et al. 1990; BLAUHUT et al. 1991). A recent meta-analysis reviewed 46 randomized clinical trials published between 1985 and 1998 involving the use of aprotinin in CPB. Compared to placebo the mean reduction of blood losses with high-dose aprotinin was 53%, with low-dose aprotinin 35% (MUNOZ et al. 1999).

Aspirin therapy is associated with increased risk of postoperative bleeding, which has been reported to increase the likelihood of repeat operation for bleeding (BASHEIN et al. 1991). Studies have been conducted to assess the efficacy of aprotinin in this group of patients. MURKIN et al. (1994) studied 54 patients with preoperative aspirin ingestion. They found a reduction of postoperative bleeding tendency from 1710 ml in the control group to 906 ml in the aprotinin group. The blood transfusion requirement was reduced from 8 units in the control group to 4.1 units in the aprotinin group. Similar results were found by others (KLEIN et al. 1998; IVERT et al. 1998; BIDSTRUP et al. 2000).

The analysis of the first 3 years' use of aprotinin in the German Heart Centre Munich (DIETRICH et al. 1992) compared the results of 902 patients treated with aprotinin to 882 control patients. The 1784 patients mostly underwent primary coronary artery bypass graft (CABG) procedures (61%), valve replacement (31%), or combined procedures (8%). The postoperative blood loss was reduced by 35% in the aprotinin group compared to the control group (678 ml vs 1037 ml). The allogeneic blood requirement was reduced by 53% (942 ml vs 1999 ml). The results in primary operations and combined procedures repeat operations were comparable in this study. In a further study these authors also compared the effect of a high-dose vs a low-dose regimen on blood coagulation activation markers, fibrinolytic parameters, and postoperative blood loss in a randomized, double-blind trial of 230 patients undergoing cardiac surgery. The high-dose was significantly more effective than the low-dose regimen in attenuating fibrinolysis and reducing the bleeding tendency, but not in reducing the F1+2 prothrombin fragments. In the opinion of the authors, a high-dose therapy is superior to low-dose aprotinin in cardiac surgery (Dietrich et al. 1998).

These initial results in patients undergoing repeat cardiac surgery were confirmed by others. Levy et al. (1995) studied 126 patients undergoing repeat CABG surgery. They found a reduction of postoperative bleeding tendency from 1700 ml in the control group to 900 ml in the aprotinin group. Accordingly, the transfusion requirement was reduced by 79% in the aprotinin group compared to the control patients. Lemmer et al. (1994), also studying repeat CABG patients, found a reduction in blood loss of 38% and a reduction of allogeneic blood requirement of 79% by the use of aprotinin.

A meta-analysis of 52 randomized trials published between 1985 and 1998 involving the use of EACA ($n = 9$) or of aprotinin ($n = 46$) in CPB revealed that total blood loss was reduced by 53% by high-dose aprotinin and 35% by either low-dose aprotinin or EACA. Transfusion requirements were significantly reduced by all three treatment schedules. The need for re-exploration because of bleeding was significantly reduced only by high-dose aprotinin (Fig. 3) (Munoz et al. 1999).

The concomitant reduction in blood transfusion requirements depends on the transfusion policy in the given hospital. Aprotinin is effective in attenuating bleeding in patients under aspirin therapy (Klein et al. 1998; Ivert et al. 1998; Bidstrup et al. 2000). It seems even more effective in redo surgery and longer lasting operations.

4. Miscellaneous Uses

a) Orthotopic Liver Transplantation (OLT) and Elective Liver Resection

Severe bleeding is common during OLT and several studies have shown that aprotinin is a potent hemostatic agent in this setting (Neuhaus et al. 1989). Most of these trials have used an open design and results have been compared with historical controls (reviewed by Garcia-Huete et al. 1997). In a comparative study (Soilleux et al. 1995), low-dose aprotinin (500000 KIU bolus and 150000 KIU/h) was as effective as high-dose (2 Mio KIU bolus and 500000 KIU/h), but there was no placebo group in this study. Therefore, it was

Outcome	Drug	# Studies	Odds Ratio and 95% Confidence Interval

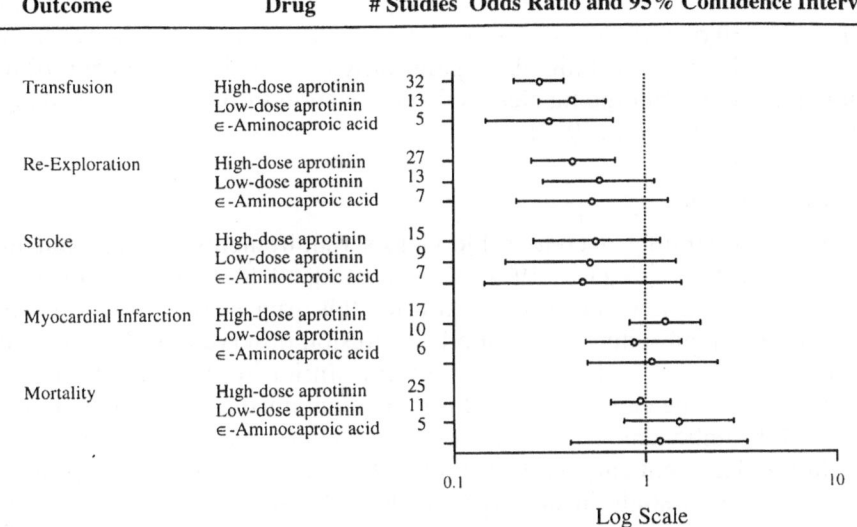

Fig. 3. Meta-analysis of randomized studies in cardiac surgery. Odds ratios and 95% confidence intervals are given for high-dose aprotinin, low-dose aprotinin, and EACA for various study parameters. Note the logarithmic scale on the abscissa. From MUNOZ et al. (1999) (corrected figure) with permission

impossible to conclude that small dose aprotinin was more effective than placebo. A more recent study has demonstrated that low-dose aprotinin controlled hyperfibrinolysis with a concomitant reduction in transfusion of blood products (MARCEL et al. 1996). This study was double-blind but unfortunately the number of patients was not large enough to give a valid answer to the question as far as the hemostatic efficacy of aprotinin is concerned. Another randomized trial in 80 patients concluded that high-dose aprotinin was not useful in reducing bleeding and blood product requirements (GARCIA-HUETE et al. 1997). However, in this study no attenuation of D-dimer increases was observed in the aprotinin group, a most unusual finding. Recently, a randomized, double-blind, placebo-controlled study has clearly confirmed that aprotinin is a useful adjunct in OLT and greatly reduces blood losses. In the EMSALT study 137 patients undergoing primary OLT received high-dose aprotinin ($n = 46$; 2 M KIU as an initial 20 min. infusion loading dose before and during induction of anesthesia, followed by 1 M KIU/h until 2 h after graft perfusion, supplemented by another 1 M KIU 30 min before graft reperfusion), "regular-dose" aprotinin as used in open-heart surgery ($n = 43$; loading dose of 2 M KIU, followed by 0.5 M KIU/h until 2 h after graft perfusion), or placebo infusions ($n = 48$). Mean total blood loss in the placebo group was 5050 ml, in the regular aprotinin group it amounted to 2825 ml (a reduction of 44%; $p = 0.04$) and in the high-dose aprotinin 2030 ml (reduction of 60%; $p = 0.02$) (PORTE et al. 2000). Thromboembolic events occurred in 2 patients in the high-dose group, none in the regular-dose group and in two in the placebo group. Mortality at 30 days did not differ between the three groups (6.5%, 4.7% and 8.3% respectively). Aprotinin also significantly reduced blood loss and transfusion in patients undergoing elective liver resection through a subcostal incision (LENTSCHENER et al. 1997).

b) Orthopedic Surgery

Aprotinin moderately decreases blood loss requirements during total hip replacement (JANSSENS et al. 1994; MURKIN et al. 1995). One or two packed red cells units per patient can be saved when this drug is used. However, with regard to the small reduction in blood transfusion observed in these patients as compared to the real efficacy of autotransfusion techniques, the use of aprotinin cannot be generalized in this indication. Furthermore, the potential increased thrombogenic risk in addition to the allergic risk limits its use after total hip replacement surgery. In contrast, the use of high doses of aprotinin in a double blind study in high risk septic and cancer patients undergoing pelvic and hip surgery has proved to be very effective in reducing significantly the need for blood transfusion as compared to a placebo group (CAPDEVILA et al. 1998).

Therefore, it seems clear that aprotinin should only be recommended for potential hemorrhagic orthopedic surgery. The same recommendation could also be proposed for cardiac surgery or liver transplant.

5. Side Effects of Aprotinin Therapy

Concerns exist about possible side effects of high-dose aprotinin therapy in open heart surgery: increased risk of graft occlusion and myocardial infarction, renal dysfunction, and allergic reactions to aprotinin. The first clinical study of the use of aprotinin in the United States indicated that aprotinin therapy may be associated with an increased incidence of myocardial infarction and graft occlusion (COSGROVE et al. 1992). Though there was no statistically significant difference in the incidence of myocardial infarctions, the concerns were based on the fact that post mortem examinations showed increased graft occlusion in patients treated with aprotinin. Four randomized, double-blind, placebo-controlled trials investigated graft patency by noninvasive methods. Using ultrafast computed tomography, LEMMER et al. (1994) found a patency rate of 92% in the aprotinin group compared to 95% in the placebo group. They found a trend without statistical significance toward lower vein and internal mammary artery (IMA) graft patency in aprotinin recipients. BIDSTRUP et al. (1993b), using magnetic resonance imaging in 90 patients undergoing primary CABG, could not demonstrate any differences in graft patency between aprotinin and control patients. These results were confirmed (HAVEL et al. 1994; KALANGOS et al. 1994). Recent large multicenter studies also demonstrated the effectiveness of high-dose aprotinin in reducing postoperative bleeding tendency without increasing the risk of myocardial infarction in repeat cardiac surgery (Levy et al. 1995). The largest study addressing this issue is the international, multicenter randomized IMAGE trial in 870 patients undergoing primary coronary surgery with CPB. Graft angiography was attempted in all patients a mean of 11 days after surgery. In 796 assessable patients aprotinin reduced thoracic drainage volume by 43% ($p < 0.0001$) and requirements for red blood cell administration by 49% ($p < 0.0001$). After adjustment for risk factors (female gender, lack of prior aspirin therapy, small and poor distal vessel quality) the aprotinin versus placebo risk ratio for graft occlusion was 1.05 (90% confidence interval 0.6–1.8). Aprotin did not affect the occurrence of myocardial infarction (aprotinin 2.9%, placebo 3.8%) or mortality (1.4% vs. 1.6 %, respectively) (Alderman et al. 1998).

Aprotinin is reabsorbed in the proximal tubules of the kidney. Therefore, the possibility exists that kidney function might be affected by the use of aprotinin. It could be shown that the excretion of α_1-microglobulin is increased in aprotinin-treated patients (Feindt et al. 1995; Fraedrich et al. 1989). However, large studies (Bidstrup et al. 1993a; Dietrich et al. 1992) did not find any association between aprotinin treatment and impaired postoperative renal failure. On the other hand, Sundt et al. (1993) reported a high incidence of renal failure in 20 patients undergoing operations of aortic aneurysms in hypothermic cardiac arrest. This was not a controlled study and used a historical control group. These results could not be confirmed in pediatric patients undergoing correction of congenital heart disease in deep hypothermic circulatory arrest (Dietrich et al. 1993). Therefore, it is conceivable that aprotinin causes transient renal dysfunction, which is clinically not significant.

Since aprotinin is a polypeptide derived from bovine lungs, it possesses antigenic properties (Weipert et al. 1997). Therefore, the possibility of an adverse reaction to this agent exists, especially in patients reexposed to this drug. Allergic reactions after reexposure have been described (Dewachter et al. 1993; Diefenbach et al. 1995; Schulze et al. 1993). The incidence of hypersensitivity reactions in one study was 2.8% in 248 patients re-exposed to aprotinin (Dietrich 1997). A time dependency for the risk of adverse reactions exists: the shorter the time interval between the two exposures the higher the risk of a reaction. Therefore, the drug should not be given within 6 months of the last exposure. It is advisable to delay the first bolus injection of aprotinin until the surgeon is ready to commence CPB. Under these precautions a reexposure to aprotinin, after an interval of more than 6 months, in patients with high risk of bleeding seems justifiable (Dietrich 1998).

Aprotinin is well tolerated by the majority of patients. The main side effect is the risk of an anaphylactic reaction in patients sensitized by prior exposure to this drug.

D. Conclusion

The efficacy of antifibrinolytic drugs has now been widely documented. Major differences do exist between agents with regard to the pharmacokinetics, the pharmacodynamics, the hemostatic potency, and the side effects. These differences have to be considered when the physician has to choose one molecule rather than another. However, aprotinin probably has the highest benefit/risk ratio and should therefore be preferred in many clinical situations; its only drawback is its high price. New fields are about to be investigated with these products. Antifibrinolytic agents, and especially aprotinin, will be useful tools in the near future to control the bleeding risk and reduce the transfusion requirements in orthopedic, urologic, vascular, gynecologic surgery, or neurosurgery.

List of Abbreviations and Acronyms

ACT   activated clotting time
APTT   activated partial thromboplastin time
ATIII   antithrombin III
CABG   coronary artery bypass graft
CPB   cardiopulmonary bypass
EACA   ε-aminocaproic acid
EMSALT   European Multicentre Study on Aprotinin in Liver
            Transplantation
HIV   human immunodeficiency virus
IMA   internal mammary artery
IMAGE   International Multicenter Aprotinin Graft patency Experience
KIU   kallikrein inhibitory units
OLT   orthotopic liver transplantation
TA   tranexamic acid
tPA   tissue-type plasminogen activator

References

Adelman B, Rizk A, Hanners E (1988) Plasminogen interactions with platelets in plasma. Blood 72:1530–1535
Alderman EL, Levy JH, Rich JB, Nili M, Vidne B, Schaff H, Uretzky G, Pettersson G, Thiis JJ, Hantler CB, Chaitman B, Nadel A, on bahalf of the IMAGE Investigators (1998) Analyses of coronary graft patency after aprotinin use: results from the International Multicenter Aprotinin Graft Patency Experience (IMAGE) trial. J Thorac Cardiovasc Surg 116:716–730
Almer S, Andersson T, Strom M (1992) Pharmacokinetics of tranexamic acid in patients with ulcerative colitis and in healthy volunteers after the single instillation of 2g rectally. J Clin Pharmacol 32:49–54
Arom KV, Emery RW (1994) Decreased postoperative drainage with addition of ε-aminocaproic acid before cardiopulmonary bypass. Ann Thorac Surg 57:1108–1113
Åstedt B (1987) Clinical pharmacology of tranexamic acid. Scand J Gastroenterol 22 [Suppl 137]:22–25
Avvisati G, Wouter ten Cate J, Büller HR, Mandelli F (1989) Tranexamic acid for control of haemorrhage in acute promyelocytic leukaemia. Lancet 2:122–124

Bashein G, Nessly ML, Rice AL, Counts RB, Misbach GA (1991) Preoperative aspirin therapy and reoperation for bleeding after coronary artery bypass surgery. Arch Intern Med 151:89–93

Bennett-Guerrero E, Sorohan JG, Canada AT, Ayuso L, Newman MF, Reves JG, Mythen MG (1997) ε-Aminocaproic acid plasma levels during cardiopulmonary bypass. Anesth Analg 85:248–251

Bergin JJ (1966) The complications of therapy with epsilon-aminocaproic acid. Med Clin North Am 50:1669–1678

Bidstrup BP, Harrison J, Royston D, Taylor KM, Treasure T (1993a) Aprotinin therapy in cardiac operations: a report on use in 41 cardiac centers in the United Kingdom. Ann Thorac Surg 55:971–976

Bidstrup BP, Hunt BJ, Sheikh S, Parratt RN, Bidstrup JM, Sapsford RN (2000) Amelioration of the bleeding tendency of preoperative aspirin after aortocoronary bypass grafting. Ann Thorac Surg 69:541–547

Bidstrup BP, Royston D, Sapsford RN, Taylor KM (1989) Reduction in blood loss and blood use after cardiopulmonary bypass with high dose aprotinin (Trasylol). J Thorac Cardiovasc Surg 97:364–372

Bidstrup BP, Underwood SR, Sapsford RN, Streets EM (1993b) Effect of aprotinin (Trasylol) on aorta-coronary bypass graft patency. J Thorac Cardiovasc Surg 105:147–153

Blauhut B, Gross C, Necek S, Doran JE, Späth P, Lundsgaard-Hansen P (1991) Effects of high-dose aprotinin on blood loss, platelet function, fibrinolysis, complement, and renal function after cardiopulmonary bypass. J Thorac Cardiovasc Surg 101: 958–967

Blombäck M, Johansson G, Johnsson H, Swedenborg J, Wabo E (1989) Surgery in patients with von Willebrand's disease. Br J Surg 76:398–400

Boughenou F, Madi-Jebara S, Massonnet-Castel S, Benmosbah L, Carpentier A, Cousin MT (1995) Antifibrinolytiques et prévention du saignement en chirurgie cardiaque valvulaire. Comparaison de l'acide tranexamique à l'aprotinine à haute dose. Arch Mal Coeur Vaiss 88:363–370

Boylan JF, Klinck JR, Sandler AN, Arellano R, Greig PD, Nierenberg H, Roger SL, Glynn MFX (1996) Tranexamic acid reduces blood loss, transfusion requirements, and coagulation factor use in primary orthotopic liver transplantation. Anesthesiology 85:1043–1048

Butterworth J, James RL, Lin Y, Prielipp RC, Hudspeth AS (1999) Pharmacokinetics of ε-aminocaproic acid in patients undergoing aortocoronary bypass surgery. Anesthesiology 90:1624–1635

Capdevila X, Calvet Y, Biboulet P, Biron C, Rubenovitch J, d'Athis F (1998) Aprotinin decreases blood loss and homologous transfusions in patients undergoing major orthopedic surgery. Anesthesiology 88:50–57

Chabbat J, Porte P, Tellier M, Steinbuch M (1993) Aprotinin is a competitive inhibitor of the factor VIIa-tissue factor complex. Thromb Res 71:205–215

Charytan C, Purtilo D (1969) Glomerular capillary thrombosis and acute renal failure after epsilon-amino caproic acid therapy. N Engl J Med 280:1102–1104

Chauvet J, Acher R (1967) The reactive site of the basic trypsin inhibitor of pancreas. Role of lysine 15. J Biol Chem 242:4274–4275

Colucci M, Semeraro N, Montemurro P, Chiumarulo P, Triggiani R, Morrone LF, Schena FP (1991) Urinary procoagulant and fibrinolytic activity in human glomerulonephritis. Relationship with renal function. Kidney Int 39:1213–1217

Cosgrove DM 3rd, Heric B, Lytle BW, Taylor PC, Novoa R, Golding LA, Stewart RW, McCarthy PM, Loop FD (1992) Aprotinin therapy for reoperative myocardial revascularization: a placebo-controlled study. Ann Thorac Surg 54:1031–1038

Crookston KP, Marsh CL, Chandler WL (2000) A kinetic model of the circulatory regulation of tissue plasminogen activator during orthotopic liver transplantation. Blood Coagul Fibrinolysis 11:79–88

Daily PO, Lamphere JA, Dembitsky WP, Adamson RM, Dans NF (1994) Effect of prophylactic epsilon-aminocaproic acid on blood loss and transfusion requirements in patients undergoing first-time coronary artery bypass grafting. A randomized, prospective, double-blind study. J Thorac Cardiovasc Surg 108:99–108

Davies D, Howell DA (1977) Tranexamic acid and arterial thrombosis [letter]. Lancet 1:49

De Boer WA, Koolen MGJ, Roos CM, ten Cate JW (1991) Tranexamic acid treatment of hemothorax in two patients with malignant mesothelioma. Chest 100:847–848

Deans R, Noel LP, Clarke WN (1992) Oral administration of tranexamic acid in the management of traumatic hyphema in children. Can J Ophthalmol 27:181–183

Del Rossi AJ, Cernaianu AC, Botros S, Lemole GM, Moore R (1989) Prophylactic treatment of postperfusion bleeding using EACA. Chest 96:27–30

Despotis GJ, Joist JH, Joiner-Maier D, Alsoufiev AL, Triantafillou AN, Goodnough LT, Santoro SA, Lappas DG (1995) Effect of aprotinin on activated clotting time, whole blood and plasma heparin measurements. Ann Thorac Surg 59:106–111

Dewachter P, Mouton C, Masson C, Gueant JL, Haberer JP (1993) Anaphylactic reaction to aprotinin during cardiac surgery [letter]. Anaesthesia 48:1110–1111

Diefenbach C, Abel M, Limpers B, Lynch J, Ruskowski H, Jugert FK, Buzello W (1995) Fatal anaphylactic shock after aprotinin reexposure in cardiac surgery. Anesth Analg 80:830–831

Dietrich W (1996) Reducing thrombin formation during cardiopulmonary bypass: is there a benefit of the additional anticoagulant action of aprotinin? J Cardiovasc Pharmacol 27 (Suppl 1):S50–S57

Dietrich W (1998) Incidence of hypersensitivity reactions. Ann Thorac Surg 65 [Suppl]:S60–S74

Dietrich W, Barankay A, Hahnel C, Richter JA (1992) High-dose aprotinin in cardiac surgery: three years' experience in 1784 patients. J Cardiothorac Vasc Anesth 6:324–327

Dietrich W, Dilthey G, Spannagl M, Jochum M, Braun SL, Richter JA (1995) Influence of high-dose aprotinin on anticoagulation, heparin requirement, and celite- and kaolin-activated clotting time in heparin-pretreated patients undergoing open-heart surgery. A double-blind, placebo-controlled study. Anesthesiology 83:679–689

Dietrich W, Jochum M (1995) Effect of celite and kaolin on activated clotting time in the presence of aprotinin: activated clotting time is reduced by binding of aprotinin to kaolin. J Thorac Cardiovasc Surg 109:177–178

Dietrich W, Mössinger H, Spannagl M, Jochum M, Wendt P, Barankay A, Meisner H, Richter JA (1993) Hemostatic activation during cardiopulmonary bypass with different aprotinin dosages in pediatric patients having cardiac operations. J Thorac Cardiovasc Surg 105:712–720

Dietrich W, Schöpf K, Spannagl M, Jochum M, Braun SL, Meisner H (1998) Influence of high- and low-dose aprotinin on activation of hemostasis in open heart operations. Ann Thorac Surg 65:70–77

Dietrich W, Spannagl M, Jochum M, Wendt P, Schramm W, Barankay A, Sebening F, Richter JA (1990) Influence of high-dose aprotinin treatment on blood loss and coagulation patterns in patients undergoing myocardial revascularization. Anesthesiology 73:1119–1126

Edmunds LH Jr (1993) Blood-surface interactions during cardiopulmonary bypass. J Card Surg 8:404–410

Endo Y, Nishimura S, Miura A (1988) Deep-vein thrombosis induced by tranexamic acid in idiopathic thrombocytopenic purpura [letter]. JAMA 259:3561–3562

Eriksson O, Kjellman H, Pilbrant A, Schannong M (1974) Pharmacokinetics of tranexamic acid after intravenous administration to normal volunteers. Eur J Clin Pharmacol 7:375–380

Feindt P, Volkmer I, Seyfert U, Huwer H, Kalweit G, Gams E (1993) Activated clotting time, anticoagulation, use of heparin, and thrombin activation during extracorporeal circulation: changes under aprotinin therapy. Thorac Cardiovasc Surg 41:9–15

Feindt PR, Walcher S, Volkmer I, Keller HE, Straub U, Huwer H, Seyfert UT, Petzold T, Gams E (1995) Effects of high-dose aprotinin on renal function in aortocoronary bypass grafting. Ann Thorac Surg 60:1076–1080

Fraedrich G, Weber C, Bernard C, Hettwer A, Schlosser V (1989) Reduction of blood ·transfusion requirement in open heart surgery by administration of high doses of aprotinin – Preliminary results. Thorac Cardiovasc Surg 37:89–91

Fritz H, Wunderer G (1983) Biochemistry and applications of aprotinin, the kallikrein inhibitor from bovine organs. Arzneimittelforschung 33:479–494

Garcia-Huete L, Domenech P, Sabaté A, Martinez-Brotons F, Jaurrieta E, Figueras J (1997) The prophylactic effect of aprotinin on intraoperative bleeding in liver transplantation: a randomized clinical study. Hepatology 26:1143–1148

Gomes MMR, McGoon DC (1970) Bleeding patterns after open-heart surgery. J Thorac Cardiovasc Surg 60:87–97

Gralnick HR, Greipp P (1971) Thrombosis with epsilon aminocaproic acid therapy. Am J Clin Pathol 56:151–154

Harker LA (1986) Bleeding after cardiopulmonary bypass. N Engl J Med 314:1446–1448

Harker LA, Hanson SR, Runge MS (1995) Thrombin hypothesis of thrombus generation and vascular lesion formation. Am J Cardiol 75 [Suppl B]:B12–B17

Harker LA, Malpass TW, Branson HE, Hessel EA,II, Slichter SJ (1980) Mechanism of abnormal bleeding in patients undergoing cardiopulmonary bypass: acquired transient platelet dysfunction associated with selective α-granule release. Blood 56:824–834

Hauert J, Nicoloso G, Schleuning WD, Bachmann F, Schapira M (1989) Plasminogen activators in dextran sulfate-activated euglobulin fractions: a molecular analysis of factor XII- and prekallikrein- dependent fibrinolysis. Blood 73:994–999

Havel M, Grabenwöger F, Schneider J, Laufer G, Wollenek G, Owen A, Simon P, Teufelsbauer H, Wolner E (1994) Aprotinin does not decrease early graft patency after coronary artery bypass grafting despite reducing postoperative bleeding and use of donated blood. J Thorac Cardiovasc Surg 107:807–810

Havel M, Teufelsbauer H, Knobl P, Dalmatiner R, Jaksch P, Zwolfer W, Muller M, Vukovich T (1991) Effect of intraoperative aprotinin administration on postoperative bleeding in patients undergoing cardiopulmonary bypass operation. J Thorac Cardiovasc Surg 101:968–972

Hiippala S, Strid L, Wennerstrand M, Arvela V, Mäntylä S, Ylinen J, Niemelä H (1995) Tranexamic acid (Cyklokapron) reduces perioperative blood loss associated with total knee arthroplasty. Br J Anaesth 74:534–537

Hoffman EP, Koo AH (1979) Cerebral thrombosis associated with amicar. Radiology 131:687–689

Horrow JC, Hlavacek J, Strong MD, Collier W, Brodsky I, Goldman SM, Goel IP (1990) Prophylactic tranexamic acid decreases bleeding after cardiac operations. J Thorac Cardiovasc Surg 99:70–74

Horrow JC, Van Riper DF, Strong MD, Brodsky I, Parmet JL (1991) Hemostatic effects of tranexamic acid and desmopressin during cardiac surgery. Circulation 84:2063–2070

Horrow JC, Van Riper DF, Strong MD, Grunewald KE, Parmet JL (1995) The dose-response relationship of tranexamic acid. Anesthesiology 82:383–392

Hoylaerts M, Lijnen HR, Collen D (1981) Studies on the mechanism of the antifibrinolytic action of tranexamic acid. Biochim Biophys Acta 673:75–85

Huber R, Kukla D, Rühlmann A, Steigemann W (1972) Pancreatic trypsin inhibitor (Kunitz). I. Structure and function. Cold Spring Harb Symp Quant Biol 36:141–148

Ichinose A, Fujikawa K, Suyama T (1986) The activation of pro-urokinase by plasma kallikrein and its inactivation by thrombin. J Biol Chem 261:3486–3489

Ivert T, Intonti M, Stain-Malmgren R, Dumitrescu A, Blombäck M (1998) Effects of aprotinin during cardiopulmonary bypass in patients treated with acetylsalicylic acid. Scand Cardiovasc J 32:289–295

Janssens M, Joris J, David JL, Lemaire R, Lamy M (1994) High-dose aprotinin reduces blood loss in patients undergoing total hip replacement surgery. Anesthesiology 80:23–29

Jerndal T, Frisen M (1976) Tranexamic acid (AMCA) and late hyphaema. A double blind study in cataract surgery. Acta Ophthalmol 54:417–429

Kalangos A, Tayyareci G, Pretre R, Didio P, Sezerman O (1994) Influence of aprotinin on early graft thrombosis in patients undergoing myocardial revascularization. Eur J Cardiothorac Surg 8:651–656

Kang Y, Lewis JH, Navalgund A, Russell MW, Bontempo FA, Niren LS, Starzl TE (1987) Epsilon-aminocaproic acid for treatment of fibrinolysis during liver transplantation. Anesthesiology 66:766–773

Kanthou C, Parry G, Wijelath E, Kakkar VV, Demoliou-Mason C (1992) Thrombin-induced proliferation and expression of platelet-derived growth factor-A chain in human vascular smooth muscle cells. FEBS Lett 314:143–148

Karski JM, Dowd NP, Joiner R, Carroll J, Peniston C, Bailey K, Glynn MFX, Teasdale SJ, Cheng DCH (1998) The effect of three different doses of tranexamic acid on blood loss after cardiac surgery with mild systemic hypothermia (32°C). J Cardiothorac Vasc Anesth 12:642–646

Karski JM, Teasdale SJ, Norman P, Carroll J, VanKessel K, Wong P, Glynn MFX (1995) Prevention of bleeding after cardiopulmonary bypass with high-dose tranexamic acid. Double-blind, randomized clinical trial. J Thorac Cardiovasc Surg 110:835–842

Karski JM, Teasdale SJ, Norman PH, Carroll JA, Weisel RD, Glynn MFX (1993) Prevention of postbypass bleeding with tranexamic acid and ε-aminocaproic acid. J Cardiothorac Vasc Anesth 7:431–435

Kassel B (1970) Bovine trypsin-kallikrein inhibitor. In: Perlman GE, Lorand L (eds) Methods in enzymology, 19. Academic Press, New York London, pp 844–852

Kawasuji M, Ueyama K, Sakakibara N, Tedoriya T, Matsunaga Y, Misaki T, Watanabe Y (1993) Effect of low-dose aprotinin on coagulation and fibrinolysis in cardiopulmonary bypass. Ann Thorac Surg 55:1205–1209

Khan MMH, Gikakis N, Miyamoto S, Rao AK, Cooper SL, Edmunds LH,Jr., Colman RW (1999) Aprotinin inhibits thrombin formation and monocyte tissue factor in simulated cardiopulmonary bypass. Ann Thorac Surg 68:473–478

Kirklin JK, Westaby S, Blackstone EH, Kirklin JW, Chenoweth DE, Pacifico AD (1983) Complement and the damaging effects of cardiopulmonary bypass. J Thorac Cardiovasc Surg 86:845–857

Klein M, Keith PR, Dauben H-P, Schulte HD, Beckmann H, Mayer G, Elert O, Gams E (1998) Aprotinin counterbalances an increased risk of peri-operative hemorrhage in CABG patients pre-treated with Aspirin. Eur J Cardiothorac Surg 14:360–366

Kraut E, Frey E, Werle E (1930) Über die Inaktivierung des Kallikreins. Hoppe-Seylers Z Physiol Chem 192:1–21

Kunitz M, Northrop JH (1936) Isolation from beef pancreas of crystalline trypsinogen, trypsin, a trypsin inhibitor and an inhibitor trypsin compound. J Gen Physiol 19:991–1007

Landymore RW, Murphy JT, Lummis H, Carter C (1997) The use of low-dose aprotinin, epsilon-aminocaproic acid or tranexamic acid for prevention of mediastinal bleeding in patients receiving aspirin before coronary artery bypass operations [letter]. Eur J Cardiothorac Surg 11:798–800

Lemmer JH Jr, Stanford W, Bonney SL, Breen JF, Chomka EV, Eldredge WJ, Holt WW, Karp RB, Laub GW, Lipton MJ, Schaff HV, Tatooles CJ, Rumberger JA (1994) Aprotinin for coronary bypass operations: efficacy, safety, and influence on early saphenous vein graft patency. A multicenter, randomized, double-blind, placebo-controlled study. J Thorac Cardiovasc Surg 107:543–551

Lentschener C, Benhamou D, Mercier FJ, Boyer-Neumann C, Naveau S, Smadja C, Wolf M, Franco D (1997) Aprotinin reduces blood loss in patients undergoing elective liver resection. Anesth Analg 84:875–881

Lethagen S, Björlin G (1991) Effect of tranexamic acid on platelet function in normal volunteers [letter]. Eur J Haematol 47:77–78

Levy JH, Bailey JM, Salmenperä M (1994) Pharmacokinetics of aprotinin in preoperative cardiac surgical patients. Anesthesiology 80:1013–1018

Levy JH, Pifarre R, Schaff HV, Horrow JC, Albus R, Spiess B, Rosengart TK, Murray J, Clark RE, Smith P, Nadel A, Bonney SL, Kleinfield R (1995) A multicenter, double-blind, placebo-controlled trial of aprotinin for reducing blood loss and the requirement for donor-blood transfusion in patients undergoing repeat coronary artery bypass grafting. Circulation 92:2236–2244

Lottenberg R, Sjak-Shie N, Fazleabas AT, Roberts RM (1988) Aprotinin inhibits urokinase but not tissue-type plasminogen activator. Thromb Res 49:549–556

Lu H, Soria C, Commin P-L, Soria J, Piwnica A, Schumann F, Regnier O, Legrand Y, Caen JP (1991) Hemostasis in patients undergoing extracorporeal circulation: the effect of aprotinin (Trasylol). Thromb Haemost 66:633–637

Marcel RJ, Stegall WC, Suit CT, Arnold JC, Vera RL, Ramsay MA, O'Donnell MB, Swygert TH, Hein HA, Whitten CW (1996) Continuous small-dose aprotinin controls fibrinolysis during orthotopic liver transplantation. Anesth Analg 82:1122–1125

Marx G, Pokar H, Reuter H, Doering V, Tilsner V (1991) The effects of aprotinin on hemostatic function during cardiac surgery. J Cardiothorac Anesth 5:467–474

Matsuzaki K, Matsui K, Tanoue Y, Nagano I, Haraguchi N, Tatewaki H (1999) Antifibrinolytic therapy with tranexamic acid in cardiac operations. Cardiovasc Surg 7:195–199

McClure PD, Izsak J (1974) The use of epsilon-aminocaproic acid to reduce bleeding during cardiac bypass in children with congenital heart disease. Anesthesiology 40:604–608

McKenna R, Bachmann F, Whittaker B, Gilson JR, Weinberg M Jr (1975) The hemostatic mechanism after open-heart surgery. II. Frequency of abnormal platelet functions during and after extracorporeal circulation. J Thorac Cardiovasc Surg 70:298–308

McNicol GP, Fletcher AP, Alkjaersig N, Sherry S (1962) The absorption, distribution and excretion of ε-aminocaproic acid following oral or intravenous administration to man. J Lab Clin Med 59:15–24

Meijers JCM, Oudijk E-JD, Mosnier LO, Bos R, Bouma BN, Nieuwenhuis HK, Fijnheer R (2000) Reduced activity of TAFI (thrombin-activatable fibrinolysis inhibitor) in acute promyelocytic leukaemia. Br J Haematol 108:518–523

Menell JS, Cesarman GM, Jacovina AT, McLaughlin MA, Lev EA, Hajjar KA (1999) Annexin II and bleeding in acute promyelocytic leukemia. N Engl J Med 340:994–1004

Menichetti A, Tritapepe L, Ruvolo G, Speziale G, Cogliati A, Di Giovanni C, Pacilli M, Criniti A (1996) Changes in coagulation patterns, blood loss and blood use after cardiopulmonary bypass: aprotinin vs tranexamic acid vs epsilon aminocaproic acid. J Cardiovasc Surg 37:401–407

Mezzano D, Panes O, Muñoz J, Pais E, Tagle R, González F, Mezzano S, Barriga F, Pereira J (1999) Tranexamic acid inhibits fibrinolysis, shortens the bleeding time and improves platelet function in patients with chronic renal failure. Thromb Haemost 82:1250–1254

Midell AI, Hallman GL, Bloodwell RD, Beall AC Jr, Yashar JJ, Cooley DA (1971) Epsilon-aminocaproic acid for bleeding after cardiopulmonary bypass. Ann Thorac Surg 11:577–582

Mohr R, Goor DA, Lusky A, Lavee J (1992) Aprotinin prevents cardiopulmonary bypass-induced platelet dysfunction. A scanning electron microscope study. Circulation 86 [Suppl II]:II-405–II-409

Munoz JJ, Birkmeyer NJO, Birkmeyer JD, O'Connor GT, Dacey LJ (1999) Is ε-aminocaproic acid as effective as aprotinin in reducing bleeding with cardiac surgery? A meta-analysis. Circulation 99:81–89

Murkin JM, Lux J, Shannon NA, Guiraudon GM, Menkis AH, McKenzie FN, Novick RJ (1994) Aprotinin significantly decreases bleeding and transfusion requirements in patients receiving aspirin and undergoing cardiac operations. J Thorac Cardio-vasc Surg 107:554–561

Murkin JM, Shannon NA, Bourne RB, Rorabeck CH, Cruickshank M, Wyile G (1995) Aprotinin decreases blood loss in patients undergoing revision or bilateral total hip arthroplasty. Anesth Analg 80:343–348

Neuhaus P, Bechstein WO, Lefèbre B, Blumhardt G, Slama K (1989) Effect of apro-tinin on intraoperative bleeding and fibrinolysis in liver transplantation. Lancet 2:924–925

Ogston D (1984) Antifibrinolytic drugs: chemistry, pharmacology and clinical usage. Wiley, New York, pp 1–180

Okamoto S, Nakajima T, Okamoto U (1959) A suppressing effect of epsilon-amino-n-caproic acid on the bleeding of dogs, produced with the activation of plasmin in the circulating blood. Keio J Med 8:247–66

Orchard MA, Goodchild CS, Prentice CRM, Davies JA, Benoit SE, Creighton-Kemsford LJ, Gaffney PJ, Michelson AD (1993) Aprotinin reduces cardiopul-monary bypass-induced blood loss and inhibits fibrinolysis without influencing platelets. Br J Haematol 85:533–541

Parsons MR, Merritt DR, Ramsay RC (1988) Retinal artery occlusion associated with tranexamic acid therapy. Am J Ophthalmol 105:688–689

Penta de Peppo A, Pierri MD, Scafuri A, De Paulis R, Colantuono G, Caprara E, Tomai F, Chiariello L (1995) Intraoperative antifibrinolysis and blood-saving techniques in cardiac surgery. Prospective trial of 3 antifibrinolytic drugs. Tex Heart Inst J 22:231–236

Popov-Cenic S, Murday H, Kirchhoff PG, Hack G, Freynes J (1985) Anlage und zusam-menfassende Ergebnisse einer klinischen Doppelblindstudie bei aorto-koronaren Bypass-Operationen. In: Dudziak R, Kirchhoff PG, Reuter HD, Schumann F (eds) Proteolyse und Proteinaseninhibition in der Herz- und Gefäßchirurgie. Schat-tauer, Stuttgart New York, pp 171–186

Porte RJ, Molenaar IQ, Begliomini B, Groenland TH, Januszkiewicz A, Lindgren L, Palareti G, Hermans J, Terpstra OT, for the EMSALT Study Group (2000) Apro-tinin and transfusion requirements in orthotopic liver transplantation: a multi-centre randomised double-blind study. Lancet 355:1303–1309

Pugh SC, Wielogorski AK (1995) A comparison of the effects of tranexamic acid and low-dose aprotinin on blood loss and homologous blood usage in patients under-going cardiac surgery. J Cardiothorac Vasc Anesth 9:240–244

Quereshi A, Lamont J, Burke P, Grace P, Bouchier-Hayes D (1992) Aprotinin: the ideal anti-coagulant? Eur J Vasc Surg 6:317–320

Rahmani B, Jahadi HR (1999) Comparison of tranexamic acid and prednisolone in the treatment of traumatic hyphema. A randomized clinical trial. Ophthalmology 106:375–379

Rahmani B, Jahadi HR, Rajaeefard A (1999) An analysis of risk for secondary hem-orrhage in traumatic hyphema. Ophthalmology 106:380–385

Ratnoff OD (1969) Epsilon aminocaproic acid – a dangerous weapon. N Engl J Med 280:1124–1125

Rinder CS, Bohnert J, Rinder HM, Mitchell J, Ault K, Hillman R (1991) Platelet activation and aggregation during cardiopulmonary bypass. Anesthesiology 75:388–393

Robert S, Wagner BKJ, Boulanger M, Richer M (1996) Aprotinin. Ann Pharmacother 30:372–380

Roos YBWEM, Vermeulen M, Rinkel GJE, Algra A, van Gijn J (1998) Systematic review of antifibrinolytic treatment in aneurysmal subarachnoid haemorrhage. J Neurol Neurosurg Psychiatry 65:942–943

Royston D, Bidstrup BP, Taylor KM, Sapsford RN (1987) Effect of aprotinin on need for blood transfusion after repeat open-heart surgery. Lancet 2:1289–1291

Royston D (1992) High-dose aprotinin therapy: A review of the first five years' experience. J Cardiothorac Anesth 6:76–100

Rydin E, Lundberg PO (1976) Tranexamic acid and intracranial thrombosis. Lancet 2:49

Rühlmann A, Kukla D, Schwager P, Bartels K, Huber R (1973) Structure of the complex formed by bovine trypsin and bovine pancreatic trypsin inhibitor. Crystal structure determination and stereochemistry of the contact region. J Mol Biol 77:417–436

Saussine M, Delpech S, Allien M, Grolleau D, Daures MF, Coulon P, Chaptal PA (1985) Saignement après circulation extracorporelle et acide epsilon amino-caproique. Ann Fr Anesth Reanim 4:403–405

Schisano G (1978) The use of antifibrinolytic drugs in aneurysmal subarachnoid hemorrhage. Surg Neurol 10:217–222

Schulze K, Graeter T, Schaps D, Hausen B (1993) Severe anaphylactic shock due to repeated application of aprotinin in patients following intrathoracic aortic replacement. Eur J Cardiothorac Surg 7:495–496

Sharifi R, Lee M, Ray P, Millner SN, Dupont PF (1986) Safety and efficacy of intravesical aminocaproic acid for bleeding after transurethral resection of prostate. Urology 27:214–219

Sindet-Pedersen S, Gram J, Jespersen J (1987) Characterization of plasminogen activators in unstimulated and stimulated human whole saliva. J Dent Res 66:1199–1203

Sindet-Pedersen S, Ramström G, Bernvil S, Blombäck M (1989) Hemostatic effect of tranexamic acid mouthwash in anticoagulant-treated patients undergoing oral surgery. N Engl J Med 320:840–843

Smith RB, Riach P, Kaufman JJ (1984) Epsilon aminocaproic acid and the control of post-prostatectomy bleeding: a prospective double-blind study. J Urol 131:1093–1095

Soilleux H, Gillon M-C, Mirand A, Daibes M, Leballe F, Ecoffey C (1995) Comparative effects of small and large aprotinin doses on bleeding during orthotopic liver transplantation. Anesth Analg 80:349–352

Sonntag VKH, Stein BM (1974) Arteriopathic complications during treatment of subarachnoid hemorrhage with epsilon-aminocaproic acid. J Neurosurg 40:480–485

Soslau G, Horrow J, Brodsky I (1991) Effect of tranexamic acid on platelet ADP during extracorporeal circulation. Am J Hematol 38:113–119

Spannagl M, Dietrich W, Beck A, Schramm W (1994) High dose aprotinin reduces prothrombin and fibrinogen conversion in patients undergoing extracorporeal circulation for myocardial revascularization. Thromb Haemost 72:159–160

Stern NS, Catone GA (1975) Primary fibrinolysis after oral surgery. J Oral Surg 33:49–52

Sterns LP, Lillehei CW (1967) Effect of epsilon aminocaproic acid upon blood loss following open-heart surgery: an analysis of 340 patients. Can J Surg 10:304–307

Sundt TM III, Kouchoukos NT, Saffitz JE, Murphy SF, Wareing TH, Stahl DJ (1993) Renal dysfunction and intravascular coagulation with aprotinin and hypothermic circulatory arrest. Ann Thorac Surg 55:1418–1424

Svanberg L, Åstedt B, Nilsson IM (1980) Abruptio placentae – treatment with the fibrinolytic inhibitor tranexamic acid. Acta Obstet Gynecol Scand 59:127–130

Taby O, Chabbat J, Steinbuch M (1990) Inhibition of activated protein C by aprotinin and the use of the insolubilized inhibitor for its purification. Thromb Res 59:27–35

Takada A, Sugawara Y, Takada Y (1989) Enhancement of the activation of Glu-plasminogen by urokinase in the simultaneous presence of tranexamic acid or fibrin. Haemostasis 1:26–31

Takada A, Takada Y, Mori T, Sakaguchi S (1990) Prevention of severe bleeding by tranexamic acid in a patient with disseminated intravascular coagulation. Thromb Res 58:101–108

Tanaka K, Takao M, Yada I, Yuasa H, Kusagawa M, Deguchi K (1989) Alterations in coagulation and fibrinolysis associated with cardiopulmonary bypass during open heart surgery. J Cardiothorac Anesth 3:181–188

Thorsen S, Müllertz S (1974) Rate of activation and electrophoretic mobility of unmodified and partially degraded plasminogen. Effects of 6-aminohexanoic acid and related compounds. Scand J Clin Lab Invest 34:167–176

Tice DA, Worth MH, Clauss RH, Reed GH (1964) The inhibition by Trasylol of fibrinolytic activity associated with cardiovascular operations. Surg Gynaecol Obst 119: 71–74

Troianos CA, Desai M, Stypula RW, Lucas DM, Pasqual RT, Newfeld ML, Pellegrini RV, Mathie TB (1995) The effect of prophylactic administration of epsilon aminocaproic acid on mediastinal bleeding after cardiopulmonary bypass. Anesth Analg 80:SCA 121

Troianos CA, Sypula RW, Lucas DM, D'Amico F, Mathie TB, Desai M, Pasqual RT, Pellegrini RV, Newfeld ML (1999) The effect of prophylactic ε-aminocaproic acid on bleeding, transfusions, platelet function, and fibrinolysis during coronary artery bypass grafting. Anesthesiology 91:430–435

van Itterbeek H, Vermylen J, Verstraete M (1968) High obstruction of urine flow as a complication of the treatment with fibrinolysis inhibitors of haematuria in haemophiliacs. Acta Haematol 39:237–242

Van Oeveren W, Jansen NJG, Bidstrup BP, Royston D, Westaby S, Neuhof H, Wildevuur CRH (1987) Effects of aprotinin on haemostatic mechanisms during cardiopulmonary bypass. Ann Thorac Surg 44:640–645

Van Oeveren W, Harder MP, Roozendaal KJ, Eijsman L, Wildevuur CRH (1990) Aprotinin protects platelets against the initial effect of cardiopulmonary bypass. J Thorac Cardiovasc Surg 99:788–797

Vander Salm TJ, Ansell JE, Okike ON, Marsicano TH, Lew R, Stephenson WP, Rooney K (1988) The role of epsilon-aminocaproic acid in reducing bleeding after cardiac operation: a double-blind randomized study. J Thorac Cardiovasc Surg 95:538–540

Verstraete M (1985) Clinical application of inhibitors of fibrinolysis. Drugs 29:236–261

Vincent JP, Lazdunski M (1972) Trypsin-pancreatic trypsin inhibitor association. Dynamics of the interaction and role of disulfide bridges. Biochemistry 11: 2967–2977

Vinnicombe J, Shuttleworth KED (1966) Aminocaproic acid in the control of haemorrhage after prostatectomy. Safety of aminocaproic acid – a controlled trial. Lancet 1:232–234

von Holstein CCSS, Eriksson SBS, Kallén R (1987) Tranexamic acid as an aid to reducing blood transfusion requirements in gastric and duodenal bleeding. Br Med J (Clin Res Ed) 294:7–10

Wachtfogel YT, Kucich U, Hack CE, Gluszko P, Niewiarowski S, Colman RW, Edmunds LH Jr (1993) Aprotinin inhibits the contact, neutrophil, and platelet activation systems during simulated extracorporeal perfusion. J Thorac Cardiovasc Surg 106:1–10

Weipert J, Meisner H, Jochum M, Dietrich W (1997) Long-term follow-up of aprotinin-specific immunoglobulin G antibodies after cardiac operations. J Thorac Cardiovasc Surg 114:676–678

Weitz JI, Hudoba M, Massel D, Maraganore J, Hirsh J (1990) Clot-bound thrombin is protected from inhibition by heparin- antithrombin III but is susceptible to inactivation by antithrombin III-independent inhibitors. J Clin Invest 86:385–391

Westaby S (1993) Aprotinin in perspective. Ann Thorac Surg 55:1033–1041

White A, O'Reilly BF (1988) Oral tranexamic acid in the management of epistaxis. Clin Otolaryngol 13:11–16

Williamson R, Eggleston DJ (1988) DDAVP and EACA used for minor oral surgery in von Willebrand disease. Aust Dent J 33:32–36

Winters KJ, Santoro SA, Miletich JP, Eisenberg PR (1991) Relative importance of thrombin compared with plasmin-mediated platelet activation in response to plasminogen activation with streptokinase. Circulation 84:1552–1560

Woo KS, Tse LKK, Woo JLF, Vallance-Owen J (1989) Massive pulmonary thromboembolism after tranexamic acid antifibrinolytic therapy. Br J Clin Pract 43:465–466

Wymenga LF, van der Boon WJ (1998) Obstruction of the renal pelvis due to an insoluble blood clot after epsilon-aminocaproic acid therapy: resolution with intraureteral streptokinase instillations. J Urol 159:490–492

Subject Index

A-23187 535
abciximab 308
acetylcholine 535
– tPA secretion, regulation 541
acetylsalicylic acid *see* aspirin
actin 31
activated partial thromboplastin time
  (aPTT) 174, 188, 297–298
– aprotinin effect 570
acute myocardial infarction
– aprotinin use, incidence 574
– aspirin, effect, ISIS-2 11
– bleeding complications 185–186
– clinical studies
– – anisoylated lys-plasmin(ogen)
    streptokinase activator complex
    (APSAC) 189–193
– – conjunctive therapy 296–309
– – lanoteplase 495–496
– – monteplase 494
– – pamiteplase 501–502
– – pro-urokinase 241–245
– – pro-urokinase, bolus
    administration 248
– – staphylokinase 439–440
– – streptokinase 177–183
– – streptokinase vs. tPA 261–262,
    264–280
– – tenecteplase (TNK-tPA) 505–508
– – tPA 216–221
– – urokinase 239–241
– contraindications for thrombolytic
  therapy 187 (table)
– delay to treatment 183–185
– insulin resistance, PAI-1 antigen and
  activity 9
– PAI-1, risk factor 9, 118

– plasma viscosity 175
– platelets, role 11
– thrombolytic treatment 13–15, 209,
  210 (fig.)
– – indications 209, 211–212 (tables)
– – experimental coronary thrombosis
    model 215
– – first administration to patients
    215–216
acute phase response
– PAI-1 118–119
– proteins 7, 297
ACUTE study 309
ADMIT trial 363
adrenaline, intravenous injection 57
adrenergic agents, tPA secretion,
  regulation 538–539
Aggrastat 306
Agkistrodon contortrix 512
AIMS study 189, 289
α_1-microglobulin 575
α_2-antiplasmin 4 (fig.), 69 (table),
  111–114, 174, 177, 209, 210 (fig.), 425
– consumption 10
– deficiencies 73, 114
– gene and protein structure 111–113
– molar concentration 5
– plasmin, inhibition of 6, 112 (fig.),
  113–114
– plasmin-SAC complex, inhibition by
  432–433
– target enzyme specificity 113–114
α_2-macroglobulin
– fibrinolytic system 4 (fig.), 30 (fig.),
  123
– scavenger inhibitor 5, 123
α-enolase 31, 144 (table)